Global Forest Resources Assessment 2000

Main report

FAO FORESTRY PAPER

140

Food and Agriculture Organization of the United Nations

Rome, 2001

The designations employed and the presentation of material in this
information product do not imply the expression of any opinion
whatsoever on the part of the Food and Agriculture Organization
of the United Nations concerning the legal status of any country,
territory, city or area or of its authorities, or concerning the
delimitation of its frontiers or boundaries.

ISBN 92-5-104642-5

All rights reserved. Reproduction and dissemination of material in this
information product for educational or other non-commercial purposes are
authorized without any prior written permission from the copyright holders
provided the source is fully acknowledged. Reproduction of material in this
information product for resale or other commercial purposes is prohibited without
written permission of the copyright holders. Applications for such permission
should be addressed to the Chief, Publishing and Multimedia Service,
Information Division, FAO, Viale delle Terme di Caracalla, 00100 Rome, Italy or
by e-mail to copyright@fao.org

© **FAO 2001**

Foreword

Humanity stands at a crucial point in its development. Never before have the Earth's ecosystems been so greatly affected by our presence. Large areas of the world's forests, which have served in the subsistence and advancement of humankind, have been converted to other uses or severely degraded. While substantial areas of productive forest remain, there is now widespread recognition that the resource is not infinite, and that its wise and sustainable use is needed for our survival. Forests are also increasingly appreciated for their aesthetic, recreational and spiritual values, which frequently conflict with purely economic objectives.

From the vantage point of the new millennium, we have the opportunity to reflect on the current condition of our planet's resources and to look carefully at the events contributing to the present situation. The Global Forest Resources Assessment 2000 (FRA 2000) provides such a perspective on the world's forests through an appraisal of their state in the year 2000, and changes since the 1980s. The assessment is a key source of factual information on forests for use by national institutions and international fora such as the Convention on Biological Diversity, the United Nations Framework Convention on Climate Change and the Convention to Combat Desertification in seeking solutions to environmental concerns.

FRA 2000 was the most comprehensive and technologically advanced assessment in FAO's 50-year history. It relied on the active participation of partners and member countries around the world. The thematic content is broader than ever before, covering forest area status and change, biological diversity, timber volume and forest biomass, non-wood forest products, trees outside the forest, forest fires and other topical issues. For the first time, comparable trend information on tropical deforestation from two successive assessment periods has been obtained through the use of statistical sampling and satellite remote sensing.

The assessment employed state-of-the-art information management systems, Internet technology and geographic information systems. One tangible benefit from their use has been the ability of FAO to release a large body of information to the general public as soon as it became available. In fact, more information is now available on the FAO Web site than could conceivably be published in the main report. But the assessment was not driven by technology. Instead, the technology was applied selectively as a complement to more conventional data gathering means.

FAO considers FRA 2000 a major achievement. However, its ultimate value will be determined by its ability to motivate the world community to take firm actions that result in the wise and sustainable use of our world's forests. Criteria and indicators of sustainable forest management provide guidance to forest users and managers on what needs to be accomplished. Yet the practical implementation of these principles must be worked out by a relatively diverse group of stakeholders, with different motivations, aspirations and needs. Therefore it is essential that decision-makers be fully involved in the process and exercise leadership in seeking solutions. Their decisions in the coming years will be difficult and the consequences far-reaching.

M. Hosny El-Lakany
Assistant Director-General
Forestry Department

Preface

The Global Forest Resources Assessment 2000 (FRA 2000) was the most comprehensive since FAO first reported on forest resources 50 years ago.

There are two possible approaches to a global assessment of forest resources. One approach is to collect data at the field level and to aggregate information upward to the country, regional and global levels. The other approach is to look down from above, either literally by using satellite remote sensing, or figuratively through global studies. FRA 2000 was based on the bottom-up approach, but supplemented by global level verification. The backbone of FRA 2000 is the data, information and knowledge provided by countries. However, because of inconsistencies in data quality and availability, country information was verified and supplemented with "top down" studies and remote sensing analysis using the latest technology. Countries were then invited to review and comment on the outcome of the combined global analysis. The result was a forest assessment of unprecedented scope and participation.

FRA 2000 emphasized collaboration and transparency. The assessment was based on the assumption that country participation in all phases of the process was the best way to ensure that countries would have a sense of ownership of the data and the results of the assessment, and would thus be inclined to use the data in the development and implementation of policies and programmes to improve the management of their forest resources. National experts reviewed and verified country data. Where countries lacked the capability to carry out their own assessments, training and assistance were provided to build national capacity. Regional workshops were held to improve data quality and to build capacity through South-South cooperation. Leading technical experts were called upon to develop methodologies and to assist with the analysis. Partnerships were formed with leading institutions to take advantage of their comparative advantages. Of particular importance, the United Nations Economic Commission for Europe (UNECE) served as the focal point for information about industrialized countries.

The mandate for FRA 2000 was established by the FAO Committee on Forestry (COFO) and the Intergovernmental Panel on Forests (IPF). At the request of both bodies, efforts were made to broaden the parameters included in the assessment to provide a comprehensive view of a broad range of forest resources.

FRA 2000 thus compiled and analysed a wide range of information about the extent, composition, protection and utilization of forests for each country. Special attention was given to estimating the rate of change of forest resources and to documenting the factors implicated in these changes. The assessment also included an independent pan-tropical remote sensing survey of forest cover change. A set of global maps of forest cover and ecological zones was prepared using the remote sensing data. The world's forests were classified into 20 ecological zones, subsets of the broader tropical, subtropical, temperate and boreal domains.

This publication constitutes the principal report of FRA 2000. The main findings on forest area and area change are presented in Part I, Chapter 1. Part I also presents the results of studies on wood volume and biomass, plantations and other key parameters studied in FRA 2000 including trees outside the forest, biological diversity, forest management, forests in protected areas, forest fires, wood supply and non-wood forest products.

Part II presents findings organized by geographic region and subregion. More detailed data by country are posted in the country profiles on the FAO Forestry Web site: www.fao.org/forestry

Part III describes the methodologies and processes underpinning the assessment. It includes chapters describing the framework for obtaining country information; the methodology used in the pan-tropical remote sensing survey of forest cover change; and the mapping processes used to obtain the global maps of forest cover and ecological zones. Also described is the development of a comprehensive forestry information system (FORIS) which was created to assemble and disseminate the FRA 2000 results. This system is integrated with other FAO databases and is accessible on the Internet.

Part IV summarizes the conclusions of the assessment, reviews the process and presents recommendations for future efforts.

Finally, detailed appendices provide terms and definitions and comprehensive tables of the global statistics presented by country and region. Also included are a listing of other FRA 2000 publications, a summary of earlier forest resources assessments, and a comparison of the results of FRA 1990 and FRA 2000.

Findings for each country are available in country pages on the FAO Forestry Web site, which are updated as new information becomes available. FRA Working Papers documenting the assumptions underlying the assessment and references to the original sources for all material are also available on the Web site.

Acknowledgements

FRA 2000 represented a major effort of the FAO Forestry Department, member countries, donors, partners and individual experts. Countries provided the basic data for the assessment in the form of technical reports and analyses. Many staff members contributed to the interpretation of satellite imagery and the technical work of assessing forest and ecological conditions. Several countries contributed donations to the assessment, including Austria, Denmark, Finland, Italy, Japan, Sweden, Switzerland and the United Kingdom. Valuable in-kind contributions were provided by Sweden, India and the United States. The United Nations Economic Commission for Europe (UNECE) led the assessment component for industrialized countries.

Important partners included the Australian Bureau of Rural Sciences; the Autonomous National University of Mexico; the Bishop Museum; the Brazilian Institute of Environment and Natural Resources; the Canadian Forest Service; CIRAD-Forêt; the Department of Remote Sensing and Landscape Information Systems, University of Freiburg, Germany; the EROS Data Center of the United States Geological Survey; the Faculty of Agricultural and Forest Sciences of the University of Florence, Italy; the Federal Agency for Nature Conservation, Germany; the Federal Research Institute for Forestry and Wood Products, Germany; the Forest National Corporation of the Sudan; the Forest Survey of India; the Foundation for American Friendship; the Institute of Remote Sensing Applications, China; the International Institute of Tropical Forestry; the International Institute for Applied Systems Analysis; the Italian Overseas Institute; Michigan State University, Tropical Rain Forest Information Center; the United States National Aeronautics and Space Administration; the Regional Community Forestry Training Center for Asia and the Pacific; Rutgers, the State University of New Jersey; the Tropical Agriculture Research and Higher Education Center; the Tropical Science Center; the United States Department of Agriculture, Forest Service; the University of Maryland; and the the United Nations Environment Programme World Conservation Monitoring Centre.

FAO is grateful for the support of all countries, organizations and experts inside and outside the organization that made the assessment possible.

Individual contributors to FRA 2000 are listed in Appendix 1. Editing and production of the report were managed by Andrea Perlis, and formatting and layout were done by Linda Ransom.

Contents

Foreword iii
Preface v
Acknowledgements vii
Abbreviations xi
List of tables xv
List of figures xvii
Executive summary xxi

PART I: GLOBAL ISSUES
1. Forest area and area change 1
2. Wood volume and woody biomass 17
3. Forest plantations 23
4. Trees outside the forest 39
5. Biological diversity 45
6. Forest management 51
7. Forests in protected areas 61
8. Fires 65
9. Wood supply 73
10. Non-wood forest products 81

PART II: FOREST RESOURCES BY REGION
11. Africa 101
12. Africa: ecological zones 103
13. North Africa 109
14. West Africa 115
15. Central Africa 121
16. East Africa 127
17. Southern Africa 133
18. Africa – small islands 139
19. Asia 143
20. Asia: ecological zones 145
21. West Asia 157
22. Central Asia 163
23. South Asia 167
24. Southeast Asia 173
25. East Asia 179
26. Europe 185
27. Europe: ecological zones 187
28. Northern Europe 197
29. Central Europe 203
30. Southern Europe 211
31. Belarus, Republic of Moldova, Russian Federation and Ukraine 217
32. North and Central America and the Caribbean 223
33. North and Central America: ecological zones 225
34. North America, excluding Mexico 235
35. Central America and Mexico 243

 36. Caribbean 249
 37. Oceania 255
 38. Oceania: ecological zones 257
 39. Australia and New Zealand 265
 40. Other Oceania 271
 41. South America 279
 42. South America: ecological zones 281
 43. Tropical South America 287
 44. Non-tropical South America 293

PART III: PROCESSES AND METHODOLOGIES
 45. Framework for implementation and country participation 299
 46. Pan-tropical survey of forest cover changes 1980-2000 305
 47. Global mapping 321
 48. Forestry information system development 333

PART IV: CONCLUSIONS AND RECOMMENDATIONS
 49. Conclusions 343
 50. Process review of FRA 2000 349
 51. Recommendations for future assessments 355

APPENDICES
 Appendix 1. FRA 2000 contributors 361
 Appendix 2. Terms and definitions 363
 Appendix 3. Global tables 371
 Appendix 4. FRA 2000 publications 445
 Appendix 5. Authors by chapter 449
 Appendix 6. Earlier global assessments 451

Index of geographic names 457
Index of botanical names 465

Abbreviations

000	thousand
APO	Associate Professional Officer
ATFP	American Tree Farm Program
ATO	African Timber Organization
AVHRR	Advanced Very High Resolution Radiometer
BEF	Biomass Expansion Factor
BRS	Bureau of Rural Sciences
CARPE	Central African Regional Program for the Environment.
CATIE	Tropical Agricultural Research and Higher Education Center
CBD	Convention on Biological Diversity
CCAD	Comisión Centroamericana de Ambiente y Desarrollo
CFS	Canadian Forest Service
CIFOR	Centre for International Forestry Research
CIRAD	Department of Remote Sensing and Landscape Information Systems
CIS	Commonwealth of Independent States
CITES	International Trade in Endangered Species of Wild Fauna and Flora
COFO	Committee on Forestry (FAO)
CPF	Collaborative Partnership on Forests
CSA	Canadian Standards Association
DBH	diameter at breast height
DEM	digital elevation model
DFAs	Regional Initiative for the Development and Implementation of National Level Criteria and Indicators for the Sustainable Management of Dry Forests in Asia
DZAf	Dry-Zone Africa Process on Criteria and Indicators for Sustainable Forest Management
EZ	ecological zone
EDC	EROS Data Center of the United States Geological Survey
ENGREF	École Nationale du Génie Rural des Eaux et des Forêts
ENSO	El Niño – Southern Oscillation
FAO	Food and Agriculture Organization of the United Nations
FFRI	Finnish Forest Research Institute
FLB	Forest Liaison Bureau
FNC	Forest National Corporation
FORIS	Forestry Information System
FRA	Forest Resources Assessment Programme of FAO
FSC	Forest Stewardship Council
GEZ	global ecological zone
GFFA	Global Forest Fire Assessment
GFIS	Global Forest Information Service
GIS	Geographic Information System
GNP	gross national product
GFSM	Global Fibre Supply Model
GLCCD	Global Land Cover Characteristics Database
GT	Green Tag (USA)

IGBP	International Geosphere and Biosphere Programme
ILO	International Labour Organisation
IPGRI	International Plant Genetic Resources Institute
IIASA	International Institute for Applied Systems Analysis
IFF	Intergovernmental Forum on Forests
IPCC	Intergovernmental Panel on Climate Change
IPF	Intergovernmental Panel on Forests
ISO	International Organization for Standardization
ITTO	International Tropical Timber Organization
IUCN	World Conservation Union
LACFC	Latin American and Caribbean Forestry Commission
LEP	Lepaterique Process of Central America on Criteria and Indicators for Sustainable Forest Management
MAI	mean annual volume increment (m^3/ha)
MCPFE	Ministerial Conference for the Protection of Forests in Europe
MON	Montreal Process on Criteria and Indicators for the Conservation and Sustainable Management of Temperate and Boreal Forests
NASA	United States National Aeronautics and Space Administration
NOAA	United States National Oceanic and Atmospheric Administration
NE	Near East Process on Criteria and Indicators for Sustainable Forest Management
NDVI	Normalized Difference Vegetation Index
NGO	non-governmental organization
NWFP	non-wood forest products
PEFC	Pan-European Forest Certification
PRA	participatory rural appraisal
RFC	Regional Forestry Commission (FAO)
RRA	rapid rural appraisal
SFI	Sustainable Forest Initiative Program, American Forest and Paper Association
SFM	sustainable forest management
SMS	Selective Management System
TROPICS	Tree Growth and Permanent Plot Information System
TARA	Tarapoto Proposal of Criteria and Indicators for Sustainability of the Amazon Forest
TBFRA	Temperate and Boreal Forests Resources Assessment 2000
TOF	trees outside the forest
TROPIS	Tree Growth and Permanent Plot Information System
TSA	Tropical South America
UNAM	Autonomous University of Mexico
UNCCD	United Nations Convention to Combat Desertification in those Countries Experiencing Serious Drought and/or Desertification, particularly in Africa
UNCED	United Nations Conference on Environment and Development
USDA	United States Department of Agriculture
UNECE	United Nations Economic Commission for Europe
UNESCO	United Nations Educational, Scientific and Cultural Organization
UNEP	United Nations Environment Programme
UNFCCC	United Nations Framework Convention on Climate Change
UNFF	United Nations Forum on Forests
UNFPA	United Nations Population Fund
USAID	United States Agency for International Development
USGS	United States Geological Survey

WCMC	World Conservation Monitoring Centre
WD	volume-weighted average wood density
WEC	World Energy Council
WFC	World Forestry Congress
WWF	World Wide Fund for Nature

List of tables

Table 1-1	Forest area information availability and quality by region	3
Table 1-2	Forest area by region 2000	5
Table 1-3	Remote sensing survey: estimates of forest area by region and at pan-tropical level in 2000	6
Table 1-4	Remote sensing survey: annual deforestation and net forest area changes during the period 1990-2000 by region and at pan-tropical level	6
Table 1-5	Comparison of forest area and forest area change estimates from the remote sensing survey with country data	8
Table 1-6	Forest area changes 1990-2000 in tropical and non-tropical areas	9
Table 1-7	Expert estimates on distribution of the reported plantation establishment over reforestation and afforestation for major plantation countries	10
Table 1-8	Tropical deforestation studies in science journals, categorized by primary information source: trends over time	14
Table 1-9	Correlation coefficients (r) between forest cover change rate and selected variables at country level	15
Table 2-1	Global carbon stocks in vegetation and top 1 m of soils	18
Table 2-2	Wood density applied for tropical tree species	19
Table 2-3	Forest volume and above-ground biomass by region	19
Table 2-4	Changes in volume and above-ground woody biomass 1990-2000 for the tropics and non-tropics	21
Table 3-1	Annual plantation rates and plantation areas by region and species group	25
Table 3-2	Regional plantation areas by purpose and ownership	26
Table 3-3	Plantation purpose and ownership by reported area for the ten largest plantation development countries	27
Table 3-4	Forest plantation purpose trends by region, 1980-2000	28
Table 3-5	Mean annual increments for selected species used in industrial forest plantations	31
Table 3-6	Characteristics of valuable hardwoods used in tropical areas	34
Table 3-7	Areas and production of non-industrial forest plantations in selected developing countries by region	35
Table 5-1	Data availability by species group	47
Table 6-1	Regional overview of number of countries involved in criteria and indicator processes	55
Table 6-2	Regional overview of size of forest area certified	58
Table 7-1	IUCN categories for protected areas as used in FRA 2000	62
Table 7-2	Forests in protected areas, based on global protected area map developed for FAO by UNEP-WCMC	63
Table 7-3	Forests in protected areas by ecological domain	63
Table 9-1	Correction factors for closed and open/fragmented forest by geographic region and ecological zone	74
Table 9-2	Forest area at varying distances from major transportation infrastructure, and accessibility for wood supply, by region	76
Table 10-1	Main categories of NWFP on which data have been collected	82
Table 10-2	Workshops on NWFP within the FRA 2000 framework	83
Table 10-3	Main NWFP of Africa	84
Table 10-4	Production and exports of gum arabic in Africa	85
Table 10-5	Main NWFP of South America	89
Table 10-6	Main NWFP of the Caribbean	93
Table 10-7	Main NWFP in Europe with major producing countries	94
Table 10-8	Main NWFP of North America	95
Table 11-1	Africa: forest resources by subregion	102
Table 12-1	Africa: extent of ecological zones	104
Table 12-2	Africa: proportion of forest by ecological zone	104
Table 13-1	North Africa: forest resources and management	110
Table 14-1	West Africa: forest resources and management	116
Table 15-1	Central Africa: forest resources and management	122
Table 16-1	East Africa: forest resources and management	128

Table 17-1	Southern Africa: forest resources and management	135
Table 18-1	Small islands: forest resources and management	140
Table 19-1	Asia: forest resources by subregion	144
Table 20-1	Asia: extent of ecological zones	146
Table 20-2	Asia: proportion of forest by ecological zone	146
Table 21-1	West Asia: forest resources and management	158
Table 22-1	Central Asia: forest resources and management	164
Table 23-1	South Asia: forest resources and management	168
Table 24-1	Southeast Asia: forest resources and management	174
Table 25-1	East Asia: forest resources and management	181
Table 26-1	Europe: forest resources by subregion	186
Table 27-1	Europe: extent of ecological zones	189
Table 27-2	Europe: proportion of forest by ecological zone	189
Table 28-1	Northern Europe: forest resources and management	198
Table 29-1	Central Europe: forest resources and management	204
Table 30-1	Southern Europe: forest resources and management	212
Table 31-1	Belarus, Republic of Moldova, Russian Federation and Ukraine: forest resources and management	218
Table 32-1	North and Central America: forest resources by subregion	224
Table 33-1	North and Central America: extent of ecological zones	226
Table 33-2	North and Central America: proportion of forest by ecological zone	226
Table 34-1:	North America excluding Mexico: forest resources and management	236
Table 35-1	Central America and Mexico: forest resources and management	244
Table 36-1	Caribbean: forest resources and management	250
Table 37-1	Oceania: forest resources by subregion	256
Table 38-1	Oceania: extent of ecological zones	258
Table 38-2	Oceania: proportion of forest by ecological zone	258
Table 39-1	Australia and New Zealand: forest resources and management	266
Table 40-1	Other Oceania: forest resources and management	272
Table 41-1	South America: forest resources by subregion	280
Table 42-1	South America: extent of ecological zones	282
Table 42-2	South America: proportion of forest by ecological zone	282
Table 43-1	Tropical South America: forest resources and management	288
Table 44-1	Non-tropical South America: forest resources and management	294
Table 45-1	Trust Funds	304
Table 46-1	Land cover classification used for the survey	307
Table 46-2	Area transition matrixes for the period 1990-2000 at pan-tropical level	311
Table 46-3	Area transition matrixes for the period 1990-2000 by region	311
Table 46-4	Estimates of forest area by region and at pan-tropical level in 2000	313
Table 46-5	Annual deforestation and net forest area change during the period 1990-2000 by region and at pan-tropical level	313
Table 46-6	Annual deforestation and net forest area change during the period 1990-2000 by ecological zone	313
Table 46-7	Comparison of forest area and forest area change estimates from the remote sensing survey with those from country data	317
Table 47-1	FRA 2000 global land cover map legend, definitions and representative land cover types	324
Table 47-2	Ecological zone breakdown used in FRA 2000	326
Table 47-3	Source maps used for the delineation of FAO global ecological zones	328
Table 47-4	Distribution of forests by ecological zone, 2000	329
Table 47-5	International and national data for protected areas	330
Table 48-1	Currently identified categories and subjects in the FAO Forestry country profiles	334
Table 49-1	Forest area changes 1990-2000 in tropical and non-tropical areas	344

List of figures

Figure 1-1	Schematic illustration of the two principal forest edges in the "forest" definition	2
Figure 1-2	Seven basic change processes for forests	3
Figure 1-3	Processes and outputs related to forest area and area change	4
Figure 1-4	Distribution of the world's forests by major ecological zone	5
Figure 1-5	Proportion of forest by country	7
Figure 1-6	Countries and forests with high rates of net forest area change 1990-2000	7
Figure 1-7	Percentage of total area change by individual change processes at regional and pan-tropical level, 1990-2000	8
Figure 1-8	Forest area changes 1990-2000	9
Figure 2-1	Above-ground woody biomass by country	20
Figure 2-2	Distribution of above-ground woody biomass among regions	20
Figure 2-3	Volume and biomass for countries with largest forest area	21
Figure 3-1	Distribution of forest plantation area by region	25
Figure 3-2	Distribution of plantation areas by genus	25
Figure 3-3	Distribution of annual planting area	26
Figure 3-4	Distribution of forest plantations end-use, worldwide	26
Figure 3-5	Ownership of industrial forest plantations, worldwide	26
Figure 3-6	Ownership of non-industrial forest plantations, worldwide	26
Figure 3-7	Leaders in forest plantation development – percentage of area	27
Figure 3-8	End-use of forest plantations, top ten countries	27
Figure 3-9	Ownership, industrial plantations, top ten countries	27
Figure 3-10	Plantation areas by genus, Asia	29
Figure 3-11	Plantation areas by genus, North and Central America	29
Figure 3-12	Plantation areas by genus, Africa	29
Figure 3-13	Plantation areas by genus, Oceania	30
Figure 3-14	Plantation areas by genus, South America	30
Figure 3-15	Predicted contribution of plantation wood to regional wood supply	30
Figure 5-1	Endangered species (all seven species groups) against forest area change for countries with more than 1 million hectares of forest	48
Figure 6-1	Geographical coverage of nine criteria and indicator processes	54
Figure 7-1	Forests in protected areas	62
Figure 7-2	Proportion of forest in protected areas in industrialized countries: comparison of results from UNECE/FAO (2000) and FRA 2000 protected areas map	64
Figure 8-1	Global availability of wildfire data	66
Figure 9-1	Proportion of forests within reach from major transportation infrastructure and its accessibility for wood supply	77
Figure 9-2	Proportion of forest areas available for wood supply within reach of land and complementary water transportation infrastructure in tropical South America	78
Figure 11-1	Africa: subregional division used in this report	101
Figure 12-1	Africa: ecological zones	103
Figure 13-1	North Africa: forest cover map	109
Figure 13-2	North Africa: natural forest and forest plantation areas 2000 and net area changes 1990-2000	111
Figure 14-1	West Africa: forest cover map	115
Figure 14-2	West Africa: natural forest and plantation areas 2000 and net area change 1990-2000	117
Figure 15-1	Central Africa: forest cover map	121
Figure 15-2	Central Africa: natural forest and plantation areas 2000 and net area change 1990-2000	123
Figure 16-1	East Africa: forest cover map	127
Figure 16-2	East Africa: natural forest and plantation areas 2000 and net area change 1990-2000	129
Figure 17-1	Southern Africa: forest cover map	133
Figure 17-2	Southern Africa: natural forest and plantation areas 2000 and net area change 1990-2000	136
Figure 18-1	Small islands: forest cover map	139
Figure 18-2	Natural forest and plantation areas 2000 and net area change 1990-2000	141
Figure 19-1	Asia: subregional division used in this report	143

Figure 20-1	Asia: ecological zones	145
Figure 21-1	West Asia: forest cover map	157
Figure 21-2	West Asia: natural forest and forest plantation areas 2000 and net area changes 1990-2000	159
Figure 22-1	Central Asia: forest cover map	163
Figure 22-2	Central Asia: natural forest and forest plantation areas 2000 and net area changes 1990-2000	166
Figure 23-1	South Asia: forest cover map	167
Figure 23-2	South Asia: natural forest and forest plantation areas 2000 and net area changes 1990-2000	168
Figure 24-1	Southeast Asia: forest cover map	173
Figure 24-2	Southeast Asia: natural forest and forest plantation areas 2000 and net area changes 1990-2000	175
Figure 25-1	East Asia: forest cover map	179
Figure 25-2	East Asia: natural forest and forest plantation areas 2000 and net area changes 1990-2000	180
Figure 26-1	Europe: subregional division used in this report	185
Figure 27-1	Northern, central and southern Europe: ecological zones	187
Figure 27-2	Belarus, Republic of Moldova, Russian Federation and Ukraine: ecological zones	188
Figure 28-1	Northern Europe: forest cover map	197
Figure 28-2	Northern Europe: natural forest and forest plantation areas 2000 and net area changes 1990-2000	199
Figure 29-1	Central Europe: forest cover map	203
Figure 29-2	Central Europe: natural forest and forest plantation areas 2000 and net area changes 1990-2000	205
Figure 30-1	Southern Europe: forest cover map	211
Figure 30-2	Southern Europe: natural forest and forest plantation areas 2000 and net area changes 1990-2000	213
Figure 31-1	Belarus, Republic of Moldova, Russian Federation and Ukraine: forest cover map	217
Figure 31-2	Belarus, Republic of Moldova, Russian Federation and Ukraine: natural forest and forest plantation areas 2000 and net changes 1990-2000	219
Figure 32-1	North and Central America: subregional division used in this report	223
Figure 33-1	North and Central America: ecological zones	225
Figure 34-1	North America excluding Mexico: forest cover map	235
Figure 34-2	North America excluding Mexico: natural forest and forest plantation areas in 2000 and net changes 1990-2000	237
Figure 35-1	Central America and Mexico: forest cover map	243
Figure 35-2	Central America and Mexico: natural forest and forest plantation areas in 2000 and net area changes 1990-2000	245
Figure 36-1	Caribbean subregion: Forest cover map	249
Figure 36-2	Caribbean: natural forest and forest plantation areas 2000 and net area changes 1990-2000	252
Figure 37-1	Oceania: Subregional division used in this report	255
Figure 38-1	Oceania: ecological zones	257
Figure 39-1	Australia and New Zealand: forest cover map	265
Figure 39-2	Australia and New Zealand: natural forest and plantation areas 2000 and net area changes 1990-2000	266
Figure 40-1	Other Oceania: forest cover map	271
Figure 40-2	Other Oceania: natural forest and forest plantation areas 2000 and net area changes 1990-2000	273
Figure 41-1	South America: subregional division used in this report	279
Figure 42-1	South America: ecological zones	281
Figure 43-1	Tropical South America: forest cover map	287
Figure 43-2	South America: natural forest and forest plantation areas 2000 and net area changes 1990-2000	290
Figure 44-1	Non-tropical South America: forest cover map	293
Figure 44-2	Non-tropical South America: natural forest and forest plantation areas 2000 and net area changes 1990-2000	295
Figure 45-1	Process of developing forest cover estimates using country information	301
Figure 46-1	Remote sensing survey processes	306

List of figures

Figure 46-2   Distribution of sampling units in the pan-tropical remote sensing survey 307
Figure 46-3   Temporal distribution of satellite images used for the survey 309
Figure 46-4   Illustration of standardization to reference years 309
Figure 46-5   Results for a sampling unit in Zimbabwe: raster maps based on dot-grid registrations 312
Figure 46-6   Summary of net changes during the period 1990-2000 by land cover classes by region 313
Figure 46-7   Distribution of the forest by ecological zone in 2000 (f3 definition) 314
Figure 46-8   Net forest area changes by region and at pan-tropical level, 1980-1990 and 1990-2000; annual deforestation rate by region and at pan-tropical level, 1980-1990 and 1990-2000 314
Figure 46-9   Net forest area change by ecological zone, 1980-1990 and 1990-2000; annual forest area change by ecological zone, 1980-1990 and 1990-2000 315
Figure 46-10 Percentage of total area change by individual change processes at regional and pan-tropical level for the period 1990-2000 316
Figure 46-11 Forest area in 2000 (left) and net forest area change (right) - comparison between country data and remote sensing survey estimates 316
Figure 47-1   FRA 2000 global forest cover map 322
Figure 47-2   FRA 2000 global ecological zone map 323
Figure 48-1   FAO Forestry homepage, from which FORIS contents are reached 333
Figure 48-2   Navigation to FAO Forestry country profiles 334
Figure 48-3   Example of an FAO Forestry country profile: summary page for Angola 335
Figure 48-4   Example of footer that appears on all FORIS country profile pages; indicating the data owner and providing a link for giving feedback to this person 336
Figure 48-5   Example of user input screen to FORIS, using a standard Web browser 336
Figure 48-6   Main process for assembly of country information and subprocess for producing estimates and outputs 338
Figure 48-7   Example (Mozambique) of extrapolation of forest area to 2000 based on two area states with reference years 1972 and 1991 339
Figure 51-1   Forestry knowledge management at the local, national or international level 355

Executive summary

The Global Forest Resources Assessment 2000 (FRA 2000) provides a comprehensive and up-to-date view of the world's forest resources at the end of the second millennium. It is the result of the collective efforts of the countries of the world. This major undertaking was based primarily on information provided by the countries, supplemented by state-of-the-art technology to verify and analyse the information and to make the results accessible to the world through the Internet.

The FRA 2000 process emphasized collaboration and transparency. Special efforts were made to transfer technology to, and increase the capability of, countries that lack adequate capacity to assess their own forest resources. Extensive consultations were carried out with experts and partnerships were forged with leading institutions from both developing and industrialized countries.

The information and knowledge provided by countries constitutes the backbone of FRA 2000. Of the 213 countries and areas represented in the assessment, 160 participated actively in the information gathering and analysis. Countries that participated fully in the assessment are perhaps best able to appreciate its importance in supporting the development of policies and programmes aimed at the management, conservation and sustainable development of their forest resources.

PROCESSES
FRA 2000 used the following approaches:
- full participation by countries in the collection, analysis and validation of data;
- capacity building at the national, subregional and regional levels;
- a remote sensing survey of forest resources in tropical countries;
- global mapping of forest cover and ecological zones;
- development of an integrated forest information system;
- participation of internationally recognized experts in all phases;
- partnerships with leading international institutions;
- comprehensive worldwide information dissemination using print media and the Internet (www.fao.org/forestry).

PARAMETERS
Based on guidance from the FAO Committee on Forestry (COFO), the Intergovernmental Panel on Forests (IPF) and the Expert Consultation on FRA 2000 (Kotka III), the following parameters were included in the assessment:
- forest area and change in forest area;
- wood volume and above-ground woody biomass;
- forest plantations;
- trees outside the forest;
- biological diversity;
- areas under forest management;
- area of forests in protected areas;
- number and extent of forest fires;
- wood supply and removals;
- non-wood forest products;
- classification and mapping of forest by ecological zones.

The most comprehensive data possible were collected at the country level and summarized by subregion, by region and globally. In some instances, data were not available for all countries. Complementary detailed information on most of the FRA 2000 subjects can be found on the Internet at www.fao.org/forestry.

FINDINGS
In the main report of FRA 2000 the key findings are presented under the following headings:
- Global perspectives;
- Forest resources by region;
- Processes and methodologies;
- Conclusions and recommendations;
- Global tables;
- Global maps.

Forest area
For the definition of forest, FRA 2000 adopted a threshold of 10 percent minimum crown cover. The definition includes both natural forests and forest plantations. It excludes stands of trees established primarily for agricultural production (e.g. fruit tree plantations).

Based on the consensus recommendation of the Intergovernmental Panel on Forests (IPF) in 1997, this same definition was used for all countries in FRA 2000. In FRA 1980 and FRA 1990, the 10 percent threshold was used for

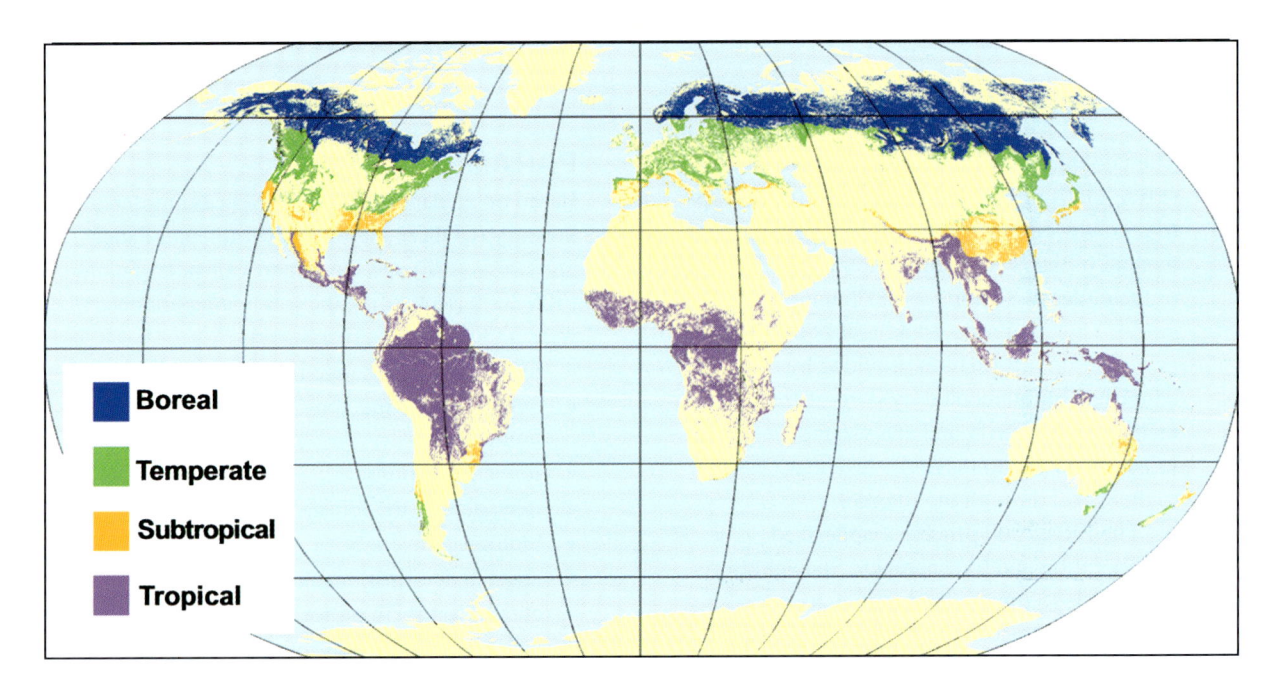

Distribution of the world's forests by major ecological zone

developing countries, but for the industrialized countries a threshold of 20 percent was used.

Using the FRA 2000 global definition of forests and new baseline information, it was estimated that the world's forest cover at the year 2000 was about 3.9 billion hectares, or approximately 0.6 ha per capita. About 95 percent of the forest cover was in natural forest and 5 percent in forest plantations. Using a combination of new global maps and statistical data, FRA 2000 also estimated the distribution of forest area by ecological zones: 47 percent is in the tropics, 33 percent in the boreal zone, 11 percent in temperate areas and 9 percent in the subtropics.

The uniform application of one forest definition had a significant impact on the global findings for the year 2000. The estimated forest area was 400 million hectares greater than the corresponding global figure reported for 1995; the change in definition particularly influenced the forest area estimates for Australia and the Russian Federation, where large areas of forest have between 10 and 20 percent canopy cover.

Another factor leading to the upward revision of forest cover since FRA 1990 was improved information from more recent national inventories which generated higher area estimates for forests in some countries. In other cases, more detailed breakdown of forest classes in the inventory reports facilitated an improved classification of national results into FRA 2000 global standards.

For comparison with the results of the 1990 assessment, the 1990 area was adjusted to the 2000 definition. Details will be presented in a forthcoming FRA Working Paper.

Changes in forest area 1990-2000

The major components of forest area change are categorized as deforestation, afforestation and natural expansion of forests into previously non-forested areas.

Deforestation is the conversion of forest to another land use or the long-term reduction of the tree canopy cover below the minimum 10 percent threshold.

Afforestation is the establishment of forest plantations in areas not previously in forest, and denotes a change from non-forest to forest. It differs from reforestation, which is the establishment of forests (through planting, seeding or other means) after a temporary loss of the forest cover. Areas under reforestation are classified as forest since the forest is actively regenerating.

Annual gross and net changes in forest area, 1990 to 2000 (million hectares per year)

Domain	Deforestation	Increase in forest area	Net change in forest area
Tropics	-14.2	+1.9	-12.3
Non-tropics	-0.4	+3.3	+2.9
World	-14.6	+5.2	-9.4

Executive summary

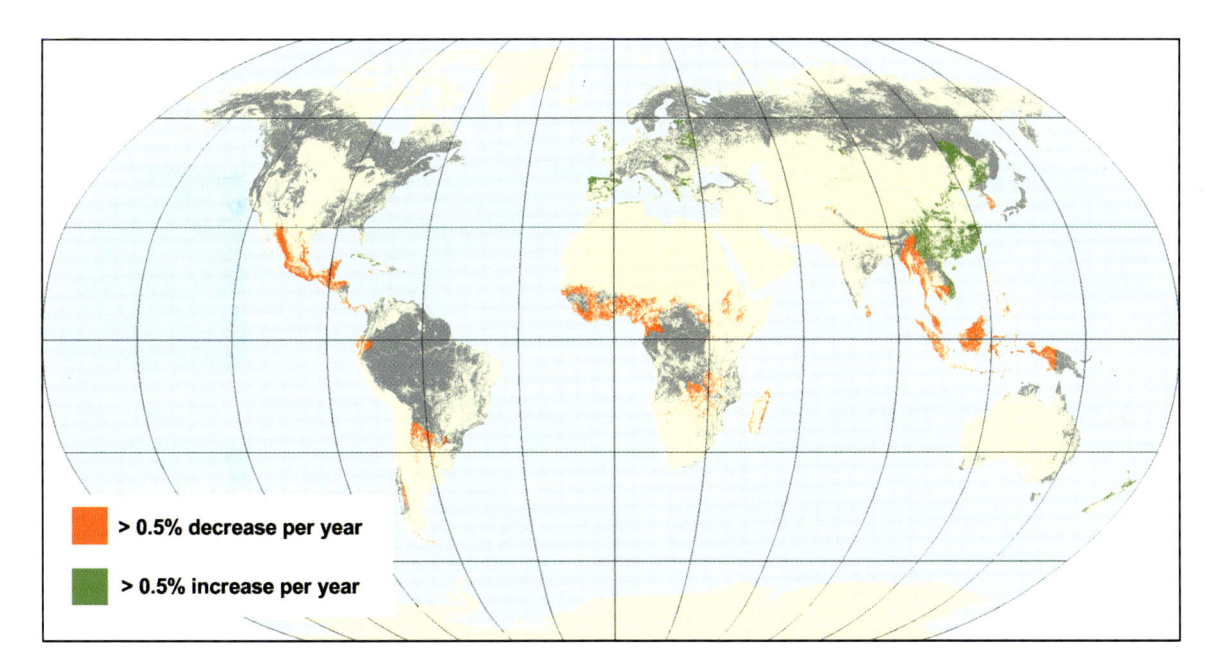

Countries and forests with high rates of net forest area change 1990-2000

Natural expansion of forests refers to the expansion of forest through natural succession on to previously non-forested lands, usually abandoned farmland.

After adjustment of the 1990 forest area to the same definition and baseline information used for FRA 2000, the net global change in total forests was calculated as the sum of deforestation (a negative change) and the gain in forest cover due to the establishment of forest plantations (afforestation) and natural expansion of forests on previously unforested lands.

Deforestation in the 1990s was estimated at 14.6 million hectares per year. The figure represents the balance of annual losses of natural forests (estimated at 16.1 million hectares per year or 0.42 percent per year) minus the area of natural forest that was replaced through reforestation with forest plantations (1.5 million hectares per year), since plantations are considered as a type of forest.

Expressed in another way, during the 1990s the world lost 4.2 percent of its natural forests, but it gained 1.8 percent through reforestation (with plantations), afforestation, and the natural expansion of forests, resulting in a net reduction of 2.4 percent over the ten-year period.

The worldwide gain in forest cover totalled 5.2 million hectares per year, the aggregate of afforestation (1.6 million hectares per year) and natural expansion of forests (3.6 million hectares per year).

Thus the net global change in forest area between 1990 and 2000 was estimated as -9.4 million hectares per year: the sum of -14.6 million hectares of deforestation and 5.2 million hectares of gain in forest cover. The global change (-0.22 percent per year) represents an area about the size of Portugal. The estimated net loss of forests for the 1990s as a whole was 94 million hectares – an area larger than Venezuela.

In addition to the analysis of statistical data from countries, which provided the core information, FRA 2000 included a pan-tropical remote sensing based statistical survey which covered 87 percent of the forests in tropical developing countries. This study provided the first consistent methodology for assessing forest change between two assessment periods. The remote sensing survey revealed that the deforestation process in the tropics is dominated by direct conversions of forest to agriculture. Statistical results from the study showed a slight decrease in the rate of forest loss, from 9.2 million hectares per year in the 1980s to 8.6 million hectares per year in the 1990s. However, this difference fell within the margin of error for the estimates.

Statistics from the country studies showed a similar pattern to those of the remote sensing survey, with slight reductions in overall net forest loss between the 1980s and 1990s. Overall, however, the loss of natural forests is still high in the tropics, and increases in plantation establishment and the natural expansion of forests have not been compensating for the losses incurred.

Worldwide changes in forests – gains and losses (million hectares per year), 1990-2000

Domain	Natural forest					Forest plantations			Total forest
	Losses			Gains	Net change	Gains		Net change	Net change
	Deforestation (to other land use)	Conversion to forest plantations	Total loss	Natural expansion		Conversion from natural forest (reforestation)	Afforestation		
Tropical	-14.2	-1	-15.2	+1	-14.2	+1	+0.9	+1.9	-12.3
Non-tropical	-0.4	-0.5	-0.9	+2.6	+1.7	+0.5	+0.7	+1.2	+2.9
Global	-14.6	-1.5	-16.1	+3.6	-12.5	+1.5	+1.6	+3.1	-9.4

Wood volume and biomass

Wood volume and woody biomass levels are important indicators of the potential of forests to provide wood and to sequester carbon. Total standing wood volume (m^3) and above-ground woody biomass (tonnes) in forests were estimated for 166 countries, representing 99 percent of the world's forest area. The world total standing volume in the year 2000 was 386 billion cubic metres of wood. The global total above-ground woody biomass was 422 billion tonnes, of which more than one-third was located in South America (with about 27 percent in Brazil alone). The worldwide average above-ground woody biomass in forests was 109 tonnes per hectare. South America had the highest average biomass per hectare at 128 tonnes. Countries with the greatest standing volume per hectare include many in Central America (such as Guatemala with 355 m^3 per hectare) and Central Europe (such as Austria with 286 m^3 per hectare), the former having high-volume tropical rain forests and the latter having temperate forests that have been managed to achieve high stocking levels.

Forest plantations

Forest plantations are defined as "forest stands established by planting and/or seeding in the process of afforestation or reforestation...". Because of their increasing significance as a supply of fibre for wood industries, rubber (*Hevea* spp.) plantations were included as forest plantations for the first time. Despite the high losses of the world's natural forests at the global level, new forest plantation areas are being established at the reported rate of 4.5 million hectares per year, with Asia and South America accounting for more new plantations than the other regions. About 70 percent of new plantations, or 3.1 million hectares per year, are considered to be successfully established. Of the estimated 187 million hectares of plantations worldwide, Asia had by far the largest area, accounting for 62 percent of the world total. In terms of composition, *Pinus* (20 percent) and *Eucalyptus* (10 percent) remain the dominant genera worldwide, although the diversity of species planted was found to be increasing. Industrial plantations (producing wood or fibre for supply to wood processing industries) accounted for 48 percent of the global forest plantation estate and non-industrial plantations (e.g. for provision of fuelwood or soil and water protection) for 26 percent. The purpose of the remaining 26 percent was unspecified.

The extent of plantations in industrialized countries was less clear than in developing countries. Many industrialized countries make no distinction between planted and natural forests in their inventories.

FRA 2000 identified the ten countries with the largest plantation development programmes (as reported by percentage of the global plantation area) as China, 24 percent; India, 18 percent; the Russian Federation, 9 percent; the United States, 9 percent; Japan, 6 percent; Indonesia, 5 percent; Brazil, 3 percent; Thailand, 3 percent; Ukraine, 2 percent and the Islamic Republic of Iran, 1 percent. These countries account for 80 percent of the global forest plantation area.

Trees outside the forest

FRA 2000 was the first of FAO's global assessments that attempted to consider trees outside the forest (TOF) – defined as trees on land not classified as forest or other wooded land. Despite the fact that TOF often play an important role in the livelihoods of the rural population, especially of women, they are often overlooked, both in forest resource assessments and in policy and decision-making processes. The consequent scarcity of information made it impossible to draw conclusions on the resource. Complicating the collection of data was the fact that neither traditional forest inventories nor modern remote sensing technology are very useful for conducting a quantitative assessment of TOF. Most of the information on trees

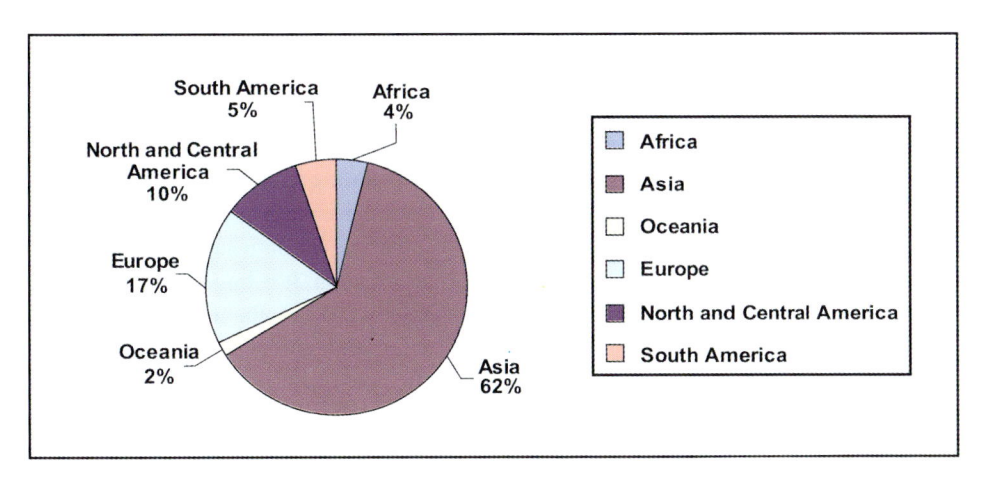

Distribution of forest plantations by region

outside the forest is site specific and scattered among different institutions and sectors, including informal sectors. The major contributions of FRA 2000 to expanding knowledge of this resource are case studies and reviews of methodologies that will be useful in future assessments, which will help to raise the awareness of the significance of TOF, especially to the lives of the rural population.

Biological diversity

FRA 2000 provides information with relevance for a number of indicators of forest biological diversity, principally new maps and detailed descriptions of forest ecological zones that are more comprehensive than those of any previous assessment. New maps of forest cover provide updated knowledge about forest fragmentation and related indicators of forest health and diversity. In addition, studies on endangered forest species and on effects on spatial attributes of forests which may influence biological diversity were carried out in the context of FRA 2000.

Forest management

Initiatives to promote sustainable forest management have stimulated many countries to implement forest management plans. FRA 2000 did not undertake a comprehensive assessment of all indicators of forest management, since most countries have only recently started to assess and monitor criteria and indicators for sustainable forest management. It would be advantageous for future global assessments to include more indicators. However, FRA 2000 did ask countries to report on forest areas under management plans. At least 123 million hectares of tropical forests are now reportedly subject to management plans, as are 89 percent of the forests in industrialized countries. However, monitoring is needed to assess implementation of these plans.

Protected forests

At the global level, 12.4 percent of the world's forests were estimated to be in protected areas according to the categories defined by the World Conservation Union (IUCN). This estimate was obtained by overlaying the new FRA 2000 forest cover map and a map of protected areas prepared for FAO by the UNEP World Conservation Monitoring Centre (UNEP-WCMC). The statistics for area of forest under protection obtained through this method were different from, and generally lower than, the areas reported by countries. Clarifying definitions and improving methods for data capture would help future efforts in this area.

Forest fires

FRA 2000 undertook a comprehensive study of forest fires during the 1990s. While statistics were available for fewer than 50 countries (none in Africa) a number of qualitative assessments were carried out on a national basis and published on the FAO Web site. In those countries where long-term data are available, the evidence indicates an increase in wildfires in the 1990-2000 period compared with most of the previous decades in the second half of the twentieth century, although available records and qualitative assessments show that the 1980-1990 period may have been equally severe. The climate phenomenon known as El Niño was implicated as a major contributing factor to the severe forest fires in the 1990s (as well as the 1980s). El Niño provoked severe droughts in generally humid or temperate areas, enhancing the potential for devastating fires.

Fire continued to be used as a major tool for land clearing and as a management tool for pasture and browse improvement in a number of developing countries. These uses need to be considered in statistics related to forest wildfires.

Wood supply

Using a combination of global databases, statistical information and GIS technology, it was estimated that 51 percent of the world's forests are within 10 km of major transportation infrastructure and potentially accessible for wood supply. This proportion increased to 75 percent for forests within 40 km from transportation infrastructure. The highest accessibility was found in subtropical forests (73 percent within 10 km of transportation infrastructure) and the lowest accessibility was found in boreal forests (34 percent within 10 km of transport).

Information on wood removals and harvesting was analysed for all major industrialized countries. Because very few tropical countries reported this information, a special study was carried out for 43 tropical countries which account for approximately 90 percent of the world's tropical forest resources. It was found that timber harvesting occurred at a wide range of intensities, between about 1 and 34 m^3 per hectare per year. There was very little evidence of implementation of low-impact logging or other model harvesting practices in the tropics.

Non-wood forest products

In many countries, especially the world's poorest countries, non-wood forest products (NWFP) are a critical component of food security and an important source of income. FRA 2000 represents the most comprehensive assessment of NWFP to date. Data were collected at the national level and validated through a series of subregional workshops. Historically, Asia is the only region where much information has been collected and reflected in national accounts, mainly because of the relatively high level of use of NWFP throughout the region. In Asia NWFP have long been an important part of national and local economies.

STRENGTHS AND WEAKNESSES OF THE FRA 2000 APPROACH

Reliance on country information

One of the greatest strengths of FRA 2000 was its reliance on the participation of individual countries, for both supply and analysis of information. It is hoped that this approach will greatly increase the likelihood that the countries will use the information to make and implement effective forest policies, and that demand for forest-related information will lead to further capacity building. While countries firmly support this approach, it has sometimes been criticized on the grounds that country information may be inaccurate or biased. FAO has addressed such concerns related to information quality by the use where possible of primary technical documents as sources of statistical information for the assessment, rather than quoted, subjective or secondary sources. Unfortunately, many countries still lack reliable primary technical information at the national level. This is a potential problem, but it is believed that the strengths of country involvement greatly outweigh the disadvantages. The goal of future assessments will be to further strengthen country capabilities and participation. In this way, FAO intends to improve the information quality as well as to assist developing countries in their inventories.

Remote sensing data compared with national inventories

The potential of remote sensing data to contribute to assessments of changes in forest cover over large land areas was demonstrated by the FRA 2000 pan-tropical remote sensing survey and the global maps. More intensive coverage would have been better than the 10 percent sample used for the pan-tropical survey component of FRA 2000, but resources were lacking to carry out a more intensive survey. In addition, there are limits to the potential of remote sensing for assessing key parameters other than forest area change, and full access to remote sensing technology is out of reach for many developing countries. FAO plans to continue to use country information combined with remote sensing in future assessments, but also to emphasize field observations as a means of gathering broad and representative information.

Change in definitions

As requested by the IPF, FRA 2000 used a new definition of forest which resulted in an upward revision of global forest cover compared with recent assessments. However, the continued use of different definitions in the developing and industrialized countries would have perpetuated the incompatibility in the two sets of estimates. The previously published FRA 1980 and 1990 figures cannot be directly compared to FRA 2000 results. However, the data from the earlier assessments were adjusted to make it possible to estimate area changes between 1990 and 2000. In addition, the remote sensing survey does give compatible change information for the tropics for the periods 1980-1990 and 1990-2000.

Inclusion of forest plantations in forest area

FRA 2000 has included plantations in the statistical estimates for forest area. This is not intended to imply that plantations are equivalent to natural forests. Great care has been taken to keep the statistics for natural and planted forests separate so that readers can draw the conclusions they feel are relevant for their needs.

FUTURE DIRECTION

FRA 2000 aimed to expand the scope of global forest assessments to include new parameters in order to shed light on environmental and social services of forests in addition to traditional measures of forest cover and timber volume. Progress was made in the assessment of a number of parameters, such as biomass, availability for wood production and non-wood forest products; but it was not possible to meet all of the demands. Part IV of the report discusses areas where potential improvements might be made in future assessments. Among the most important are the following.

- Further emphasis should be placed on capacity building to improve national forest assessments. The best way to improve global assessments is to improve national assessments.
- Collaboration with key partners should be increased to make a better use of scarce resources.
- Country participation and the role of the FAO Regional Forestry Commissions should be increased in planning and implementing all phases of future assessments, from data collection through analysis.
- FAO will seek to continue to identify new parameters whose assessment may contribute to sustainable forest management and to develop practical ways for assessing them.
- FAO should work towards reducing the interval between successive assessments, or towards establishing rolling regional assessments.
- Continued improvement is needed in developing standardized definitions and categories for collecting information, which will help to broaden the capacity of forest resources assessment to respond to various needs and uses.
- Continued work is needed to improve methodologies and techniques for data collection and analysis, particularly for systematic field observations of broad sets of forest parameters complemented by remote sensing technology. Only through first-hand and representative observations can a solid basis for forest policy processes be established.
- An expert consultation in March 2000 recommended the establishment of a standing global team of experts who will serve in an advisory capacity on key policy and process issues related to global assessments.

For the future, any individual, organization or country that develops more reliable or current information is encouraged to contribute it as soon as it is available so that it can be used to strengthen the next global assessment.

Part I: Global issues

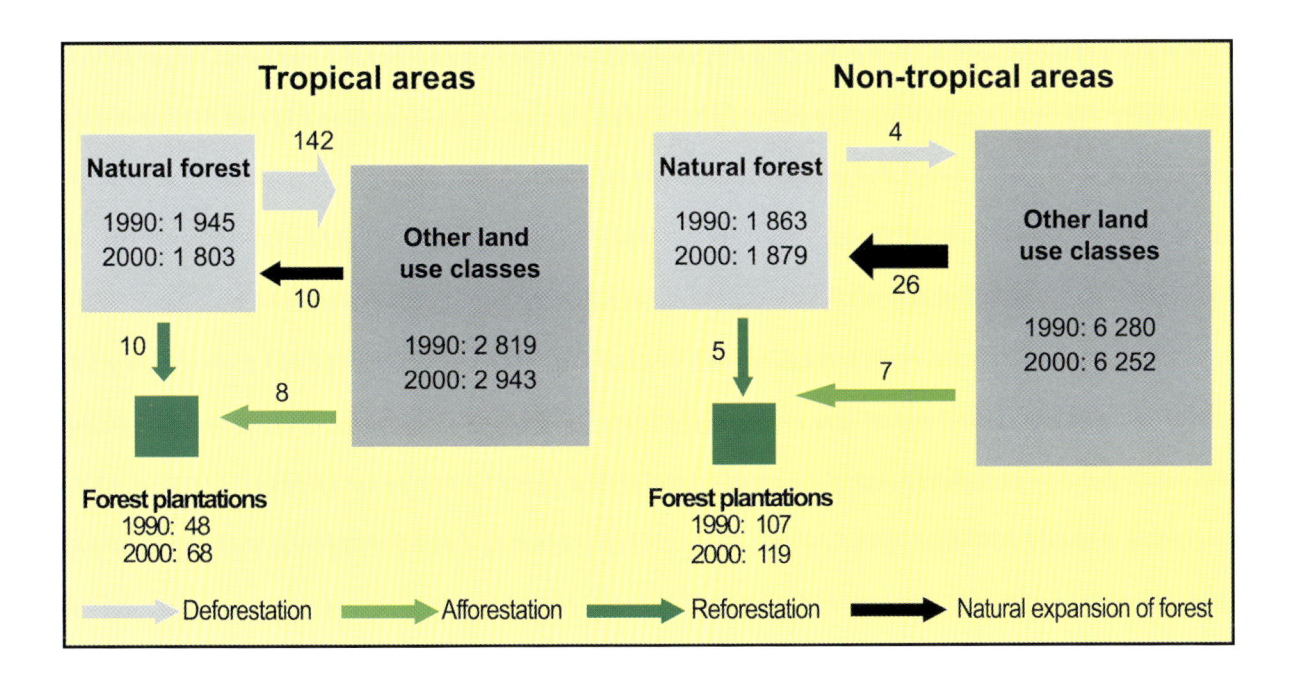

Forest area changes 1990-2000 (million hectares)

Chapter 1

Forest area and area change

ABSTRACT

Forest area and area change was a major theme in FRA 2000. Estimates were based on a comprehensive analysis of the latest forest inventory data available for each country. To support the country findings a systematic pan-tropical remote sensing survey provided estimates at the regional and pan-tropical levels. A detailed study of forest plantations was included in the analysis. A number of qualitative studies were carried out to enrich the knowledge on forest area change. This was the most comprehensive survey of forest area and area change at the global level to date, revealing extensive and detailed findings, but also considerable information gaps, particularly in Africa. The findings indicate that the world's forests covered 3 869 million hectares in 2000, about 30 percent of the world's land area. The net change in forest area was -9.4 million hectares per year, representing the difference between a deforestation rate of 14.6 million hectares per year of natural forests and an expansion of 5.2 million hectares per year of natural forests and forest plantations. In addition 1.5 million hectares per year of natural forests were converted to forest plantations. Most of the forest losses were in the tropics. The rate of net change was slightly lower in the 1990s compared to the 1980s, due to a higher estimated rate of forest expansion in the 1990s. A survey of scientific literature showed that the subject is rich in publications, but also that conclusions are not representative of all forests. Country studies point at land use rights as a common main determinant behind land use change (deforestation), and the remote sensing survey indicated that direct conversions of forests to permanent agriculture were more prominent than shifting agriculture in forest change processes.

INTRODUCTION

The core of FRA 2000 is the estimate of forest area and changes in forest area over time. The work has led to new knowledge about the dynamics of the world's forests. Forest area is an easily understood baseline parameter that provides the first indication of the relative importance of forests in a country or region. Estimates of forest area changes may also provide a clue to the demand for land for other uses and environmental pressures on forest ecosystems.

The main limitation of the emphasis on forest area is that this is not necessarily a good qualitative indicator of the health of a forest ecosystem. For example, environmental values such as critical areas for biological diversity may be concentrated in small and scattered areas or where forests are interwoven with other land uses. Such values may not be well assessed through spatial area-based classifications of disturbance. Social values may be derived from complex interactions and synergies between agriculture and forestry at the local level, and they may not be highly correlated to the absolute extent of the forest. Economic values are more dependent on variables such as productivity, volume, species composition, accessibility, demand, non-wood forest products and regulations than on the overall forest area and area change.

Nevertheless, forest area and area change remains a basic theme for global forest resources assessments. FRA 2000 has made considerable efforts to record and explain the extent of forests and the area transition to and from forests. An effort has been made to show clearly where the source data are weak and where the transformation into global classes was particularly difficult. In addition to the area estimates, qualitative studies of the change processes have been made through literature reviews.

This chapter synthesizes work that is described in more detail in other chapters and in the tables in the Appendix; a large number of FRA Working Papers; and conclusions reached at numerous workshops and meetings during the FRA 2000 process.

TERMS AND DEFINITIONS

Global assessments of forest resources coordinated by FAO have a long history. Over time, basic definitions have been developed that are generally accepted by participating countries and are well known to experts of forest

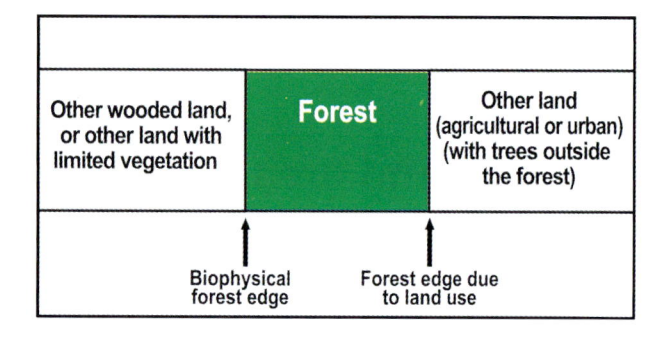

Figure 1-1. Schematic illustration of the two principal forest edges in the "forest" definition

inventories and assessments. It is nonetheless important to continue to develop and refine a core set of terms in light of changing technology and information requirements of countries, which are the primary users of the assessments. Common terms are needed to produce consistent and comparable estimates for each country. On the other hand, there is no single definition that fits well everywhere, owing to the greatly varying conditions throughout the world. Definitions of forests and forestry must extend over and take into consideration not only the climatic range from boreal to tropical ecological zones, but also economic variations from countries where recreational values rank among the highest priorities, to locations where collection of fuelwood and non-wood forest products is part of daily life. Global definitions by necessity involve compromises.

International terms and definitions are not static, but follow the general development of international processes. For example, the importance of forests as carbon sinks was not widely discussed several decades ago, yet this issue is now at the top of the international political agenda. New terms are introduced as subjects enter into the international debate and old terms may need to be modified to better serve the current requirements for information. While this is a desirable evolution, it is also important to keep definitions consistent over time. Forests change relatively slowly, and it is necessary to compare estimates several decades apart to establish reliable trends. For this purpose, FRA 2000 has tried to maintain a globally homogeneous set of definitions that allows comparisons with earlier global forest resources assessments.

The most important exception to this rule was a change in the definition of "forest" for industrialized countries in order to adopt a single common definition for all countries. In FRA 1990, a 20 percent canopy cover threshold was used in countries with temperate and boreal forests for which data were collected by ECE/FAO. In contrast, 10 percent canopy cover was used as the minimum definition of forest in tropical countries. This distinction dated back many years, based on differences in forest inventories in different regions. During discussions of forest resource assessments in the Intergovernmental Panel on Forests (IPF), consensus emerged to use 10 percent minimum canopy cover as the common definition for forests in all countries in FRA 2000. As a result, FRA 2000 provided globally homogeneous data on forest cover. However, comparisons with earlier assessments required considerable work and extrapolation of previous data, particularly in dry subtropical zones and boreal zones where the extent of sparsely stocked forests with between 10 and 20 percent canopy cover is considerable.

The definitions used in FRA 2000 are found in Appendix 2. It is important to note that "forest" is defined both by the presence of trees and by the absence of other land uses regardless of the legal status of the land. In other words, "forest" is a combination of a land use classification and a land cover classification. There is some disagreement about this approach in the scientific community, but a majority of experts on forest assessments agree that such an approach is necessary if the FRA 2000 results are to be of optimal use to policy-makers. One practical outcome of this approach is that interpreters of remote sensing images must have knowledge of the situation on the ground; it is possible to interpret land cover from space, but it is not so simple to identify land use.

In estimating forest change, there are two principal kinds of forest edges. The first is the edge between a forest and areas where the climatic conditions are too harsh to support forest vegetation, such as the northern edge of a boreal forest or the edge between a forest and a desert. The second is the edge between a forest and areas where other land uses are practised, including agriculture, urban land use or infrastructure (Figure 1-1). The first edge is strictly derived from biophysical properties, whereas the second is subject to other considerations.

One question is, what degree of other land uses can be allowed without disqualifying the "forest" classification? Obviously some grazing can occur in what is called a forest, as well as collection of non-wood forest products. However, when other land uses dominate, the land use classification would not be forest. This raises questions about which products and purposes should be included in the "forest" classification. Stands of rubber trees and oil palms are included, whereas fruit orchards and agroforestry areas are not. National park areas are included, whereas urban parks are not. It is important to document how the definitions were applied in each case to enable future comparisons.

In addition to land use classification, area change processes were central to the assessment. These change processes were defined by seven terms for which the distinctions between terms were clear and the set of terms as a whole covered all possible changes. The seven identified change processes (Figure 1-2) can be grouped into land use changes (deforestation, afforestation, expansion of natural forests) and internal changes within the forest class (reforestation, regeneration of natural forests, degradation, improvement).

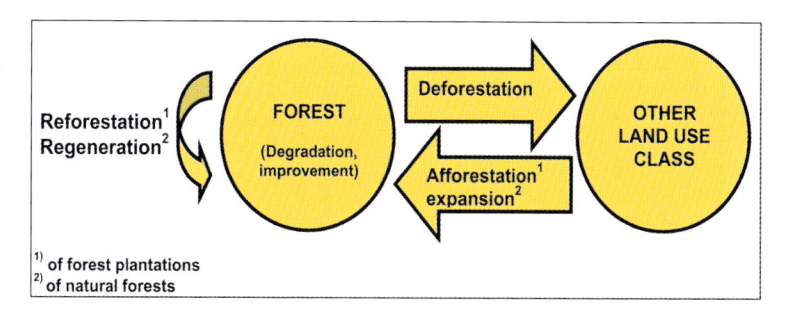

Figure 1-2. Seven basic change processes for forests

METHODS

Several different approaches were used to assess the extent of forest and its development over time. In Figure 1-3 the main processes and outputs are shown, together with a note on where further information can be found. For area statistics, FRA 2000 generated information at three scales – country (based on surveys of national inventory and mapping reports), region (FRA 2000 remote sensing survey) and world (FRA 2000 global mapping). For the estimates of area and area change, only country- and regional-level information was used, as the global forest map did not provide sufficient precision. The global-level information was used to derive relational data such as the distribution of forests by ecological zones.

In addition to the area statistics, narrative descriptions on woody vegetation and forest plantations were developed for many countries to accompany and enrich the area statistics. Descriptions of forest resources – including forest area and area change – are elaborated for all countries and presented for each subregion in Part II of this report. Finally, qualitative studies and literature reviews were carried out to deepen the knowledge on factors underlying forest changes.

FRA 2000 considered all available documents that contained primary country-level information on forest area, forest area change and forest plantations. Requests for information to all countries were followed up by in-country assignments in most developing countries, workshops and meetings involving more than 100 countries and a final validation of results by country correspondents.

In many countries, primary information on forest area was not available or was not reliable. Other countries lacked a time series of forest area information. In these instances, FRA 2000 had to rely on secondary information and/or expert estimates. Table 1-1 summarizes the information

Table 1-1. Forest area information availability and quality by region

Region	Reference year for latest available area data (area weighted)	Expert estimate	General mapping	Detailed mapping	Field survey	Yes	No	High	Medium	Low	References reviewed (No.)
Africa	1991	24	6	5	10	35	21	11	12	13	547
Asia	1995	14	6	9	1	28	14	11	14	3	284
Europe	1997				3	38		32	6		44
North and Central America	1995	15	2	11		22	9	21	1		304
Oceania	1992	12		5		6	13	5	1		85
South America	1991	4		10		11	3	8	2	1	280
World	**1994**					**140**	**60**	**88**	**36**	**17**	**1 544**

Source: Appendix 3, Table 2.
Note: Source data for industrialized countries were not classified. Number of reviewed references are sums from Table 2 in Appendix 3 and some references were therefore double counted. For industrialized countries, the reference UNECE/FAO (2000) was used.

availability for forest area. The world average reference year for source data is 1994, with considerably older dates in some developing countries. A high proportion of developing countries had to rely on expert opinion for the latest area estimates. Furthermore, fewer than half of all countries have time series information with high compatibility between the observations.

The FRA 2000 pan-tropical survey of forest cover changes was used to complement and validate the results of the country surveys for tropical countries. Building on the methodology and experiences from FRA 1990, the survey covered a representative sample of tropical forests over the period 1980 to 2000. The survey is described in detail in Chapter 46. Two types of qualitative studies of forest area change were carried out. First, a review was made of all available documentation within the country, including grey literature, combined with interviews of key people who are responsible for national inventories and assessments. These surveys were documented in detailed country reports and annotated bibliographies on forest change processes. Second, an exhaustive survey was carried out of scientific literature (peer-reviewed papers published in journals of science) covering aspects of tropical deforestation. The

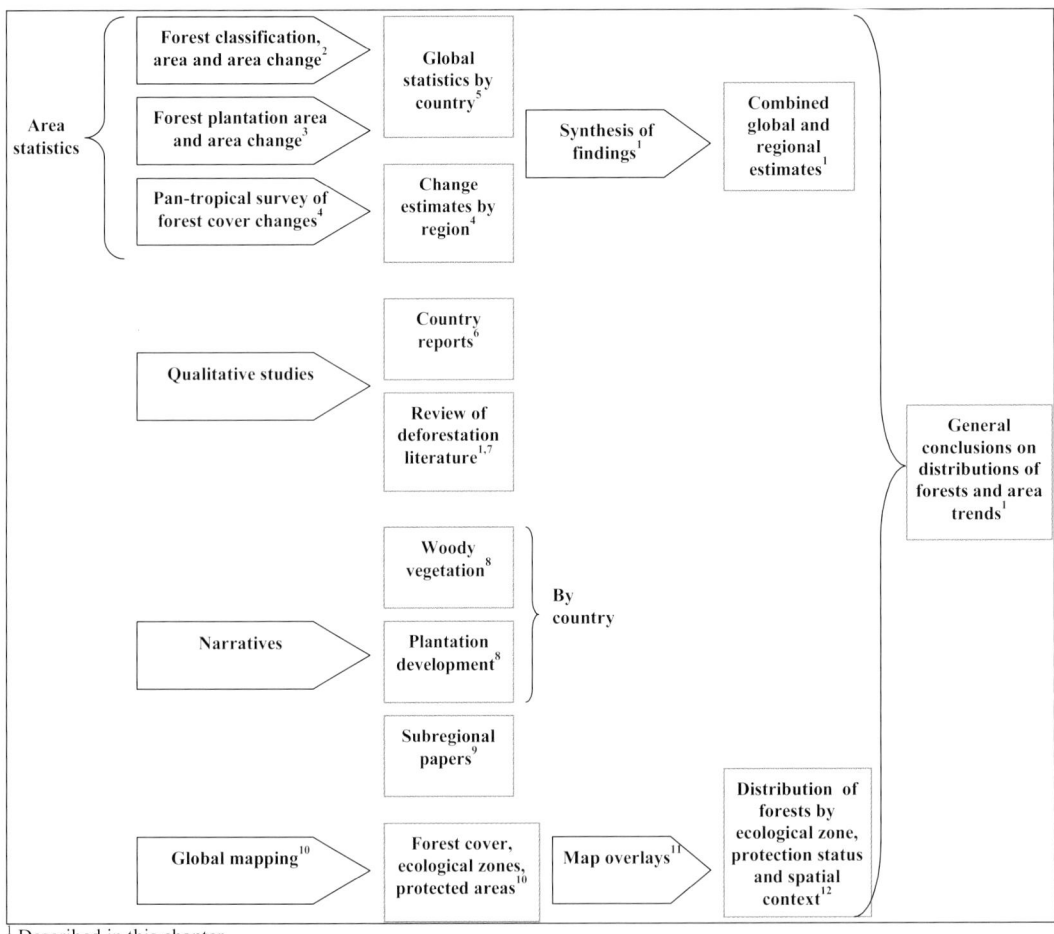

[1] Described in this chapter
[2] See Chapter 48
[3] See Chapter 3
[4] See Chapter 46
[5] See Appendix 3, Table 2-6
[6] See Appendix 3, Table 16
[7] See FRA Working Paper No. 27 (www.fao.org/forestry/fo/fra/index.jsp)
[8] See country narratives in the FAO Forestry Web site country profiles (www.fao.org/forestry/fo/country/nav_world.jsp)
[9] See Part II of this report
[10] See Chapter 47
[11] See Chapter 7 and Chapter 10
[12] See Appendix 3, Table 14-15 as well as tables and graphs in subregional chapters in Part II of this report

Figure 1-3. Processes and outputs related to forest area and area change

Table 1-2. Forest area by region 2000

Region	Land area	Total forest (natural forests and forest plantations)				Natural forest	Forest plantation
	million ha	Million ha	% of land area	% of all forests	Net change 1990-2000 million ha/year	million ha	million ha
Africa	2 978	650	22	17	-5.3	642	8
Asia	3 085	548	18	14	-0.4	432	116
Europe	2 260	1 039	46	27	0.9	1 007	32
North and Central America	2 137	549	26	14	-0.6	532	18
Oceania	849	198	23	5	-0.4	194	3
South America	1 755	886	51	23	-3.7	875	10
WORLD TOTAL	**13 064**	**3 869**	**30**	**100**	**-9.4**	**3 682**	**187**

Note: Changes are the sums of reported changes by country.
Source: Appendix 3, Tables 3, 4 and 6.

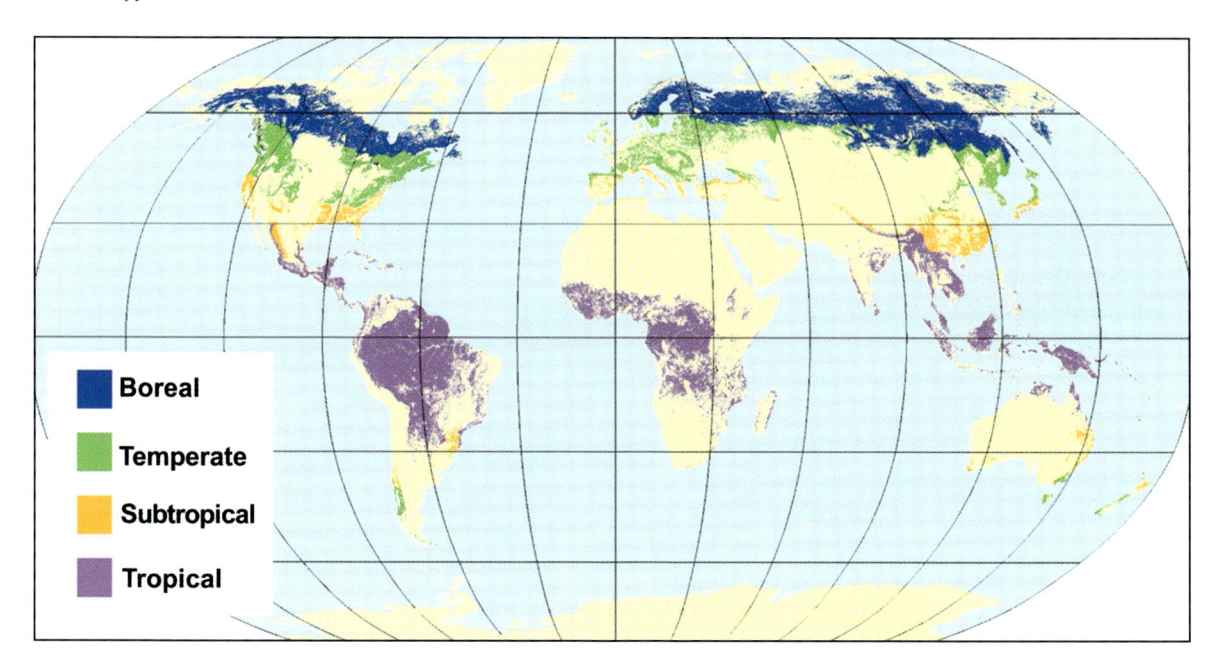

Figure 1-4. Distribution of the world's forests by major ecological zone

survey was carried out by Rutgers University (United States) and recorded the geographic extent, methodology applied and conclusions drawn on factors behind deforestation for each reference, followed by an analysis of spatial and temporal patterns.

Global maps of forest cover, ecological zones and protected areas were developed by using low-resolution satellite imagery and geographic information system (GIS) technology (see Chapter 47 and Figure 1-4). This technology was not used to estimate forest area or area change owing to limitations in the imagery. However, by overlaying country boundaries, country estimates were made for the proportional distribution of forests by ecological zone; protected forest areas; and the proportion of forests available for wood supply.

RESULTS

The tables in Appendix 3 display statistics for 213 countries and areas. In addition, the FAO Forestry Web site includes comprehensive country profiles where the results, methods and background material are presented in detail for each country (FAO 2001a). The global tables in Appendix 3 may also be downloaded from the Web site. Table 1-2 shows the distribution of forests by region and the estimated annual net change. Figure 1-5 shows the proportion of forest by country, and Figure 1-6 indicates where the net change rates are greatest.

Tables 3 and 4 in Appendix 3 display the estimates of forest area in 2000 and estimates of the change in forest area 1990-2000. Table 5 in Appendix 3 shows the distribution over global land use classes at the latest reference year. The

FAO estimates are based on national reports and extrapolated to 2000. Countries validated the FAO estimates.

The FRA 2000 area and area change estimates were independent estimates; they were not based on results of earlier global assessments and were not based on models. Rather, they were estimates for each individual country based on the best available country data.

More than 600 forest types occurring in the national reports were transformed into global classes. In most cases, this reclassification was straightforward. Some classes, particularly in dry ecological zones, were difficult to transform as the results are sensitive to the method of analysis applied. Australia and Angola are examples where the forest edge towards drier woodlands was difficult to determine from national classifications. The reclassifications from national classes are documented in the FAO Forestry country profiles (FAO 2001a).

Forest plantation area (Table 1-2) was estimated separately. As plantations constitute only five percent of total forest, they are often misrepresented in national-level mapping surveys. More detailed reports on the plantation estate were analysed to provide a better picture of the resource. The separate results for plantations were incorporated in the overall national forest area estimates. One effect of this approach was that for some countries there was a discrepency between the national forest area data in which plantations were sometimes included as one mapping class, and the separate plantation area estimate (see Appendix 3, Table 6 for more detailed FRA 2000 results).

Table 1-3 shows the forest area estimates for tropical forests based on the pan-tropical remote sensing survey, including estimated standard errors (since resource limitations did not allow a 100 percent sample). Table 1-4 indicates the estimated deforestation and net change rates for the studied regions, with accompanying standard errors. It should be noted that no statistically significant difference between the periods 1980-1990 and 1990-2000 was observed for any region.

Table 1-5 compares the remote sensing survey findings with the country data obtained from national reports. There was a statistically significant difference in the estimates of forest area for each region, but this was a consistent discrepancy as the remote sensing survey showed a lower estimate for all regions. More interestingly, the change estimates from the remote sensing survey and the country data

Table 1-3. Remote sensing survey: estimates of forest area by region and at pan-tropical level in 2000

Region	Forest area			
	Million ha		%	
	Estimate	SE	Estimate	SE
Africa	519	37	42	3
Latin America	780	49	63	4
Asia	272	23	45	4
Pan-tropical	**1 571**	**66**	**51**	**2**

Note: SE = Standard error of the mean. that the figures are related to the surveyed area, representing about 90 percent of the total forest land in the pan-tropical region. These estimates refer to the most inclusive definition of forest (f3) as defined in Chapter 46.

Table 1-4. Remote sensing survey: annual deforestation and net forest area changes during the period 1990-2000 by region and at pan-tropical level

Region	Annual deforestation *million ha/year*	Annual net forest area change *million ha/year*		Annual rate of net forest area change *%/year*	
	Estimate	Estimate	SE	Estimate	SE
Africa	-2.3	-2.1	0.4	-0.34	0.06
Asia	-2.5	-2.3	0.6	-0.79	0.20
Latin America	-4.4	-4.2	1.1	-0.51	0.15
Pan-tropical	**-9.2**	**-8.6**	**1.3**	**-0.52**	**0.08**

Note: SE = Standard error of the mean.

correspond well for Latin America and Asia, whereas for Africa the difference is very large. The likely reason was poor inventory information for many African countries and, as a consequence, an apparent exaggeration of deforestation for a few countries (for example, the Sudan and Zambia).

Figure 1-7 shows the contribution of different change processes to the overall change in forest area. Direct conversion of forests to permanent agriculture or other land uses was much more prevalent than gradual intensification of shifting agriculture. Large-scale conversions dominate in Latin America, whereas direct conversion of forests into small-scale agriculture dominates in Africa. In Asia, intensified shifting agriculture practices accounted for a larger share of the overall changes, including migration into new areas as well as gradual change of existing areas towards more permanent agriculture.

The estimates of change in forest area are of potential significance to forest policy-makers, so they are described in more detail here. An increase or decrease in total forest area does not necessarily correspond to qualitative changes of the forest. FRA 2000 therefore attempted to identify the type of forest change – what is the new land use that replaced former forest land? Where there was an increase in forest land, what was the previous land use?

Forest area and area change 7

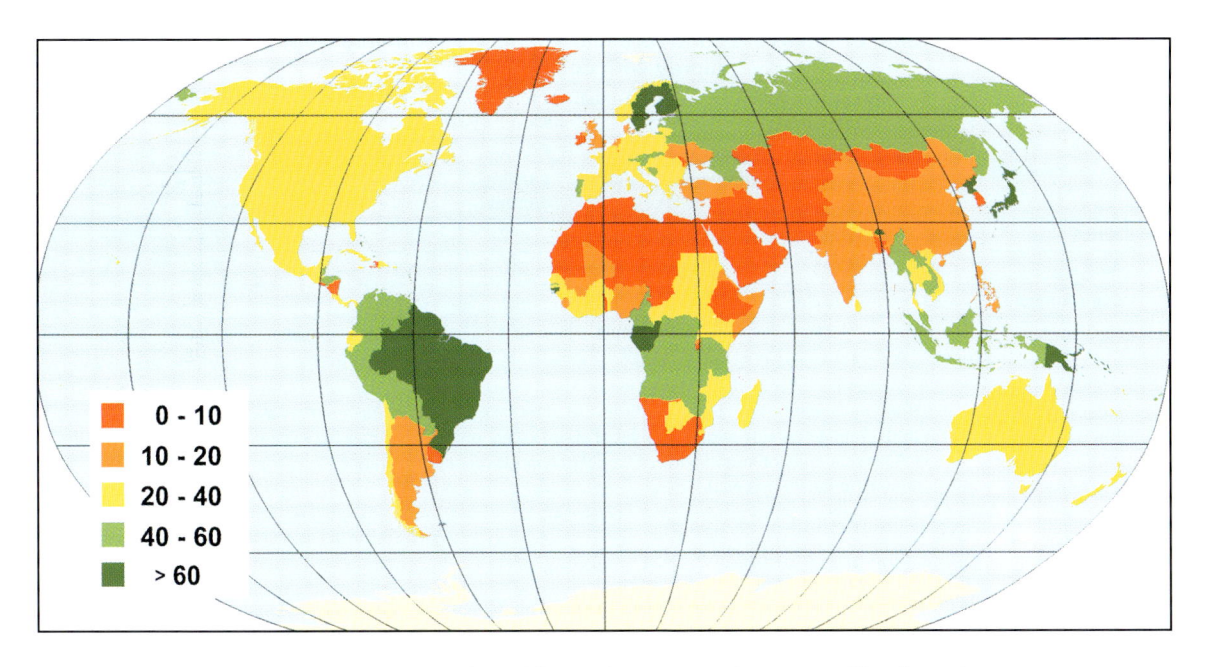

Figure 1-5. Proportion of forest by country (percent of land area)

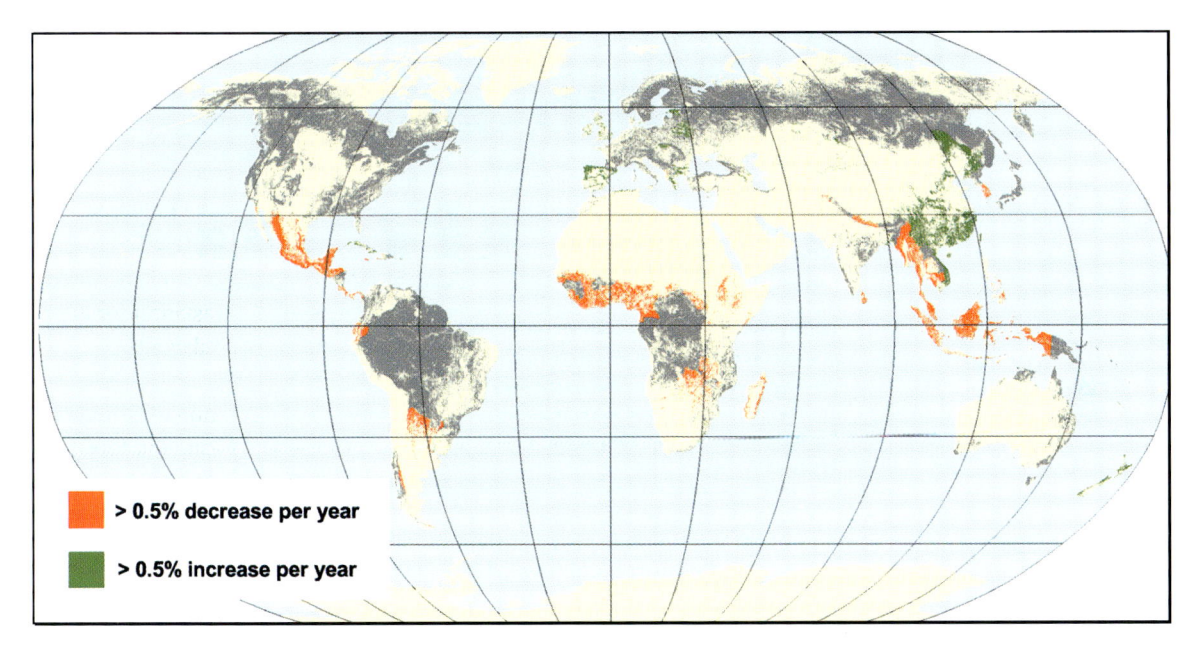

Figure 1-6. Countries and forests with high rates of net forest area change 1990-2000

One important qualitative change is the conversion of natural forest to forest plantations. This change may have implications for biological diversity as well as the future productivity and use of the forest. Providing only one reference figure (e.g. either the overall net forest area change rate or the deforestation rate) would only give a partial picture of the forest area dynamics. It is therefore important to account for each of the change processes separately, to the extent possible.

Estimates were made for the conversion of natural forests to plantations for the major domains (tropics and non-tropics). Other qualitative changes (reforestation of forest plantations, regeneration of natural forests, forest degradation and forest improvement) are very important for forest policy development and forestry planning; however, the available statistics are not comprehensive for enough countries to make definitive estimates. The review of available studies and literature showed many statements that forests are being degraded; however, it was not possible to make objective estimates of the extent or severity of these changes for most countries because of data limitations. Qualitative changes are reported in FRA 2000 country reports and briefs, but it was not possible to derive globally valid statistics.

Table 1-5. Comparison of forest area and forest area change estimates from the remote sensing survey with country data

Region	Forest area 2000 million ha			Annual net forest area change million ha/year			Annual forest area change rate %/year		
	Country data	Remote sensing survey	Significant difference	Country data	Remote-sensing survey	Significant difference	Country data	Remote-sensing survey	Significant difference
Africa	622	484	**	-5.2	-2.2	**	-0.77	-0.43	**
Asia	289	224	**	-2.4	-2.0	n.s.	-0.78	-0.84	n.s.
Latin America	892	767	*	-4.4	-4.1	n.s.	-0.45	-0.51	n.s.
Pan-tropical	1 803	1 475	***	-12.0	-8.3	**	-0.62	-0.54	n.s.

Note: Only the results from the countries included in the remote sensing survey were compiled to obtain the country data given in the table. The remote sensing estimates refer to the f2 definition of forest (see Chapter 47), that which most closely corresponds to the definition used in compiling the country data. The hypothesis tested in the table is that the country data value is the true value of the sampled population of the remote sensing survey. Level of significance of the difference between country data and remote sensing estimates: *** = 99.9 percent level of significance, ** = 99 percent level of significance, * = 95 percent level of significance, n.s. = not significant at the 95 percent level.

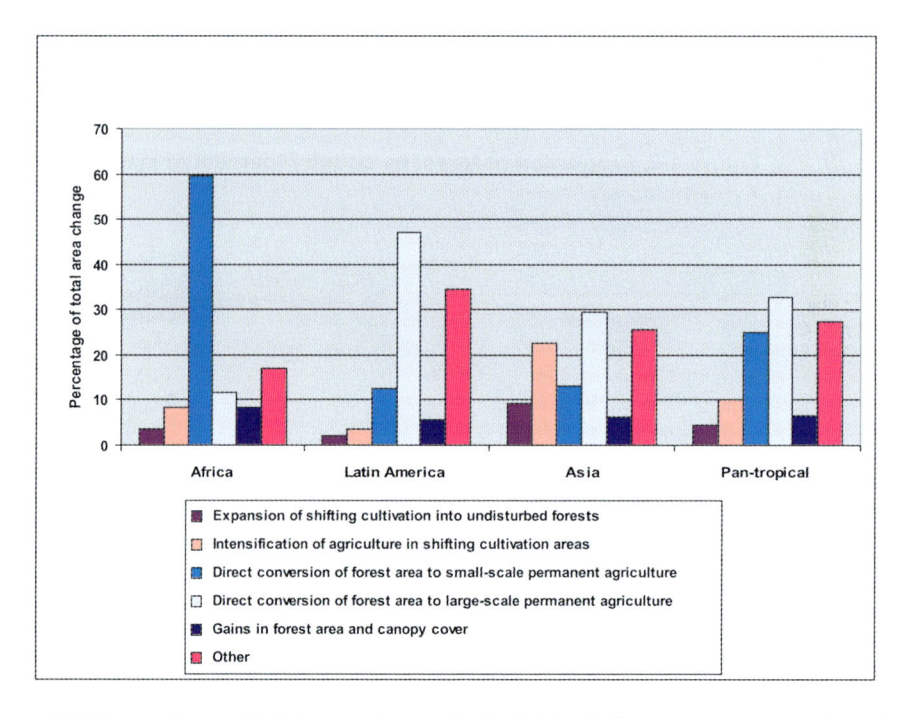

Figure 1-7. Percentage of total area change by individual change processes at regional and pan-tropical level, 1990-2000

The FRA 2000 estimates of forest change therefore focused on the transformation to and from natural forests and forest plantations. Three separate studies were aggregated:
- area statistics of total forest area and area changes for each country;
- area statistics on forest plantation area and area changes for each country;
- results of the FRA 2000 remote sensing survey of area changes in the tropics.

These studies brought different strengths and weaknesses to synthetic analysis. Taking them together, and building on the strengths of each study, FAO developed reliable estimates for the tropics and non-tropics (Table 1-6, Figure 1-8). The steps and assumptions in this analysis are described below.

The starting point and base statistics were the estimated annual net change of forest area 1990-2000 for each country. For convenience, entire countries were assumed to lie either inside or outside the tropical domain. Tropical countries included the following subregions, as defined in this report: West, Central, East and Southern Africa (except South Africa), Central America and Mexico, Tropical South America, the Caribbean, South and Southeast Asia, and Other Oceania. The sum of net changes in the tropical and non-tropical domains appears in the last column in Table 1-6. Before this column was added to Table 1-6, however, two systematic errors in the material were considered.

First, when the results of the FRA 2000 remote sensing survey were compared with

Table 1-6. Forest area changes 1990-2000 in tropical and non-tropical areas
(million hectares per year)

Domain	Natural forest					Forest plantations				Total forest
	Losses			Gains	Net change	Gains		Net change	Net change
	Deforestation (to other land use)	Conversion to forest plantations	Total loss	Natural expansion		Conversion from natural forest (reforestation)	Afforestation		
Tropical	-14.2	-1.0	-15.2	+1	-14.2	+1	+0.9	+1.9	-12.3
Non-tropical	-0.4	-0.5	-0.9	+2.6	+1.7	+0.5	+0.7	+1.2	+2.9
Global	-14.6	-1.5	-16.1	+3.6	-12.5	+1.5	+1.6	+3.1	-9.4

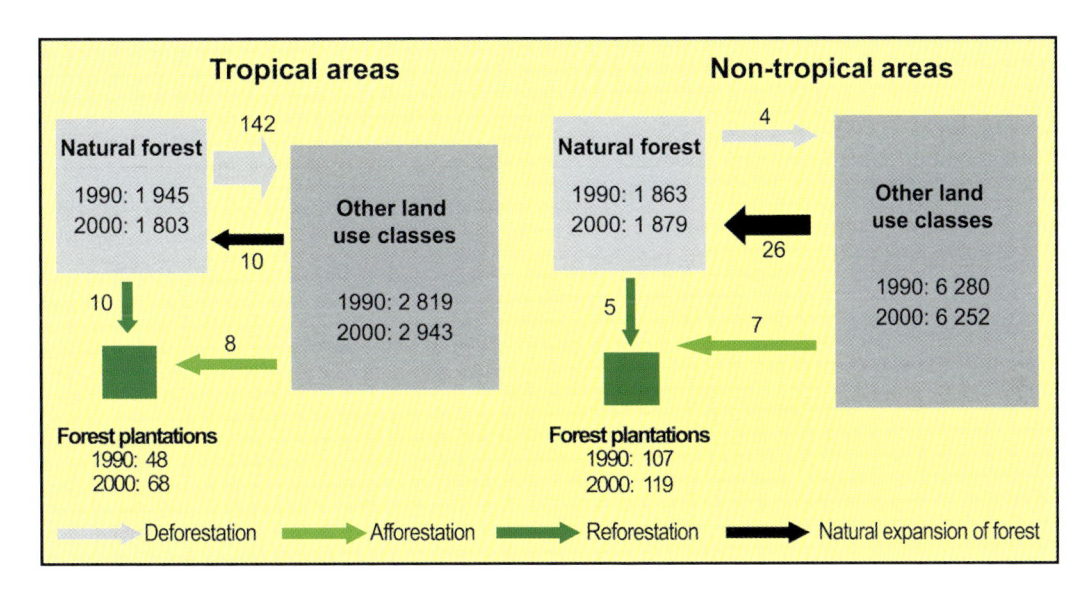

Figure 1-8. Forest area changes 1990-2000 (million hectares)

country data, a relatively good correlation was found for Asia and Latin America; but the negative area change for Africa appeared to be considerably overestimated in the national reports, possibly by as much as 3 million hectares annually for tropical Africa (-5.2 million hectares per year in the national reports and -2.2 million hectares per year in the remote sensing survey; see Table 1-5). It was known that country data were very weak for most African countries, and in some cases the reported change rate seemed very high (e.g. for the Sudan and Zambia). The discrepancy in area change estimates for Africa was secured at 99 percent confidence level, so it was necessary to make adjustments. Bearing in mind that large parts of Africa have dry forest types for which changes are not easily detected in satellite images, it is reasonable not to adopt the remote sensing survey results uncritically. It was assumed that the remote sensing survey and national report estimates were equally reliable for the region as a whole. Thus, the average of the remote sensing survey estimate and the national report estimate of forest area change to constitute a valid estimate for tropical Africa as a whole. This estimate of annual negative change for tropical Africa is about 1.5 million hectares lower than that of the national reports.

Second, the plantation establishment was considered to be exaggerated in the national reports in relation to the actual success rate for the 1990s as a whole. Based on many field studies, it was assumed that only 70 percent of the reported plantation establishment was successful. Overall, the successful plantation expansion rate was 1.4 million hectares less than what the country reports suggested. This adjustment of plantation estimates is consistent with past FAO analyses used since 1995 (FAO 1995).

It is noted that the above two calibrations were of roughly the same magnitude (-1.5 and -1.4) and as they were in different directions, the combined effect on the totals was minimal. For simplicity, it was therefore assumed that they were equal and that the net change rates in Appendix 3 could be adopted directly.

New forest plantations are established either on non-forest land (afforestation) or on land

Table 1-7. Expert estimates on distribution of the reported plantation establishment over reforestation and afforestation for major plantation countries

Country	Domain	Reforestation as % of reported plantation establishment	Afforestation as % of reported plantation establishment
Argentina	Non-tropical	50	50
Brazil	Tropical	75	25
China	Non-tropical	40	60
India	Tropical	50	50
Indonesia	Tropical	90	10
Thailand	Tropical	25	75
United States	Non-tropical	0	100

where they replace natural forest (reforestation). As these represent different change processes, it was important to quantify the proportion of each, for the tropical and non-tropical domains as a whole. This was done by expert opinion on the proportion of afforestation and reforestation for seven countries reporting major plantation establishments (80 percent of the world total) (Table 1-7). These proportions were considered valid throughout the respective domains and were extrapolated to the entire plantation establishment area. It was further assumed that no significant plantation areas were lost to other land uses or to natural forests. For tropical countries, the afforestation rate was estimated at 0.9 million hectares per year and the reforestation rate at 1 million hectares per year; for non-tropical countries the results were 0.7 and 0.5 million hectares per year respectively, (see Gains under Forest plantations in Table 1-6).

Based on the above estimates for plantation change rates, the net changes of natural forest area were calculated in Table 1-6 as a net loss of 14.2 million hectares per year for the tropics and a net gain of 1.7 million hectares per year for the non-tropics.

The next step in the analysis was to distinguish between positive and negative changes within the natural forest. This was done in two different ways. For the non-tropics, the expansion of natural forests was calculated as the sum of all positive changes at the country level, an annual increase of 2.6 million hectares per year. This can be considered a very conservative estimate, as there may be local changes which are not reflected in country totals.

For the tropics, it was less relevant to use the country statistics, as the majority of countries had large negative net changes which would effectively disguise the expansion of natural forests. Instead, results from the remote sensing survey were used. The remote sensing survey indicated a total expansion of forests of 0.56 million hectares per year. It could be expected that the method underestimates expansion of forests, as this is a slow process which may be difficult to capture in remote sensing interpretation. Furthermore, the remote sensing survey had a minimum interpretation unit of 25 ha, which means that expansion of small-scale forest formations (0.5 to 25 ha) was not accounted for. In addition, a large proportion of land with (currently) low proportion of forest cover was not included in the sampled population. Finally, some tropical countries, notably Zimbabwe and Madagascar, were not included in the survey. These factors all suggested that the expansion in the tropics was higher than the estimated 0.56 million hectares per year. It was thus assumed that tropical natural forests expanded by about 1 million hectares per year (notwithstanding the substantial losses in other areas which are accounted for in the next paragraph).

Given the expansion rates for natural forests, and given that the conversion from natural forests into plantations was already known from the plantation analysis, the deforestation of natural forest could be calculated by subtracting these changes from the net change of natural forests. This gave an estimate of the global deforestation rate of about 14.6 million hectares per year, of which 14.2 million hectares per year occurred in the tropics.

These results (Table 1-6, Figure 1-8) represent the overall conclusions by FRA 2000 with respect to the change in forest area. These estimates were based on thorough analysis of three independent and original data sets (country data, forest plantation data and remote sensing survey data). An error estimate for the combined results is not possible to obtain, as the errors for national estimates are generally not known. However, building on the results of the remote sensing survey (about 15 percent sampling error on tropical forest area change estimates) and the study on reliability made by UNECE/FAO (2000), it can be concluded that the estimates have a high precision. More importantly, combining the three studies made it possible to eliminate some systematic errors in the material.

COMPARISON WITH EARLIER GLOBAL ASSESSMENTS

FRA 2000 was the first global forest assessment to use a common definition for all forests worldwide. Previous assessments used a minimum canopy cover threshold of 10 percent for developing countries and 20 percent for industrialized countries to define forests, based in part on past forest inventory practices in the two domains. When the results of FRA 1990 were reviewed, a number of experts suggested that the next global assessment should use a common forest definition for all regions. Following a consensus recommendation of the IPF, it was decided that FRA 2000 would use the 10 percent minimum canopy threshold for all countries. (That is, when observed from above, at least 10 percent of the land area is covered by forest canopy. "Other wooded land" has a canopy cover between 5 and 10 percent).

As mentioned above, the FRA 2000 area and change estimates were not based on results of earlier assessments. The data for state of forest resources from FRA 1990 were reviewed within the framework of the present assessment for purposes of comparison with the year 2000 data. To ensure comparability, the original 1990 data were adjusted, taking into consideration the following:

- availability of new national forest inventory data which improved the estimates for 1990;
- adjustment of existing 1990 data to the FRA 2000 definitions, and improved reclassification of national vegetation categories in accordance with these definitions;
- adjustment based on other new reliable data and information which had not been available in 1990;
- redefined political country boundaries.

The change in definition for non-tropical forests was the major reason that estimated global forest area for 2000 is 400 million hectares higher than the interim estimate for 1995 which used the FRA 1990 definition (FAO 1997). The effect was most significant for Australia and the Russian Federation. The estimate for Australia's forest area in 2000 was 155 million hectares, compared with 41 million hectares in 1995, in part because the 2000 estimate included large expanses of sparsely stocked forests with canopy cover between 10 and 20 percent that previously had been classified as other wooded land. For similar reasons, the estimate for the Russian Federation is 850 million hectares in 2000, compared with 764 million hectares in 1995.

Forest inventories conducted after 1990 resulted in different figures for a number of countries (including 47 developing countries) than were previously reported, and the inclusion of these results has also contributed to the higher estimate for 2000. In other countries (including 19 developing countries) a more detailed breakdown of forest classes in national inventory reports facilitated an improved reclassification of national results into FRA 2000 forest classes; the new estimates include as forest some areas previously classified as other wooded land. Further details are documented in a FRA Working Paper (in preparation) which reports on the results of an analysis of forest cover change estimates made in FRA 1990 (changes 1980-1990) and in FRA 2000 (changes 1990-2000).

It was difficult to create time series based directly on the forest area estimates in the different assessments because of variations in definitions and information quality as explained in the previous paragraphs. However, it was possible to compare the area change estimates for the 1980s and 1990s with due consideration to the effect of variations between assessment methodologies. The comparison showed that estimated net loss of forest (i.e. the balance of the loss of natural forest and the gain in forest area through afforestation and natural expansion of forest) was lower in the 1990s than in the 1980s. The net annual change in forest area was reported to be -9.4 million hectares for the 1990-2000 period (this report), -11.3 million hectares in the 1990-1995 period (FAO 1997) and -13.0 million hectares in 1980-1990 (FAO 1995b).

There is higher confidence in the 1990-2000 change estimates than in the earlier estimates. Nonetheless, if the effects of differences in definitions, methodologies and updating of national forest inventories are taken into consideration, some general conclusions can be made regarding deforestation over the past 20 years.

The change of forest definition for industrialized countries, while notably increasing global estimates of forest cover, did not greatly affect the estimated rate of change of global forest area. The change in definition had the greatest impact on the forest area of Australia and the Russian Federation, where conversions of forest to other land uses were relatively small on a global scale and thus did not significantly alter worldwide change rates. For most other industrialized countries, the revised 1990 national forest area figures (based on FRA 2000 definitions, methodologies and new data) showed a high degree of consistency and comparability with the 1990 figures of the previous two assessments. The three assessments used essentially the same definition for natural forest for developing countries. Although new estimates at the national level were not always comparable with earlier assessments, they did not significantly affect the estimates of global change rates. The new definition for plantations (which allowed the inclusion of rubber tree plantations) affected the forest area figure for a few tropical countries, but without significant effect on the world forest area change rate. It should also be noted that the three assessments used the same methodology to assess forest area change in the industrialized countries.

The findings of the FRA 2000 pan-tropical remote sensing survey supported the results of the country-based assessment. The survey indicated a net rate of change for tropical forests that was slightly lower in the 1990s than in the 1980s, but the difference was not statistically significant. The survey's findings on forest cover change in the 1980s and 1990s, which are completely compatible with one another, confirm a continued high rate of forest loss in the tropics during the 1990s. This result fits well with the results of the country assessment, as net gains in forest area are reported for the non-tropical countries as a whole while net losses are occurring in the tropics.

In conclusion, after analysis of the estimates of present and previous assessments, FRA 2000 pointed to a lower rate of net loss of forests worldwide in the 1990s than in the 1980s, owing mainly to a higher rate of natural expansion of forest area. At the same time, the worldwide loss of natural forests has continued at roughly comparable high levels over the past 20 years.

ON THE RELIABILITY OF THE FOREST AREA ESTIMATES

The nature of FRA 2000 statistics made it difficult to calculate confidence intervals for most estimates, with the exception of the remote sensing survey. At the country level very few countries, including developed countries, can derive statistically controlled confidence intervals for both forest area and forest area change. For some countries, the results were based on expert estimates using first-hand knowledge of the country but limited field data. For most countries, detailed field inventories provided reliable results, but often the results could not be compared with other inventories using comparable definitions.

For the survey of industrialized countries, UNECE/FAO (2000) addressed the precision issue in an attempt to estimate indirectly the standard error for some key variables including forest area but not area change. The main conclusion was that the precision was high, at ±3 percent "likely range" for forest area. The same general conclusion is valid for the global estimates in this report. In general, the country data quality is roughly comparable between the industrialized and developing countries. In both domains, highly reliable national forest inventories were uncommon and most country information on forest area was derived from aggregated land classifications. Furthermore, reclassification to global classes caused similar difficulties in both domains.

For the tropical domain, the remote sensing survey provided a unique possibility to calibrate for systematic errors in the country estimates. As discussed above, the extent of forests showed a good correlation. For area change, however, only two regions had a good match, and remote sensing estimates for Africa were greatly different from the aggregate country estimates. It was concluded that country data for Africa overestimate deforestation for the region as a whole, and a calibration was applied in the above global estimates (Table 1-6). Furthermore, systematic overestimations of plantation establishment were adjusted for, although this calibration was based on an expert estimate.

In conclusion, it appears that the precision at the global, tropical and non-tropical levels is good for estimates of area and area change, but that systematic errors may still distort the overall picture. Two major systematic errors were adjusted for as described above, but others may still be hidden in the material. For example, secondary forests in South America are often excluded from area change statistics and would contribute to a lower net change rate in this region if accounted for. As another example, the extent of forest plantations in Europe may be larger than reported, as the option "semi-natural" is not given in the global classification scheme.

RESULTS FROM QUALITATIVE STUDIES

The qualitative studies undertaken by FRA 2000 were extensively documented in FRA Working Papers (listed in Appendix 4 and available on the FAO Forestry Web site). It was generally not difficult to establish a good understanding of the important factors affecting land use change in a local context, where climatic, cultural, policy and economic parameters are reasonably constant. It was considerably more difficult to generalize the findings to an international level.

Although specific land use practices varied considerably among regions, a common finding was that rules that govern the right to use the land and its products tend to be correlated with the management of the land and the tendency of forest land to be converted to other uses. For example, rights to land use were often established after *de facto* conversion to agriculture, creating a strong incentive to encroach on forests.

Deforestation has been a popular research subject in the past decade. A survey of scientific papers (FAO 2000b) found over 1 200 published papers on tropical deforestation since 1980, of which 825 contained findings related to deforestation processes and were included in the analysis. While it is not possible to conclude that this bulk of research papers can describe all deforestation processes, nonetheless the material represents a significant input to international discussions and negotiations. It is therefore important to understand the extent to which this research is representative of the global situation.

The accumulated information on tropical deforestation studies showed in a recognizable pattern. Throughout the 1980s, as concern about deforestation grew, the number of publications increased – from 8 in 1980 to 41 in 1989. Since 1990 the rate of publications has remained relatively constant, between 45 and 60 publications per year. For purposes of discussion, the number of published papers is taken as an indicator, regardless of the area covered by the studies or the originality of their data or analysis.

Half of the studies were published since 1992. Almost one-third of the publications on tropical deforestation had no clear geographical reference point; they discussed a particular aspect of the problem in an abstract way or they undertook a global analysis of the problem. Slightly more than two-thirds of the studies had a clear geographical point of reference, but they were distributed unevenly across countries. As a generalization, easily accessed countries were represented more often than others.

The research methods also changed during the past two decades as summarized in Table 1-8. General studies drawing on secondary sources and first-hand accounts by field researchers predominate in the early publications about the problem. The methodological patterns changed from the 1980s to the 1990s. Funding for studies employing remote sensing and surveys have increased in frequency, while first-hand accounts of deforestation processes have declined in number. The number of studies based exclusively on secondary sources has also declined somewhat, although they continued to account for about 46 percent of the published research in the 1990s.

The study observed some trends in the causes of deforestation prevailing in the literature – noting that deforestation may have been defined in different ways in different publications. In the 1990s, deforestation was as frequently attributed to logging, plantation expansion, smallholder agriculture, road building, population increase, and demands for fuelwood as in earlier studies. In publications on Latin America, factors such as incentives to create or expand cattle ranches and government colonization projects are no longer as frequently cited. More publications in the 1990s cited the expansion of markets through the growth in urban populations, improvements in transportation, and the search for raw materials in more remote settings as causes of deforestation. The increased level of foreign debt was suggested as a source of pressure to develop export crops at the expense of forest area. It is not clear if these factors have increased, or if the recent emphasis on globalization has perhaps stimulated more research and writing on these topics.

The qualitative studies of forest change and deforestation carried out within the framework of FRA 2000, including detailed country studies, provided an interesting overview of the knowledge of forest change processes. The results provided useful insights for countries where studies were carried out, and some distinct geographic and temporal pan-tropical patterns emerged. However, perhaps a more important conclusion was that while many studies on changes in forest area were made over the past several decades, the studies were not well coordinated and were not necessarily representative of the global situation. It is therefore difficult for analysts to draw valid conclusions from the literature and to use existing results to develop policies that address forest change. The high proportion of studies that used secondary information indicated that the knowledge on forest dynamics may not be proportional to the number of published scientific papers.

Table 1-8. Tropical deforestation studies in science journals, categorized by primary information source: trends over time

Data reference year	Remote sensing %	Survey %	Field observation %	Secondary source %	Total %	Total No. of studies
Pre-1980	8	8	39	46	100	88
1980s	8	5	30	57	100	276
1990s	17	15	20	47	100	332
Total	12	9	27	52	100	696

Source: FAO 2000b.
Note: Survey refers to household surveys and similar approaches.

A HISTORICAL PERSPECTIVE ON FOREST AREA AND DEFORESTATION

Clearing of forests to yield higher returns from land has a long history. Most studies estimate that about half of the Earth's land area was covered by forests 8 000 years ago, as opposed to 30 percent today (e.g. Ball 2001). Historically, deforestation has been much greater in temperate regions than in the tropics. Allowing that long-term changes in forest area are influenced by climatic fluctuations as well as by the actions of humans, the rate of deforestation since the introduction of agriculture might be estimated at about a quarter of a million hectares per year over the long term. However, much higher rates have been experienced in certain areas in the short term: for example, deforestation rates during the westward expansion in the United States in the late 1800s were roughly comparable to deforestation rates in the tropics today.

In this historical perspective it is obvious that increasing human population has been correlated with a negative impact on the extent of forests. Agriculture has expanded and replaced vast tracts of forests in all parts of the world to meet the demand for food and fibre. In some cases forests were removed primarily for wood products and the land was not reforested. Agricultural expansion has shifted between regions over time, following the general developments of civilizations, economies and increasing populations. It is still common in developing countries. The hypothesis that population growth *per se* drives deforestation through the demand for new agricultural land has also prevailed in many current papers and reports addressing deforestation. However, it has also been demonstrated in the United States and elsewhere in the twentieth century that population growth does not necessarily cause forest loss, especially if the rate of improvement in agricultural productivity is greater than the rate of population growth.

When agriculture dominated the economy of the now industrialized countries, governments commonly stimulated the clearing of forests for agricultural use as a means of economic development or as a means of providing a livelihood for poorer people. Wood was treated as a resource to be exploited, and forests were often viewed more as a nuisance than a treasure. Only recently has deforestation become a negative concept, first in countries where industrial forest products became important and the supply of raw material was threatened, and later in all countries as awareness of environmental issues increased and as the importance of forests for sustainable development and food security was better understood.

In recent decades, the rate of forest conversion has been particularly high in the tropics. FRA 2000 estimates tropical deforestation at 14.2 million hectares per year during 1990-2000, which means that almost 1 percent of the tropical forest is being lost each year. At the same time, the world's population has increased faster than ever before, but the direct link to deforestation and demand for agricultural land seems to have become less obvious. As the economies of most countries have grown, the relative importance of the agricultural sector has decreased. Most countries have experienced large-scale migration to cities. According to the United Nations Population Fund (UNFPA 2001), the population growth of urban areas greatly outpaces non-urban areas. Globally, only 13 percent of population growth is in rural areas, and rural populations are declining in most developed countries.

Table 1-9 shows relatively weak correlations at the national level between forest area change rate, demographic parameters and gross national product (GNP) per capita. The table suggests that the decision to abandon the population-driven deforestation model used in FRA 1990 was correct; but also that the forest change processes are too complex to be completely explained by any single indicator.

Table 1-9. Correlation coefficients (r) between forest cover change rate and selected variables at country level

Variable	Population density	Population change rate	Population, rural proportion	GNP/caput	Forest area change rate
Population density		-0.09	0.00	0.12	-0.04
Population change rate	-0.09		0.31	-0.36	-0.26
Population, rural prop.	0.00	0.31		-0.59	-0.38
GNP/caput	0.12	-0.36	-0.59		0.21
Forest area change rate	-0.04	-0.26	-0.38	0.21	

Source: Appendix 3.
Note: All data at national level and unweighted.

From the FRA 2000 findings, it appears that the expansion of agriculture was less prevalently associated with intensified shifting agriculture than with direct transformation of forest into permanent agriculture (or other land uses) at both large and small scales. This implies that economic and policy factors other than subsistence farming are more important in the deforestation processes.

On the positive side, many developing countries are trying to adopt policies to sustainably manage natural forests. For example, many countries are committed to monitoring progress towards sustainable forest management by national criteria and indicators. Numerous countries with substantial forest resources are trying to implement national forest programmes. This is a major development since the early 1990s and the United Nations Conference on Environment and Development (UNCED). Many industrialized countries actually experienced an increase in forest area in the 1990s, suggesting a positive link between development and the capability of a country to maintain or regain forest cover. In developed countries there is an increasing tendency for marginal lands to be valued more highly for forest goods and services than to be maintained for agriculture.

In conclusion, the changes in forest area observed for the period 1990-2000 were substantial, with a continued high rate of deforestation in the tropics. The net change, however, was lower than in the previous decade because of increased expansion of forests, primarily in non-tropical areas. The direct link to population growth and shifting cultivation earlier used to explain deforestation seems to be less valid in the most recent decade. However, demand for agricultural land remains the major driving force leading to deforestation. Factors related to land use rights seemed to determine forest area changes, as well as the general level of economic development, agricultural productivity and urbanization.

BIBLIOGRAPHY

Ball, J.B. 2001. *Global forest resources; history and dynamics. The forests handbook.* Oxford, Blackwell Science.

FAO. 1995a. *Forest Resources Assessment 1990 – Tropical forest plantation resources.* FAO Forestry Paper No. 128. Rome.

FAO. 1995b. *Forest Resources Assessment 1990 – Global synthesis.* FAO Forestry Paper No. 124. Rome.

FAO. 1997. *State of the World's Forests 1997.* Rome.

FAO. 2000a. *On definitions of forest and forest change.* FRA Working Paper No. 33. Rome.

FAO. 2000b. *Tropical deforestation literature: geographical and historical patterns in the availability of information and the analysis of causes.* FRA Working Paper No. 27. Rome.

FAO. 2001a. Forestry country profiles. www.fao.org/forestry/fo/country/nav_world.jsp

FAO. 2001b. Forest Resources Assessment homepage. www.fao.org/forestry/fo/fra/index.jsp

Sweden. National Board of Forestry (NBF). 2000. *Statistical Yearbook of Forestry 2000.* Jönköping, Sweden. www.svo.se/statistik

Swedish University of Agricultural Sciences. 2000. *Skogsdata 2000.*, Umeå, Sweden.

UNECE/FAO. 2000 *Forest Resources of Europe, CIS, North America, Australia, Japan and New Zealand: contribution to the global Forest Resources Assessment 2000.* Geneva Timber and Forest Study Papers No. 17. New York and Geneva, UN. www.unece.org/trade/timber/fra/pdf/contents.htm

United Nations Population Fund (UNFPA). 2001. *Demographic, economic and social indicators.* www.unfpa.org/swp/2000/english/indicators/indicators2.html

Chapter 2

Wood volume and woody biomass

ABSTRACT

Wood volume and above-ground woody biomass in forests were estimated for each country in 2000. Changes of these parameters during the 1990s were estimated at the global level. Information to support estimates of forest volume and woody biomass were not satisfactorily available for many countries, particularly in the tropics. This meant that assumptions and extrapolations had to be used. The year 2000 estimate for the global volume of forests was 386 billion cubic meters and the estimate for worldwide above-ground woody biomass was 422 billion tonnes. Results showed that the wood volume increased by 2 percent during the 1990s, largely because of increment in temperate and boreal forests. At the same time the above-ground woody biomass decreased by about 1.5 percent. A simultaneous increase of volume and decrease of woody biomass was possible because tropical forests were lost that contained considerably more biomass in relation to stem volume, compared to boreal forests where gains were recorded.

INTRODUCTION

Wood volume and woody biomass levels are important indicators of the potential of forests to provide wood and to sequester carbon. Wood is needed as a construction material, for pulp and paper manufacture, for fuel and energy, and for a wide variety of other uses. Because living forests trap and hold large amounts of carbon in their woody biomass they have also been indentified as potentially important regulators of the world's climate. Conversely, forests also may be a source of emissions when forests are burned or when wood from trees and other organic matter decomposes. releasing carbon dioxide back into the atmosphere.

The role of forests as major terrestrial sinks (and sources) of carbon dioxide has received significant additional attention since the adoption of the 1997 Kyoto Protocol to the United Nations Framework Convention on Climate Change (UNFCCC). Data on the carbon content in forest ecosystems has been estimated by the Intergovernmental Panel on Climate Change (IPCC 2000) (Table 2-1) but is far from complete and uncertainty is large. Determination of the amount of carbon in the woody biomass was not possible without a number of assumptions and uncertainties. National forest inventories, which are carried out in many countries, can be an important source of data and information about net productivity and biomass, but these often use different inventory methodologies and are not widely available or aggregated at the regional or global levels (GTOS 2000, 2001).

Volume and biomass statistics were among the most important parameters for FRA 2000. Statistics were compiled from country information sources following standard terms and definitions. For FRA 2000, volume is defined as the "stem volume of all living trees more than 10 cm diameter at breast height (or above buttresses if these are higher), over bark measured from stump to top of bole" (FAO 1998a).[1] (The definition excludes branches.) This term is referred to as "volume over bark" (VOB). "Above-ground woody biomass" was estimated for the assessment, defined as "The above ground mass of the woody part (stem, bark, branches, twigs) of trees, alive or dead, shrubs and bushes, excluding stumps and roots, foliage, flowers and seeds" (FAO 1998a). While total woody biomass would provide a more comprehensive measure of a forest ecosystem's capacity to sequester carbon, the algorithms needed to convert forest volume data to total woody biomass are lacking for much of the world's forests.

Volume and biomass data were available for most of the industrialized countries. Many of these countries also had statistics on growing stock, increment, felling and natural losses. However, reliable national-level data on volume and biomass in developing countries were not widely available. In those countries, most of which were in the tropics, volume estimates had to be based on local inventories or on inventories

---

[1] For the industrialized countries, trees down to 0 cm diameter were included.

Table 2-1. Global carbon stocks in vegetation and top 1 m of soils

Biome	Area	Global carbon stocks (Gt C)		
	million km²	Vegetation	Soils	Total
Tropical forests	17.6	212	216	428
Temperate forests	10.4	59	100	159
Boreal forests	13.7	88	471	559
Tropical savannahs	22.5	66	264	330
Temperate grasslands	12.5	9	295	304
Deserts and semideserts	45.5	8	191	199
Tundra	9.5	6	121	127
Wetlands	3.5	15	225	240
Croplands	16.0	3	128	131
Total	**151.2**	**466**	**2 011**	**2 477**

Source: IPCC (2000). Note that definitions used may differ from those in FRA 2000.

that only covered certain aspects of the forests, such as the commercial timber volume, or that were limited to only a few species (see e.g. IBAMA 1997; Malleux 1975). Throughout the world, inventories rarely employed the same standards, terms and definitions applied by FAO for volume measurements.

Biomass studies in developing countries were even less common than inventories of timber volume. Relevant exceptions were the national studies on biomass for many of the Central American countries focusing on the amount of carbon sequestered by the forests (USAID 1998). In other instances, biomass assessments for fuelwood production provided the baseline data (Banze *et al.* 1993).

Global studies encountered during the assessment (mentioned in FAO 1997) include those by Reichle (1981), Brown and Lugo (1982) and Olson *et al.* (1983). However, they were not appropriate for FRA 2000 since the study sites were often not representative of the population of interest (Brown and Lugo 1992). Consequently, their results could not be successfully extrapolated to the global level.

METHODS

Volume per hectare

For the developing countries, estimations of volume by hectare were based on inventory reports containing volume data for the various national forest types. In cases where the reported minimum diameter at breast height (DBH) was larger or smaller than 10 cm, data were adjusted. Stem volume of missing DBH classes was estimated either through regression equations established between DBH classes and the corresponding volume (when data were sufficient) or by using a volume expansion factor (VEF)

(FAO 1997; FAO 1998b). The VEF was used in situations where the volume per hectare was reported for DBH larger than the threshold of 10 cm and regression analysis could not be applied.

Various VEFs had to be used to match the wide range of volume data coming from the inventory reports. Differences in data composition were frequently due to the range of species and the type of forest being inventoried. For example, the minimum DBH in the inventories ranged from 5 cm to more than 50 cm (CIRAD 1991; Hammermaster and Saunders undated). Timber producing countries in humid tropical areas often estimated only volume for DBH classes larger than 30 or 40 cm. Conversely, in dry regions of Africa, a minimum DBH of 7 to 10 cm was used (Chakanga and Selanniemi 1999; CIRAD 1991; Saket *et al.* 1999). Volume data from most Asian countries were reported for a minimum DBH of 10 cm, and some countries from the humid tropical regions (Indonesia, Bangladesh, Brunei) reported minimum DBH of 20 to 50 cm. Volume data that included stems and branches required additional adjustments for the biomass calculations. In such cases, the volume of branches was excluded by using the ratio of 46 percent branches to 53 percent stem found by Saket (1994).

In many countries, only local inventories were available, which frequently focused on high-volume forests of interest for exploitation. In these areas, additional data adjustments had to be made since direct extrapolation from high-volume forests to all forests in a country would lead to overestimations. For a small number of countries where information on volume was not available, estimates were made using collateral information including the global ecological zone and forest cover maps combined with data from neighbouring countries that have similar

Table 2-2. Wood density applied for tropical tree species (tonnes of oven-dry biomass per cubic metre green volume)

Tropical region	Mean	Common range
Africa	0.56	0.50-0.79
America	0.60	0.50-0.69
Asia	0.57	0.40-0.69

Source: FAO (1997).

ecological and socio-economic conditions. Thus, the volume per hectare for all national forest types could be estimated.

Industrialized countries reported volume statistics as documented in UNECE/FAO (2000).

Biomass per hectare

For developing countries, biomass per hectare was calculated for each national forest type based on the volume statistics (VOB per hectare) and information on wood density (Table 2-2), and by expanding the volume to take into account the biomass of other above-ground components as follows (See also FAO 1998b).

Total forest biomass (t/ha) = VOB * WD * BEF,
where:

VOB = volume over bark (m^3 per hectare),
WD = volume-weighted average wood density (tonnes of oven dry biomass per cubic metre green volume),
BEF = biomass expansion factor (ratio of above-ground oven-dry biomass of trees to oven-dry biomass of inventoried volume).

Industrialized countries reported biomass statistics as documented in UNECE/FAO (2000).

Total volume and biomass

Total volume and biomass for each developing country were obtained by multipying the estimated volume and biomass per hectare with the forest area for each national forest type, and then adding the results for the various forest types into national totals. This means that the area distribution of forest types was an important component of the total volume and biomass estimates. The FRA 2000 documentation of national forest types, their areas and correspondence with global classes was therefore essential (see Chapter 1).

Industrialized countries reported total volume and biomass as documented in UNECE/FAO (2000).

Changes 1990-2000

Changes in forest volume and biomass occur in two different ways. First, areas that are transformed into forests (through afforestation or natural expansion) or deforested represent changes to the overall stock of forest volume and biomass. Second, the balance between increment, natural losses and fellings affects the volume and biomass per hectare within the forest. Seen over the long term, the latter can be used to indicate degradation (decreasing volume per hectare) or improvement (increasing volume per hectare) of the forests.

Volume and biomass changes resulting from forest area changes were estimated by country by multiplying the net forest area change 1990-2000 (see Chapter 1) with the average standing volume and biomass per unit area for the country as a whole. The results by country were added for the tropics and non-tropics as a whole.

Changes within the forests could only be estimated for industrialized countries (UNECE/FAO 2000), representing temperate and boreal forests, about 40 percent of the world forest area. For the remaining area, no comprehensive data for change estimates were available.

RESULTS

Wood volume 2000

The global volume of growing stock was estimated at 386 billion cubic metres in 2000. The

Table 2-3. Forest volume and above-ground biomass by region

Region	Forest area	Volume		Biomass	
		by area	total	by area	total
	million ha	m^3/ha	Gm^3	t/ha	Gt
Africa	650	72	46	109	71
Asia	548	63	35	82	45
Oceania	198	55	11	64	13
Europe	1 039	112	116	59	61
North and Central America	549	123	67	95	52
South America	886	125	111	203	180
Total	3 869	100	386	109	422

Source. Appendix 3, Table 7.

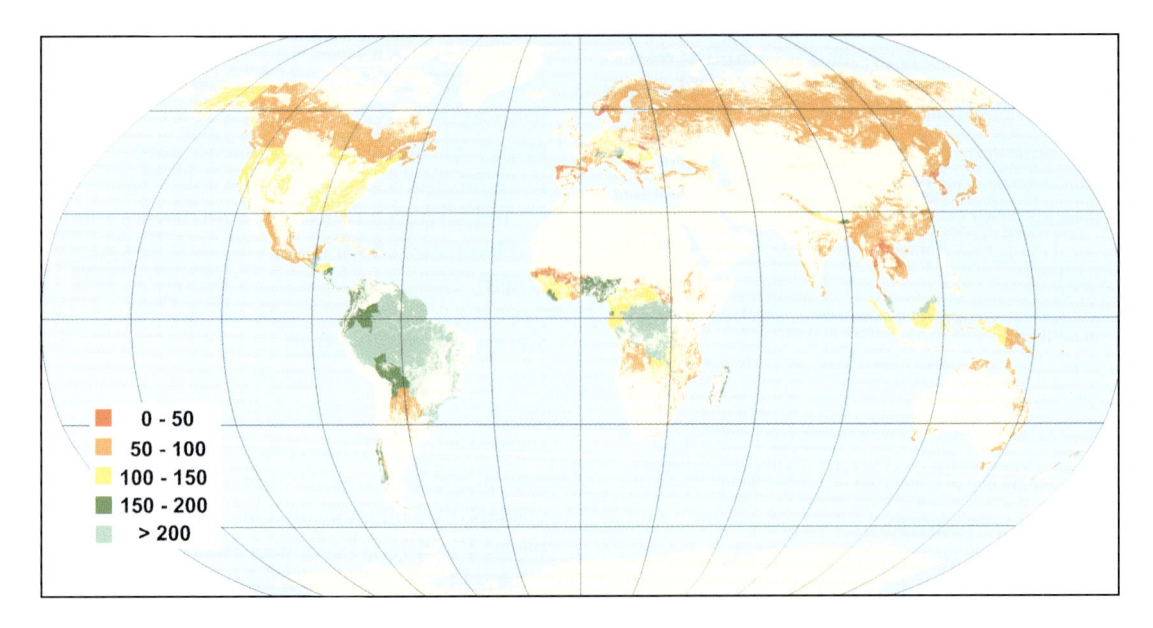

Figure 2-1. Above-ground woody biomass by country (tonnes/ha)

regions with the largest volume were Europe (including the Russian Federation) with 30 percent (116 billion cubic metres) and South America with 29 percent (111 billion cubic metres) (Table 2-3). Oceania shows the lowest growing stock with 11 billion cubic metres or 3 percent of the global volume. Estimates for each country are found in Appendix 3, Table 7.

Woody biomass 2000

The global estimate for above-ground woody biomass was 422 billion tonnes. The region with the largest quantity of biomass was South America with 43 percent of the world total or 180 billion tonnes. Brazil alone accounted for 27 percent of the world's above-ground woody biomass. Africa had the second largest quantity with 17 percent of the world total, or 71 billion tonnes. The other regions together accounted for 40 percent of the global above-ground biomass (Figure 2-1, Figure 2-2, Figure 2-3). Estimates for each country are found in Appendix 3, Table 7.

Changes 1990-2000

Changes for the 1990s, related to the transformation of areas to and from forests, were estimated at -9 billion cubic metres, corresponding to -16 billion tonnes of woody biomass. The losses were mainly in the tropics, whereas the non-tropics had an increase of volume and biomass (Table 2-4).

Changes within the forests were only known for industrialized countries, which reported an aggregated increase of 18 billion cubic metres of wood for the 1990s, or just over 1 m^3 per hectare per year, corresponding to 9 billion tonnes of woody biomass. These numbers represent the changes in temperate and boreal forests, about 40 percent of the total forest area (Table 2-4) (UNECE/FAO 2000).

Total changes for the 1990s, i.e. the sum of area-related changes and known within-forest changes, amounted to a volume increase of 9 billion cubic metres, or 2 percent. This corresponds to a decrease of 7 billion tonnes of woody biomass, or 1.5 percent (Table 2-4). An increase of volume and at the same time a decrease of woody biomass is possible because tropical forests contain considerably more biomass in relation to stem volume than boreal forests.

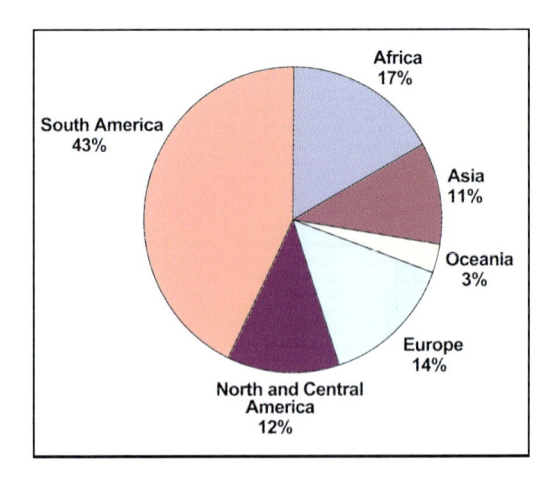

Figure 2-2. Distribution of above-ground woody biomass among regions

DISCUSSION

Problems of comparability of national data and reliability of aggregated results arose mainly because of differences in the national systems of nomenclature (i.e. measurement rules and definitions) and differences in the reference period(s). Differences in definitions and measurement rules were made comparable by harmonizing and standardizing the national nomenclature and data sets. Data from developing countries are highly variable in terms of quality and spatial, thematic and temporal resolution. Results of the assessments for the temperate/boreal countries were more complete as, in addition to the growing stock, they generally included a comprehensive analysis of increment, natural losses, felling and removals.

One component of forest volume and biomass change was the transformation of areas to and from forests. The average stocking level was used to estimate these flows. This is a simplification, as the gained forest area will only over a longer period develop into well-stocked forests.

Furthermore, and on the other hand, forests that are converted to other land uses may already to some extent have been degraded to lower levels of volume and biomass. Finally, the conversion of forests will not generally result in a completely treeless landscape. Without supporting data, it was reasonably assumed that the areas in transition involve forests at an average stocking level.

Forest volume and biomass stocks also changed within the forest as a balance of increment, natural losses and fellings. These factors were quantified only for industrialized countries. For remaining areas it was not possible to support assumptions on the changes during the 1990s. On the one hand, degradation occurs, for example in tropical forests, and reduces stocking levels. On the other, net increment occurs for example in large secondary forest formations.

The overall change estimates are thus not complete, as the within-forest development is not known for 60 percent of the forest area, including all tropical forests. The balance of increment,

Table 2-4. Changes in volume and above-ground woody biomass 1990-2000 for the tropics and non-tropics

Domain	Changes 1990-2000 as a result of forest area change			Changes 1990-2000 within the forest		Total change 1990-2000		Totals, 2000	
	Area	Volume	Biomass	Volume	Biomass	Volume	Biomass	Volume	Biomass
	million ha	Gm^3	Gt	Gm^3	Gt	Gm^3	Gt	Gm^3	Gt
Tropics	-123	-12	-18	n.a.[1]	n.a.[1]	-12	-18	179	282
Non-tropics	+29	+3	+2	+18[2]	+9[2]	+21	+11	207	140
Total	-94	-9	-16	+18	+9	+9	-7	386	422

[1] No data existed for estimating the balance of increment, losses and fellings within tropical forests.
[2] Refers to balance of increment, losses and fellings in industrialized countries as reported in UNECE/FAO (2000); changes within other non-tropical forests are not known.

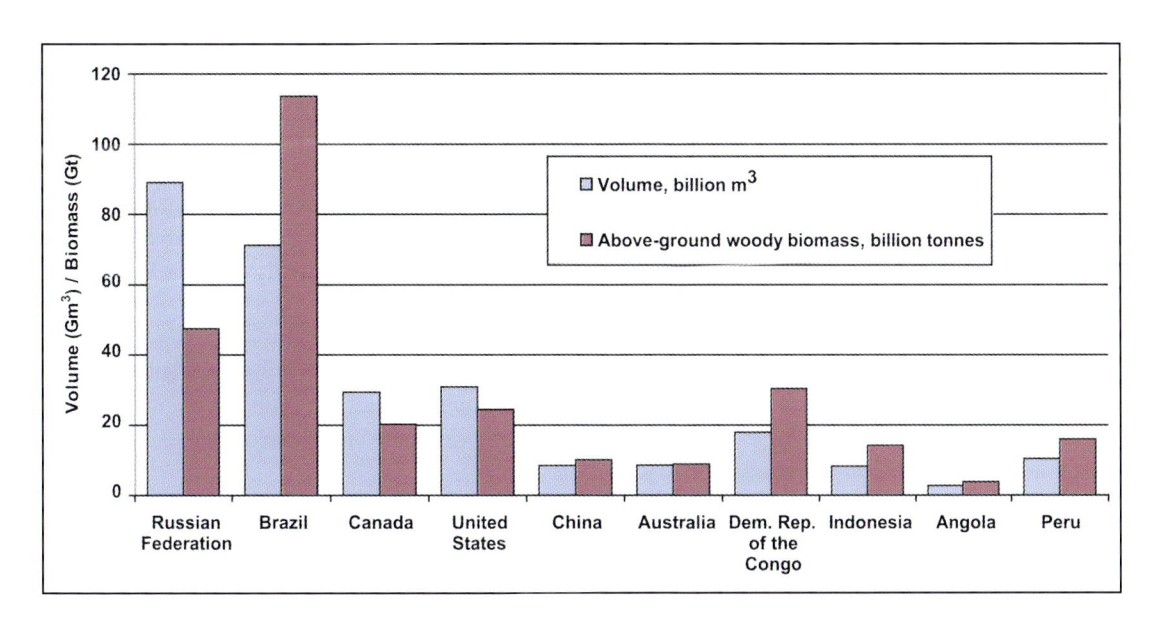

Figure 2-3. Volume and biomass for countries with largest forest area

natural losses and fellings could not be estimated for these areas, and sufficient expert knowledge did not exist to make reasonable assumptions. It is likely that changes in both directions were significant, but no reliable knowledge existed to judge the relative magnitude of positive and negative changes. At the same time, the increase of volume in temperate and boreal forests was well documented and large enough to affect the overall balance of volume and biomass worldwide.

BIBLIOGRAPHY

Banze, C.J., Monjane, M.J. & Matusse, R.V. 1993. *Avaliação de Biomassa Lenhosa nas Áreas de Maputo/Corredor do Limpopo, Corredor de Beira e Nampula/Corredor de Nacala*. Maputo, Mozambique, Ministry of Agriculture and Fisheries, National Directorate of Forestry and Wildlife (DNFFB).

Brown, S. & Lugo, A.E. 1982. The storage and production of organic matter in tropical forests and their role in the global carbon cycle. *Biotropica*, 14: 161-187.

Chakanga, M. & Selanniemi, T. 1999. *Forest inventory report of Caprivi State Forest*. Windhoek, Namibia, Namibia Finland Forestry Programme, National Forest Inventory Sub-component.

CIRAD. 1991. *Projet d'inventaire des ressources ligneuses (PIRL)*. Mali.

FAO. 1997. *Estimating biomass and biomass change of tropical forests – a primer*, FAO Forestry Paper No. 134. Rome.

FAO. 1998a. *FRA 2000 terms and definitions*. FRA Working Paper No. 1. Rome. www.fao.org/forestry/fo/fra/index.jsp

FAO. 1998b. *FRA 2000 Guidelines for assessment in tropical and sub-tropical countries*. FRA Working Paper No. 2. Rome. www.fao.org/forestry/fo/fra/index.jsp

Global Terrestrial Observing System (GTOS). 2000. *Terrestrial Carbon Observation Synthesis Workshop*. Ottawa, Canada, 8-11 February 2000. GTOS-23. www.fao.org/gtos/gtospub/pub23.htm

GTOS. 2001. *IGOS-P Carbon Cycle Observation Theme: Terrestrial and Atmospheric Components*. October, 2000 (revised February 2001). GTOS 25. www.fao.org/gtos/gtospub/pub25.htm

Hammermaster, E.T. & Saunders, J.C. Undated. *Forest resources and vegetation mapping of Papua New Guinea*.

Instituto Brasileiro do Meio Ambiente e dos Recursos Naturais Renováveis (IBAMA). 1997. *Diagnóstico e avaliação do sector florestal Brasileiro*. Brasilia, Brazil.

Intergovernmental Panel on Climate Change (IPCC). 2000. *Land use, land use change and forestry. A special report*. Cambridge, UK, Cambridge University Press.

Malleux, J. 1975. *Mapa Florestal del Peru, Memoria explicativa*. Lima, Peru, Universidad Nacional Agraria la Molina, Departemento de Manejo Forestal.

Olson, J.S., Watts, J.A. & Allison, L.J. 1983. *Carbon in live vegetation of major world ecosystems*. DOE/NBB-0037. Springfield, Virginia, USA, National Technical Information Service, US Department of Commerce.

Reichle (Editor). 1981. *Dynamic properties of forest ecosystems*. International Biological Programme No. 23. Cambridge, UK, Cambridge University Press.

Saket, M., Monjane, M. & Dos Ánjos, A. 1999. *Results of the detailed forest inventory in the districts of Marromeu, Cheringoma and Muanza*. Maputo, Mozambique, DNFFB.

Saket, M. 1994. *Report on the updating of the exploratory National Forest Inventory*. Maputo, Mozambique, DNFFB.

UNECE/FAO. 2000. *Forest Resources of Europe, CIS, North America, Australia, Japan and New Zealand: contribution to the global Forest Resources Assessment 2000*. Geneva Timber and Forest Study Papers No. 17. New York and Geneva, UN. www.unece.org/trade/timber/fra/pdf/contents.htm

United States Agency for International Development (USAID) & Programa Ambiental Regional para Centro América (PROARCA)/Central America Protected Areas System (CAPAS). 1998. *Estimación de la cantidad de carbono almacenado y captado (masa aérea) en los bosques*. Guatemala, Central American Commission on Environment and Development (CCAD).

Chapter 3
Forest plantations

ABSTRACT

Forest plantations covered 187 million hectares in 2000, of which Asia accounted for 62 percent. The forest plantation area represents a significant increase from the 1995 estimate of 124 million hectares. The reported new annual planting rate is 4.5 million hectares globally, with Asia and South America accounting for 89 percent. About 3 million hectares are estimated to be successful. Globally, half the forest plantation estate is for industrial end-use, one-quarter for non-industrial end-use and one-quarter not specified. Globally, the main fast-growing, short-rotation species are in the genera *Eucalyptus* and *Acacia*. Pines and other coniferous species are the main medium-rotation utility species, primarily in the temperate and boreal zones.

The potential for forest plantations to partially meet demand from natural forests for wood and fibre for industrial uses is increasing. Although accounting for only 5 percent of global forest cover, forest plantations were estimated in the year 2000 to supply about 35 percent of global roundwood. This figure is anticipated to increase to 44 percent by 2020. In some countries forest plantation production already contributes the majority of industrial wood supply. There is increasing interest in development of forest plantations as carbon sinks; however, failure to resolve international debates on legal instruments, mechanisms and monitoring remains a serious constraint.

In developing countries about one-third of the total plantation estate was primarily grown for woodfuel in 1995 – although it should be noted that planted trees on farmland, in villages and homesteads and along roads and waterways contribute significantly to fuelwood supplies, enabling the demand to be met in most instances.

INTRODUCTION

New forest plantation areas were reported as being established globally at the rate of 4.5 million hectares per year, with Asia and South America accounting for more new plantations than the other regions. Of plantations established, about 3 million hectares per year were estimated as being successful. Of the estimated 187 million hectares of plantations worldwide in 2000, Asia had by far the largest area. In terms of composition, *Pinus* spp. (20 percent) and *Eucalyptus* spp. (10 percent) remain dominant worldwide, although the overall diversity of species planted was shown to be increasing. Industrial plantations account for 48 percent, non-industrial plantations for 26 percent and plantations for unspecified use for 26 percent of the global forest plantation estate.

The results of the plantation assessment were the first global estimates with a uniform definition of forest plantations and can therefore not be directly compared to previous estimates. FRA 2000 country statistics on plantations may also differ from those reported in prior FAO publications (FAO 1981; FAO 1995), partly because of changes in definitions. Countries participated directly in the assessment, providing technical documentation and supporting analysis and validating the results generated by FAO. Several experts around the world were enlisted to provide detailed information on various aspects of the plantation situation in the form of special studies.

CONCEPTS AND DEFINITIONS

Between the extremes of afforestation and unaided natural regeneration of natural forests, there is a range of forest conditions in which human interventions occur. European forests have long traditions of human intervention in site preparation, tree establishment, silviculture and protection; yet these are not always defined as forest plantations. The traditional forest plantation concept tends to be applied to single species, uniform planting densities and even age classes. Terms such as "natural forest under management" or "assisted natural regeneration" are applied to stands of indigenous species in more heterogeneous management mechanisms in Europe and other industrialized temperate and boreal countries.

In FRA 2000 "forest plantations" are defined as those forest stands established by planting or/and seeding in the process of afforestation or reforestation. They are either of introduced or indigenous species which meet a minimum area requirement of 0.5 ha; tree crown cover of at least 10 percent of the land cover; and total height of adult trees above 5 m.

In country responses, terms such as "human made forest" or "artificial forest" were considered synonyms for forest plantations as defined in FRA 2000. Because of their increasing significance as a supply of fibre to the wood industries sector, rubber (*Hevea brasiliensis*) plantations were included as forest plantation resources.

METHODS

The area of existing forest plantations would ideally all have been derived from statistically designed inventories of forest plantations or statistics for planted areas reported by planting agencies or appearing in national reports. However, information also comes from many other sources including nursery production, seedling distribution and estimates derived from the goals of planting programmes. The vast range of agencies, industries and non-governmental organizations within countries engaged in planting programmes made the comprehensive collection of all relevant source documents a major logistical exercise. For FRA 2000, over 800 source documents were analysed to derive the forest plantation estimates. In most developing countries a national clearinghouse for collecting information on plantations is either lacking or ineffective owing to the enormity of the task and limited resources.

Data collection

To retrieve the source documents for the plantation study, FAO made formal requests to all developing countries, some of which contributed the necessary materials. Most of the reports were collected directly by FAO staff during FRA 2000 workshops and visits to national ministries. For consistency FRA 2000 prepared guidelines and questionnaires for the collection of forest plantation statistics in which the objectives, scope, definitions, sources of data and templates for specific data collection were supplied to each country. Parameters requested included:

- total estimated forest plantation area, 2000;
- annual area of new plantations;
- species groups: broadleaf (including *Hevea* spp.), conifer, non-forest like, African oil palm (*Elaeis guineensis*), coconut palm (*Cocos nucifera*) bamboo or unspecified;
- purpose and end-use objective of forest plantations: industrial (producing wood or fibre for supply to wood processing industry) or non-industrial (fuelwood, soil and water protection);
- ownership: public, private, other (e.g. traditional, customary) or unspecified.

Other data requested in the guidelines, which proved difficult for countries to provide by species group, included age class distribution; end-use by forest product (industrial plantations); growth and yield (mean annual increment); standing volumes; and rotation lengths. Despite the absence of these data, FRA 2000 is the most comprehensive forest plantation resources assessment that has been carried out.

In previous assessments of forest plantation resources, plantation data were available up to the reference year for most countries, since the reporting followed the reference year. In FRA 2000, the reference year was 2000, so if data were not available to that date, then existing area and annual planting data were used to extrapolate the necessary information. For the few countries that have no data sets since 1990, the rate of planting in preceding years and future planting programmes were considered in projections to the year 2000.

FAO also enlisted the assistance of several experts around the world to make specific technical contributions on the forest plantation situation in the 1990s. These studies constituted an important part of the global results as well, and complemented the country information.

Analysis and interpretation

The quantity and quality of forest plantation data provided is dependent upon the capacity of the national forest inventory systems to collect and analyse data and to adjust the information to conform with global and regional reporting parameters. In many developing countries there is a lack of institutional capacity to carry out periodic national forest inventories, so data can be incomplete, inconsistent, outdated and of variable reliability. Because of this, it was necessary to derive and in some instances to verify forest plantation statistics through desk research using available country reports. All sources of country data were referenced and made available in a transparent manner. In addition, regional and national focal persons were appointed to assist in the forest plantation data collection, to ensure that the latest data were available and to maintain coordination and communication between FRA 2000, FAO regional offices and each participating country. On completion of the data sets, a formal verification process was undertaken with each participating country.

RESULTS

Regional forest plantation areas, species and annual plantings

The annual plantation rates and plantation areas by regions and species groups are summarized in Table 3-1.

According to global forest plantation area distribution, as depicted in Figure 3-1, Asia accounts for 62 percent of the total; Europe, 17 percent; North and Central America, 9 percent; South America, 6 percent; Africa, 4 percent; and Oceania, less than 2 percent.

Globally, broadleaves make up 40 percent of forest plantation area with *Eucalyptus* the principal genus. Coniferous species make up 31 percent of which *Pinus* is the principal genus (Figure 3-2).

In FRA 2000 the global rate of new planting was estimated at 4.5 million hectares per year.

Asia accounted for 79 percent and South America for 11 percent (Figure 3-3).

Purpose and ownership within the global forest plantation estate

Purpose and ownership of forest plantations vary markedly among regions (Table 3-2). Industrial plantations provide the raw material for wood processing for commercial purposes, including timber for construction, panel products and furniture, and pulpwood for paper. In contrast, non-industrial plantations are aimed for example at supplying fuelwood, providing soil and water conservation, wind protection, biological diversity conservation and other non-commercial purposes.

In many countries, particularly in the developing world, the end purpose of the plantations is not clearly defined at the outset. In some of these cases, valuable tree resources are established which coincidentally match future needs. However, in other cases the lack of planning may result in plantations that have little commercial value and a low potential for local use.

Globally, 48 percent of the forest plantation estate is for industrial end-use; 26 percent for

Table 3-1. Annual plantation rates and plantation areas by region and species group

Region	Total area	Annual rate	Plantation areas by species groups (000 ha)							
	000 ha	*000 ha/yr*	Acacia	Eucalyptus	Hevea	Tectona	Other broadleaf	Pinus	Other conifer	Unspecified
Africa	8 036	194	345	1 799	573	207	902	1 648	578	1 985
Asia	115 847	3 500	7 964	10 994	9 058	5 409	31 556	15 532	19 968	15 365
Europe	32 015	5	-	-	-	-	15	-	-	32 000
North and Central America	17 533	234	-	198	52	76	383	15 440	88	1 297
Oceania	3 201	50	8	33	20	7	101	73	10	2 948
South America	10 455	509	-	4 836	183	18	599	4 699	98	23
WORLD TOTAL	**187 086**	**4 493**	**8 317**	**17 860**	**9 885**	**5 716**	**33 556**	**37 391**	**20 743**	**53 618**

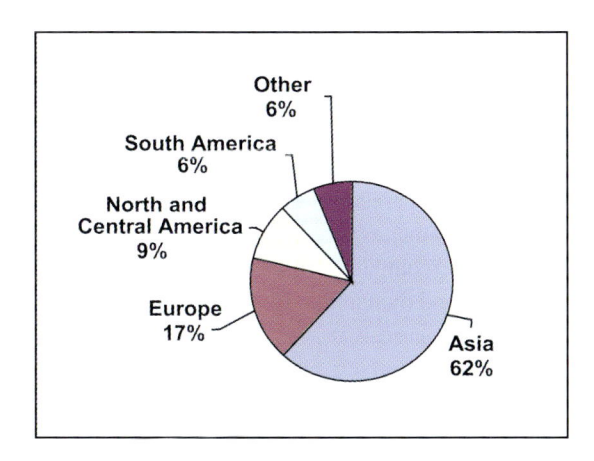

Figure 3-1. Distribution of forest plantation area by region

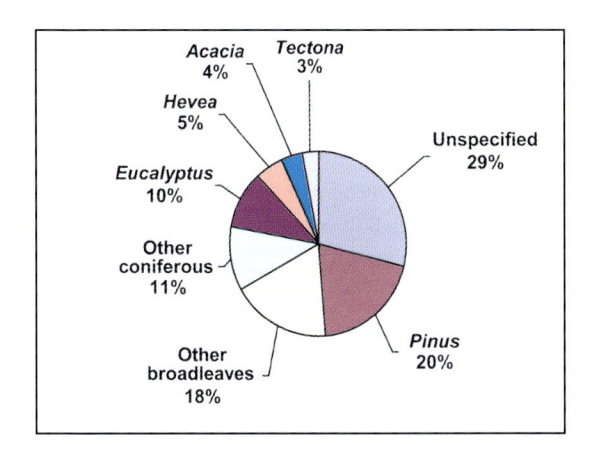

Figure 3-2. Distribution of plantation areas by genus

Table 3-2. Regional plantation areas by purpose and ownership

Region	Total area	Industrial purpose (000 ha)					Non-industrial purpose (000 ha)					Purpose unspec.
		Public	Private	Other	Unspec.	Sub-total	Public	Private	Other	Unspec.	Subtotal	
Africa	8 036	1 770	1 161	51	410	3 392	2 035	297	611	330	3 273	1 371
Asia	115 847	25 798	5 973	27 032	-	58 803	17 177	17 268	9 145	72	43 662	13 381
Europe	32 015	-	-	-	569	569	9	6	-	-	15	31 431
North and Central America	17 533	1 446	15 172	118	39	16 775	362	58	16	35	471	287
Oceania	3 201	151	14	-	24	189	2	3		19	24	2 987
South America	10 455	1 061	3 557		4 827	9 445	251	528	-	225	1 004	6
WORLD TOTAL	187 086	30 226	25 876	27 202	5 871	89 175	19 836	18 161	9 772	680	48 449	49 463

Source: FRA 2000

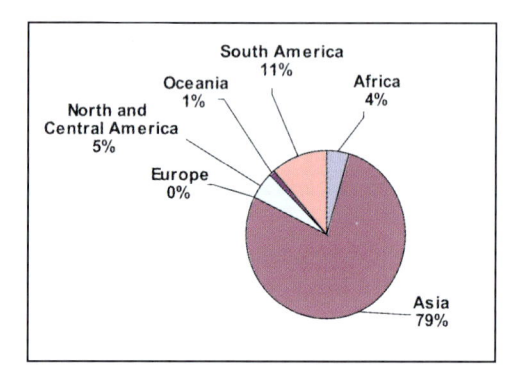

Figure 3-3. Distribution of annual planting area

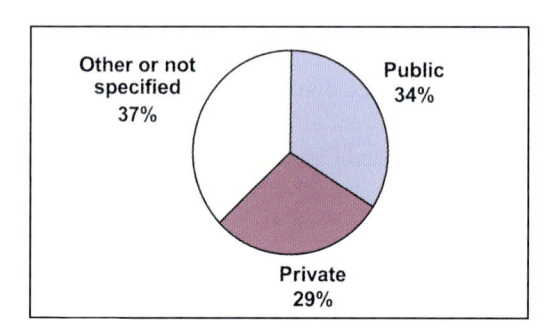

Figure 3-5. Ownership of industrial forest plantations, worldwide

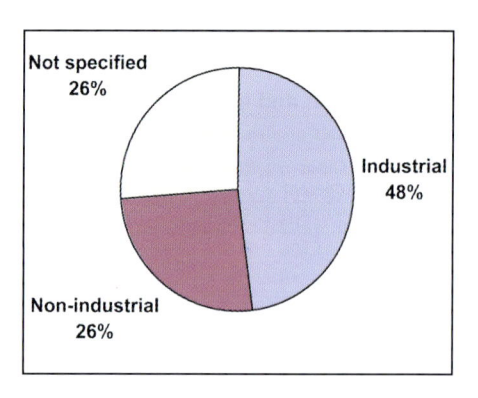

Figure 3-4. Distribution of forest plantations end-use, worldwide

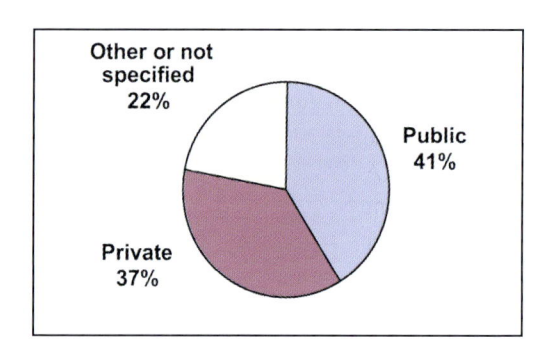

Figure 3-6. Ownership of non-industrial forest plantations, worldwide

non-industrial (fuelwood, soil and water, other); and 26 percent is not specified (Figure 3-4).

Globally, industrial plantations are 34 percent publicly owned, 29 percent privately owned and 37 percent other or unspecified (Figure 3-5). Within non-industrial plantations, 41 percent are publicly owned, 37 percent are privately owned and 22 percent are other or unspecified (Figure 3-6).

Leaders in forest plantation development (top ten countries by area)

As detailed in Table 3-3, the ten countries with the largest forest plantation development account for 79 percent of the global forest plantation development area. Six of these countries, accounting for 56 percent of global forest plantations, are in Asia.

The top ten countries according to area are China, 24 percent; India, 17 percent; the Russian Federation, 9 percent; the United States, 9 percent; Japan, 6 percent; Indonesia, 5 percent; Brazil, 3 percent; Thailand, 3 percent; Ukraine, 2 percent and the Islamic Republic of Iran, 1 percent (Figure 3-7).

Within the top ten, an estimated 52 percent of forest plantations are grown for industrial purposes to supply raw material for industry; 26 percent for non-industrial uses (fuelwood, soil and water protection, biodiversity conservation);

Table 3-3. Plantation purpose and ownership by reported area for the ten largest plantation development countries

Country	Total area	Industrial purpose 000 ha					Non-industrial purpose 000 ha					Unspecified purpose
	000 ha	Public	Private	Other	Unspecified	Subtotal	Public	Private	Other	Unspecified	Subtotal	
China	45 083	10 182	-	26 994		37 176	102	-	7 805	-	7 907	-
India	32 578	8 258	3 749	-	-	12 007	11 370	8 641	560	-	20 571	-
Russian Federation	17 340	n.a.	n.a.	n.a.	n.a.	n.a.	n.a.	n.a.	n.a.	n.a.	n.a.	17 340
United States	16 238	1 185	15 053	-	-	16 238	-	-	-	-	-	-
Japan	10 682	n.a.	n.a.	n.a.	n.a.	n.a.	n.a.	n.a.	n.a.	n.a.	n.a.	10 682
Indonesia	9 871	4 531	1 228	-	-	5 759	358	3 754	-	-	4 112	-
Brazil	4 982	-	-	4 802	-	4 802	-	-	180	-	180	-
Thailand	4 920	850	314	-	-	1 164	1 219	2 537	-	-	3 756	-
Ukraine	4 425	n.a.	n.a.	n.a.	n.a.		n.a.	n.a.	n.a.	n.a.	n.a.	4 425
Islamic Republic of Iran	2 284	241	-	-	-	241	1 938	105	-	-	2 043	-
Top 10 Total	148 403	25 247	20 344	31 796	-	77 387	14 987	15 037	8 545	-	38 569	32 447
Top 10 %	79%					87%					80%	66%
WORLD TOTAL	187 086	30 226	25 876	27 202	5 871	89 175	19 836	18 161	9 772	680	48 449	49 463

Source: FRA 2000

and the purpose was not specified in 22 percent (Figure 3-8). The industrial forest estate in these top ten countries was owned publicly, 33 percent; privately, 26 percent; and other or unspecified, 41 percent (Figure 3-9).

SELECTED GLOBAL TRENDS, 1980-2000

Comparisons

FRA 2000 country statistics on plantations may differ from those reported in prior FAO publications (FAO 1981; FAO 1991), partly because of changes in definitions. For example, rubber (*Hevea* spp.) plantations were not previously considered as forest plantations but are included in FRA 2000 plantation data. Previous assessments also used regional reduction factors to indicate the successful proportion of plantations remaining after establishment. The FRA 2000 assessment applied reduction factors according to the best available data from each country independently. There have also been changes in the information base from which the estimates were derived. The statistics now include data from many industrialized countries, none of which were included in the prior global assessment reports. Despite these differences, comparison of FRA results from each decade allows analysis of some trends including planting rates, genera, areas and purpose (end-use).

Global forest plantation estate

The global forest plantation estate has increased from 17.8 million hectares in 1980 and 43.6 million hectares in 1990 to 187 million hectares in 2000 (Table 3-4).

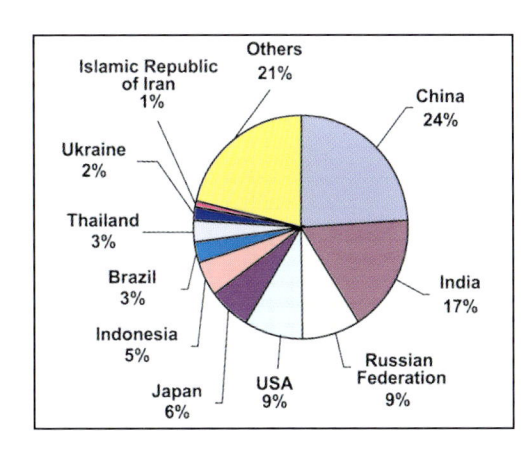

Figure 3-7. Leaders in forest plantation development – percentage of area

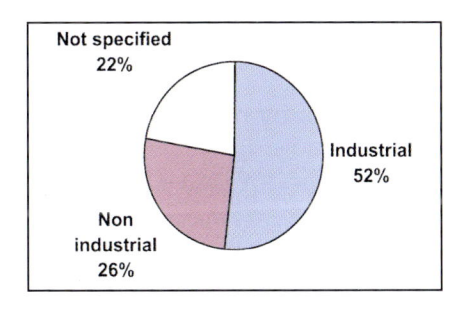

Figure 3-8. End-use of forest plantations, top ten countries

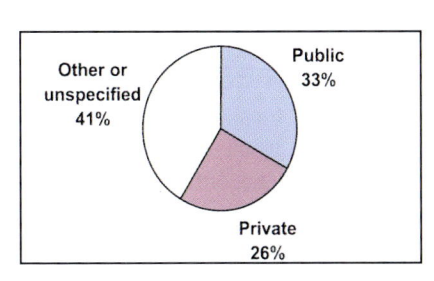

Figure 3-9. Ownership, industrial plantations, top ten countries

Table 3-4. Forest plantation purpose trends by region, 1980-2000

Region	Plantation area by purpose 000 ha			
	Total	Industrial	Non-industrial	Unspecified
2000				
Africa	8 036	3 392	3 273	1 371
Asia	115 847	58 803	43 662	13 381
Oceania	3 201	189	24	2 987
Europe	32 015	569	15	31 431
North and Central America	17 533	16 775	471	287
South America	10 455	9 446	1 004	6
GLOBAL TOTAL	**187 087**	**89 175**	**48 449**	**49 463**
1990				
Africa	2 990	1 366	1 623	
Asia	31 775	8 991	23 119	
Oceania	189	167	22	
Europe				
North and Central America	691	457	234	
South America	7 946	4 645	3 301	
GLOBAL TOTAL	**43 590**	**15 625**	**28 300**	
1980				
Africa	1 713	939	780	
Asia	11 088	3 487	7 601	
Oceania	88	41	47	
Europe				
North and Central America	287	272	15	
South America	4 604	2 261	2 348	
GLOBAL TOTAL	**17 779**	**7 000**	**10 791**	

Source: FAO 1981, 1995, 2000

Although in 2000, 26 percent of plantations continued to be for unspecified purpose, there was a significant increase in plantations for industrial purposes in the past decade: from 39 percent in 1980 and 36 percent in 1990 to 48 percent in 2000. There has been a corresponding decrease in forest plantations for non-industrial purposes.

Species trends by region – a graphic illustration

Species trends from FRA 1980, FRA 1990 and FRA 2000 are graphically illustrated by region in Figures 3-10 to 3-15 (FAO 1981; FAO 1995). The graphics are not to scale but illustrate relative growth within the region over the period and show trends in species used.

IMPACTS OF THE FOREST PLANTATION ESTATE

The potential for forest plantations to partially meet demand for wood and fibre for industrial uses is increasing. According to FRA 2000, the global forest plantation area accounts for only 5 percent of global forest cover and the industrial forest plantation estate for less than 3 percent. However, as an indication only, forest plantations were estimated in the year 2000 to supply about 35 percent of global roundwood and an increase to 44 percent anticipated by 2020 (ABARE and Jaakko Pöyry 1999) (Figure 3-15). If plantation development is targeted at the most appropriate ecological zones and if sustainable forest management principles are applied, forest plantations can provide a critical substitute for natural forest raw material supply. In several countries industrial wood production from forest plantations has significantly substituted for wood supply from natural forest resources. Forest plantations in New Zealand met 99 percent of the country's needs for industrial roundwood in 1997; the corresponding figure in Chile was 84 percent, Brazil 62 percent and Zambia and Zimbabwe 50 percent each. This substitution by forest plantations may help reduce logging pressure on natural forests in areas in which unsustainable harvesting of wood is a major cause of forest degradation and where logging roads facilitate access that may lead to deforestation.

Forest plantations

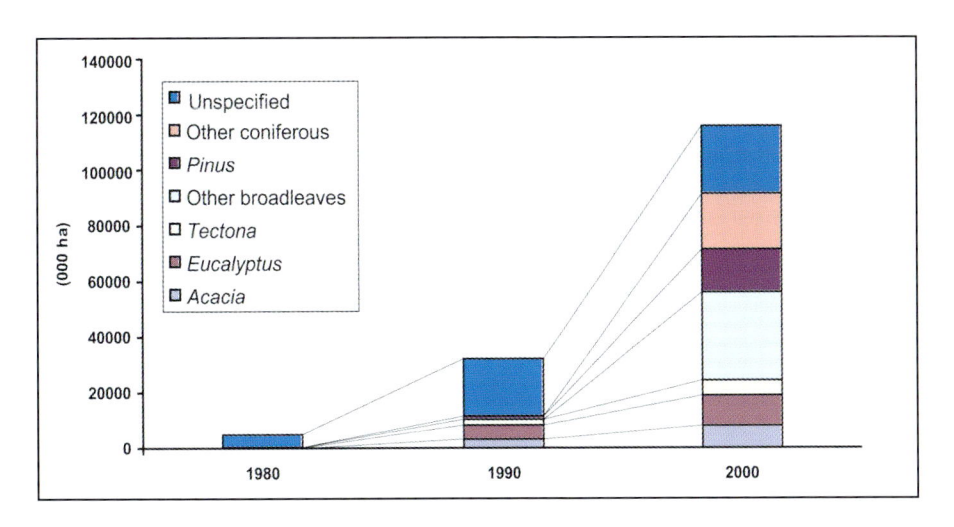

Figure 3-10. Plantation areas by genus, Asia

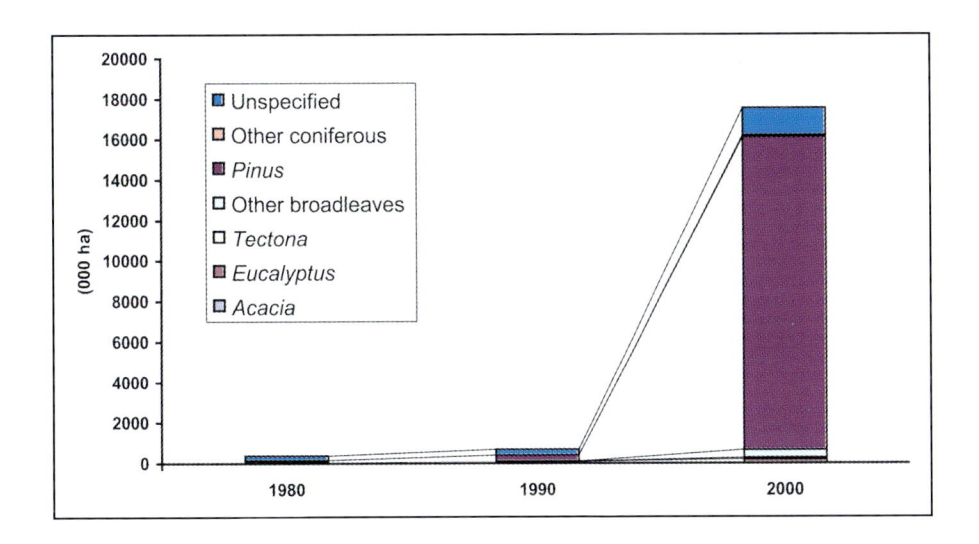

Figure 3-11. Plantation areas by genus, North and Central America

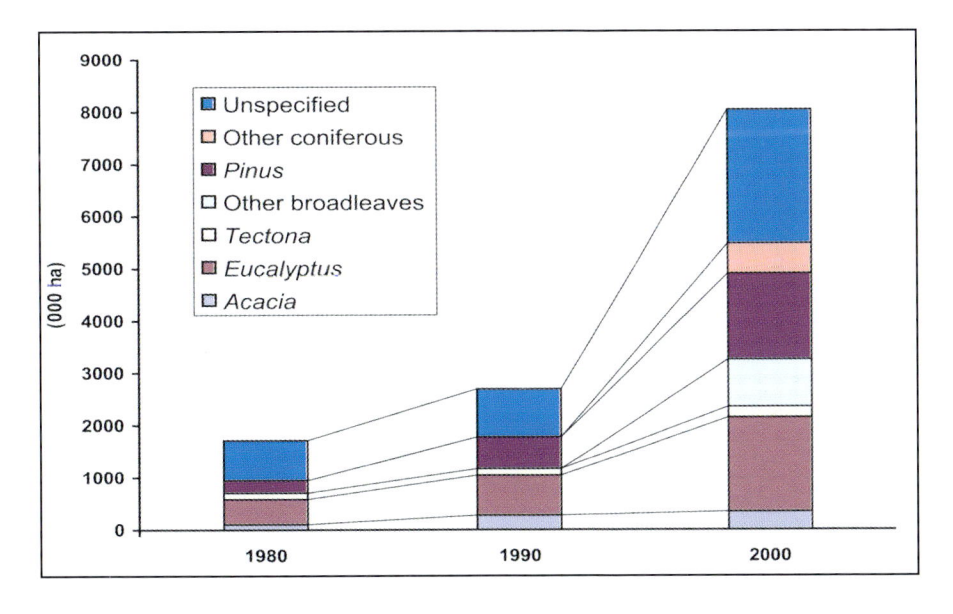

Figure 3-12. Plantation areas by genus, Africa

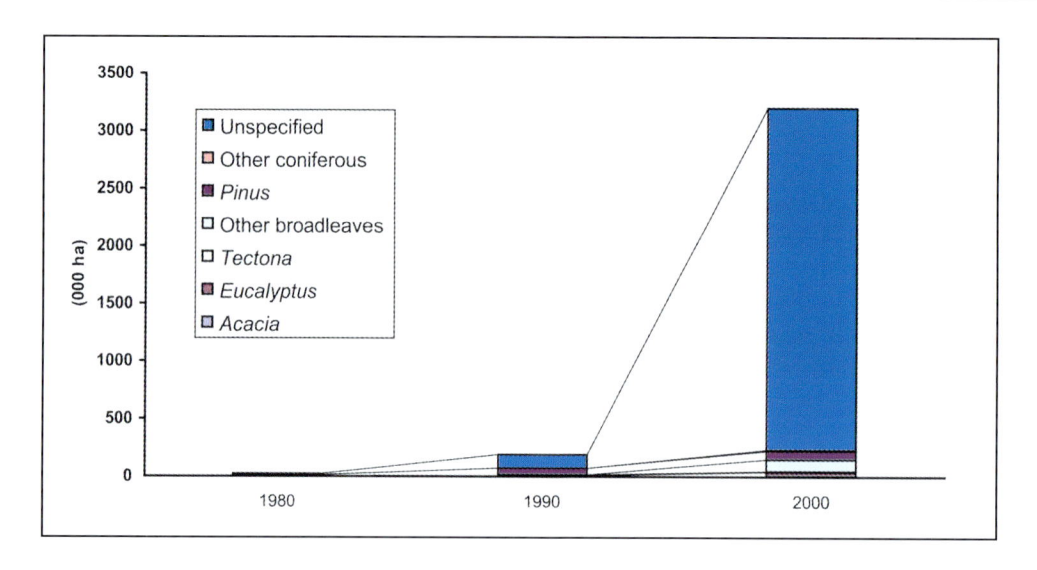

Figure 3-13. Plantation areas by genus, Oceania

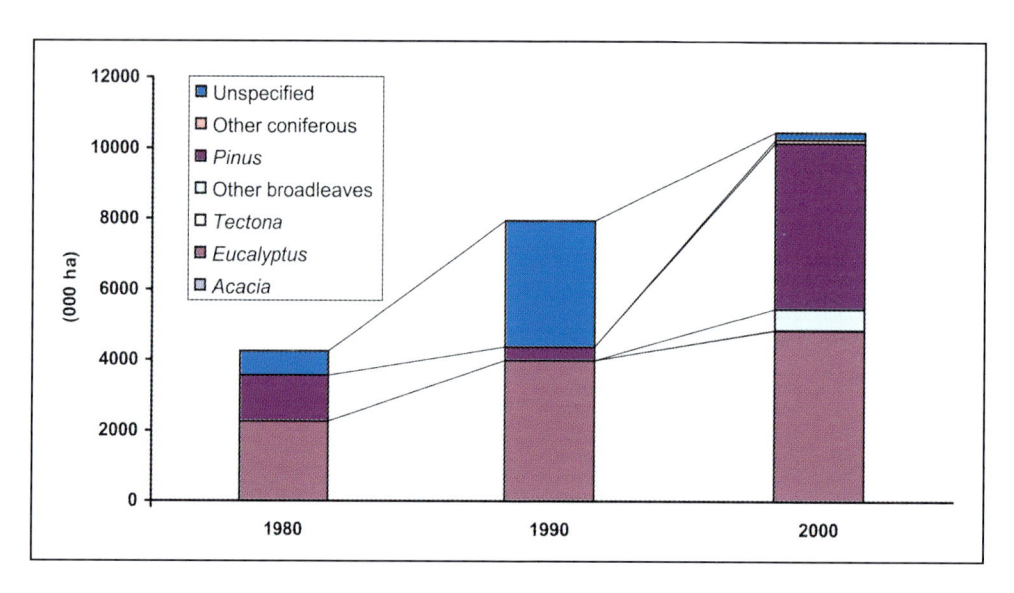

Figure 3-14. Plantation areas by genus, South America

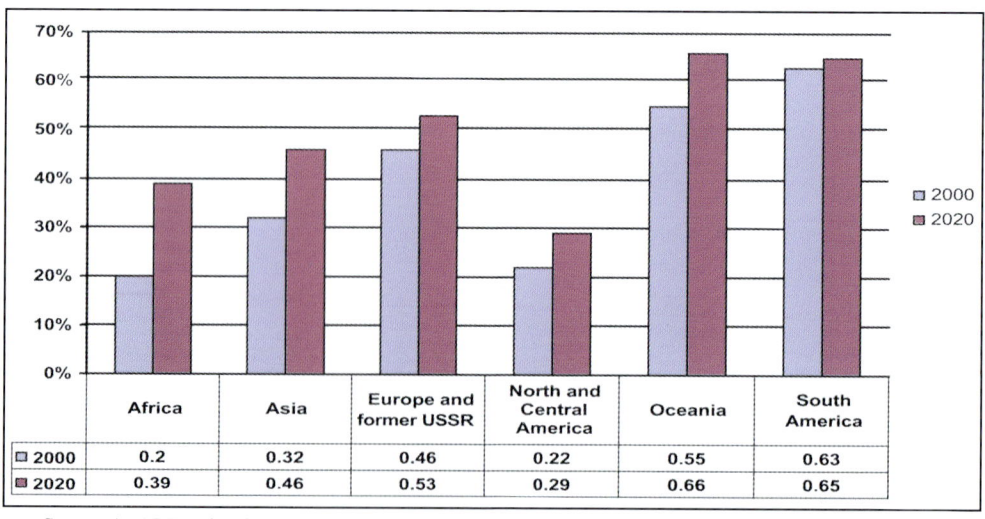

	Africa	Asia	Europe and former USSR	North and Central America	Oceania	South America
2000	0.2	0.32	0.46	0.22	0.55	0.63
2020	0.39	0.46	0.53	0.29	0.66	0.65

Source: ABARE and Jaakko Pöyry 1999

Figure 3-15. Predicted contribution of plantation wood to regional wood supply

Table 3-5. Mean annual increments for selected species used in industrial forest plantations*

Species	MAI
	m³/ha/yr
Eucalyptus	
E. deglupta	14-50
E. globulus	10-40
E. grandis	15-50
E. saligna	10-55
E. camaldulensis	15-30
E. urophylla	20-60
E. robusta	10-40
Pinus	
P. caribaea var. caribaea	10-28
P. caribaea var. hondurensis	20-50
P. patula	8-40
P. radiata	12-35
P. oocarpa	10-40
Other species	
Araucaria angustifolia	8-24
Araucaria cunninghamii	10-18
Gmelina arborea	12-50
Swietenia macrophylla	7-11
Tectona grandis	6-18
Casuarina equisetifolia	6-20
Casuarina junghuhniana	7-11
Cupressus lusitanica	8-40
Cordia alliodora	10-20
Leucaena leucocephala	30-55
Acacia auriculiformis	6-20
Acacia mearnsii	14-25
Terminalia superba	10-14
Terminalia ivorensis	8-17
Dalbergia sissoo	5-8

Source: Webb *et al.* 1984; Wadsworth 1997.
* Some promising trials are included.

Forest plantations also provide additional non-wood forest products, from the trees planted or from other elements of the ecosystem that they help to create. They contribute environmental, social and economic benefits. Forest plantations are used in combating desertification, absorbing carbon to offset carbon emissions, protecting soil and water, rehabilitating lands exhausted from other land uses, providing rural employment and, if planned effectively, diversifying the rural landscape and maintaining biodiversity.

Not all forest plantation development has positive economic, environmental, social or cultural impacts. Without adequate planning and without appropriate management, forest plantations may be grown in the wrong sites, with the wrong species/provenances, by the wrong growers, for the wrong reasons. Examples exist where natural forests have been cleared to establish forest plantation development or where customary owners of traditional lands may have been alienated from their sources of food, medicine and livelihoods. In some instances poor site/species matching and inadequate silviculture have resulted in poor growth, hygiene, volume yields and economic returns. In other instances, changes in soil and water status have caused problems for local communities. Land use conflicts can occur between forest plantation development and other sectors, particularly the agricultural sector.

The negative impacts of forest plantations can draw the focus away from the fact that forest plantation resources are totally renewable and can be economically, socially, culturally and environmentally sustainable with prudent planning, management, utilization and marketing.

SELECTED FOREST PLANTATION TOPICS

Mean annual volume increment (MAI) of select industrial species

For plantation planning and modelling, data from FRA 2000 need to be supplemented by growth and yield information. Average growth rates of frequently planted species are summarized in Table 3-5.

On average *Eucalyptus* and *Pinus* species, which dominate industrial plantations in developing countries, have similar MAIs of 10 to 20 m³ per hectare per year. However, many of the popular species of both genera frequently achieve much faster growth rates. Thus *Eucalyptus grandis,* which is the most widely planted *Eucalyptus* species, can achieve 40 to 50 m³ per hectare per year and in very exceptional conditions with advanced tree improvement 100 m³ per hectare per year. Other widely planted tropical hardwoods including *Casuarina equisetifolia, Casuarina junghuhniana, Tectona grandis* and *Dalbergia sissoo* have MAIs of less than 15 m³ per hectare per year and frequently under 10 m³ per hectare per year (FAO 2001h).

Climate and site have a very large impact on growth rates. The humid tropics and more fertile sites are more conducive to higher growth rates than locations with long dry seasons or infertile or degraded soil. Teak on many sites in India, for example, frequently has an MAI of 4 to 8 m³ per hectare per year, partly because of drought combined with poor soils. Some species such as *Gmelina arborea* and some of the *Eucalyptus* species are very site sensitive. *Pinus* spp., in contrast, generally tolerate adverse conditions better and are more flexible with respect to site.

Both tree breeding and silviculture have improved growth rates. Good examples are *Eucalyptus grandis* and *E. urophylla* in Brazil and *Pinus radiata* in some countries of the Southern Hemisphere. Advanced silviculture typically includes improved nursery and establishment techniques such as good site preparation, weed control and judicious use of fertilizer. It has been suggested that growth of teak (*Tectona grandis*), for example, could be doubled in Kerala, India and Bangladesh, and increased sixfold in Indonesia by adopting these practices. With coppice species productivity varies with rotation, the first and second coppice rotations usually being more productive than the seedling one.

The growth patterns vary among species. For example, very fast growing species such as *Gmelina arborea* can reach a peak MAI in less than 10 years, while *Pinus caribaea* var. *hondurensis* grown in Trinidad reaches maximum MAI at about 25 years and *P. radiata* at over 40 years. With *Cupressus lusitanica* in Costa Rica, the MAI maximum is reached at about 30 years (FAO 2001h).

Rotation lengths can reflect both end-use and economics. Many fast-growing *Eucalyptus*, *Acacia* and *Casuarina* species and *Gmelina arborea* are grown on short rotations of under 15 years as they are used primarily for pulp or woodfuel. Usual rotations in Kenya for *E. grandis* are 6 years for domestic woodfuel, 7 to 8 years for telephone poles and 10 to 12 years for industrial woodfuel. In Brazil this species is largely grown for pulp or charcoal on 5 to 10 year rotations. Species being grown for high-value sawlogs usually have longer rotations; teak (*Tectona grandis*) is grown on 50 to 70 year rotations and high-value conifers such as *Araucaria angustifolia* on 40 year rotations. Generally pines are grown on medium-length rotations of 20 to 30 years, unless grown solely for pulpwood, when shorter rotations may be adopted.

Modelling of growth, rotations, harvest yields and product mix by species is important for decision-making in forest management. One of the major obstacles to model development for planners and managers is a lack of suitable data. Data can come from a range of sources, including temporary and permanent sample plots and experiments. Experiments and protocols for obtaining data need to be carefully designed so that reliable information is obtained over the complete range of conditions to which the model is to apply. Tree Growth and Permanent Plot Information System (TROPIS), sponsored by the Centre for International Forestry Research (CIFOR), seeks to coordinate and improve access to tree growth information.

A growth and yield model developed for *Pinus elliottii* plantations in coastal Zululand, South Africa, predicts height, basal area, total stem and merchantable volume and stocking, with age on a stand level for harvest planning. In New Zealand several simulation models have been developed for *P. radiata* which predict similar variables but also include wood quality, harvesting and marketing aspects, which make it possible to link the silvicultural options to industrial use.

Sustaining productivity

It is possible not only to sustain but also to increase productivity in successive rotations. This requires clear definition of the end-use objective for forest plantation development and a holistic view in their management. There is a need to integrate strategies for genetic improvement programmes, nursery practices, site and species/provenance matching, appropriate silviculture (site preparation, establishment, weeding, fertilizing, pruning, thinning), forest protection and harvesting practices with prudent management. New Zealand and the southern United States have shown that substantial gains can be made by adopting this holistic approach. In developing countries where resources may be constrained, highly technical solutions may not be essential but it is critical to get the fundamentals correct: careful species and provenance choice, good nursery stock, site preparation, planting techniques, weed control and, less frequently, fertilizer inputs. Once fast-growing, uniform plantations have been established, later silvicultural tending may become increasingly important, depending on the end-use objective (FAO 2001e).

Current evidence suggests that plantation production can be sustainable if foresters implement prudent genetic and silvicultural tree improvement programmes and sound management practices (Evans 1999). There has been, however, limited long-term research on the subject; there are few definitive studies, limited to few species. In one of the most promising studies, with *Pinus patula* in Swaziland grown intensively on about 15 year rotations, site productivity was maintained or increased over three rotations. The question of declining growth in teak (*Tectona grandis*) plantations in Indonesia and India remains unclear (FAO 2001b).

How forest plantations are managed affects the chemical and physical properties of the soils and site. However, only recently have long-term studies been undertaken to evaluate these critical factors or processes. The methods adopted for site preparation (ripping, ploughing, scarifying, bedding, windrowing, controlled burning), establishment (manual, mechanical), weeding (manual, chemical, mechanical), fertilizer application, pruning and thinning (manual and mechanical, for commercial to waste), forest protection and harvesting (manual, mechanical, clear-fell or selection) all affect the pool of nutrients in the ecosystem. Interference with the drainage, litter and recycling of organic matter and change in the physical conditions of soils during these operations are critical to long-term sustainability. Because of litter recycling and the rapid development of tree roots, plantations are used for rehabilitation of fragile and degraded lands prone to soil erosion and excessive water runoff. Tree plantations often have higher evapotranspiration rates than grassland or agricultural crops, and thus change the hydrology of the site. This can be either beneficial (for example by reducing salinity problems in some dryland conditions) or detrimental (if it reduces water required for other uses) (FAO 2001b).

The rare studies of changes of productivity between rotations have concluded that negative changes have primarily been due to inappropriate or inadequate management practices or weed invasions rather than a result of the plantations themselves.

Burning and excessive cultivation in site preparation, soil compaction from mechanical operations, inappropriate harvesting techniques and poor forest protection can contribute to loss of nutrients and soil erosion, with a resultant loss in productivity of forest plantation sites. This cannot be addressed solely by addition of fertilizer, but by the adoption of the whole range of tree improvement, silviculture, protection and harvesting techniques in an integrated forest management strategy.

Valuable hardwood plantations

Long-rotation, slow-growing but valuable hardwood species have special technical properties, such as strength, natural durability, hardness and easy machining, and appearance (grain, figure, texture, colour and other aesthetic qualities) that make them suitable for high-value end-uses such as furniture. These high-grade hardwoods contrast with short-rotation, fast-growing, lesser-quality woods used for woodfuel, pulpwood or reconstituted products and less demanding building timbers. In tropical countries teak (*Tectona grandis*), mahogany (*Swietenia* spp.) and rosewood (*Dalbergia* spp.) are the main hardwood plantation species, while in temperate countries oak (*Quercus* spp.), ash (*Fraxinus* spp.), cherry (*Prunus* spp.), walnut (*Juglans* spp.), tulipwood (*Jacaranda* spp.) and hard maple (*Acer* spp.) predominate.

Because many valuable hardwood species are difficult to establish because of their ecological requirements or disease or insect susceptibility, focus has been on the easier species to grow, including teak (*Tectona grandis*), Indian rosewood (*Dalbergia sissoo*) and mahogany (*Swietenia macrophylla*). In 1995 the global areas of these species were 2 254 000, 626 000 and 151 000 ha, respectively. They accounted for about 10 percent of total hardwood plantations in the tropics. More than 90 percent of teak plantations were located in Asia, mainly in Indonesia, India, Thailand, Bangladesh, Myanmar and Sri Lanka. About 95 percent of rosewood plantations are located in India and Pakistan. The largest mahogany (*Swietenia macrophylla*) plantations are located in Indonesia and Fiji, which together make up about 80 percent of the established area (FAO 2001g). A summary of the main characteristics of valuable hardwood species commonly grown in tropical areas is given in Table 3-6.

The market preference for large piece sizes, slow growth and very long rotation lengths (e.g. 50 to 70 years for teak) combine to reduce the attractiveness for commercial investment in these species. This is only partially counteracted by their value. The low return on capital investment, coupled with the long wait period for this return, has made it difficult to interest private investors without supportive, secure and stable government policies.

Table 3-6. Characteristics of valuable hardwoods used in tropical areas

Use categories	Desirable wood properties	Main end-uses	Matching valuable hardwood species	Comments
Decorative timbers	Appearance, consistent quality, dimensional stability, durability, good machining, staining and finishing properties	Quality furniture and interior joinery	*Tieghemella* spp.; *Entandophragma cylindricum, Chorophora* spp., *Aucoumea klaineana, Afrormosia* spp., *Entandophragma utile, Mansonia* spp., *Lovoa* spp., *Khaya* spp., *Swietenia* spp., *Dalbergia* spp., *Aningeria* spp.	Highest value, competition from temperate hardwoods and MDF
High to very high-density timbers	Appearance, strength, high natural durability, availability in large sizes	Principally in construction	*Dipterocarpus* spp., *Lophira* spp., *Chlorophora* spp., *Ocotea rodiaei*	Small share of total tropical timber use
Low to medium-density utility timbers	Appearance, clear grain, natural durability, good machining properties	External joinery, shop fittings, medium-priced furniture	*Shorea* spp, *Hevea brasiliensis, Terminalia* spp., *Heritiera* spp.	Most commonly used, prone to competition from substitutes

Source: Based on FAO 1991.

As markets demand a continuity of supply, plantations need to be on a sustainable scale within a region. Some of the less common species are not known in the marketplace. Other potential market problems are that the timber may be wrongly associated with tropical deforestation and changing fashions that often occur with decorative timbers. Niche marketing is important for valuable hardwoods.

Projections for supplies of timber from existing valuable hardwood plantations indicate that because of the age class distribution and long rotations there will not be a significant increase in supply in the next 20 years (FAO 2001g).

Future promotion of quality hardwood plantations needs to emphasize choice of species with versatile end-uses, market research and development to hold on to niche markets and maintained high standards from production to marketing. Careful site selection, use of high-quality planting materials of superior genetic origin and good silviculture are important. Planting programmes should be economically viable, environmentally appropriate and socially desirable. Incentives may also be necessary to stimulate private investment because of the long rotations.

Even though valuable hardwood plantations have the potential to reduce the pressure on natural forests, they will not prevent deforestation resulting from agricultural encroachment. The supply of large quantities of high-value timber could perhaps undermine the value of natural forest stands and so lead to more rapid destruction. Hence it is advisable, where possible, to manage plantations and forest resources and forest products in a complementary manner.

Plantations and wood energy

Woodfuels from plantations or natural or semi-natural forests are particularly important in developing countries, providing about 15 percent of their total energy demand (WEC 1999). Woodfuel provides about 7 percent of energy demand for the world as a whole and in industrialized countries only 2 percent. Woodfuel provides more than 70 percent of energy needs in 34 developing countries and more than 90 percent in 13 countries (including 11 in Africa). Woodfuel makes up about 80 percent of total wood use in developing countries and about 89 percent in Africa (FAO 2001f).

The prediction of a woodfuel crisis in developing countries in the 1980s was based largely on looking at supply and demand from forest plantations and natural forests. The reaction to the expected woodfuel crisis was to plant trees for this purpose, often in the form of traditional plantations. Many programme failures resulted from lack of appreciation for the complexities of bioenergy supply and demand, failure to take into account social aspects and people's needs and poor programme structures. The importance of planted trees on farmland, in villages and homesteads and along roads and waterways as a source of woodfuel supply was underestimated.

Rural communities harvest stems, branches, stumps, twigs, leaves and litter for woodfuels in chronic woodfuel supply areas. In these instances the nutrient recycling process is broken, resulting in degradation of forest plantation sites. In many rural communities in developing countries, woodfuel is considered a public free good, to be foraged from public natural and plantation forests. Often women and children collect the woodfuel at little or no cost. As a result, the growing of private forest plantations specifically for woodfuel, in which development costs and rotation cycles are involved, can be a foreign concept.

Asian studies show that forest-based supply can range from 13 percent in the Philippines to as high as 73 percent in Nepal. In many countries less than 50 percent of fuelwood is from forests.

Globally, non-industrial forest plantations in 1995 were estimated to cover about 20 million hectares (FAO 2000). This was almost 17 percent of the world's total plantation area in 1995. A significant proportion of these plantations were planted for woodfuel, and 98 percent were in developing countries. These plantation figures do not account for trees planted outside the forest on farms or in villages, etc., nor do they consider plantations that were considered agricultural plantations, such as *Hevea* or palm plantations.

In developing countries about one-third of the total plantation estate was grown primarily for woodfuel in 1995 (Table 3-7). Three-quarters of these plantations were in Asia (excluding Japan), where they accounted for 60 percent of total plantation production. In Latin America more than half of plantation production went to woodfuel; in Africa and Oceania a larger proportion of plantation production was as industrial wood. However, plantations, in general, provided only a small proportion of total woodfuel used. Uruguay is an interesting exception (FAO 2001f).

Production of woodfuel from plantations currently makes only a small contribution to energy requirements, although it is very important in some localities and countries. Plantations currently supply 5 percent of woodfuel. Production from these non-industrial plantations is likely to double over the next 20 years, even with little expansion in area, because the age class distribution is heavily concentrated in young plantations. In an optimistic scenario where planting continues at the same rate as in the past ten years and then gradually declines, a 350 percent increase in woodfuel production would be anticipated by 2020. By-products from wood-

Table 3-7. Areas and production of non-industrial forest plantations in selected developing countries by region

Region	Area of woodfuel estate[a] 000 ha	% of total plantation estate	1995 estimates Plantation woodfuel[d] million m^3	% of plantation production	% of total woodfuel use[b]	2020 Predicted woodfuel[c] million m^3
Africa	**2 154**	**37**	**12.2**	**34**	**3**	**20.6**
Ethiopia	135	88	1.5	93	3	1.6
Madagascar	122	52	1.5	84	16	1.7
Sudan	233	78	1.1	76	7	3.2
Asia[d]	**15 090**	**33**	**53.8**	**60**	**5**	**334.8**
China	3 854	18	5.5	20	2	56.7
India	8 308	67	30.2	92	11	137.7
Indonesia	399	13	4.2	52	5	8.2
Oceania[e]	**14**	**10**	**<0.1**	**12**	**<1**	
Latin America	**3 123**	**35**	**20.4**	**55**	**8**	**47.0**
Brazil	1 946	47	12.6	51	12	25.1
Peru	210	72	1.5	70	9	3.6
Uruguay	232	67	2.1	71	95	5.9
Developing countries	**20 380**	**33**	**86.4**	**47**	**5**	**302.4**

[a] Assumes non-industrial plantations are primarily for woodfuel.
[b] Based on estimates in WEC (1999) and FAO (2000).
[c] Scenario 3 from FAO (2000) – for 10 years new planted area same as recent years, followed by a gradual decline – an optimistic estimate.
[d] Asia includes Turkey but excludes Japan.
[e] Oceania excludes Australia and New Zealand.

using industries will also contribute to increased fuelwood supply. The situation is less positive in Africa, where for a few countries declines are projected in plantation-based woodfuel production (FAO 2001f).

New sources of fibre

Since FRA 1990, advances in wood utilization technology have resulted in increasing importance of new sources of fibre – rubber (*Hevea brasiliensis*), coconut palm (*Cocos nucifera*) and African oil palm (*Elaeis guineensis*) – especially in the Southeast Asian subregion. These species account for 9.7, 12.0 and 6.0 million hectares of plantations, respectively. All grow in the humid tropics. In terms of plantation area, Asia has 92 percent of the world's rubber, 86 percent of the world's coconut palm and 78 percent of the world's African oil palm. Indonesia, Thailand and Malaysia have almost three-quarters of the rubber plantations; Indonesia and the Philippines have about half the coconut resources; and Malaysia has 55 percent of the oil palm resource. All three species are grown principally for other products rather than wood, so when overmature they are available for fibre-based industries at minimal cost (FAO 2001c).

Rubberwood is harvested when latex productivity declines (beyond 30 years) and yields 100 m^3 per hectare of roundwood, but recovery for lumber is only 25 to 45 percent because of poor form and small size. Most of the planted stands in Southeast Asia are owned by smallholders and are geographically dispersed, with poor accessibility and poor-quality stems. Currently the major proportion of industrially utilized rubberwood comes from large-scale plantations. Quality furniture, parquet, panelling, reconstituted panels, general utility timber and woodfuel, including charcoal, are made from rubberwood. However, the rubberwood must be processed within days of harvesting to minimize sapstain attack. The most developed downstream industries are in Malaysia, where the production of sawn rubberwood timber rose from 88 000 m^3 in 1990 to 137 000 m^3 in 1997 and medium density fibreboard (MDF) production from rubberwood reached 1.16 million cubic metres per annum by 1999. Exports of rubberwood furniture have grown from about US$74 million in 1991 to US$683 million in 1998. Rubberwood has become a substitute for light tropical forest hardwoods. Its acceptance as a sustainable plantation-grown, environmentally friendly timber has given it wide appeal (FAO 2001c).

Coconut palms are harvested as the copra yields decline (beyond 60 years) and yield 90 m^3 per hectare of coconut wood. Coconut palm has variable properties and is intrinsically difficult for conversion but can yield a relatively low-cost, general-utility timber for construction, panelling, stairs, door jambs, furniture, flooring and power poles. In 1993 Indonesia had 65 million cubic metres of overmature coconut stems which needed disposal before replanting. There is increasing interest in this raw material in European and North American markets. It is unlikely to replace conventional timber, but likely to find its way into niche markets. It will continue to be used as a low-cost construction timber (FAO 2001c).

Oil palm plantations are harvested for fibre beyond the 25 to 30 year rotations and yield about 235 m^3 per hectare. It is estimated that over 1.6 billion cubic metres of fibre will be available in the years to come from established resources in Southeast Asia. From 1996 to 1999 the area increased by 18 percent. In Malaysia the area has increased by 3 million hectares in the past 30 years.

Most oil palm plantations (unlike rubber and coconut) in the main growing countries, Malaysia and Indonesia, are managed by plantation companies or cooperatives. Oil palm by-products such as kernel shells, pressed fibres and empty fruit bunches are currently used in heat generation at the extraction plants. Water in the stems can reach five times the weight of dry matter. The high moisture content as well as the high amounts of parenchyma tissue rich in sugar and starches make conversion into quality forest products a challenge. An MDF plant in Malaysia is currently being planned to utilize oil palm stems (FAO 2001c).

Plantation substitutes for natural forest products

With growing concerns about the status and loss of natural forests, the rapid expansion of protected areas and large areas of forest unavailable for wood supply, plantations are increasingly expected to provide substitutes for products from natural forests, particularly in Asia and the Pacific.

In Asia and the Pacific it is estimated that 52 percent of natural forests are not available for wood harvest because they are inaccessible or uneconomic to exploit. Of the unavailable forest in the region, it is estimated that about 38 percent is legally reserved. In addition, logging bans have been imposed on large areas of natural forest covering about 10 million hectares. The reasons for these bans vary but were related to deforestation and forest degradation causing environmental problems in Thailand, the Philippines and China and to conservation requirements in Sri Lanka and New Zealand (FAO 2001a).

As a result of the net effect of deforestation and removal of natural forests from wood production, some areas in the Asia and the Pacific region have wood deficits and roundwood harvesting is exceeding sustainable levels of cut. The worst affected areas are South Asia and insular Southeast Asia, with continental Southeast Asia also under strain. In contrast, New Zealand has surplus plantation wood available for export.

Of six examples studied in the Asia and the Pacific region, New Zealand is more than self-sufficient in wood production based on plantations. In China and Viet Nam, the importance of plantations will increase as planted resources mature. There have been serious problems with implementing plantation development programmes in Sri Lanka, the Philippines and Thailand. In Sri Lanka, India and elsewhere in the tropics, trees outside the forest are playing a critical role in roundwood and woodfuel supply (FAO 2001a).

Most countries in the region are becoming importers of wood, with imports expected to rise. Sometimes logging bans have shifted the problem to other countries. Problems with acquiring large areas of land in some countries make it difficult to implement industrial plantations. In the Philippines, Thailand and Viet Nam there have been social conflicts with local indigenous people or between traditional forest use and development, as well as between the rich and the poor. Sometimes incentives and the development of social forestry programmes are being used to help resolve such problems.

While it is clear that plantations will have an increasingly significant role in substituting products from natural forests, the impact will be felt on a case-by-case basis as governments and investors determine where and how plantations can be technically, economically and socially feasible as well as environmentally friendly. In the near term, plantations in Asia and the Pacific can make a contribution but cannot replace harvests from natural forests. It is likely that both in the region and globally the current pace of industrial plantation development will barely keep pace with losses from deforestation and transfer of natural forests to protected status. While it would be theoretically possible, actual plantation development is at present not sufficient to offset both growing consumption and declining harvest from natural forests (FAO 2001a).

Plantations and carbon sequestration

In the past ten years, the development of forest plantations as carbon offsets has evolved towards a market mechanism, although an organized market with carbon prices defined according to supply and demand forces is still a long way off. The adoption of the Kyoto Protocol in 1997 triggered a strong increase in investment in plantations as carbon sinks, although the legal and policy instruments and guidelines for management are still debated. A number of countries have already prepared themselves for the additional funding for the establishment of human-made forests. The 1997 Costa Rica national programme was the first to establish tradeable securities of carbon sinks that could be used to offset emissions and the first to utilize independent certification insurance.

To date, greenhouse gas mitigation funding covers about 4 million hectares of forest plantations worldwide (FAO 2001d). The recognition of afforestation and reforestation as the only eligible land use, land use change and forestry activities under the Clean Development Mechanism of the Kyoto Protocol, as agreed in Bonn during the second part of the Sixth Conference of the Parties to UNFCCC in July 2001, will lead to a steep increase in forest plantation establishment in developing countries. The sink decision of the Bonn Agreement is expected to funnel additional funds into forest activities in developing countries and thus to strengthen the international efforts in this field. However, it will also require a monitoring and verification system to ensure that these plantations will not be established at the expense of the local population or efforts to conserve biological diversity. Thus the decisions taken in Bonn to make the Kyoto Protocol ratifiable will also bear new challenges for forest plantation development.

CONCLUSIONS

New forest plantation areas are reported to be increasing globally at the rate of 4.5 million hectares per year, but net areas may be much less. Asia and South America account for more new plantation development than other regions. The

Asian region has the largest areas in forest plantations.

Broadleaf species account for 40 percent of forest plantations, coniferous species 31 percent and unspecified species 29 percent.

Industrial plantations account for 48 percent and non-industrial 26 percent of global forest plantations. Industrial plantation resources are dominated by China, India and the United States, while non-industrial plantation resources are dominated by China, India, Thailand and Indonesia. Forest plantation ownership in both industrial and non-industrial plantations is evenly balanced between public and private.

Data on forest plantations remain weak, however, for detailed analysis.

Forest plantations can provide critical environmental, social and economic benefits. Sound forest plantation management, tree improvement and silviculture can sustain and/or enhance productivity of forest plantations. To do so, however, it is important that forest plantations be managed in accordance with a defined end-use objective.

Forest plantations provide a critical substitute for raw material supply from natural forests, including industrial roundwood and fuelwood. In addition, non-forest species such as rubber (*Hevea brasiliensis*), coconut (*Cocos nucifera*) and oil palm (*Elaeis guineensi*) are becoming important sources of wood and fibre. Finally, there is increasing potential for forest plantation investment to offest carbon emissions.

BIBLIOGRAPHY

Australian Bureau of Agriculture and Resource Economics (ABARE) and Jaakko Pöyry Consulting. 1999. *Global outlook for plantations*. ABARE Research Report 99.9. Canberra, ABARE.

Evans, J. 1999. *Sustainability of forest plantations – the evidence*. Issues Paper. London, Department for International Development.

FAO. 1981. *Forest Resources Assessment Project 1980*. Rome.

FAO. 1991. *High value markets for tropical sawnwood, plywood and veneer in the European Community,* by R.J. Cooper. Rome.

FAO. 1995. *Forest Resources Assessment 1990. Tropical forest plantation resources*. FAO Forestry Paper No. 128. Rome.

FAO. 2000. *The global outlook for future wood supplies from forest plantations,* by C. Brown. FAO Working Paper GFPOS/WP/03. Rome.

FAO. 2001a. *Role of forest plantations as substitutes for natural forests in wood supply – lessons learned from the Asia-Pacific region,* by T. Waggener. Forest plantations thematic paper series. Rome. (unpublished)

FAO. 2001b. *Biological sustainability of productivity in successive rotations,* by J. Evans. Forest plantations thematic paper series. Rome. (unpublished)

FAO. 2001c. *Non-forest tree plantations,* by W. Killmann. Forest plantations thematic paper series. Rome. (unpublished)

FAO. 2001d. *Plantations and greenhouse gas mitigation: a short review,* by P. Moura Costa & L. Aukland. Forest plantations thematic paper series. Rome. (unpublished)

FAO. 2001e. *Plantation productivity,* by W. Libby. Forest plantations thematic paper series. Rome. (unpublished)

FAO. 2001f. *Plantations and wood energy,* by D. Mead. Forest plantations thematic paper series. Rome. (unpublished)

FAO. 2001g. *Promotion of valuable hardwood plantations in the tropics. A global overview,* by F.K. Odoom. Forest plantations thematic paper series. Rome. (unpublished)

FAO. 2001h. *Mean annual volume increment of selected industrial forest plantation species,* by L. Ugalde & O. Perez. Forest plantations thematic paper series. Rome. (unpublished)

Wadsworth, F.H. 1997. *Forest production for tropical America*. USDA Forest Service Agriculture Handbook 710. Washington, DC, USDA Forest Service.

Webb, D.B., Wood, P.J., Smith, J.P. & Henman, G.S. 1984. *A guide to species selection for tropical and subtropical plantations*. Tropical Forestry Papers No. 15. Oxford, UK, Commonwealth Forestry Institute.

World Energy Council (WEC). 1999. *The challenge of rural energy poverty in developing countries*. London, FAO/WEC.

Chapter 4
Trees outside the forest

ABSTRACT

The world has billions of trees that are not included in the FRA 2000 definitions of "forest" and "other wooded land". Trees outside the forest (TOF) include trees in cities, on farms, along roads and in many other locations which are by definition not a forest. All trees make a contribution to the environment and to the social and economic well-being of humankind. This chapter briefly describes the importance of trees outside the forest and some of the issues related to their assessment. FRA 2000 did not attempt a comprehensive global assessment of TOF, nor has such an assessment ever been carried out. However, many studies have been made of TOF for specific countries or land areas, often with an emphasis on their economic contributions. The chapter provides a summary of selected studies and discusses the practical and conceptual difficulties related to a comprehensive global assessment. Suggestions are made for improvements that might be made in future assessments.

INTRODUCTION

The significance of trees outside the forest (TOF) can be observed in several contexts. In countries with low forest cover, TOF resources constitute the main source of wood and non-wood "forest" products, even though trees may be so scattered that the maps produced by FRA 2000 indicate that no forests exist. Trees are found on agricultural lands, in densely populated areas, in fruit-tree plantations and in home gardens, which often cover a large proportion of the land. In urban areas trees provide important aesthetic and environmental services in addition to providing shade and greatly increasing the livability of cities. Communities, farmers and herders who do not have access to forests diversify their production and protect their land by maintaining various tree systems on their farms.

Deforestation has been mapped and quantified, but very little is known about the fate of land formerly under forest; forest clearing is often followed by the establishment of production systems of which trees are an integral part. Not much is known about the dynamics of trees on farmlands and their corresponding contribution to the production of wood and other products and services. Similarly, little is known about changes in tree cover in fields and urban systems. Knowledge of trees outside the forest comes mostly from local studies on agroforestry, sylvipastoralism and urban, social, community or rural forestry.

This widespread and multipurpose resource, familiar to farmers but poorly defined by managers and mostly absent from official statistics and development policies, needs to be better assessed and known. Growing populations, shrinking forests and degraded ecosystems all suggest that trees outside the forest are destined to play a larger local and global role in meeting the challenges of resource sustainability, poverty reduction and food security. Trees outside the forest relieve the pressure on forest resources, conserve farmland, boost agricultural productivity, blunt the harmful impact of urban growth on the environment, increase food supplies, provide income and in general make valuable contributions to food security.

FRA 2000 did not undertake a global assessment of trees outside the forest, mainly because of resource limitations; nor has there ever been a comprehensive global assessment of trees outside the forest and their products. However, a number of studies have been carried out for specific sectors or geographic areas, often with an emphasis on their economic contributions. This chapter provides a summary of selected studies and discusses the practical and conceptual difficulties related to a comprehensive global assessment.

The chapter responds to the concern expressed by the Expert Consultation on Forest Resouces Assessment 2000 (Kotka III) regarding the lack of information on TOF (Finnish Forest Research Institute, 1996). For more information, national case studies and working papers can be found on the FRA Web page. An FAO Conservation Guide on trees outside the forest will be published at the end of 2001.

DEFINITION OF TREES OUTSIDE THE FOREST

Trees outside the forest are defined by default, as all trees excluded from the definition of forest and other wooded lands (see Appendix 2). Trees outside the forest are located on "other lands",[2] mostly on farmlands and built-up areas, both in rural and urban areas. A large number of TOF consist of planted or domesticated trees. TOF include trees in agroforestry systems, orchards and small woodlots. They may grow in meadows, pastoral areas and on farms, or along rivers, canals and roadsides, or in towns, gardens and parks. Some of the land use systems include alley cropping and shifting cultivation, permanent tree cover crops (e.g. coffee, cocoa), windbreaks, hedgerows, home gardens and fruit-tree plantations.

Classification of trees outside the forest presents certain difficulties. There are existing classifications for agroforestry, but none applicable to all trees outside the forest (Kleinn 2000). For practical reasons, the FRA 2000 definition of "forest" combines aspects of both land cover and land use. This approach creates difficulties not only for classification of forest, but also for classification of TOF.

In a study on data-gathering on TOF in Latin America (Kleinn *et al* 1999), where classification was primarily based on land use criteria, separating land use and land cover aspects was found to be a main source of misinterpretation. There was a possibility of confounding coffee plantations and trees in pasture with forest, given their high density. This clearly shows some of the problems involved in establishing a simple and reliable *a posteriori* classification.

In France, the National Forest Inventory (IFN) and Teruti Land Use Study[3] have begun to attempt coordinating classifications of trees outside the forest. The objective is eventually to use the annual Teruti data to update the IFN ten-year data, with a single national nomenclature as a possible end result (IFN 2000).

FUNCTIONS AND CHALLENGES

In industrialized countries, farmers list shade and shelter, soil protection and improvement of the landscape and rural environment as their main reasons for growing trees (Auclair *et al.* 2000). In the tropics, farmers grow woody species for food security and subsistence. Trees outside the forest are a major source of food (Bergeret and Ribot 1990). Livestock fodder produced by TOF can be a matter of life and death in semi-arid or mountainous areas.

Fuelwood remains the prime source of energy in developing countries, representing up to 81 percent of the wood harvest (FAO 1999). In contrast, in the industrialized countries, fuelwood accounts for less than 10 percent of total fuel consumption (FAO 1998). Very few studies have reported on overall fuelwood output from stands and single trees outside the forest, but agroforestry systems and orchards are known to provide a large part of the resource.

Trees outside the forest are an important source of non-wood forest products,[4] described in more detail in Chapter 11.

Trees outside the forest have an important ecological role. Planted trees and shrubs in fields help to check runoff and erosion and control flooding, as well as helping to purify water and protect against wind. Trees lining rivers and streams help to maintain biodiversity, providing spawning beds for fish and shellfish and shade which reduces eutrophication.

The unique role of trees in soil protection and conservation, checking wind and water erosion and maintaining soil fertility is universally acknowledged. Also important are the cumulative benefits of trees on smallholdings to soil and water conservation, in particular in the larger context of mountain watershed management; their positive impact on climate; and their role in buffering the effects of desertification and drought.

RESULTS OF SELECTED STUDIES

In spite of the limits of data at the regional or global level, a number of local initiatives have been carried out. The approaches of the various studies differ in accordance with the purpose and scale of the analysis. Few studies use methods resembling the conventional forest inventory. Many studies rely on existing literature or

---

[2] "Other lands" include farmlands (including pasture and meadows), built areas (including human settlements and infrastructure), bare lands (including oasis) and snow and ice.
[3] The Teruti survey of the Central Bureau of Statistical Surveys and Studies was initiated by the French Ministry of Agriculture in 1981; it monitors changes in TOF areas and wooded areas.

[4] Non-wood forest products (NWFP) are products of biological origin other than wood derived from forests, wooded lands and trees outside forests.

estimates drawn from surveys and interviews. The quantification of products is often based on different parameters, such as estimates of global output, marketed output, observed or potential productivity or economic value. Thus the reliability of the results is uncertain.

The following are the results of some national initiatives that have assessed trees outside the forest (FAO 2001).

In Kerala, the most densely inhabited state of India, a study estimated that of the total annual production of 14.6 million cubic metres of wood in the state, about 83 percent was from homesteads (house compounds and farmlands), 10 percent from estates (plantations of rubber, cardamom, coffee and tea) and only about 7 percent from forest areas (26.6 percent of the state area is under forest cover (FSI 1998). Trees outside the forest met about 90 percent of the fuelwood requirements of the state. Fuel from coconut trees alone, including both wood and non-wood materials (pruned and fallen), constituted about 70 percent of the total fuelwood supply (Krishnakutty 1990).

A study in Haryana State in India, an intensively cultivated state with about 3.8 percent of its area classified as forest land but only about 2 percent under actual forest cover (FSI 1998), showed that farm forestry (trees along farm bonds and in small patches up to 0.1 ha) accounted for 41.2 percent of the total growing stock of wood. Multiple tree rows along roads and canals accounted for 13 percent and 9.6 percent, respectively; village woodlots for 24 percent; and block plantations of less than 0.1 ha for 10.6 percent (FSI 2000).

In Morocco, where forest cover is less than 5 percent of the land cover and other wooded lands only 7 percent, nearly 20 percent of the land may be occupied by trees outside the forest, namely as wooded pasture (84 percent) and fruit-tree plantations (12 percent) (Rosaceae, citrus, olives trees, palm trees, walnut trees, fig trees, almond trees). Fruit production has an important place in the national economy (MADRPM 2000). It is noteworthy that even when a forest is largely destroyed, the carob (*Ceratonia siliqua*) is one of the few species traditionally conserved, as it is highly appreciated by farmers for multiple purposes, providing both fodder and income from the sale of its fruit for export. However, there are no reliable data on the distribution and potential of this "forest" resource, which is of interest for farmers, herders, concessionaires and the government and distributed on agricultural and forest lands.

In the Sudan, the National Forest Inventory has undertaken a national land use inventory to provide area and volume statistics for planning at the subnational and national levels (FAO 1995). The inventory was designed to provide preliminary estimates regarding products other than the traditional fuelwood and timber, such as the amount of gum, fruit or nuts that can be collected and the distribution of non-wood species of interest (Glen 2000).

In Costa Rica, the Tropical Agricultural Research and Higher Education Center (CATIE), in collaboration with Freiburg University, Germany, is developing a regional methodology for Central America to assess tree resources outside the forest. A mix of satellite remote sensing, aerial photos and ground sampling is used to address the complexity of the resource (number of species, distribution and structure) and to allow dynamic monitoring of resources at the national and regional levels (Kleinn *et al.* 1999). Kleinn *et al* (1999) studied data-gathering on TOF in eight Latin American countries (Brazil, Costa Rica, Colombia, Guatemala, Haiti, Honduras and Peru). None of these countries had established a database and the search for information was multisectoral. The statistics on land cover and land use gave some idea of the relative importance of trees outside the forest.

In Kenya, extensive tree planting on farmlands was promoted in the 1970s and 1980s, with land tenure security as a major incentive. There is an increasing trend of tree cover and species diversification on privately owned farms (Kiyiapi2000). Assuming that the present rate of increase in tree planting will continue, it was estimated that farms produced about 9.4 million cubic metres of wood in 2000 and will produce about 17.8 million cubic metres in 2020. Their share of the total wood produced in the medium- and high-potential districts was projected to increase to 80 percent in 2020 (FDK 1994). Njenga *et al.* (1999) indicate that tree crops contributed 18 to 51 percent of the total household income at the farm level. Indeed, while natural stands of trees have declined there has been a corresponding increase in tree planting in much of the densely populated plateaus of Kenya. As natural forests are reduced or become inaccessible, agroforestry systems help people to diversify production and income and to protect themselves from shortages of fuel and wood (FDK 1994).

In Bangladesh, natural forest formations cover less than 6 percent of the country and the population growth rate is extremely high. An inventory of homestead/village forests in the country (FAO 1981; Douglas 1981; Hammermaster 1982, quoted in Singh 2000) indicated that trees outside the forest constitute a vital resource for local populations, providing food, fodder and fuelwood. The sampling method was based on dual village/household sampling with an agro-ecological and administrative sampling base. Rural Bangladesh was divided into six major regions considered as agro-ecological strata, each subdivided into *thanas* (administrative entities, subdistricts). The households making up the sampling units were chosen at random from a number of villages. The inventory sampled data on palm trees and cane as well as trees, bamboo and thickets. The results, expressed per stratum and per inhabitant, provide volumetric data for fuelwood and sawnwood, and species data for total amounts under and over 20 cm. This inventory was apparently the first to nationwide assessment of trees outside the classified forests in Bangladesh.

METHODS AND TOOLS FOR FUTURE ASSESSMENTS

A priority challenge of future assessments is to know the state and dynamics of all tree resources both in and outside the forest. A country embarking on a planning exercise cannot confine itself solely to the trees within its forests, especially when its wood resources appear to be insufficient.

The choice of tools and methods used to describe or assess trees outside the forest depends on the scale of analysis, kind of data and degree of exactitude desired. The tools used are not generally specific or new; rather, they are combined and implemented in original ways. The inventory in Bangladesh described above is one of the numerous examples of methods developed for gathering data on TOF. The Bangladesh study gives evidence of the adaptations needed to bring conventional forest inventory procedures in line with the specific nature of this resource.

In some aspects – e.g. structure, spatial distribution and extent of area cover – trees outside the forest are more difficult to assess than forest formations. The assessment of TOF does not lend itself to the potential cost savings associated with expanded uses of remote sensing technology. Remote sensing by satellite presents more difficulties for assessing TOF resources than for assessing attributes such as forest area. However, satellite data do allow a region to be stratified on the basis of ecological criteria and land cover, providing the basis for a good working document for more specific work in the future.

The most commonly used remote sensing technology for TOF resources is aerial photography, which can be used to describe spatial distribution and to distinguish TOF cover classifications, providing the appropriate scale is chosen. However, high costs prohibit widespread use of aerial photography for TOF assessments in most countries. The new 1 m resolution satellite sensors represent a possible future alternative to aerial photography.

Some TOF field inventories are modelled on forest inventory methods and keep to biological and physical criteria; others emphasize social aspects, choosing villages as the sampling units. For measurements on the ground, sampling arrangements designed for forest stands may not be the most effective arrangements for trees. Less traditional sampling plans which would theoretically be better suited to this resource should be tested on various categories of TOF, especially those covering fairly large areas.

Studies of the social and economic benefits or impacts of TOF often rely on household surveys, interviews or standardized appraisals such as rapid or participatory rural appraisal.

The integration of the last two approaches – biophysical inventory and socio-economic analysis – is not simple and calls for caution given the great variety of social situations that are only meaningful in the local context.

Environmental benefits or impacts of TOF might be indirectly assessed by linking measurable indicators, such as the number and type of trees, with environmental variables such as water quality or erosion. In an urban setting, tree cover might have direct impact on the ambient temperature. Measuring the environmental impact of tree management is an issue for all natural resource planning or management operations.

Assessment of trees outside the forest requires geographical, ecological, biophysical, social and economic data. However, this implies that an important amount of information will have to be carefully processed. The diversity of end-uses for this information, including land use planning and analysis based on inventory, will need to be considered in data assembly and processing and in the presentation of results.

It is important to know the status of trees outside the forest at any given moment, but it is even more essential to be able to trace patterns of change over time in the same area. The two most commonly used approaches have been comparison of aerial photos taken at sufficiently long intervals, and surveys among villagers/managers combined with field inventories.

Some countries, such as France and the United Kingdom, have undertaken periodic inventories based on the establishment of permanent plots linked to permanent forest inventories. However, the high cost of this type of operation limits the number of countries able to adopt it. India and Bangladesh are now experimenting with options for the future.

The current trend towards decentralized authority in land use planning suggests the importance of carrying out assessments at the local level, where the geographical, historical and socio-economic context is relatively harmonious. A minimum number of common rules concerning methods and arrangements is necessary, however, if the data are to be comparable at the country level. Certainly, the technical side of assessing trees outside the forest is complex, and more research is needed to better pinpoint the resource.

CONCLUSIONS

Trees outside the forest are increasingly recognized by policy-makers, planners and managers as an essential component of sustainable development. This ancient resource has been part of the daily context and culture of rural populations and in many cases TOF resources are critical to food security. However, much work and discussion will be needed before trees growing in non-forest areas can be considered an integral part of planning and development policies.

One important need is a consensus working definition that can be adapted with time and circumstances to fit the rapidly changing economic, ecological, social and cultural context of this resource. This would facilitate the work of framing laws that are neither sectoral nor contradictory, incorporating rights of ownership, use and access for land and trees. In many countries there is a need for more secure land tenure and user rights for trees growing outside forest areas, especially for less empowered sectors of the population, including women.

Databases on trees outside the forest, although fairly substantial in some countries, remain fragmented, diffuse, sometimes empirical and often sectoral. Advances are needed in practical approaches to the use of knowledge to assess the true contribution of TOF resources to economic needs, social demand and ecosystem maintenance. Inventories and assessments of TOF resources based on reliable and accessible methods are essential to effective land use planning. Strategies to promote and support trees outside forest areas need to address the importance of sustainability as they seek to maintain traditional advantages to populations while expanding opportunities for new benefits from this resource.

Moves towards devolution and empowering people in the management of local resources should enhance and promote the conservation and sustainable use of trees outside the forest. FAO is committed to improving the assessment of trees outside the forest in future global assessments, and to assisting member countries in building their capacity to assess TOF resources effectively and to use this knowledge to evolve and implement effective sustainable development policies and programmes.

BIBLIOGRAPHY

Auclair, D., Prinsley, R., & Davis, S. 2000. *Trees on farms in industrialised countries: silvicultural, environmental and economics issues.* Kuala Lumpur, Malaysia, IUFRO.

Bellefontaine, R. & Ichaou, A. 1999. Pour une gestion reproductive des espaces sylvo-pastoraux des zones à climats chauds et secs, une règle d'or: l'O.S.R. – orienter, simplifier, mais surtout régénérer. Réseau International des Arbres Tropicaux, *Le Flamboyant*, 51: 18-21.

Bergeret, A. & Ribot, J. 1990. *L'arbre nourricier en pays sahélien.* Paris, Editions de la Maison des Sciences de l'Homme.

FAO. 1995. *Forest products consumption in the Sudan*. Final Report. Forest Development project FAO/GCP/SUD/047/NET. Rome.

FAO. 1998. Wood energy situation and trends. Paper prepared for the World Energy Council (WEC) as a contribution to the study on energy in developing countries, presented at the 1998 WEC Congress, Houston, Texas, USA.

FAO. 1999. *State of the world's forests 1999.* Rome.

FAO. 2001. *Arbres hors forêts – vers une meilleure prise en compte.* FAO Conservation Guide No. 36. Rome. (in press)

Finnish Forest Research Institute. 1996. *Expert Consultation on Global Forest Resource Assessment 2000. Kotka III.* Proceedings of FAO Expert Consultation on Global Forest Resources Assessment 2000 in cooperation with ECE and UNEP with the Support of the Government of Finland (KOTKA III). Kotka, Finland, 10-14 June 1996. Eds. Nyyssonen, A. & Ahti, A. Research Papers No. 620. Helsinki, Finland.

Forest Department of Kenya (FDK). 1994. *Kenya Forestry Master Plan.* Nairobi, Ministry of Environment and Natural Resources.

Forest Survey of India (FSI). 1998. *State of forest report 1997.* Dehra Dun.

FSI. 2000. *Trees outside the forest resource of Haryana, Forest Survey of India, Northern Zone, Shimla.* Dehra Dun (in press).

Glen, W.M. 2000. *Trees outside the forest assessment in the Sudan: a contribution to the Forest Resources Assessment 2000 report.* FAO, Rome (draft paper).

Inventaire Forestier National (IFN). 2000. *Les arbres hors forêts en France – une ressource inféodée à l'agriculture. A case study in contribution to the FAO Forest Resource Assessment Programme.* FAO, Rome.

Kleinn, C. 2000. On large-area inventory and assessment of trees outside the forest. *Unasylva*, 200(51): 3-10.

Kleinn, C., Baker, C.G., Botero, J., Bolivar, D., Girón, L., *et al.* 1999. *Compilation of information on trees outside the forest: a contribution to the Forest Resource Assessment 2000 of FAO. Regional study for Latin America (including also Haiti).* CATIE-FAO.

Kiyiapi, J.L. 2000. Forest cover type, habitat diversity and anthropgenic influences on forest ecosystems adjoining the Maasai Mara Game Reserve, Kenya. *In* M. Hansen & T. Burk (eds.), *Integrated tools for natural resources inventories in the 21st Century*, p. 296-304. Proceedings of the IUFRO Conference, August 16-20, 1998. USDA, North Central Research Station, General Technical Report NC-212. Boise, Idaho, USA.

Krishnakutty, C.N. 1990. *Demand and supply of wood in Kerala and their future trends.* Kerala Forestry Research Institute (KEFRI) Research Report 67. Peechi, Kerala, KEFRI.

Morocco. Ministère de l'agriculture, du développement rural et des pêches maritimes (MADRPM). 2000. *Stratégie des filière de production végétale à l'horizon 2020.* Vol. II. National Symposium on Agriculture and Rural Development, 19-20 July 2000. Rabat.

Njenga, A., Wamicha, W.N. & van Eckert, M. 1999. Role of trees in small holder farming systems of Kenya: results from high, medium and low potential areas in Kenya. *Proceedings of a Workshop on Off-Forest Tree Resources of Africa*, 12-16 July, Arusha, United Republic of Tanzania. Nairobi, Kenya, African Academy of Sciences.

Singh, C.D. 2000. *Valuation and evaluation of trees outside the forest (TOF) of Bangladesh. Draft paper for a regional study for Asia and Pacific in contribution to the Forest Resource Assessment (FRA) 2000.* FAO, Rome.

Chapter 5
Biological diversity

ABSTRACT

Biological diversity denotes the variety of life forms, the ecological roles they perform and the genetic diversity they contain (FAO 1989). While recognizing the complexity of the issue, the Expert Consultation on Global Forest Resources Assessment 2000 ("Kotka III") recommended that FRA 2000 address key indicators that might contribute towards a better understanding of the status and trends in forest biological diversity, including information on forests by ecological zones; protection status; naturalness; and fragmentation. In other chapters and in the global tables, this report provides information related to these indicators. Two studies carried out within the framework of FRA 2000 are summarized in the present chapter. These address the number of forest-occurring ferns, palms, trees, amphibia, reptiles, birds and mammals by country; and the spatial attributes of forests that define one aspect of "naturalness", applicable at the global level. Conceptual difficulties related to the assessment of biological diversity in forests at the global level are also addressed.

INTRODUCTION

It is generally accepted that the conservation of forest biological diversity, at the levels of ecosystems, landscapes, species, populations, individuals and genes, is essential to sustain the health and vitality of forest ecosystems, thereby safeguarding their productive, protective, social and environmental functions.

The greatest threat to forests and their diversity is conversion to other land uses. Increasing pressure from human populations and aspirations for higher standards of living, without due concern to the sustainability of the resources underpinning such developments, heighten these concerns. While some land use changes are inevitable, it is important that such changes be planned and managed to address complementary goals. Concerns for biological and genetic conservation should be major components of land use planning and forest management strategices (Soulé and Sanjayan 1998; Wilcox 1990; FAO 1995; FAO/IPGRI/DFSC 2001; FAO 2001).

Biological diversity is intensively discussed at policy levels and within the global scientific community, and it is the focus of attention of many international and national non-governmental organizations. A number of dedicated journals directly address biological diversity and related issues. At the international level, many organizations and agencies address biological diversity in their programmes (FAO 2001c). Forest biological diversity is a concern of the Collaborative Partnership on Forests (CPF) chaired by FAO. The Secretariat of the Convention on Biological Diversity (CBD) plays a lead role within the CPF on this issue.

Guidance was sought from the FAO Expert Consultation on Global Forest Resources Assessment 2000 (Kotka III) regarding the extent to which FRA 2000 might address this important issue (Finnish Forest Research Institute, 1996). The Kotka III report included the following recommendation on biological diversity:

> The meeting recognized the conceptual and practical difficulties of directly measuring biological diversity, but noted that considerable progress in understanding the situation and trends for biological diversity in the world's forests could be made by including in the global framework questions on the following:
> - naturalness (breakdown into natural forest, semi-natural forest and plantations);
> - protection status (using IUCN categories to improve comparability);
> - fragmentation;
> - better information on forests by ecological zone.

Extensive reporting on these indicators, including protected areas, is provided in other chapters of this report. In addition, FRA 2000 carried out two studies on specific aspects of forest biological diversity. The results of these studies are summarized in this chapter. This chapter also discusses some of the conceptual and practical difficulties related to the assessment of forest biological diversity at the global level.

For further discussion of the status of efforts to assess forest biological diversity at the global level, refer to "Assessments of forest-based biological diversity" in the *State of the World's Forests 1999* (FAO 1999).

ASSESSING FOREST BIOLOGICAL DIVERSITY

The goal of conserving biological diversity is to ensure that variability and variation will continue to be present and can dynamically develop and evolve both through natural processes and through the direct or indirect intervention or influence of humans (Eriksson *et al.* 1993; FAO 1989; FAO 2001c). The values derived from biological diversity are associated with different scales. These include ecosystems, landscapes, species, populations, individuals and genes. Varying and complex interactions exist between all of these levels (see Namkoong 1986; FAO 2000; Sigaud *et al.* 2000). In implementing a conservation strategy, it is important to specify which level of diversity is discussed and to identify the ultimate aim of the strategy (Eriksson *et al.* 1993; Palmberg-Lerche 1999; FAO/IPGRI/DFSC 2001).

Because biological diversity encompasses the complexity of all life forms, its assessment and monitoring are only possible for specific aspects or particular, defined goals. There is no single, objective measure of biological diversity, only complementary measures appropriate for specified and, by necessity, restricted purposes (Norton 1994; Williams 1999). The use of "indicator species" as a surrogate in biological diversity assessment is a common approach.

A number of major challenges must be faced in designing an assessment of global forest biological diversity. These are not unique to biological diversity, but are general inventory problems for variables in which target parameters are complex and highly variable.

First, the complexity and variation of forest biological diversity at the global level must be expressed in a simplified, uniform and easily understood set of variables that represent the major values of forest biological diversity. Such a set of variables must, by necessity, be based on generalizations that use indirect (surrogate) measures, typically in the form of indicators which are based on the general (qualitative) condition of the forest and the likely development following management events or natural developments (Thuresson *et al.* 1999).

Second, the inherently local nature of variations in biological diversity requires that data be inventoried on a sample plot basis and then generalized into broader spatial representations for reporting purposes. World maps indicating diversity at ecosystem or species levels can only indicate spatial variations at large scales, perhaps 10 km and larger. Summarized tables with national-level statistics on such variables will be much less detailed. Even if a good set of indicators is identified, part of the meaning may be lost when data are interpreted as an average over larger areas. This problem can, in theory, be at least partly remedied by reporting on local variations of specific indicators rather than averages, but this leads to other problems: assessing local variations is complex and very costly, and the results become more abstract, difficult to comprehend and difficult to incorporate in policy processes.

RESULTS OF FRA 2000 STUDIES

Given the above difficulties and severe data limitations, two studies were implemented within the framework FRA 2000 by the United Nations Environment Programme (UNEP) and the World Conservation Monitoring Centre (WCMC), (FAO 2001a; FAO 2001b). The first study reviewed and documented the number of forest-occurring species by country, and the proportion of these that were considered to be endangered according to the classification and definition of IUCN (see below). The second study addressed indicators of the spatial attributes and integrity of forests that might be applied at the global level and that define one aspect of "naturalness".

Study on endangered forest species

A desk study of endangered forest-living species was made to provide a generalized estimate of the national importance of forests as habitats for biological diversity, at the ecosystem and species levels (FAO 2001a). It was recognized from the outset that obtaining accurate data would be difficult; therefore specific groups of species were selected for review based on the anticipated availability of data.

The study was principally designed to make use of existing data from the databases at UNEP-WCMC, which supported a number of published documents, including the *IUCN Red List of Threatened Animals* (IUCN 1996), the *IUCN Red List of Threatened Plants* (IUCN 1997) and the *World List of Threatened Trees* (IUCN 2000). The categories used and criteria for establishing endangered status for species are fully documented in the source publications. This

Table 5-1. Data availability by species group

Group	All species occurring in country				Forest-occurring species			
	All species		Endemic species		All species		Endemic species	
	Total[1]	Endangered[1]	Total	Endangered[1]	Total	Endangered	Total	Endangered[1]
Ferns	Good	Good	Limited	Good	Good	Good	Limited	Good
Palms	Good	Good	Good	Good	Good	Good	Good	Good
Trees	No data	Good[2]	Limited	Good	No data	Good[2]	Limited	Good
Amphibia	Good	Good	Partial	Good	No data	No data	No data	Good
Reptiles	Good	Good	Partial	Good	No data	No data	No data	Good
Birds	Good	Good	Partial	Good	No data	No data	No data	Good
Mammals	Good	Good	Partial	Good	No data	No data	No data	Good

[1] Column included in presentation of global statistics (Appendix 3, Table 13).
[2] For most countries.

information is also available on the Internet (www.unep-wcmc.org).

Seven categories of species were selected for review. However, even using this procedure, data were only partially available, as indicated in Table 5-1.

Total number of species by country. In general, estimates of number of species by taxonomic group, by country, were available in the literature. To a large extent this information was also already recorded in the UNEP-WCMC species database. This database was updated as part of the FRA 2000 study. An exception to overall availability of data was information related to the category "trees", for which data on country (and global) totals were not available. "Trees" is not a readily definable group and data will have to be gathered on a species-by-species (or genus by genus) basis, in the absence of an authoritative, global world list of trees. Currently the only possibility is to estimate figures based on national floras, where these exist. Such work was beyond the scope of the FRA 2000 study.

Total of forest-occurring species by country. Reliable data were only available for two relatively small groups, palms and ferns. Figures for all forest-occurring species in these groups by country were not available from the literature nor in the UNEP-WCMC species databases.

Endemic species by country. Some information on endemic animal species, i.e. species occurring in one single country, was available in the UNEP-WCMC database of endemic animal species. Complementary information from country-based reports was added, where available. Palms, once again, proved to be an exception, with good data available.

Endangered species per country, aggregated at global level. Good data were available for all categories, based on the UNEP-WCMC threatened species databases.

Forest-occurring endangered species per country, aggregated at global level. Good data were available for the category "plants". Data were not available for animal species. As reported for the above categories, the task of identifying the occurrence of threatened animal species in forests, on a species-by-species basis, was beyond the scope of the present study.

Endangered endemic species by country. Data for this subset were available in the UNEP-WCMC databases.

Endangered forest-occurring endemic species by country. In this subset, all threatened endemic species that occurred in one single country, in forest ecosystems, were identified.

The results, by country, are shown in (Appendix 3, Table 13) Information is given in those columns in which reliable data were available for the above-mentioned groups of species. Figure 5-1 displays the total number of endangered, country endemic and forest-occurring species against the forest area change for countries with more than 1 million hectares of forest.

The FRA 2000 questionnaire sent by UNECE/FAO to industrialized temperate and boreal zone countries included a request for information on endangered forest-occurring species. The outcome is not directly comparable with the findings of the global study reported above, and the results are therefore not included in this chapter. However, this complementary information can be found in UNECE/FAO (2000).

Study on spatial indicators

The methodological study for spatial indicators of forest biodiversity (FAO 2001b) was carried out

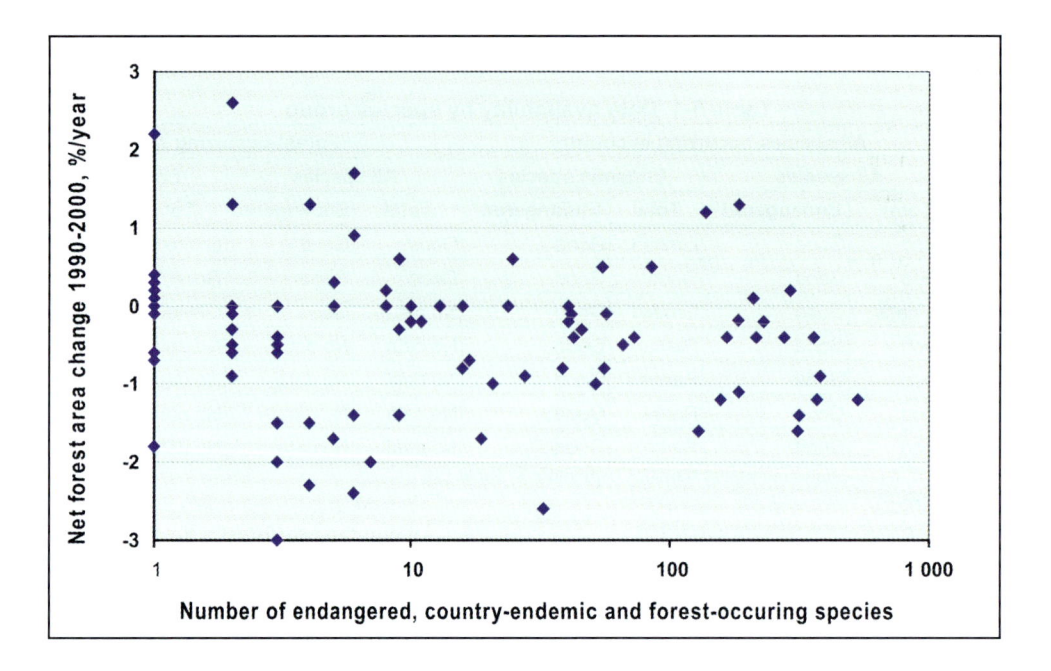

Figure 5-1. Endangered species (all seven species groups) against forest area change for countries with more than 1 million hectares of forest

for FAO by UNEP-WCMC as a contribution to FRA 2000. It was based on the assumption that deforestation and forest fragmentation have a negative impact on biological diversity by altering the spatial configuration of forests. The study considered the possibility of monitoring likely impacts of such forest disturbance on biological diversity by considering the following parameters and making the listed, general assumptions.

- *Area effects*: Forest area was assessed, based on the assumption that patches of forest that are reduced in size will support only subsets of the species found in larger, continuous forest areas, and that these species are more vulnerable to loss because of their relatively small population size.
- *Edge and gradient effects*: The increase in areas where forests interface with non-forest ecosystems was assessed, based on the assumption that this will negatively affect environmental variables and biotic interactions.
- *Isolation effects*: The isolation of populations of a given species from other populations of the same species was assessed, based on the assumption that this will reduce geneflow (genetic exchange) among populations.[5]

Forest configuration and spatial integrity at broad geographic scales were assessed using geographic information system technology (GIS) and considering the following:

- *Patch size*: the area of each contiguous unit of forest cover;
- *Spatially weighted forest density*: the percent of cells within a given radius that are occupied by forest;
- *Connectivity*: distance between isolated forest fragments and larger forest areas.

The study concluded that GIS can potentially be useful in monitoring changes in these spatial indicators over time, using repeatable algorithms. However, considerable conceptual work remains to be done to link the parameters to actual impacts on biological diversity.

CONCLUSIONS

While a number of generally agreed indicators of changes in forest area, structure and composition can be assessed, there is no accepted methodolgy for directly linking these changes to their impacts on forest biological diversity in forest ecosystems, landscapes, species, populations and genes. This is especially evident when information is aggregated at the global level. Compounding this problem is the lack of agreement at national and local levels regarding the extent to which these linkages are relevant and scientifically sound, and the extent to which comprehensive assessments

---

[5] Such reduction in geneflow may, however, have either positive or negative effects from a genetic point of view, depending on the size of the fragments in which isolated populations develop, variation in the original fragment, and the biology, breeding systems and overall patterns of variation of the species concerned.

are technically possible and economically feasible.

The study on spatial indicators concluded that baseline assessments and monitoring of spatial integrity and naturalness would advance the state of knowledge about forest biological diversity. There is a need to monitor trends, not only in forest quantity, but also in forest quality with respect to biological diversity.

It is suggested that future action focus on the further development of and support to the testing and implementation of indicators related to each of the globally accepted criteria for sustainable forest management (FAO 2001d). In such action the level or levels of diversity targeted for conservation must be clearly specified (ecosystems, landscapes, species, genes), and action must be accompanied by regular monitoring to assess progress towards stated objectives.

Information on the status and trends of the world's forests is of basic importance to assessing the status and trends of forest biological diversity. The FRA 2000 studies described above were aimed to further contribute to this issue. However, it is recognized that the value of information on endangered species has some serious limitations in this regard, compounded by the lack of basic data. The use of spatial information is, from a technological point of view, a feasible approach for monitoring and modelling, but its relevance for the assessment of status and trends in forest biological diversity remains to be determined.

BIBLIOGRAPHY

Eriksson, G., Namkoong, G. & Roberds, J.H. 1993. Dynamic gene conservation for uncertain futures. *Forest Ecology and Management,* 62: 15-37.

FAO. 1989. *Plant genetic resources: their conservation in situ for human use.* Document prepared in collaboration with UNESCO, UNEP and IUCN. FAO, Rome.

FAO. 1993. *Conservation of genetic resources in tropical forest management: principles and concepts.* FAO Forestry Paper No. 107. Based on the work of R.H. Kemp, with scientific review by G. Namkoong and F. Wadsworth. FAO, Rome.

FAO. 1998. *Management of forest genetic resources: some thoughts on options and opportunities,* by C. Palmberg-Lerche. Forest Genetic Resources No. 26. Rome.

FAO. 1999. *State of the world's forests 1999.* Rome.

FAO. 2000. *Management of forest genetic resources: their conservation, enhancement and sustainable utilization.* Forestry Department Information Note. Rome. www.fao.org/forestry/FODA/Infonote/en/t-fgr-e.stm

FAO. 2001a. *Forest occurring species of conservation concern: review of status of information for FRA 2000.* FRA Working Paper No. 53. www.fao.org/forestry/fo/fra/index.jsp

FAO. 2001b. *Assessing forest integrity and naturalness in relation to biodiversity.* FRA Working Paper No. 54. www.fao.org/forestry/fo/fra/index.jsp

FAO. 2001c. *International action in the management of forest genetic resources: status and challenges.* Based on the work of C. Palmberg-Lerche. Forest Genetic Resources Working Papers No. 1. Rome.

FAO. 2001d. *Criteria and indicators for sustainable forest management: a compendium.* Paper compiled by F. Castañeda, C. Palmberg-Lerche and P. Vuorinen. Forest Mangement Working Papers No. 5. Rome.

FAO. 2001e. *Conservation of forest biological diversity and forest genetic resources,* by C. Palmberg-Lerche. Forest Genetic Resources No. 29. Rome (In press).

FAO/IPGRI/DFSC. 2001. *Conservation and management of forest genetic resources.* Volume 2: *Forest genetic resources conservation and management in managed natural forests and protected areas (in situ).* Rome, International Plant Genetic Resources Institute.

Finnish Forest Research Institute. 1996. *Expert consultation on Global Forest Resource Assessment 2000. Kotka III.* Proceedings of FAO Expert Consultation on Global Forest Resources Assessment 2000 in cooperation with ECE and UNEP with the Support of the Government of Finland (KOTKA III). Kotka, Finland, 10-14 June 1996. Eds. Nyyssonen, A. & Ahti, A. Research Papers No. 620. Helsinki. Finland.

IUCN. 1996. *The IUCN red list of threatened animals.* Gland, Switzerland.

IUCN. 1997. *The IUCN red list of threatened plants for ferns and palms.* Gland, Switzerland.

IUCN. 2000. *The world list of threatened trees.* Gland, Switzerland.

Libby, W.J. 1987. Genetic resources and variation. In: *Forest trees in improving vegetatively propagated crops*. New York. Academic Press.

Namkoong, G. 1986. Genetics and the forests of the future. *Unasylva*, 38(152): 2-18.

Norton. 1994. On what we should save: the role of culture in determining conservation targets. In: *Systematics and conservation evaluation*, p 23-29. Eds. Forey, P.L., Humphries, C.J. & Vane-Wright, R.I. Oxford, UK, Oxford University Press.

Palmberg-Lerche, C. 1999. Conservation and management of forest genetic resources. *Journal of Tropical Forestry Research,* 11(1): 286-302.

Sigaud, P., Palmberg-Lerche, C. & Hald, S. 2000. International action in the management of forest genetic resources: status and challenges. In: B. Krishnapillay *et al.* (Eds): *Forests and society: the role of research.* XXI IUFRO World Conference, Kuala Lumpur, Malaysia, August 2000. Vol. I.

Soulé, M.E. and Sanjayan, M.A. 1998. Conservation targets: do they help? *Science,* 279: 2060-2061.

Thuresson, T., Drakenberg, B. & Ter-Gazaryan, K. 1999. *A sample based forest resource assessment of the forests possible for exploitation in Armenia.* Jönköping, Sweden. National Board of Forestry.

UNECE/FAO. 2000. *Forest resources of Europe, CIS, North America, Australia, Japan and New Zealand: contribution to the global Forest Resources Assessment 2000.* Geneva Timber and Forest Study Papers No. 17. New York and Geneva, UN. www.unece.org/trade/timber/fra/pdf/contents.htm

Wilcox, B.A. 1990. *Requirements for the establishment of a global network of in situ conservation areas for plants and animals.* Rome, FAO. (unpublished)

Wilcox, B.A. 1995. Tropical forest resources and biodiversity: the risks of forest loss and degradation. *Unasylva*, 46(181): 43-49.

Williams, P.H. 1999. Key sites for conservation: area selection methods for biodiversity. In: G.M. Mace, A. Balmford and J.R. Ginsberg (Eds). *Conservation in a changing world: integrating processes into priorities for action.* Cambridge, UK, Cambridge University Press.

Chapter 6
Forest management

ABSTRACT

This chapter provides information on status and trends in forest management by reporting on three selected national-level forest management indicators: whether the country is a member of an international initiative to develop and implement criteria and indicators for sustainable forest management; the area of forest covered by a management plan in each country; and the area of forest certified for sustainable forest management in each country. As of 2000, 149 countries were involved in nine different criteria and indicator processes. The information supplied on areas of forests under management indicate that 89 percent of forests in industrialized countries are being managed "according to a formal or informal management plan". National statistics on forest management plans were not available for many developing countries; preliminary estimates showed that at least 123 million hectares, or about 6 percent of the total forest area, were covered by a "formal, nationally approved forest management plan covering a period of at least five years". The area of certified forests worldwide at the end of 2000 was estimated to be about 80 million hectares, or about 2 percent of total forest area. Most certified forests are located in temperate, industrialized countries. Reliable information on longer-term trends of forest management worldwide is not readily available. The FAO Forest Resources Assessments of 1980 and 1990 and a study undertaken by the International Tropical Timber Organization (ITTO) in 1988 provide useful points of reference. In summary, the situation as regards forest management has improved in most regions during the period 1990-2000. Future global forest assessments should provide much improved data, as more countries begin to monitor indicators for sustainable forest management.

INTRODUCTION

Developments in forest management over the past decade have focused on progress towards sustainable forest management, an approach that balances environmental, socio-cultural and economic objectives of management in line with the Forest Principles[6] agreed at the United Nations Conference on Environment and Development (UNCED) in 1992.

These efforts have stimulated changes in forest policy and legislation and in forest management practices in many countries. Public participation in forest management has increased in many countries. Broader approaches to forest management, such as ecosystem management and landscape management, are becoming more widely accepted and implemented. These approaches recognize the dynamism of ecological and social systems, the benefits of adaptive management, and the importance of collaborative decision making. Integrated strategies for forest conservation, in which conservation of forest resources and biological diversity entails management both inside and outside forest protected areas, are increasingly being developed.

On the international level, efforts to encourage sustainable forest management include initiatives to achieve a common understanding of the concept through the development of criteria and of indicators[7] by which sustainability of forest management can be assessed, monitored and reported at national and local levels. In some countries, model and demonstration forests have been established to demonstrate sustainable management in practice for a variety of forest types and management objectives.

As regards production forests, countries are moving towards broader management objectives. Initiatives established in the past decade included the International Tropical Timber Organization

---

[6] The full title is the Non-Legally Binding Authoritative Statement on Principles for a Global Consensus on the Management, Conservation and Sustainable Development of All Types of Forest.

[7] **Criteria** define the essential elements or principles against which sustainability of forest management is judged, with due consideration paid to the environmental, economic and socio-cultural roles of forests and forest ecosystems. Each criterion is defined by quantitative or qualitative **indicators**, which are measured and monitored regularly to determine the effects of forest management interventions over time.

(ITTO) Year 2000 Objective, which promoted sustainable forest management in countries that produce and consume tropical timber. A number of regional and national forest harvesting codes were also developed. Certification of forest products, a market-based mechanism devised to encourage the sustainable management of forests, has recently received considerable attention.

Despite these indications that there may be cause for cautious optimism, reliable information on status and trends in forest management worldwide is not readily available. Few attempts have been made in the past to estimate the extent of sustainable forest management in the world as a whole. Given the number of countries and the wide variety of forest types, local conditions and management objectives, this is, perhaps, not surprising. Previous attempts have, as a consequence, focused on specific regions and on specific management objectives and definitions of sustainable forest management.

Past studies which provide useful points of reference include the FAO Forest Resources Assessments of 1980 and 1990 and a study undertaken by ITTO in 1998. A recent assessment of progress towards the ITTO Year 2000 Objective provides valuable qualitative information on the forest management status in all ITTO producer and consumer countries.

Past studies on tropical forests

FRA 1980. The FAO/UNEP assessment of tropical resources in 1980 covered 76 countries in tropical America, Africa and Asia. The area of productive, closed natural tropical forest (comprising broad-leaved, coniferous and bamboo forest) was estimated at 886 million hectares of which an estimated 42 million (4.7 percent) were subject to intensive management for wood production.[8] Almost 80 percent of the area that was intensively managed was located in only one country (India). An additional 169 million hectares (19.1 percent) were subject to harvesting without intensive management and the remaining 674 million hectares (76 percent) were classified as "undisturbed".

No estimate was provided for the management status of "unproductive" closed forests, totalling 315 million hectares of forest unavailable for wood production for physical or legal (including protected status) reasons. Nor was an estimate provided for the management status of open forest formations. (FAO/UNEP1982; FAO 1988).

FRA 1990 did not report information on forest management in tropical countries.

ITTO 1988. In 1988, a study by ITTO (Poore *et al.* 1989) reviewed the status of forest management in 17 of the then 18 producer country members and concluded that the total area of natural forest under sustained-yield management for timber production[9] was about 1 million hectares[10] out of a potential productive forest of about 690 million hectares, or approximately 0.1 percent. However, the study also concluded that large areas *nearly* met the criteria for sustainable management.

Both of these studies were limited to production forests in the tropics and reported the area under management in percentage of the potential production forest area, not of the actual area subject to timber harvesting. A large area of the potentially productive forest area was, in fact, classified as "undisturbed" and therefore not in need of being managed for wood production.

Past studies on temperate and boreal forests

FRA 1980. As part of the UNECE/FAO assessment of forest resources of Europe, the Soviet Union and North America, a questionnaire was sent out to 32 countries of the UNECE in December 1981. In addition to providing information on forest cover, 24 countries in Europe (including Cyprus, Israel and the Soviet Union) reported on the area of closed forest being managed according to a forest management plan and on the size of the forest area without a plan but subject to controls relating to management or use. All of these countries reported that all their closed forests were subject to either a management plan or some form of control of management or use. The total area of closed forest

---

[8] Defined as follows: "The concept of intensive management is used here in a restricted way and implies not only the strict and controlled application of harvesting regulations but also silvicultural treatments and protection against fires and diseases." (FAO 1988, FAO/UNEP 1982)

[9] Defined in this study as follows: "Management should be practised on an operational rather than experimental scale and should include the essential tools of management (objectives, felling cycles, working plans, yield control and prediction, sample plots, protection, logging concessions, short-term forest licences, roads, boundaries, costings, annual records and the organization of silvicultural work). Management might be at any level of intensity provided that objectives were clearly specified so that one could assess whether they were being attained; and that there was proven performance (indications that the next crop would be satisfactory and that sufficient natural regeneration exists for the following crop)". (Poore *et al.* 1989; Poore 1990)

[10] When comparing the figure to the above, it should be noted that India was not included in the ITTO study.

in the 24 reporting European countries was estimated at 142 million hectares, of which 83 million hectares, or 59 percent, were reportedly managed according to a management plan. The Soviet Union reported that all of its closed forest area, equalling 792 million hectares, was being managed in accordance with a plan.

FRA 1990. Thirty-four countries were covered in the UNECE/FAO 1990 assessment of forest resources of the temperate zones, and 26 of these (23 European countries, Canada, the United States and Australia) provided information on their forest management status. The total forest area of these 26 industrialized countries in the temperate/boreal zone was estimated at 626 million hectares, of which 347 million hectares, or 56 percent, were considered to be under active management.[11] In Europe, according to the information provided by 23 European countries, the total forest area was estimated at 129 million hectares, of which 92 million hectares, or 71 percent, were reported as being under active management. Although changes in definitions make direct comparisons difficult, a general trend of increase in the percentage of the area under management between 1980 and 1990 was noticeable.

Progress towards the ITTO Year 2000 Objective

This study, undertaken in 2000 and covering all producer and consumer country members of ITTO, assessed progress towards achieving sustainable management of tropical forests and trade in tropical timber from sustainably managed resources (the ITTO Year 2000 Objective). It did not provide quantitative information on the area of forest under sustainable forest management, but recorded a very considerable improvement over the situation recorded in ITTO producer countries in 1988. The study concluded, however, that only six producer countries (Ghana, Guyana, Indonesia, Malaysia, Cameroon and Myanmar) appeared to have established all the conditions that make it likely that they can manage their forest management units sustainably. All consumer countries were found to be committed to sustainable forest management and most European countries were reported as considering that their forests would meet the criteria for sustainable forest management (Poore and Thang 2000).

The present study

The present study does not attempt to estimate the total area of forests under sustainable forest management worldwide, since this would entail extensive field visits to provide a reliable estimate and since discussions on what constitutes sustainable forest management are still ongoing. Rather, it includes information on the following selected indicators demonstrating countries' commitment to working towards sustainable forest management:

- whether the country is currently engaged in an international initiative to develop and implement criteria and indicators for sustainable forest management (one measure of political commitment to the concept of sustainable forest management);
- the area of forest covered by forest management plans (irrespective of management objective);
- the area of forest certified as being under sustainable forest management (applicable primarily to production forests).

Three separate studies provide information on other important indicators: the area of forest classified as protected area (see Chapter 7), the area under approved forest harvesting schemes and the area of forest available for woody supply (see Chapter 9).

Many other relevant and important indicators of sustainable forest management exist but have not been included in the present study because of lack of adequate and comparable information. Efforts will be made to collate information on additional indicators for future reporting.

METHODS

Representatives of the nine ecoregional processes on criteria and indicators provided the information on the number of countries that were participating in these processes during a meeting held in Rome in November 2000.

To obtain updated information on the area of forests under management plans, the topic was specifically included in FRA 2000. The FRA 2000 *Guidelines for assessments in tropical and subtropical countries*, which were sent to all developing countries, included a table for recording the area of forest subject to a forest

---

[11] Defined as "Forest and other wooded land that is managed according to a professionally prepared plan or is otherwise under a recognized form of management applied regularly over a long period (five years or more)".

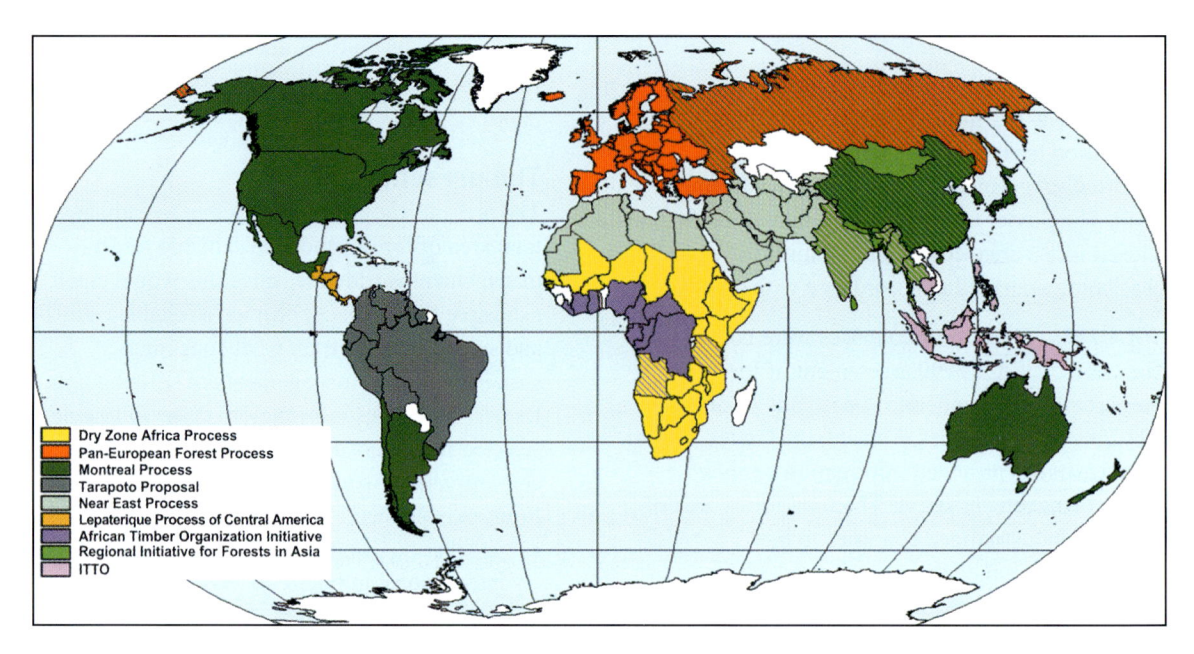

Figure 6-1. Geographical coverage of nine criteria and indicator processes

management plan[12] according to national forest type classification and main management objective (production, conservation, other) where possible. For the industrialized temperate/boreal countries, information was requested on area of managed forest[13] broken down according to ownership status rather than main management objective. No distinction was made among forest types.

As the area of forest subject to a forest management plan is not necessarily consistent with the area of forests being managed sustainably, the information provided by FRA 2000 was supplemented with information provided on areas under certification. This information, which is confined to areas being managed for wood supply, was compiled through a desk study from a variety of mostly Web-based sources.

RESULTS

Criteria and indicators

At the end of December 2000, 149 countries were members of one or more of the following nine initiatives: the Pan-European, Montreal, Tarapoto, Dry-Zone Africa, Near East, Central America/Lepaterique and Dry-Forest Asia processes and/or initiatives, and action taken by ITTO and the African Timber Organization (ATO).[14] Refer to Table 6-1 for a regional overview and to Appendix 3 for details. Figure 6-1 illustrates the geographical coverage of the nine processes.

These initiatives and processes are conceptually similar in objectives and overall approach, although differing in detail. National-level criteria of sustainable forest management focus on the following globally agreed elements: extent of forest resources; biological diversity; forest health and vitality; productive functions of forests; protective functions of forests; socio-economic benefits and needs; legal, policy and institutional framework. The indicators vary widely among initiatives owing to differences in forest types and environmental, social, political and cultural conditions.

National-level criteria and indicators are being complemented by the development and implementation of criteria and indicators defined at the forest management unit level in a number of the processes as well as by other actors such as NGOs and the private sector.

The degree of implementation of criteria and indicators at the national level varies considerably. In many cases, action is limited by the lack of trained personnel or institutional

---

[12] Defined in this context as: "The area of forest which is managed for various purposes (conservation, production, other) in accordance with a formal, nationally approved, management plan over a sufficiently long period (five years or more)".

[13] The term "managed" as applied to forest and other wooded land being defined as: "Forest and other wooded land which is managed in accordance with a formal or an informal plan applied regularly over a sufficiently long period (5 years or more). The management operations include the tasks to be accomplished in individual forest stands (e.g. compartments) during the given period".

[14] At least two additional countries (Cuba and the Lao People's Democratic Republic) have developed national-level criteria and indicators for sustainable forest management without being members of any of the aforementioned ecoregional processes.

Table 6-1. Regional overview of number of countries involved in criteria and indicator processes

Region	Total number reported in FRA 2000	Member of one or more criteria and indicators processes	International/ecoregional processes
Africa	56	46	Near East, Dry Zone Africa, ATO, ITTO
Asia	49	36	Near East, Dry-Forest Asia, ITTO, Montreal, Pan-European
Oceania	20	5	Montreal, ITTO
Europe	41	40	Pan-European, Montreal, Near East
North and Central America	34	11	Lepaterique; Montreal, ITTO
South America	14	11	Tarapoto, ITTO, Montreal
WORLD TOTAL	**213**	**149***	

* Includes four countries being invited to join the Pan-European process as of December 2000 (Bosnia and Herzegovina, Georgia, San Marino and Yugoslavia). Belgium and Luxembourg are counted as two countries in this table although reported on jointly in the main table.

capacity for collecting and analysing information and for following up the development and implementation of improved management prescriptions based on the information obtained.

As most of the nine processes have begun only in the past few years, it is anticipated that future global assessments will be able to gain significant information on a number of forest management indicators.

Forest management plans

Eighty-three countries, including all major industrialized countries,[15] provided national-level information on forest management as part of FRA 2000 reporting. An additional 14 countries supplied comparable information to FAO's Latin American and Caribbean Forestry Commission in 2000. National figures are, however, still missing from a fairly large number of developing countries, including many of the larger countries in Africa and some key countries in Asia. Data are also missing from many smaller countries in Oceania and the Caribbean. An attempt has been made to obtain information from auxiliary sources, but, as can be seen from the results presented in Appendix 3, information is still lacking from a number of countries.

Disparities in replies from industrialized countries, which are not entirely explicable by differences in national situations, suggest that there is a lack of uniformity in the way in which the definition of forest area managed has been interpreted and applied – notably in the distinction between management for wood supply only and for all forest functions, and between management according to approved management plans and less formal forms of management.[16] There was also uncertainty as to whether areas where a decision has been made not to manage the area at all had always been included in the "managed" category, as was recommended by UNECE/FAO (2000), so figures may not be directly comparable between countries.

The definition used for developing countries, on the other hand, was limited to forest areas subject to a formal and nationally approved forest management plan. This definition appeared to cause fewer difficulties in interpretation and application but precludes a direct comparison of results with those from industrialized countries.

In summary, the results indicate that 89 percent of the forests in industrialized countries (accounting for 45 percent of the total forest area in the world, most of which is located in the temperate and boreal zones) were subject to a formal or informal management plan. National figures are still missing from a fairly large number of developing countries, including many of the larger countries in Africa and some key countries in Asia. Nevertheless, results obtained so far indicated that of a total forest area of 2 139 million hectares in non-industrialized countries, at least 123 million hectares, or about 6 percent, were covered by a formal, nationally approved forest management plan with a duration of at least five years.

In analysing the results provided in the table, it is important to keep in mind that the total area reported to be subject to a forest management

---

[15] Europe (including Cyprus, Turkey and Israel which are listed under Asia in Appendix 3, Table 9), Commonwealth of Independent States (CIS), Canada, the United States, Japan, Australia and New Zealand.

[16] Examples of possible deviations from the definition are found in Azerbaijan, where it is assumed that because all forests and other wooded lands are State-owned, they are therefore managed. In Australia, the data on managed forest is limited to areas managed for wood production.

plan is not necessarily equivalent to the total area of forest under sustainable forest management. The present study does not indicate whether the plan is appropriate, being implemented as planned or having the intended effects. Some areas reported as being covered by a management plan may thus not be under sustainable forest management. Conversely, many areas may be under sustainable forest management without the existence of a formal management plan. Furthermore, remote areas with lack of access or very limited human use may not require a management plan or management activities to achieve a management objective of being safeguarded for the future.

Qualitative information on status and trends in silviculture and forest management is presented for selected countries in the country profiles on the FAO Forestry Web pages (www.fao.org/forestry/fo/country).

Regional overview. In Africa, only seven countries provided national-level information, representing less than 3 percent of the total forest area in the region. Only two tropical moist forest countries were included in this list. The percentage of the forest area under a formal, nationally approved forest management plan in these seven countries in 2000 ranged from 2 to 78 percent, with the total area subject to management plans equalling 15 percent of the combined forest area in these countries. Partial figures were obtained from an additional seven countries. All available figures added up to 5.5 million hectares of forests under management plans, equivalent to only 0.8 percent of the total forest area of Africa. Efforts are currently under way to obtain clarification from a number of countries and to supplement the existing information with data from other sources.

In Asia, national-level information was provided by 21 of the 49 countries and areas reported on, accounting for 30 percent of the total forest cover as information was not obtained from the two countries in the region with the largest forest areas. The percentage of the forest area under a formal, nationally approved forest management plan in these 21 countries in 2000 ranged from 23 to 100 percent.[17] It should be kept in mind that the countries of the Commonwealth of Independent States (CIS), where figures ranged from 82 to 100 percent, reported on the area subject to management with a formal or informal management plan and included areas where a decision had been made not to manage the area at all.[18] The total area in the region reported as being subject to forest management plans (including partial results from two countries, and with the two different definitions used kept in mind) equalled 134 million hectares or 24 percent of the total forest area.

In Oceania, only three of the 20 countries and areas provided national data on the area of forest managed. However, these three countries (Australia, New Zealand and Papua New Guinea) together accounted for 98 percent of the total forest area in the region. The figure for Papua New Guinea was limited to formal forest management plans, whereas the definition used for Australia and New Zealand included areas under informal plans and areas where a decision had been made not to manage the area at all. Including the partial results from Solomon Islands, and with the use of two different definitions kept in mind, the total forest area under management plans was 167 million hectares or 84 percent of the total forest area in the region.

In Europe, 39 countries and areas provided national information on areas of forest managed, including areas subject to informal management plans and areas where a decision had been taken not to manage the area at all. With the exception of Italy, which reported only on areas subject to a formal, nationally approved management plan, the figures ranged from 33 to 100 percent of the total forest area in each country. Quite a large number of countries (19) reported that all their forests were managed according to the definition above, including the Russian Federation, which alone accounted for 82 percent of the total forest area in this region. Looking at the region as a whole, 98 percent of the total forest area was reported as being managed.

Thirteen of the 34 countries and areas reported in North and Central America provided national information on the area of forest under management.[19] The total forest area covered by these countries equalled 99 percent of the combined forest area of the region. Canada and the United States, accounting for 86 percent of the total forest area in this region, used the definition that includes areas under informal forest

---

[17] The percentage for the Philippines is above 100 percent as the area figure represents "forest lands", parts of which are not defined as forests according to the FRA 2000 definition.

[18] The exception being Georgia, which does not include areas classified as "undisturbed" as being managed.
[19] Information was especially lacking from many smaller Caribbean countries with a limited extent of forests.

management plans and areas where a decision had been taken not to manage the area at all. The remaining countries reported on areas subject to a formal management plan. Most of these excluded forest plantations from their reporting to the meeting of the FAO Latin American and Caribbean Forestry Commission in 2000. The reported figures ranged from 2 percent of the total forest area to 74 percent. The total area reported as being under management plans in the region (including partial data and with the two different definitions in mind) equalled 310 million hectares or 56 percent of the total forest area.

Eleven of the 14 countries and areas reported in South America provided information on the size of the forest area subject to a formal management plan. Most of them included only natural forests in their report to the meeting of the FAO Latin American and Caribbean Forestry Commission in 2000. The area subject to a forest management plan varied between 0.1 and 25 percent of the total forest area in each country. For the region as a whole, 26 million hectares, or 3 percent, of the total forest area was reportedly subject to a formal management plan. Given that the countries that reported accounted for 94 percent of the combined forest area in the region, these figures may seem low. However, it should be kept in mind that many countries in this region have large expanses of forests that are located in remote areas with lack of access or with very limited human use, which may not require a management plan. It is also uncertain whether all countries included protected forest areas in their reporting on areas covered by forest management plans.

Certification

Criteria and indicators provide a means to measure, assess, monitor and demonstrate progress towards achieving sustainability of forest management in a given country or in a specified forest area over a period of time. Certification, on the other hand, is an instrument used to confirm the achievement of certain predefined minimum standards of forest management in a given forest area at a given point in time. Certification is essentially a marketing tool, used by forest owners who perceive an economic benefit from undergoing the certification process.

A number of international, regional and national forest certification schemes now exist, focusing primarily on forests managed for timber production purposes. The volume of timber covered by these schemes, while increasing, is still relatively low. Depending on how the term "area certified" is defined, the area of certified forests worldwide at the end of 2000 was estimated at 81 million hectares or about 2 percent of the total forest area.

Whereas certification implies that an area is well or sustainably managed, the total area of well-managed forest is not limited to certified areas. Many uncertified forests, including both those managed primarily for wood production and those with other management objectives, may also be under sound management.

While some important wood-producing countries in the tropics have forest areas certified under existing certification schemes or are in the process of developing new schemes, most certified forests are located in a limited number of temperate, industrialized countries. At the end of 2000, about 92 percent of all certified forests worldwide were located in the United States, Finland, Sweden, Norway, Canada, Germany and Poland. At the same time, only four countries with tropical moist forests (Bolivia, Brazil, Guatemala and Mexico) had more than 100 000 ha of certified forests, for a combined total of 1.8 million hectares.

Table 6-2 provides a regional overview. A breakdown by country is provided in Appendix 3. Note that only selected schemes have been included in these tables. Areas certified according to the ISO 14001 Environmental Management System Standard are, for instance, not included in these figures since this standard was not developed specifically to ascertain whether sustainable forest management is being undertaken. Some national certification schemes may also have been left out by oversight. If estimates for these areas were added,[20] the world under certification would be about 110 million hectares (2.8 percent of total forest area).

Some forest certification schemes are now being extended to include certain non-wood forest products (NWFP). This is proving to be more complex than certifying for timber, as the same forest area may have to be assessed for one or more NWFP, which may have different requirements. It is possible, for example, that a forest is managed for timber in a sustainable way while its NWFP resources are being overharvested, and *vice versa*.

---

[20] 27 million hectares in Canada have been certified under ISO 14001 – not counting those areas that have been certified by more than one scheme – and more than 300 000 ha in New Zealand.

Table 6-2. Regional overview of size of forest area certified

Region	Forest area certified
	000 ha
Africa	974
Asia	158
Oceania	410
Europe	46 708
North and Central America	30 916
South America	1 551
WORLD TOTAL	**80 717**

Comparison with previous estimates

A direct comparison with previous estimates of the forest management status in tropical and temperate/boreal regions is not possible because of differences in definitions used. However, it is worth noting that in 1980 an estimated 42 million hectares of forest in 76 tropical countries were reported to be subject to "intensive management for wood production purposes". In 2000, information received so far indicates that at least 117 million hectares[21] of forests in these countries were covered by a formal, nationally approved forest management plan of a duration of more than five years. Most, but not all, of these forests were managed for wood production purposes. A reported 2.2 million hectares of forests in these countries had obtained forest certification by third parties by the end of 2000.

The ITTO study referred to earlier estimated that in 1988 a maximum of 1 million hectares of forest in 17 tropical timber producing countries were being managed sustainably for wood production purposes (Poore *et al.* 1989). Judging from the area under management plans and/or certified in the same 17 countries in 2000, a considerably larger area may now be under sustainable management for wood production purposes. Currently, more than 35 million hectares of forests in these countries are covered by a formal forest management plan, and 1.7 million hectares of forests have been certified by third parties. A considerably larger area is likely to be eligible for certification or under sustainable management for purposes other than timber production. As a case in point, six tropical countries (Cameroon, Ghana, Guyana, Indonesia, Malaysia and Myanmar), with a combined forest area of 206 million hectares, appear to have established all the conditions needed to enable them to manage their forests sustainably in the near future (Poore and Thang 2000).

[21] National data missing from some countries.

The situation in temperate and boreal forests appears to have remained stable or to have improved over the past 20 years. In the early 1980s, all areas classified as closed forests in the former Soviet Union were reported as being "managed according to a forest management plan" (UNECE/FAO 1985). In 2000 the Russian Federation and most of the States of the CIS reported that all forests were being "managed according to a formal or informal plan". Nineteen countries in Europe[22] provided information on the forest management situation in the early 1980s, 1990 and 2000 (UNECE/FAO 1985; UNECE/FAO 1992; UNECE/FAO 2000). The proportion of closed forests "managed according to a forest management plan" in 1980 was 64 percent; in 1990, the proportion of forests "under active management" was 71 percent; and in 2000, 95 percent of the forest area was reported to be "managed in accordance with a formal or informal management plan".

The percentage of the forest area under management in Canada and the United States increased from 60 and 41 percent respectively in 1990 to 71 and 56 percent respectively in 2000.

CONCLUSIONS

A total of 149 countries are currently members of one or more of nine regional or ecoregional initiatives to develop and implement criteria and indicators for sustainable forest management. All of these were established within the last ten years.

One of the indicators identified in these initiatives is the extent of the forest area managed according to a management plan and/or for specific management objectives.[23]

All industrialized countries (accounting for 45 percent of the total forest area in the world, most of it in the temperate and boreal zones) have reported on the area of forest managed as part of FRA 2000 reporting. The results indicate that 89 percent of the forests in these countries are being managed subject to a formal or informal management plan. National figures are still

[22] Including Turkey, which is listed under Asia in Appendix 3, Table 9.
[23] The formulation of this indicator varies between initiatives. Most processes includes the extent of forest area (area or percent) subject to a forest management plan, the exceptions being the Montreal Process and the Tarapoto Proposal, which do not specify a management plan *per se* but rather the percentage of forest area managed for specific objectives. In the Tarapoto proposal, the existence of a forest management plan is, however, one of the indicators at the forest management unit level. The Pan-European Forest Process uses the expression "managed according to a management plan or management guidelines".

missing from a fairly large number of developing countries, including many of the larger countries in Africa and some key countries in Asia. Nevertheless, results obtained so far show that of a total forest area of 2 139 million hectares in non-industrialized countries, at least 123 million hectares, or about 6 percent of the total forest area, are covered by a formal, nationally approved forest management plan with a duration of at least five years.

It must be emphasized that the total area reported to be subject to a formal or informal forest management plan is not necessarily equivalent to the total area of forest under sustainable forest management. The present study does not indicate whether the plan is appropriate, being implemented as planned or having the intended effects. Some areas reported as being covered by a management plan may, therefore, not be sustainably managed, while other areas not currently under a formal management plan may be.

The use of different definitions makes it difficult to compare the situation between industrialized countries and developing countries and to derive a global total of forests under management plans. In addition, some industrialized countries interpreted the definitions in different ways. Moreover, many developing countries did not include forests in protected areas in the area under management, and some countries excluded plantations. These problems suggest a need for further refinement and consistency of approaches in future reporting on the area of forest under management plans.

One way of demonstrating that a particular forest is being managed sustainably for wood production purposes is through the act of third-party certification. A number of international, regional and national forest certification schemes now exist. Depending on how the term "area certified" is defined, the area of certified forests worldwide at the end of 2000 was estimated at around 81 million hectares or about 2 percent of the total forest area. About 92 percent of these forests were located in seven temperate industrialized countries (the United States, Finland, Sweden, Norway, Canada, Germany and Poland). Only four countries with tropical moist forests (Bolivia, Brazil, Guatemala and Mexico) were listed as having more than 100 000 ha of certified forests, for a combined total of 1.8 million hectares.

Whereas certification implies that an area is well or sustainably managed for wood production, the total area of well-managed forest is not limited to certified areas. Many uncertified forests, including both those managed primarily for wood production and those with other management objectives, may also be under sound management.

Notwithstanding the inherent difficulties in comparing FRA 2000 results with those from previous studies because of differences in definitions used and countries included, there are, however, indications that, overall, the situation as regards forest management has improved in most regions over the past 20 years.

BIBLIOGRAPHY

American Forest and Paper Association. 2000a. Report to the joint UNECE Timber Committee & FAO European Forestry Commission. Rome, October 2000.

American Forest and Paper Association. 2000b. Summary of key enhancements to the SFI program in 2000. www.afandpa.org/forestry/SFI/3pagesummary1.pdf

American Tree Farm System. 2000. Web site. www.treefarmsystem.org

Argow, K. 2001. Personal communication: areas under green tag certification.

Canadian Sustainable Forestry Certification Coalition. 2000. Certification status in Canada, 21 December 2000. www.sfms.com/decade.htm

FAO. 1988. *An interim report on the state of forest resources in the developing countries.* Miscellaneous paper FO:MISC/88/7. Rome.

FAO/UNEP. 1982. *Tropical forest resources.* FAO Forestry Paper No. 30. Rome.

Forest Stewardship Council (FSC). 2000. *Forests certified by FSC-accredited certification bodies.* DOC. 5.3.3, 31 December 2000. http://fscoax.org/html/5-3-3.html

National Woodland Owners Association. 2000. *What is green tag forestry?* www.nationalwoodlands.org/green_tag.htm

New Zealand Forest Industries Council. 2000. *Information note on certification.*

Pan European Forest Certification Council (PEFCC). 2000. *PEFCC Newsletter* No. 4, December. www.pefc.org/news.htm

Poore, M.E.D. 1990. Sustainability in the tropical forest. *Journal of the Institute of Wood Science,* 12(2):103-106.

Poore, D., Burgess, P., Palmer, J., Rietbergen, R. & Synnott, T. 1989. *No timber without trees – sustainability in the tropical forest – A study for ITTO*. London, Earthscan Publications.

Poore, D. & Thang, H.C. 2000. *Review of progress towards the year 2000 objective*. Report presented at the 28th Session of the ITTC held on 24-30 May 2000, Lima, Peru. ITTC(XXVIII)/9/Rev.2. Yokohama, Japan, ITTO.

Singh, H. 2001. Personal communication: areas under certification and audit in Malaysia

UNECE/FAO. 1983. *The forest resources of the ECE Region (Europe, the USSR, North America)*. Geneva and Rome.

UNECE/FAO. 1992. *The forest resources of the temperate zones. The UNECE/FAO 1990 Forest Resources Assessment*. Volume 1. General Forest Resource Information. New York, UN.

UNECE/FAO. 2000. Forest resources of Europe, CIS, North America, Australia, Japan and New Zealand: contribution to the global Forest Resources Assessment 2000. Geneva Timber and Forest Study Papers No. 17. NewYork and Geneva, UN. www.unece.org/trade/timber/fra/pdf/contents.htm

Chapter 7

Forests in protected areas

ABSTRACT
This chapter summarizes the results of two initiatives to assess the status of protected forest areas as of the year 2000. The UNEP World Conservation Monitoring Centre (UNEP-WCMC) prepared an updated map of protected forest areas for FAO based on detailed surveys by the World Conservation Union (IUCN) and using IUCN protected area management categories. In addition, industrialized countries submitted statistics on protected forest areas in response to a questionnaire prepared by UNECE/FAO. At the global level, the FAO/UNEP-WCMC mapping project indicates that 12.4 percent of the world's forest area is in protected areas as classified by IUCN. However, there are sometimes discrepancies in statistics reported by different agencies within the same country. Continuous improvement is needed in the assessment approaches used by responsible international organizations and by countries.

INTRODUCTION
In the past two decades, many countries have set aside considerable portions of their forests as national parks or under other categories of protected conservation. The effectiveness of protection and the level of development activity allowed within protected areas have varied considerably. Some countries have suggested that most or all of their forests fall under the protected area status as a consequence of general forestry legislation, but a number of others have held to more traditional views of protection and have included in their reporting only the legally designated protected areas which met international standards.

FRA 2000 prepared an updated report on the protection status of forests at the end of the second millennium.

METHODS
The FRA 2000 assessment of forests in protected areas was based on a new global map of protected forest areas developed for FAO in collaboration with the UNEP World Conservation Monitoring Centre (UNEP-WCMC). In addition, industrialized countries reported on protected forest areas in response to questionnaires prepared by UNECE/FAO.

UNEP-WCMC maintains a global digitized spatially referenced database of protected areas which are classified according to the categories established by the World Conservation Union (IUCN) (see Table 7-1) (UNEP-WCMC 2001). The database is updated periodically through the use of questionnaires circulated by IUCN to national and subnational land management agencies throughout the world. The raw data in the UNEP–WCMC database includes all land under protected management status, not just forest land. Hence, the UNEP-WCMC global protected areas map was overlaid with the new FRA 2000 global forest cover map to arrive at an updated global protected forests map (Figure 7-1) showing the locations of forests in protected areas.

A major technical difficulty in this process arose because some of the geographic reference points in the UNEP-WCMC database are single points rather than the actual shape of the protected area. It was necessary to project a circular shape at the appropriate scale of the actual area at the reference point for the site. As a result, a reasonably accurate intermediate map and associated statistics were generated, but the representation for a given protected area will not be accurate. The map indicates only the cross-tabulation of the forest cover and protected area maps, not the actual protection status of the forest.

The map was overlaid on a country boundary map to generate statistics on the proportion of forests in protected areas for each country. Statistics were not generated for countries and areas smaller than 2 500 km² since the classification accuracy would likely be low for relatively small areas.

The same protected forests map was then overlaid on the FRA 2000 global ecological zones map, and statistics on the proportion of forests inside protected areas were generated for each ecological zone.

Table 7-1. IUCN categories for protected areas as used in FRA 2000

Category	Definition
I - Strict nature reserve/ wilderness area	**Protected area managed mainly for science or wilderness protection.** These areas possess some outstanding ecosystems, features and/or species of flora and fauna of national scientific importance, or they are representative of particular natural areas. They often contain fragile ecosystems or life forms, areas of important biological or geological diversity, or areas of particular importance to the conservation of genetic resources. Public access is generally not permitted. Natural processes are allowed to take place in the absence of any direct human interference, tourism and recreation. Ecological processes may include natural acts that alter the ecological system or physiographic features, such as naturally occurring fires, natural succession, insect or disease outbreaks, storms, earthquakes and the like, but necessarily excluding man-induced disturbances.
II – National park	**Protected area managed mainly for ecosystem protection and recreation.** National parks are relatively large areas, which contain representative samples of major natural regions, features or scenery, where plant and animal species, geomorphological sites, and habitats are of special scientific, educational and recreational interest. The area is managed and developed so as to sustain recreation and educational activities on a controlled basis. The area and visitors' use are managed at a level which maintains the area in a natural or semi-natural state.
III - Natural monument	**Protected area managed mainly for conservation of specific natural features.** This category normally contains one or more natural features of outstanding national interest being protected because of their uniqueness or rarity. Size is not of great importance. The areas should be managed to remain relatively free of human disturbance, although they may have recreational and touristic value.
IV – Habitat/species management area	**Protected area managed mainly for conservation through management intervention.** The areas covered may consist of nesting areas of colonial bird species, marshes or lakes, estuaries, forest or grassland habitats, or fish spawning or seagrass feeding beds for marine animals. The production of harvestable renewable resources may play a secondary role in the management of the area. The area may require habitat manipulation (mowing, sheep or cattle grazing, etc.).
V – Protected landscape/seascape	**Protected areas managed mainly for landscape/seascape conservation and recreation.** The diversity of areas falling into this category is very large. They include those whose landscapes possess special aesthetic qualities which are a result of the interaction of man and land or water, traditional practices associated with agriculture, grazing and fishing being dominant; and those that are primarily natural areas, such as coastline, lake or river shores, hilly or mountainous terrains, managed intensively by man for recreation and tourism.
VI – Managed resource protection area	**Protected area managed for the sustainable use of natural ecosystems.** Normally covers extensive and relatively isolated and uninhabited areas having difficult access, or regions that are relatively sparsely populated but are under considerable pressure for colonization or greater utilization.

Source: McNeely and Miller 1984.

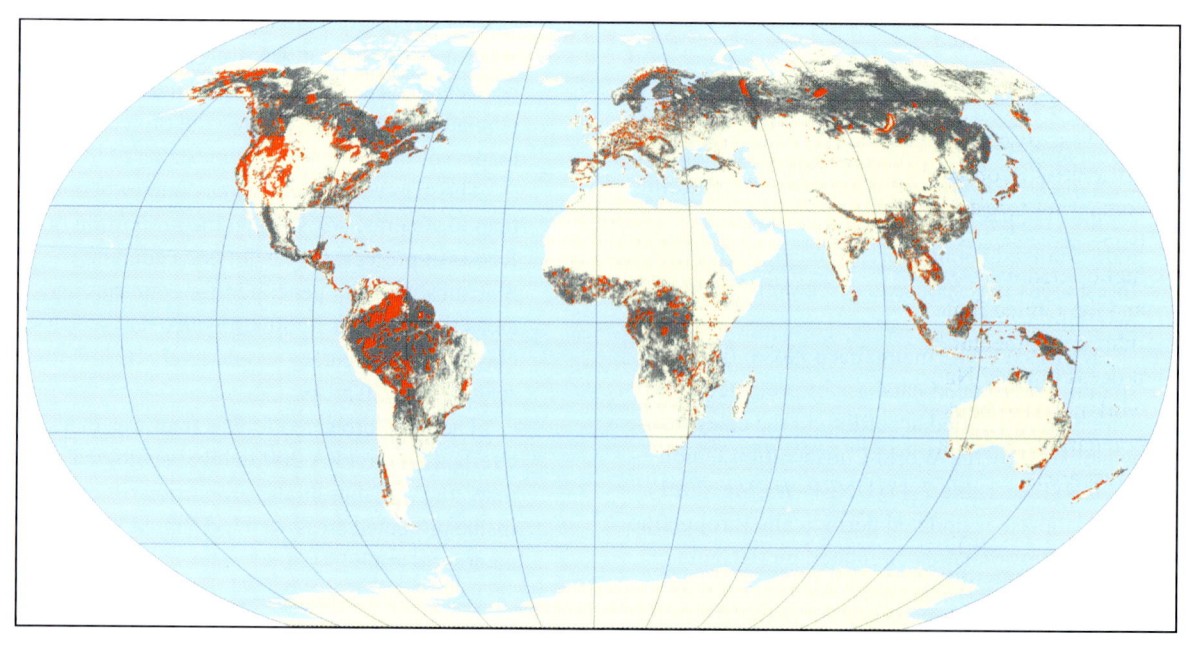

Figure 7-1. Forests in protected areas (in red; other forests in green)

RESULTS

The total extent of forests in protected areas was estimated at 479 million hectares, which is equivalent to 12.4 percent of the world's forest area. As shown in Table 7-2, the two regions of the Americas have a higher proportion of forests in protected areas than other regions. A relatively small proportion, 5.0 percent, of European forests are protected. However, this low figure is explained by the fact that the region's forest area is dominated by the vast forest areas in Siberia, Russian Federation, which for the most part are not officially protected. The results by country are listed in Appendix 3, Table 9 and can be found in the country profiles on the FAO Web site, www.fao.org/forestry.

Results by ecological domain indicate that tropical and temperate forests have the highest proportion of forest in protected areas, whereas only 5 percent of boreal forests are located in protected areas (Table 7-3).

Results from country responses to the UNECE/FAO questionnaires (UNECE/FAO 2001) are listed in Appendix 3. However, comparison of the country responses and results from the global maps shows considerable discrepancies (Figure 7-2). Several countries have interpreted the IUCN categories more broadly in the FRA 2000 questionnaires than in the IUCN surveys. In particular, in response to the UNECE/FAO survey, some countries reported all of their forest area as "protected" because they have national legislation regarding the management or protection of all forests. Other agencies within these same countries, however, did not report all forests as "protected" when they responded to separate requests from IUCN for information on protected areas.

These discrepancies highlight continuing difficulties in obtaining a consistent approach for comparing forest areas that countries report as being protected. Some of these difficulties were highlighted in an expert meeting hosted by Brazil and the United States in March 1999 as part of the Intergovernmental Forum on Forests (IFF) process (United Nations 1999). Clearly, more work needs to be done to improve the national comparability of statistics and maps on protected forest areas.

CONCLUSIONS

The proportion of forest under protection is of particular interest to many governments and to

Table 7-2. Forests in protected areas, based on global protected area map developed for FAO by UNEP-WCMC

Region	Forest area 2000	Forest in protected areas	Proportion of forest in protected areas
	million ha	million ha	%
Africa	650	76	11.7
Asia	548	50	9.1
Oceania	198	23	11.7
Europe	1 039	51	5.0
North and Central America	549	111	20.2
South America	886	168	19.0
Total	3 869	479	12.4

Note: Numbers have been calibrated to the total forest area as reported in Chapter 1.

Table 7-3. Forests in protected areas by ecological domain

Ecological domain	Forest area 2000	Forest in protected areas	Proportion of forest in protected areas
	million ha	million ha	%
Tropical	1997	304	15.2
Subtropical	370	42	11.3
Temperate	507	83	16.3
Boreal	995	49	5.0
Total	3 869	479	12.4

Note: Ecological domains according to FRA 2000 global ecological zone map (Chapter 47). Numbers have been proportionally calibrated to the total forest area.

civil society. FRA 2000 estimated that in 2000, 12.4 percent of the total global forest area fell in the protected area categories defined by the IUCN, as mapped in the global protected areas map.

The map is comprehensive from a global perspective, although there are gaps for individual countries. Some countries did not release the spatial extent of their protected areas, which made the overlay analysis difficult and required the use of approximations using buffered point data to represent land areas.

The discrepancy between results from the global map analysis and the areas reported by national FRA 2000 correspondents is interesting. Obviously, the interpretation of the IUCN classification and its implementation in the national context vary among countries. It is then not surprising that the definitions of protected areas are still being discussed at the international level. It can be expected that as the international discussions on forest protection continue, the standards for designating and reporting protected

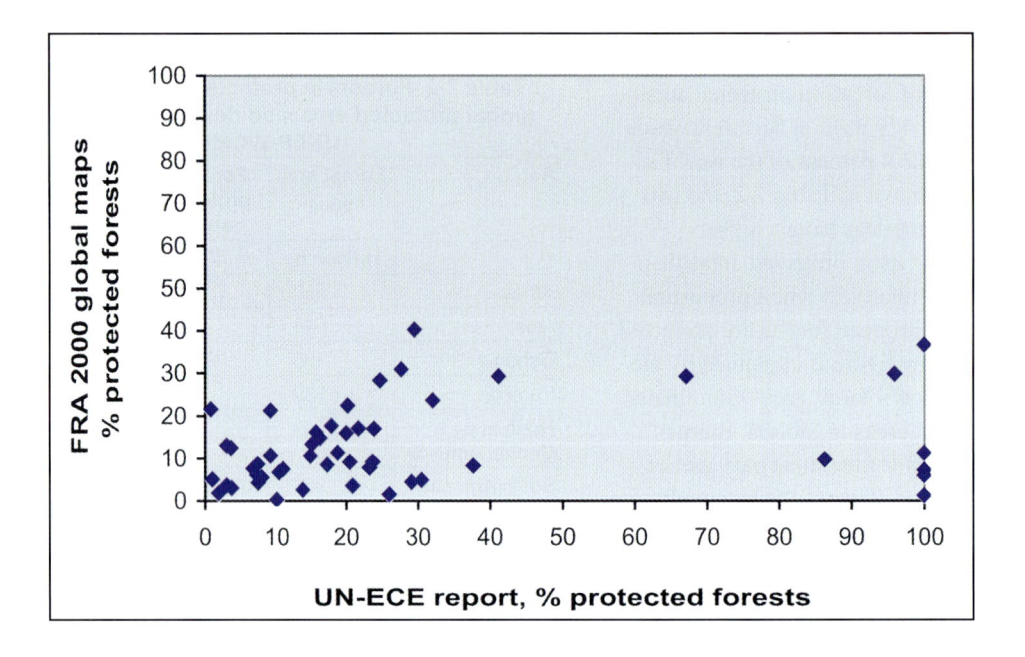

Figure 7-2. Proportion of forest in protected areas in industrialized countries: comparison of results from UNECE/FAO (2000) and FRA 2000 protected areas map

areas will converge and the correlation among different ways of reporting statistics will increase. However, there is considerable work to be done by FAO, UNEP-WCMC, IUCN and national agencies to ensure improvement in the comparability of assessments of protected forests.

At the global level, the proportion of forests in protected areas estimated in FRA 2000 exceeds 10 percent, a figure that has been suggested as a minimum target for protected forest areas. However, it should be noted that statistics at the global level may not be representative of the protection afforded to forests in different ecological zones or in different countries. It should also be noted that varying levels of protection are included in the six IUCN categories, and that not all legally protected forests are effectively managed.

Continuous improvement of the baseline information on protected areas is essential to monitor national commitments to nature conservation. This would also provide a framework for monitoring the status of forest ecosystems within protected areas.

BIBLIOGRAPHY

McNeely, J.A. & Miller, K.R., eds. 1984. Categories, objectives and criteria for protected areas. In: *National parks, conservation and development: the role of protected areas in sustaining society.* Washington, DC, IUCN/Smithsonian Press.

UNEP-WCMC. 2001. *Protected areas information, 1996. Global protected areas summary statistics.* www.unep-wcmc.uk.org

UNECE/FAO. 2000. *Forest resources of Europe, CIS, North America, Australia, Japan and New Zealand: contribution to the global Forest Resources Assessment 2000.* Geneva Timber and Forest Study Papers No. 17. New York and Geneva, UN. www.unece.org/trade/timber/fra/pdf/contents.htm

United Nations. 1999. *Report on the international expert meeting on protected forest areas,* 15-19 March 1999, San Juan, Puerto Rico.

Chapter 8
Fires

ABSTRACT

The 1990s were marked by periods of severe drought, setting the stage for devastating wildfires in practically every corner of the world. The widespread public, media and political attention focused on these wildfires caused decision-makers and resource management agencies to concentrate on policies and practices that could reduce the flammability and vulnerability of wildland ecosystems in the future. An FAO sponsored Meeting on Public Policies Affecting Forest Fires brought together 71 participants from 33 countries and 13 international organizations in October 1998 to develop recommendations to strengthen the fire management capacity of FAO member countries. Participants indicated that a global fire information system was needed to provide immediate access to real-time data and information on fires.

Based on the need for this global fire information system, FRA 2000 initiated a Global Fire Assessment for the 1990s. FAO requested member countries to complete a Fire Management Country Profile which highlighted essential information and data. This chapter summarizes the results of the compilation of fire management information for FAO's six geographical regions (FAO 2001).

Following the summary of regional fire management highlights, several conclusions are drawn characterizing the global fire situation in the 1990s. Policy-makers are beginning to realize that continued emphasis only on emergency response will not prevent large and damaging fires in the future. Emergency preparedness and response programmes must be coupled with better land use policies and practices. Active work towards sustainable forestry practices with community involvement is an important strategy for better conservation of natural resources together with reduced impacts of wildfires.

INTRODUCTION

Severe forest fires around the world gained international attention during the 1990s. Millions of hectares burned in 1997 and 1998 and smoke blanketed large regions of the Amazon Basin, Central America, Mexico and Southeast Asia, disrupting air and sea navigation and causing serious public health problems. Significant losses of forest vegetation and biomass resulted. Ecosystems generally not subject to fires, such as the Amazon rain forest in Brazil and the cloud forest of Chiapas in Mexico, sustained considerable damage. The global wildfire situation in 1999-2000 was again serious, although on a smaller scale. Fires were widespread in Indonesia in 1999 and 2000, but not on a scale comparable to 1997-1998. The major fires of 2000 occurred in Ethiopia, the eastern Mediterranean and the western United States.

These were "headline news" fires, but widespread fires in many places of the world do not reach the international press. Hundreds of thousands to millions of hectares burn annually in fire-adapted ecosystems in dry west Africa, large areas of Africa south of the equator, central Asia, southern South America and Australia. For example, during the 2000 fire season as much as 200 million hectares south of the equator in Africa (including savannahs and grasslands) may have burned. Prevention and control of these recurring and widespread fires could reduce adverse impacts on ecosystems and the livelihoods of local people.

Comprehensive national, regional or global statistics on wildland fires are not available that would allow a reliable and precise comparison of global fire occurrence in the 1980s and 1990s. However, some general observations can be made. Both decades experienced high annual variability in regional and national fire occurrence and impacts. El Niño episodes, such as in 1982-1983 and 1997-1998, were the most important climatic factor affecting area burned and fire impacts in both decades. In these years most of tropical Asia, Africa, the Americas and Oceania experienced extreme wildfire situations. During 1997-1998, the number of land-clearing fires and other escaped fire situations increased in the equatorial forest regions of Southeast Asia and South America.

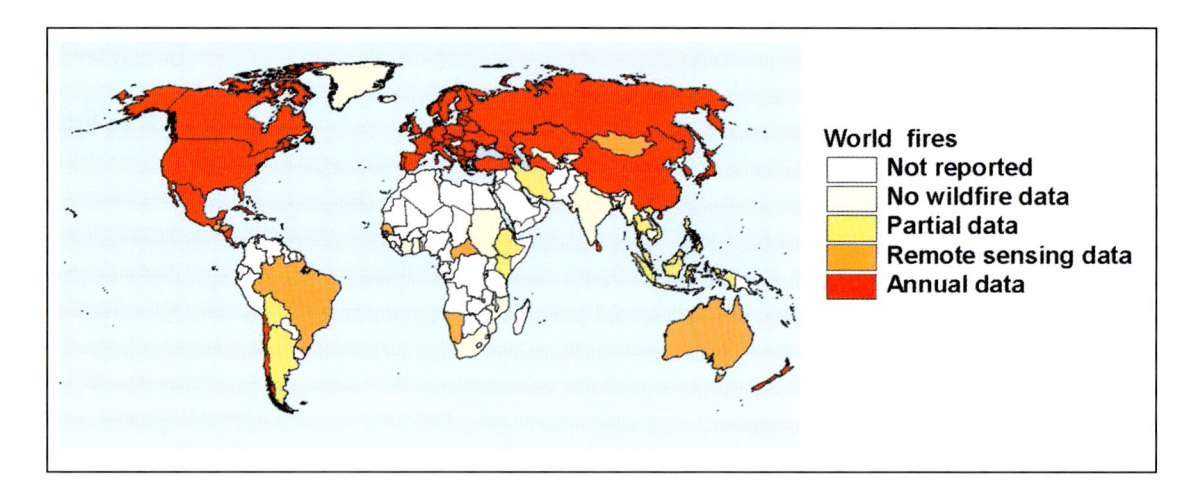

Figure 8-1. Global availability of wildfire data

The northern temperate and boreal forest zones also experienced extremely dry years in both decades. Central eastern Asia was affected most severely in 1987, particularly Siberia in the Russian Federation and northeastern China. The Russian Far East was also severely affected during the 1998 drought.

Statistical evidence from Canada suggests that there has been an increasing trend in area burned starting in the early 1980s and continuing into the 1990s. Wildland fire statistics for National Forests in the United States show a similar increase from the mid-1980s onwards. However, a change in fire response strategies in Canada and unnatural fuel accumulations in the United States as a result of long-term fire exclusion help to explain some of these changes.

In summary, there was no worldwide trend during the past two decades. Some areas suffered more fires because of increasing land use intensity. Other regions have become more susceptible to larger and more damaging fires as a result of long-term fire exclusion. Another important consideration is the large areas of degraded forests and other wooded lands that have been converted to grassland and shrubland through repeated fires. These lands are much more prone to frequent burning, which also prevents a return to tree cover.

ASSESSMENT METHODS

Fire data are compiled for industrialized countries and published by UNECE/FAO as Forest Fire Statistics every two years. However, as global data are not available, FAO member countries were requested to complete a standard questionnaire on forest fire data. Unfortunately this met with little success, so a standardized fire profile was developed which enabled countries to complete thematic information even in the absence of numeric data. These profiles, completed by 47 countries, describe how fires affect people and natural resources and how the countries are organized to manage fires. The profiles are displayed on FAO's Web pages and aggregated in *Global forest fire assessment: 1990-2000,* an FRA Working Paper (FAO 2001).

The fire situation in the six geographical regions, Africa, Asia, Europe, Oceania, North and Central America and the Caribbean and South America, is summarized in FAO (2001). Fire management highlights from the six regions are presented in the following section.

A map was prepared depicting the availability of global fire data based on data collected by the UNECE and by the submission of data by countries to the Global Fire Assessment (Figure 8-1).

RESULTS

Africa

Africa is often referred to as the "fire continent" owing to the regular and widespread occurrence of wildland fires. This description is equally pertinent to southern, western and eastern Africa where the savannah biome is a major plant community. Africa is highly prone to lightning storms and has a fire climate with both dry and wet periods, where fires can burn the fuels produced and accumulated during the wet, rainy period. Although lightning was the primary ignition source of fires in the Africa savannahs in the past, the situation today is one where humans have become more important than lightning as a source of ignition. Africa has the most extensive area of tropical savannah in the world, characterized by a grassy understorey that becomes extremely flammable during the dry season.

Most wildland fires in Africa burn in fire-adapted ecosystems. A recent research report by the United States National Aeronautics and Space Administration (NASA) indicated that about 130 million hectares of savannahs and grasslands burn annually in Africa south of the equator (for comparison, the country of South Africa covers an area of 122 million hectares). The heaviest burning is concentrated in the moist subtropical belt which includes Angola, the southern Congo, Zambia, northern Mozambique and southern United Republic of Tanzania. During the 2000 fire season, the area burned south of the equator may have reached more than 200 million hectares.

A case study from the Central African Republic showed that in the second half of the 1990s just over 43 percent of Sudanian savannahs (equal to 8.6 million hectares) and 58 percent of Guineo-Congolian/Sudanian savannahs (equal to about 62 million hectares) burned.

In Ethiopia, the delayed onset of the rainy season and increasing land use pressure resulted in an extreme wildfire season in early 2000. Land conversion burning and escaping fires led to large-scale wildfires, particularly in the montane south. The government called for international assistance and a coalition of countries (Germany, South Africa, Canada and the United States) responded. By the end of the dry season in April 2000, however, more than 100 000 ha of montane forests had been severely affected or destroyed by fire.

Fire is an important danger to forests of the North African countries. In Morocco, the number of annual fires has increased from 150 to 200, and the annual area burned has increased from 2 000 to 3 100 ha, since the 1970s.

Asia

The Asian region suffered extreme wildfire and smoke episodes during the 1990s. Insular Southeast Asia was most affected by several El Niño – Southern Oscillation (ENSO) events in the 1990s, particularly during the extreme ENSO in 1997-1998. Extended droughts favoured the application of land use fires, forest conversion burning (use of fire in land use change) and extended wildfire situations. The fires have caused impoverishment or destruction of primary and secondary equatorial rain forest ecosystems over large areas. Indonesia was the main source of smoke-haze that affected the entire region for almost one year and affected the health of more than 100 million people living in the region.

Continental South and Southeast Asia continued to experience extended wildfires in the seasonal (deciduous) forests, e.g. monsoon forests and forest savannahs. Human-induced wildfires in the deciduous forests have been common through history. As a traditional element of forest utilization, especially for improving grazing conditions (silvopastoral land use) or for improving productivity or facilitating harvest of non-wood forest products, these fires partially represent prescribed burning systems. However, many of the fires are not contained and tend to escape as extended wildfires.

In Central Asia, the region most challenged by fire is between the steppe and southern boreal forests. Steppe fires exert a tremendous pressure on the adjoining forests. Political and socio-economic changes in Mongolia during the 1990s were the major reasons for a dramatic increase in wildfire occurrence. Campfires set by inexperienced cattle herders and collectors of non-wood forest products as well as an increase in other forest uses because of deteriorating economic conditions are the main causes of escaping wildfires. Very serious fire seasons affected forested and steppe lands on 10.2 million hectares in 1996 (including 2.36 million hectares of forest) and 12.4 million hectares in 1997 (including 2.71 million hectares of forest). Forests are most seriously affected by fire in the transition zone between steppe and the montane-boreal dense forest.

In China, the main fire regions are in Inner Mongolia (with fire features similar to Mongolia), the montane-boreal forest in the northeast and the southern tropical forests. Advanced fire management systems, including the use of remote sensing for detecting and monitoring fires, are in place. In early 1999 a severe spring drought affected the whole of central Asia and led to widespread forest and steppe fires.

Seasonally dry forests in the continental Southeast Asian countries exhibited typical seasonal burning patterns affecting several million hectares in 1999-2000. Although most of these forests are fire-adapted, fire protection contributes to increased productivity, soil conservation and reduction of erosion, runoff and subsequent flooding. After the 1997-1998 fire and smoke episode in Southeast Asia, Thailand established a self-funded Forest Fire Management Centre which will serve as a centre of excellence in fire management training and research for the Association of Southeast Asian Nations (ASEAN) region.

In insular Southeast Asia the total land area affected by fire in 1997-1998 as a result of escaped fires during the El Niño drought was about 9.7 million hectares, 6.5 million hectares in Kalimantan, Indonesia, alone. After the end of the drought in 1998 the situation stabilized. Average to above-average long-term rainfall was recorded in the critical areas of the Indonesian archipelago, particularly in Sumatra and Kalimantan, and increased public awareness and law enforcement somewhat reduced the large-scale use of fire in forest conversion. Consequently, there were fewer wildfires during 1999-2000. Still, nearly 23 000 fire events were detected through the use of satellite remote sensing. Most of them were small land use fires except for a 14 000 ha fire in the coastal wetlands to the east of Palembang, South Sumatra, which lasted for over three months.

Europe

Fire is the most important natural threat to forests and wooded areas of the Mediterranean basin. It destroys many more trees than all other natural calamities: parasite attacks, insects, extreme wind events, frost, etc. Mediterranean countries have a relatively long dry season, lasting between one and three months on the French and Italian coasts in the north of the Mediterranean, and more than seven months on the Libyan and Egyptian coasts in the south.

Today, the average annual number of forest fires throughout the Mediterranean basin is close to 50 000, i.e. twice as many as during the 1970s. It is not easy to form an accurate picture of the overall increase, however, owing to the varying databases. In the countries where data have been available since the 1950s, a large increase in the number of forest fires can be observed from the beginning of the 1970s: Spain (from 1 900 to 8 000), Italy (from 3 000 to 10 500), Greece (from 700 to 1 100) and Turkey (from 600 to 1 400). Only former Yugoslavia deviates from the general trend (from 900 to 800).

The average annual accumulated area burned by wildfires in the Mediterranean countries is approximately 600 000 ha, almost twice the annual average during the 1970s. The trend observed is, however, much less uniform than for fire numbers. A worsening situation is clearly observed in Greece (from 12 000 to 39 000 ha), Italy (from 43 000 to 118 000 ha), Spain (from 50 000 to 208 000 ha) and former Yugoslavia (from 5 000 to 13 000 ha). The situation in Portugal has also worsened, although its statistical series starts later. In Cyprus, no apparent trend emerges from the statistics, but some years present a very high maximum (e.g. 1974). Finally, the total burnt area has remained relatively stable in Croatia, France, Israel and Turkey.

Unlike other parts of the world, where a large percentage of fires are of natural origin (lightning), the Mediterranean basin is marked by a prevalence of human-caused fires. Natural causes represent only a small percentage of all fires (from 1 to 5 percent, depending on the country), probably because of the absence of climatic phenomena such as dry storms.

The 1999 and 2000 fire seasons in the Russian Federation were less critical than in 1998, when 4.27 million hectares of forest and other land under fire protection were affected by fire. In 1999, the area burned was 752 000 ha and 1.14 million hectares were burned up to September 2000. The future of fire management in the Russian Federation depends on final institutional arrangements at the federal and regional levels. The European Union is sponsoring a technical cooperation project to improve fire information and response and an exchange of fire management specialists is continuing with the United States.

In September 1996 the FAO/UNECE/ILO Team of Specialists on Forest Fire called for a regional Baltic action plan concerning collaboration in forest fire protection and proposed a first regional conference. This proposal was submitted to the government of Poland. The government responded positively and hosted the First Baltic Conference on Forest Fires in Radom-Katowice in May 1998. The meeting was attended by scientists, managers and representatives from administrations of the host country (Poland), the Baltic States (Estonia, Latvia, Lithuania), the Nordic countries (Denmark, Finland, Norway, Sweden), Germany and the Russian Federation.

At the conference the establishment of pan-Baltic programmes and exchange mechanisms encompassing fire research, fire management training, the use of prescribed fire (in forestry, nature conservation and landscape management) and mutual fire emergency assistance were proposed. The conference participants agreed to develop a concerted regional Baltic Forest Fire Action Plan within the framework of the Baltic 21 Action Programme.

The fire problem zones in the countries bordering the southern Baltic Sea (Estonia, Germany, Latvia, Lithuania, Poland) and Belarus are dominated by pine forests which are favoured by the continental climate.

The western European countries bordering the Atlantic Ocean, the English Channel and the North Sea have fewer wildfire problems than the central-eastern countries of Europe. They only occasionally experience large wildfires. For example, statistical data for the United Kingdom show an average annual area burned of 428 ha between 1980 and 1996. Wildfire risk in the region of the Alps and southeastern Europe (non-Mediterranean) is determined by the characteristics of either mountain mixed deciduous-conifer forest or lowland broad-leaved forest. Both in Austria (average area burned annually between 1980 and 1996: 105 ha) and in Switzerland (average area burned annually in the same period: 407 ha) a high proportion of forest fires is caused by lightning, mainly at higher elevations. In 1994 in Austria and Switzerland 27 and 33 percent, respectively, of all fire starts were caused by lightning.

Oceania

The Oceania region is dominated by Australia, a fire-prone continent with a large variety of vegetation types and fire regimes. Fire plays a major role in the ecology of most vegetation types, and humans have had to both live with and learn to manage fire. Most of Australia's vegetation formations are fire-adapted and many are fire-dependent for regeneration. The majority of Australian wildfires are ignited accidentally or purposely by humans, although lightning is important, especially in remote areas. About 115 000 and 230 000 fires per year were depicted by satellite remote sensing during the fire seasons 1998-1999 and 1999-2000. After the end of the 1997-1998 El Niño drought, fire activity in Australia and New Zealand returned to normal. Fire statistics from New Zealand show that between 1989 and 1999 an average of 6 322 ha of forests burned annually.

North and Central America and the Caribbean

Together, Canada and the United States cover nearly 18.8 million square kilometers, about 14 percent of the world's land area. The two countries share one of the longest common borders in the world, creating numerous opportunities for transboundary cooperation in fire management. Mexico has a forested area of 141.7 million hectares (according to the national definition of forest), of which 56.8 million hectares are temperate and tropical forests and 58.4 million hectares are zones with arid and semi-arid vegetation. The common border between Mexico and the United States is also long, about 3 200 km, providing many opportunities for international cooperation during fire emergencies.

International and regional cooperation in fire management increased significantly during the past decade. Under the North American Forestry Commission, a Fire Management Working Group brings together specialists from Canada, the United States and Mexico to work on common problems. The Northeast Fire Compact between Canada and the northeastern United States has been in place for many years; and a Northwest Compact is a recent development sharing firefighting resources both ways across the border between Canada and the United States. There is also a Great Lakes Compact which specifies fire management cooperation along the central portion of the international border.

Agreements for sharing resources also exist along the border between Mexico and the United States. Central America has been especially proactive in developing cooperative efforts among all countries in this area. Central American countries meet periodically to establish common fire management policies and strategies to help each other.

Large-scale fires throughout North America, Central America and the Caribbean in 1998-2000 clearly indicated that public policies and practices, as well as prolonged drought, contributed to the severity of fire impacts. In the United States, for example, a policy emphasis on fire exclusion over many decades has led to the build-up of unnatural accumulations of fuels within fire-dependent ecosystems. Fires that now occur burn at much higher intensity and are more difficult to control. In Central American pine forests, fire is part of the silvicultural practice.

Intense drought conditions in the western United States in 2000 contributed to wildfires that burned about 2.5 million hectares of forests and grasslands. Montana, Idaho and Oregon were declared national disaster areas and the National Guard, Army, and Marines were called into action. In an unprecedented move, firefighting personnel were requested from Canada and Mexico and from as far away as Australia and New Zealand. The firefighting effort cost the United States about US$1 billion, but it was really the onset of fall rains that quelled the fires.

Mexico experienced seven consecutive years of drought from 1994 to 2000. In 1998, El Niño conditions created the most difficult wildfire season in Mexico's history. Mexico had 14 445 wildfires affecting 849 632 ha, the largest area ever burned in a single season. Seventy-two people died in fire control activities which involved the military, state governments, many federal agencies and volunteers. Mexico received support from the United States Government in the form of equipment, technical support and financial resources.

A review of fire conditions in Mexico and Central America indicates that the number of fires is often related to traditional burning for land clearing and agriculture. Firefighters are overwhelmed by the number of fires during the dry season.

South America

Fire as a land use tool is deeply rooted in the culture, society and traditions of most countries in the region. Fire has been used to prepare agricultural lands for crops or grazing, to open impenetrable lands to new agricultural uses, to facilitate hunting or to maintain an open landscape.

Without exception, country fire officials throughout the Southern Hemisphere believe that uncontrolled wildfire is fast emerging as a major concern. This was a recurring theme in the presentations at the first South American Seminar on Control of Forest Fires, held in Belo Horizonte, Brazil, in 1998. The continuing use of fire in land use practices, population pressures and a decrease in the economic stature of many of the people in the region are primary causes for the increase in wildfire problems.

The exact scope of the problem is difficult to determine. Fire statistics in many cases are non-existent, significantly incomplete or misleading. There is not a common understanding or definition of what constitutes a wildland fire. A review of available statistics suggests that 50 to 95 percent of wildfire starts in the region are the result of agricultural burns or land-clearing burns escaping control. Agricultural burning has been occurring for so many centuries that vast quantities of smoke or many hectares on fire evoke little concern. Satellite imagery cannot differentiate the unmanaged and uncontrolled wildfires from controlled burns. During the early months of 1998, satellite imagery heightened government and international awareness regarding the vast number of "hot spots" in the region.

The province of La Pampa in central Argentina, experienced an unusual fire season in 1993. Fires burned 1 227 440 ha of grassland and shrublands, with great economic loss. This was four times the annual average. In 1994, 25 firefighters died in a rangeland fire in the coastal area of northeastern Patagonia. During the 1995-1996 season, large wildfires affected the Patagonian/Andean region in general and, in particular, the oldest National Park in the country. In response to public concern, the Federal Government established a National Fire Management Plan. In 1999, large fires affected the central and southern areas of the country. One of the oldest pine plantations in Patagonia was lost, causing a great impact on the community. Two fatalities occurred in two different fires. The Mesopotamic region had an unusually critical fire situation in 2000. The fires affected large areas of pastureland and eucalyptus and pine plantations.

Serious wildfires occurred in Brazil during the 1990s. For example, in 1998 almost 20 percent of the State of Roraima burned. Economic losses due to yearly fires in Amazonia are high and smoke contributes to serious respiratory health problems. Fires have also led to interruptions in the electrical energy supply and the closure of airports as well as contributing to the loss of biological diversity.

In Chile, in comparison to the 1980s, fires increased by 13 percent in the 1990s, from an average of 4 800 to 5 530 per year. Nevertheless, the average fire size dropped from 11.3 to 9.1 ha as a result of improved strategies, organizational methods and cooperation among firefighting partners.

Droughts in 1992, 1993, 1997 and 1998 in Chile caused enormous damage to the environment and losses of facilities and miscellaneous structures. During the 1997-1998 fire season, fire behaviour was extreme in the deep south, and in 1998-1999 the central part of the country was affected. The latter was the most difficult fire season in Chile's history, with 6 830 fires and 101 691 ha burned. The "La Rufina" fire alone burned 25 400 ha, 14 houses, cattle and power lines among other losses.

CONCLUSIONS

Policy-makers are beginning to realize that continued emphasis only on emergency response will not prevent large and damaging fires in the future. Emergency preparedness and response programmes must be coupled with better land use policies and practices. Actively working towards sustainable forestry practices with community involvement is an important strategy for better conservation of natural resources and reduced impacts of wildfires.

Between 1998 and 2000, several international initiatives related to sustainable development and wildland fire prevention, preparedness, management and response were started or continued. Many countries are starting to develop policies and practices to improve their institutional capacity to prevent, prepare for and combat forest fires. At the same time, it should be remembered that fire is one of the natural forces that has influenced plant communities over time and as a natural process it serves an important function in maintaining the health of certain ecosystems. Consequently, the traditional view of fire as a destructive agent requiring immediate suppression has given way to the view that fire can and should be used to meet land management goals under specific ecological conditions.

Reviewing the global fire situation in the 1990s, it is possible to conclude as follows.

Many countries are now starting to develop policies and practices to improve their institutional capacity to prevent, prepare for, and combat forest fires. The Ministries of Environment and Agriculture in Mexico, for example, have collaborated since the disastrous 1998 fire season to reduce the threat of agricultural burning to forests.

In Brazil, measures have been taken to stress fire prevention programmes and to train farmers in burning practices that will better control fires used in agriculture.

Strategies are being developed in the United States to determine the extent to which tree thinning, timber harvest and prescribed burning could restore forest health and reduce fire hazards.

Wildfires during drought years continue to cause serious impacts to natural resources, public health, transportation, navigation and air quality over large areas. Tropical rain forests and cloud forests which typically do not burn on a large scale were devastated by wildfires during the 1990s.

Many countries and regions have well-developed systems for documenting, reporting and evaluating wildfire statistics. However, many fire statistics do not provide sufficient information on the damaging and beneficial effects of wildland fires.

Satellite systems have been used effectively to map active fires and burned areas, especially in remote areas where other damage assessment capabilities are not available.

Some countries still do not have a system in place to report annually number of fires and area burned in a well-maintained database, often because other issues such as food security and poverty are more pressing.

Even those countries supporting highly financed fire management organizations are not exempt from the ravages of wildfires in drought years. When wildland fuels have accumulated to high levels, no amount of firefighting resources can make a notable difference until the weather becomes more moderate (as observed in the United States in the 2000 and 2001 fire season).

Uncontrolled use of fire for forest conversion, agriculture and pastoral purposes continues to cause a serious loss of forest resources, especially in tropical areas.

Some countries are beginning to realize that intersectoral coordination of land use policies and practices is an essential element in reducing wildfire losses. In some cases, sustainable land use practices and the participation of local communities in integrated forest fire management systems are being employed to reduce resource losses from wildfires.

In some countries, volunteer rural fire brigades are successful in responding quickly and efficiently to wildfires within their home range, and residents are taking more responsibility to ensure that homes will survive wildfires.

Although prescribed burning is used in many countries to reduce wildfire hazards and achieve resource benefits, other countries have prohibitions against the use of prescribed fire.

Fire ecology principles and fire regime classification systems are being used effectively as an integral part of resource management and fire management planning.

Fire research scientists have been conducting cooperative research projects on a global scale to improve understanding of fire behaviour, fire effects, fire emissions, climate change and public health.

In numerous cases, intersectoral and international cooperation in helping to lessen the impact of wildfires on people, property and natural resources reached unprecedented levels in the 1990s.

Institutions like the Global Fire Monitoring Center have been instrumental in bringing the world's fire situation to the attention of a global audience via the Internet.

In addition to statistics on forest management indicators, qualitative information on status and trends in silviculture and forest management has been collected through a literature review and is presented in Forestry country profiles on the FAO Web page (www.fao.org/forestry/fo/country/index.jsp). Profiles currently exist for 20 countries in Asia and are under preparation for 25 countries and territories in the Caribbean; 13 countries in Central and South America; 10 countries in Central Africa and 22 countries and territories in Oceania.

BIBLIOGRAPHY

FAO. 2001. *Global forest fire assessment 1990-2000.* FRA Working Paper No. 55. www.fao.org/forestry/fo/fra/index.jsp

Chapter 9
Wood supply

ABSTRACT

Studies were made on the extent of forests theoretically accessible for industrial wood supply and on actual harvesting and wood removals by country. The study on accessibility was carried out using global maps and geographic information system (GIS) technology. A spatial model was developed to estimate forest area within varying distances from major transportation infrastructure, excluding protected areas and forest areas considered to be above an economically exploitable altitude. Adjustments were made for closed and open/fragmented forest classifications, differences among geographic regions and ecological zones, and distortions due to map projection. It was estimated that 51 percent of the world's forests lie within 10 km, and 75 percent within 40 km, of major transportation infrastructure. Results are presented by region and ecological domain. Boreal and tropical forests are more remote than other forests. In some regions, notably North America, protected areas represent a significant accessibility limitation. Protected areas have, however, a minor impact on accessibility at the global level. For the study on extent and intensity of forest harvesting, information was analysed for 43 tropical countries, representing approximately 90 percent of the world's tropical forests. About 11 million hectares of tropical forests were harvested annually, with intensity ranging from 1 to 34 m^3 per hectare. Detailed information on removals and harvesting was assembled for industrialized countries. Excluding the Russian Federation, over 70 percent of the increment was harvested in these countries. Data are reported by country in Appendix 3.

INTRODUCTION

The world's forests present a large potential for industrial exploitation. Information on forest areas accessible for wood supply is important for land use planning, for development of forest industries and from a policy perspective. At the same time, accessibility can be seen as a potential threat to degradation of forest ecosystems. Two studies on wood supply were done in FRA 2000 – one to estimate the theoretical accessibility of forests for industrial wood supply, and one to collect information on areas and volumes actually harvested.

The study on accessibility was based on the FRA 2000 global maps on forest cover, protected areas and ecological zones. Because it was based on coarse-resolution data, the study is not useful at the local level; the study was intended as a reference for policy-makers at the national and international levels.

The concepts used in the study were developed in previous studies, including the Global Fibre Supply Model (GFSM) (FAO 1998) and the Temperate and Boreal Forest Resources Assessment (TBFRA) (UNECE/FAO 2000). Both earlier studies included a compilation of recent forest inventory statistics. GFSM focused on access to global industrial fibre sources including non-wood fibre. Projections were made from inventory data to estimate the current situation as well as to forecast accessibility of raw material. The UNECE/FAO study was based on the results of a comprehensive survey of industrialized countries with temperate and boreal forests. Both studies reached similar conclusions to FRA 2000, with some variations for some geographic regions and countries.

The study on actual harvesting and removals was mainly based on information obtained from national reports. Most previous assessments of forest harvesting and removals have been restricted to case studies of individual operations, providing a mosaic of information on several topics under specific conditions. For FRA 2000 a more comprehensive overview was requested of the extent of forest harvesting schemes and intensity of harvesting by country.

METHODS

Accessibility

The study was based on an analysis of global thematic maps of forest cover, protected areas and global ecological zones produced for FRA 2000. The mapping processes are described in detail in Chapter 47. The analysis was carried out with ArcView 3.2 software using grid themes with cell

Table 9-1. Correction factors for closed and open/fragmented forest by geographic region and ecological zone

Geographic region/ecological zone	Closed forests	Open/fragmented forests
World except North and Central America		
Dry tropical and subtropical zones	3.0	2.0
All other ecological zones	1.0	0.5
North and Central America		
All ecological zones except temperate and boreal United States and Canada	1.0	0.5
Temperate and boreal United States and Canada	0.7	0.3

size of 2 × 2 km^2 in the Robinson world map projection with central meridian 0. The resulting figures were post-processed to correct area distortions due to the map projection.

The global ecological zone map (FAO 2000a) was developed by several institutions, including the United States Geological Survey (USGS) EROS Data Center (EDC); the Ecological Laboratory of Toulouse (LET), France; and the UNEP World Conservation Monitoring Centre (UNEP-WCMC). The base for the classification of the ecological zones was the Köppen-Trewartha system and the data set that describes five domains based on temperature: tropical, subtropical, temperate, boreal and polar, with a further division into a second level of 20 global ecological zones, as recognized by FRA 2000. Data are presented at 1 km ground resolution.

The forest cover map (FAO 2000b) was produced by FAO and EDC and comprises woody vegetation cover divided into three classes: closed forest, open/fragmented forest and other wooded lands. The data set was based mainly on remote sensing material and presented at a 1 km ground resolution.

The protected areas map (UNEP-WCMC 2000) was produced for FAO by UNEP-WCMC and consists of both international and national protected areas at 1 km ground resolution.

A digital elevation model was derived from two topographic data sets developed by EDC, HYDRO1k (EDC 1996b) and GTOPO30 (EDC 1996a), both with the resolution of 30 arc-second corresponding to approximately 1 km ground resolution.

As standard map data several themes were employed from the ESRI *Digital chart of the world* (ESRI 1995). The political theme was employed for country boundaries, the road and railroad themes were used to describe land transport infrastructure, and the inland water drainage theme was employed to describe water transport infrastructure.

The data sets were geometrically adjusted and all grid themes were resampled to the cell size of 2 × 2 km^2 at the extension of the world in Robinson map projection to achieve a usable overlay of all themes and a feasible level of detail to perform the map analysis.

The following assumptions were made regarding the availability of forest for wood supply.

- Forest in protected areas was considered not available for wood supply.
- Forest above a certain altitude was considered not accessible. The limit for economically feasible forest exploitation was set at 3 000 m altitude in the tropical domain, 2 500 m in the subtropical domain, 2 000 m in the temperate domain and 1 000 m in the boreal domain.
- Accessibility was determined by distances to major transportation infrastructure. Major roads and railroads from the ESRI *Digital chart of the world* (ESRI 1995) were used with the assumption that they represent infrastructure suitable for wood transportation. In tropical South America major rivers below 400 m altitude were considered as additional transportation routes. Rivers are used for wood transportation also in other parts of the world, but the road and railroad infrastructure already covers major parts of these regions. From the major transportation routes it is understood that forests are accessible through minor forest roads and streams.

Two major adjustments were applied to the area figures from the map analysis. First, adjustments were made for closed forest and open/fragmented forests in each geographic region and ecological zone. The correction factors were determined manually (Table 9-1), based on comparisons between the forest areas by region in the global map and the forest areas given in country reports. Dry forests tended to be underrepresented in the remote sensing results used to produce the forest cover map. Tropical and subtropical dry forests were therefore given a higher correction factor. The class open/fragmented forest in the forest cover map partly represents mosaics of forests, and the area figures for this class must therefore be adjusted downwards. In North and Central America the forest areas were overrepresented, especially in the United States and Canada, and therefore lower correction factors were used.

Second, adjustments were made for area distortions caused by the map projection. For practical reasons, all maps were processed in the Robinson map projection (Robinson 1963), which is one of the most frequently used projections for global maps and presents a "true" shape of the world. The projection presents a uniform scale only within latitudes ±38° and the scale error becomes greater as latitude increases. The areas derived from the Robinson map were calibrated for each country by calculating the relation between the areas in Robinson map projection and the areas in Lambert Azimuthal Equal Area map projection (a map projection that displays geographical data by continent at a fairly true scale). The calibrated statistics contained corrected proportions of areas.

Harvesting and removals

Tropical countries. In-depth analysis on removals and harvesting was carried out for 43 selected tropical countries where detailed country references were available. The countries included in the study represented the forested tropical countries of Africa (19 countries), Asia including Oceania (10 countries) and America (14 countries) (FAO 2000c). The selected countries have in common that their forest cover exceeds 5 percent of the total land area or exceeds 1 million hectares. Collectively, the 43 countries accounted for approximately 90 percent of the world's tropical forests.

Information was gathered both through an extensive literature review and through statistical modelling, supplemented by country visits to Gabon, Suriname and Papua New Guinea. The literature research focused on data published in government reports and on documentation where the original source could be determined. Analysed data included only commercial and legal utilization of the timber resource from natural broad-leaved forest in the tropics. Other forms of utilization of the resource (e.g. fuelwood, illegal or unrecorded harvesting) that might affect forest conditions in individual countries are mentioned in descriptive country profiles published in the FAO forestry country profiles (FAO 2001).

Industrialized countries. Industrialized countries generally have substantial information on harvesting which it was possible to collect for FRA 2000. Countries responded to detailed questionnaires covering several aspects of harvesting and removals, including breakdown by species and balance among increments, natural losses and fellings. A full account of the methods and results is found in UNECE/FAO (2000).

RESULTS

Accessibility

The results from the analysis by geographical region are presented in Table 9-2. These results are broken down also by ecological domain in Figure 9-1.

In Europe, access to wood supply in the vast areas of boreal forests is to a great extent limited by the absence of major transportation infrastructure. In contrast, in the European temperate and subtropical forests, the major limiting factor is protected area status.

Most of the forests in South America are found in the tropical region. Remoteness from roads and railroads in the Amazon basin implies the use of rivers as potential transportation routes. Water transportation improves the accessibility of wood supply areas by almost 50 percent (Figure 9-2), but accessibility still remains limited. The protection status of some forest areas is also a factor limiting access.

In North and Central America vast areas of boreal forests offer low accessibility for wood supply because of an undeveloped transportation infrastructure, while access to temperate and subtropical wood supply is relatively restricted by protected areas.

The tropical forests in Africa are generally closer to infrastructure than tropical forests in Asia and South America.

Accessibility to wood supply areas in Asia is mainly restricted by protected areas and altitude limits, and to a lesser extent by remoteness.

Tropical forest is the main forest type in Oceania, and its accessibility for wood supply is restricted by lack of transportation infrastructure. Protected areas are the major limiting factor for access to the region's subtropical and temperate forests.

Globally, 51 percent of the world's forests considered available for wood supply were found to be located within 10 km of major transportation infrastructure. Fourteen percent of the world's forests were considered unavailable for wood

Table 9-2. Forest area at varying distances from major transportation infrastructure, and accessibility for wood supply, by region

Forest accessibility, by geographical region	Total forest area *million ha*	Cumulative forest areas at varying distances from major transportation infrastructure *million ha*					
		10 km	20 km	30 km	40 km	50 km	Unlimited[a]
Africa	650						
Inaccessible forest		n.a.	n.a.	n.a.	n.a.	n.a.	n.a.
Forest in protected areas		33	53	62	65	67	69
Forest available for wood supply		422	533	562	572	576	581
Proportion of forests available for wood supply[b]		65%	82%	86%	88%	89%	89%
Asia	548						
Inaccessible forest		18	24	25	26	26	26
Forest in protected areas		34	46	51	53	54	59
Forest available for wood supply		344	412	430	439	444	462
Proportion of forest available for wood supply[b]		63%	75%	79%	80%	81%	84%
Oceania	198						
Inaccessible forest		n.a.	n.a.	n.a.	n.a.	n.a.	n.a.
Forests in protected areas		10	15	17	18	18	21
Forest available for wood supply		110	141	153	159	164	177
Proportion of forest available for wood supply[b]		56%	71%	77%	81%	83%	90%
Europe	1 039						
Inaccessible forest		12	22	28	34	39	56
Forest in protected areas		18	23	26	29	31	37
Forest available for wood supply		518	657	727	776	813	946
Proportion of forest available for wood supply[b]		50%	63%	70%	75%	78%	91%
North and Central America	549						
Inaccessible forest		11	15	17	19	20	24
Forest in protected areas		61	82	88	92	94	101
Forest available for wood supply		248	309	335	351	363	424
Proportion of forest available for wood supply[b]		45%	56%	61%	64%	66%	77%
South America	886						
Inaccessible forest		1	2	2	2	2	2
Forests in protected areas		24	41	53	62	68	141
Forest available for wood supply		333	475	554	608	644	742
Proportion of forest available for wood supply[b]		38%	54%	63%	69%	73%	84%
World	3 869						
Inaccessible forest		41	62	73	81	87	109
Forests in protected areas		180	261	297	318	333	428
Forest available for wood supply		1 976	2 527	2 761	2 906	3 004	3 332
Proportion of forest available for wood supply[b]		51%	65%	71%	75%	78%	86%

[a] Unlimited distance implies that all forests are within economic reach.
[b] Proportion of total forest in the region that is accessible for wood supply and within reach of transportation infrastructure.

supply because they were located in protected areas or in areas above accessible altitude.

The results from the analysis by ecological zones (Figure 9-1) indicate that the main limit to accessibility for wood supply in tropical forests is remoteness from transportation infrastructure, especially in the Amazon region.

The subtropical forests are relatively accessible through transportation infrastructure. About 10 percent are not accessible because of protected area status, and about 6 percent of the forests lie above the altitude limits set for the study.

Most of the temperate forests are accessible by transportation infrastructure. About 15 percent are excluded because of protected status.

The world's boreal forests are to a great extent limited for wood supply because of remoteness and to some extent also because of altitude and protected status.

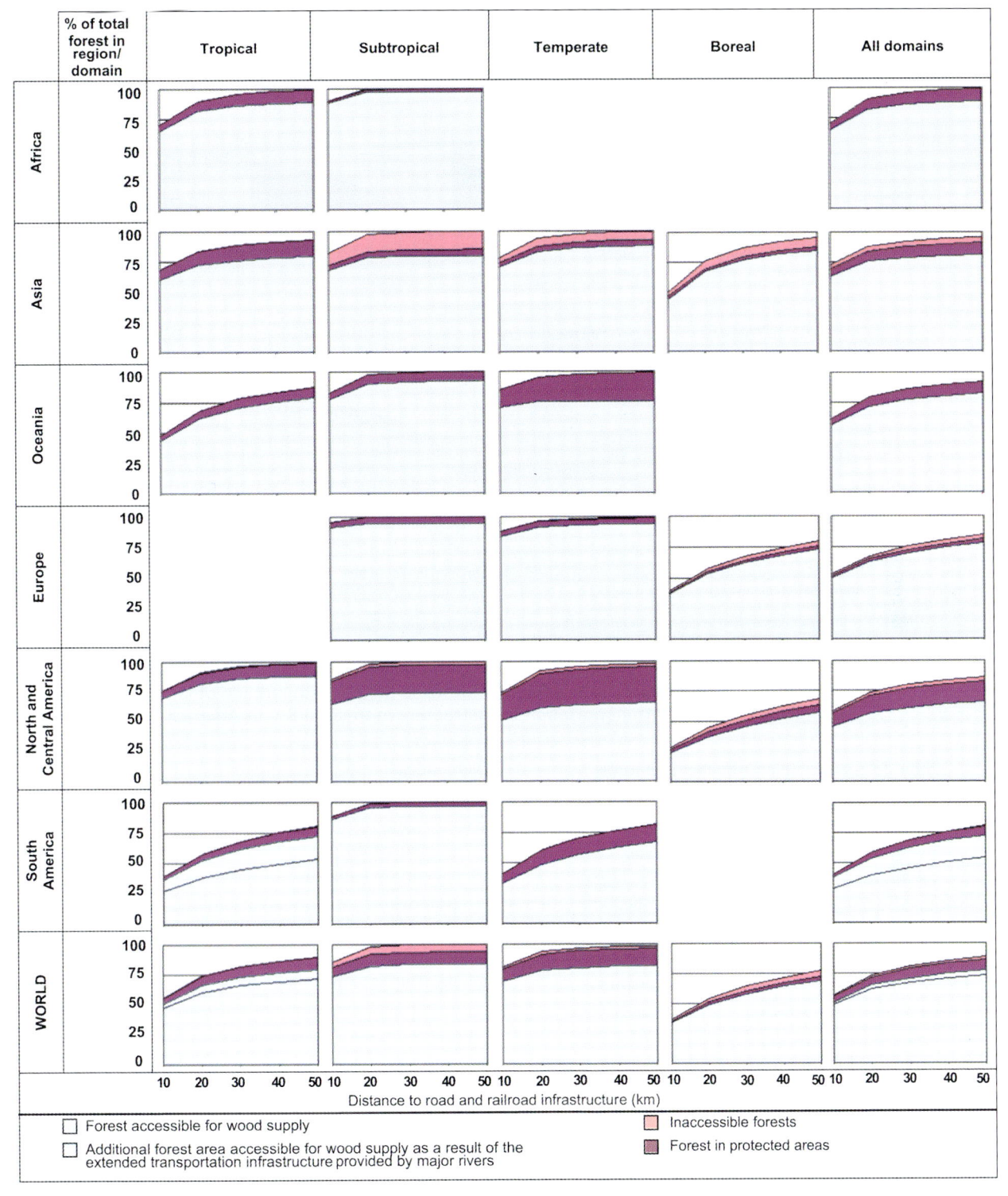

Wood supply

Figure 9-1. Proportion of forests within reach from major transportation infrastructure and its accessibility for wood supply

Notes: Boreal forest also includes forest in the polar domain. Major rivers were only considered as transportation infrastructure in tropical South America (see assumptions). The area of forest by ecological zone accounts for all forests, in contrast with the analysis in Chapter 2 where countries as a whole were considered tropical.

Harvesting and removals

Tropical countries. Country information on harvested area, harvesting intensity and volume harvested for the 43 studied tropical countries is presented in Appendix 3, Table 10.

In Africa, of the 5.9 million hectares under timber harvesting schemes, 3.3 million hectares were harvested annually. The harvesting intensity was highly variable in the countries, ranging from 1 m^3 per hectare in Zambia to 13 m^3 per hectare

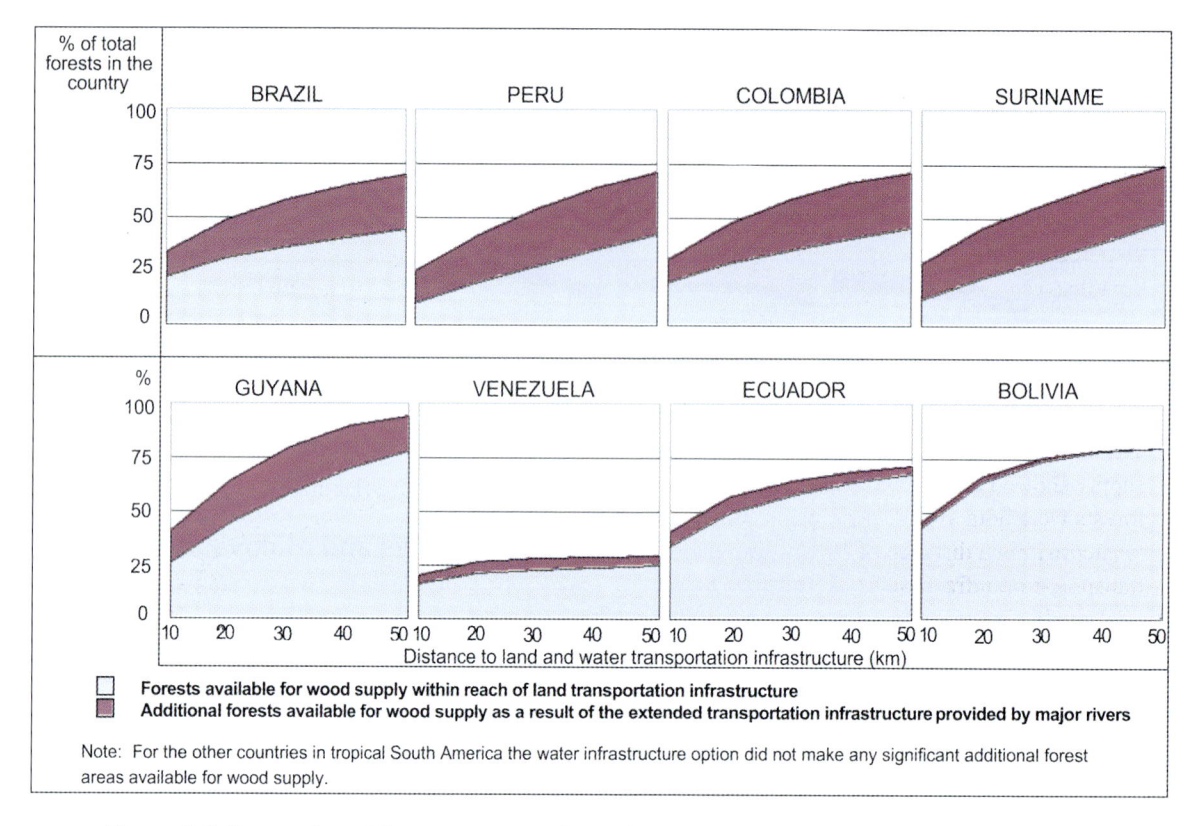

Figure 9-2. Proportion of forest areas available for wood supply within reach of land and complementary water transportation infrastructure in tropical South America

in Gabon. In Asia and Oceania, the total forest area under a harvesting scheme in 2000 amounted to 27.3 million hectares, of which 6.2 million hectares were actually harvested each year. The harvesting intensity in most countries was higher than in Africa and ranged from 5 m^3 per hectare in Myanmar to 23 m^3 per hectare in Viet Nam. In tropical America, a total of 16.7 million hectares were under a timber harvesting scheme, while some 1.9 million ha were harvested annually. The harvesting intensity ranged from 1 m^3 per hectare in Bolivia to 34 m^3 per hectare in Brazil.

Industrialized countries. As reported in TBFRA (UNECE/FAO 2000), annual fellings in the temperate and boreal domains were 1 632 million cubic metres in the mid-1990s. Of this, over half, 922 million cubic metres, was in the two countries of North America, and another 28 percent in Europe (465 million cubic metres). The Russian Federation, which accounts for 30 percent of the region's net annual increment, accounted for only 9 percent of its fellings. The contrast between growth and harvest in the Russian Federation was linked to the country's economic, social and infrastructure problems related to the process of economic transition. In the mid-1980s, Russian fellings and removals were at least three times higher than at the end of the 1990s. This dramatic decline in fellings of the world's largest forest resource has significant consequences for global wood supply and the global carbon balance.

Removals from temperate and boreal forests were 1 260 million cubic metres, implying considerable harvest losses for the region as a whole. Taking into account only forest land, removals were 88 percent of fellings in Europe, 74 percent in the Commonwealth of Independent States (CIS) and 90 percent in North America. Much of this variation may be explained by differences in conditions and practices among regions.

For the temperate and boreal domains as a whole, fellings of growing stock accounted for 53 percent of net annual increment. There are, of course, rather wide differences between regions. In general, a larger portion of the increment is harvested in those regions with powerful forest industries. Thus the proportion is 79 percent in North America, 72 percent in the Nordic countries and 63 percent in central-western and northwestern Europe. In the CIS it is only 17 percent, while in other countries of the temperate and boreal domain it is 52 percent.

If the data are broken down by species, it is apparent that coniferous stands are used much more intensively than broadleaves. The felling/increment ratio was 62 percent for conifers and 42 percent for broadleaves. For Europe, the ratio was 68 percent for conifers and 56 percent for broadleaves. In North America, the ratio was 98 percent for conifers and 54 percent for broadleaves.

CONCLUSIONS

Accessibility

The study on accessibility of forests for wood supply showed that 51 percent of the world's forests were both available for wood supply and located within a distance of 10 km from major transportation infrastructure. Fourteen percent of total world forests were considered unavailable owing to their location either in protected areas or above accessible altitude.

Boreal and tropical forests are in general more remote than the forests in other regions. This suggests that additional areas for wood supply will depend on the development of roads and railroads in such remote areas.

Regionally, protected areas represent a significant limitation for access to wood supply areas, for example in the subtropical and temperate forests in the United States and Western Europe. Note that some protected areas were not included in this study because of missing information on spatial extension (particularly for some European countries), in which case the forest area accessible for wood supply may have been overestimated.

It should be noted that there are some differences in the findings of this study when compared with the UNECE/FAO (2000) study of industrialized countries, based on differences in assumptions and in methodology.

A major strength of this study is that the same model was used for the entire world, enabling global comparisons. The model used is flexible for changes in input data, which makes it possible to update the results as new data become available. Map analyses were made at ground resolution of 2×2 km^2, which resulted in high-quality output. The original data held a resolution of 1×1 km^2 but it was not feasible to maintain that level in the digital map material because of higher processing requirements. All map sources were based on recent detailed remote sensing techniques and inventories and validated by various institutions and international experts.

Some protected areas in the UNEP-WCMC database are only recorded as point data without further spatial specification; these were not included in the analysis, since geographic representation is essential for the GIS analysis. If point data had been included in the analysis as circle areas (as was done in the protected area statistics elsewhere in FRA 2000), the total forest area under protection would have been about 20 percent higher. On the other hand, some forests in protected areas may be available for wood supply under certain conditions that comply with the management practices allowed in the protected area. These potentially offsetting factors are not further elaborated in this global study.

Harvesting and removals

Incomplete data on timber removals and harvesting in tropical countries made it difficult to reach global or regional conclusions. Studies carried out under FRA 2000 in tropical countries found a very wide range of harvesting practices and intensities. The information assembled for FRA 2000, however, constitutes a first attempt to establish a baseline by country.

For industrialized countries, more data were available. However, some inherent quality problems were almost impossible to resolve, and must be kept in mind when the data are used. Data were mostly supplied by national forest inventories and based on measurements in the forest, taken over the life cycle of the forest inventory. They are therefore not necessarily comparable with the annual data on removals published at the national level and in the *Timber Bulletin* and the *FAO Yearbook of Forest Products*. The latter are often estimates based on parameters that are more easily measurable on an annual basis, such as inputs of raw material to the forest industries. From time to time these estimates for annual felling are calibrated against the forest inventory data. For this reason, TBFRA data should not be directly compared to annual removals data published elsewhere, including other data from UNECE/FAO.

If the 10 km distance to infrastructure is taken as a reference, about half or just under 2 billion hectares of the world's forests are accessible for wood supply, distributed fairly evenly between tropical/subtropical and temperate/boreal domains. In the tropical domain, about 11 million hectares were harvested annually, which represents about 1 percent of the accessible area. Area data for removals were not reported for the temperate and boreal domain, but (excluding the

Russian Federation) the removals are above 70 percent of the increment, indicating a higher intensity of wood extraction for industrial purposes than in the tropical domain.

BIBLIOGRAPHY

EROS Data Center (EDC). 1996a. *GTOPO30.* United States Geological Survey. http://edcdaac.usgs.gov/gtopo30/gtopo30.html

EDC. 1996b. *HYDRO1k.* United States Geological Survey. http://edcdaac.usgs.gov/gtopo30/hydro/index.html

ESRI. 1995. *Digital chart of the world.* http://gisstore.esri.com/acb/showdetl.cfm?&Product_ID=313

FAO. 1998. Global fibre supply model. Rome.

FAO. 2000a. *Global ecological zone map.* Rome. www.fao.org/forestry/fo/fra/index.jsp

FAO. 2000b. *Forest cover map.* Rome. www.fao.org/forestry/fo/fra/index.jsp

FAO. 2000c. *Environmental impact assessment related to forest utilization.* FAO/ Government Cooperative Programme/Austria. Project Report. (unpublished)

FAO. 2001. *Reduced impact logging in tropical forests: literature synthesis, analysis and prototype statistical framework.* Working Paper FOP/08. Rome.

Robinson, A., *et al.* 1978. *Elements of cartography.* New York, John Wiley. 4th edition.

UNECE/FAO. 2000. *Forest resources of Europe, CIS, North America, Australia, Japan and New Zealand: contribution to the global Forest Resources Assessment 2000.* Geneva Timber and Forest Study Papers No. 17. New York and Geneva, UN. www.unece.org/trade/timber/fra/pdf/contents.htm

UNEP-WCMC. 2000. *WCMC protected areas map.* www.wcmc.org.uk/protected_areas/data

Chapter 10
Non-wood forest products

ABSTRACT
Non-wood forest products (NWFP) are a major source of food and income. However, few countries monitor their NWFP systematically, so an accurate global assessment is difficult. This chapter provides a summary of NWFP for which data have been collected and describes the most important NWFP in each region, with estimates of economic value where available. Some of the major problems associated with collecting and analysing data on NWFP are discussed, and suggestions for improving this situation are advanced.

INTRODUCTION
Non-wood forest products (NWFP)[24] play an important role in the daily life and well-being of millions of people worldwide. NWFP include products from forests, from other wooded land and from trees outside the forest. Rural and poor people in particular depend on these products as sources of food, fodder, medicines, gums, resins and construction materials. Traded products contribute to the fulfilment of daily needs and provide employment as well as income, particularly for rural people and especially women. Internationally traded products, such as bamboo, rattan, cork, gums, aromatic oils and medicinal plants, contribute to economic development. However, most NWFP are used for subsistence and in support of small-scale, household-based enterprises.

Despite their real and potential importance, national institutions do not carry out regular monitoring of the resources or evaluation of the socio-economic contribution of NWFP as they do for timber and agricultural products. In the *FAO Yearbook of Forest Products*, for example, statistical data on products such as cork, tannins, bamboo and various oils were covered for the period 1954 to 1971 only. Today, countries that monitor NWFP utilization at the national level remain the exception.

The past decade has witnessed greatly increased interest and activities concerning NWFP, especially with regard to their social and economic role. Numerous ongoing projects promote NWFP use and commercialization as a means of improving the well-being of rural populations and while conserving existing forests.

Countries are increasingly encouraged to monitor their forest resources, including attributes such as biological diversity and NWFP and their use. Although significant advances in research on both the socio-economics and the biology of NWFP have taken place in the last few years, assessment of NWFP and the resources that provide them is still a difficult task. This difficulty is party attributable to the multitude and variety of products; the many uses at local, national and international levels; the multiplicity of disciplines and interests of different ministries and agencies involved in NWFP assessment and development; the fact that many NWFP are used or marketed outside traditional economic structures; and the lack of common terminology and units of measurement.

METHODS
Monitoring and evaluation of the entire variety of forest resources providing NWFP in a given country are not technically or economically feasible. Thus, the approach used for FRA 2000 was to identify and describe products of national relevance for which monitoring and evaluation are most urgently needed. Highlighted are products widely used on national markets or gathered for export. The selection of relevant products should help countries to focus their first efforts on improved data collection for major NWFP.

In order to evaluate the socio-economic importance of NWFP utilization, available information for each country was reviewed and

---
[24] There are a variety of definitions for non-wood forest products (NWFP) and the related terms non-timber forest products (NTFP) and non-wood goods and services (NWGS) corresponding to different perceptions and different needs. For the purposes of this paper, the following definition of NWFP is used: "Non-wood forest products are goods of biological origin other than wood, derived from forests, other wooded lands and trees outside the forests" (FAO 1999e).

compiled in a standard format. Key information on products and their resources and economic value was collected and aggregated at the national level. The aim is to assist the national institutions of FAO member countries in collecting, compiling and analysing relevant data and national-level statistics on NWFP for improved policy formulation.

Specific preparatory activities for the collation of country-based data on NWFP were started by FAO as part of FRA 2000. The difficulty of collecting globally comparable information on non-wood goods and forest services, which are often site-specific and highly diverse in their characteristics, was recognized by the Expert Consultation on Forest Resources Assessment 2000 (Kotka III) in 1996. These difficulties were further confirmed when countries were requested to report on their NWFP; developed and developing countries alike found it very difficult to provide comprehensive and accurate information.

A globally applicable standard classification system for NWFP does not exist. However, NWFP can be classified in many different ways: according to end-use (medicine, food, drinks, utensils, etc.), by the part used (roots, leaves, bark, etc.) or in accordance with major international classification systems such as the *Harmonized Commodity Description and Coding System* developed under the auspices of the Customs Cooperation Council. For the aims of this project, NWFP were mainly classified according to their end-use (Table 10-1).

A standard reporting format for collecting data by country utilization was developed to cover the following key information requirements:

- the relative importance of selected NWFP and the status of NWFP statistics in the country;
- major uses of NWFP (subsistence, trade and cultural values with production/trade figures);
- the scientific, trade and local names (and part used) of the species;
- resource base, management systems and harvesting methods (e.g. cultivated or gathered from wild origins in natural forest, from plantations or agroforestry systems) and impact of the present utilization on the resource base;
- resource access and property rights;
- recent trends in utilization (decreasing, stable or increasing).

Table 10-1. Main categories of NWFP on which data have been collected

Plant products		Animals and animal products	
Categories	Description	Categories	Description
Food	Vegetable foodstuffs and beverages provided by fruits, nuts, seeds, roots, mushrooms, etc.	Living animals	Mainly vertebrates such as mammals, birds, reptiles kept/bought as pets.
Fodder	Animal and bee fodder provided by leaves, fruits, etc.	Honey, beeswax	Products provided by bees
Medicines	Medicinal plants (e.g. leaves, bark, roots) used in traditional medicine and/or for pharmaceutical companies.	Bushmeat	Meat provided by vertebrates, mainly mammals.
Perfumes and cosmetics	Aromatic plants providing essential (volatile) oils and other products used for cosmetic purposes.	Other edible animal products	Mainly edible invertebrates such as insects (e.g. caterpillars) and other "secondary" products of animals (e.g. eggs, nests)
Dying and tanning	Plant material (bark and leaves) providing tannins and other plant parts (especially leaves and fruits) used as colorants.	Hides, skins for trophies	Hide and skin of animals used for various purposes.
Utensils, handicrafts and construction materials	Heterogeneous group of products including thatch, bamboo, rattan, wrapping leaves, fibres.	Medicine	Entire animals or parts of animals such as various organs used for medicinal purposes.
Ornamentals	Entire plants (e.g. orchids) and parts of the plants (e.g. pots made from roots) used for ornamental purposes.	Colourants	Entire animals or parts of animals such as various organs used as colourants.
Exudates	Substances such as gums (water soluble), resins (water insoluble) and latex (milky or clear juice), released from plants by exudation.	Other non-edible animal products	e.g. bones used as tools

Table 10-2. Workshops on NWFP within the FRA 2000 framework

Region	Countries represented
East Africa (Kenya)	Ethiopia, Eritrea, Kenya, United Republic of Tanzania, Somalia, Sudan, Uganda
Southern Africa (Zimbabwe)	Angola, Botswana, Lesotho, Malawi, Mozambique, Namibia, South Africa, Swaziland, Zambia, Zimbabwe
Central Africa (Gabon)	Burundi, Cameroon, Central African Republic, Chad, Congo, Gabon, Equatorial Guinea, Madagascar, Rwanda
West Africa (Côte d'Ivoire)	Benin, Burkina Faso, Cape Verde, Côte d'Ivoire, Gambia, Ghana, Guinea, Guinea-Bissau, Liberia, Mali, Niger, Nigeria, Senegal, Sierra Leone, Togo
Insular East Africa (Madagascar)	Comoros, Mauritius, Seychelles, Madagascar
Caribbean (Trinidad and Tobago)	Antigua & Barbuda, Bahamas, Barbados, Belize, Cuba, Dominica, Dominican Republic, Grenada, Guyana, Haiti, Jamaica, St. Lucia, St. Vincent & the Grenadines, St. Kitts & Nevis, Suriname, Trinidad & Tobago
Near East (Lebanon)	Iran, Tunisia, Saudi Arabia, Lebanon, Cyprus, Syria, Jordan, Turkey, Sudan
Central America (Costa Rica)	Costa Rica, Honduras, Guatemala, El Salvador, Nicaragua, Panama

If available, a qualitative assessment of the importance of services from forests (e.g. grazing, recreation, tourism, environmental services) was also sought. Based on the above format, country profiles include a standardized text that provides qualitative and quantitative information on NWFP and a standardized summary table that provides available quantitative information. The format remained rather flexible across countries and regions because of the inherent variability of information available on NWFP. Country profiles also include references to the source documents where the data were found as well as key contact sources in each country. Country profiles are found on the FAO Forestry Internet site (www.fao.org/forestry).

The main sources of data consulted were country reports to regional consultations on NWFP held in Africa, Latin America and Asia; documents in the FAO series of publications on NWFP; country reports to the regional Forestry Commissions; and project reports. In addition, in-country studies were commissioned in selected countries under the UNECE/FAO Partnership Programme in order to collect data available within the country. Finally, data validation was done in regional workshops by national experts. Eight regional workshops for data validation were held between October 1998 and March 2000 (Table 10-2).

The draft country profiles were discussed with country representatives during these workshops to validate available information and add missing data. No validation workshop was held for countries in Asia, as data validation was done by comparing country results with those from two previous workshops held in Asia (1992, 1994).

For Europe, North America, Australia, Japan and New Zealand, the UNECE/FAO Timber Section in Geneva conducted a study on non-wood goods and forest services. Data for this study were collected from officially designated national correspondents by means of a questionnaire. While the UNECE/FAO study for temperate and boreal countries also reports on services provided by forest lands, including aesthetic, cultural, historic, spiritual and scientific values, it was not possible to report on these services for countries in the other regions.

Subregional and regional syntheses were compiled based on the country profiles. All documents will eventually become available both on the FAO Web site and as printed working papers.

RESULTS

Africa

The most important NWFP for the different African subregions, i.e. North, West, Central, East, insular East and southern Africa, are medicinal plants, edible products (mainly edible plants, mushrooms, bushmeat and bee products) and fodder (see Table 10-3). Products of relevance for specific subregions are exudates (East and West Africa), cork and aromatic plants (North Africa), ornamental plants and living animals (insular East Africa) and rattan (Central Africa). NWFP are collected in all kinds of habitats, whether in closed or open forests, woodlands (e.g. miombo woodlands in East and southern Africa) or shrublands (mainly in arid zones). Many products (e.g. shea butter) are derived from trees outside the forest located in agricultural fields, fallow areas or home gardens. Plantations have been established for species providing high-value products, mainly traded on the world market, such as *Acacia senegal* or *Cinchona* spp.

Medicinal plants are of major importance for all African regions, both for their use in traditional medicine and for trade. In Africa, a large percentage of the population depends on medicinal plants for health care. The number of species used is not known; in Ethiopia, for example, 600 plant species are documented as being used in traditional medicine. This important role is underlined by the high ratio of traditional healers to Western-trained medical doctors, estimated to be 92:1 in Ghana (Kwahu District) and 149:1 in Nigeria (Benin City).

Medicinal plants used in traditional medicine are either collected directly by the user or sold in local markets. In addition, medicinal plants are traded on the world market. The most important African countries exporting medicinal plants (including plants from cultivated sources) are Egypt and Morocco. Important internationally traded species include *Thymus* spp., *Laurus nobilis, Rosmarinus officinalis* (North Africa), *Prunus africana* (East, southern and Central Africa), *Warburgia salutaris* (East and southern Africa) and *Harpagophytum procumbens* and *Harpagophytum zeyheri* (southern Africa).

NWFP provide important foodstuffs, in particular during the "hungry season" and in marginalized areas. Important edible plants include fruits (e.g. *Irvingia gabonensis, Elaeis guineensis)*, nuts (e.g. *Vitellaria paradoxa*), seeds *(e.g. Cola acuminata)*, vegetables (*Gnetum africanum)*, bark (e.g. *Garcinia* sp.*)*, roots (e.g. *Dioscorea* sp.) and spices (e.g. *Piper guineense)*. Mushrooms such as *Cantharellus* spp. and *Boletus* spp. are mainly collected in East and southern Africa. Bushmeat is an important edible product, in particular in the humid parts of Central and West Africa. Species hunted include antelopes, gazelles, monkeys, wild boar and porcupines. Honey and beeswax are of major importance in East and southern Africa. Ethiopia, one of the major producing countries in Africa, exported 3 000 tonnes of honey and 270 tonnes of beeswax annually between 1984 and 1994.

Table 10-3. Main NWFP of Africa

Subregion	Main NWFP	Selected national statistical data available	Reference
North Africa	Cork, medicinal plants, aromatic plants, fodder	Algeria: Annual cork (Quercus suber) production of 6 000 tonnes exploited from 460 000 ha of cork forests	NCQC 2000
		Morocco: Annual export of 6 850 tonnes of medicinal plants worth US$12.85 million in 1992-1995	Lange & Mladenova 1997
		Egypt: Annual export of 11 250 tonnes of medicinal plants worth US$12.35 million in 1992-1995	Lange & Mladenova 1997
		Tunisia: Annual production of 10 000 tonnes of *Pinus halepensis* seeds	El Adab 1993
East Africa	Exudates, medicinal plants, bee products	*Eritrea:* Export of 49 tonnes of gum arabic (*Acacia senegal*) and 543 tonnes of olibanum (*Boswellia papyrifera*) in 1997	Eritrea Ministry of Agriculture 1998
		Ethiopia: Annual honey production of 20 000 tonnes in 1976-1983 and annual production of gum arabic of 375 tonnes in 1988-1994	FAO 1998b
		Tanzania: Export of 756 tonnes of *Cinchona* sp. bark, worth US$258 000 in 1991	Chihongo 1992
Insular East Africa	Edible plants, medicinal plants, ornamental plants, living animals	*Madagascar:* Export of 300 tonnes of *Prunus africana* bark worth US$1.4 million in 1993	Walter 1996
Southern Africa	Edible plants, medicinal plants, bee products, fodder	*Namibia:* Annual export of 600 tonnes of *Harpagophytum* spp. worth US$1.5-2 million in 1998	FAO 1998a
		Zambia: Honey production of 90 tonnes and beeswax production of 29 tonnes worth US$170 000 and US$74 000, respectively, in 1992	Zambia MENR 1997; Njovu 1993
Central Africa	Edible plants, medicinal plants, bushmeat, rattan	*Cameroon:* Annual export of 600 tonnes of *Gnetum* spp. leaves worth US$2.9 million	Shiembo 1999
		Rwanda: Production of 23 000 tonnes of honey in 1998	FAO 1999a
West Africa	Edible plants, medicinal plants, bushmeat, fodder	*Burkina Faso:* Annual export of 14 200 tonnes of shea butter (*Vitellaria paradoxa*) worth US$2.4 million in 1984-1990	Zida & Kolongo 1991
		Guinea: Annual use of more than 100 million chewing sticks (*Lophira lanceolata*)	Camara 1991
		Liberia: Annual use of 100 000 tonnes of bushmeat for subsistence purposes	FAO 1997b

Table 10-4. Production and exports of gum arabic in Africa

Country	Year	Annual production *tonnes*	Annual exports *tonnes*	Reference
Chad	1997/98	not available	10 000–15 000	FAO 1999b
Eritrea	1997	not available	49	Eritrea Ministry of Agriculture 1998
Ethiopia	1988-94	250-300 (*Acacia senegal*)	not available	Chikamai 1997
		50-100 (*Acacia seyal*)	not available	Chikamai 1997
Ghana	1988-94	<10	not available	Chikamai 1997
Kenya	1988-94	200-500	not available	Chikamai 1997
Mali	1989	293	not available	FAO 1991
Niger	1970s	not available	300	Niger Ministère de l'Hydraulique et de l'Environnement 1998
Nigeria	not available	4 000-10 000 tonnes	not available	Nour 1995
Senegal	1990s	not available	500–800	Senegal MDRH 1993
Sudan	1994	22 735 (*Acacia senegal*)	18 339 (*Acacia senegal*)	FAO 1995b
	1994	11 049 (*Acacia seyal*)	4 396 (*Acacia seyal*)	FAO 1995b
Tanzania	1994	1 000	500	Makonda & Ishengoma 1997

Fodder is of great importance in the arid and semi-arid zones. Fodder is mainly provided from tree leaves, shrubs and bushes such as *Acacia tortilis* (Zimbabwe), *Khaya senegalensis*, *Faidherbia albida* and *Balanites aegyptiaca* (all West Africa). Forage plays an essential role in animal-based production systems; in the Niger, for example, tree forage contributes 25 percent of the fodder supply for ruminants during the dry season.

Exudates are another group of products of major importance for sub-Saharan Africa. Important products include gum arabic (*Acacia senegal, Acacia seyal*) (Table 10-4) as well as resins such as olibanum (*Boswellia papyrifera*), myrrh (*Commiphora myrrha*) and opopanax (*Commiphora* spp.). These products are mainly provided by three East African countries, the Sudan (gum arabic, olibanum), Ethiopia (olibanum) and Somalia (myrrh, opopanax).

In insular East Africa, ornamental plants and living animals are of major importance. Major ornamental plants are *Trochetia boutoniana* in Mauritius and *Cyathea* sp. (fern tree), *Ficus* sp., various orchids and aquatic plants in Madagascar. In 1993, 300 000 individual plants of the aquatic plant *Aponogeton* sp., worth US$70 000, were exported from Madagascar. The most valuable Malagasy animals in trade are reptiles and amphibians (e.g. *Mantella aurantiaca*); their annual export value reached US$700 000 in 1990-1995.

Cork and aromatic plants are important in North Africa. Thirty-three percent of the world's cork forests (*Quercus suber*) are located in North Africa, i.e. Algeria, Morocco and Tunisia. However, this region only contributes 9 percent (30 000 tonnes) of the world cork production of 350 000 tonnes. In particular, Algeria has low cork production (2 percent of world production) in spite of its extensive resource, making up some 21 percent of the world's cork forests. Aromatic plants such as *Thymus* sp., *Rosmarinus officinalis*, *Acacia farnesiana* and *Eucalyptus* spp. are important products of Egypt, Morocco and Tunisia. In Tunisia, for example, the export of essential oils reached 230 tonnes worth US$3.2 million in 1996.

Depletion of habitat and/or overexploitation are the main threats to the resources providing NWFP. Overexploitation has been documented for species such as *Acacia farnesiana, Cyathea* spp, *Cycas thouarsii, Gnetum africanum, Podocarpus* sp., *Prunus africana, Warburgia salutaris* and *Xylopia aethiopica* as well as for some species of rattans, orchids, reptiles, birds, frogs, lemurs and primates. Some of these species (e.g. *Prunus africana*) are included in the annexes of the Convention on International Trade in Endangered Species (CITES). Non-wood forest products provide an important source of income for women. In Morocco, for example, extraction of edible oils from the argan tree, *Argania spinosa*, is mainly carried out by women.

Asia

Asia is by far the world's largest producer and consumer of NWFP, not only because of its population size but even more because of the traditional use of a vast variety of different products for food, shelter and cultural needs. NWFP have been vitally important to forest-dwellers and rural communities for centuries. Local people collect, process and market bamboo, rattan, resins, fruits, honey, mushrooms, gums, nuts, tubers, edible leaves, bushmeat, lac, oil seeds, essential oils, medicinal herbs and tanning materials. Both rural and increasingly urban communities (both affluent and poor, but for different products) draw upon forests for a variety of needs.

Asia is unique in that most countries in the region have included data on production and trade of major NWFP in their national statistics for many decades and have developed their own nationally applicable definitions, terminology and classifications for their "minor forest produce".[25] The types and the relative importance of the listed products change from country to country, but the most important products at the regional level are rattan, bamboo, medicinal and aromatic plants, spices, herbs, resins, mushrooms, forest fruits and nuts, vegetables and leaves and fodder. In addition, the Philippines, Indonesia and Malaysia include assessments of NWFP resources in their national forest inventories. These NWFP resources include rattan, bamboo, resin and essential oil-producing species like sandalwood (*Santalum* spp.) and agarwood (*Aquilaria* spp.), as well as some palm species such as *Nypa fruticans*, *Oncosperma* spp. and *Metroxylon* spp. (sago).

China and India are by far the world's largest producers and consumers of various NWFP. China produces and processes more wild products than any other country in the world. There is growing interest worldwide in its natural foodstuffs, traditional medicines and herbs and in its handicrafts, made mainly from rattan and bamboo. Thus, China dominates world trade in NWFP (estimated at US$11 billion in 1994). It is closely followed by India and then by Indonesia, Viet Nam, Malaysia, the Philippines and Thailand.

By subregion, medicinal plants are of major importance in continental Asia, particularly for the higher-elevation regions of Nepal, Bhutan, northern India and Pakistan and southwestern China. High-value medicinal plants include *Nardostachys jatamansi*, *Dioscorea deltoidea* and *Swertia chirayta*. In the drier regions in continental and south Asia, fodder is the main NWFP.

The rich forests of insular and Southeast Asia have traditionally been a major source of a wide variety of non-wood forest products. Those for which there is significant production and trade include bamboo and rattan, medicines and herbs (*Ephedra* sp., *Anamirta cocculus*, *Cinnamomum camphora*), essential oils (*Styrax* spp., *Pogostomon cablin*, *Cassia* spp., *Citronella* sp.), spices, sandalwood, fruits and resins (naval stores, copal).

Rattan is the most important internationally traded NWFP in the world. At the local level, it is of critical importance as a primary, supplementary and/or emergency source of income in rural areas. There are approximately 600 species of rattans, of which some 10 percent are commercially used for industrial processing (mainly furniture making). Key genera are *Calamus, Daemonorops, Korthalsia* and *Plectocomia*. Indonesia hosts the bulk of the world's rattan resources (by both volume and number of species) and is the largest supplier of cane, with an estimated annual production of 570 000 tonnes.

However, Asian rattan resources are being depleted through overexploitation and loss of forest habitat. Only Indonesia, the Philippines, Malaysia and, to a lesser extent, the Lao People's Democratic Republic and Papua New Guinea, still have some significant rattan resources left. In the Philippines, the latest national forest inventory data of 1988 show an available growing stock of approximately 4 500 million linear metres of rattan (all species combined). However, no follow-up rattan inventory has been made and it is presumed that most of the commercial species have been cut. The total area of rattan plantations in the Philippines is estimated at between 6 000 and 11 000 ha.

In the Peninsular Malaysian Permanent Forest Reserves the 1992 National Forest Inventory estimated 32.7 million total rattan plants (irrespective of age) of which the most abundant (about 37 percent) were *Korthalsia* species. Of *Calamus* species, *C. manan* is the most abundant with around 5.9 million clumps. The rattan plantation area is estimated at around 30 000 ha. In some of the traditional rattan producing countries, such as China, India, Thailand, Sri Lanka, Bangladesh, Nepal, Myanmar, Viet Nam and Cambodia, the long-term sustainability of the

---

[25] In China, for example, " all crops obtained from trees, including walnuts, apples and grapes" are by law under the Ministry of Forestry and included in the country's forest products statistics.

rattan processing industry has been undermined by the depletion of rattan stocks in natural forests. Although some smallholder rattan gardens exist, investment in industrial-scale rattan plantations is presently negligible, resulting in an insecure future supply.

Bamboo is by far the most commonly used NWFP in Asia. There are more than 500 species. Although international trade in bamboo products is still of lesser importance than trade in rattan or medicinal plants, it has dramatically increased in the last decade. Unlike rattan, bamboo is moving out of the craft industry phase and now provides raw material for industrial products (shoots, construction poles, panelling and flooring products, pulp). This has important repercussions for the bamboo resource base. Bamboo is increasingly becoming a domesticated crop grown by farmers. Harvesting of bamboo in forests is still important in Myanmar and the Lao People's Democratic Republic and in remote mountain forests in northern India, central China and Viet Nam.

China has the largest area of bamboo forests, with an estimated area of 7 to 17 million hectares (depending on what "bamboo forest" is defined to include – from dispersed bamboo in degraded natural forests to full-scale plantations), mostly of *Phyllostachys* and *Dendrocalamus* spp. Annual production of bamboo poles ranges from 6 to 7 million tonnes (one-third of total known world production). Average value of world trade in bambooware is on the order of US$36.2 million. China (US$20 million in 1992) and Thailand are the main suppliers; Malaysia, Myanmar, the Republic of Korea, Indonesia, Viet Nam, the Philippines and Bangladesh are minor exporters. Bamboo shoots supply a rapidly expanding and fashionable export market, with China the major world producer and exporter (1.6 million tonnes of fresh shoots in 1999) followed by Thailand, with minor quantities from Indonesia, Viet Nam and Malaysia. Most bamboo shoots are produced on farms.

Since ancient times, forest-gathered medicinal plants have been a key component of the traditional health systems of the region, and they still are today. Most countries maintain and have legalized a dual system of providing both "Western medicine" and traditional health care (Aryuveda, Jamu and others). Traditional health care systems in the region recognize a long list of about 4 000 medicinal plants of commercial importance. Some species have become active ingredients in Western medicine, resulting in growing demand and trade. This demand has led to overharvesting of several species to the point that some species have been listed as endangered by the Convention on International Trade in Endangered Species of Wild Fauna and Flora (CITES). It is estimated that three-quarters of the total production is still gathered from wild sources. However, domestication and production of medicinals in home gardens is rapidly increasing. Total world trade in medicinal plants in 1992 was of the magnitude of US$171 million. China is the biggest producer as well as exporter of medicinal plants, accounting for 30 percent of world trade (by value) in 1991, followed by the Republic of Korea, the United States, India and Chile. Singapore and Hong Kong are the main re-exporters in Asia.

The extensive pine forests in the region provide the resource for the collection of pine-related products such as resins, seeds and mushrooms. China and Indonesia dominate the world's production of oleoresins (naval stores) from all sources (largely *Pinus* spp.), which ranges between 1.1 and 1.2 million tonnes annually. China has emerged as the world's largest producer of resin, with an annual production level of nearly 400 000 tonnes. Pine nuts (seeds of *Pinus gerardiana*, *P. pinea*, *P. koraiensis* and *P. cembra*) are an important product with a growing and high-value market, particularly in developed countries. Seeds of chilghoza pine (*Pinus gerardiana*) are produced and exported by India, Afghanistan and Pakistan. China is the world's largest producer and exporter of *Pinus koraiensis* seeds – one of the larger-seeded species – as well as seeds of *Pinus cembra,* the Siberian equivalent of the edible seeds from the European *Pinus pinea*. Production levels vary greatly from year to year.

Wild edible mushrooms, particularly morels belonging to the genus *Morchella*, are another product of considerable economic and commercial significance. Morels are prized for culinary use, particularly as a gourmet food. Morels grow naturally in the temperate forests of India, Pakistan, Afghanistan, China, Nepal and Bhutan. Total world production is estimated at 150 tonnes. Pakistan and India are the major producers, each producing and exporting about 50 tonnes of dry morels annually (equivalent to 500 tonnes of fresh morels). Total world trade in morels is on the order of US$50 million to US$60 million. China is a major producer and exporter of other wild mushroom species. The Chinese black auricular fungus (*Auricularia auricula*) is well known for its quality, and

1 000 tonnes are exported annually, earning US$8 million. The annual production of *Tremella fuciformis* often reaches 1 000 tonnes, a third of which is exported. The annual harvest of shiitake mushrooms (*Lentinus edodes*) is about 120 000 tonnes, accounting for 38.3 percent of world production. China is the second largest producer in the world with annual exports of over 1 000 tonnes of dried mushrooms, valued at US$20 million.

Asia is also the world's leading producer of several essential oils. Total world trade in raw essential oils exceeds US$1 billion per year, but the major share comes from cultivated sources. Major wild sources of essential oils in the region include sandalwood (*Santalum* spp.), agarwood (*Aquilaria* spp.), tung oil (*Aleurites* spp.) and eucalypt oils. China, Indonesia, Thailand, India and Viet Nam are the major suppliers of these oils.

Spices, condiments and culinary herbs are another important group of products (although most now come from domesticated sources) that constitute a significant component of world trade. Indonesia is the largest world producer of nutmeg and mace and accounts for three-quarters of world production and export. Indonesia produced 15 800 tonnes of nutmeg during 1990. World trade in cinnamon is between 7 500 and 10 000 tonnes annually. Sri Lanka contributes 80 to 90 percent, most of the balance coming from Seychelles and Madagascar. The world trade in cassia is on the order of 20 000 to 25 000 tonnes annually, of which Indonesia accounts for two-thirds and China most of the remainder. Minor producers include Viet Nam and India. About 2 000 to 3 000 tonnes of cassia bark are exported from Viet Nam annually. Europe, the United States and Japan are the major markets.

Products of lesser importance include sago, illipe nuts, bird nests, karaya gum, kapok and shellac. Sago is starch obtained from the stem of the sago palm (*Metroxylon* spp.). Indonesia is the major producing and exporting country. During 1991, Indonesia exported 10 108 tonnes of sago flour and meal to Japan, Hong Kong and Singapore, valued at US$2.32 million. Malaysia also produces small volumes.

Illipe nut is the commercial name for the winged fruits produced by about 20 different species of *Shorea* trees. The seeds from these fruits contain an oil whose chemical and physical properties are remarkably similar to those of cocoa butter. Large quantities of illipe nuts are collected and sold to be used in the manufacture of chocolate, soap and cosmetics. Indonesia dominates world trade in illipe nuts, exporting about 15 000 tonnes annually, worth about US$8 million.

Bird nests are built by two species of cave-dwelling swiftlets, *Collocalia fuciphaga* and *C. maxima,* in Malaysia and Thailand. These are collected for sale to the Chinese market at home and abroad. Malaysia is the major producer and exporter of bird nests. Malaysian exports during 1991 totalled 18.6 tonnes, mainly to Hong Kong, Singapore, Japan and China (Taiwan), valued at around US$1 million.

Karaya gum, also known as Indian tragacanth, is obtained from tapping trees of the genus *Sterculia*. India is the only major producer. Total world production is about 5 500 tonnes per annum.

Kapok is a mass of silky fibres in the fruit of the ceiba tree (*Ceiba pentandra*), used as a filling for mattresses, life preservers and sleeping bags and as insulation. The tree grows in many South Asian countries (as well as on the Pacific islands and in Africa and Central America). Thailand and Indonesia are the main suppliers in the world trade. Japan, China, the European Union and the United States are the major markets. During 1992 the total value of world trade was about US$11 million, of which about 66 percent was contributed by Thailand and 16 percent by Indonesia.

Thailand and India dominate world trade in shellac, each exporting, on an average, about 6 000 tonnes per annum. Shellac is produced from lac, a gummy substanced produced as a protective covering by *Coccus lacca,* a scale insect that feeds on certain trees in India and southern Asia. Viet Nam's annual exports average around 300 tonnes. China produces about 3 000 tonnes.

South America

The most important NWFP in South America are edible products (food and drinks such as Brazil nuts, fruits and palm hearts, palm wines, mushrooms and maté), resins, latexes and essential oils (pine resins, natural rubber and eucalyptus oil), medicinal plants, fibres and construction materials (palm fibres, bamboo), fodder, colourants and tannins (see Table 10-5).

In the Amazon region, the most well-known edible products, with a considerable domestic, regional and international market, are Brazil nuts and palm hearts. Brazil nuts are still collected almost entirely from wild sources of *Bertholletia excelsa* in Bolivia, Brazil and Peru and are a

Table 10-5. Main NWFP of South America

Subregion	Main NWFP	Selected national statistical data available	Reference
Tropical South America (Bolivia, Brazil, Colombia, Ecuador, Paraguay, Peru, Venezuela)	Brazil nuts	World export of Brazil nuts US$30 million (20 percent from Brazil, 75 percent from Bolivia and 5 percent from Peru) 20 000 tonnes of shelled Brazil nuts entering the world trade in 1999: *Brazil*: 7 800 tonnes, *Bolivia* 10 000 tonnes and *Peru* 2 200 tonnes in 1999	Consejo Nacional de la Castaña, Bolivia (in Collison *et al.* 2001)
	Natural rubber	*Bolivia:* Exports of 3 981 tonnes in 1983; decreased to 150 tonnes in 1995 and are now practically non-existent, mainly because of low price and substitution of synthetic rubber *Brazil*: production of 53 000 tonnes of natural rubber in 1997. Area cultivated 58 715 ha in 1994 *Peru*: Estimated area is 1.4 million ha. Rubber extraction has practically disappeared due to the competitiveness of Malaysian plantations and synthetic substitutes	Banco Central de Bolivia and Inst. Nac. de Estadistica (in Wende 2001) IBGE 1998c (in FAO 1999d) Rios Torre 2001
	Palm hearts	*Brazil*: production of 20 653 tonnes in 1995 *Bolivia*: value of exports in 1997: US$12 355 420	IBGE 1998 (in FAO1999d)
	Edible oil	*Brazil*: production of 76 000 tonnes of babassu oil (*Orbignya phalerata*) in 1997	FAO 1999d
	Colourants	*Peru*: estimated export of Annatto seeds 4 000 tonnes in the mid-1990s *Peru*: production of 500 tonnes of cochineal in 1993; export of 77 tonnes of carmine valued at US$6 700 000 in 1993	FAO 1995c FAO 1995c
	Tannins	*Peru* : 2 900 tonnes of *Caesalpinia spinosa* in 1999 for a value of US$2.5 million	Rios Torre 2001
	Resins, Copaiba	*Venezuela:* production of 7000 tonnes of crude resin from *P. caribaea* in mid-1990s *Brazil*: in the mid-1990s, production was around 60 000-65 000 tonnes. Most of the processed products are consumed domestically, but significant quantities are exported (Brazil exported 13 500 tonnes of rosin and 3 000 tonnes of turpentine in 1993) *Brazil*: production of 72 tonnes of copaiba oil in 1995	FAO 1995e FAO 1995 IBGE 1998 (in FAO 1999d)
	Medicinal plants	*Peru*: *Uncaria tomentosa* (Cat's claw) 535 tonnes in 1999 *Chile:* Quillay (*Quillaja saponaria*) exports of 872 tonnes in 1997 for a value of US$3 700	Rios Torre 2001 Campos Roasio 1998
	Fibres	*Ecuador*: Exports of Panama hats made with the fibres of *Carludovica palmata* US$4.6 million in 1992. 2 000 shops for the manufacture of Panama hats *Chile*: wicker (*Salix viminalis*) export of 942 tonnes in 1997 for a value of US$803 000	FAO 1996 Campos Roasio 1998
	Vegetable ivory	*Ecuador*: export of 327 tonnes of tagua nuts (*Phytelephas* spp.) valued at US$2 408 400 in 1992	FAO 1995c
	Fodder	*Peru*: estimated 1 750 tonnes of *Prosopis pallida* fruits	Rios Torre 2001
Non-tropical South America (Argentina, Chile, Uruguay)	Food and drinks	*Argentina*: the value of maté production in 1998: US$80 million; exports of maté over US$28 million in 1998. *Chile*: Exports of mushrooms (all species, dried, salted and frozen) US$7 million in 1997	Resico 2001 Campos Roasio 1998
	Essential oils	*Chile*: in 1997, 14 tonnes of eucalyptus oil were produced for a value of US$44 000. Exports of Chilean hazelnut (*Gevuina avellana*) in 1997: 5.7 tonnes for a value of US$92 000. Exports of *Rosa moschata* in 1997 83 tonnes for a value of US$1 million	Campos Roasio 1998
	Resins	*Argentina*: Pine resin production 19 904 tonnes, rosin and derivatives 256 840 tonnes and turpentine 2 985 tonnes in 1996	DRFN (in FAO 1999d)
	Medicinal plants	*Chile*: exports of 1 400 tonnes of boldo leaves for a value of US$810 000 in 1996	Campos Roasio 1998
	Tannins	*Argentina*: Production of 283 908 tonnes of quebracho colorado in 1996	Argentina MDSMA 2001

major component of the extractive economies of these countries. While they represent only a small percentage of the world edible nut trade, they bring considerable revenue to the producing countries. Bolivia is the largest world exporter of Brazil nuts.

zThe production of palm hearts is concentrated mainly in the tropical areas of Brazil, Bolivia, Colombia, Venezuela, Guyana and Peru. Palm hearts are extracted from wild stands of *Euterpe oleracea* and *Euterpe precatoria* or from cultivated palm species like *Bactris gasipaes*. In the Amazon region the fruits of these palms also play an important role in food and drink.

Other important palm species (at both the subsistence and commercial level) from which edible seeds and industrial oils are produced include *Orbignya phalerata*, *Mauritia flexuosa* and *Jessenia bataua*. Several tree species such as *Platonia insignis*, *Myrciaria dubia*, *Theobroma grandiflorum* and *Couepia longipendula* also produce edible fruits or nuts of local importance. Seeds of *Araucaria angustifolia* (Argentina and southern Brazil) and *Araucaria araucana* (Argentina and Chile) are commonly used for food.

In Argentina, Uruguay, Paraguay and southern Brazil, the leaves of *Ilex paraguariensis* are used to brew maté, an extremely popular tea-like beverage. This plant has been moved from its natural habitat in forest ecosystems (in the Alto Paraná region, Alto Uruguay region and northeastern Argentina) into large-scale plantations, especially in Argentina and Brazil.

Mushrooms (particularly *Boletus luteus* and *Lactarius deliciosus*), growing mainly in plantations of *Pinus radiata*, are a major item in domestic and export markets, for example in Chile.

Rosewood (*Aniba rosaeodora*), andiroba (*Carapa* spp.) and sassafras (*Ocotea pretiosa*) are essential oil-producing species with commercial value. Chile is an important producer and exporter of eucalyptus oil (from *Eucalyptus globulus* and other *Eucalyptus* species). Chilean hazelnut (*Gevuina avellana*) and musk rose (*Rosa moschata* – cultivated) are other oil-producing species. Cumaru (*Dypterix odorata*) is commercially exploited in Brazil as a flavouring agent.

Latex extracted from *Hevea brasiliensis*, indigenous to the Amazon region, is the basis for the production of natural rubber. Other exudates from tropical South America are jatobá (*Hymenaea courbaril*), maçaranduba (*Manilkara huberi*), sorva (*Couma* spp.), balata (*Manilkara bidentata*) and balsamo (*Myroxylon balsamum*). Copaiba (*Copaifera* spp.) and dragon's blood (*Croton draconoides*) are used in medicine. Gum brea (*Cercidium australe*) is used in Argentina for various industrial applications. A hard vegetable wax is obtained from the seeds of the carnaúba palm (*Copernicia prunifera*) in Brazil.

Pine resin is extracted from various *Pinus* species. The main products derived from pine resin are rosin and turpentine used in the manufacture of adhesives, paper sizing agents, printing inks, as a solvent for paints and varnishes, as a cleaning agent and for other purposes. Brazil, Argentina (*Pinus elliottii*) and Venezuela (*Pinus caribaea*) are commercial producers and exporter of pine resin. Brazil is the biggest producer of gum naval stores in South America.

The region has a long tradition of medicine based on plants. One of the legacies of the South American people is the bark of *Cinchona* species, the source of the antimalarial drug quinine. World production of quinine bark is approximately 8 000 to 10 000 tonnes per year. Important producer countries in South America are Brazil, Bolivia and Colombia. Cat's claw (the bark of *Uncaria tomentosa*) contains alkaloids and acid glycosides, several of which have immunostimulatory, anti-inflammatory and antimutagenic properties. Quillay (*Quillaja saponaria*) is used for the extraction of saponine (mainly from the bark), which has many applications in the drug and cosmetic industries. Chile is the most important producer of quillay in South America. Boldo (*Peumus boldus* or *Boldoa fragrans*) is an endemic tree that grows in the semi-arid regions of Chile. Boldine, the active substance extracted from the leaves, is used in medicine for its analgesic, diuretic and antirheumatic properties.

Annatto (obtained from dried seeds of *Bixa orellana*) and cochineal (from the insect *Dactilopius coccus*, feeding on certain cactus species) are sources of natural colourants. Peru is the main producer of both. A red colourant (carmine) is produced from the extract of cochineal. In Peru, production of annatto is export oriented and is very heavily dependent upon the harvesting of wild trees. By contrast, Brazil produces annatto to meet local demand of several thousand tonnes annually. Supplies are mainly dependent upon small farmers.

Peru is the world's largest producer of tara fruits (*Caesalpinia spinosa*) for the extraction of tannins (80 percent of world production). Production is mainly from natural stands but in part from agroforestry systems. Peru is the Andean country with the largest *Caesalpinia* forests, followed by Bolivia and to a lesser extent Chile, Ecuador and Colombia. Quebracho colorado (*Schinopsis* spp.) is a source of tannin in Argentina and Paraguay.

Fibres include the leaves of the palm *Carludovica palmata,* used for the production of Panama hats in Ecuador. *Attalea funifera* and *Leopoldina piassaba* are sources of fibre in Brazil. *L. piassaba* is also harvested on a small scale in Venezuela and Colombia. *Heteropsis* spp. are exploited for their aerial roots in the Brazilian Amazon. In non-tropical South America (particularly Chile), the young branches of *Salix viminalis* are split and woven for the production of furniture, baskets and other household items.

Bamboos are largely used in construction, furniture and handicrafts in Ecuador, Colombia and Venezuela, with *Guadua angustifolia* and *Chusquea* spp. used in the Andean regions of Ecuador and Chile.

In South America, large areas are under a cover of shrubs and low tree species – for example, the campo cerrado and caatinga of central-eastern and northeastern Brazil, the chaco in Argentina, Paraguay and Bolivia, and the arid coastal areas of Peru and Chile. In these areas the most important economic activity is often livestock raising, with livestock feeding almost exclusively on the fruits and leaves of these plants. In the arid zones of Chile large areas are under the cover of trees of *Prosopis tamarugo* and *Prosopis chilensis*. In Peru there are approximately 1.4 million hectares of dry woodlands, predominantly covered with *Prosopis pallida*, used for fodder and for the extraction of algarobina (a cocoa substitute) from the pods.

The nuts of the tagua palm (*Phytelephas* spp.) in northern South America produce a kind of vegetable ivory which is carved for handicrafts and made into buttons. The production of other forest seeds is also important.

In terms of forest management, there is very little experience in South America with management of NWFP or with integrated management of forests for timber and NWFP. Trials have been conducted for some species, (for example, *Uncaria tomentosa* and palm hearts in Peru). For some species subject to high extraction pressure, governments have set up regulations to reduce the ecological impact (for example, guidelines for use of *Araucaria araucana* in Argentina, measures to regulate felling of *Ocotea pretiosa* trees in Brazil and a ban on export of raw bark of *Uncaria tomentosa* in Peru). However, most harvesting is done opportunistically and often in a predatory manner. The result is that wild populations of various species are threatened by overexploitation and habitat destruction. Species for which overharvesting is documented include *Jubaea chilensis*, *Araucaria araucana* (listed in Appendix 1 of CITES), *Uncaria tomentosa* and *Guadua angustifolia*.

Over the years there has been a general reduction of the proportion of South American NWFP in the international markets, as shown by trade statistics for commercial products (for example latexes, gums, resins). The Brazilian government agency for statistics, Instituto Brasileiro de Geografia e Estatistica (IBGE), surveyed the production of some 34 products based on their past economic importance. In 1980, 11 of these had economic value (i.e. production value higher than US$200 000) and the total production value was US$160.2 million. By 1995, the number of products had decreased to six and the production value had dropped to US$65.4 million (ITTO 1998). The decline, in many cases, can be ascribed to competition from synthetic substitutes or products from domesticated sources, but in some cases it is caused by the degradation of the natural resource base.

On the other hand, some other products have seen a sharp rise in demand. In Bolivia, for example, in the past ten years recorded palm heart extraction increased from 11 to 4 185 tonnes. Expectations about the medicinal properties of *Uncaria tomentosa* have given rise to a boom in the production of bark in recent years.

Central America

The subregion that includes Central America and Mexico is endowed with rich and diverse forests, ranging from cloud forest to temperate hardwood and conifer forests to moist tropical high forests. As a result, the subregion has a wide variety of plant and animal species, providing a large number of different types of NWFP. The most important and common products in all countries of the subregion are medicinal plants, wild fruits, latex and handicrafts and utensils made with fibres of several plant species. Of local and national importance are ornamental plants (Guatemala, Costa Rica), fodder (El Salvador), fauna products (Nicaragua), pine resin (Honduras) and construction materials (Belize, Panama).

Medicinal plants are by far the most important. Some products, such as zarzaparilla roots (*Smilax* spp.), were already exported in large quantities in the seventeenth century to Spain. Costa Rica is the largest producer, with a yearly production of 170 tonnes of several species, but with a growing share now of cultivated origin (at present 50 percent). Major medicinal products are sen seeds (*Caesalpinia pulcherrima*), zarzaparilla roots and balsamo (*Myroxylon balsamum*). In Guatemala, the main species are calahuala (*Polypodium* spp.), with a yearly production of 50 tonnes of which 30 tonnes are exported (at a value of US$140 000), and yerba de toro (*Tridax procumbens*) with annual exports of 15 tonnes (US$90 000). In Honduras the major species is *Polypodium aureum*, with yearly export value of US$110 000.

The forests of the region contain more than a hundred tree and palm species with edible fruits, e.g. cohune (*Attalea cohune*) and pejibaye palms (*Bactris gasipaes*) and tropical forest trees such as anono (*Annona* spp.), guabo (*Inga* spp.), zapotillo (*Couepia polyandra*) and caimito (*Chrysophyllum cainito*). Costa Rica exports around 36 tonnes of zapote (*Pouteria sapota*) yearly. In El Salvador, flour is made from seeds of ojushte (*Brosimum alicastrum*) and from pito seeds (*Erythrina berteroana*); between 3 and 16 tonnes of the latter are exported per year.

Chicle is a major product of the region's tropical lowland forests. It is a latex tapped from the sapodilla tree (*Manilkara zapota*) and is used for making chewing gum. Sapodilla is most frequent in Guatemala (Petén) and Belize, where tree densities in the forests vary from 24 to 40 trees per hectare. The high tree density is an indication of the tree's use in pre-Columbian times, when the Olmec and Maya collected the latex or possibly managed stands for local consumption and export. Only trees with diameter at breast height (DBH) greater than 30 cm can be tapped by law. Chicle production in Guatemala was about 1 000 tonnes per year from 1940 until the 1970s, but has now dropped to some 500 tonnes (valued at US$2 million in 1998) because of deforestation and habitat degradation.

A wide variety of plant species are used for handicrafts and construction materials, mainly palms such as *Desmoncus* sp., *Sabal* spp., palma chonga (*Astrocaryum* spp.) and bellota (*Carludovica palmata*). These palms provide leaves, fibres and canes comparable to rattan.

Export of cane furniture from Nicaragua amounts to US$5.7 million per year. Other handicrafts include hats made from pita palm leaves (*Cardulovica palmata*), pine needle baskets (*Pinus oocarpa*) and bamboo products. An important handicraft in the region is sculptures and mouldings made from small pieces of timber species such as conacaste (*Enterolobium cyclocarpum*) and cedro (*Cedrela odorata*) and from vegetable ivory (*Phytelephas seemannii*).

Other NWFP of national importance include honey (with, for example, production from *Apis mellifera* of 200 tonnes per year, valued at US$3.5 million, in El Salvador); bushmeat from paca (*Agouti paca*); birds; iguanas (*Iguana iguana*) and garrobo (*Ctenosaura similis*), including eggs and live animals (with, for example, approximately 350 000 green iguanas exported from El Salvador in 1997, valued at US$1 million, although increasingly of reared origin) and pine resin products (particularly in Honduras, with annual export value of around US$2 million). Forage from forest land is also reported to be very important, although no quantitative data are available.

Caribbean

The most important NWFP of the Caribbean[26] are medicinal and aromatic plants, edible products (mainly fruits, mushrooms, bushmeat and bee products) and construction materials, utensils and handicrafts (see Table 10-6).

Medicinal plants are mainly used by rural communities. In Grenada, over 80 percent of the population uses herbal medicines. Important aromatic plants include candlewood *(Amyris balsamifera)*, citronella (*Cymbopogon citratus*), rosewood (*Aniba rosaeodora*), sassafras (*Ocotea pretiosa*), common hazel (*Gevuina* spp.), vetiver (*Vetiveria zizanioides*) and *Eucalyptus* spp. Grenada is the world's second largest producer of essential oils derived from the seeds of the nutmeg tree, *Myristica fragrans*. Some 25 percent of the world production comes from Grenada, contributing around 40 percent of the country's export revenue. However, nutmeg exports

---

[26] The Caribbean subregion here includes large islands (Cuba, Dominican Republic, Haiti, Jamaica, Puerto Rico, Trinidad and Tobago), small islands (Antigua and Barbuda, Aruba, Bahamas, Barbados, British Virgin Islands, Cayman Islands, Dominica, Grenada, Guadeloupe, Montserrat, Saint Christopher and Nevis, Santa Lucia, Saint Kitts and Nevis, Saint Pierre and Miquelon, Saint Vincent and the Grenadines, United States Virgin Islands) and continental countries (Belize, Guyana, Suriname).

Table 10-6. Main NWFP of the Caribbean

Subregion	Main NWFP	Selected national statistical data available	Reference
Caribbean	Medicinal plants, aromatic plants, edible products, construction materials tools and handicrafts	*Guyana*: Export of 1 456 tonnes of palm hearts (*Euterpe oleracea*) worth US$1 965 978	van Andel 1998
		Suriname: Annual export of fruits of *Astrocaryum segregatum* (awara) and *Astrocaryum maripa* (maripa) worth US$1 740 in 1996-2000	FAO 2001a
		Grenada: Production of 2 347 tonnes and export of 1 863 tonnes of nutmeg (*Myristica fragans*) in 1993	FAO 1994b

declined by nearly 50 percent from 3 362 tonnes in 1986 to 1 863 tonnes in 1993 because of decreased world demand for raw nutmeg and competition from other producing countries.

Important edible products are fruits such as maripa (*Astrocaryum maripa*) and awara (*Astrocaryum segregatum*) in Suriname and balata (*Manilkara bidentata*), hog plum (*Spondias mombin*) and serrette (*Brysonima coriacea*) in Trinidad and Tobago.

The heart of the manicole palm (*Euterpe oleracea*) is one of the most important products in Guyana and the principal source of income for Amerindian communities in the coastal wetlands. Annual production rose from 942 tonnes in 1993 to 1 648 tonnes in 1995, with export revenue of US$2 million. Other countries, including Cuba and Trinidad and Tobago, also cultivate this species.

Important faunal foodstuffs are honey and bushmeat. Beekeeping is an important activity in the Dominican Republic and Cuba. In Suriname, the dependence of indigenous people and urban inhabitants on wildlife species for protein threatens many species.

Construction materials, utensils and handicrafts are another important group of NWFP in the Caribbean. In Guyana, aerial roots of nibi (*Heteropsis flexuosa*) are used for the manufacture of furniture, while roots of kufa (*Clusia* spp.) are used as household items. For families in the lower Pomeroon Basin, nibi harvesting is the most important source of income. In Santa Lucia, latanier (*Cocothrinax barbadensis*) is used in broom production. Latanier is sold in rural and urban areas but faces competition from imported plastic brooms. In Jamaica, jippi jappa (*Carludovica palmata*) is the principal source of material from the forest for making hats, bags, table mats, etc. In addition, strips of the rose apple (*Eugenia jambos*) are used to make baskets and hampers. Bamboo (*Bambusa vulgaris*) is an important product used in Grenada as scaffolding during construction and as a raw material in the production of different handicrafts. Some villages are dependent in income from these handicrafts. There is concern about the supply of bamboo because of high demand.

The great expansion of the tourist sector has increased the consumption of palm leaves for thatch. In the Dominican Republic, for example, "palma cana" (*Sabal umbraculifera*) is used for thatch for both temporary and permanent structures. In Trinidad and Tobago, *Sabal mauritiiformis, Maximiliana caribea* and *Manicaria saccifera* are used for thatching.

Europe

As reported in UNECE/FAO (2000) there are, in general, few reliable and systematically collected data on NWFP production for most European countries. National potentials, quantities and value by product category and the volumes traded or consumed are poorly known and/or documented in national forest statistics. A few countries maintain some regular statistics on NWFP for which harvesting permits are issued by the forest authorities, such as mushrooms, berries, game meat and hunting. Key NWFP on which data are reported include, in order of importance, Christmas trees (including production from plantations on farms and from cutting in forests), mushrooms, berries and game meat. A few countries also report on decorative foliage, cork, pine resin, herbal plants, honey and nuts (particularly chestnuts, acorns, hazelnuts and stone nuts). For nuts, herbal plants and honey, reported data on total country production include significant outputs from agricultural lands and are usually reported in agricultural statistics. An overview of the main European NWFP for which data are available is presented in Table 10-7.

Table 10-7. Main NWFP in Europe with major producing countries

Product	Main producing countries (reference year)[1]	Quantity *thousand tonnes*	Value *million US$*
Christmas trees[2]	Germany	20.0[3]	235.3
	France (1985-1995)	5.6[3]	75.4
	United Kingdom (1995)	3.0[3]	66.7
	Denmark (1996)	7.0[3]	24.1
Mushrooms and truffles (all species combined)[4]	Czech Republic	23.9	39.1
	Belarus (1995-1996)	10.1	15.2
	Sweden	8.5	31.8
	France	8.2	107.7
	Finland (1996)	6.0	14.1
	Italy (1995)	2.4	44.7
Fruits and berries (all species combined)[4]	Albania	60.0	114.0
	Finland (1996)	40.0	67.1
	Norway (1994-1996)	25.0	45.3
	Czech Republic	22.7	39.2
Game meat	Sweden	17.1	76.1
	Poland (1995)	8.1	
	Finland (1996)	7.9	64.0
	Czech Republic (1990-1994)	6.8	
	Norway (1994-1996)	6.6	66.5
Cork	Portugal (1995)	135.0	145.3
	Spain[5]	110.0	
	Albania	18.1	7.2
	Italy (1995)	10.4	7.2
Pine resin	Portugal (1995)	40.0	80.7
Decorative foliage	Albania	198.5	143.0
	Switzerland (1996)	117.5[6]	
	Denmark	25.0	49.8

Source: UNECE/FAO 2000, except as noted.
[1] The reference year (or range for average) of production and value data is given when available.
[2] The value of Christmas trees was based on net national income in Denmark and retail prices in France, Germany and the United Kingdom.
[3] Million trees.
[4] Reported values for both mushrooms and berries are income to collectors (Finland) or estimated market prices.
[5] Not reported in UNECE/FAO 2000. *Source:* FAO 2001c.
[6] Thousand m^3.

North America

In Canada and the United States, a great variety of NWFP are gathered, mainly for personal use, and their collection is widespread among rural populations. However, only a few products are included in national forest product statistics (UNECE/FAO 2000).

In Canada, reported products include Christmas trees, pelts and maple syrup. Reported data from the United States cover five products (Christmas trees, mushrooms, pelts, maple syrup and commercial fish catch). Even so, the records on mushrooms refer to only four major species from among 25 to 30 species that are commercially used. There are no data for some widely consumed products such as game meat and berries.

Major NWFP of the region are foodstuffs and forest plants for ornamental purposes (Table 10-8), but there is little information on production sources, the number of collectors, volume or value. Moreover, reported figures vary considerably from year to year. For example, in Canada, the reported value of maple syrup production varied from US$59.1 million in 1992 to US$44.9 million in 1993.

Commercial demand for mushrooms and berries is increasing throughout the region. In the Canadian province of British Columbia, 35 mushroom species are now commercially harvested. In the United States, data on mushrooms were available from only one region of the country, the Pacific Northwest, where

Table 10-8. Main NWFP of North America

Product	United States Quantity (reference year)	United States Value million US$	Canada Quantity	Canada Value million US$
Christmas trees *(million trees)*	35[1] (1993-1996)		4.5 (1997)	48.6[2]
Mushrooms/truffles		41.1		
Decorative foliage		128.5		
Pelts *(million units)*	5.7 (1995)	40.6	1.3 (1995-1996)	
Maple syrup[3] *(million litres)*	6.2 (1991)	39.3	15.3 (1995)	44.9 (1993)

Source: UNECE/FAO 2000, except as noted.
[1] Not reported in UNECE/FAO 2000. *Source:* FAO 1999c. Most trees are harvested for domestic use.
[2] Wholesale value.
[3] *Source:* FAO 2001c.

commercial harvest is done largely for export markets in Asia and Europe.

No data at the country level are available on the production of medicinal and herbal plants collected in the region for personal and/or commercial use. Medicinals are collected mainly on forest lands, but a growing share is now being produced through farming.

The Pacific Northwest region of the United States has a significant industry based on processing decorative forest foliage. About one-quarter of its production is for export to Europe.

Hunting, game meat and animal trophies provide significant income to both private forest owners and public land management agencies in the region. Canada produces the world's largest number of pelts, and the United States ranks third (after the Russian Federation). In both countries the data reported are for total harvest, which includes species that are not associated with forests. Reported value is the price received by the trapper.

Sport fishing in the region is very popular, although it is difficult to separate fish harvest occurring in forests. The reported harvest in the United States is restricted to salmon species, which spend part of their life in forest environments. The United States salmon harvest in 1995 was 517 000 tonnes, valued at US$521 million.

CONCLUSION

Data collection for this study confirmed that there is a serious lack of quantitative data at the national level on non-wood forest products and even less on the resources that provide them, with the exception of Asia where there is a tradition of national collection of information on NWFP resources and consumption. Information is scarce and often mixed with agricultural production statistics. Statistical data, where they exist at all, are mostly limited to selected internationally traded products and, in this case, data are usually limited to export quantities. Information on the resource base and on subsistence use of NWFP is non-existent, mainly because of the multitude of products used by local people and the technical difficulty and high cost of measuring and reporting on them.

Even when data exist, they are seldom based on recurrent, statistically designed surveys and inventories, and it is therefore difficult to assess the reliability of the information. For example, even in Asia much of the information is based on national inventories up to ten years out of date. A similar problem exists for the economic value associated with the products because value can be calculated at different stages of production and processing. The data obtained from traditional forestry institutions responsible for the forest resources often differ from the trade data reported by customs agencies.

National level data on the resources and on production and trade (quantities and values) of major products are essential to assess the full contribution of the forest sector to the economy of the country, and for forest management and policy development. In some cases NWFP resource and product information is available on a national basis, but in most cases, the information is available only for parts of the country. Therefore, extrapolation is necessary but difficult.

Because of the factors described above, as well as the lack of internationally agreed-upon terminology, concepts and clear definitions, statistical data on NWFP resources and production are not usually comparable among or even within countries or regions. Therefore, regional and global aggregation of production and value is very difficult. A classification system with unified terminology and measurements is needed.

Most of the products are extracted from natural stands in various types of forest and woodland ecosystems. However, among the current issues of global resource monitoring is the lack of management of non-wood resources. For products in high demand, this often leads to unsustainable harvest levels and the potential endangerment or extinction of the species. This has serious socio-economic implications for people dependent on the availability of these resources. Some important products, such as bamboo, are evolving into farmed crops, while others, such as many medicinal plants, are becoming endangered because of deforestation and/or overharvesting. The use of synthetic substitutes has made many others, such as guta percha, balata, sorva, copal and piassaba fibres, obsolete.

BIBLIOGRAPHY

Argentina. Ministerio de Desarrollo Social y Medio Ambiente (MDSMA). 2001. www.medioambiente.gov.ar

Biodiversity and Development (BIODEV). 1994. *Study of the collection of wild specimen of Malagasy plants and animals destined for export (with emphasis on regions chosen for future projects development).* Study IIb for the TRADEM project. Antananarivo, Madagascar.

Camara, A.K. 1991. *Guinée, séminaire sur les statistiques en Afrique.* Thiès, Sénégal, FAO.

Campos Roasio, J. 1998. *Productos forestales no madereros en Chile.* Serie Forestal 10. Santiago, Chile, FAO Regional Office for Latin America and the Caribbean.

Chemli, R. 1997. Medicinal, aromatic and culinary plants of Tunisian flora. In *Proceedings of the International Expert Meeting on Medicinal, Culinary and Aromatic Plants in the Near East.* Cairo, Egypt 19-21 May 1997. FAO Regional Office for the Near East.

Chihongo, A.W. 1993. Pilot country study on NWFP for Tanzania. In *Non-wood forest products: a regional expert consultation for English-speaking African countries,* 17-22 October 1993. Arusha, Tanzania, Commonwealth Science Council and FAO.

Chikamai, B. 1997. Production, markets and quality control of gum arabic in Africa: Findings and recommendations from an FAO project. In J.O. Mugah, B.N. Chikamai & E. Casadei, eds. *Conservation, management and utilization of plant gums, resins and essential oils.* Proceedings of a regional conference for Africa held in Nairobi, Kenya, 6-10 October 1997.

Collinson, C., Burnett, D. & Agreda, V. 2000. *Economic viability of Brazil nut trading in Peru.* Chatham Maritime, Kent, UK, Natural Resources Institute, University of Grenwich.

Cunningham, A.B. 1993. *African medicinal plants.* People and Plants Working Paper No. 1. Paris, WWF/UNESCO/Royal Botanic Gardens Kew.

Cunningham, M., Cunningham, A.B. & Schippmann, U. 1997. *Trade in Prunus africana and the implementation of CITES.* Bonn, German Federal Agency for Nature Conservation.

El Adab, A. 1993. Les produits forestiers et leur importance dans l'economie Tunisienne. *Séminaire sur les Produits de la Forêt Méditerranéenne.* Florence, Italy, 20-24 September 1990.

Eritrea. Ministry of Agriculture. 1998. *Forestry data report on Eritrea.* In EC/FAO/UNEP. *Proceedings of Subregional Workshop on Forestry Statistics – IGAD region.* Rome.

FAO. 1991. Proceedings of the 10th World Forestry Congress, September 1991, Paris. Rome.

FAO. 1994a. *Non-wood forest products in Asia.* RAPA Series No. 28. Bangkok, FAO Regional Office for Asia and the Pacific.

FAO. 1994b. *Nutmeg and derivates.* Working paper FO:MISC/94/7. Rome.

FAO. 1995a. *Trade restrictions affecting international trade in non-wood forest products.* FAO Non-wood Forest Products No. 8. Rome.

FAO. 1995b. *Gums, resins and latexes of plant origin,* by J.J.W. Coppen. FAO Non-wood Forest Products No. 6. Rome.

FAO. 1995c. *Memoria: consulta de expertos sobre productos forestales no madereros para América Latina y el Caribe.* Serie Forestal No.1. Santiago, Chile, FAO Regional Office for Latin America and the Caribbean.

FAO. 1995d. *Natural colourants and dyestuffs,* by C.L. Green. FAO Non-wood Forest Products No. 4. Rome.

FAO. 1995e. *Gum naval stores: turpentine and rosin from pine resin,* by J.J.W. Coppen &

G.A. Hone. FAO Non-wood Forest Products No. 2. Rome, FAO.

FAO. 1996. *Desarrollo de productos forestales no madereros en América Latina y el Caribe*, by C. Chandrasekharan, T. Frisk & J.C. Roasio. Serie Forestal N. 10. Santiago, Chile, FAO Regional Office for Latin America and the Caribbean.

FAO. 1997a. *Assistance à l'exportation des produits forestiers*. Vol. 1. Projet TCP/MAG/6611. Antananarivo, Madagascar.

FAO. 1997b. *Wildlife and food security in Africa*, by Y. Ntiamoa-Baidu. FAO Conservation Guide No. 33. Rome.

FAO. 1998a. *Non-wood forest products of Namibia*, by J. Haiwla. EC-FAO Partnership Programme, Project GCP/INT/679/EC. Rome.

FAO. 1998b. *Non-wood forest products of Ethiopia*, by G. Deffar. EC-FAO Partnership Programme, Project GCP/INT/679/EC. Rome.

FAO. 1998c. *Non wood forest products from conifers*. FAO Non-wood Forest Products No. 12. Rome.

FAO. 1999a. *Les données statistiques sur les PFNL au Rwanda*, by A. Murekezi. EC-FAO Partnership Programme, Project GCP/INT/679/EC. Rome.

FAO. 1999b. *Statistiques sur les PFNL: Tchad*, by A.M. Haggar. EC-FAO Partnership Programme, Project GCP/INT/679/EC. Rome.

FAO. 1999c. *Preliminary consideration for the development of statistics on non-wood forest products in Suriname*, by P.R. Peneux. EC-FAO Partnership Programme, GCP/INT/679/EC. Rome/Paramaribo.

FAO. 1999d. *Non-wood forest products study for Mexico, Cuba and South America*. FRA Working Paper No. 11. Rome. (draft)

FAO. 1999e. Towards a harmonized definition of non-wood forest products. *Unasylva*, 198: 63-64.

FAO. 2000a. *Proceedings of subregional workshop on data collection and outlook effort for forestry in the Caribbean, Port-of-Spain,* Trinidad & Tobago. 21-25 February 2000. EC-FAO Partnership Programme, CP/INT/679/EC. Rome.

FAO. 2000b. *Situación forestal en la región de América Latina y el Caribe: Periodo 1998-1999.* 21st session of the Latin American and Caribbean Forestry Commission, Santa Fe de Bogotá, Colombia, 4-8 September 2000. www.fao.org/documents

FAO. 2000c. *Statistical data on non-wood forest products in Trinidad and Tobago,* by L. Kisto. EC-FAO Partnership Programme, GCP/INY/679.

FAO. 2000d. *Evaluación de los recursos forestales no madereros en América Central.* FRA Working Paper No. 22. Rome.

FAO. 2001a. *Pilot study on data collection and analysis related to non-wood forest products in Suriname,* by C.M. Rahan-Chin. EC-FAO Partnership Programme GCP/INT/679/EC. Rome/Paramaribo (draft report).

FAO. 2001b. *Non-wood forest products of Africa – a regional and national overview*, by S. Walter. Working Paper FOPW/01/1. Rome.

FAO. 2001c. *Non-wood products from broad-leaved trees*. FAO Non-wood Forest Products. Rome. (in press).

International Tropical Timber Organization (ITTO). 1998. *Non-wood tropical forest products: processing, collection and trade.* ITTO project PD 143/91 vers 2(I). Executive summary of the technical report.

Lange, D. & Mladenova, M. 1997. Bulgarian model for regulating the trade in plant material for medicinal and other purposes. In *Medicinal plants for forest conservation and health care*. FAO Non-Wood Forest Products Series No. 11. Rome.

Makonda, F.B.S. & Ishengoma, R.C. 1997. Indigenous knowledge and utilization potentials of selected gum, resin and oil plant species of Tanzania. In J.O. Mugah, B.N. Chikamai & E. Casadei, eds. *Conservation, management and utilization of plant gums, resins and essential oils.* Proceedings of a regional conference for Africa, 6-10 October, Nairobi.

Natural Cork Quality Council (NCQC). 2000. *The cork industry.* www.corkqc.com, viewed 30/08/2000.

Natural Resources Conservation Authority (NRCA). 2001. *Animal species of Jamaica protected under the Wildlife Protection Act.* Kingston, Jamaica. www.nrca.org/biodiversity/species

Niger. Ministère de L'Hydraulique et de l'Environnement. 1998. *Rapport national du Niger sur les ressources génétiques forestières.* Niamey.

Njovu, F.C. 1993. Non-wood forest products Zambia. A country pilot study for the expert consultation for English speaking African countries. In *Non-wood forest products: a*

regional expert consultation for English-speaking African countries, 17-22 October 1993. Arusha, Tanzania, Commonwealth Science Council and FAO.

Nour, H.O.A. 1995. *Quality control of gum arabic.* Mission report. Khartoum.

Resico, C. 2001. *Compilación y análisis sobre los productos forestales no madereros – Argentina.* Proyecto información y análisis para el manejo forestal sostenible: integrando esfuerzos nacionales e internacionales en 13 paises tropicales en America Latina. GCP/RLA/133/EC. Santiago, Chile, FAO Regional Office for Latin America and the Caribbean.

Rios Torre, M. 2001. *Compilación y análisis sobre los productos forestales no madereros – Perú.* Proyecto Información y análisis para el manejo forestal sostenible: integrando esfuerzos nacionales e internacionales en 13 paises tropicales en America Latina. GCP/RLA/133/EC. Santiago, Chile, FAO Regional Office for Latin America and the Caribbean.

Senegal. Ministère du développement rural et de l'hydraulique (MDRH). 1993. *Plan d'Action Forestier.* Dakar.

Shiembo, P.N. 1999. The sustainability of eru *Gnetum africanum* and *Gnetum buchholzianum*: over-exploited NWFP from the forests of Central Africa. In T.C.H. Sunderland, L.E. Clark & P. Vantomme (eds.). *The NWFP of Central Africa: current research issues and prospects for conservation and development.* Rome, FAO.

Sunderland, T. & Tako, C.T. 1999. *The exploitation of* Prunus africana *on the island of Bioko, Equatorial Guinea.* Report for the People and Plants Initiative, WWF-Germany and the IUCN/SSC Medicinal Plant Specialist Group. www.ggcg.st/bioko/bioko_prunus.htm

Tratado de Cooperación Amazónica (TCA). 1996. *Frutales y hortalizas promisorios de la Amazonia.* Lima.

UNECE. 1998. *Non-wood goods and services of the forests.* Geneva Timber and Forest Study Papers No. 15. Geneva.

UNECE/FAO. 2000. *Forest resources of Europe, CIS, North America, Australia, Japan and New Zealand – Main report.* Geneva.

van Andel, T. 1998. Commercial exploitation of non-timber forest products in the Northwest district of Guyana. *Caribbean Journal of Agriculture and Natural Resources*, 2(1): 15-28

van Andel, T.R. 2000. *Non-timber forest products of the north-west district of Guyana. part 1 and 2.* Tropenbos – Guyana Series 8a/8b. Georgetown, Guyana, Tropenbos Guyana Programme.

Walter, S. 1996. *Moeglichkeiten und Grenzen der Nutzung von Nicht-Holz Waldprodukten in Madagaskar, dargestellt am Beispiel des Naturreservates Zahamena.* Thesis. University of Giessen, Germany.

Wende, L. 2001. *Compilación y análisis sobre los productos forestales no madereros – Bolivia.* Proyecto Información y análisis para el manejo forestal sostenible: integrando esfuerzos nacionales e internacionales en 13 paises tropicales en America Latina. GCP/RLA/133/EC. Santiago, Chile, FAO Regional Office for Latin America and the Caribbean.

WWF. Undated. *Forest harvest: an overview of non-timber forest products in the Mediterranean region*, by Y. Moussouris & P. Regato. Rome, WWF Mediterranean Programme.

Zambia. Ministry of Environment and Natural Resources (MENR). 1997. *Zambia Forestry Action Plan.* Volume II – Challenges and opportunities for Development. Lusaka

Zida, O.B. & Kolongo, S.L. 1991. Main NWFP in North Africa. Seminar on Forestry Statistics in Africa. Thiès, Sénégal. FAO.

Part II: Forest resources by region

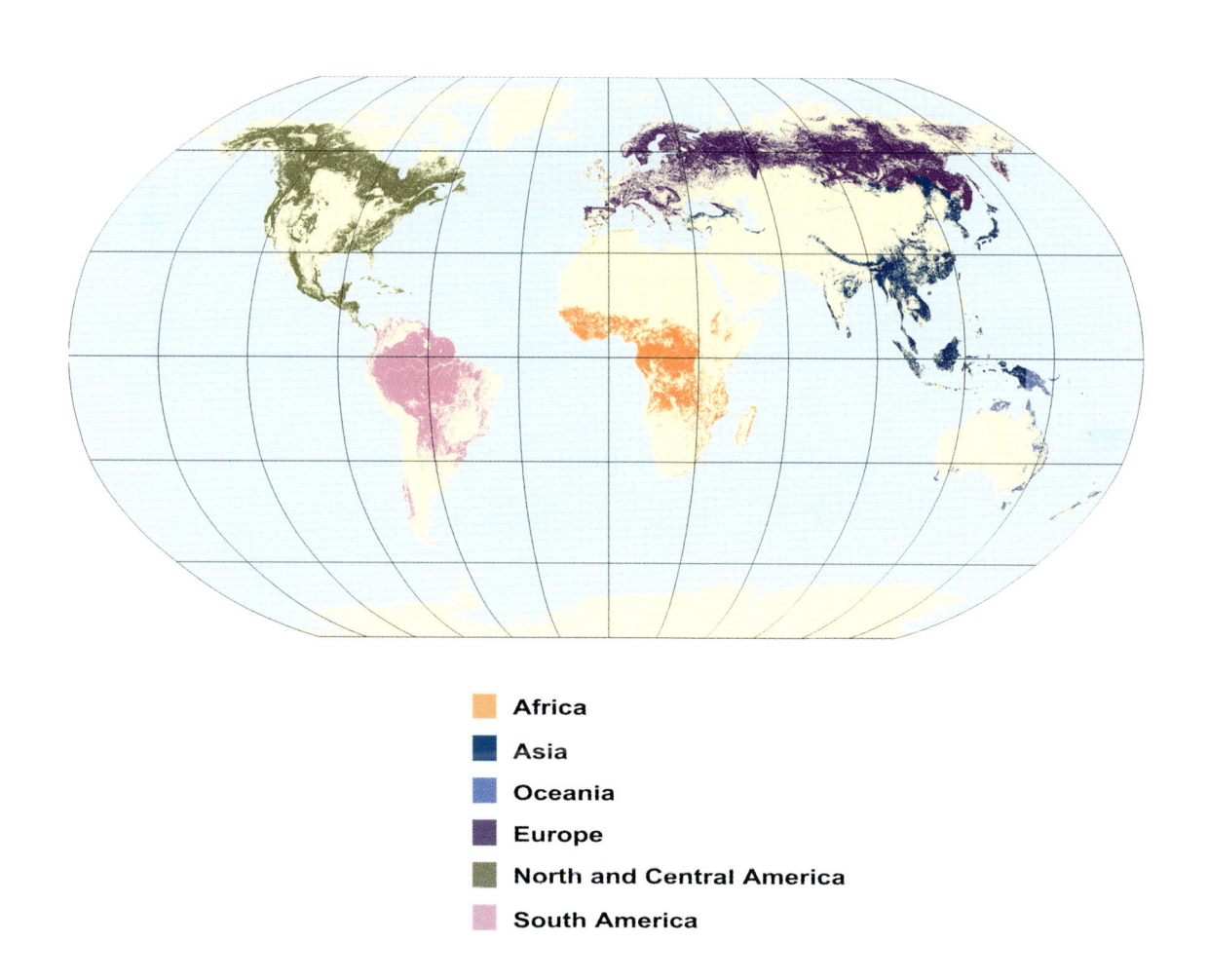

Africa
Asia
Oceania
Europe
North and Central America
South America

Chapter 11
Africa

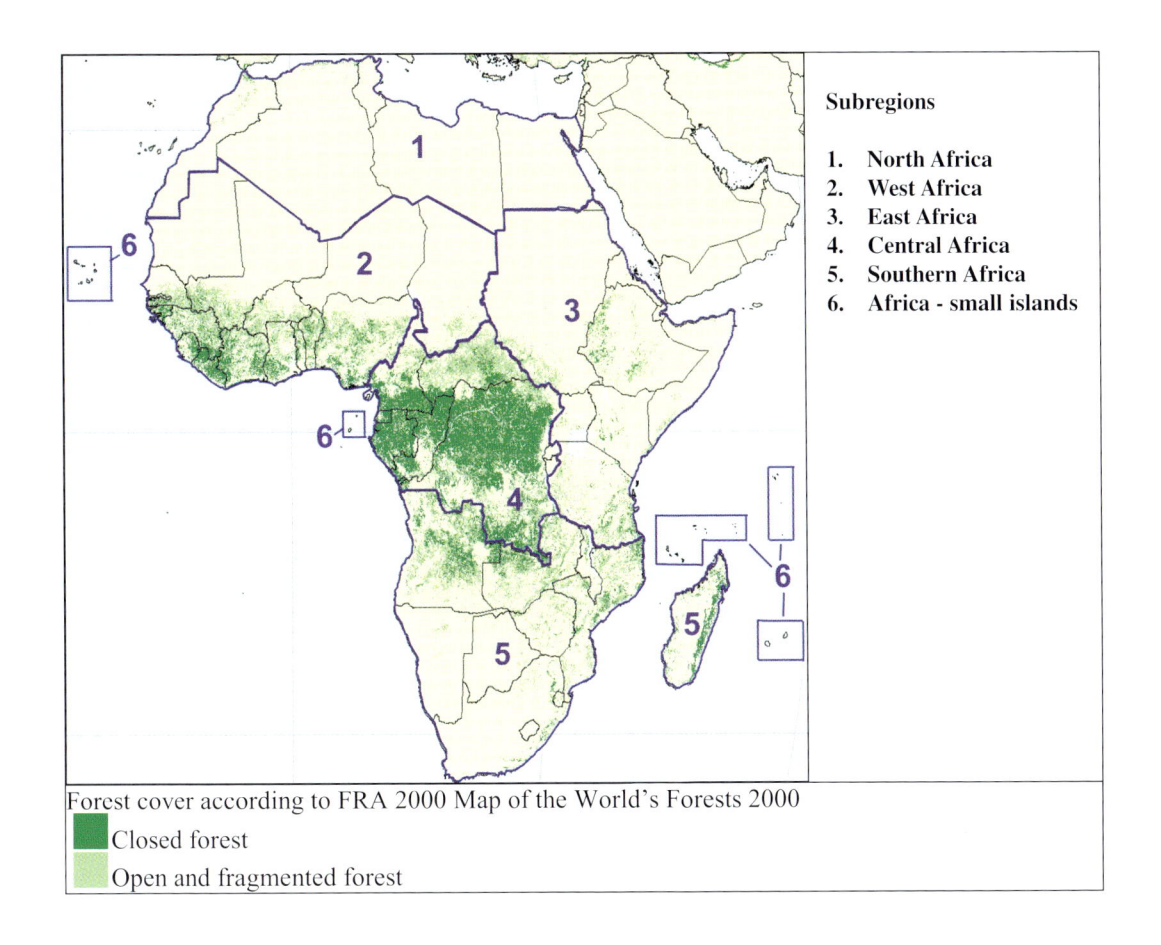

Figure 11-1. Africa: subregional division used in this report

Africa (see Figure 11-1[27] and Table 11-1) contains about 650 million hectares of forests, corresponding to 17 percent of the world total. African forests amount to 0.85 ha per capita, which is close to the world average. Almost all forests are located in the tropical ecological domain, and Africa has about one-quarter of all tropical rain forests. Only 1 percent of the forest area is classified as forest plantations. The net change of forest area is the highest among the world's regions, with an annual net loss, based on country reports, estimated at -5.3 million hectares annually, corresponding to -0.78 percent annually.

---

[27] The division into subregions was made only to facilitate the reporting at a condensed geographical level and does not reflect any opinion or political consideration in the selection of countries. The graphical presentation of country areas does not convey any opinion of FAO as to the extent of countries or status of any national boundaries.
www.fao.org/forestry/fo/country/nav_world.jsp

Table 11-1. Africa: forest resources by subregion

Subregion	Land area	Forest area 2000 - Natural forest	Forest plantation	Total forest 000 ha	Total forest %	Total forest ha/capita	Area change 1990-2000 (total forest) 000 ha/year	Area change 1990-2000 (total forest) %	Volume and above-ground biomass (total forest) m³/ha	Volume and above-ground biomass (total forest) t/ha
	000 ha	000 ha	000 ha	000 ha	%	ha/capita	000 ha/year	%	m³/ha	t/ha
Central Africa	403 298	227 377	634	228 011	56.5	2.6	-852	-0.4	127	194
East Africa	590 078	134 132	1 291	135 423	23.0	0.7	-1 357	-1.0	28	38
North Africa	601 265	4 569	1 693	6 262	1.0	n.s.	33	0.5	32	51
Southern Africa	649 213	192 253	2 601	194 854	30.0	1.6	-1 741	-0.9	42	72
West Africa	733 359	83 369	1 710	85 079	11.6	0.4	-1 351	-1.5	61	84
Africa – small islands	1 181	130	107	237	20.1	0.1	4	1.9	88	121
Total Africa	2 978 394	641 830	8 036	649 866	21.8	0.8	-5 262	-0.8	72	109
TOTAL WORLD	13 063 900	3 682 722	186 733	3 869 455	29.6	0.6	-9 391	-0.2	100	109

Source: Appendix 3, Tables 3, 4, 6 and 7.

Chapter 12
Africa: ecological zones

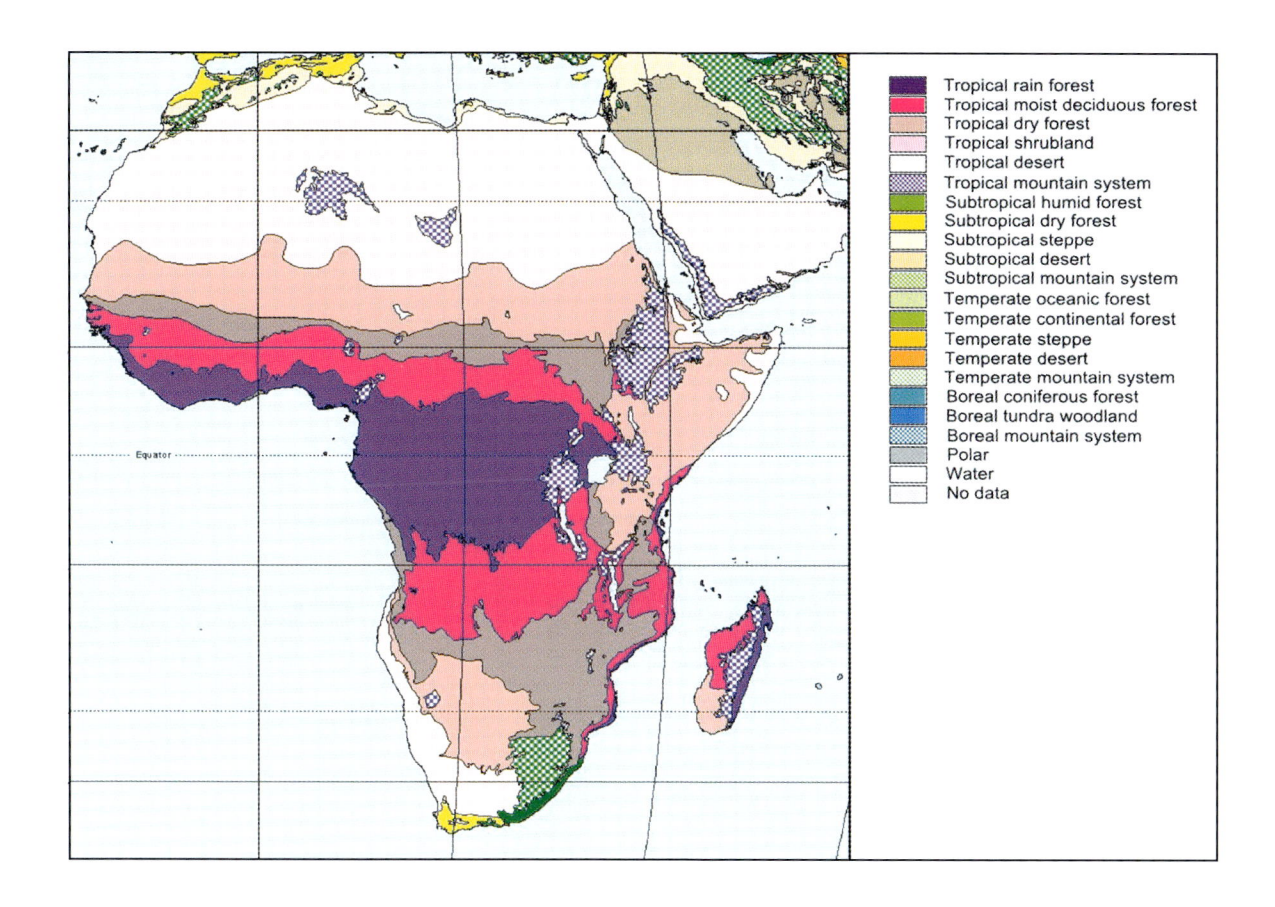

Figure 12-1. Africa: ecological zones

Figure 12-1 shows the ecological zones of Africa, as identified and mapped by FRA 2000. Table 12-1 contains area statistics for the zones by subregion, and Table 12-2 indicates the proportion of forest in each zone by subregion.

TROPICAL RAIN FOREST

This zone covers the central part of Africa on both sides of the equator as well as the southeastern coast. The climate is more or less tropical. Rainfall ranges from 1 000 mm to more than 2 000 mm per year. If there is a dry season, it does not exceed three to four months and always occurs in winter. Temperature is always high, generally more than 20°C, except in the mountains.

The greater part of the zone was formerly covered with rain forests and swamp forests. Today, little undisturbed rain forest remains and secondary grassland and various stages of forest regrowth are extensive. Compared to the rain forests of South America and Asia, African forests are relatively poor floristically.

The most extensive formation is the Guineo-Congolian lowland rain forest, concentrated in the Congo Basin. It is a tall, dense forest, more than 30 m high with emergents up to 60 m and several strata. Some species are deciduous but the forest as a whole is evergreen or semi-evergreen. The large trees include *Entandrophragma* spp., *Guarea cedrata*, *Guarea thompsonii*, *Lovoa trichilioides*, *Maranthes glabra*, *Parkia bicolor*, *Pericopsis elata* and *Petersianthus macrocarpus*. Small patches of moist evergreen and semi-evergreen rain forest occur with a single dominant, usually *Brachystegia laurentii*, *Cynometra alexandri*, *Gilbertiodendron dewevrei*, *Julbernardia seretii* or *Michelsonia microphylla*, all Leguminosae.

Table 12-1. Africa: extent of ecological zones

Subregion	Tropical						Subtropical					Temperate					Boreal			
	Rain forest	Moist	Dry	Shrub	Desert	Mountain	Humid	Dry	Steppe	Desert	Mountain	Oceanic	Continental	Steppe	Desert	Mountain	Coniferous	Tundra	Mountain	Polar
Central Africa	291	112	13	1		19														
East Africa	21	68	79	268	103	76														
North Africa					497	20		26	48		11									
Southern Africa	26	187	192	106	76	22	8	8			31									
West Africa	70	106	86	226	222	10														
Total Africa	409	473	370	601	898	147	8	35	48		42									
TOTAL WORLD	1468	1117	755	839	1192	459	471	156	491	674	490	182	726	593	552	729	865	407	632	564

Note: Data derived from an overlay of FRA 2000 global maps of forest cover and ecological zones. The subregion Africa – small islands is not included in the table because of incomplete information.

Table 12-2. Africa: proportion of forest by ecological zone

Subregion	Tropical						Subtropical					Temperate					Boreal			
	Rain forest	Moist	Dry	Shrub	Desert	Mountain	Humid	Dry	Steppe	Desert	Mountain	Oceanic	Continental	Steppe	Desert	Mountain	Coniferous	Tundra	Mountain	Polar
Central Africa	65	44	74			23														
East Africa	6	15	32	5		9														
North Africa					23	7														
Southern Africa	34	28	42	7		15	16	7			3									
West Africa	47	35	74	1		6														
Total Africa	57	31	48	4		11	16	19			4									
TOTAL WORLD	69	31	64	7	0	26	31	45	9	2	20	25	34	4	1	26	66	26	50	2

Note: Data derived from an overlay of FRA 2000 global maps of forest cover and ecological zones. The subregion Africa – small islands is not included in the table because of incomplete information.

The rain forest of Madagascar is 25 to 30 m tall, without large emergent trees but very rich in species. It is evergreen and grows up to 800 to 1 000 m altitude. Some of the important families represented in the upper canopy are Euphorbiaceae, Rubiaceae, Araliaceae, Ebenaceae (*Diospyros* spp.), Sapindaceae, Burseraceae (*Canarium* spp.), Anacardiaceae, Elaeocarpaceae (*Echinocarpus* spp.), Lauraceae, Guttiferae, Myrtaceae, Malpighiaceae and the conspicuous giant monocot the traveller's palm (*Ravenala madagascariensis*).

The drier periphery of the zone has transitional forest types. In West Africa these evergreen or semi-evergreen forests include *Afzelia africana, Aningeria altissima, Aningeria robusta, Chrysophyllum perpulchrum, Cola gigantea, Khaya grandifolia* and *Mansonia altissima*. Other important species are *Triplochiton scleroxylon, Celtis mildbraedii,* *Holoptelea grandis, Sterculia* spp., *Trilepisium madagascariense* and *Chlorophora excelsa*.

Mangroves extend along the muddy, sheltered coasts of the Gulf of Guinea from Angola to Senegal. They include *Rhizophora racemosa, Rhizophora harrisonii, Rhizophora mangle, Avicennia africana, Avicennia nitida, Laguncularia racemosa* and *Acrostichum aureum*.

TROPICAL MOIST DECIDUOUS FOREST

This zone lies on the Great African Plateau to the south of the Guineo-Congolian Basin, mostly at an altitude of 900 to 1 000 m but in some places up to 1 500 m, as well as along the southeastern coast of Africa and in the central part of Madagascar. The dry season is always pronounced, lasting up to six months. There is a single rainy season, in summer, but there is pronounced regional variation. Annual rainfall for the zone varies between 800 and 1 500 mm, but can reach 2 000 mm locally.

Dry evergreen forest is widely distributed on Kalahari sands, featuring species of *Marquesia, Berlinia* and *Laurea*. Semi-evergreen forest of the Guineo-Congolian type is mainly confined to Angola. On the eastern coastal plain, forest is the climax but has been largely replaced by wooded grassland and cultivation.

Everywhere else the most characteristic vegetation is woodland – wetter Zambezian miombo woodland to the south and Sudanian woodland to the north. Zambezian woodlands are characterized by several species of *Brachystegia* (*B. floribunda, B. glaberrima, B. taxifolia, B. wangermeeana, B. spiciformis, B. longifolia, B. utilis*) with canopy heights sometimes up to 30 m. Associated species include *Marquesia macroura, Pterocarpus* spp., *Julbernardia* spp. and *Isoberlinia* spp. Sudanian woodlands, generally lower, are characterized by several species of *Acacia* and by *Isoberlinia doka*. Other characteristic species include *Acacia dudgeoni, Acacia gourmaensis, Antidesma venosum, Faurea saligna, Lophira lanceolata, Maprounea africana, Maranthes polyandra, Monotes kerstingii, Ochna afzelii, Ochna schweinfurthiana, Protea madiensis, Terminalia glaucescens* and *Uapaca togoensis*.

In Madagascar, the primary vegetation is a dry deciduous forest or thicket, but the most extensive vegetation is now secondary grassland. Nevertheless, some areas of forest remain, especially along the coast, with *Dalbergia* spp. on lateritic soils; *Tamarindus indica* on sandy soils; and *Adansonia* spp. and *Bathiaea* sp. on calcareous plateaus.

Mangroves occur along sheltered coasts of the Indian Ocean, dominated by *Rhizophora mucronata, Avicennia marina* and *Sonneratia alba*. Other tree and shrub species include *Ceriops tagal, Bruguiera gymnorrhiza* and *Xylocarpus obovatus*.

TROPICAL DRY FOREST

Farther from the equator and the wet southeastern coast, rainfall decreases and the dry season is always long six to seven months. Rainfall varies between 500 and 1 000 mm. Temperature is always high, with the mean temperature of the coldest month about 20°C. Similar conditions are found in Ghana (Accra) and Angola (Cabinda).

Woodland is the predominant vegetation type under these drier conditions. In the Zambezian region there is drier miombo, mopane (*Colophospermum mopane*) woodland or Sudanian woodland in the southern valleys and depressions and scrub woodland in the southern lowlands with *Acacia caffra, Acacia davyi* and *Acacia luederitzii*. In the Sudanian region, woodland species include *Acacia albida, Acacia macrostachya* and *Acacia nilotica*. In the Sudan, woodland species typically include *Anogeissus leiocarpus* and various species of *Combretum*. Where cultivation is possible, most of the land is bush fallow. Near Accra, Ghana, some patches of dry semi-evergreen forest with *Diospyros abyssinica* and *Millettia thonningii* remain. In Cabinda, Angola, the prevalent vegetation is wooded grassland with *Adansonia digitata* and many individuals of *Anacardium occidentale* and *Mangifera indica*, two introduced trees. A conspicuous tree of this zone is the baobab (*Adansonia digitata*) with its bizarre big trunk.

TROPICAL SHRUBLAND

In the Sahelian zone, the Kalahari and the southwestern part of Madagascar, rainfall becomes lower while temperatures are still high. Rainfall is always less than 1 000 mm and reaches scarcely 200 mm in the drier parts. The mean temperature of the coldest month is generally more than 20°C, except in the Kalahari where temperatures are lower (to 10°C). Even though Somalia lies across the Equator the climate is semi-arid to arid, with annual rainfall between 400 and 750 mm and very high temperatures.

In these very dry areas, spontaneous vegetation is generally pseudo-steppe, scrub woodland or thicket. In the Sahelian zone, wooded grassland (mainly with *Anogeissus* and *Acacia* species) is located in the south and semi-desert grassland in the north. Somalia has predominately deciduous shrubland and thicket with *Acacia* and *Commiphora* species. In the Kalahari, stunted scrub woodland with acacia (*Acacia karroo*) and shrub pseudo-steppe forms the landscape. In Madagascar, some dry deciduous forest still occurs in the northern part of the zone but the most characteristic vegetation type in the western part is deciduous thicket with Didiereaceae.

TROPICAL MOUNTAIN SYSTEMS

The main mountain systems are the Cameroon highlands, the mountains of Kenya, the Kivu ridge and the Ethiopian highlands. Some lower and isolated hills occur, such as the Fouta Djalon, Jos and Mandara Plateaus in West Africa, Hoggar in the Sahara and Windhoek Mountain in southern Africa. Madagascar has a high central range.

The climate is similar to that of the surrounding lowlands but with lower temperatures and, often, higher rainfall. Above 800 to 1 200 m, temperature decreases and vegetation changes, defining submontane, montane and high-elevation ecofloristic zones.

The vegetation is extremely diverse and varies with climate. On most mountains the lowermost vegetation is forest. Between the lowland forest and the rather different (in physiognomy and flora) montane forest, there is a submontane transition zone. In many places, however, fire and cultivation have destroyed the vegetation of this transition zone. Montane forest, generally above 1 500 to 2 000 m, is lower in structure than lowland and submontane forests. At the upper part of the montane level is an Ericaceous belt followed, above 3 000 m, by alpine vegetation.

In western Africa, on the Kivu ridge or the wetter slopes of the Ethiopian highlands and East African mountains, the trees of the upper stratum are 25 to 45 m tall with middle and lower layers. Characteristic species include *Aningeria adolfi-fredrici, Chrysophyllum gorungosanum, Cola greenwayi, Diospyros abyssinica, Drypetes gerrardii, Olea capensis, Podocarpus latifolius, Prunus africana, Syzigium guineense* subsp. *afromontanum* and *Xymalos monospora*.

Bamboo (*Arundinaria alpina*) forest or thicket occurs between 2 300 and 3 000 m on most of the high mountains in East Africa and sporadically on some of the mountains of Cameroon.

In Madagascar, the original vegetation in the mountains was moist montane forest with *Tambourissa* and *Weinmannia* species, sclerophyllous montane forest with *Dicoryphe* and *Tina* species on the eastern slopes and drier, "tapia" (*Uapaca bojeri*) forest on the western slopes. These forests have been replaced over extensive areas by secondary grassland.

In other areas shrubland and thicket is the prevalent vegetation.

SUBTROPICAL HUMID FOREST

This zone is restricted to a narrow zone along the east coast of southern Africa, roughly between 25° and 34°S. It has moderately high and well-distributed rainfall and, except in the extreme south, is frost free. Annual rainfall is 800 to 1 200 mm and the mean temperature of the coldest month is 7° to 15°C. Mean annual temperatures diminish from 22°C in the north to 17°C in the south. Further inland, climate changes rapidly over short distances.

In most of the zone the natural vegetation is evergreen or semi-evergreen forest, the most luxuriant stands approaching rain forest stature and structure. The canopy varies in height from 10 to 30 m. About 120 species occur, although more than 30 are not usually present in any one stand. Endemic species include *Atalaya natalensis, Anastrabe integerrima, Beilschmiedia natalensis, Brachylaena uniflora, Cola natalensis, Commiphora harveyi, Cordia caffra, Diospyros inhacaensis* and *Manilkara concolor*. Today, where the original vegetation has not been completely replaced, land cover often consists of a mosaic of forest, scrub forest, bushland, thicket and secondary grasslands. Where rainfall is too low to support forest, the most widespread climax vegetation is evergreen and semi-evergreen bushland and thicket.

SUBTROPICAL DRY FOREST

This zone includes parts of North Africa and South Africa with a Mediterranean climate. There is a pronounced dry season in summer. Most of the rainfall (400 to 1 000 mm per year) occurs in winter although in the eastern regions of South Africa it is more evenly distributed (subtropical humid). The annual temperature varies but the mean temperature of the coldest month, in the lowlands, is always more than 7°C.

In northern Africa, the climax vegetation is forest, with *Quercus suber, Quercus faginea, Quercus ilex* and *Pinus pinaster* in the most humid parts under marine influence and *Tetraclinis articulata, Q. ilex* and *Pinus halepensis* in more continental situations. In many places, as a result of degradation by overgrazing these forests have been replaced by scrub.

In South Africa, the prevalent vegetation of this zone is fynbos, sclerophyllous shrublands 1 to 4 m high, with the main shrub genera *Protea, Cliffortia, Muraltia, Leucospermum, Restio, Erica* and *Serruria*. The only tree species, silver tree (*Leucadendron argenteum*), is found on the slopes of Table Mountain.

SUBTROPICAL STEPPE

This transitional belt lies in the Marrakech and Agadir Basins in Morocco and the lower inland plateaus in Algeria and Tunisia. Rainfall varies from 200 to 500 mm with a long dry hot season of 6 to 11 months. The mean temperature of the coldest month is always more than 7°C. Vegetation in this zone is a tree pseudo-steppe with *Acacia gummifera, Ziziphus lotus* and *Pistacia atlantica*. In Morocco (Sous) the typical vegetation type is *Argania* spp. forest.

SUBTROPICAL MOUNTAIN SYSTEMS

In northern Africa, the Atlas Mountains dominate the landscape and extend over 3 000 km. Their altitude reaches 1 500 m in Tunisia, 2 500 m in Algeria and 4 165 m in Morocco. The Rif Atlas experiences a humid climate because of proximity to the Atlantic Ocean. Rainfall approaches 1 000 mm, with a short summer drought. Further inland, the dry season is always pronounced and the climate becomes semi-arid to the south.

In South Africa, the largest highland area is the Highveld region, more than 1 000 m in altitude, bordered by the Drakensberg, reaching more than 3 000 m. The mountain ranges in the Cape region also belong to this ecological zone. The climate is humid with a tropical regime. Rainfall varies from 500 to 1 100 mm with a short winter dry season. Winter temperatures are only somewhat low, more than 7°C up to 1 500 m. In the northern Atlas Ranges, the lower slopes are covered by mixed forest with deciduous oaks or *Quercus ilex* associated with *Pinus pinaster* or *P. halepensis*. Above 1 600 m this forest gives way to *Cedrus atlantica* forest. In the southern, drier ranges is *Juniperus thurifera* forest.

In southern Africa an evergreen montane forest with *Podocarpus* and *Apodytes* species grows on the Drakensberg slopes. In the Cape region, a forest with conditions resembling those of temperate forest, *Podocarpus* spp., *Ocotea* spp. and *Olea capensis,* grows on the slopes of the Outeniekwaberge, facing the sea.

BIBLIOGRAPHY

Ecological Laboratory of Toulouse (LET). 2000. *Ecofloristic zones and global ecological zoning of Africa, South America and Tropical Asia,* by M.F. Bellan. Rome, FAO.

Hamilton, A. 1989. *African forests*. In H. Lieth & M.J.A. Werger (editors). *Tropical rain forest ecosystems: biogeographical and ecological studies. Ecosystems of the world*, Vol. 14b. Amsterdam, Elsevier.

Walter, H. 1985. *Vegetation of the Earth and ecological systems of the geo-biosphere.* Third, revised and enlarged edition. Berlin, Springer–Verlag.

White, F. 1983. *The vegetation of Africa – a descriptive memoir to accompany the UNESCO/AETFAT/UNSO vegetation map of Africa*. Natural Resources Research, No. 20. Paris, UNESCO.

Chapter 13
North Africa

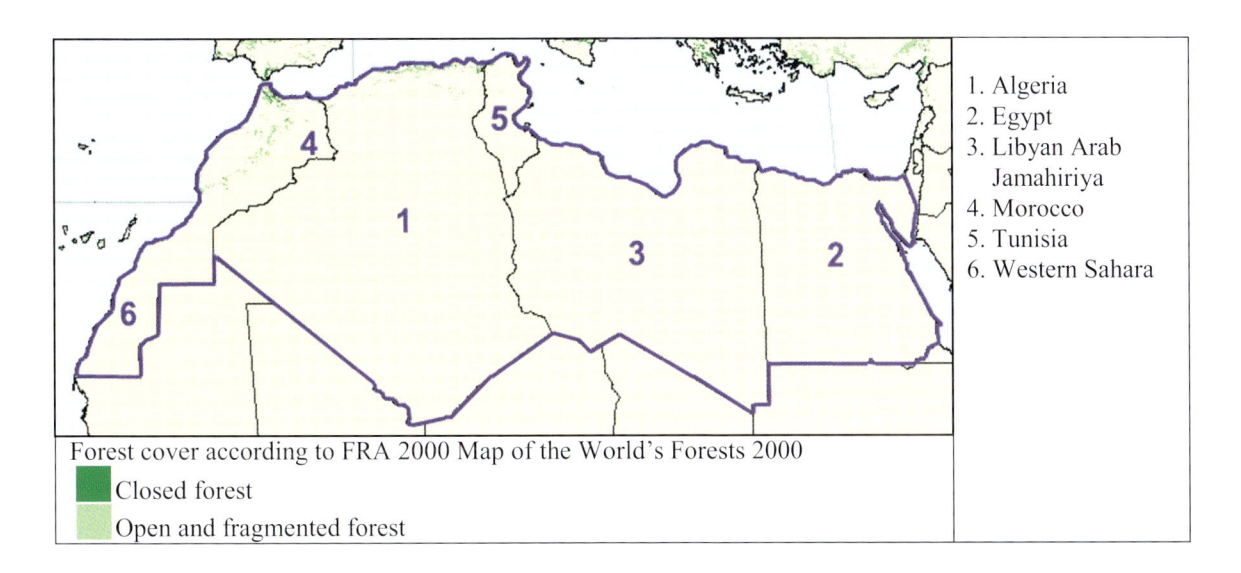

Figure 13-1. North Africa: forest cover map

Legend:
- 1. Algeria
- 2. Egypt
- 3. Libyan Arab Jamahiriya
- 4. Morocco
- 5. Tunisia
- 6. Western Sahara

Forest cover according to FRA 2000 Map of the World's Forests 2000
- Closed forest
- Open and fragmented forest

The subregion is bordered by the Atlantic Ocean in the west, the Red Sea in the east and the Mediterranean Sea to the north and includes Algeria, Egypt, the Libyan Arab Jamahiriya, Morocco, Tunisia and Western Sahara[28]. The area is 6 million square kilometres, of which 94 percent is in the desert ecosystems of the North African Sahara. The forest cover in this subregion is among the lowest in the world at around 1 percent of the land surface (Figure 13-1).

The subregion is characterized, in general, by a hot and dry to very dry climate. Its northern part falls under the temperate influence of the Mediterranean, while the central and southern regions are desert. Owing to the latitude range from 19° to 37° N and altitude of up to 4 165 m in the High Atlas of Morocco, the rainfall regime is quite variable. The average annual precipitation is below 100 mm in the Sahara but as high as 1 500 mm in the regions of Ain Draham and Djebel El Ghorra in Tunisia and up to 2 000 mm in the mountains of Morocco. However, less than 10 percent of the subregion receives more than 300 mm per year. A hot, dry sirocco wind blowing north from the Sahara is frequent during the summer season, bringing blinding sand and dust storms to the coastal regions.

In the past, under the combined effects of severe climate, growing populations and lack of adequate land use planning, the forest cover suffered from large-scale deforestation. Clearing of forests and the use of fire for cultivation and grazing reduced the cover to patchy relics as compared to the reported cover present during previous centuries. Overgrazing, fires (particularly in Algeria) and droughts continue to hamper efforts to conserve and develop forests. In the absence of adequate forest cover in most of the region, the process of desertification has continued, critically affecting fragile ecosystems as well as the economy.

FOREST RESOURCES

Algeria, Morocco and Tunisia prepared national forest inventories in 1982, 1995 and 1996, respectively. Morocco's national forest inventory included Western Sahara (Morocco AEFCS 1996d). The Algeria data set is obsolete. Information used by FRA 2000 was generated from a countrywide inquiry led by a local consultant. It was a simple updating of the 1982 inventory on the basis of local knowledge. Tunisia and Algeria have started updating their forest inventories. Their new inventories involve comparable mapping methodologies and sampling designs. The sampling schemes are, however,

---
[28] For more details by country, see www.fao.org/forestry

Table 13-1. North Africa: forest resources and management

Country/area	Land area	Forest area 2000					Area change 1990-2000 (total forest)		Volume and above-ground biomass (total forest)		Forest under management plan	
		Natural forest	Forest plantation	Total forest								
	000 ha	000 ha	000 ha	000 ha	%	ha/capita	000 ha/year	%	m^3/ha	t/ha	000 ha	%
Algeria	238 174	1 427	718	2 145	0.9	0.1	27	1.3	44	75	597	28
Egypt	99 545	0	72	72	0.1	n.s.	2	3.3	108	106	-	-
Libyan Arab Jamahiriya	175 954	190	168	358	0.2	0.1	5	1.4	14	20	-	-
Morocco	44 630	2 491	534	3 025	6.8	0.1	-1	n.s.	27	41	-	-
Tunisia	16 362	308	202	510	3.1	0.1	1	0.2	18	27	400	78
Western Sahara	26 600	152	-	152	0.6	0.5	n.s.	n.s.	18	59	-	-
Total North Africa	601 265	4 569	1 693	6 262	1.0	n.s.	33	0.5	32	51	-	-
Total Africa	2 978 394	641 830	8 036	649 866	21.8	0.8	-5 262	-0.8	72	109	-	-
TOTAL WORLD	13 063 900	3 682 722	186 733	3 869 455	29.6	0.6	-9 391	-0.2	100	109	-	-

Source: Appendix 3, Tables 3, 4, 6, 7 and 9.

based on independent sets of temporary plots. Results from Egypt and the Libyan Arab Jamahiriya were produced from secondary sources.

Morocco, Algeria and Tunisia account for 91 percent of the forest cover in the subregion, although the land area of these countries accounts for less than 50 percent of the total land area of the subregion. The subregional forest cover is about 1 percent of the continent's forest area and about 0.16 percent of the world forest area, although the subregion's total land area accounts for 20 percent of Africa and 4.5 percent of the world (Table 13-1, Figure 13-2).

The extent of natural forest cover is closely correlated with annual precipitation. Natural forest is thus more abundant in a 100 to 200 km zone in the north of Tunisia, Algeria and Morocco where annual rainfall ranges between 300 and 2 000 mm. It decreases, becomes scarce or even completely disappears as annual rainfall decreases towards the south and the east of the subregion.

These findings are based on FAO definitions of forests and trees. At the national level, however, other vegetation components are reported as part of forest cover. Shrub formations of garrigue and maquis, without a tree layer, are widespread. They consist of two main groups of species. The first includes shrubby species that, under all edaphic and climatic conditions, remain below tree size when mature. Among these species are *Arbutus unedo, Alnus glutinosa, Calycotome villosa, Myrtus communis, Prunus avium* and *Rosmarinus officinalis*. The second group consists of species that are dwarfed because of soil and or climate unsuitability. Among these are *Pinus halepensis, Quercus suber, Quercus ilex, Quercus coccifera, Olea europaea, Pistacia lentiscus* and *Ceratonia siliqua*. The area of garrigue and maquis is estimated at 1 249 640 ha in Morocco (Morocco AEFCS 1996d), 1 662 000 ha in Algeria (Ikermoud 2000) and 328 000 ha in Tunisia (Selmi 2000). The steppes of *Stipa tenacissima* (alfa), which is an herbaceous ecological succession of garrigue from pine forest, are reported as part of the forest domain in Tunisia, Algeria and Morocco.

Except for Morocco, forest cover change is positive in all countries of the subregion. Egypt shows the highest change rate of 3.3 percent followed by the Libyan Arab Jamahiriya (1.4 percent), Algeria (1.3 percent) and Tunisia (0.2 percent). The positive change of forest cover in this region is mainly the result of tree planting efforts and also of policies oriented towards resource conservation. The high rate of change in Egypt is due to the fact that the amount of forest cover is very small and any tree planting makes a significant difference.

In terms of area, Algeria reported the largest planting programme. An average of 29 411 ha are planted every year, followed by Tunisia with 4 500 ha, Libyan Arab Jamahiriya with 1 100 ha and Egypt with 100 ha. Morocco reported an average annual planting of 40 ha. The area of planted forests was estimated at 1 693 000 ha and accounts for about 27 percent of the total forest cover of the subregion.

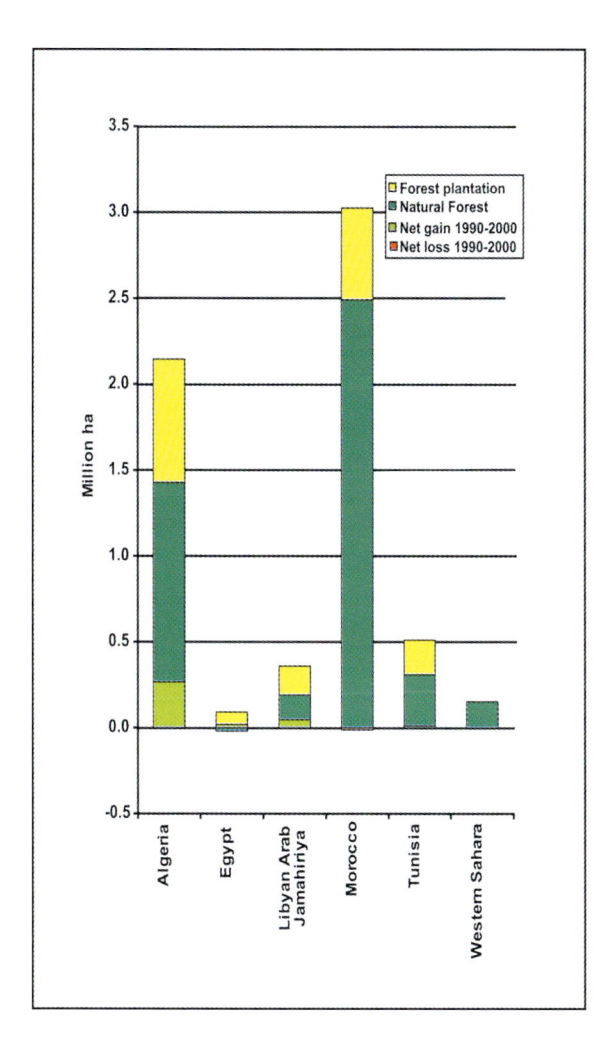

Figure 13-2. North Africa: natural forest and forest plantation areas 2000 and net area changes 1990-2000

The largest amount of woody biomass is found in Algeria, which accounts for 50 percent of the total biomass in the subregion. Algeria is followed by Morocco with 38 percent, Tunisia 4 percent and the rest of the countries with 7 percent. The relatively high biomass in Algeria is due to the high stocking level of forest plantations.

There are no systematic studies on biodiversity. The information available was produced from national forest inventories or from spatially limited vegetation community and wildlife surveys. Despite the droughts and aridity that characterize the area, this subregion has conserved an important part of its original fauna and flora. Tunisia's flora, for instance, is still rich at 2 200 species (Selmi 2000). Among the endemic vegetation species in North Africa, 20 are found only in Tunisia (Tunisia DGF 1997).

Morocco's relief and diversity of climate have favoured a great diversity of ecosystems, which means an appreciable floristic wealth. Over 4 200 species and subspecies have been recorded, of which 800 are endemic. In Algeria, the various bio-climatic conditions, ranging from Saharan in the south to humid in the north, has favoured a rich flora; 3 300 vegetation species have been recorded of which 640 are threatened and 256 are endemic (Algeria DGF 2000).

FOREST MANAGEMENT AND USES

Formal management of forests in the North African countries started gradually in the early 1950s. Since then important achievements have been made in putting a substantial part of the resources under management plans. Only two of the six countries in North Africa provided national-level information for FRA 2000 on the forest area covered by a formal, nationally approved forest management plan (Table 13-1). Algeria reported that 597 000 ha or 28 percent of its forest area was covered by a formal management plan, whereas Tunisia reported that 400 000 ha or 78 percent of its forest area was covered by such a plan. Auxiliary reference sources indicate that a large percentage of the forest area of Morocco (about 80 percent) was also under management (Morocco AEFCS 1996c) although no information was provided for FRA 2000. Egypt and the Libyan Arab Jamahiriya did not provide information on the status of forest management for FRA 2000.

In Tunisia, management planning covers the productive forests. Because of their high environmental, social and economic value, maquis and garrigue are also planned for future management. Among the existing plans, 50 percent need updating (Tunisia DGF 1997).

Algeria's achievements in forest management planning are notable. Plans essentially cover productive forests of *Pinus halepensis, Pinus pinaster, Quercus faginea, Quercus afares, Quercus ilex* and *Quercus suber*. Priority for management planning and implementation is given to *P. halepensis* because of its environmental and economic importance. For other species, particularly *Q. suber, Q. faginea* and *Q. afares,* there are delays in implementing management plans (Tunisia DGF 2000).

In Morocco, priority has been given to natural forests where stands are composed of species that have high social and economic value. Management has been extended to various formations of *Cedrus atlantica, Pinus* spp., *Q. suber* and a number of other broadleaf and coniferous species (Morocco AEFCS 1997).

Since the main function of forest cover is protection of soil against erosion and the landscape from further degradation, efforts have been employed to establish protected areas in national parks and natural reserves. Tunisia has established eight national parks covering about 200 000 ha, of which 12 percent are composed of various forest formations. The national parks were designed to protect relics of forests or threatened wildlife and vegetation species. Therefore, they cover a wide spectrum of ecosystems (Tunisia DGF 1997).

Algeria's protected areas system, excluding the desert parks of Ahagar and Tassili in the south, extends over an area of 250 000 ha, of which 113 000 ha are covered by various forest formations and 59 000 ha by maquis. As in Tunisia, the protected areas include a large array of ecosystems of particular interest for their biodiversity (Algeria DGF 2000).

The biodiversity of Morocco is among the highest in the Mediterranean basin. In order to protect this national heritage Morocco has identified a network of protected areas composed of ten national parks and 146 reserves (Morocco AEFCS 1996b). This protected area system harbours an important array of ecosystems. The forest area in the national parks is estimated at about 120 000 ha.

Algeria's forest resources are mainly State-owned with only 8.7 percent belonging to private entities (Ikermoud 2000). In Tunisia, the private sector owns about 5.2 percent of the forest cover, all of it in plantations (Selmi 2000). Privately owned forest in Morocco was estimated at 2.9 percent of the total forest cover. All the private forests are planted. Information on ownership is not available from Egypt or the Libyan Arab Jamahiriya.

Forest fires are a great threat to forest resources despite efforts to limit their negative impact. In Algeria, the number of fires recorded in the forest domain varies from year to year. The lowest number recorded during the past 15 years was 562 fires and the largest was 2 322, with an average of 1 256 (Ikermoud 2000). The average area affected by fire each year over the same period was estimated at 37 917 ha or 1.8 percent of the national forest cover. Over the same period Tunisia recorded 134 fires with an average impact of 1 783 ha per year (Selmi 2000), which accounts for 0.4 percent of the nation's forest cover. No information is available on forest fires for the other countries in the subregion, but in view of the social, economic and environmental similarities among the countries, the impact of forest fires is likely to be analogous to that in Algeria and Tunisia.

CONCLUSIONS AND ISSUES

Assessment of forest cover and change for the North African countries was not straightforward. National definitions and classification systems differed widely from those used by FRA 2000. Close collaboration with Tunisia and Algeria permitted national experts to reclassify their national classes into the global classification system. Information on the forest cover of Morocco is the most recent (Morocco AEFCS 1996d). Results of its national forest inventory were published in 1996. This included information on the forest cover of Western Sahara, which was extracted. The existing sets of data from all countries were produced from single inventories without information on change over time. As Tunisia is more advanced in updating its inventory, preliminary results were used and gave preliminary trends.

Reported data on forest plantations are sometimes misleading. They often include enrichment planting in naturally regenerated stands or shrubby species such as *Atriplex* spp., *Acacia* spp., *Calligonum comosum*, *Prosopis juliflora*, *Opuntia ficus-indica* and *Parkinsonia aculeata* used as fodder, for dune fixation or for soil stabilization (FAO undated).

Forest resources have been recognized by the countries in the subregion as important economic, social and environmental assets. Algeria, Morocco and Tunisia, the most forested countries, deploy considerable effort in the conservation, development and exploitation of their resources on a sustainable basis through improved legislation, sustainable management and implementation of challenging development programmes. As a result, effects of desertification are minimized despite unfavourable natural and social conditions and the sector's productivity has improved significantly. Many products are extracted from the forests including timber and other wood and non-wood products. The contribution of the forestry sector to the national economies and to meeting the needs of rural people in these countries is appreciable. In Morocco, for instance, the contribution of the forestry sector to the national economy is estimated at 10 percent of the agricultural gross domestic product if all resource uses are considered (Morocco AEFCS 1997).

In addition to timber and fuelwood, cork produced from *Quercus suber* bark generates important income. Morocco has 366 000 ha, Algeria 230 000 ha and Tunisia 46 000 ha of this species, producing about 15 000 tonnes (Morocco AEFCS 1996c), 9 600 tonnes (Ikermoud 2000) and 8 100 tonnes (Tunisia DGF 1997), respectively.

Forest ecosystems in these countries have many functions, not only economic, sometimes discordant. Under climatic conditions that are sometimes extremely severe the forest is expected to perform multiple functions to produce various wood and non-wood products for household consumption and industrial processing for the local market or even export, to protect biodiversity, to conserve soil and water and to combat desertification (Algeria DGF 2000; Morocco DGF 1997).

Forestry and wildlife legislation varies considerably among the countries of the subregion. It was recently revised in Tunisia and Morocco, where new concepts of local population involvement, incentives for tree planting and the condition that "the forest domain should not be reduced" were introduced. Although legislation has succeeded in reducing the rate of deforestation in some countries and in halting it in others, the forest cover in the subregion continues to diminish because of fires and particularly overgrazing. The North African countries, despite their recent social and economic development, still have large rural populations that graze livestock. A substantial part of the livestock pasture is in the forest, which has led in places to severe degradation because of a complete failure of natural regeneration (Algeria DGF 2000; Tunisia DGF 1997; Morocco AEFCS 1996a).

Increasing pressure on resources by people in the subregion, in addition to the severe climate and low soil fertility, has rendered ecosystems even more fragile, and in some places their renewal is jeopardized. Natural forest, where it is under the strict control of local foresters enforcing appropriate legislation, is protected from significant conversion to other land uses. The problem that still remains, which can deeply affect the resource, is the general degradation of forest cover and biodiversity over time. Desertification is also progressing northwards, making the recovery of vegetation on cleared and abandoned land impossible without human assistance through soil preparation, use of fertilizer and watering during regeneration (Algeria DGF 2000; Tunisia DGF 1997; Morocco AEFCS 1996a).

BIBLIOGRAPHY

Algeria. Direction Générale des Forêts (DGF). 2000. *Etude prospective du secteur forestier en Algérie*. Algiers.

FAO. Undated. *Ressources forestières de la Libye*. Working Paper. Rome. (unpublished).

Ikermoud, M. 2000. *Evaluation des ressources forestières nationales*. Algiers, Algeria, DGF.

Morocco. Administration des Eaux et Forêts et de la Conservation des Sols (AEFCS). 1996a. *Colloque national sur la forêt, rapport des modules*. Rabat.

Morocco AEFCS. 1996b. *National parks and natural reserves of Morocco*. Rabat.

Morocco AEFCS. 1996c. *Maroc*. Rabat.

Morocco AEFCS. 1996d. *Rapport final, inventaire des ressources forestières du Maroc*. Rabat.

Morocco AEFCS. 1997. 17ème session du comité CFFSA/CEF/CFPO des questions forestières méditerranéennes, *Silva Mediterranea*, rapport national. Rabat.

Morocco AEFCS. Undated. *Apperçu sur le Maroc forestier*. Rabat.

Selmi, K. 2000. *Tunisia, rapport sur les ressources forestières en Tunisie pour le FRA 2000*. Tunis, Tunisia, DGF.

Tunisia. Direction Générale des Forêts (DGF). 1995. *Résultas du premier inventaire forestier national en Tunisie*. Tunis.

Tunisia DGF. 1997. *Plan directeur national des ressources forestières et pastoral*. Tunis.

Chapter 14
West Africa

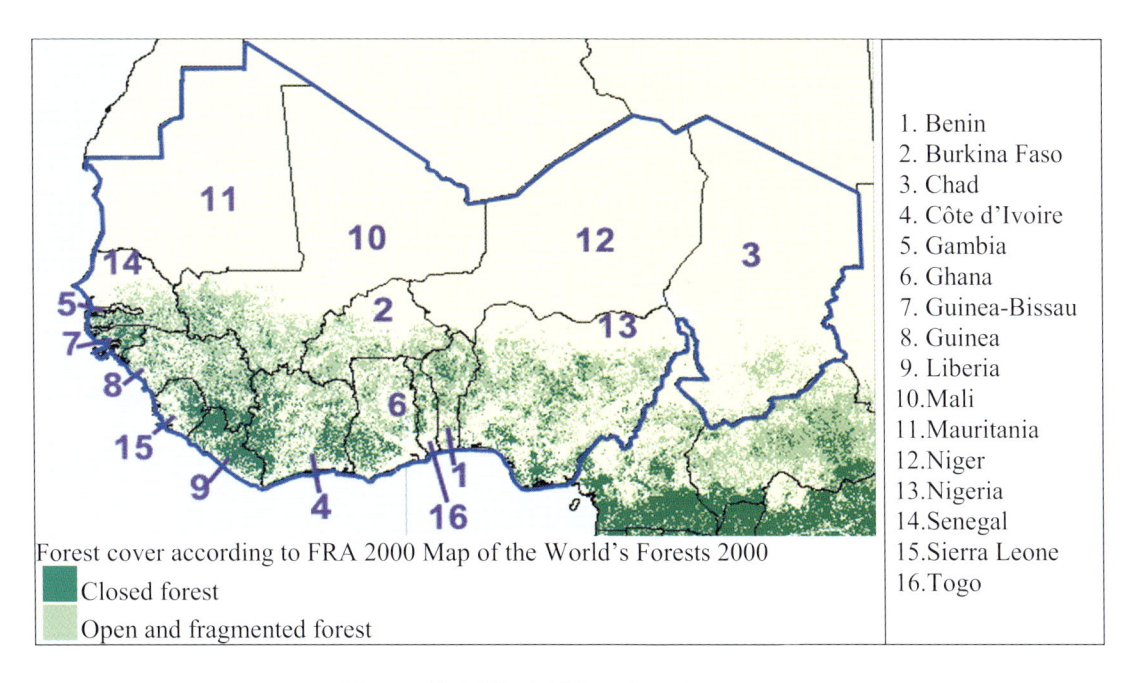

Figure 14-1. West Africa: forest cover map

West Africa includes 16 countries distributed along a climatic gradient from the Sahel region in the north to the Guineo-Congolese zone in the south (Figure 14-1)[29]. This subregion supports a wide range of natural vegetation which includes tropical humid forests, dry forests and savannah. Tropical humid forests can be divided into tropical rain forests and tropical deciduous forests. Tropical rain forests form a belt from the eastern border of Sierra Leone all the way to Ghana. They gradually dissipate near the Volta River; thus, they continue from eastern Benin through southern Nigeria. The tropical deciduous forests lie along the fringes of the tropical rain forests. The dry forest band stretches from northern Nigeria and Chad to Senegal. Dryer climate zones are also characterized by woodlands (tree and shrub savannah, parklands and bush fallows). The West African dry regions correspond to the transition zone of the Sahel as well as the regional centre of Sudanese endemism (Bellefontaine *et al.* 2000).

Humid regions belong to the Guineo-Congolese endemism centre (IUCN 1996). However, West African rain forests are less biodiverse than the central African ones and the endemism is relatively low (IUCN 1996). Nevertheless, Côte d'Ivoire, Ghana and Nigeria are among the 50 most biodiverse countries in the world (WCMC 1994). For example, Nigeria includes about 4 600 plant species, of which approximately 200 are endemic.

Chad, Mali, Mauritania and the Niger are by far the largest countries of West Africa, with a total land area accounting for almost 65 percent of the subregion, although mostly desert. Indeed, the whole forest cover of these four countries represents only 6 percent of their total land area. By contrast, the Gambia and Guinea-Bissau are the smallest but the most forested countries of West Africa.

FOREST RESOURCES

Forest resource knowledge and information quality vary by country. For most West African countries, information and data on forest resources and areas are dated, obsolete and/or partial. Indeed, only a few countries carried out an evaluation of their forest resources at the national level during the 1990s (Benin, Burkina Faso, Guinea-Bissau, the Gambia, Nigeria). Other West African countries made earlier national forest

---

[29] For more details by country, see www.fao.org/forestry

Table 14-1. West Africa: forest resources and management

Country/area	Land area	Forest area 2000					Area change 1990-2000 (total forest)		Volume and above-ground biomass (total forest)		Forest under management plan	
		Natural forest	Forest plantation	Total forest								
	000 ha	000 ha	000 ha	000 ha	%	ha/capita	000 ha/year	%	m³/ha	t/ha	000 ha	%
Benin	11 063	2 538	112	2 650	24.0	0.4	-70	-2.3	140	195	-	-
Burkina Faso	27 360	7 023	67	7 089	25.9	0.6	-15	-0.2	10	16	694	10
Chad	125 920	12 678	14	12 692	10.1	1.7	-82	-0.6	11	16	-	-
Côte d'Ivoire	31 800	6 933	184	7 117	22.4	0.5	-265	-3.1	133	130	1 387	19
Gambia	1 000	479	2	481	48.1	0.4	4	1.0	13	22	-	-
Ghana	22 754	6 259	76	6 335	27.8	0.3	-120	-1.7	49	88	-	-
Guinea	24 572	6 904	25	6 929	28.2	0.9	-35	-0.5	117	114	112*	n.ap.
Guinea-Bissau	3 612	2 186	2	2 187	60.5	1.8	-22	-0.9	19	20	-	-
Liberia	11 137	3 363	119	3 481	31.3	1.2	-76	-2.0	201	196	-	-
Mali	122 019	13 172	15	13 186	10.8	1.2	-99	-0.7	22	31	-	-
Mauritania	102 522	293	25	317	0.3	0.1	-10	-2.7	4	6	-	-
Niger	126 670	1 256	73	1 328	1.0	0.1	-62	-3.7	3	4	-	-
Nigeria	91 077	12 824	693	13 517	14.8	0.1	-398	-2.6	82	184	832*	n.ap.
Senegal	19 252	5 942	263	6 205	32.2	0.7	-45	-0.7	31	30	-	-
Sierra Leone	7 162	1 049	6	1 055	14.7	0.2	-36	-2.9	143	139	-	-
Togo	5 439	472	38	510	9.4	0.1	-21	-3.4	92	155	12	2
Total West Africa	733 359	83 369	1 710	85 079	11.6	0.4	-1 351	-1.5	61	84	-	-
Total Africa	2 978 394	641 830	8 036	649 866	21.8	0.8	-5 262	-0.8	72	109	-	-
TOTAL WORLD	13 063 900	3 682 722	186 733	3 869 455	29.6	0.6	-9 391	-0.2	100	109	-	-

Source: Appendix 3, Tables 3, 4, 6, 7 and 9.
*Partial result only. National figure not available.

assessments (Senegal, 1985; Sierra Leone, 1986; Chad, 1988; Togo, 1975; Liberia, 1981). The remaining West African countries have undertaken partial assessments covering only parts of their national forests. Consequently, the forest areas for some West African countries presented in Table 14-1 are based on national expert estimates (Chad, Ghana, Liberia, Mauritania, the Niger, Sierra Leone, Togo). A workshop was organized in Yamoussoukro, Côte d'Ivoire on data collection for this subregion in 1999 with the participation of all West African countries except Chad (FAO 2000).

West African countries have limited forest resources (approximately 11 percent of the total land area) because of the climate (countries of the Sahelo-Sudanese zone), large populations (e.g. Nigeria, Benin, Togo), agricultural clearing or long-term export of wood products (e.g. Côte d'Ivoire). Therefore, the forests of this subregion represent only 13 percent of the total forest cover of the continent and 2 percent of the world forest area. Guinea-Bissau is by far the most forested country with 60 percent of its land area covered by forests. Mauritania and the Niger, on the other hand, are the least forested countries (0.3 and 1.0 percent of their total land area, respectively)

because of dry climatic conditions. West Africa has a high annual negative rate of forest area change (-1.5 percent on average) compared to the whole of Africa (-0.78 percent). In terms of area, Nigeria and Côte d'Ivoire have by far the greatest negative annual loss of forest cover. The Niger has the highest annual deforestation rate (Table 14-1, Figure 14-2).

Forest plantations in West Africa account for more than 20 percent of all African plantations. However, the statistics on planted forests are not reliable in several countries because of lack of inventories, frequent fires, lack of maintenance and/or uncontrolled clearing (e.g. Guinea, Ghana, Liberia, Chad). In the humid part of West Africa, countries have significant areas of forest plantations, mostly for industrial purposes (e.g. Côte d'Ivoire, Benin, Nigeria). Nevertheless, plantations for timber production, which are expensive and difficult to manage, have been insufficient to compensate for the extensive exploitation of natural forests. In addition, the area of plantations for high-grade hardwood timber similar to that which is extracted from the humid forests is insufficient to have any impact on the supply of such timber in the foreseeable future (FAO 2000). In dry parts of the subregion,

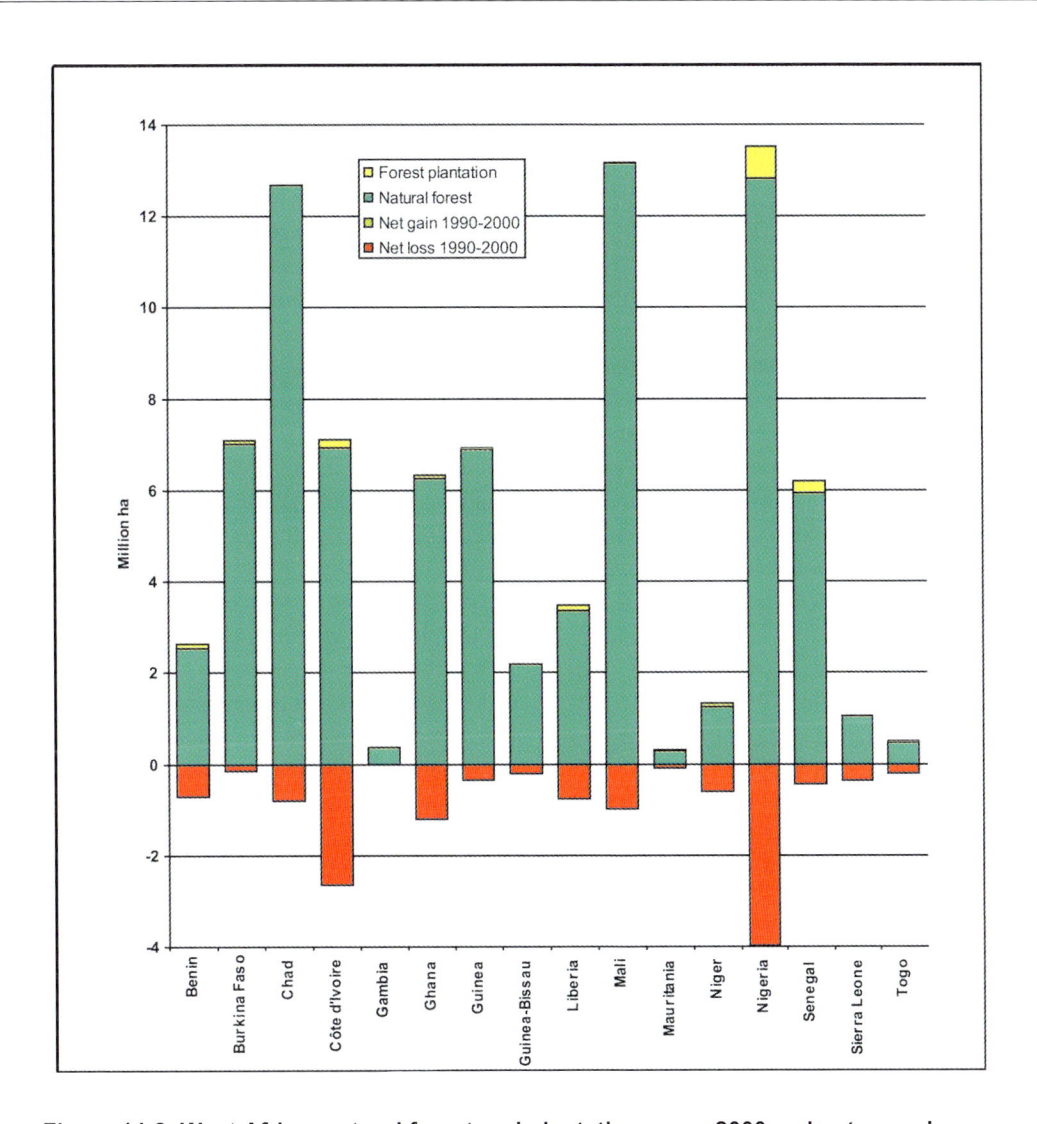

Figure 14-2. West Africa: natural forest and plantation areas 2000 and net area change 1990-2000

forest plantation areas are less important and are mainly non-industrial (with the exception of Senegal). Many plantations have been established to try to stop, or even to reverse, the desertification process, which is the main ecological problem of numerous countries with dry climates (the Niger, Chad, Mali, Nigeria, Burkina Faso, Mauritania, Senegal) (FAO 2000).

The total volume of West African forests is estimated at approximately 5 billion cubic metres over bark, which is 11 percent of the volume of all African forests. The volume and biomass estimates for most countries are based on existing forest inventories. In humid zones, volume assessment is focused on timber volume. In dry zones, volume assessment usually includes the whole ligneous biomass, including trunks and branches, for fuelwood consumption. Maximal production of natural vegetation in West Africa was estimated to vary from 0.1 to 2.75 m^3 per hectare per year according to rainfall and vegetation type (Bellefontaine *et al.* 2000).

Wood provided by trees outside the forest is extremely important in this subregion. Indeed, the sparse forest cover of most West African countries makes this material very valuable, notably in dry zones where a large part of fuelwood is harvested outside the forest. Jensen (1995) estimated that the volume in fallows and sparse trees on agricultural lands constitutes approximately 30 percent of the wood resources in Burkina Faso and 19 percent in the Gambia.

FOREST MANAGEMENT AND USES

Only three of the 16 countries in West Africa provided national-level information on the forest area covered by a formal, nationally approved forest management plan (Table 14-1). Of these countries, Togo had the lowest percentage (2 percent) and Côte d'Ivoire the highest (19 percent). Partial figures were available from Nigeria (lowland rain forests only) indicating that at least 832 000 hectares (or 6 percent) of the total forest area of the country was covered by a management plan. Information was lacking for the remaining countries, including Ghana, which according to a recent ITTO study (ITTO 2000) appeared to have established all the conditions that make it likely that the country can manage its forest management units sustainably.

Decentralization has started in a majority of West African countries, clarifying the role and ownership of resources. Nevertheless, land tenure is sometimes very complex because of overlapping land tenure rights and uses. This is particularly notable in savannah regions and even more in Sahelian zones where forest, pastoral and agricultural domains overlap (Bellefontaine et al. 2000).

Natural forest exploitation and management has a long history in the humid part of West Africa. A number of different systems of tropical silviculture have been tried in the past to maximize yield (e.g. tropical shelterwood, modified selection, etc.). These silvicultural techniques have not always been successful for both ecological and managerial reasons (FAO 2000; Dupuy et al. 1999). In all the countries of the subregion with tropical humid forests, government forestry departments control the right to exploit timber. Regulations specify the logging methods and the most appropriate logging systems. Private timber companies or individuals are awarded concessions by the government and issued contracts that spell out the regulations and procedures to be followed, including in some cases restocking and post harvest operations. However, monitoring and control by government are often lacking owing to limited resources. For forest plantations, agreements and contracts are set up to manage their exploitation and to prevent conflicts (FAO 2000).

In dry zones, a number of pilot projects are currently in progress or have been completed to assess the consequences of increased public participation in forest management. During the 1980s, numerous projects were undertaken with limited participation. Since then, considerations of land tenure, user and interest groups and problems of conflicting uses have led to increased decentralized management of natural resources for the benefit of local people (Dupuy et al. 1999). In addition, local participation has slowly increased in reforestation programmes. The decentralization process is illustrated by the Energie II project in the Niger, whose main objective is sustainable forest management for fuelwood utilization. The project is based on the transfer of management responsibility for renewable natural resources (but not the property) from the State to the local population (Bellefontaine et al. 2000).

Large quantities of wood energy are consumed in the subregion. Fuelwood is thought to constitute 85 percent of the total energy consumption in these countries, but there is no reliable information on wood trade and consumption (FAO 2000). The highest consumption is found in Burkina Faso and the Gambia. High population density has led to overexploitation of forests for fuelwood in the dry forests of the Niger, Nigeria, Togo and Benin, where this resource is becoming increasingly scarce, leading to occasional shortages (Bellefontaine et al., 2000).

Non-wood forest products are important to local people, but there are few statistics except for commercially marketable products. Information is available on some products (such as gum arabic in Chad) because their export contributes to the national budget. Trees are also an important source of fodder in dry zones.

Many countries have created coordinating agencies for environmental management. Decentralization of government functions in environmental planning is also now taking place (Benin, the Gambia, Ghana). Some countries have formulated new legislation on environmental and natural resource management as well as establishing monitoring and regulatory systems (e.g. Côte d'Ivoire). For example, Ghana has prepared environmental impact assessment guidelines (FAO 2000). Several NGOs are working on sustainable conservation of biodiversity in protected areas. In 1997, the World Conservation Monitoring Centre estimated that there are approximately 128 legally protected areas in West Africa (WCMC 1997).

CONCLUSIONS AND ISSUES

Most West African countries have defined, or are currently defining, new forestry policies that include the concept of sustainable forest management. Nevertheless, most countries have insufficient financial and material means to implement these policies properly (FAO 2000). In general, forestry programmes are poorly funded. Hence, forestry institutions in most West African countries are ill-equipped to implement their functions. For most of these countries, forest resource information is generally unreliable, relatively outdated and in need of revision. Many countries do have the administrative and technical capacity to carry out forest plantation work. However, maintenance and commercialization of plantations are also impacted by the lack of financial means (FAO 2000).

Pressure on forest resources in West Africa results from multiple factors, including rapid population growth, economic development, poverty and government policies (lack of decentralization in some countries, lack of adequate information on forests, poor project implementation, etc.). Failure to recognize the legal rights of indigenous peoples and other traditional communities in their territories can also lead to deforestation. This, coupled with the absence of land security, often creates a situation of open access where no person or community is responsible for the land. Conflicts in some West African countries (Liberia, Sierra Leone) have led to the destruction of forests and infrastructure as well as the settlement of refugees in forested areas. Urban population growth generally leads to deforestation in the immediate vicinity because of forest exploitation for fuelwood, building materials and land for settlement (FAO 2000; Bellefontaine *et al.* 2000).

The main causes of deforestation are agriculture coupled with poor farming practices (shifting cultivation and cash crops), logging (poor logging practices, poor concessionaire agreements, etc.), and other land uses such as urban development and mining. Forests have been widely overexploited for timber in the subregion (FAO 2000). In humid forests, replacement by cash crops and tree plantations is one of the main causes of deforestation. Large areas of tropical rain forest have been cleared to plant cash crops such as cocoa, coffee and rubber (e.g. Côte d'Ivoire). Forest fires are considered one of the greatest constraints to conservation and sustainable forest management. Uncontrolled fires, in conjunction with shifting cultivation, result in poor herbaceous vegetation dominated by species such as *Panicum maximum* and *Imperata cylindrica* (Louppe *et al.* 1995).

Mangroves are also under increasing pressure from economic development in coastal zones, conversion into agricultural lands (rice fields) and fuelwood collection for coastal cities (FAO 2000).

In dry zones, the scarcity of fertile soil for agriculture increases the pressure on forested lands. In addition to immediate and permanent conversion to agriculture, a progressive degradation because of short forest fallow also takes place. Hence, the preservation of fertility is no longer assured. Continuing overgrazing in some areas in addition to uncontrolled fires accelerates soil degradation processes. Fuelwood scarcity and forest degradation occur with high populations and insufficient forest area. Clearing and fuelwood collection can easily exceed the regenerative capacity of the ecosystem. This is the widespread "fuelwood crisis" characteristic of numerous Sahel countries (Mauritania, Senegal, Mali, Burkina Faso, the Niger, Chad). The rate of tree cutting for fuel is increasing. However, most countries have developed energy policies and regional cooperation in the energy sector is being improved through the creation of an African Energy Commission (FAO 2000).

Climate is a major natural factor that reinforces the effects of human activities on the environment, particularly in tropical dry zones where severe droughts are frequent and soil quality is poor. Desertification is the major ecological issue for countries in the southern Sahara. Considerable effort has been expended to stop and even reverse this trend, including reforestation with exotic species, green belt plantations and agroforestry development. In addition, there have been important advances in the development of high-yield crop varieties, research in agroforestry systems (since the 1970s) to improve productivity and sustainability and the beginning of research into low-cost modification of shifting cultivation. There has also been a great deal of scientific research on nitrogen-fixing trees, which are important in the conservation of soil fertility. The utilization of these trees has led to spectacular results in sand dune fixation in Senegal. In addition, there is regional collaboration on practical and consistent regulations for protection of forests, to deal with the negative effects of conflicts on the environment and land-related issues such as desertification (Bellefontaine *et al.* 2000; FAO 2000).

Forestry projects in West Africa currently embrace concepts of management of renewable resources by integrating multiple land uses with the participation of local people. Participation has increased during the last few years but still needs to be improved. Other important issues include improving forest staff training, conducting educational programmes for the public on sustainable agricultural practices and agroforestry techniques, promoting alternative energy sources and energy saving techniques and improving the use of wood products (FAO 2000).

BIBLIOGRAPHY

Bellefontaine, R., Gaston, A. & Petrucci, Y. 2000. *Management of natural forests of dry tropical zones.* FAO Conservation Guide No. 32. Rome.

Dupuy, B., Maître, H.-F. & Amsallem, I. 1999. *Techniques de gestion des écosystèmes forestiers tropicaux: état de l'art.* Working Paper FAO/FPIRS/05 prepared for the World Bank forest policy implementation review and strategy. Montpellier, France, Cirad-Forêt.

FAO. 2000. *Actes de l'atelier sous-régional sur les statistiques forestières et perspectives pour le secteur forestier en Afrique/FOSA sous région ECOWAS.* Yamoussoukro, Côte d'Ivoire, 13-18 December 1999. FAO, Rome.

ITTO. 2000. *Review of progress towards the year 2000 objective,* by D. Poore and Tang Hooi Chiew. Report presented at the 28th session of the ITTC, 24-30 May 2000, Lima, Peru. ITTC(XXVIII)/9/Rev.2.

IUCN. 1996. *Atlas pour la conservation des forêts tropicales d'Afrique.* J.-P. de Monza, éd.

Jensen, A.M. 1995. Evaluation des données sur les ressources ligneuses au Burkina Faso, Gambie, Mali, Niger et Sénégal. In: *Examen des politiques, stratégies et programmes du secteur des énergies traditionnelles.* World Bank. 2nd version.

Louppe, D., Ouattara, N. & Coulibaly A. 1995. The effects of bush fires on vegetation: the Aubréville fire plots after 60 years. *Commonwealth Forestry Review.* 74(4): 288-292.

WCMC. 1994. *Priorities for conserving global species richness and endemism.* WCMC Biodiversity Series No. 3. World Conservation Press.

WCMC. 1997. *United Nations List of Protected Areas 1997.* www.wcmc.org.uk/protected_areas/data/un_97_list.html

Chapter 15
Central Africa

Figure 15-1 legend:
1. Burundi
2. Cameroon
3. Central African Republic
4. Congo
5. Democratic Republic of the Congo
6. Gabon
7. Equatorial Guinea
8. Rwanda

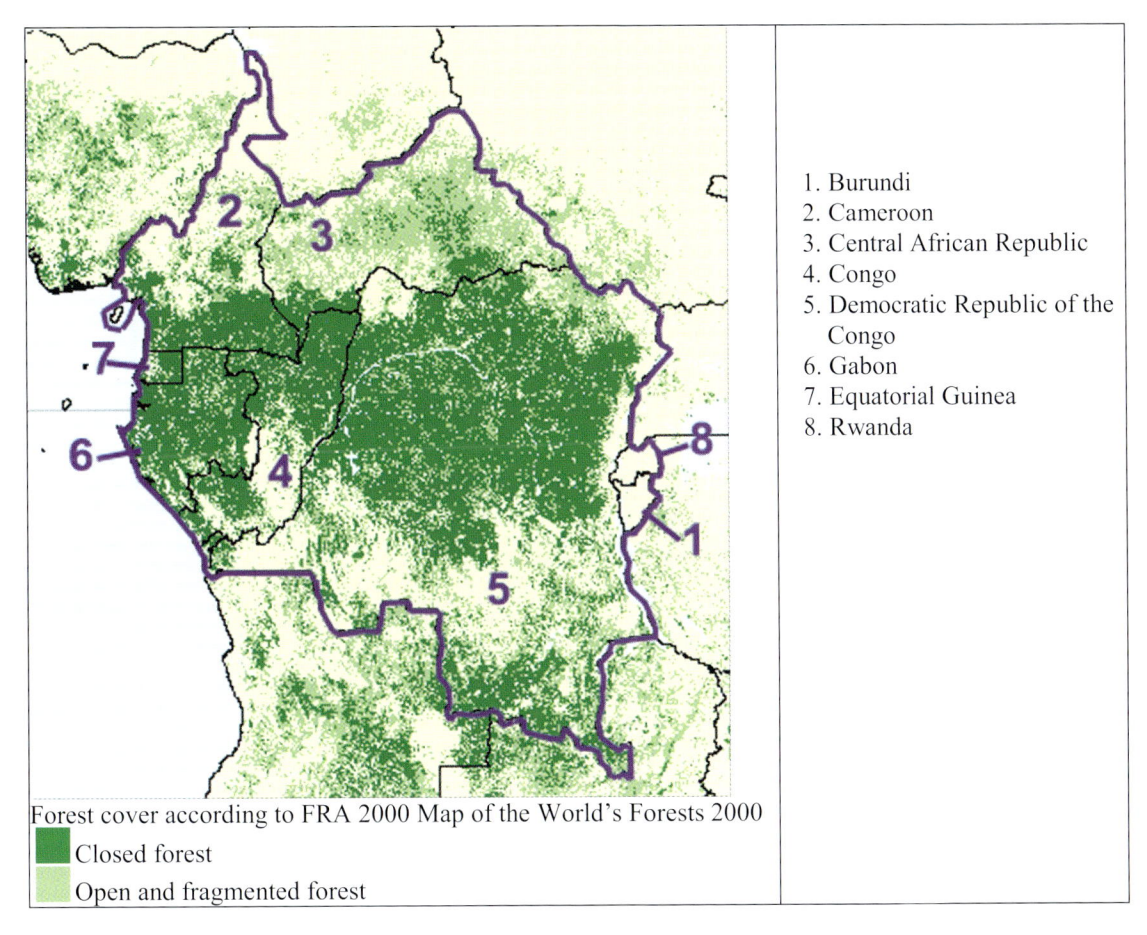

Forest cover according to FRA 2000 Map of the World's Forests 2000
- Closed forest
- Open and fragmented forest

Figure 15-1. Central Africa: forest cover map

Central Africa[30] is an important forested subregion with approximately 57 percent of its area covered with natural forests. Central Africa contains the largest remaining contiguous expanse of moist tropical forest on the African continent and the second largest in the world (after the Amazon forest). This quasi-uniform forest cover encompasses Gabon, Equatorial Guinea, the Congo, the majority of Cameroon and the Democratic Republic of the Congo as well as a small part of the Central African Republic (Figure 15-1). The Democratic Republic of the Congo is by far the largest country of this subregion, with more than 226 million hectares of land. Burundi and Rwanda are among the smallest countries of central Africa and the continent. An important characteristic of this subregion is the zonal climate distribution that induces a gradient of ecosystems and hence biodiversity. The lowland evergreen broadleaf rain forest (including swamp forests localized for the greater part in the eastern Congo and the western Democratic Republic of the Congo) and the semi-deciduous broadleaf forest dominate this subregion and count among the richest in Africa. The montane forests (Rwanda, Burundi, Cameroon and the Democratic Republic of the Congo) are of lower biodiversity but often have a greater number of endemic species (IUCN 1996). Central Africa also includes dry forests in the northern Central African Republic and Cameroon.

Central Africa is rich in natural resources, has played a large part in history and continues to play a role as a reservoir for the export of raw materials to the industrialized nations. In particular, wood and, more recently, petroleum

---
[30] For more details by country, see www.fao.org/forestry

Table 15-1. Central Africa: forest resources and management

| Country/area | Land area | Forest area 2000 ||||| Area change 1990-2000 (total forest) || Volume and above-ground biomass (total forest) || Forest under management plan ||
| | | Natural forest | Forest plantation | Total forest |||||||||
	000 ha	000 ha	000 ha	000 ha	%	ha/capita	000 ha/year	%	m³/ha	t/ha	000 ha	%
Burundi	2 568	21	73	94	3.7	n.s.	-15	-9.0	110	187	-	-
Cameroon	46 540	23 778	80	23 858	51.3	1.6	-222	-0.9	135	131	-	-
Central African Republic	62 297	22 903	4	22 907	36.8	6.5	-30	-0.1	85	113	269*	n.ap.
Congo	34 150	21 977	83	22 060	64.6	7.7	-17	-0.1	132	213	-	-
Dem. Rep. of the Congo	226 705	135 110	97	135 207	59.6	2.7	-532	-0.4	133	225	-	-
Gabon	25 767	21 790	36	21 826	84.7	18.2	-10	n.s.	128	137	-	-
Equatorial Guinea	2 805	1 752	-	1 752	62.5	4.0	-11	-0.6	93	158	-	-
Rwanda	2 466	46	261	307	12.4	n.s.	-15	-3.9	110	187	-	-
Total Central Africa	403 298	227 377	634	228 011	56.5	2.6	-852	-0.4	127	194	-	-
Total Africa	2 978 394	641 830	8 036	649 866	21.8	0.8	-5 262	-0.8	72	109	-	-
TOTAL WORLD	13 063 900	3 682 722	186 733	3 869 455	29.6	0.6	-9 391	-0.2	100	109	-	-

Source: Appendix 3, Tables 3, 4, 6, 7 and 9.
*Partial result only. National figure not available.

are the main exports. The uses of the forest are multiple, including non-wood forest products collection, and vary from low-impact harvesting to high-intensity commercial logging. Central Africa is not a uniform political or socio-economic entity: more than 70 percent of the population in central Africa is rural, although Gabon and the Congo are the most urbanized. Population densities in certain regions are among the lowest in Africa. However, Rwanda and Burundi are very densely populated, with 90 percent of their population living in rural conditions. In general, central African countries are among the poorest in the world, with the exception of Gabon (FAO 2000).

FOREST RESOURCES

Forest resource knowledge is relatively low and most of the central African forest inventories cover only part of the productive forested domain (Cameroon, the Congo, Gabon, Rwanda and the Central African Republic). At the national level, the information regarding forest areas is obsolete where it exists at all and needs to be updated. The last national forest inventory of Burundi dates to 1976 and that of the Democratic Republic of the Congo to 1982. The most recent national-level data are those of Equatorial Guinea (1992). Consequently, the figures presented in Table 15-1 are, for the most part, based on national expert estimates. A workshop was also organized in Gabon in 1999 on data collection for this subregion with participation from all central African countries (FAO 2000).

Central African forests represent the second largest area of rain forest in the world and constitute 35 percent of the African forest area as well as approximately 6 percent of the world forest cover. The Democratic Republic of the Congo contains more than 60 percent of the subregion's forest area. Gabon is the most forested country with 85 percent of its total land area covered by forests. Burundi and Rwanda have the lowest proportion of forest cover (4 and 12 percent, respectively). Despite the lack of accurate statistics, it is clear that the forests of the Congo basin have experienced relatively low annual rates of clearing compared to other tropical forests and compared to the whole of Africa. Nevertheless, they have been subjected to progressive degradation that is difficult to estimate. Burundi and Rwanda have the highest annual negative rates of forest area change while the Congo, the Central African Republic and Gabon present rates lower than or equal to -0.1 percent a year (Table 15-1, Figure 15-2). The largest areas cleared each year are found in the Democratic Republic of the Congo and Cameroon.

Because central Africa contains such a large forest resource, reforestation efforts have been minimal. Also, these efforts have consisted primarily of commercial plantations rather than reforestation of logged-over or degraded areas. Approximately 634 000 ha of plantations have been established in central Africa with varying degrees of success. Many plantations in Cameroon, Gabon and the Democratic Republic of the Congo have failed owing to lack of maintenance and poor management. Accurate statistics on plantation and reforestation rates are also lacking. Indeed, certain countries have stopped state control, causing national plantation expertise to wane slowly. More than half of the plantation area is located in Burundi and Rwanda because of an extensive plantation programme instituted between 1975 and the early 1990s (FAO 2000).

The quantity and quality of forest resources available represents considerable potential. In fact, the total volume of central African forests represents more than 60 percent of the total African volume and 7 percent of the entire world volume. Central African forest volume is estimated at 47 billion cubic metres over bark, which corresponds to an average of 127 m^3 per hectare. In terms of biomass, the estimate is more than 44 billion tonnes because of the high wood density and a high percentage of branches that averages 194 tonnes per hectare. Central African forests come close to constituting two-thirds of the forest biomass reserves on the continent. Volume and biomass estimates for most central African countries are extracted from existing forest inventories (Equatorial Guinea, Cameroon, the Central African Republic and the Democratic Republic of the Congo). For the remaining countries, they are based on expert estimates (Burundi, Rwanda and Gabon) or extrapolation from nearby countries having comparable ecological characteristics (the Congo).

Biodiversity is exceptional in central Africa and the level of endemism is high. The Democratic Republic of the Congo, for example, contains more than 11 000 plant species, of which more than 30 percent are endemic. More than 1 100 species of birds and 400 species of mammals are also found there (these last two figures are the highest in Africa) (Tchatat 1999). The central African dense forests have important timber potential owing, in part, to high-value commercial species, notably "redwood" species belonging mainly to the Meliaceae family. The main commercial species are, among others, okoumé (*Aucoumea klaineana*), limba (*Terminalia superba*), tiama (*Entandrophragma angolense*) and sapelli (*Entandrophragma cylindricum*). Nevertheless, removals are uniformly less than increment. Indeed, the rain forests, in spite of their species variety and the abundance of big trees, contain only a limited number of commercially marketable species and

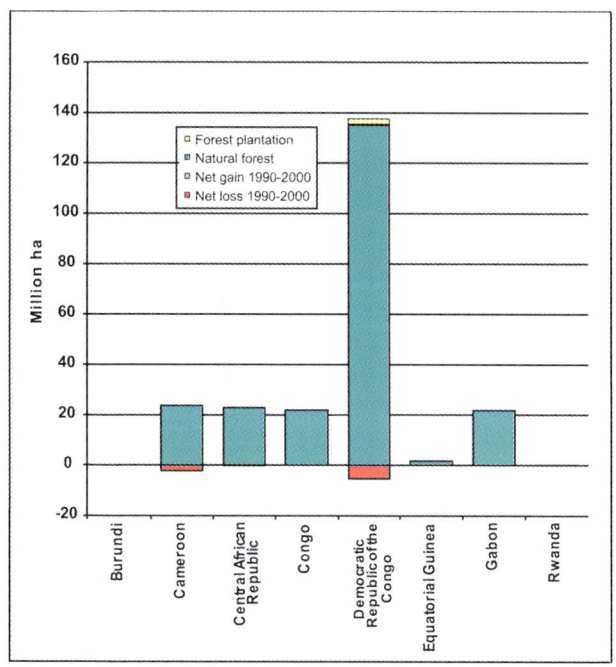

Figure 15-2. Central Africa: natural forest and plantation areas 2000 and net area change 1990-2000

the exploitable trees are scattered. These two factors, combined with poor accessibility (lack of roads) and timber transportation problems make the harvest selective and much lower than the potentially exploitable volume. Also, depending on market conditions, the concessionaires often limit themselves to only the highest quality timber, which is mainly located in closed forests (Dupuy *et al*. 1999). Nevertheless, forest overexploitation increases with the needs of populations, which is the case in the montane forests of Burundi and Rwanda (FAO 2000).

Wood from trees outside the forest is also important, notably where the natural forests are limited, as in Rwanda and Burundi, where agroforestry systems and small private plantations are encouraged to provide forest products (FAO 2000).

FOREST MANAGEMENT AND USES

None of the countries in central Africa provided information on the forest area covered by a formal, nationally approved forest management plan (Table 15-1). Nevertheless, significant efforts have begun to establish the framework for field-level implementation of sustainable forest management practices in the subregion (FAO 2000). A recent ITTO study (Poore and Thang 2000) thus reported that Cameroon is one of only six ITTO tropical producer countries which appeared to have established all the conditions that make it likely that they can manage their forest management units sustainably.

All of the central African countries have adopted strategies and forest action plans that take into account their specific needs. Some of these policies are very recent (Gabon, Cameroon and the Central African Republic). Countries have also modified their forest laws and management regulations. However, some countries have delayed their execution because of political disturbances, economic difficulties or violent civil crises. Furthermore, the current technical, financial, political and institutional conditions are not favourable in most of the countries (FAO 2000; Dupuy *et al.* 1999).

Currently, most central African forests belong to the State, although some countries have maintained traditional land tenure rights (Gabon, Cameroon and the Central African Republic, for example). Forest management is entrusted to a public forestry department and forests are classified as production, protection or nature reserves according to their characteristics. Generally, the forestry administration is charged with the execution of forest conservation, reforestation and exploitation activities as well as forest inventories and the preparation and implementation of management plans. However, many forestry administrations in central Africa lack the resources needed to implement their functions effectively and administer large areas of forest at the national level (FAO 2000; CARPE 1996).

Forest management for timber exploitation is focused on the demarcation of concession areas and control of harvested volumes. Production forests are generally awarded to timber companies or individuals (i.e. concessionaires) under more or less long-term concession agreements (temporary harvest permits). In Gabon, a resource inventory and a forest management plan proposal are compulsory before any exploitation. In the Congo and Cameroon, the national forest estate has been divided into forest management units, each having (in principle) a sufficient area to feed an independent wood industry under coordinated resource use and management plans. Various projects have established sustainable management strategies for forest resources. Pilot projects for sustainable forest production also exist in Cameroon and the Central African Republic. Paper production is almost non-existent in the region. The exact contribution of the forestry sector to the state economy is not generally defined in the available statistics. Nevertheless, it is apparent that Burundi and Rwanda have insufficient forest resources to meet internal needs and imported products must supplement national production.

In central Africa, 65 million persons live inside or near forests (Aubé 1996) and depend on them for energy, food, medicines, etc. As elsewhere in Africa, forests are the main source of domestic energy. Around 80 percent of the population of Cameroon and the Democratic Republic of the Congo use fuelwood for their domestic energy needs. In Gabon and Rwanda, 80 to 94 percent of the total fuel consumption comes from woody biomass. However, in spite of its importance, few data are available because of the informal character of fuelwood collection. Non-wood forest products are also important in the life of the local people and are widely used. Game holds an essential place in the Congo basin (Tchatat 1999). There is little information on markets and consumption patterns. The few statistics available are from isolated studies.

Central African countries have legally established large forest areas under protection. Certain zones have remarkable plant and animal diversity and are generally protected (e.g. the national parks of Dja in Cameroon and Dzanga-Ndoki in the Central African Republic). There are a number of protected area projects managed by regional and national offices. There is considerable legislation relating to protected areas at national levels but a large percentage of it is out of date and, in many cases, sufficient resources and mechanisms to ensure effective implementation are lacking (FAO 2000; Fotso 1996). In 1997, the World Conservation Monitoring Centre estimated that there are about 83 legally protected areas covering about 5 percent of the total subregion lands (WCMC 1997).

CONCLUSIONS AND ISSUES

Information about the forest resources of the eight central African countries is mostly based on national expert estimates. Forest inventory data are often unreliable, dated, obsolete, partial or unavailable. At present, data collection in central Africa is mostly done as part of forest management activities. Significant improvement in statistical data collection and analysis at the national level is needed for a better knowledge of forest resources.

All central African countries have adopted sustainable forest management policies. However, their implementation is generally poor because of lack of resources and institutional weaknesses. In addition, for some of these countries (Burundi, Rwanda, the Congo and the Democratic Republic of the Congo), political and social crises during the last decade have had negative effects on forest sustainability. Nevertheless, significant efforts have been undertaken by national scientific research units in each country to improve the technical and economical management of production forests (FAO 2000).

There are multiple causes of deforestation in central Africa. Some are direct (agriculture, urbanization, mining, etc.), others are indirect, such as socio-economic factors (population pressure, poverty, international market fluctuations, etc.) or political factors (political instability, etc.). The principal causes of deforestation in dense forests are agriculture (shifting cultivation and cash crops) and fuelwood harvesting, mostly in the high population density zones. Shifting cultivation can lead to drastic forest resource degradation if it is not managed in a sustainable way. The Democratic Republic of the Congo, Cameroon, Burundi and Rwanda have the highest rates of rural population increase in the subregion. As a consequence, agriculture is one of the main causes of deforestation in these countries as opposed to Gabon and the Congo, which are more urbanized (CARPE 1996). Natural resources in peri-urban zones are subject to high pressure from urban area expansion and utilization for fuelwood and building materials (FAO 2000).

Commercial logging is selective in the highly forested countries of the subregion and leads mainly to forest degradation rather than deforestation. Degradation can lead to the depletion of commercial species in the short term. After several harvests, the dense forest is often degraded into an open forest sensitive to fire (although fires are usually more important on other wooded lands). In addition, construction of logging roads encourages people to settle and convert forested lands into agriculture (FAO 2000; Dupuy et al. 1999).

Migrant populations, because of economic, social or political reasons, have destroyed forests through settlement, uncontrolled logging and fire. This critical situation can lead to the destruction of infrastructure and to overall instability of the forestry sector. This was the case in Burundi and Rwanda during the last decade where most productive lands were converted to agriculture. Efforts to reforest degraded or clear-cut areas have begun in those two countries as well as the promotion of agroforestry practices (FAO 2000).

Popular participation in forest management planning and implementation has increased in central African countries. Some other important issues include strengthening forestry training institutions, conducting conservation awareness programmes for the public and carrying out long-term ecological research on the value of services provided by forests.

BIBLIOGRAPHY

Aubé, J. 1996. *Étude pour favoriser le développement des produits forestiers non ligneux dans le cadre du Central African Regional Program for the Environment (CARPE).* Washington, DC, Forestry Support Program, USAID. http://carpe.umd.edu/Products/

Central African Regional Program for the Environment (CARPE). 1996. *CARPE workshop, Libreville, Gabon.* Washington, DC, USAID. http://carpe.umd.edu/

Dupuy, B., Maître, H.-F. & Amsallem, I. 1999. *Techniques de gestion des écosystèmes forestiers tropicaux: état de l'art.* Working paper FAO/FPIRS/05 prepared for the World Bank forest policy implementation. Review and Strategy. FAO, Rome/ Montpellier, France, Cirad Forêt.

FAO. 2000. *Collecte et analyse de données pour l'aménagement durable des forêts – joindre les efforts nationaux et internationaux.* Proceedings of subregional workshop on forestry statistics. EC-FAO Partnership Programme GCP/INT/679/EC, subregional workshop for Congo Basin countries, Lambarene, Gabon, 27 September – 1 October 1999. Rome. www.fao.org/forestry/fon/fons/outlook/Africa/ACP/Lamb/Lamb-17.htm

Fotso, C. 1996. *Problématique de la conservation de la biodiversité en Afrique centrale.* Cameroun, Conservation et utilisation rationnelle des Ecosystèmes Forestiers d'Afrique Centrale (ECOFAC). CARPE Libreville. USAID. http://carpe.umd.edu/products/

IUCN. 1996. *Atlas pour la conservation des forêts tropicales d'Afrique,* ed. J.-P. de Monza. Paris.

Poore, D. & Thang, H.C. 2000. *Review of progress towards the year 2000 objective.*

Report presented at the 28th Session of the International Tropical Timber Council ITTC(XXVIII)/9/Rev. 2, 24-30 May 2000, Lima, Peru. Yokohama, Japan, ITTO.

Tchatat, M. 1999. *Produits forestiers autres que le bois d'œuvre (PFAB): place dans l'aménagement durable des forêts denses humides d'Afrique centrale.* Projet régional de capitalisation et transfert des recherches sur les écosystèmes forestiers de l'Afrique humide. Série FORAFRI. Document 18.

WCMC. 1997. *United Nations List of Protected Areas 1997.* www.wcmc.org.uk/protected_areas/data/un_97_list.html

Chapter 16
East Africa

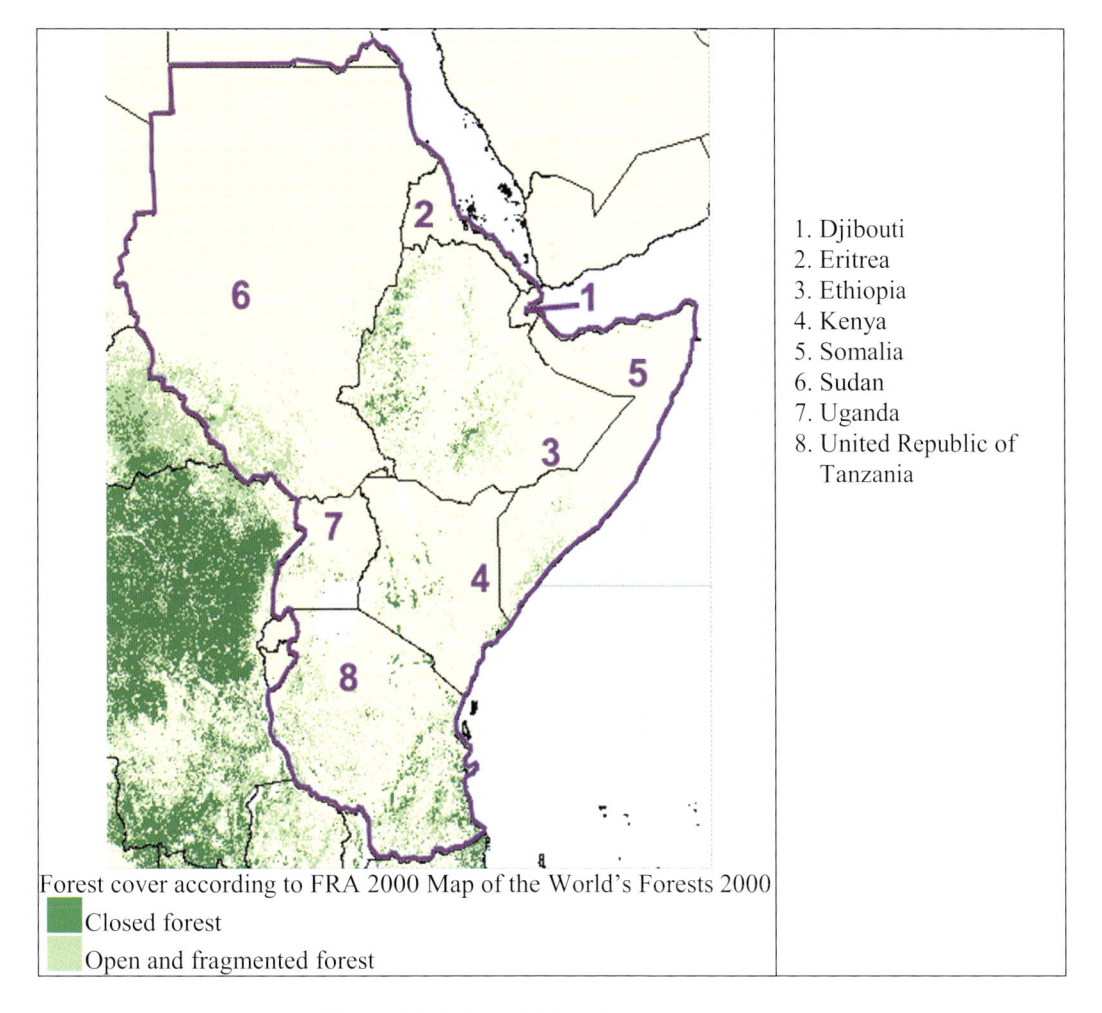

Figure 16-1. East Africa: forest cover map

Legend:
1. Djibouti
2. Eritrea
3. Ethiopia
4. Kenya
5. Somalia
6. Sudan
7. Uganda
8. United Republic of Tanzania

Forest cover according to FRA 2000 Map of the World's Forests 2000
- Closed forest
- Open and fragmented forest

The subregion of East Africa lies between 21° north latitude and 11° south latitude. The Tropic of Cancer crosses southern Egypt near its border with the Sudan. With eight countries (Djibouti, Eritrea, Ethiopia, Kenya, Somalia, the Sudan, Uganda and the United Republic of Tanzania),[31] East Africa covers a land area of 5.9 million square kilometres. The Sudan, with a land area of 2.4 million square kilometres, is the largest country in Africa. The subregion is bordered by the Red Sea and the Indian Ocean on the east (Figure 16-1).

East Africa is a relatively dry area strongly influenced by the Sahara Desert. Desert covers more than 1 million square kilometres, including all of the northern Sudan. The climate is characterized by high temperatures and low precipitation (less than 200 mm). Very arid and semi-arid climates are also found in Somalia, Djibouti and along the coast of Eritrea, with annual rainfall ranging between 400 and 750 mm. Most of Ethiopia and the mountains of Kenya have montane climates with higher rainfall and lower temperatures. Uganda and the coast of the United Republic of Tanzania are mostly characterized by a very humid climate with high temperatures and a very short dry season. The rest of Tanzania, Kenya and Uganda have typical tropical climates with a long dry season.

East Africa has suffered from many social problems. The Sudan is involved in a civil war in the southern part the country and Ethiopia, Eritrea

---
[31] For more details by country, see www.fao.org/forestry

Table 16-1. East Africa: forest resources and management

| Country/area | Land area | Forest area 2000 ||| Area change 1990-2000 (total forest) || Volume and above-ground biomass (total forest) || Forest under management plan ||
| | | Natural forest | Forest plantation | Total forest |||||||| |
	000 ha	000 ha	000 ha	000 ha	%	ha/capita	000 ha/year	%	m³/ha	t/ha	000 ha	%
Djibouti	2 317	6	-	6	0.3	n.s.	n.s.	n.s.	21	46	-	-
Eritrea	11 759	1 563	22	1 585	13.5	0.4	-5	-0.3	23	32	-	-
Ethiopia	110 430	4 377	216	4 593	4.2	0.1	-40	-0.8	56	79	112	2
Kenya	56 915	16 865	232	17 096	30.0	0.6	-93	-0.5	35	48	120*	n.ap.
Somalia	62 734	7 512	3	7 515	12.0	0.8	-77	-1.0	18	26	-	-
Sudan	237 600	60 986	641	61 627	25.9	2.1	-959	-1.4	9	12	-	-
Uganda	19 964	4 147	43	4 190	21.0	0.2	-91	-2.0	133	163	-	-
United Republic of Tanzania	88 359	38 676	135	38 811	43.9	1.2	-91	-0.2	43	60	-	-
Total East Africa	590 078	134 132	1 291	135 423	23.0	0.7	-1 357	-1.0	28	38	-	-
Total Africa	2 978 394	641 830	8 036	649 866	21.8	0.8	-5 262	-0.8	72	109	-	-
TOTAL WORLD	13 063 900	3 682 722	186 733	3 869 455	29.6	0.6	-9 391	-0.2	100	109	-	-

Source: Appendix 3, Tables 3, 4, 6, 7 and 9.
*Partial result only. National figure not available.

and Somalia have been devastated by war. Much of the population of Rwanda crossed the border to seek refuge in Tanzania and Uganda. Refugees from Somalia are in Ethiopia and Kenya. The effects of war, combined with the severe climate, have placed increased pressure on the land and have had a heavy impact through deforestation. Fires are a major problem. Desertification has increased, especially in the Sudan where 13 of its 26 states have been declared "affected by desertification" by the UN Convention for Combating Desertification (El Hassan and Mohamed 1999).

FOREST RESOURCES

Because of the very difficult social situation, there is little information on the forest resources of East Africa. Only three countries have relatively new data. Eritrea has recent forest cover mapping with a reference year of 1997 (FAO 1997). Tanzania has recently completed a land cover/forest mapping project with a reference year of 1995 (United Republic of Tanzania HTS 1997). Uganda has a biomass inventory dated 1992 (Uganda FD 1996). The Sudan has a partial forest inventory that covers the "gum belt" regions (Sudan FNC 2000). The other countries (Djibouti, Ethiopia, Kenya and Somalia) have old and fragmented information. Estimates by local experts provided the data for these countries (Bekele 2000; Ndambiri and Kahuki, 2000). The total forest area assessed by forest inventories is less than 0.5 million square kilometres and corresponds to about a third of the total forest cover. The largest area of the forest cover is in the Sudan, with 46 percent, followed by Tanzania with 29 percent and Kenya with 13 percent. The remaining 12 percent is located in the rest of the subregion. The forest area of the subregion accounts for 21 percent of the total forest area of Africa and 4 percent of the world's forests (Figure 16-2, Table 16-1).

Natural forests in East Africa total 134 million hectares. Uganda has the highest deforestation rate, but the largest area of deforestation occurs in the Sudan, where it is estimated that almost 1 million hectares are deforested annually.

The heavy deforestation occurring in the subregion is not balanced by tree planting. The primary use of wood in East Africa is for fuel. Despite successive wars the Government of Eritrea has set up a programme to protect natural forests by permanent and temporary closures of areas of natural vegetation, replanting of indigenous species and increases in the areas of plantations (FAO 1997). The tree planting programme focuses to a large extent on planting *Acacia senegal* for the production of gum arabic. The most recent tree planting efforts in Ethiopia were in the 1970s when significant areas of eucalyptus plantations were established. Today most of these areas are degraded. The Ethiopian Forestry Action Plan of 1994 recommended a serious programme of tree planting within the next 20 years. Kenya established significant areas of plantations during the 1970s and 1980s (Kenya MENR 1994), but the area planted declined in the 1990s. Because of civil war almost no tree

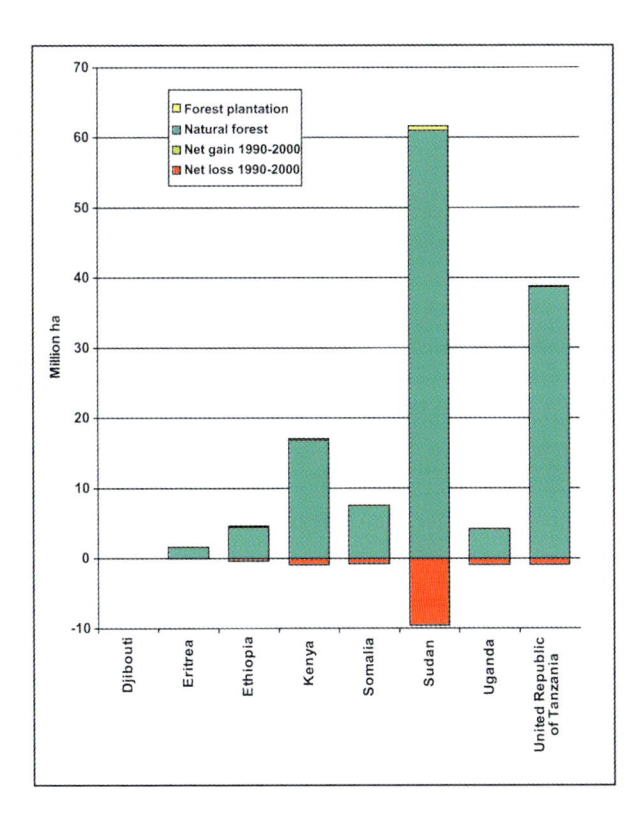

Figure 16-2. East Africa: natural forest and plantation areas 2000 and net area change 1990-2000

planting activity has been reported in Somalia. The Sudan has the largest area of plantations with significant areas of *Acacia senegal* and *A. nilotica* (Sudan FNC 2000). Plantations in Tanzania are estimated to be 0.3 percent of the total forest area and this is predicted to increase in the future. The most important product is fuelwood. To meet increasing demand and secure its sustainable resources, Tanzania has recently revised its forestry strategic plan. Forests supply about 90 percent of the energy demand in Uganda. However, plantations only account for 1 percent of the total forest area.

FOREST MANAGEMENT AND USES

Information on forest management is generally lacking for East Africa. Ethiopia was the only country in the subregion that provided national-level information to FRA 2000 on the forest area covered by a formal, nationally approved forest management plan (Table 16-1), while Kenya provided partial information (plantations only). Although not reported for FRA 2000, some forests in natural reserves and national parks are also covered by management plans in several East African countries including Kenya, Tanzania and Uganda.

In several countries, the available forests cannot meet the increasing demand for fuelwood. At the same time, countries of this region recognize that indigenous forests are able to provide a variety of valuable products when well managed. Biodiversity is regarded as a potential source of income, especially when wildlife is considered.

Although the Sudan has had forest legislation for a long time, forest management is not well established. A forest policy was issued with the objective of reserving 20 percent of the area of the country as forests under sustainable management at the end of the 1980s. At the beginning of the 1990s, about 4 percent of the total forest area had been reserved under a presidential decree, but none of the areas were reported as under management for FRA 2000 (Sudan FNC 2001).

Ethiopia is mainly an agricultural country with limited forest cover. The indigenous forest is still shrinking owing to rapid deforestation. Forestry activities are currently being reorganized (Bekele 2000). The government is moving towards a federal system and, in the future, the regions will carry out forest management activities. During this transition period little activity has taken place.

Eritrea has no tradition of formal forest management. A forest policy has been developed since independence and the major effort has been to try to reduce the degradation of the country's resources by planting trees on mountains and escarpments and along roadsides.

Forest management activities are almost non-existent in Somalia. There is little information on fuelwood requirements and most areas are under strong pressure.

In Kenya, the forest areas under management are mainly industrial forest plantations (Kenya MENR and FINNIDA 1992) and some indigenous forests in protected areas, although the government recognizes the role of these forests in agriculture and livestock management and their key role in sustaining wildlife. The tourism industry in this country is highly dependent on wildlife and provides a major contribution to the country's income.

Fuelwood is a major use in Tanzania, but the forests are also a source of income from non-wood forest products (honey, tannins, gum arabic, etc.) and tourism. According to the National Forest Policy (United Republic of Tanzania MNRT 1998), about a quarter of the forest area is devoted to national parks, forest reserves and game reserves. All these areas are reportedly under management although no information was provided for FRA 2000 on the area of forest management plans. In the latest revision of the National Forest Policy, published in 1998, the Ministry of Natural Resources and Tourism declared that sustainable management of the resources is a major issue to be addressed and that it will also try to promote sustainable forest management outside forest conservation areas. These forests are subject to conversion to other uses such as shifting cultivation and grazing and also suffer from degradation due to repeated forest fires.

Deforestation is a major problem in Uganda. Forests are rapidly decreasing even though the country is the most humid and wet in the subregion. The main reason is conversion to agriculture and clearing for fuel. In 1992 the government tried to address the problem with a National Tree Planting Programme (Uganda 1998). The programme was also assisted by NGOs and the private sector in afforestation and reforestation programmes under agroforestry practices, in peri-urban plantations and in private woodlots. The government and NGOs also agreed to promote the use of energy-efficient technologies.

CONCLUSIONS AND ISSUES

Assessment of forests was not straightforward for this region. Most of the work was carried out in close cooperation with local experts who supplied information and local knowledge. The results highlight a situation of progressive degradation and reduction of East African forests due to social conditions created by war, population pressure and the limited potential area of forests. Wars also increased poverty in the area and were a disincentive to donors and investments.

Forest fires are a major problem in most of the region and, unfortunately, although the countries are conscious of the consequences, there are almost no programmes to monitor and control them.

Desertification is progressively affecting the hot, dry areas of the Sudan, Eritrea, Ethiopia, Djibouti and Somalia. Forest resources are seriously threatened by droughts and adverse human activities such as grazing, fires and shifting cultivation. Afforestation and reforestation programmes are badly needed. Degraded areas around large settlements require immediate action. Programmes are also needed to substitute other sources of energy, at least for industrial needs.

Adequate forest policies need to be developed and applied. There is a strong need for sustainable forest management, especially in countries dependent on forests for fuelwood, timber, non-wood forest products and tourism. Countries with small or diminishing forest resources, such as Eritrea, Somalia and the Sudan, need forest policies to support tree planting and forest management.

BIBLIOGRAPHY

Bekele, M. 2000. Ethiopia submission to FRA 2000.

Centre de coopération internationale en recherche agronomique pour le développement – Institut d'élevage et de médecine vétérinaire des pays tropicaux (CIRAD-IEMVT). 1991. *Carte de la végétation et des ressources pastorales. 1/250 000.* Institut de la carte internationale de la végétation (1987).

El Hassan, H.M. & Mohamed, Y. 1999. Personal communication.

Ethiopia. Ministry of Natural Resources Development and Environmental Protection (MNRDEP). 1994. Ethiopian Forestry Action Program (EFAP).

FAO. 1993. *Forest plantation inventory and management planning – Kenya – project findings and recommendations.* FO:DP/KEN/86/052. Terminal report. Nairobi.

FAO. 1997. *Support to forestry and wildlife sub-sector. Pre-investment study.* TCP/ERI/6721. Rome.

Getachew, E. 1999. Assessment of fuelwood resources in *Acacia* woodlands in the rift valley of Ethiopia. Towards the development of planning tools for sustainable management. Doctoral thesis. Swedish University of Agricultural Sciences, Umeaa, Sweden.

Hawkes, M.D. 1991. *Lower Shabelle Region woodland inventory starter kit – Forestry development and strengthening of the forestry department.* GCP/SOM/042/FIN. Rome, FAO.

Kenya. Ministry of Environment and Natural Resources (MENR) & Department for International Development Cooperation, Finland (FINNIDA). 1992. *Kenya Forestry Master Plan.* First incomplete draft.

Kenya. MENR. 1994. *Kenya Forestry Master Plan.* Kenya.

Ndambiri, J.K. & Kahuki, C.D. 2000. Kenya submission to FRA 2000.

Sudan. Forest National Corporation (FNC). 2000. Country submission to FRA 2000.

Sudan. FNC. 2001. Summary brief on the forestry sector in the Sudan.

Uganda. Forest Department, Ministry of National Resources (FD, MNR). 1996. *The national biomass study (NBS)*. Kampala. www.imul.com/forestry/forestry.html

Uganda. 1998. Country report on assessment of the intergovernmental panel on forest proposal.

United Republic of Tanzania. Hunting Technical Services (HTS). 1997. *Forest resources mapping project. National Reconnaissance Level Land Use and Natural Resources Mapping Project*. Dar es Salaam, United Republic of Tanzania, Ministry of Natural Resources and Tourism (MNRT).

United Republic of Tanzania – Ministry of Natural Resources and Tourism (MNRT). 1998. *National Forest Policy*. Dar es Salaam.

Chapter 17
Southern Africa

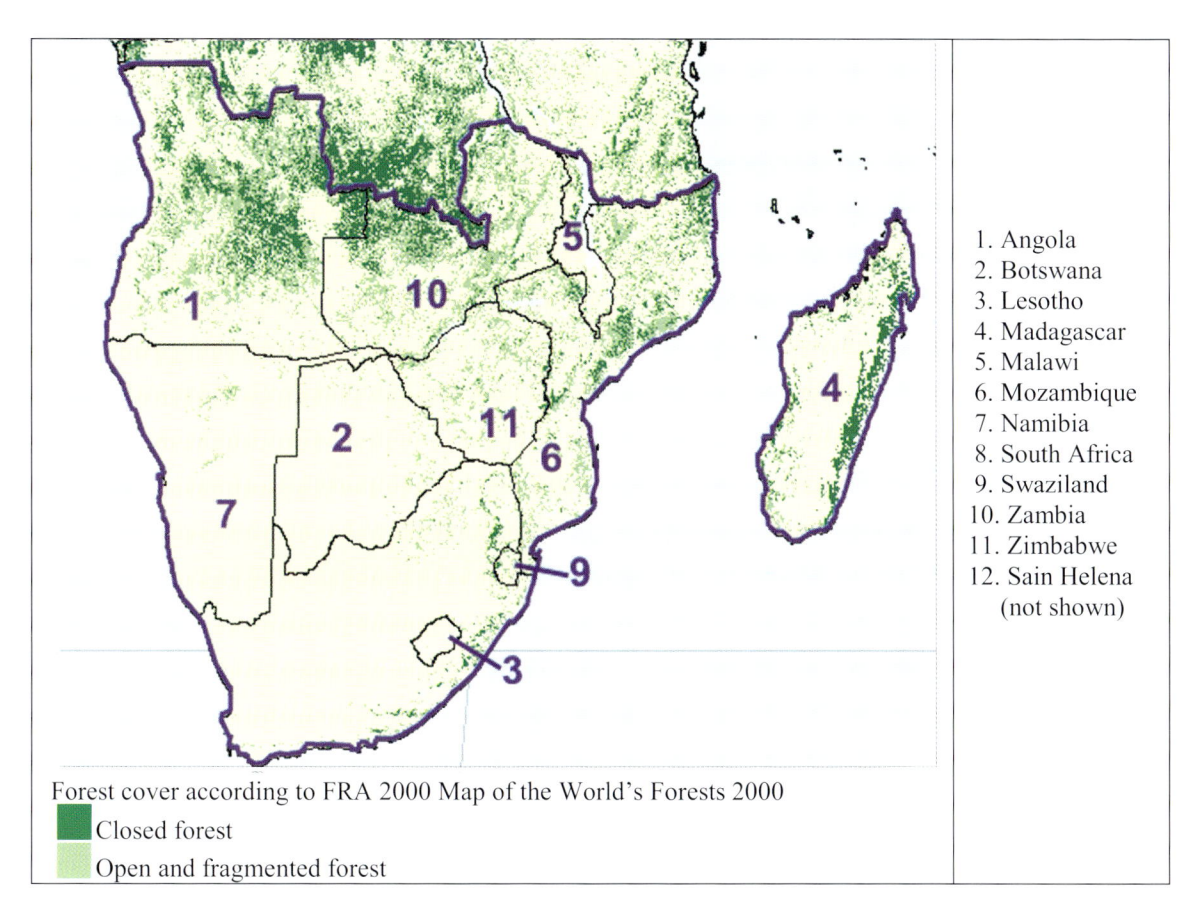

Forest cover according to FRA 2000 Map of the World's Forests 2000
- Closed forest
- Open and fragmented forest

1. Angola
2. Botswana
3. Lesotho
4. Madagascar
5. Malawi
6. Mozambique
7. Namibia
8. South Africa
9. Swaziland
10. Zambia
11. Zimbabwe
12. Sain Helena (not shown)

Figure 17-1. Southern Africa: forest cover map

The subregion is bordered by the Indian Ocean in the east and the Atlantic-Indian basin in the south and includes the countries of Angola, Botswana, Lesotho, Madagascar, Malawi, Mozambique, Namibia, Saint Helena, South Africa, Swaziland, Zambia and Zimbabwe.[32] The total area is 6.49 million square kilometres, a substantial part of which belongs to the Kalahari Desert ecosystems. Despite the vast area of the Kalahari Desert, the forest cover in this subregion is moderately high at around 30 percent of the countries' total land area (Figure 17-1).

The subregion is characterized by different climatic conditions depending on location. On the southern side of South Africa the climate is warm-temperate humid (Gelgenhuys 1993). It becomes subtropical north of the Cape region, in Lesotho and southern Mozambique and then tropical in the remainder of the subregion. In the western strip, from Angola to north of the Cape region, the dry climate of the tropical Kalahari Desert dominates. The rainfall regimes also vary greatly. The average annual precipitation is very low and limited to a few rainy weeks in the desert area to more than 2 200 mm in the mountains of Tsaratanana in Madagascar (Madagascar ONE 1997) and Gurue and Chimanimani in Mozambique (Chidumayo 1997).

Economic, social and environmental functions of the forest cover and resources vary greatly among countries. Angola, Madagascar, Mozambique and Zambia have the greatest timber production capacities from natural forests. The ecological conditions in Namibia, Lesotho, Swaziland, Botswana and South Africa do not favour timber-producing natural forests. In countries such as Malawi and Zimbabwe, natural forests of high timber potential were largely

---

[32] For more details by country, see www.fao.org/forestry

eliminated through clearing for agriculture, fuelwood and pole collection, infrastructure development and overstocking of domestic animals. The forestry sector continues to be a huge reservoir providing an array of goods and services vital to the livelihood of local populations in all countries. The proportion of the population in these countries living in rural areas is still very high and people rely to a large extent on forest resources for shelter, food, energy, construction material, employment and other products for domestic consumption as well as for trade (Howell and Convery 1999).

To alleviate the lack of natural forest cover in countries such as South Africa and Swaziland, considerable effort was undertaken to create and maintain artificial forests that are nowadays highly productive.

FOREST RESOURCES

Lesotho, Madagascar, Malawi, Mozambique, Namibia and Swaziland have relatively recent data from national forest surveys prepared during the 1990s. The national forest inventory of Mozambique was an updating of the first one carried out in 1980. It is the most complete in terms of maps and statistics and was produced from successive inventories at ten-year intervals. Data sets for Angola, Saint Helena and Zambia were produced from surveys done in 1983, 1980 and 1978, respectively. Botswana, South Africa and Zimbabwe do not have information from inventories with national coverage. Lesotho, Malawi and Swaziland prepared detailed forest maps of their forests.

In this subregion only Mozambique and Swaziland have complete updated and historical data which facilitate the estimation of forest cover and forest cover change for 2000. In Botswana, the only information is the soil map, which is of limited usefulness for forest cover estimation. The description of vegetation for each soil type is vague and inaccurate (De Wit and Bekker 1990). Angola's baseline information covers the entire country, but it is generally based on crude secondary sources of low reliability (Horsten 1983). In South Africa, information on natural forests is incomplete and refers merely to broad classes of vegetation, including shrub and forest formations, whose reclassification into the FAO classes is not straightforward for lack of clear definitions. It was produced from the National Land Cover Survey (South Africa DWAF 2000). The same source provides better information on plantations. It gives total planted areas, their distribution by zone and species and the volume of various products extracted.

Namibia and South Africa are among the countries with the least forest cover in the subregion. This is because they include a substantial part of the Kalahari Desert where woody vegetation is sparse whenever it exists. Lesotho and Saint Helena have the least cover. The lack of forest cover in Lesotho is reportedly the result of unfavourable natural factors for forest development (climate and poor soils) combined with excessive use of the limited resource by rural people for fuelwood and construction material and the demand for land for other uses (Lesotho Forestry Division 1996).

Angola's forest cover accounts for 36 percent of the subregion's total forest area, followed by Mozambique and Zambia with 16 percent each (Table 17-1). These three countries total about 68 percent of the forest cover while their land area makes up only 43 percent of the subregion. The subregional forest cover is about 30 percent of the continent's forest area. Compared to the world, it is about 5 percent, although the total land area amounts to only 22 percent of Africa and 5 percent of the world.

The extent of the natural forest cover is closely correlated with the level of annual precipitation. Natural forest is more abundant and developed in areas where precipitation is more than 400 mm per year. Below this amount, woody vegetation tends to be shrubs and bushes. On the eastern side of Madagascar, and to a lesser extent at some sites in the north of Angola, where rainfall is high and frequent all year round, moist and tropical rain forests are common.

Forests are also widespread across the tropical dry regions where miombo, mopane and *Acacia* woodlands are dominant. According to Chidumayo (1997) woodland is an ecological designation for stands of trees in relatively dry regions with pronounced seasonal effects and distinct physiognomic and structural characteristics. Miombo woodland is the most extensive vegetation type and covers a substantial area of Angola, Malawi, Mozambique, Zambia and Zimbabwe. It extends north to the United Republic of Tanzania and the Democratic Republic of the Congo. Miombo woodland is dominated by the presence of leguminous trees of the genera *Brachystegia, Isoberlinia* and *Julbernardia,* which are associated with others such as *Uapaca kirkiana, Acacia* spp., *Afzelia quanzensis, Pericopsis angolensis, Bauhinia* spp., *Burkea africana* and *Combretum* spp. Mopane

Table 17-1. Southern Africa: forest resources and management

Country/area	Land area	Forest area 2000 Natural forest	Forest plan-tation	Total forest			Area change 1990-2000 (total forest)		Volume and above-ground biomass (total forest)		Forest under management plan	
	000 ha	000 ha	000 ha	000 ha	%	ha/capita	000 ha/year	%	m^3/ha	t/ha	000 ha	%
Angola	124 670	69 615	141	69 756	56.0	5.6	-124	-0.2	39	54	-	-
Botswana	56 673	12 426	1	12 427	21.9	7.8	-118	-0.9	45	63	-	-
Lesotho	3 035	0	14	14	0.5	n.s.	n.s.	n.s.	34	34	n.s.	2
Madagascar	58 154	11 378	350	11 727	20.2	0.8	-117	-0.9	114	194	-	-
Malawi	9 409	2 450	112	2 562	27.2	0.2	-71	-2.4	103	143	-	-
Mozambique	78 409	30 551	50	30 601	39.0	1.6	-64	-0.2	25	55	-	-
Namibia	82 329	8 040	0	8 040	9.8	4.7	-73	-0.9	7	12	54*	n.ap.
Saint Helena	31	0	2	2	6.5	0.3	n.s.	n.s.	-	-	-	-
South Africa	121 758	7 363	1 554	8 917	7.3	0.2	-8	-0.1	49	81	828*	n.ap.
Swaziland	1 721	362	161	522	30.3	0.5	6	1.2	39	115	-	-
Zambia	74 339	31 171	75	31 246	42.0	3.5	-851	-2.4	43	104	-	-
Zimbabwe	38 685	18 899	141	19 040	49.2	1.7	-320	-1.5	40	56	92*	n.ap.
Total Southern Africa	**649 213**	**192 253**	**2 601**	**194 854**	**30.0**	**1.6**	**-1 741**	**-0.9**	**42**	**72**	**-**	**-**
Total Africa	**2 978 394**	**641 830**	**8 036**	**649 866**	**21.8**	**0.8**	**-5 262**	**-0.8**	**72**	**109**	**-**	**-**
TOTAL WORLD	**13 063 900**	**3 682 722**	**186 733**	**3 869 455**	**29.6**	**0.6**	**-9 391**	**-0.2**	**100**	**109**	**-**	**-**

Source: Appendix 3, Tables 3, 4, 6, 7 and 9.
*Partial result only. National figure not available.

woodlands occupy areas of low rainfall and high temperature from Inhambane and Tete provinces in Mozambique to north of Namibia and southern Angola and large areas of Zimbabwe and Botswana. The main species in this formation are *Colophospermum mopane* associated with a wide range of other species such as *Adansonia digitata* and *Sclerocarya birrea* (Chidumayo 1997; White 1983). *Acacia* woodland is common in various parts of the Zambezian phytoregion where the rainfall is low and the soil is suitable. Along the coast from Cabo Delgado to the Cape of Good Hope, woody vegetation is characterized by coastal forests with different floristic, structural and physiognomic properties from the woodland types (White 1983). Dry montane forest occurs in small patches at higher elevations.

Mangroves are very common along the coast of the tropical regions with a large concentration in Mozambique, where the area was estimated in 1990 at 396 000 ha (Saket 1994a), in Madagascar, estimated at 332 000 ha (WCMC 2000) and in Angola, where there are an estimated 28 000 ha (Horsten 1983).

With the exception of Swaziland, where the forest cover has changed positively over the last ten years, all other countries show various levels of deforestation (Table 17-1, Figure 17-2). Angola, Mozambique and South Africa show relatively low deforestation rates of 0.2 to 0.1 percent per year. Rates of deforestation depend on the combined effects of many factors in relation to development and conservation policies, reigning ecological conditions, the fragility of ecosystems and the social environment such as the size and economy of the rural population, employment opportunities for rural people, intensity of use, etc. South Africa (South Africa DWAF 2000) reported that 46 percent of its national population is rural and 9.2 million rural dwellers live in and around the forests and woodlands. About 31 percent of the national population use wood as their main energy source. Commercialization of forest products was reported to be increasing as more people seek additional opportunities for cash income. The causes of deforestation in South Africa seem to be mostly from unsustainable exploitation. In Angola and Mozambique, a positive effect on natural forests of the long-running civil wars was reported by various studies. Vast areas in these countries became totally inaccessible for security reasons as local people fled to large settlements and safe corridors, thus creating conditions for recovery of the forest vegetation. Since the restoration of peace in Mozambique in 1992, signs of accelerated deforestation have become more visible everywhere as a result of the return of refugees to their homelands. The unbalanced distribution of people prevails in Angola and, while deforestation is very high in some safe zones, it does not occur in the large unsafe areas (Saket 1994a).

The largest deforestation rates in the subregion occur in Malawi and Zambia, where forests are being converted into agriculture and settlement areas (Zambia MENR 1998). In absolute terms, Malawi loses 71 000 ha per year and Zambia 851 000 ha. The losses in Zambia amount to 49 percent of the deforestation in the subregion. Zambia is losing 14 times more forest per person than Malawi. The lower per capita rate in Malawi, despite its low living standard, has to do with the establishment of national conservation policies for natural resources that have already become scarce (FAO 1999).

The final group of countries, composed of Botswana, Madagascar, Namibia and Zimbabwe, has moderate deforestation. When deforestation is expressed in terms of area, Zimbabwe has the highest rate with 320 000 ha per year followed by Botswana and Madagascar. Expressing deforestation as area per person, Botswana takes the lead with 0.08 ha followed by Namibia with 0.05 ha and then Zimbabwe with 0.03 ha.

South Africa reported the largest planting programme with 1 554 000 ha, or 1.3 percent of its national land area. Swaziland's total plantation area of 160 500 ha amounts to 9.3 percent of its land area. Madagascar, Angola and Malawi spent moderate efforts in tree planting. Most of the forest plantations in these countries are for industrial purposes such as wood pulp.

In terms of woody biomass, Angola accounts for almost 27 percent of the subregional total, followed by Zambia with 23 percent. Angola, Madagascar, Mozambique and Zambia have 78.1 percent of the biomass while the total land area accounts for less than 52 percent of the subregion. Madagascar shows the highest tonnage of biomass per hectare. Malawi and Swaziland also have relatively high biomass per hectare owing to well-stocked plantations.

FOREST MANAGEMENT AND USES

Lesotho was the only country in southern Africa that provided national-level information to FRA 2000 on the forest area covered by a formal, nationally approved forest management plan (Table 17-1). Partial information was available from three countries (Namibia, South Africa and Zimbabwe) in the form of the forest area which had obtained third party certification by the end of 2000. For South Africa, this area corresponds to 9 percent of its total forest area. Madagascar

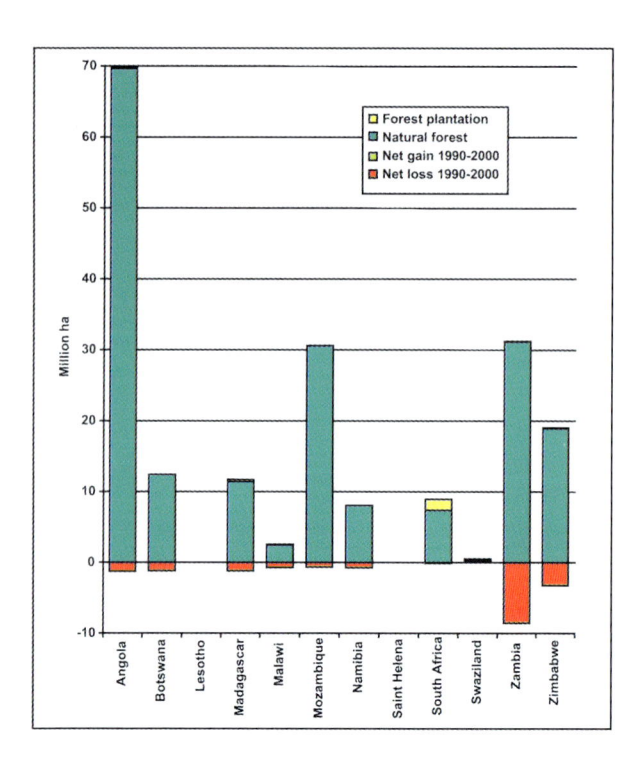

Figure 17-2. Southern Africa: natural forest and plantation areas 2000 and net area change 1990-2000

reportedly had 397 000 ha of natural forest under management (Madagascar MEF 1999), although no information was provided for FRA 2000.

Mozambique prepared its first forest management plan in 1999 as a model for a 45 000 ha timber concession. Application of management planning policy to productive natural forest is a national concern that is reflected in the newly revised legislation but remains dependent on factors external to the forestry sector and on a number of practical concerns (Saket 1999a). South Africa's management planning covers mainly plantations (South Africa DWAF 2000). Woodland and natural forests are still largely unmanaged. Zimbabwe has recently initiated some work on woodlands and a technical note on woodland management in Zimbabwe was prepared in 1992 (Hofstad 1992).

Zambia's management policy is described in its Forestry Action Plan, which recognizes the need for sustainable forest management if the nation is to be able to curb the considerable deforestation every year, to sustain timber production and secure protection of biological diversity and watersheds. In Malawi, the annual consumption of fuelwood was estimated at 6.4 million cubic metres against an annual growth of 5.3 million cubic metres (as of 1995). The deficit was thus 1.1 million cubic metres that had to be taken from growing stock, further depleting the resource (Moyo *et al.* 1993).

In South Africa, 95 percent of the protected areas do not have complete inventories of all groups of fauna and flora (South Africa DWAF 2000). According to White (1983), the South African ecosystems constitute a reservoir of biological diversity. In South Africa, the contribution of the forest to biodiversity includes 40 to 71 mammals, 106 birds, 649 woody plants and 649 herbaceous plants. Sixteen percent of the mammals and 13 percent of the birds are rare and endangered (South Africa DWAF 2000). There are at least 8 500 plant species in the Zambezian phytoregion of which 4 600 are endemic, more than 7 000 species in the Cape phytoregion with about half of them endemic, 3 500 in the Karoo-Namib phytoregion with more than half endemic, 3000 in the Inhambane-Zanzibar phytoregion with several hundreds endemic and about 3 000 in the Kalahari-Highveld phytoregion but with only a few endemic. The southern African countries have moved to protected areas of various types under such designations as national parks, game reserves and forest reserves (White 1983).

The forestry resources of the southern African countries are predominantly in State ownership. In Mozambique, for instance, all the lands, and thus the resources, are owned by the State. The situation is similar in Zambia and Angola. In the other countries, the private sector has limited access to ownership of forest resources (Saket 1994a).

Owing to a climate characterized, in most of the subregion, by pronounced wet and dry seasons, high temperatures and low air humidity and frequent droughts, the vegetation consists of open to relatively close-canopy deciduous forests, thickets or shrubs with an abundant grass layer. The long dry season, the loss of tree foliage and the accumulation of abundant dry material on the ground from leaf litter, dry grass and fallen dead branches create optimal conditions for intensive fires each year from May to October. In Mozambique, for instance, 40 percent of the country is burnt by fire every year and more than 80 percent of the area affected is forested (Saket 1999b). If an average of 7 tonnes per hectare of dry biomass (leaf litter, grass and dead branches) are burnt, the total biomass consumed by fire in Mozambique alone amounts about 157 million tonnes (Chidumayo 1997).

Forest fires are mostly caused by humans for various reasons such as improving visibility for hunting, facilitating timber exploitation, clearing land for agriculture, protecting households, opening land for settlement and charcoal making. Sometimes people set fires for no obvious reason. Fires are also frequently set during honey collection or cooking or accidentally by cigarettes. Forest fires started naturally (e.g. by lightning) are rare.

CONCLUSIONS AND ISSUES

Assessment of forest cover and change for the southern African countries was not straightforward. National definitions of forests and trees and the classification systems differed widely from those used in FRA 2000. However, close collaboration with these countries permitted local experts to work with the FRA team to reclassify the national vegetation, land use and land cover classes into the global classification system.

The quality of information on the status of forest cover in southern Africa is governed by the economic, social and environmental importance of the resources. In countries where natural forests do not produce timber and the priority of the subsector in national policies is very low, little information is available. Existing data in some countries are frequently generated from land use, soil or other thematic mapping work that pays little attention to the forest cover. The low timber potential of the forests in the subregion, compared to the tropical rain forest, also contributes to the scarcity of up-to-date information, even in the most forested countries. The only country in the subregion that has carried out two successive national inventories in the 1980s and 1990s is Mozambique, but further updating is not yet programmed. In some countries wildlife development projects were the origin of information on forests within these limited protected areas.

Land clearing for agriculture, forest fires and overexploitation for fuelwood and timber are reported as the major causes of deforestation in all countries. Zambia has the highest deforestation rate. In Mozambique, pressure on resources is mounting since the return of refugees following restoration of peace in 1992. Forestry legislation is usually outdated and practical implementation is still very limited in many countries for a number of reasons, including lack of operational funds, understaffing, insufficient training of technical staff and weak policing systems (Saket 1999b).

Pastures are widely overused, with subsequent soil erosion and desertification in large parts of the subregion where rainfall is low and irregularly distributed over the year. In countries where timber is exploited under licences for selected species and quantities without regard to proper silviculture, the forests have been deeply degraded or even stripped of a number of their most valuable species and their biodiversity adversely affected (Saket 1994a; Saket 1999b). Deforestation and degradation of natural forests may in some areas be attributable to international demand for tropical timber and to increasing demand for fuelwood. The loss of forest cover is contributing to soil erosion, causing water pollution and siltation of rivers and dams.

BIBLIOGRAPHY

Angola. Ministère de l'agriculture et du développement rural (MADR). 1994. *Rapport de la mission de consultation pour le sous-secteur forestier.* Luanda, Institut de développement forestier (IDF).

Chidumayo, E.N. 1997. *Miombo ecology and management, an introduction.* Stockholm, Environment Institute.

De Wit, P.V. & Bekker, R. 1990. *Explanatory note of the land system map of Botswana.* Gaborone, Soil Mapping and Advisory Services.

FAO. 1999. *State of the World's Forests 1999.* Rome.

Gelgenhuys, C.J. 1993. *Composition and dynamic of plant communities in the Southern Cape Forests.* Pretoria, Council for Scientific and Industrial Research (CSIR).

Hofstad, O. 1992. Technical note on woodland management in Zimbabwe. Harare, Forestry Commission.

Horsten, F. 1983. *Madeira, uma análise da situação actual.* Luanda, Sector de Divulgação e Informação, Direcção Nacional da Conservação da Natureza (DNCN), Ministerio da Agricultura.

Howell, D. & Convery, I. 1999. Socio-economic study on communities in future concession areas: recommendations for equitable and sustainable management. Maputo, Direcção Nacional de Florestas e Fauna Bravia (DNFFB).

Lesotho. Forestry Division. 1996. *Lesotho national forestry action plan.* Programme document. Maseru.

Madagascar. Ministère des eaux et des forêts (MEF). 1999. Rapport national sur le secteur forestier malgache, by Randriama Ampianina, V. & Razafiharison, A. *Proceedings of subregional workshop on forestry statistics.* EC-FAO Partnership Programme GCP/INT/679/EC, subregional workshop for Congo Basin countries, Lambarene, Gabon, 27 September-1 October 1999. Rome, FAO. www.fao.org/forestry/fon/fons

Madagascar. Office national pour l'environnement (ONE). 1997. *Bulletin statistiques, environnement.* Antananarivo.

Moyo, S.P., O'Keefe, P.O. & Sill, M. 1993. *The Southern African environment, profiles of the SADC Countries.* London, Earthscan Publications.

Saket, M. 1994a. *Report on the updating of the exploratory national forest inventory.* Maputo, DNFFB.

Saket, M. 1994b. *Study for the determination of the rate of deforestation of the mangrove vegetation in Mozambique.* Maputo, DNFFB.

Saket, M. 1999a. *Management plan for the timber concession area in Maciambose, province of Sofala.* Maputo, Direcção Nacional de Florestas e Fauna Bravia (DNFFB).

Saket, M. 1999b. *Tendencies of forest fires in Mozambique.* Maputo, DNFFB.

South Africa. Department of Water Affairs and Forestry (DWAF). 2000. *Report on the state of the forests in South Africa.* Pretoria, Department of Water Affairs and Forestry.

White, F. 1983. *Vegetation of Africa – a descriptive memoir to accompany the Unesco/AETFAT/UNSO vegetation map of Africa.* Natural Resources Research Report XX. Paris, UNESCO.

World Conservation Monitoring Centre (WCMC). 2000. *Forest and protected areas, Madagascar.* www.latinsynergy.org/mad_map.htm

Zambia. Ministry of Environment and Natural Resources (MENR). 1998. *Zambia forestry action plan, volume 1 – Executive summary.* Lusaka, MENR.

Chapter 18
Africa – small islands

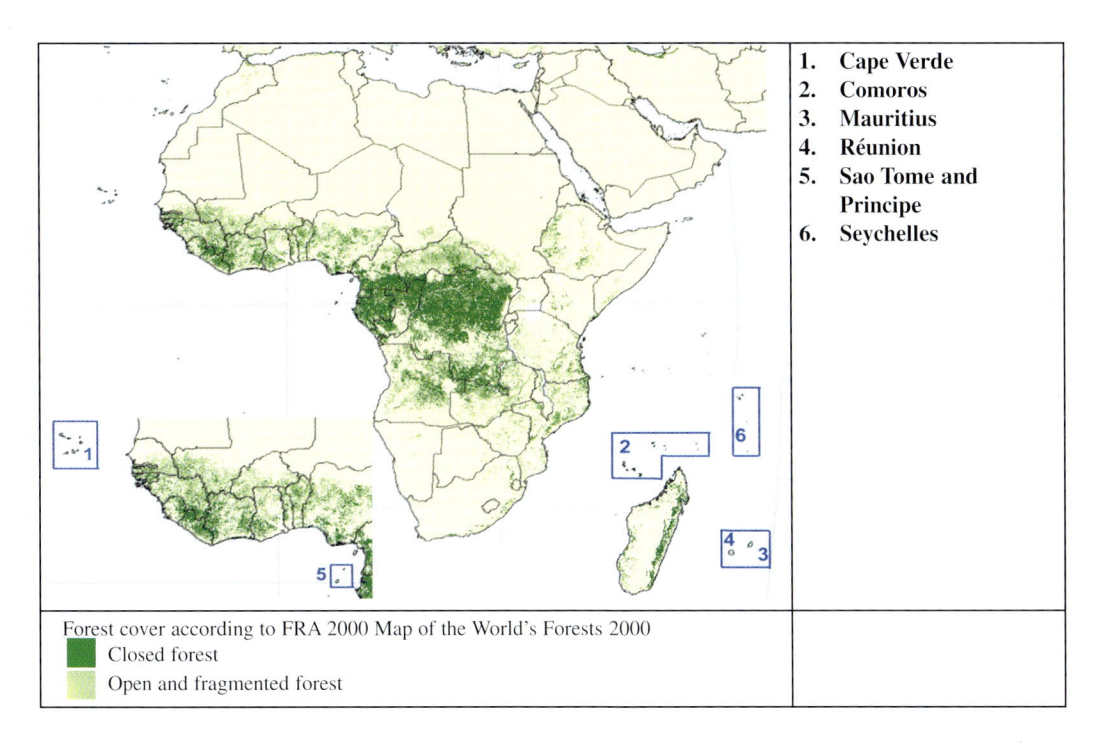

1. Cape Verde
2. Comoros
3. Mauritius
4. Réunion
5. Sao Tome and Principe
6. Seychelles

Forest cover according to FRA 2000 Map of the World's Forests 2000
- Closed forest
- Open and fragmented forest

Figure 18-1. Small islands: forest cover map

The countries included in this subregion are the islands of Cape Verde, Comoros, Mauritius, Réunion, Sao Tome and Principe and Seychelles (Figure 18-1).[33] The total land area is 1 181 million hectares. Forests cover 21 percent of this area.

The Cape Verde islands originally had extensive dry savannah woodland cover, but most of it was cleared for agriculture and, combined with an arid climate and steep terrain, the result has been widespread soil erosion and desertification. However, the archipelago can be divided into four broad ecological zones (arid, semi-arid, subhumid and humid) according to altitude and average annual rainfall, which ranges from 200 mm in arid coastal zones to over 1 000 mm in humid high zones. Much of the forest cover is relatively immature agroforestry plantations utilizing species such as *Prosopis juliflora, Leucaena leucocephala* and *Jatropha curcas.*

The Comoros islands were originally heavily forested but most of the lowland forest has been cleared for agriculture. The upland forests on Mohéli have been the least degraded, with remnant tropical forests on upland slopes and cloud forest above 600 m. Grande Comore is more degraded, with eruptions from the Kartala volcano also affecting the vegetation. Anjouan has remnant rain forest in Forêt de Moya. All three islands have areas of mangroves and strand vegetation. The climate is basically governed by the winds – the Indian monsoon (northeasterly) during the hot wet season and the trade winds (southerly and southeasterly) during the cool dry season.

The forests of Mauritius originally extended across much of the island with lowland moist evergreen forests in the south and east, moist upland forests in the centre of the island, dry palm forests in the north and dry savannah forests in the west. A high proportion of State-owned natural forest is protected. A fairly extensive reforestation programme constitutes the major forestry activity.

Réunion was originally covered with thick tropical forest. Much of it has been converted to agricultural uses. Common species include *Cryptomeria* spp. and a variety of palms. A high

---
[33] For more details by country, see www.fao.org/forestry

Table 18-1. Small islands: forest resources and management

Country/area	Land area	Forest area 2000					Area change 1990-2000 (total forest)		Volume and above-ground biomass (total forest)		Forest under management plan	
		Natural forest	Forest plantation	Total forest								
	000 ha	000 ha	000 ha	000 ha	%	ha/capita	000 ha/year	%	m³/ha	t/ha	000 ha	%
Cape Verde	403	0	85	85	21.1	0.2	5	9.3	83	127	-	-
Comoros	186	6	2	8	4.3	n.s.	n.s.	-4.3	60	65	-	-
Mauritius	202	3	13	16	7.9	n.s.	n.s.	-0.6	88	95	-	-
Réunion	250	68	3	71	28.4	0.1	-1	-0.8	115	160	-	-
Sao Tome and Principe	95	27	-	27	28.3	0.2	n.s.	n.s.	108	116	-	-
Seychelles	45	25	5	30	66.7	0.4	n.s.	n.s.	29	49	-	-
Africa – small islands	1 181	130	107	237	20.1	0.1	4	1.9	88	121	-	-
Total Africa	2 978 394	641 830	8 036	649 866	21.8	0.8	-5 262	-0.8	72	109	-	-
TOTAL WORLD	13 063 900	3 682 722	186 733	3 869 455	29.6	0.6	-9 391	-0.2	100	109	-	-

Source: Appendix 3, Tables 3, 4, 6, 7 and 9.

proportion of Réunion's plant species are endemic. This vegetation is fragile and threatened by various human activities: wood production, tourism, secondary forest product collection and invasive exotic species.

Sao Tome and Principe is composed of two volcanic islands in the Gulf of Guinea. Moist forests cover nearly three-quarters of the total land area and can be divided into three zones: low altitude moist closed forest, moist submontane evergreen forest and closed cloud forest. The low altitude forest has been extensively cleared and is now primarily savannah-type vegetation to the north and palms and coconuts to the south. The montane tropical high forest is, however, largely intact. The cloud forest has short trees with open crowns. Sao Tome and Principe currently has no formally protected areas. Most of the remaining primary forest has survived because of inaccessibility on steep slopes in the wettest, most inhospitable parts of the island that are unsuitable for either cultivation or human habitation. The primary forest actually has very little exploitable woody resources (in terms of species and size) and is not under pressure for the collection of fuelwood.

The main areas of natural forest in the Seychelles are in the uplands of Male and Silhouette islands and in the Vallée du Mai on Praslin Island. The latter is home to the coco-de-mer palm (*Lodoicea maldivica*), unique to the Seychelles. Relict lowland forests include species such as *Calophyllum inophyllum*. The upland forests are predominantly secondary forests in inaccessible locations. The larger islands also have areas of dry palm forests unique to the Seychelles. The small coral islands are generally covered by scrub vegetation of species such as *Pemphis acidula*. Mangroves occur around the coastline of a number of islands. The Seychelles has established a relatively large planted forest estate with *Casuarina* spp. and *Albizia* spp. the primary species. The country has a large proportion of its territory in protected areas.

FOREST RESOURCES

The Seychelles has the greatest percentage of forest cover of the islands, while Réunion and Sao Tome and Principe have the largest forest area (Table 18-1). Comoros has the highest deforestation rate. The area of forest plantations is relatively high in Cape Verde and constitutes the only forest cover, providing a net increase in forest cover over the last ten years (Figure 18-2). Forest plantations for industrial purposes are found on other islands of the subregion.

FOREST MANAGEMENT AND USES

In Cape Verde, the Tropical Forest Action Plan gave direction to the integral management of the forest resources in the country. Essential characteristics are the complete involvement of the forest-dependent communities, establishment of a national forest planning exercise and a multisector approach to the implementation of the forest action plan (Cape Verde MPAR 1994). There are currently no formally protected areas.

Forests in the Comoros are important because they support more than 78 percent of the energy needs of the country. Harvesting practices have not been regulated. Most forest areas are State-owned, which then permits access to these lands by private industry. Comoros is moving towards improved forest management and protection techniques. Reform of the forest legislation and forest policies has been proposed in addition to forest inventories, delimitation of use for agriculture and forestry, silvicultural practices, forest reserves, regulation of forest exploitation and use (in both natural forests and plantations), development practices for communities and training to strengthen the forestry sector (Houssen 2000).

In Mauritius, the native forest has almost disappeared with the exception of a few inaccessible areas which have been declared as natural reserves and natural parks. The native forests have been largely converted to agriculture or plantations of fast-growing species. The forests are both State and privately owned, although the State forests have considerably more growing stock (Appanah 2000).

In Sao Tome and Principe, the natural forest cover can be divided into two categories, the ecological reserve and a commercial zone. Secondary forest is mainly a consequence of regeneration of native species on former cacao and coffee plantations. Forests are also utilized for shade in agricultural and pastoral systems. Forest exploitation is based on needs, which vary from timber production to fuelwood consumption. Tree cutting requires permission of the National Direction of Agriculture and Forest (Soto Flandes 1985).

In Seychelles, the forest sector contributes less than 0.4 percent to the national economy and is thus perceived as marginal. Forests are of considerable importance, however, for tourism. The water supply is also highly dependent on the cover provided by forests. Most of the forest area is natural, and around 45 percent has been declared national parks or conservation areas. The main forest management problems include housing encroachment, invasive exotic plant species competing with the endemic and indigenous species and prevention and control of forest fires (Vielle 2000).

CONCLUSIONS

Efforts to promote appropriate forest management are taking place in this subregion. In those countries where there are natural forests, policies and legislation to protect these areas have been developed. Plantations are increasing in all the countries, both for energy and timber supply. Forest management plans and regulation of resource use have been implemented in order to curtail illegal and destructive practices in forested areas. Agroforestry, agrosilvopastoral and protection practices as well as the training of personnel and general strengthening of the forest sector has been promoted in the last ten years.

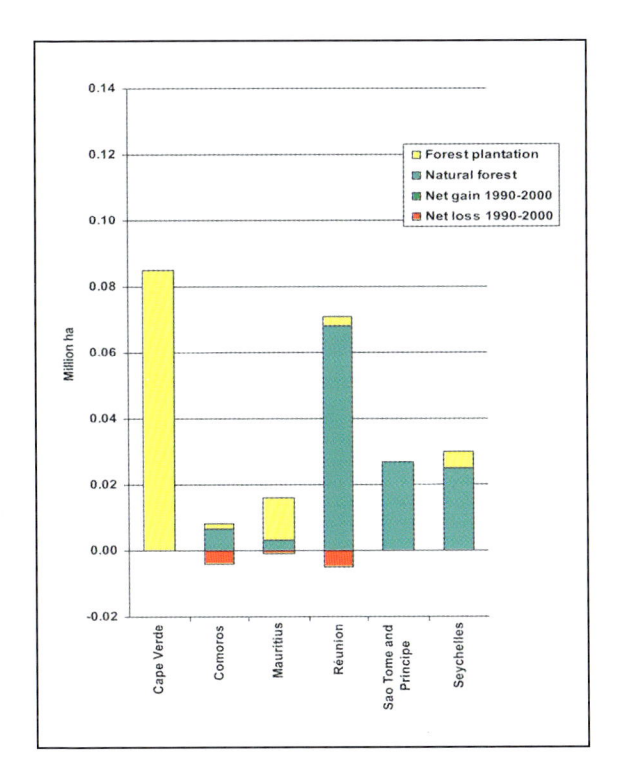

Figure 18-2. Natural forest and plantation areas 2000 and net area change 1990-2000

BIBLIOGRAPHY

Appanah, P.S. 2000. Mauritius. Rapports nationaux sur le secteur forestier. *Collection and analysis for the sustainable management of forest.* 15-18 July, Madagascar. Partnership Programme CE-FAO (1998-2000).

Cape Verde. Ministère de pêches, agriculture et animation rurale (MPAR). 1994. *Tropical Forest Action Plan Cape Verde, principal document.* Praia, Direction générale de l'agriculture, de la sylviculture et l'élevage.

Houssen, M.A. 2000. Comoros. Rapports nationaux sur le secteur forestier. *Collection and analysis for the sustainable management of forest.* 15-18 July, Madagascar. Partnership Programme CE-FAO (1998-2000).

São Tomé. Ministério da Economia. 2000. *Rapport, situation des forêts et de la faune sauvage en Sao Tomé-et-Principe.* São Tomé, Direcção das Florestas.

Soto Flandes, M. 1985. *Rapport de mission de planification forestière.* Sao Tomé-et-Principe. Rome, FAO.

Vielle, M. de Ker. 2000. Seychelles. Rapports nationaux sur le secteur forestier. *Collection and analysis for the sustainable management of forest.* 15-18 July, Madagascar. Partnership Programme CE-FAO (1998-2000).

Chapter 19
Asia

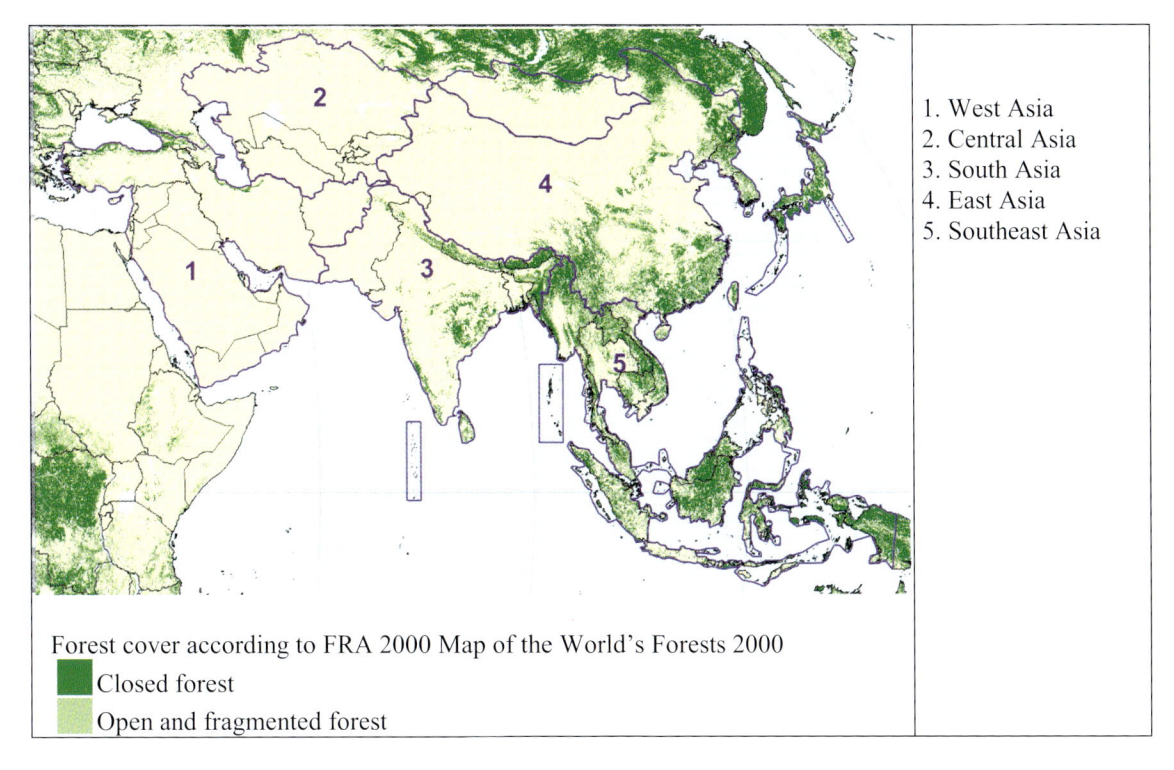

Figure 19-1. Asia: subregional division used in this report

Legend:
- 1. West Asia
- 2. Central Asia
- 3. South Asia
- 4. East Asia
- 5. Southeast Asia

Forest cover according to FRA 2000 Map of the World's Forests 2000
- Closed forest
- Open and fragmented forest

Asia (see Figure 19-1[34] and Table 19-1) as a whole contains about 548 million ha of forests which corresponds to 14 percent of the world total. Asian forests amount to 0.2 ha per capita, which is low compared to the world average. Most forests are located in the tropical ecological domain and Asia has about 21 percent of all tropical rain forests. Subtropical forests are extensive and Asia has more subtropical mountain forests than any other region and more than one third of the world total. More than 60 percent of the world's forest plantations are located in Asia. The net change of forest area is relatively low, with an annual net loss, based on country reports, estimated at 364 000 ha, corresponding to 0.2 percent annually.

---

[34] The division into subregions was made only to facilitate the reporting at a condensed geographical level and does not reflect any opinion or political consideration in the selection of countries. The graphical presentation of country areas does not convey any opinion of FAO as to the extent of countries or status of any national boundaries.

Table 19-1. Asia: forest resources by subregion

Subregion	Land area	Forest area 2000 Natural forest	Forest plantation	Total forest 000 ha	%	ha/capita	Area change 1990-2000 (total forest) 000 ha/year	%	Volume and above-ground biomass (total forest) m³/ha	t/ha
	000 ha	*000 ha*	*000 ha*	*000 ha*	*%*	*ha/capita*	*000 ha/year*	*%*	*m³/ha*	*t/ha*
Central Asia	545 407	29 536	384	29 920	5.5	0.5	208	0.7	62	40
East Asia	992 309	146 254	55 765	202 019	20.4	0.1	1 805	0.9	62	62
South Asia	412 917	42 013	34 652	76 665	18.6	0.1	-98	-0.1	49	77
Southeast Asia	436 022	191 942	19 972	211 914	48.6	0.4	-2 329	-1.0	64	109
West Asia	698 091	22 202	5 073	27 275	3.9	0.1	48	0.2	101	87
Total Asia	**3 084 746**	**431 946**	**115 847**	**547 793**	**17.8**	**0.2**	**-364**	**-0.1**	**63**	**82**
TOTAL WORLD	13 063 900	3 682 722	186 733	3 869 455	29.6	0.6	-9 391	-0.2	100	109

Chapter 20
Asia: ecological zones

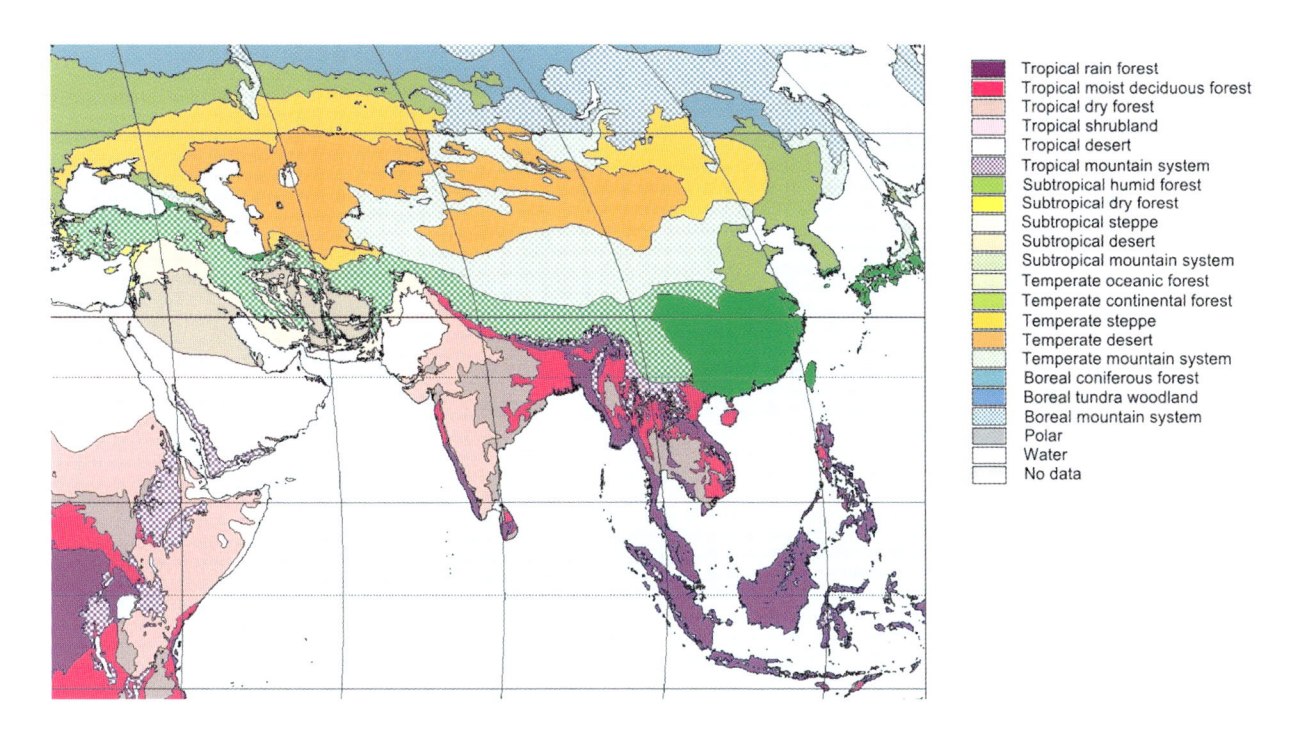

Figure 20-1. Asia: ecological zones

Figure 20-1 shows the distribution of ecological zones in Asia. Table 20-1 contains area statistics for the zones by subregion and Table 20-2 indicates the proportion of forest in each zone by subregion.

TROPICAL RAIN FOREST

This zone covers the southwestern coasts of India and Sri Lanka, Myanmar and the eastern Himalayan foothills, the coastal lowlands of Southeast Asia, the Philippines and most of the Malay Archipelago.

The western coasts of the Asian continent are very wet owing to monsoonal rains. Viet Nam and the Philippines deviate from this pattern and their eastern coasts are wet. Across the zone, annual rainfall is everywhere more than 1 000 mm and often more than 2 000 mm. There is no dry season in the equatorial regions. Everywhere else there is a short dry season, generally one to four months. Temperatures are always high.

In the wettest parts of this extensive zone the prevailing vegetation type is dense moist evergreen forest. A striking characteristic is the occurrence of Dipterocarpaceae only to the west of Wallace's Line.[35] The mangrove forests of the Ganges Delta and western New Guinea are the most extensive in the world. In the drier parts of the area, mainly in eastern Indonesia and the Himalayan foothills, semi-deciduous or moist deciduous forests occur. In the Brahmaputra valley, these are valuable sal forests (*Shorea robusta*).

The lushest rain forests are found in the Malay Archipelago, harbouring a very rich flora. Over half (220) of the world's flowering plant families are represented as well as about one-quarter of the genera (2 400), of which about 40 percent are endemic. Of 25 000 to 30 000 species about one-third are trees of more than 10 cm in diameter. Dipterocarpaceae, which are particularly diverse in genera and species, dominate rain forests west

---

[35] Wallace's Line. Imaginary line postulated by A.R. Wallace as the dividing line between Asian and Australian fauna in the Malay Archipelago. It passes between Bali and Lombok islands and between Borneo and Sulawesi, then continues south of the Philippines and north of the Hawaiian islands (*Columbia Encyclopedia* 2001).

Table 20-1. Asia: extent of ecological zones

Subregion	Total area of ecological zone (million ha)																		
	Tropical						Subtropical				Temperate					Boreal		
	Rain forest	Moist	Dry	Shrub	Desert	Mountain	Humid	Dry	Steppe	Desert	Mountain	Oceanic	Continental	Steppe	Desert	Mountain	Coniferous	Tundra	Mountain	Polar
Central Asia											1		1	135	308	99			1	
East Asia		4				8	200				140		120	65	156	307	16			
South Asia	31	58	98	119	57	12			22	5	41					5				
Southeast Asia	272	79	49	2		46	1				1									
West Asia					223	22	6	13	95	145	168	9	9	5	7					
Total Asia	303	141	146	121	280	88	208	13	116	150	351		130	210	468	418	16		1	
TOTAL WORLD	1 468	1 117	755	839	1 192	459	471	156	491	674	490	182	726	593	552	729	865	407	632	564

Note: Data derived from an overlay of FRA 2000 global maps of forest cover and ecological zones.

Table 20-2. Asia: proportion of forest by ecological zone

Subregion	Forest area as proportion of ecological zone area (percentage)																		
	Tropical						Subtropical				Temperate					Boreal		
	Rain forest	Moist	Dry	Shrub	Desert	Mountain	Humid	Dry	Steppe	Desert	Mountain	Oceanic	Continental	Steppe	Desert	Mountain	Coniferous	Tundra	Mountain	Polar
Central Asia														3		6			72	
East Asia		28				58	36				28		32	10		8	85			
South Asia	53	23	58	8		79			7		22									
Southeast Asia	55	46	79	*		56														
West Asia					49	34				5		19			26					
Total Asia	55	36	65	10		46	36	34	2		16		31	5	0	8	85		76	
TOTAL WORLD	69	31	64	7	0	26	31	45	9	2	20	25	34	4	1	26	66	26	50	2

* Estimate uncertain because of discrepances in global forest cover map.
Note: Data derived from an overlay of FRA 2000 global maps of forest cover and ecological zones.

of the Wallace Line. They contribute many (Sumatra, Malaysia), most (Borneo) or all (Philippines) of the top canopy giant trees. The main genera are *Dipterocarpus, Shorea, Dryobalanops* and *Hopea*. Other important tree families include Anacardiaceae, Ebenaceae, Leguminosae, Sapindaceae, Euphorbiaceae and Dilleniaceae. *Pometia, Canarium, Cryptocarya, Terminalia, Syzygium, Casuarina* and *Araucaria* are among the chief tree genera of forests east of the Wallace Line.

The Asian mangroves, most widely distributed in the Indonesian archipelago and the Sundarbans of Bangladesh, are richer in species than comparable formations elsewhere. Mangrove forests can reach heights of 30 to 40 m and are best developed in sheltered bays or in extensive estuaries. Conspicuous species are *Avicennia alba, A. officinalis, A. marina, Bruguiera cylindrica, B. gymnorrhiza, Ceriops decandra, Excoecaria agallocha, Rhizophora apiculata,* *R. mucronata, Sonneratia alba, S. caseolaris* and *Nipa fruticans*.

TROPICAL MOIST DECIDUOUS FOREST

This zone includes the lowlands of Sri Lanka; much of peninsular India; the hilly basin forming most of the country of Myanmar; the Red River valley and the lower foothills of the surrounding mountains in northern Viet Nam; the low plateaus on the western side of the Annamitic Range in southern Viet Nam, the Lao People's Democratic Republic and Cambodia; the plains and western foothills of the mountains in the Philippines; the low, flat, often swampy plains of the southern part of New Guinea and parts of Hainan Island and the Lezhou Peninsula in China.

Where the influence of the southwest monsoon is less, rainfall is generally between 1 000 and 2 000 mm with a dry season of three to six months. Temperatures are always high, with a mean temperature of the coldest month generally above 20°C but sometimes slightly lower, as in northern India, Myanmar or the Indochinese peninsula. In China, the southern parts of Lezhou and Hainan Island have a similar climate.

The natural vegetation is mostly deciduous or semi-deciduous forest, commonly known as monsoon forest. Many dominant trees belong to the Leguminosae, Combretaceae, Meliaceae or Verbenaceae. Dipterocarpaceae are also present, but less conspicuous than in the rain forest. Teak forests (*Tectona grandis*) in western and northern Thailand, the Lao People's Democratic Republic, Myanmar and peninsular India and sal forests (*Shorea robusta*) in eastern India and the Ganges valley are of great economic value. Tree species associated with the teak forest include *Lagerstroemia* spp., *Xylia kerrii*, *Adina cordifolia*, *Vitex* spp., *Tetrameles nudiflora*, *Afzelia xylocarpa*, *Diospyros* spp., *Sindora cochinchinensis* and *Pinus merkusii*. In the sal forests, tree species of *Dillenia*, *Terminalia*, *Adina* and *Pterospermum* are codominant. Bamboo brakes (*Dendrocalamus strictus*) are common in India and Myanmar. Extensive deciduous forests remain on hilly parts of Myanmar and some patches in northern Viet Nam on the Red River plain. In the remaining part of Indochina the zone is widely covered with deciduous dipterocarp and teak forest. In Papua New Guinea there is a different type of dry evergreen or semi-evergreen deciduous forest, characterized by species such as *Garuga floribunda*, *Protium macgregorii*, *Intsia bijuga* and *Acacia* spp. (Mimosaceae) and the presence of Myrtaceae, Proteaceae and Rutaceae.

In China, tropical moist deciduous forest is found below 700 m in basins and river valleys of the southern mountains on Hainan Island. The main species include *Heritiera parvifolia*, *Amesiodendron chinense*, *Litchi chinensis*, *Vatica hainanensis*, *Diospyros hainanensis*, *Hopea hainanensis*, *Lithocarpus fenzelianus*, *Homalium hainanensis*, *Podocarpus imbricata*. The middle layer of the rain forest often includes *Dysoxylum binectariferum*, *Sindora glabra*, *Ormosia balansae*, *Pterospermum heterophyllum*, *Gironniera subaequalis*, *Schefflera octophylla*, *Dillenia turbinata* and *Hydnocarpus hainanense*. Hill moist forest grows from 700 to 1 200 m and is composed of *Altingia obovata*, *Manglietia hainanensis*, *Michelia balansae*, *Madhuca hainanensis* and species of Fagaceae, Lauraceae, Theaceae, and Aquifoliaceae. In the central part of the island, coniferous forests grow on low mountains and hills at altitudes below 800 m. *Pinus latteri* dominates and forms second-growth pure forests or mixtures with *Liquidambar formosana*, *Chukrasia tabularis* and *Engelhardtia roxburghiana*. Mangrove forests grow along shorelines around the island except for the west coast. Tree species include *Avicennia marina*, *Rhizophora mucronata*, *R. apiculata*, *Bruguiera conjugata*, *B. cylindrica*, *Ceriops tagal*, *Sonneratia acida*, *Xylocarpus granatum* and others.

TROPICAL DRY FOREST

The zone comprises the coastal plains along the Gulf of Bengal and the northeastern part of the Deccan Plateau in India and Sri Lanka. In Myanmar, it includes the basin around Mandalay. The zone occupies the wide, flat alluvial basin of the Chao Phraya River in Thailand as well as the Korat Plateau and the Mekong River valley. In Cambodia, the area is the whole low central plain built by the lower Mekong River and the Tonle Sap. The Mekong delta in Viet Nam is part of this zone. Narrow coastal stretches also occur in southern Papua New Guinea.

These areas are sheltered from the humid winds blowing from the oceans and only partially receive, in summer, the southwest monsoon. In winter they are influenced by the dry winds of the northeast monsoon. Rainfall ranges between 1 000 and 1 500 mm, with a dry season of five to eight months. Mean temperature of the coldest month is always above 15°C, often 20°C.

Dry evergreen forest occurs on the dry eastern Coromandel Coast of India and in northern Sri Lanka. The vegetation is a stunted woody formation with *Manilkara hexandra*, *Chloroxylon swietenia*, *Albizia amara* and *Capparis zeylanica*.

Dry deciduous dipterocarp forests and woodlands are more common throughout Viet Nam, the Lao People's Democratic Republic, Cambodia and Thailand. Characteristic species include *Dipterocarpus intricatus*, *D. obtusifolius*, *D. tuberculatus*, *Pentacme siamensis* and *Shorea obtusa*. In Thailand, some of these woodlands include teak (*Tectona grandis*) and a pine species (*Pinus merkusii*).

In mixed deciduous woodlands, teak and pine occur with dipterocarps or Leguminoseae. They are found in Thailand, Myanmar, the Lao People's Democratic Republic and Viet Nam. In India, woodlands are also common but only a few dipterocarps occur, notably *Shorea robusta* and *S. talura*. In southern Papua New Guinea there are some dry deciduous forests with Myrtaceae and Eucalyptus woodland.

TROPICAL MOUNTAIN SYSTEMS

Tropical mountain systems include the eastern Himalayas; mountains stretching from Tibet to northern Indochina, the Malaysian Peninsula and the Annamitic Range; the central mountain ranges of the islands of Indonesia and the Philippines; relatively high peaks (over 2 000 m) in India and Sri Lanka; and mountains in the southwestern Arabian Peninsula.

Most tropical mountains of Asia, i.e. those reaching at least 1 500 to 2 000 m, have a wet climate. The Himalayas have a subtropical northwestern part and a tropical wet southeastern part. Nepal is a transitional region between these two areas. In all tropical mountains, between 1 000 to 1 500 m and 4 000 m, annual precipitation is more than 1 000 mm, sometimes more than 2 000 mm. There is a pronounced dry season of three to five months in the submontane zone of the eastern Himalayas, with the mean temperature of the coldest month above 15°C. Everywhere else, the dry season, if it occurs, is very short. The mean temperature of the coldest month rapidly decreases with increasing elevation. Above 4 500 to 5 000 m there is permanent snow.

The mountains in the southwestern Arabian Peninsula have a drier climate. Annual rainfall ranges from 400 mm in the lower foothills to 800 mm on the higher escarpments. There are two rainy seasons, March to April and July to September.

Forests generally cover the Himalayan slopes up to 4 000 m. Beginning around 1 000 m, tropical lowland forest is replaced by an evergreen forest with *Castanopsis, Schima, Engelhardtia* and *Lithocarpus* species and, locally, Himalayan chir pine forest (*Pinus roxburghii*). From 2 000 to 3 000 m is a belt of evergreen oak forest, followed higher up by coniferous forest (*Abies* spp., *Tsuga* spp.).

In Myanmar and Thailand, evergreen oak forests are found above 1 500 m with pine forest. An evergreen forest with Lauraceae and Fagaceae grows from 1 500 to 2 000 m in the Lao People's Democratic Republic and Viet Nam while a mixed broad-leaved/coniferous forest takes over above this elevation. Woodlands with oaks and pines also occur at high altitude. In Thailand, northern Lao People's Democratic Republic and Viet Nam, these forests have been affected by shifting cultivation and mosaics of forests and thickets predominate at lower elevations.

In Malaysia, as well as in Indonesia and the Philippines, the montane (evergreen) rain forest still covers relatively large areas. This forest is best developed between 1 400 and 2 400 m altitude and is characterized by Fagaceae (of the genera *Castanopsis, Lithocarpus* and *Nothofagus* in Papua New Guinea), Lauraceae, Juglandaceae (*Engelhardtia* spp.), Magnoliaceae (*Casuarina junghuhniana*), conifers (*Podocarpus* spp. and *Pinus* spp. in Sumatra), *Dacrydium* spp., *Araucaria* spp., *Libocedrus* spp., *Phyllocladus* spp. and others. In the subalpine zone, between 2 400 and 4 000 m, dense or discontinuous montane thickets are found. Coniferous forest containing *Araucaria* spp., *Podocarpus* spp. and/or *Libocedrus* spp. often occurs in this belt. The alpine zone extends above 4 000 m.

Mountains are the only locations on the Arabian Peninsula where forests grow. From around 1 000 to 1 500-1 800 m is *Acacia-Commiphora* deciduous scrub or savannah. From 1 500-1 800 to 2 000 m is evergreen woodland or forest with *Olea africana, Podocarpus* spp., *Olea chrysophylla, Trochonanhus comphoratus* and other species while from 2 000 to 3 000 m is coniferous forest of *Juniperus procera*.

SUBTROPICAL HUMID FOREST

This ecological zone has its main distribution in southeastern China south of the Yangtze River, the southern tip of the Republic of Korea and the southern half of Japan. There are two distinct small geographic units in the Near East, humid forests at the foot of the Caucasus Mountains extending westward along the Black Sea and in the foothills of the Talysh Mountains at the Caspian Sea.

Winters are mild to warm and summers are hot and wet. Northerly cold fronts from Siberia heavily influence winter temperatures while in summer the Pacific monsoon brings large amounts of precipitation to the region.

Annual mean temperatures in China and the Korean Peninsula range from 15° to 17°C in the northern part of the zone to around 21°C in the south and southeast. Annual precipitation varies between 800 and 1 300 mm throughout the northern region, but further south it becomes wetter, up to 1 800 mm and sometimes 2 500 mm in low mountains. Annual rainfall diminishes towards the west, away from the coast. In the northern and central parts of the zone rainfall is evenly distributed throughout the year. In the south, most of the rain falls between May and October. A dry season from November to April is distinctive. The island of Taiwan Province of China is under the strong influence of the maritime monsoon climate, with higher average temperatures and greater rainfall.

The climate in Japan is greatly influenced by the monsoon. Generally speaking, the summers are very hot and the winters rather cold with snow and frosts. Mean annual temperature is around 14° to 17°C. The yearly precipitation over most of Japan is much greater than over the continent. Mean annual precipitation ranges from around 1 200 mm to more than 2 500 mm locally with two peak rainy seasons, "Baiu" (June to July) and "Shurin" (autumn rain).

The climate of the coastal plains and lowlands south of the Black Sea and the Caspian Sea is warm-temperate with an annual average temperature around 14° to 15°C. Large amounts of precipitation fall throughout the year (1 500 to 2 000 mm, locally up to 4 000 mm). In the Colchis area the climate is mild owing to the influence of the Black Sea (yearly amplitude of the monthly average temperatures 15° to 19°C), with mild winters (average temperature of the coldest month 5° to 6°C).

Two types of woody vegetation prevail south of the Yangtze River in eastern China, pine forest and deciduous forest mixed with evergreen species. The dominant conifer here is *Pinus massoniana*. The mixed deciduous evergreen forests, a unique subtropical vegetation type, include *Quercus acutissima, Q. variabilis, Q. dentata, Q. glandulifera, Q. fabrei, Liquidambar formosana, Pistacia chinensis, Ulmus parvifolia, Zelkova schnederiana, Celtis sinensis, Dalbergia hupeana, Albizia macrophylla, Tilia miqueliana, Cyclobalanopsis glauca, C. myrsinaefolia, Castanopsis sclerophylla, C. carlesii, Lithocarpus glabra, Phoebe sheareri, Cinnamomum chekiangense, Machilus thunbergii* and *Ilex purpurea*. Bamboo stands are common in the region, with more than 20 species of *Phyllostachys* of which *P. edulis* is most common.

The western mid-latitude mountains feature conifer forests dominated by such species as *Abies chensiensis, A. fargesii, A. ernestii, Picea complanata* and *P. neoveitchii* as well *as Pinus armandii, P. henryi* and Platycladus orientalis. *Pinus tabulaeformis* and *P. bungeana* forests are distributed over western portions of the Qinling Range. Deciduous broad-leaved forests contain more than 300 woody species, the major trees including *Quercus acutissima, Q. variabilis, Q. liaotungensis, Q. aliena* var. *acuteserrata, Q. dentata, Q. glandulifera, Betula albo-sinensis* and *Toxicodendron vernicifluum*.

In the southeastern low mountain and hill region as well as the Sichuan Basin the representative vegetation is typically evergreen broad-leaved forests as well as coniferous forests. Distributed across the entire region is an evergreen broad-leaved laurel forest of *Cyclobalanopsis glauca*. There are *Castanopsis eyrei* and *C. fargesii* in the central to northern parts of the area and *C. hystrix* and *C. lamontii* from the western to eastern sides of the Nanling Mountains. Conifer forests are primarily those of *Pinus massoniana, P. taiwanensis* and *Cunninghamia lanceolata*. The region is one of the most important bamboo regions in China. There are two million hectares of *Phyllostachys edulis* in the area. Several other species from the same genus, *P. bambusoides, P. nidularis, P. mannii, P. nigra* var. *henonis* and *P. heteroclada*, also occupy a broad range.

On the Yungui Plateau in southern and southwestern China, regional evergreen broad-leaved laurel forests are similar to those of eastern areas, consisting the same genera, *Castanopsis, Lithocarpus, Cyclobalanopsis, Cinnamomum* and *Phoebe*, but often with different species. The conifer forest here is dominated by *Pinus yunnanensis*, which grows widely from 1 000 to 3 100 m, with pure stands usually from 1 600 to 2 800 m.

The forests of Taiwan Province of China are distributed along a distinct gradient from the coastal region to the high mountains. Mangrove forests occur along shallow shorelines. Southern subtropical rain forest covers low hills (below 500 m) in northern Taiwan. Major upper-storey species include *Cyclobalanopsis glauca, Castanopsis carlesii, C. kusanoi, Ficus microcarpa, Cryptocarya chinensis, Acer oblongum, Elaeocarpus japonica, Ilex rotunda* and *Engelhardtia roxburghiana*. Evergreen broad-leaved forests extend to 500 to 1 800 m slopes and include *Castanopsis kawakamii, C. fargesii, C. uraiana, Lithocarpus brevicaudatus, L. ternaticupula, L. amygdalifolius* and *Cinnamomum camphora*.

The predominant natural vegetation in Japan is evergreen broad-leaved forest of several types. The major tree species are *Machilus thunbergii* and *Castanopsis cuspidata* in coastal areas and *Cyclobalanopsis glauca, C. gilva. C. salicina,*

C. myrsnaefolia and *C. acuta* (evergreen oaks) in inland areas. Conifers, such as *Podocarpus macrophyllus*, *P. nagi* and *Torreya nucifera* also occur in these forests. At higher elevations, *Tsuga sieboldii* and *Abies firma* grow in mixture with the broad-leaved evergreen species. The medium to lower strata contain small trees and shrubs of such broad-leaved evergreen species as *Aucuba japonica*, *Damnacanthus indicus* and *Neolitsea sericea*. Secondary forests of *Pinus densiflora*, *Quercus serrata* and *Quercus acutissima* now cover large areas. Natural stands of *Pinus densiflora* are restricted to extreme habitats.

The two forests in the Near East, although of relatively small extent, are the most diverse and productive in the region. Both forests are dense broad-leaved summer-green types. The forest canopy consists of oak species (in the west *Quercus imeretina*, *Q. hartwissiana*, in the east *Q. castaneifolia*) and also *Castanea sativa*, *Pterocarya pterocarpa*, *Diospyros lotus* and *Fagus sylvatica* subsp. *orientalis*. *Zelkova carpinifolia*, *Carpinus betulus* and some *Acer* species are present in the subcanopy layer. At higher altitudes mixed hornbeam-oak forests (*Quercus iberica*, *Carpinus orientalis*, *Fagus sylvatica* subsp. *orientalis* and *Castanea sativa*) replace this vegetation. Small areas of swamp and fen forests (*Alnus barbata*, *A. subcordata*, *Pterocarya pterocarpa*) occur along riverbanks and estuaries.

SUBTROPICAL DRY FOREST

This zone is confined to the Near East and occupies a relatively narrow belt along the Mediterranean Sea and the low hills running parallel to the coast. The northern part of the Jordan-Arava Rift Valley is also included. The zone has a typical Mediterranean climate with mild, humid winters and dry, moderately hot summers. Annual rainfall ranges from around 400 to 800 mm, decreasing from north to south.

Various types of pine forest occur, with either Aleppo pine (*Pinus halepensis*), *P. brutia* or *P. pinea* as dominant species. Otherwise, the typical Mediterranean woody maquis vegetation dominates this zone. *Ceratonia-Pistacia lentiscus* maquis dominates the coastal plains up to around 200 m, while *Quercus calliprinos-Pistacia palaestina* maquis is the main vegetation from 200 to 1 000-1 200 m. Other important tree species include *Quercus ithaburensis*, *Q. infectoria*, *Q. ithaburensis*, *Q. coccifera*, *Laurus nobilis*, *Arbutus andrachne*, *Cercis siliquastrum*, *Juniperus phoenicea*, *Myrtus communis*, *Olea europaea*, *Phillyrea* spp., *Pinus halepensis* and *P. brutia*.

SUBTROPICAL STEPPE

This zone is confined to western Asia, mainly the Near East but also in Afghanistan and Pakistan. The climate is semi-arid. Annual rainfall ranges from about 200 to 500 mm and falls during winter in the Near East. Eastern Afghanistan and Pakistan receive most of their rainfall from June to September. Although differences in temperature between seasons are relatively high, winters are not severe.

The vegetation consists of low shrubs and grasses interspersed with sparse trees, particularly at wetter locations. At higher, more humid locations in the Near East a forest-steppe can be found with trees such as *Amygdalus korshinskyi*, *A. arabica*, *Acer monspessulanum*, *Pistacia atlantica*, *Pyrus bovei*, *Rhamnus palaestina* and *Crataegus aronia*. In Pakistan, the woody steppe vegetation consists of shrubs and small trees. Main species are *Acacia modesta* and *Olea cuspidata* accompanied by *Ziziphus jujuba*, *Dodonea viscosa* and others. Owing to prolonged human activity the original vegetation has been considerably altered.

SUBTROPICAL MOUNTAIN SYSTEMS

Subtropical mountain systems cover extensive areas in Asia in a nearly continuous west-east belt from the mountains and highlands of Turkey to the eastern reaches of the Himalayas in southern China.

The climate of the Near Eastern mountain systems is extremely diverse, both in temperature and rainfall. Winter precipitation is predominant, ranging from 500 to 1 400 mm. The rainy season is from around September to May or June, while the rest of the summer is dry and hot.

All along the Himalayan ranges the rainfall increases from west to east and the climatic regime changes gradually from Mediterranean to typical monsoon types. The rain also decreases from the outer to the inner parts of the ranges. At the submontane and montane levels, rainfall ranges from less than 1 000 to 1 500 mm, with at least one or two dry months even up to seven or eight. The mean temperature of the coldest month varies from around 15°C in the submontane zone to less than 10°C above 2 000 m. Snow occurs above 3 000 m, with frequent winter frost. Precipitation is 500 to 1 000 mm.

China's subtropical mountains comprise the central interior highlands and southwestern high mountains. The region has a harsh climate at high elevations but warmer, moist conditions in the medium to low mountains. Annual mean temperature ranges from 8° to 18°C in eastern areas with the January mean above 0°C and the extreme low at -20°C. Annual rainfall is 800 to 1 200 mm, up to 3 000 mm locally. A dryer and colder climate prevails towards the western higher mountain areas. In southern Tibet, mean annual temperatures in the mountains are 6° to 8°C, average in winter is 2° to 4°C and in summer around 15°C. Annual precipitation ranges from 300 to 700 mm. River basins in the south at 500 m elevation are relatively warm and moist with annual rainfall more than 1 200 mm and a distinct dry-rainy seasonal change as a result of the impact from the Indian Ocean monsoon.

Mediterranean mountain vegetation is diverse and includes dense humid forests, shrubland, forest-steppe and treeless grass steppe. The forests can be either deciduous broad-leaved or coniferous. In Lebanon and the Syrian Arab Republic a summer-green oak forest is found between 1 000 and 1 600 m altitude. The forest climax is *Quercus cerris* accompanied by *Quercus boissieri* and fragments of *Quercus libani*. In western Turkey, black pine (*Pinus nigra*) dominates this belt. From 1 500 to 2 000-2 200 m, there is a subalpine coniferous forest with cedar (*Cedrus libani*), fir (*Abies cilicica*) and juniper (*Juniperus excelsa*). Juniper forest occupies the drier areas. Above 2 200 m, alpine dwarf shrubs and meadows occur.

Forest-steppe and steppe vegetation occupy major parts of the central highlands and plateaus of Turkey and Iran. At humid locations grows a deciduous oak forest dominated by *Quercus persica* or other oak species, often in combination with juniper (*Juniperus* spp.). In the valleys there are *Fraxinus oxycarpa*, *Platanus orientalis* and *Ulmus campestris* as well as various species of *Populus*, *Salix*, *Tamarix*, etc. Tree steppe with pistachio, almond and juniper occur at sub-dry locations.

Well-developed forest grows on the higher slopes of the mountains bordering the Black and Caspian Seas. At both locations we find summer-green dense forest between approximately 800 and 2 000 m. The Hyrcanian montane forest is *Fagetea hyrcanica* with *Fagus orientalis*, accompanied by *Carpinus betulus*, *Acer insigne* and *Quercus castaneifolia*. The Euxinian montane forest is composed of deciduous broad-leaved trees and conifers with species of oak, fir and pine.

Vegetation of the northwestern and western Himalayas is extremely diverse. In southern Afghanistan, open deciduous woodland is the dominant vegetation at medium high altitudes. *Pistacia atlantica* woodland 4 to 6 m high occurs from around 1 100 to 1 800-2 000 m. Between 2 000 and 2 800 m, *Amygdalus* communities prevail (*Amygdalus* cf. *communis*, *A. kuramica* and *Fraxinus xanthoxyloides*).

In eastern Afghanistan and Pakistan, different types of west Himalayan evergreen sclerophyllous forests and woodlands occur. Woodland of *Quercus baloot* is most extensive and occurs at an altitude of around 1 300 to 2 000 m. Depending on the water supply, they are either open woodlands with stunted trees 3 to 6 m high or true forests with trees 15 m or more in height. *Quercus dilatata* and *Quercus semecarpifolia* communities are confined to the higher parts of wet mountains. The first dominates between 1 900 and 2 400 m, the latter from 2 400 to 2 900 m. Both species form rich, mesophylous forests 8 to 20 m in height.

Coniferous forests are the most extensive mountain forests. Chir pine (*Pinus roxburghii*) forests dominate the lower mountain slopes from 900 m up to 1 700 to 2 000 m altitude, accompanied by some oaks (*Quercus dilatata*) and other broad-leaved species. West of the Indus, *Pinus gerardiana* forest is found between 2 000 and 2 500 m. A dense forest of *Cedrus deodara* is found between 2 500 and 3 100 m in areas with 450 to 650 mm annual rainfall. Other trees of this forest are *Picea morinda*, *Pinus excelsa* and *Abies webbiana*. With decreasing rainfall, *Juniperus seravschanica* gradually replaces the cedar. East of the Indus, increased precipitation favours blue pine (*Pinus wallichiana*). A dense, mixed forest dominated by *Picea smithiana* and *Abies webbiana* grows in high rainfall areas (greater than 800 mm per year) between 2 900 and 3 200 m. In areas with winter rains, *Juniperus* spp. woodlands dominate at altitudes ranging from 1 500 to 3 000 m. Further eastward, under the monsoon-influenced climate, *Juniperus* woodland occurs above 3 000 m. Typical subalpine woody vegetation, ranging between 3 000 and 4 000 m altitude, is a mixture of conifers and broad-leaved low trees or shrubs. Main species are *Abies webbiana*, *Abies spectabilis*, *Betula utilis* and *Rhododendron campanulatum*.

In Azad Jammu and Kashmir, from 1 500 to 3 000 m, coniferous forests occur with *Pinus excelsa* and *Cedrus deodara*, mixed with thickets and grasslands. Above 3 000 m they give way to mixed forests and woodlands with *Betula* and *Abies*. To the east, from Himachal Pradesh to central Nepal, the submontane level from 1 000 to 2 000 m is characterized by open woodlands with *Pinus roxburghii*. Above 2 000 m, dense evergreen forests occur, with oaks or conifers (*Cedrus deodara, Picea* spp., *Pinus excelsa*), then *Abies-Quercus* forests above 3 000 m.

The alpine conifer forests of China are dominated by *Abies faberi* and *Picea complanata*, usually from 2 000 to 3 000 m. The species are often associated with *Tsuga chinensis, Picea complanata, Acer* spp., *Tilia* spp. and *Betula albo-sinensis* but form pure stands at higher altitudes, up to 4 000 m. Conifer forests that grow in pure stands on low and medium altitude mountains are *Pinus massoniana, P. yunnanensis, Cunninghamia lanceolata* and *Cupressus funebris*. Further west and at higher elevations are alpine conifer forests of highly cold-tolerant species dominated by *Picea balfouriana* and *Abies squamata*, which often form pure stands on north-facing slopes from 3 000 to 4 000 m.

Abies spectabilis and *Picea linzhiensis* are the dominant species in the alpine conifer forests of southern Tibet. The former species, also called Himalayan fir, is found in pure stands or in association with *Abies georgei* and *Picea likiangensis* in the southern part at 3 100 to 4 000 m on north-facing slopes. *Picea linzhiensis* forms pure stands in the southeast from 2 900 to 3 900 m. Conifer forests at medium elevations are dominated by *Pinus griffithii*. Southern subtropical monsoon rain forest occurs on valley lands under 500 m and is composed of *Shorea robusta, Terminalia catappa, Tetrameles nudiflora* and *Dillenia pentagyna*.

In central Taiwan Province of China, coniferous and broad-leaved mixed forests occupy mountain slopes from 1 800 m up to 3 000 m altitude. Major species include *Chamaecyparis obtusa* var. *formosana* and *C. formosensis*. Broad-leaved components include *Cyclobalanopsis stenophylloides, Trochodendron aralioides, Acer formosum* and *Sassafras randaiense* as well as lower-layer species from the genera *Eurya, Ilex, Symplocos* and *Hydrangea*. Alpine conifer forests occur in the Yushan and Bishan Mountains at elevations generally above 3 000 m with *Abies kawakamii* as the major species.

TEMPERATE CONTINENTAL FOREST

This zone includes the temperate forests of China, the Korean Peninsula and Japan. In China, the annual mean temperature varies greatly, from 2°C in the north to 14°C in the south. Climate is distinctly seasonal; winter is relatively long (four to seven months) and spring short (one to three months). In the northern part, warm summers have monthly average temperatures above 20°C in the warmest month and a growing season lasting 100 to 150 days. Annual precipitation is between 400 and 800 mm for most of the area to 1 000 mm over the southeastern part of the zone. In the southern part, mean temperature in the coldest months still falls below 0°C. Warm summers bring the average temperature up to 24°C in the warmest month except in the mountains. The growing season lasts 200 days. Annual precipitation of 600 to 1 000 mm is unevenly distributed over the year. Coastal areas experience higher rainfall, 1 000 to 1 400 mm. Similar climatic conditions prevail on the Korean Peninsula and in northern Japan.

The northern part of the zone (in northeastern China) features well-stocked *Pinus koraiensis* mixed forests on low mountains of 400 to 600 m. Associated species include *Picea jezoensis* var. *microsperma, Picea koraiensis, Abies nephrolepis, Betula platyphylla, B. costata, B. davurica, Populus davidiana, Quercus mongolica, Tilia amurensis, Acer mono, A. ukurunduense, A. tegmentosum, Ulmus davidiana* var. *japonica, Fraxinus mandshurica* and *Juglans mandshurica*. Once disturbed, the mixed forests usually degrade into *Populus davidiana* and *Betula platyphylla* second-growth forests. *Pinus koraiensis* mixed forest in Changbaishan has a similar composition but more species, adding *Abies holophylla, Pinus sylvestris* var. *sylvestriformis, P. densiflora, Taxus cuspidata, Thuja koraiensis, Fraxinus rhynchophylla* and several maple and linden species.

In contrast to the generally forested eastern part of northeastern China, the rest of the zone has little tree cover left. Pockets of natural second-growth forests exist, represented by *Pinus densiflora, P. tabulaeformis* and several deciduous oaks, including *Quercus acutissima, Q. variabilis, Q. dentata, Q. aliena, Q. serrata, Q. liaotungensis* and *Q. mongolica*. Planted species in the countryside are mostly *Populus, Salix* and *Ulmus* species and *Sophora japonica, Ginkgo biloba, Platycladus orientalis, Sabina chinensis, Paulownia fortunei, Catalpa bungei, Castanea mollissima, Diospyros kaki, Ziziphus jujuba, Toona sinensis, Ailanthus altissima* and *Robinia pseudoacacia*. Cultivated bamboo stands are scattered in the plains, mostly *Phyllostachys glauca, P. vivax, P. bambusoides* and *P. propinqua*. The region also has reported some successful agroforestry experiments using fast-growing timber species, *Populus* and *Paulownia* for instance.

The temperate forests of Japan are deciduous, summer-green, broad-leaved forests dominated by beech. The main trees are *Fagus crenata, Kalopanax septemlobus, Tilia japonica, Quercus mongolica* var. *grosseserrata, Acer mono,* etc. Moist habitats in valley bottoms and on alluvial fans support *Pterocarya rhoifolia* forests with *Ulmus laciniata, Athyrium pycnocarpon, Acer mono, Dryopteris crassirhizoma* and others. Habitats with a high water table in the lowlands of northern Honshu and Hokkaido support *Alnus japonica* forest.

TEMPERATE STEPPE

This ecological zone encompasses the vast steppes of Central Asia, occupying the eastern part of Inner Mongolia in China and central and eastern Mongolia.

The zone has a long, cold winter and a short, but warm, summer. Annual average temperatures vary between 2° and 10°C, with mean temperatures of the coldest month (January) ranging from -10° to -20°C. Mean temperature reaches 24°C in the warmest summer month. The growing season lasts 100 to 175 days. Annual rainfall ranges from 200 to 400 mm, locally up to 600 mm, and the maximum occurs during the second half of summer. Spring, as a rule, is dry.

Natural vegetation is primarily grass and shrub steppe. In some areas, pockets of woodland can be found. Tree species are represented by *Pinus tabulaeformis, P. bungeana, Picea wilsonii, P. meyeri* and *Larix principis-rupprechtii*, individually forming pure stands or sometimes admixed with *Abies nephrolepis. Populus davidiana* and *Betula platyphylla* come in from the northeast to form second-growth pure or mixed stands when spruce forests are disturbed while *Populus cathayana* is common in valleys and lowlands.

TEMPERATE MOUNTAIN SYSTEMS

The vast mountain systems of Central Asia, including the Tibetan Plateau in China and the Altai and Khangai mountain systems of Mongolia, comprise this ecological zone. The mountains of Japan also form part of the zone.

In the lower mountains of north-central China, mean annual temperature decreases from 14°C in the warmer eastern low hills to 8°C in the cooler western highlands. The difference in the July mean temperature is 20° versus 26°C between east and west, whereas January varies between 0° and -10°C. Similarly, mean annual precipitation typically averages 800 to 300 mm between east and west, most of which falls during summer. Nevertheless, this transitional region is seasonally moist enough to support monsoon vegetation.

On the Tibetan Plateau, temperature distribution generally follows elevation contour lines. Mean annual temperature goes from the 6° to 10°C range around 3 000 m, to 3° to 7°C above 4 000 m, to below -2°C above 5 000 m. Annual mean precipitation follows an east-west gradient from 800 mm on the eastern rim of the plateau to less than 50 mm in the west near the Pakistan-Afghanistan border.

The climate of the Mongolian mountain systems is characterized by widely ranging temperatures, both throughout the year and during the day. Annual precipitation ranges from approximately 200 to 600 mm, most of it falling during the second half of summer.

The transitional region of eastern China, including the Yellow Loess Plateau, has only limited natural forests, mostly in the high, inaccessible mountains. These forests are characterized by *Pinus tabulaeformis, P. bungeana, Picea wilsonii, P. meyeri* and *Larix principis-rupprechtii*, individually forming their own pure stands or sometimes mixed with a small amount of *Abies nephrolepis*. On the Yellow Loess Plateau and the surrounding areas, local residual woodlands are scattered with such similar species as *Pinus tabulaeformis* but also *P. armandii, Platycladus orientalis, Sabina chinensis, Quercus liaotungensis, Q. baronii, Populus davidiana, Betula platyphylla, Fraxinus chinensis, Toxicodendron vernicifluum* and *Zelkova sinica* as well as *Acer* and *Tilia* species.

Natural forests are better preserved in the western, higher mountains in the provinces of Gansu, Shanxi and Sichuan. Both conifer and broadleaf forests are present in these mountains. *Abies faxoniana* and *Picea asperata* dominate alpine conifer forests at 2 500 to 3 800 m. Conifer species that prefer a warmer environment, such as *Picea wilsonii, P. brachytyla, P. complanata, Tsuga chinensis* and *T. dumosa*, occupy lower elevations of 2 000 to 3 000 m or sometimes at 3 400 m, forming pure stands. Among the medium-elevation conifer forests are *Pinus tabulaeformis* and *Cupressus chengii* in pure stands. They grow from around 1 300-1 400 m to 2 100 m. *Pinus armandii* forests can extend up to 2 700 m. Deciduous broad-leaved forests are less prominent. *Betula platyphylla*, *B. albo-sinensis*, *B. utilis* and *Populus davidiana* are the most common species on the 2 600 to 3 500 m slopes, associated with *Tilia chinensis*, *Acer* spp., *Dipteronia sinensis*, *Populus cathayana* and *P. purdomii* in mixed forests.

There is a great diversity of mountain vegetation in Mongolia. The forest belt mainly contains larch forests, sometimes mixed with Siberian cedar or stone pine (*Pinus sibirica*) and spruce or fir. On sandy sediments on the lower slopes pine stands dominate and, together with larch, form the forest-steppe belt. In Mongolian-Altai the forest belt is often absent. The forest belt of the Khangai Mountains is in the range of 1 800 to 2 300 m with larch stands. Thickets of *Salix* spp. and *Potentilla fruticosa*, with occasional larch, cover the broad river valleys.

In Japan, the lower mountain zone is covered with deciduous beech forest dominated by *Fagus crenata* and *Quercus crispula*. The subalpine belt supports coniferous forests with *Abies mariesii* and/or *A. veitchii*. The altitudinal lower limit of the coniferous forests becomes gradually higher southwards, ranging from 700 m in northern Honshu to 1 500 m in central Honshu. Mixed forest of *Thuya standishii* and *Tsuga diversifolia* is present on ridges with shallow soils in the subalpine region of Honshu. *Betula ermanii* and *Alnus maximowiczii* are deciduous trees found in the subalpine and alpine regions. Mixed or pure stands are developed on boulders and shallow soils along snow valleys and on subalpine volcanic habitats. Prevailing coniferous forests on Hokkaido are dominated by *Picea jezoensis* and *Abies sachalinensis*, sometimes accompanied by *Picea glehnii*.

BOREAL CONIFEROUS FOREST

This zone is confined to the northern part of northeastern China. The zone is essentially Daxinganling (the Greater Xingan Range), a medium-altitude plateau. The zone has a rigorous climate with a long, cold winter. Mean annual temperature ranges between -1° and -6°C, the mean minimum of the coldest month is below -25°C and the extreme low is below -45°C. Soils are either permafrost or frozen for most of the year. Relatively warm summers bring a monthly mean temperature of 15°C in the warmest months with a growing season of about 90 days. Most of the annual mean precipitation of 500 mm falls during the summer season.

Forests in this zone are mostly simple, natural stands of three types. First, *Larix gmelini* is widely spread on 300 to 1 100 m slopes. It forms large, pure stands as well as mixed stands with *Betula platyphylla*, *Populus davidiana* and *Quercus mongolica*. Second, *Pinus sylvestris* var. *mongolica* forests are mostly distributed in the north between 300 to 900 m. They mostly form small pure stands. Third, *Pinus pumila* dominates on mountain tops or ridges of 1 100 to 1 400 m, forming low stands. In addition, *Pinus sibirica* forest is found in the northwestern portion of the Daxinganling. Among deciduous broad-leaved forests, *Betula platyphylla* and *Populus davidiana* grow as natural second-growth forests following disturbance of *Larix gmelini*, either in pure stands or in mixtures. *Quercus mongolica* forests are found in the south on dry, south-facing slopes below 600 m. Deciduous broad-leaved mixed forests, composed of *Populus suaveolens*, *Chosenia arbutifolia*, *Ulmus davidiana* var. *japonica* and *Salix* spp. are scattered along the Heilongjiang River and its tributaries.

BIBLIOGRAPHY

Blasco, F., Bellan, M.F. & Aizpuru, M. 1996. A vegetation map of tropical continental Asia at scale 1:5 million. *Journal of Vegetation Science.*

Columbia Encyclopedia. 2001. *The Columbia Encyclopedia.* Sixth edition. www.bartleby.com/65/wa/

Ecological Laboratory of Toulouse (LET). 2000. *Ecofloristic zones and global ecological zoning of Africa, South America and Tropical Asia*, by M.F. Bellan. Rome, FAO.

FAO. 1989. *Classification and mapping of vegetation types in tropical Asia.* FAO, Rome.

Gunin, P.D. et al. (eds). 1999. *Vegetation dynamics of Mongolia.* Geobotany 26. Dordrecht, The Netherlands, Kluwer Academic Publishers.

Numata, M. et al. (eds). 1975. *Studies in conservation of natural terrestrial ecosystems in Japan. Part 1. Vegetation and its conservation.* Japan International Biological Program (JIBP) Synthesis, Volume 8. Tokyo.

Satoo, T. 1983. Temperate broad-leaved evergreen forests of Japan. In *Temperate* *broad-leaved evergreen forests.* Ed. J.V. Ovington. Ecosystems of the World 10. Amsterdam, Elsevier.

UNESCO/FAO. 1969. *Vegetation map of the Mediterranean zone.* Explanatory notes. Arid zone research, XXX. Rome.

Whitmore, T.C. 1981. *Wallace's line and plate tectonics.* Oxford, UK, Monographs on biogeography.

Whitmore, T.C. 1989. Southeast Asian tropical forests. In *Tropical rain forest ecosystems: biogeographical and ecological studies.* Eds. H. Lieth & M.J.A. Werger. Ecosystems of the World 14b. Amsterdam, Elsevier.

Zhu, Z. 1992. *Geographic distribution of China's main forests.* Nanjing, China, Forestry University.

Zohary, M. 1973. *Geobotanical foundations of the Middle East.* Volumes 1 and 2. Stuttgart, Germany, Gustav Fischer Verlag.

Chapter 21
West Asia

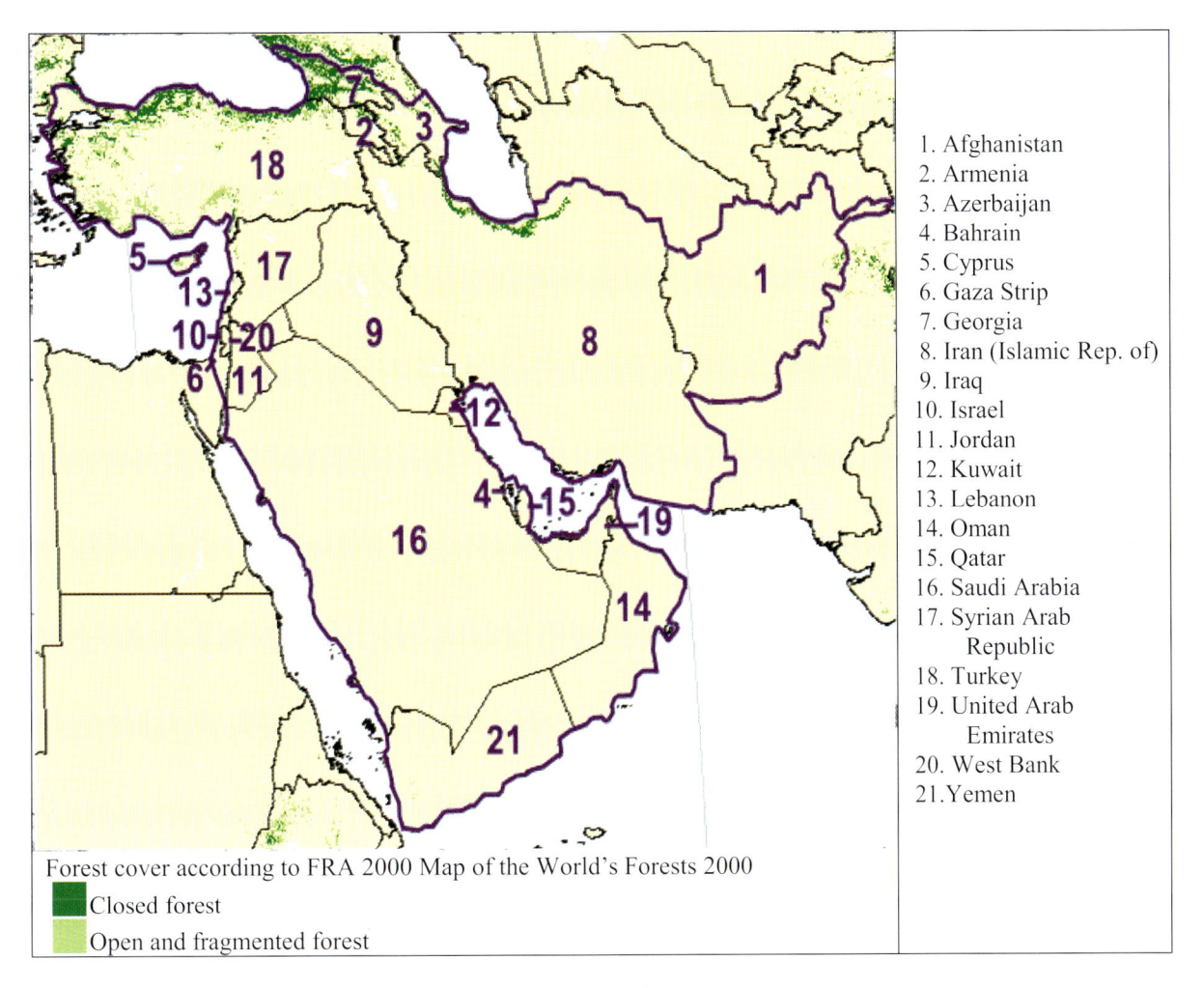

```
1. Afghanistan
2. Armenia
3. Azerbaijan
4. Bahrain
5. Cyprus
6. Gaza Strip
7. Georgia
8. Iran (Islamic Rep. of)
9. Iraq
10. Israel
11. Jordan
12. Kuwait
13. Lebanon
14. Oman
15. Qatar
16. Saudi Arabia
17. Syrian Arab Republic
18. Turkey
19. United Arab Emirates
20. West Bank
21. Yemen
```

Forest cover according to FRA 2000 Map of the World's Forests 2000
- Closed forest
- Open and fragmented forest

Figure 21-1. West Asia: forest cover map

The following countries and areas comprise the West Asia subregion: Afghanistan, Armenia, Azerbaijan, Bahrain, Cyprus, Gaza Strip, Georgia, Iran (Islamic Republic of), Iraq, Israel, Jordan, Kuwait, Lebanon, Oman, Qatar, Saudi Arabia, Syrian Arab Republic, Turkey, United Arab Emirates, West Bank and Yemen[36] (Figure 21-1).

In general, these countries and areas are among those that are forest poor, with only 3.2 percent of the total areas under forest cover and less than 1 percent of the world's forest cover. The forest area per capita is 0.1 ha, which is very low, only 15 percent of the world average.

Owing to the prevailing arid conditions of the region, forests mostly comprise open woodlands and lands with scattered trees and xerophytic shrubs. However, in the highlands of Cyprus, Turkey, the Caspian Sea, Georgia, Armenia, Azerbaijan and Afghanistan, temperate and moist forests are found.

In countries lacking natural forests, fast-growing and multipurpose tree species such as *Eucalyptus* spp., *Casuarina* spp., poplars and acacias are planted in the form of windbreaks or shelterbelts and used in agroforestry systems. In some countries that have natural forests, such plantations provide significant amounts of wood. In Turkey, 4 million cubic metres of wood per

---
[36] For more details by country, see www.fao.org/forestry

annum are produced, mainly from poplar plantations (Heywood 1997).

FOREST RESOURCES

The land area of the subregion is about 5.4 percent of the global land area. The total forest area is about 3.2 percent of the subregion's land area and less than 1 percent of the world's forests. Only six countries of the region have more than 1 million hectares of forest land. The largest area is in Turkey, with about 37.5 percent of the subregion's forests, followed by Iran, Georgia, Saudi Arabia, Azerbaijan and Afghanistan, which have about 24.5 million hectares and 89.7 percent of the total forest area in the subregion. The remaining countries have about 2.8 million hectares. Bahrain, Kuwait, Oman, Qatar and the United Arab Emirates have only plantations (Table 21-1).

Various factors external to the forestry sector have had a significant impact on forest resources. Among these are urbanization, economic changes and conflicts. Many countries of the subregion are undergoing rapid urbanization, including both seasonal and permanent migration of rural populations to urban areas. Economic difficulties in some countries have hindered efficient conservation and sustainable management of natural resources, including forests. National and regional disputes and wars have also caused serious forest resource degradation in some countries of the region such as Afghanistan, Iraq and Lebanon (FAO 1998).

The survey methods and quality of information vary among countries. Afghanistan carried out a systematic forestry inventory, published in 1993, based on remote sensing images from 1989 to 1991, with maps and technical reports. For Iran, a survey based on satellite images, aerial photos and a field survey was carried out for the Caspian forests and central Zagros in 1999. For other parts of the country a sample inventory was used. The information on forest cover for Yemen was done using satellite imagery, aerial photos and fieldwork. Data were published in 1993. For Iraq and Lebanon,

Table 21-1. West Asia: forest resources and management

Country/area	Land area	Forest area 2000					Area change 1990-2000 (total forest)		Volume and above-ground biomass (total forest)		Forest under management plan	
		Natural forest	Forest plantation	Total forest								
	000 ha	000 ha	000 ha	000 ha	%	ha/capita	000 ha/year	%	m³/ha	t/ha	000 ha	%
Afghanistan	64 958	1 351	-	1 351	2.1	0.1	n.s.	n.s.	22	27	-	-
Armenia	2 820	338	13	351	12.4	0.1	4	1.3	128	66	351	100
Azerbaijan	8 359	1 074	20	1 094	13.1	0.1	13	1.3	136	105	1 094	100
Bahrain	69	n.s.	0	n.s.	n.s.	-	n.s.	14.9	14	14	-	-
Cyprus	925	172	0	172	18.6	0.2	5	3.7	43	21	172	100
Gaza Strip	38	-	-	-	-	-	-	-	-	-	-	-
Georgia	6 831	2 788	200	2 988	43.7	0.6	n.s.	n.s.	145	97	2 438	82
Iran (Islamic Republic of)	162 201	5 015	2 284	7 299	4.5	0.1	n.s.	n.s.	86	149	-	-
Iraq	43 737	789	10	799	1.8	n.s.	n.s.	n.s.	29	28	-	-
Israel	2 062	41	91	132	6.4	n.s.	5	4.9	49	-	132	100
Jordan	8 893	41	45	86	1.0	n.s.	n.s.	n.s.	38	37	-	-
Kuwait	1 782	0	5	5	0.3	n.s.	n.s.	n.s.	3.5	21	21	-	-
Lebanon	1 024	34	2	36	3.5	n.s.	n.s.	n.s.	-0.4	23	22	-	-
Oman	21 246	0	1	1	0.0	n.s.	n.s.	n.s.	5.3	17	17	-	-
Qatar	1 100	0	1	1	0.1	n.s.	n.s.	n.s.	9.6	13	12	-	-
Saudi Arabia	214 969	1 500	4	1 504	0.7	0.1	n.s.	n.s.	12	12	-	-
Syrian Arab Republic	18 377	232	229	461	2.5	n.s.	n.s.	n.s.	29	28	-	-
Turkey	76 963	8 371	1 854	10 225	13.3	0.2	22	0.2	136	74	9 954	97
United Arab Emirates	8 360	7	314	321	3.8	0.1	8	2.8	-	-	-	-
West Bank	580	-	-	-	-	-	-	-	-	-	-	-
Yemen	52 797	449	-	449	0.9	n.s.	-9	-1.9	14	19	-	-
Total West Asia	698 091	22 202	5 073	27 275	3.9	0.1	48	0.2	101	87	-	-
Total Asia	3 084 746	431 946	115 847	547 793	17.8	0.2	-364	-0.1	63	82	-	-
TOTAL WORLD	13 063 900	3 682 722	186 733	3 869 455	29.6	0.6	-9 391	-0.2	100	109	-	-

Source: Appendix 3, Tables 3, 4, 6, 7 and 9.

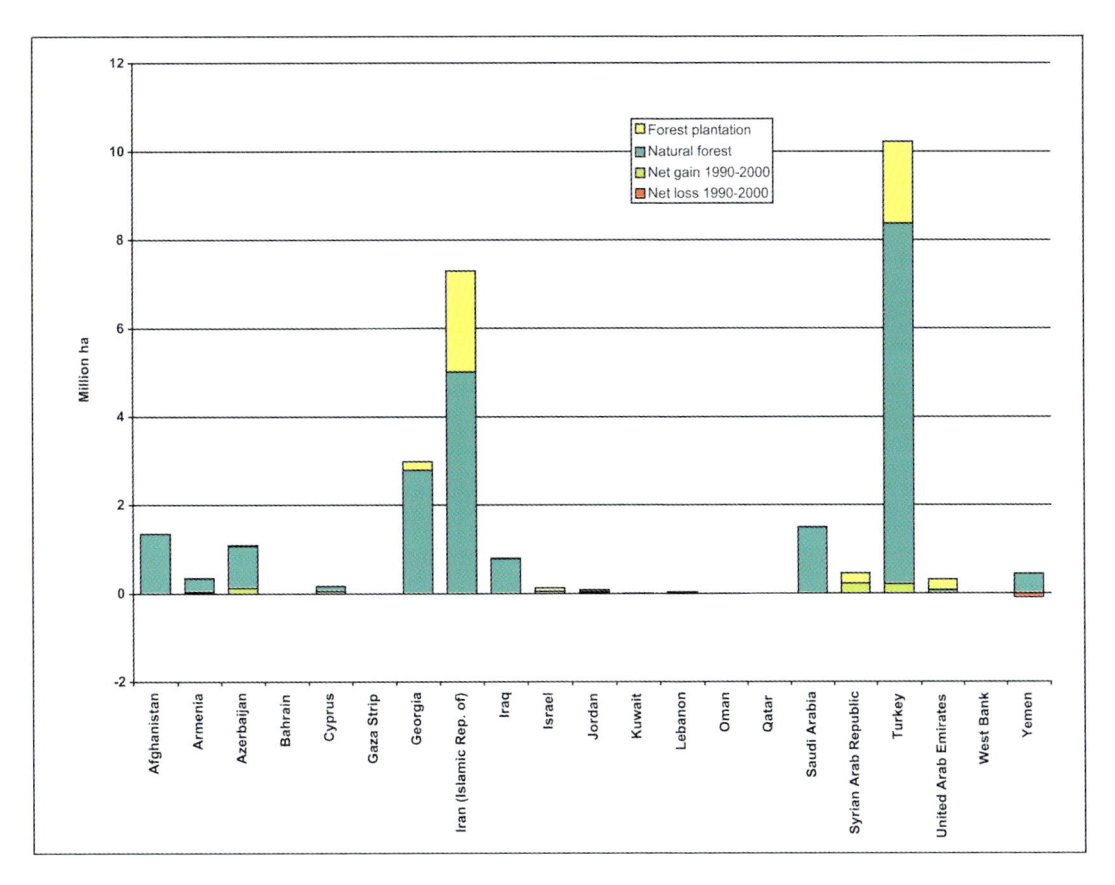

Figure 21-2. West Asia: natural forest and forest plantation areas 2000 and net area changes 1990-2000

information on forest cover is based on surveys and studies conducted prior to 1990. Estimates for Saudi Arabia are based on a 1994 inventory of the southwestern part of the country using aerial photos and fieldwork. For other parts of the country, estimates are based on annual reports and studies. Information for Jordan and the Syrian Arab Republic is based on secondary sources of annual reports and studies. For Armenia, Cyprus, Georgia and Turkey, information is based on literature review and secondary sources. For Azerbaijan and Israel, the data are based on secondary sources. Information on Bahrain, Kuwait, Oman, Qatar and the United Arab Emirates is based on records and surveys of planted areas. No information was provided for the Gaza Strip and the West Bank.

Turkey and Iran have the highest proportion of forest cover in the subregion with 37.5 percent and 26.8 percent, respectively (Table 21-1, Figure 21-2). The rates of forest area change in the region vary from country to country. Forest cover increased in Armenia, Azerbaijan, Cyprus and Turkey. The greatest increase in area was in Turkey. However, Cyprus has the largest annual rate of change. The greatest negative change in both the rate and gross area of forest cover was in Yemen. Afghanistan, Georgia, Iran, Iraq, Jordan, the Syrian Arab Republic and Saudi Arabia had no change in the gross area of forest cover.

In general, forest lands in the region are State-owned, although there is some variation among the countries regarding ownership and the rights of forest dwellers and local populations. In Turkey, the Ministry of Forestry is responsible for forestry activities. In other countries, the Forestry Departments are under the ministries of agriculture or natural resources (Heywood 1997; Duzgun and Ozu-Urlu 2000; Loubani 2000).

Forests of the region are composed of productive forests, degraded forests and eroded unproductive forests as well as some mangrove areas along the Red Sea. The predominant species are pines and oaks. According to Duzgun and Ozu-Urlu (2000), in Turkey 51 percent of the forest area is considered as productive and 49 percent unproductive degraded forests, range lands and eroded forests. About 38.8 percent of the forest area is in pines and 26 percent in oaks.

The subregion has close to 3 percent of the world's forest plantation area. Iran and Turkey have the largest area of plantations. These are established for industrial and protective purposes in addition to fuelwood and charcoal production.

Pines, *Eucalyptus* spp. and acacias are the main species. In Iran, afforestation is promoted by providing free seedlings to landowners. In Turkey, the National Afforestation and Erosion Control Mobilization law passed in 1995 increased the rate of afforestation to around 300 000 ha annually (Duzgun and Ozu-Urlo 2000). In the five Persian Gulf countries which have only plantations, the United Arab Emirates has the largest gross area. The remaining Persian Gulf countries have about 2.5 percent of the total planted area of the five countries. The annual rate of change in these countries is based on the ratio of the latest annual planted area to the total planted area. The country reports submitted for FRA 2000 are the first published data for Bahrain, Kuwait, Oman and Qatar.

The forests of Georgia and Azerbaijan have a larger wood volume and biomass than the world average while Iran has the greatest biomass per hectare. The lowest volume and biomass values are for Saudi Arabia and Yemen.

FOREST MANAGEMENT AND USES

Six of the 21 countries and areas in West Asia provided national-level information on the forest area managed (Table 21-1). They all applied the definition used by industrialized countries of forests managed in accordance with a formal or an informal plan applied regularly over a sufficiently long period (five years or more). Georgia, which had the lowest percentage (82 percent) of the forest area under management of the six countries, did not include forests classified as "undisturbed by man" in the area of forest being managed. The remaining countries appear to have followed the recommendation of including areas where a conscious decision has been made not to undertake any management interventions and reported that 100 percent of their forest area was being managed according to the above definition.

During the last decade, there has been increased interest in improvement and sustainable management of natural forests. Some countries in the region (Turkey, Cyprus, Lebanon and the Syrian Arab Republic) have started national forestry programmes. Others have undertaken action on some elements of their national strategic framework, such as policy reviews or new legislation (FAO 1998). Countries are making efforts to implement integrated programmes involving forestry, pasture, agriculture and rural development institutions and introducing participatory approaches involving forest dwellers and forest villagers in planning and management of the forests on a sustainable basis. In Iran, 130 000 ha of forests were recently transferred to cooperatives with more than 500 members (Abdollahpour 2000). Use of forests by local communities exists, but has not been quantified. However, in Turkey there are more than 17 000 villages located in or near forests that depend on the forests for their livelihood (Duzgun and Ozu-Urlo 2000).

There are estimates on wood production for Armenia, Azerbaijan, Cyprus, Georgia and Turkey (UNECE/FAO 2000). With the exception of Turkey, the contribution of the forestry sector to the gross national product (GNP) in all countries is unknown, since it is combined with agricultural production. In Turkey, the forestry sector is independent. It contributes only 0.8 percent to the GNP of the country, but this excludes indirect and intangible benefits. It is difficult to calculate the economic value of non-wood forest products since most of the products are collected directly from the forests and consumed by local people. In Turkey, the income from exports of these products is around US$80 to US$100 million. The most important non-wood forest products are fruits, nuts, medicinal plants and animal fodder (Duzgun and Ozu-Urlo 2000).

Tourism and water and soil conservation are gaining importance in the region. In Cyprus, the social benefits attributable to recreation, tourism, improvements in agricultural yields and water and soil conservation were estimated to be more than US$70 million per year, while the annual revenue from timber sales is about US$1 million (Theophanous 2000).

Fuelwood is still a major source of energy in the region. Many forests are exploited by the rural population as a source of fuelwood and charcoal for their domestic needs. The percentage of total roundwood production consumed for fuel was 98 percent in Lebanon in 1998, 97 percent in Afghanistan, 67 percent in Iraq, 66 percent in Jordan, 44 percent in Turkey, 41 percent in Saudi Arabia, 32 percent in the Syrian Arab Republic, 29 percent in Iran and 23 percent in Cyprus (FAO Forestry Web page).

General information on forest fires is available for some countries. In the Syrian Arab Republic, there were 347 forest fires during 1995 to 1999 and around 1 400 ha were burnt (Ibrahim 2000). In Turkey, about 2 000 fires occurred in the last ten years and about 12 500 ha were burnt annually (Duzgun and Ozu-Urlo 2000). Forest fire control networks have been established in Lebanon in cooperation with the French Government (Akl 2000) and in Jordan with the assistance of international agencies (Loubani 2000).

CONCLUSIONS AND ISSUES

The 21 countries and reporting areas of the subregion can be classified into 14 countries with natural forests and plantations, five countries with plantations only and two reporting units, the Gaza Strip and the West Bank, with no information.

Based on the country reports sent to FAO, there is accurate forest cover information based on satellite images for Afghanistan. Moderately reliable information based on surveys was found for Armenia, Azerbaijan, Cyprus, Georgia, Iran, Israel, Jordan, Saudi Arabia, Turkey and Yemen. For the remaining countries reported in this subregion, the data were collected from secondary sources.

The difficulty in estimating forest cover area and change for these countries is due to the lack of direct compatibility of the local definitions of forest types with the FAO definitions. Forest cover change has not been estimated for most countries owing to the lack of base data and/or owing to national over- or underestimates of forest cover in 1990.

All countries have policies for conservation and sustainable management of forest resources. They recognize the protective and environmental functions of forests, particularly the aspects related to combating desertification, the protection of watersheds and irrigated zones (FAO 1993) and their role in generating higher income and employment for rural communities.

Most forests in the subregion are State-owned. Privately owned forests account for a small percentage of the total forest area, mostly in the form of small woodlots and linear plantations. The participation of NGOs in forestry activities is still limited in most countries.

In many countries of the subregion, population growth and increased demand on forest products, overgrazing, shifting of forestlands to agricultural use and urbanization are leading to overexploitation and increased degradation of the forest resources, resulting in the inability of the forests to regenerate. The arid climate in most of these countries limits forest productivity (FAO 1993).

Although few countries in the subregion have national forest programmes, some of the actions now being undertaken in many countries include forest policy reviews, new legal instruments and review of institutions. Public participation in forest management and conservation is receiving more attention through governmental bodies, research institutes, NGOs and local communities. Other important issues include the need to identify criteria and indicators for sustainable forest management as well as quantifying the indirect benefits and services of forests and plantations.

BIBLIOGRAPHY

Abdollahpour, M. 2000. *Forest policy in Iran. Country report.* FAO regional workshop on forest policy formulation and implementation in the Near East countries, 3-6 June 2000, Cairo.

Akl, G. 2000. *Forest policy in Lebanon. Country report.* FAO regional workshop on forest policy formulation and implementation in the Near East countries, 3-6 June 2000, Cairo.

Duzgun, M. & Ozu-Urlu, E. 2000. *Forest and Forestry Policy Development in Turkey. Country report.* FAO regional workshop on forest policy formulation and implementation in the Near East countries, 3-6 June 2000, Cairo.

FAO. 1993. *Forestry policies in the Near East region: analysis and synthesis.* FAO Forestry Paper No. 111. Rome.

FAO. 1998. *Overview and opportunities for the implementation of national forest programmes in the Near East.* Damascus, Syrian Arab Republic, 6-9 December 1998. Secretariat note. Near East Forestry Commission 13th Session.

Heywood, H. 1997. The International Expert Meeting: plant resources and their diversity in the Near East, 19-21 May 1997. FAO Forestry Department. Cairo, FAO Regional Office for the Near East.

Ibrahim, H. 2000. *Forests in Syria. Country Report.* FAO regional workshop on forest policy formulation and implementation in the Near East countries, 3-6 June 2000, Cairo.

Jafari, M. & Hosseinzadeh. 1997. *Present status of afforestation in Islamic Republic of Iran.* Technical Publication No. 176. Tehran, Research Institute of Forests and Rangelands.

Loubani, M.S. 2000. *Forest policy and national forest programs in Jordan. Country report.* FAO regional workshop on forest policy formulation and implementation in the Near East countries, 3-6 June 2000, Cairo.

Theophanous, S. 2000. *Forest policy of Cyprus. Country report.* FAO Regional workshop on forest policy formulation and implementation in the Near East countries, 3-6 June 2000, Cairo.

UNECE/FAO. 2000. *Forest resources of Europe, CIS, North America, Australia, Japan and New Zealand: contribution to the global Forest Resources Assessment 2000.* Geneva Timber and Forest Study Papers 17. New York and Geneva, United Nations. www.unece.org/trade/timber/fra/pdf/contents.htm

Chapter 22
Central Asia

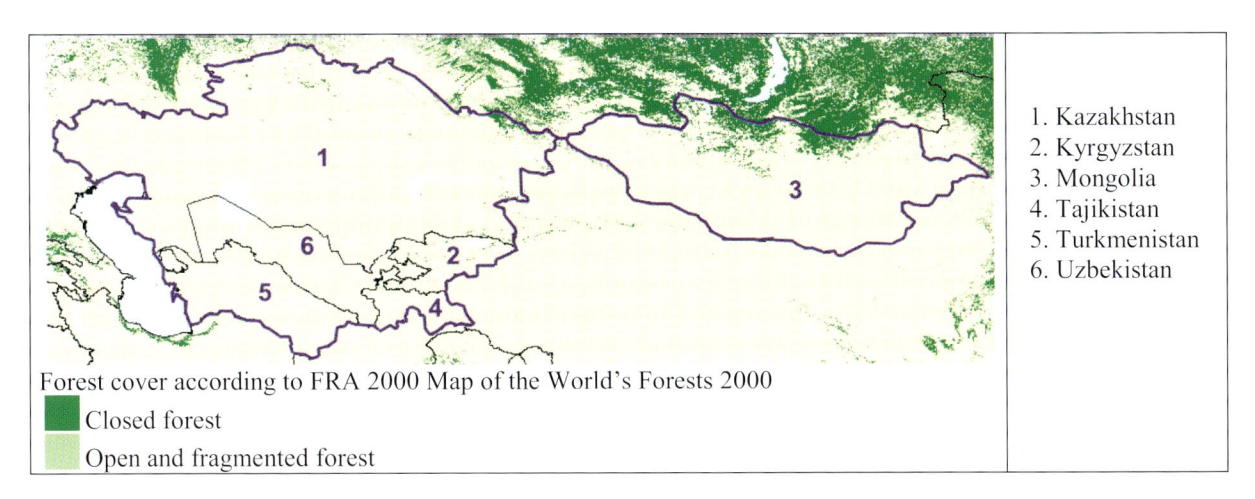

Forest cover according to FRA 2000 Map of the World's Forests 2000 ■ Closed forest □ Open and fragmented forest	1. Kazakhstan 2. Kyrgyzstan 3. Mongolia 4. Tajikistan 5. Turkmenistan 6. Uzbekistan

Figure 22-1. Central Asia: forest cover map

The countries included in this subregion are Kazakhstan, Kyrgyzstan, Mongolia, Tajikistan, Turkmenistan and Uzbekistan (Figure 22-1).[37]

Kazakhstan is a large, sparsely populated country. A significant part of the country is desert but the northern regions, where the forests are located, are ecologically similar to southern Siberia. Kyrgyzstan is a mountainous country with a predominantly agricultural economy. The forests of Mongolia are mainly located in the northern part of the country along the Russian Federation border, forming a transition zone between the Siberian taiga forest and the central Asian steppes. The taiga forests are mainly larch (*Larix sibirica*) and cedar (*Pinus cembra* var. *sibirica*), with *Pinus silvestris* and *Betula* spp. also relatively common. There are also significant areas of arid shrub land in the southern and southwestern parts of the country, mainly saxaul (*Haloxylon ammodendron*) forest. Tajikistan is a landlocked, mountainous country. All its forests are classified as not available for wood supply, most of them reserved for conservation and protection reasons. Turkmenistan is located to the east of the Caspian Sea. Its terrain consists of flat or rolling sandy desert with hills and mountains to the south. The climate is continental, with very low precipitation and extremes of temperature between summer and winter. Uzbekistan's terrain consists of flat or rolling sandy desert with broad, intensely irrigated valleys and steppes in the east.

The steppes and deserts of Kazakhstan are virtually treeless. They are drought-resistant native, although grain crops have largely supplanted native vegetation in the northern steppes. Scrub plants are common in the Qyzylqum desert. Thickets of elm (*Ulmus* spp.) poplar (*Populus* spp.), reeds and shrubs grow along the banks of rivers and lakes. Coniferous trees grow in thick forests on the mountain slopes in the extreme east and southeast.

Animal life varies by zone. The country is home to the extremely rare saiga antelope, protected by government decree. Various animals thrive in the deserts, including gazelles, rodents such as gophers, sand rats and jerboas and reptiles such as lizards and snakes. Wild boars, jackals and deer are found near the rivers and lakes. The mountains are home to ibex, lynx, wolves, wild boars and brown bears as well as the endangered snow leopard.

The eastern steppes of Mongolia are of great ecological importance because, unlike most other grasslands in Central Asia and the rest of the world, there has been relatively little modification by human use. Eastern Mongolia is also home to vast herds of migratory Mongolian gazelles that were once widespread throughout Mongolia and neighbouring areas of Russia and China but are now limited owing to the disruption of migration routes. Mongolia is divided into six basic natural

---
[37] For more details by country, see www.fao.org/forestry

Table 22-1. Central Asia: forest resources and management

Country/area	Land area	Forest area 2000					Area change 1990-2000 (total forest)		Volume and above-ground biomass (total forest)		Forest under management plan	
		Natural forest	Forest plantation	Total forest								
	000 ha	000 ha	000 ha	000 ha	%	ha/capita	000 ha/year	%	m³/ha	t/ha	000 ha	%
Kazakhstan	267 074	12 143	5	12 148	4.5	0.7	239	2.2	35	18	12 148	100
Kyrgyzstan	19 180	946	57	1 003	5.2	0.2	23	2.6	32	-	1 003	100
Mongolia	156 650	10 645	-	10 645	6.8	4.1	-60	-0.5	128	80	-	-
Tajikistan	14 087	390	10	400	2.8	0.1	2	0.5	14	10	400	100
Turkmenistan	46 992	3 743	12	3 755	8.0	0.9	n.s.	n.s.	4	3	3 755	100
Uzbekistan	41 424	1 669	300	1 969	4.8	0.1	5	0.2	6	-	1 969	100
Total Central Asia	545 407	29 536	384	29 920	5.5	0.5	208	0.7	62	40	-	-
Total Asia	3 084 746	431 946	115 847	547 793	17.8	0.2	-364	-0.1	63	82	-	-
TOTAL WORLD	13 063 900	3 682 722	186 733	3 869 455	29.6	0.6	-9 391	-0.2	100	109	-	-

Source: Appendix 3, Tables 3, 4, 6, 7 and 9.

zones differing in climate, landscape, soil, flora and fauna. Mixed coniferous forest is found on cooler, moister northern slopes while steppe vegetation predominates on other aspects. Whereas taiga species are predominant in the mountain forest steppe of the Khangai and Khentii, steppe species dominate the mountain forest steppe of the Altai. Providing habitat for species from both the steppe and the taiga, this zone has a high degree of biological diversity. Wide river valleys separate the hilly terrain characteristic of this zone. Desert steppe occupies a large band, more than 20 percent of Mongolia's area, extending across the country between the steppe and desert zones. The climate is arid with frequent droughts and an annual precipitation of 100 to 125 mm. Frequent strong winds and dust storms strongly influence the vegetation. Still, many of Mongolia's nomadic herders occupy this zone (United Nations 2001).

Wetlands are an important habitat type. Standing water covers about 15 000 km² and there are some 50 000 km of rivers. These wetlands are also extremely diverse, ranging from cold, deep ultra-oligotrophic lakes to temporary saline lakes. Many of the rivers have extensive floodplains. Threats to ecosystems are generally limited in extent and severity. However, mining, especially open cast, is on the increase and is causing soil erosion and pollution in some areas (WCMC 1992). Tajikistan, with an area of 143 100 km², is the smallest country in Central Asia. It is extremely mountainous; almost half the country lies above 3 000 m. Plant life varies by region. Vegetation on the steppes includes drought-resistant grasses and low shrubs. Vast fields of wild poppies and tulips grow on the steppes where they rise into the foothills. The mountain slopes are covered with dense forests of coniferous trees. Ancient forests are found on the lower mountain slopes. Wildlife is abundant and extremely diverse, including the endangered snow leopard (Environmental Information Systems undated).

Turkmenistan is the third largest country in Central Asia, after Kazakhstan and Mongolia. The entire central part of the country is occupied by one of the largest sand deserts in the world, the Garagum. About four-fifths of the country is steppe (semi-arid grassy plain) that is part of the southern portion of the vast Turan lowland. Freshwater resources are scant in Turkmenistan, and extensive canal systems are crucial conduits for irrigation and drinking-water. The mountain streams dissipate upon reaching the arid sands and parched clay of the Garagum, so Turkmenistan's only significant water sources are rivers that originate in other countries. Plant life is sparse in the vast, arid desert, where only drought-resistant grasses and desert scrub grow. The mountain valleys in the south support wild grapevines, fig plants and old forests of wild walnut trees. The mountain slopes are covered with dense thickets called tugai. The wildlife in the mountains includes the caracal (or Persian lynx), goats, cheetahs and snow leopards. In the desert, gazelles, foxes and wildcats thrive. In the tugai live jackals, wild boar and the rare pink deer. Reptiles are abundant and include the central Asian cobra, the desert monitor, several species of gecko and a tortoise. Migratory birds, such as ducks, geese and swans, inhabit the Caspian shore during winter.

Uzbekistan is a landlocked country. Mountains dominate the landscape in the east and northeast. To the west of the mountains, Uzbekistan is generally low in elevation. More than two-thirds of Uzbekistan's territory is covered by desert and steppe (semi-arid grassy plains). One of the largest deserts in the world, the vast, barren Qyzylqum, lies in north-central Uzbekistan and extends into Kazakhstan. In northeastern Uzbekistan, southwest of Tashkent, lies the Mirzachol desert. Uzbekistan's mixed topography provides divergent wildlife habitats. In the steppes the endangered saiga antelope can be found as well as roe deer, wolves, foxes and badgers. The desert monitor thrives in the Qyzylqum desert, along with a type of gazelle and a number of rodent species. The river deltas are home to wild boars, jackals and deer, with a variety of pink deer living in the Amu Darya delta. The endangered snow leopard lives in the eastern mountains. The mountains are also home to several types of mountain goat, including the alpine ibex (characterized by enormous back-curving horns), as well as lynx, wild boars, wolves and brown bears. Plant life is equally diverse. Drought-resistant grasses and low shrubs cover the steppes except in areas that have been cleared for crop cultivation.

FOREST RESOURCES

With the exception of Mongolia, the countries of this subregion were included in the report *Forest resources of Europe, CIS, North America, Australia, Japan and New Zealand (industrialized temperate and boreal countries)*, published as the UNECE/FAO contribution to FRA 2000 (UNECE-FAO 2000). The original data were collected at the national level on the basis of national definitions and sampling techniques. It was necessary to adjust the national data to fit the international definitions. This adjustment, while increasing the comparability and internal consistency of the international data set, reduces accuracy by introducing an additional source of error[38] (UNECE-FAO 2000).

The data for Kazakhstan, Kyrgyzstan, Tajikistan, Turkmenistan and Uzbekistan come from country submissions and are summarized in the above-cited report. Data for Mongolia were provided by the Ministry for Nature and Environment and consist of one data set. The reference year for the estimate is also uncertain although inventory work began around 1963.

Forest and other wooded land in Central Asia accounts for just 5 percent of the total land area, which is less than 1 percent of the world forest cover. Turkmenistan has the largest percentage of forest cover while Tajikistan has the lowest percentage (see Table 22-1, Figure 22-2). Plantation areas are significant in Uzbekistan. Mongolia is the only country which shows a net forest cover loss. The forests of Kyrgyzstan, Tajikistan and Uzbekistan are not available for wood supply, mainly for economic reasons, although forests are important for environmental reasons, notably soil and water protection. On the other hand, Turkmenistan reported more than 90 percent of its forest area as available for woody supply (UNECE-FAO 2000). Mongolia shows the highest production of volume and biomass in the region.

FOREST MANAGEMENT AND USES

Apart from Mongolia, all the countries in Central Asia provided national-level information on the forest area managed (Table 22-1) applying the definition used by industrialized countries of forests managed in accordance with a formal or an informal plan applied regularly over a sufficiently long period (five years or more). They all appear to have followed the recommendation of including areas where a conscious decision has been made not to undertake any management interventions and reported that 100 percent of their forest area was being managed according to the above definition.

All countries reported that forests play an important role in soil and water protection and watershed management. Collection of non-wood forest products is of importance for the local population. The demand for forest products is met by imports, mainly from the Russian Federation.

Mongolia has a relatively large land area under formal protection in a network of around 48 parks and reserves. Fires destroy significant areas of forest and steppe woodland each year. Sawmilling is the main forest industry in Mongolia.

CONCLUSIONS AND ISSUES

This subregion contains a small percentage of the world's forest area. Nevertheless, the forest functions of water and soil conservation are

---

[38] Notes about the country data are included in the above-cited report and describe the adjustment process and data quality. These notes are useful for users to make their own judgement about the quality of the data. This information on adjustments and data quality is intended to improve the credibility of the data set as a whole (UNECE-FAO 2000).

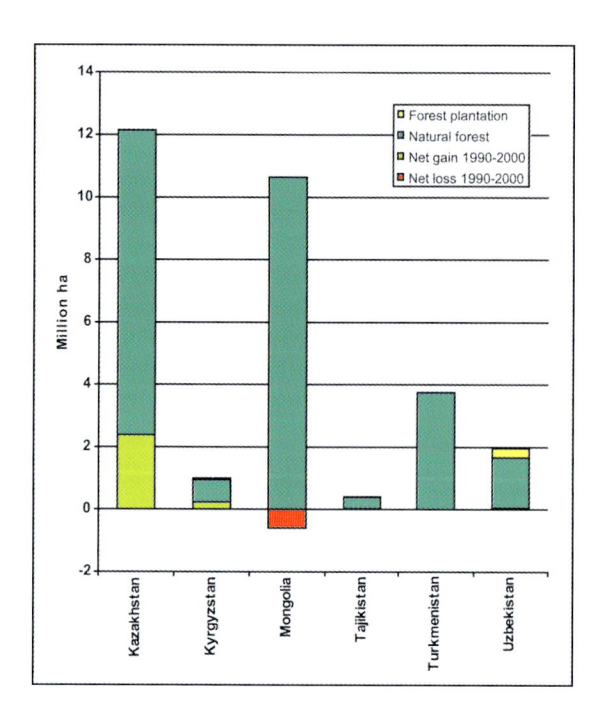

Figure 22-2. Central Asia: natural forest and forest plantation areas 2000 and net area changes 1990-2000

important for all the countries and a good reason for the protection of forested areas.

Forest degradation is of special concern in Tajikistan. In the last few years the development of new territories, the establishment of new villages and a lack of fuel have led to destructive deforestation. As a result of these processes the area in forests has diminished and, more important, the number of valuable and endemic types of plants has declined. Systems for monitoring their condition are in place which will allow the development of concrete measures for their protection and rational use. Urbanization and deterioration of the socio-economic situation have resulted in additional impact on the fauna. Three nature reserves fell within the zone of war actions. At present, the situation in these nature reserves is normal but in order to bring them to the level of 1985 to 1990 considerable effort and expenditure will be required (Environmental Information Systems undated).

There was limited information related to forest use in this subregion, which makes it difficult to predict trends as to the area and quality of the forests. Governments are making efforts to promote sustainable management of the forest areas, and the current shortage of water and increased tourism could be incentives to promote the further protection of forest resources.

BIBLIOGRAPHY

Environmental Information Systems. Undated. Republic of Tajikistan. www.grida.no/enrin/htmls

UNECE-FAO. 2000. *Forest resources of Europe, CIS, North America, Australia, Japan and New Zealand (industrialized temperate and boreal countries).* Main report. Global Forest Resources Assessment 2000. New York and Geneva.

United Nations. 2001. *Mongolian wild heritage.* www.un-mongolia.mn/wildher/desert-steppe.htm

World Conservation Monitoring Centre (WCMC). 1992. *Protected areas of the world. A review of national systems.* Mongolian People's Republic. www.wcmc.org.uk/cgi-bin/pa_paisquery.p

Chapter 23
South Asia

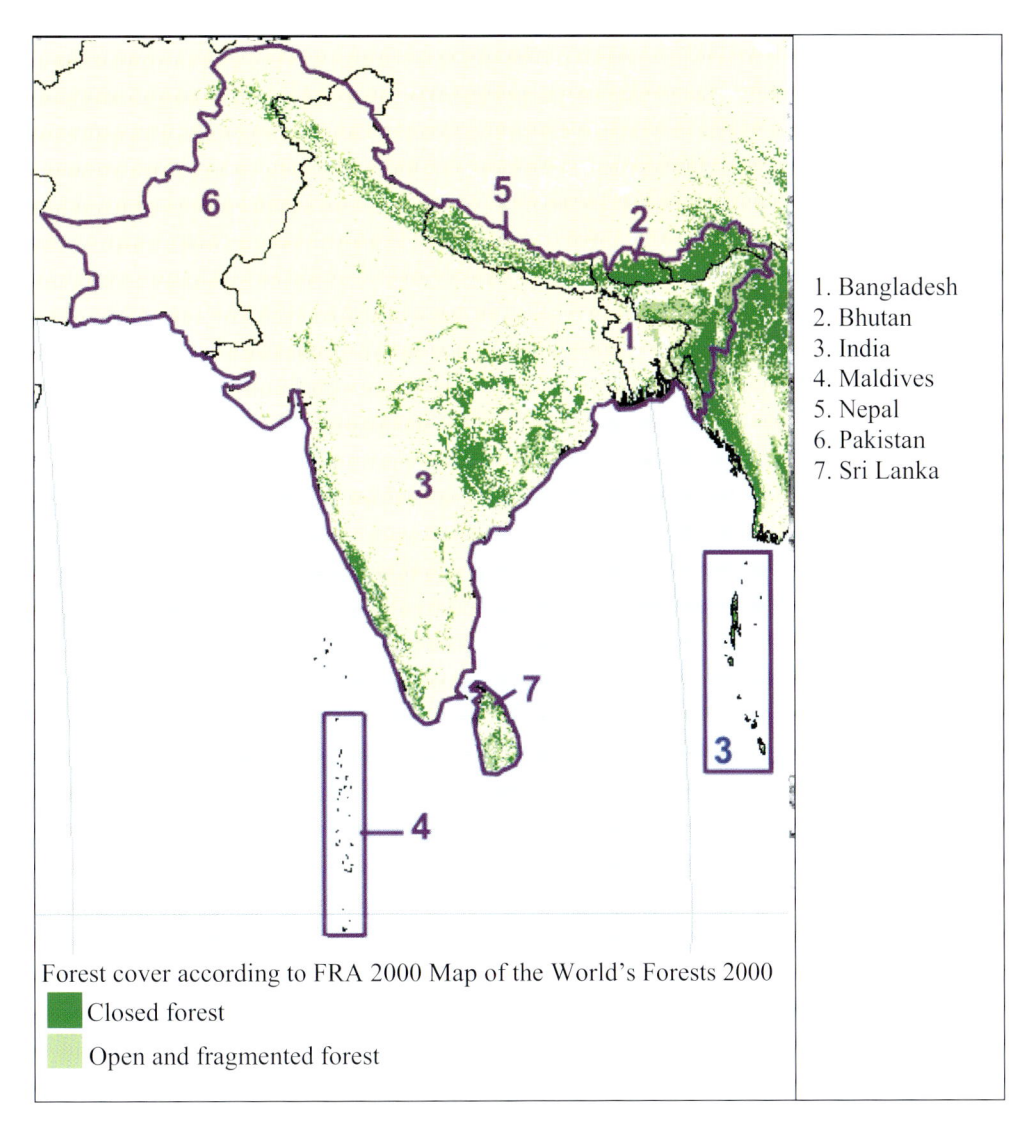

Figure 23-1. South Asia: forest cover map

Legend:
1. Bangladesh
2. Bhutan
3. India
4. Maldives
5. Nepal
6. Pakistan
7. Sri Lanka

Forest cover according to FRA 2000 Map of the World's Forests 2000
- Closed forest
- Open and fragmented forest

The South Asia subregion spans seven countries (Bangladesh, Bhutan, India, Maldives, Nepal, Pakistan and Sri Lanka).[39] The areas of these countries vary from 30 000 ha (Maldives) to 297 319 000 ha (India). The subregion is a reservoir of great biodiversity, in and outside forests, and has untapped potential to develop the use of trees outside the forest. The subregion supports about 22 percent of the global population but has only about 2 percent of the world's forests spread over about 3 percent of total land area (Figure 23-1).

National and international developments during the last decade have changed the way people and institutions in South Asia perceive and value forests and their functions. This has redefined the roles of the State and the people and is leading to new approaches to forest management, planning, monitoring and policy. Increasingly, sustainable forests and healthy ecosystems, rather than merely sustained yield, are being adopted as objectives for managing forests. People and local institutions are being viewed as part of the solution in promoting sustainable forests and ecosystems rather than as merely agents of deforestation.

---

[39] For more details by country, see www.fao.org/forestry

Table 23-1. South Asia: forest resources and management

Country/area	Land area	Forest area 2000 Natural forest	Forest plantation	Total forest			Area change 1990-2000 (total forest)		Volume and above-ground biomass (total forest)		Forest under management plan	
	000 ha	000 ha	000 ha	000 ha	%	ha/capita	000 ha/year	%	m³/ha	t/ha	000 ha	%
Bangladesh	13 017	709	625	1 334	10.2	n.s.	17	1.3	23	39	1 334	100
Bhutan	4 701	2 995	21	3 016	64.2	1.5	n.s.	n.s.	163	178	699	23
India	297 319	31 535	32 578	64 113	21.6	0.1	38	0.1	43	73	46 159	72
Maldives	30	1	-	1	3.3	n.s.	n.s.	n.s.	-	-	-	-
Nepal	14 300	3 767	133	3 900	27.3	0.2	-78	-1.8	100	109	1 010	26
Pakistan	77 087	1 381	980	2 361	3.1	n.s.	-39	-1.5	22	27	-	-
Sri Lanka	6 463	1 625	316	1 940	30.0	0.1	-35	-1.6	34	59	1 940	100
Total South Asia	412 917	42 013	34 652	76 665	18.6	0.1	-98	-0.1	49	77	-	-
Total Asia	3 084 746	431 946	115 847	547 793	17.8	0.2	-364	-0.1	63	82	-	-
TOTAL WORLD	13 063 900	3 682 722	186 733	3 869 455	29.6	0.6	-9 391	-0.2	100	109	-	-

Source: Appendix 3, Tables 3, 4, 6, 7 and 9.

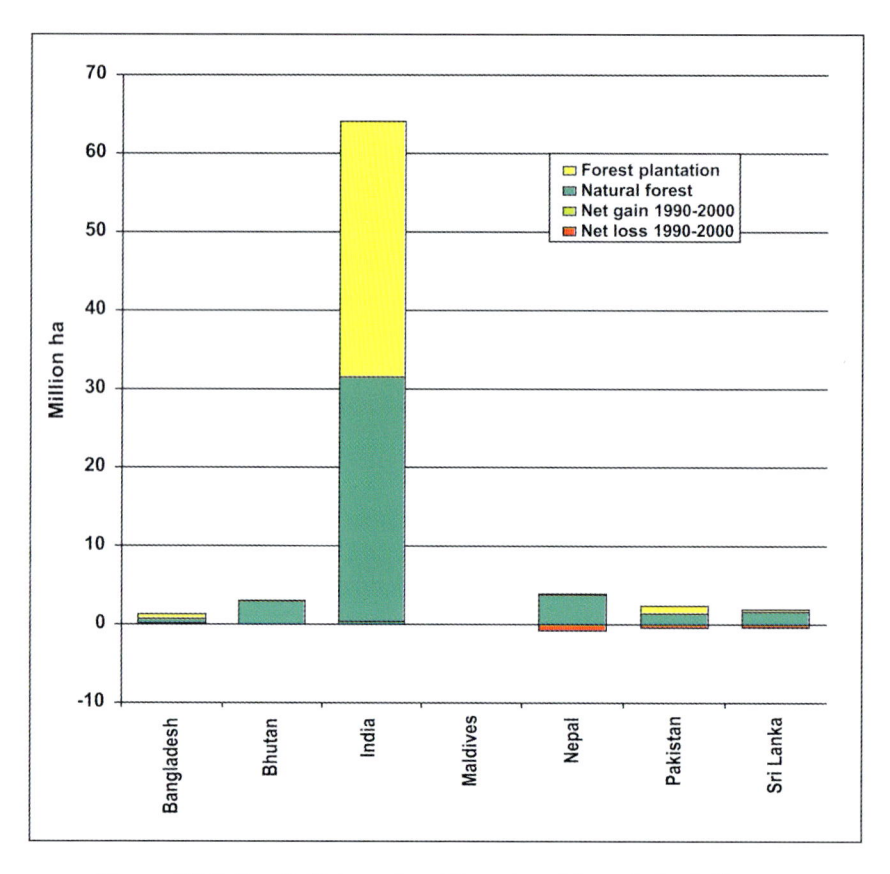

Figure 23-2. South Asia: natural forest and forest plantation areas 2000 and net area changes 1990-2000

Poverty and population pressure are the two factors most responsible for the degradation of forest resources in the subregion. Therefore, apart from control of population growth, countries of the subregion are working hard to achieve higher rates of economic growth to provide additional employment and income.

FOREST RESOURCES

FRA 2000 organized two regional workshops in the subregion – one to explain concepts, definitions and data needs and the other to compile country information, including preliminary trends, and to seek comments from the country representatives. The second workshop also reviewed the use of forest information in planning and the use of electronic networking. A strategy was developed for integrated information collection, storage and use for sustainable forest planning.

The coverage, reference year and definitions of forest assessments differ among the countries. For example, forest assessments in Bangladesh and Nepal have been done in part while those in Bhutan, Sri Lanka and India have covered the entire forested area. The forest assessment of Bhutan used in FRA 2000 utilized 1989 panchromatic SPOT 1 images. Information for Pakistan is available only for 1990. India regularly assesses its forest cover every second year for the entire country. The last assessment (1997) utilized remote sensing imagery from Indian satellites and used FAO definitions. Nepal completed its latest forest assessment over a period of ten years (1986-1996). It used three independent sets of information (Landsat TM satellite imagery for 14 Terai districts, aerial photos for 51 hill districts and the latest inventory data for the remaining ten districts). Sri Lanka utilized 1992 Landsat TM imagery supplemented by IRS-1 imagery in its last assessment, using its own set of definitions.

This subregion has a negative rate (0.13 percent per annum) of forest cover change, which is roughly double the negative rate of change for the Asia region (0.07 percent per annum) but is roughly half the negative rate (0.22 percent per annum) of change for the world (Table 23-1). The forest cover for Bhutan and Maldives has remained roughly the same during the last decade. It has increased in Bangladesh and India but has decreased for Nepal, Pakistan and Sri Lanka (Figure 23-2). The total increase in forest cover for Bangladesh is a result of plantation programmes – the natural forest cover is highly impacted and a large proportion of the forests have been significantly degraded. The maximum rate of decline is found in Nepal and the least in Pakistan. The countries with the highest proportion of forest cover are Bhutan, Sri Lanka and Nepal with 64.2 percent, 30.0 percent and 27.3 percent, respectively (FAO 2000a, b, c, d).

India has the largest area of plantations in the subregion for the production of industrial raw material and fuelwood, and Bhutan has the lowest plantation area. The subregion has made a very large commitment to plantations for the size of its land area. With only about 3 percent of the world's land area, the region has 18.5 percent of the world's plantations. Similarly, with only about 13.4 percent of the land area, the contribution of this subregion to the total plantation area in the Asia region is about 29.9 percent.

Although plantation activity in the subregion is more than a century and a half old, all the countries still need to improve the quality of their planting material, maintenance, monitoring, assessment and databases. Strategic and commercial aspects motivated plantation activity in the subregion, starting with teak (*Tectona grandis*) in 1840 in India, irrigated plantations of sheesham (*Dalbergia sissoo*) in Pakistan in 1866, teak plantations in 1871 in Bangladesh and similar plantations in Sri Lanka and Bhutan in 1947. The current level of private planting exceeds public planting, which currently focuses on satisfying social (conservation and environmental) rather than commercial needs. This has changed the landscape picture across the subregion over the last two decades. The most preferred plantation species in India, Bangladesh and Sri Lanka have been teak and eucalyptus while in Pakistan and Nepal it has been sheesham.

The average volume (49 m^3 per hectare) and biomass (77 tonnes per hectare) estimates for the subregion are slightly less than for the Asia region (63 m^3 per hectare and 82 tonnes per hectare, respectively) and much less than for the world (100 m^3 and 109 tonnes, respectively). It is noteworthy that volume (163 m^3 per hectare) and biomass (178 tonnes per hectare) of the forests in Bhutan are more than one and a half times the world average.

At the ecosystem level the forests in South Asian countries have been classified from two to 16 broad forest types. The forests of Bangladesh are classified into three broad categories based on topographic conditions: hill forests, plain sal forests (*Shorea robusta*) and littoral mangrove forests. The hill forests contain most of the productive forest areas and plain sal forests the least. Hill forests consist of seven forest types (tropical wet evergreen, tropical mixed evergreen, tropical moist deciduous, tropical open deciduous, bamboo, lowland fresh water swamp and savannah). The plain sal forests are of the tropical moist deciduous type. The mangrove littoral forests along the southern coast are of five types (fresh water mangrove, moderately saline mangrove, salt water mangrove and mangroves on rapidly accreting sand and mudflats or on low-lying offshore islands) and occupy numerous estuaries and offshore islands. Most of the original natural habitats have been lost owing to disturbance and the main undisturbed areas are confined to protected areas, where about 968 species belonging to 812 genera and 501 families have been identified.

Bhutan has seven broad natural forest types: fir, mixed conifer, blue pine (*Pinus wallichiana*), chir pine (*Pinus roxburghii*), hardwoods, broad-leaved hardwoods mixed with conifers, broad-leaved and forest scrub. The fir forests are found between 2 700 and 3 800 m. Towards the tree line (3 600 to 3 800 m) the fir forests become stunted and grade into juniper and rhododendron scrub. The mixed conifer forests occur between 2 000 and 2 700 m and occupy the largest portion of the subalpine zone. Blue pine forests occur in the temperate zone between 1 800 and 3 000 m. The chir pine forests are found at low altitude (900 to 1 800 m) under subtropical conditions. The broad-leaved hardwood forest can be divided into three subcategories: upland hardwood (2 000 to 2 900 m), lowland hardwood (1 000 to 2 000 m) and tropical hardwood (below 1 000 m). The forest scrub type includes alpine and temperate scrub occurring naturally between the limits of the tree line and barren rocks.

India has 16 broad forest types: tropical wet evergreen, tropical semi-evergreen, tropical moist deciduous, littoral and swamp, tropical dry deciduous, tropical thorn, tropical dry evergreen, subtropical broad-leaved hill, subtropical pine, subtropical dry evergreen, montane wet temperate, Himalayan moist temperate, Himalayan dry temperate, subalpine, moist alpine scrub and dry alpine scrub (Champion and Seth 1968). The tropical wet evergreen forests are found in the Western Ghats, Upper Assam and the Andamans. The tropical semi-evergreen forests occur along the western coast and in Assam, the eastern Himalaya, Orissa and the Andamans. The tropical moist deciduous forests are present in the Andamans, Uttar Pradesh, Madhya Pradesh, Gujarat, Maharashtra, Mysore and Kerala. The littoral forests are found all along the coast and the swamp forests in the deltas of the larger rivers. The tropical dry deciduous forests occur from the foot of the Himalaya to Cape Comorin except in Rajasthan, the Western Ghats and Bengal. The tropical thorn forests grow in a large strip in South Punjab, Rajasthan, the upper Gangetic Plains, the Deccan Plateau and lower peninsular India. The tropical dry evergreen forests are restricted to the Karnataka coast. The subtropical broad-leaved hill forests are limited to the lower slopes of the Himalaya in Bengal and Assam and other hill ranges such as Khasi, Nilgiri and Mahableswar. The subtropical pine forests are found between 1 000 and 1 800 m throughout the whole length of the Himalaya. The subtropical dry evergreen forests are present in the Bhabar, the Siwalik and the western Himalaya up to about 1 000 m. The montane wet temperate forests are found in Madras, Kerala, the eastern Himalayas, Bengal, Assam and Northeast India. The Himalayan moist temperate forests occur between 1 400 and 3 300 m in Indian-administered Kashmir, Himachal Pradesh, Punjab, Uttar Pradesh, Darjeeling and Sikkim. The Himalayan dry temperate forests occur in Ladakh, Lahol and Chamba. The subalpine forests are present at the upper limit of trees in the Himalaya. The moist alpine scrub occurs along the entire length of the Himalaya above 3 000 m. Dry alpine scrub vegetation is found at the uppermost limit (3 500 m) of vegetation in the Himalaya.

Maldives has two main forest types (mangrove and littoral). These forests have a pattern of salt-tolerant bushes and trees at the island edges and larger trees and coconut palms further inland. The forests at the coastal fringes mainly consist of *Pemphis acidula* and *Suriana maritima*. Inland, the low-lying, richer soils support numerous species such as *Calophyllum inophyllum* and *Hibiscus tiliaceus* that are very important to local people.

Nepal has six bioclimatic forest vegetation types (tropical, subtropical, temperate, subalpine, alpine and nival). The tropical forests are below 1 000 m and account for a total of 1 829 species of flowering plants and about 81 species of pteridophytes. Subtropical forests occur between 1 000 and 2 000 m and support more than 1 945 flowering plant species. The temperate forests are spread between 2 000 and 3 000 m and mainly support broad-leaved evergreen forest. The subalpine forests are present between 3 000 and 4 000 m and support more than 1 400 flowering plants and about 177 endemic species out of a total of 246 endemic plants in Nepal. The alpine forests occur between 4 000 and 5 000 m and are characterized by the presence of various stunted bushy shrubs. Nival vegetation is found above 5 000 m. This zone is mostly without vegetation except for some lichens on exposed rocky places.

Pakistan has four major types of forest (mangrove, coniferous, riverain and scrub). The mangrove or coastal forests are located in shallow waters along the coast near the mouth of the Indus River. The riverain forests occur in Sind and

Punjab along the banks of the Indus and other rivers. The coniferous forests can be grouped into four types: chir pine, upland hardwoods, high-level conifers and alpine. The chir pine forests, or low-level conifers, occur from a little below 900 m up to 1 650 m on the mountain slopes. The upland hardwood forests are present on mountains above 1 500 m elevation. The high-level conifers grow in the temperate zone and range in altitude from 1 650 m up to about 3 000 m. The alpine forests are present between 2 850 and 3 600 m and are a mixture of conifers and broad-leaved trees. The shrub category includes three types of forest (tropical thorn forests, subtropical dry evergreen forests and alpine scrub). The tropical thorn forests occur in the plains and are also known as desert scrub. The subtropical dry evergreen forests are present on hill slopes up to about 1 000 m. The alpine scrub is found above 3 500 m.

Sri Lanka has eight forest types. Lowland mesophyllous evergreen dipterocarp forests are common in wet zones at elevations up to 900 m. Lower montane notophyllous dipterocarp rain forests are common in the wet zone, especially at an elevation between 900 and 1 525 m. Lower montane notophyllous evergreen mixed rain forests are common at elevations between 900 and 1 370 m. Upper montane microphyllous evergreen dipterocarp rain forests are widespread above 1 525 m. Upper montane microphyllous evergreen mixed rain forests are common at elevations above 1 370 m. Lowland semi-deciduous forests are widespread in dry zone lowlands and mainly consist of deciduous species supplemented with evergreen and semi-evergreen species. Lowland semi-deciduous woodland/thorn shrub is widespread in low arid areas.

FOREST MANAGEMENT AND USES

Forest management has a long tradition in South Asia and all countries, except Pakistan and the Maldives, provided national-level information for FRA 2000 on the forest area covered by a formal, nationally approved forest management plan (Table 24-1). The figures reported by Bhutan and Nepal equalled 23 and 26 percent of their total forest area in 2000, respectively, while the area reported by India equalled 72 percent of its forest area. Bangladesh and Sri Lanka both reported that all their forests were being managed according to a formal, nationally approved forest management plan.

Problems such as the inability of forest resources to satisfy demand at the local level are ubiquitous across the subregion. The rapidly increasing use of forest resources by a fast-growing population, poverty and poor enforcement of forest regulations are the three main problems that adversely affect the forest resources of this subregion.

Forest planning and management in these countries is guided by their respective national forest policies. Countries in the subregion have increasingly recognized the importance of biodiversity contained in their forests and have set aside forests for conservation of biodiversity. The past decade has witnessed an increase in the involvement of the private sector, increased empowerment and participation of stakeholders in local forest processes, and considerable investments in poverty alleviation and promotion of alternative sources of renewable energy. Several programmes were initiated to increase the stock of trees outside the forest and forest plantations.

During the last 12 years, all the countries except Maldives have adapted new national forest policies or are in the process of doing so (Bangladesh in 1994, Bhutan in 1991, India in 1988, Nepal in 1989, Pakistan [under revision] and Sri Lanka in 1995). The general thrust of these policies is to promote participatory and people-oriented planning and management and provide a framework to address institutional inadequacies preventing the sustainable use of forest resources.

Bangladesh, Bhutan, Maldives and Sri Lanka plan forests at two (national and district) levels. Other countries, such as India, Pakistan and Nepal, do so at three levels (national, region/state and district/division). All of the countries except Maldives have a long-term plan at the national level such as a forestry master plan or national forest action plan that spans about 20 years and working/management/operational plans at the district level for a period of about 10 to 15 years. The availability of financial resources largely defines the level of implementation of these plans, which varies from country to country.

Many countries have more than 100 years of experience in raising forests. Most of the forests and planted trees on village, private and institutional lands do not, however, have management plans even though they meet most domestic requirements for forest products.

Forests are mainly owned by the State. However, all countries now realize the importance of local social institutions and capacity building in sustaining forest resources and are working to revive or establish such institutions and develop participative forest management programmes.

This new perception has not yet been able to make significant changes in the traditional use of goods and services from forests. The collection of fuelwood is still the main use since fuelwood continues to be the main source of domestic energy. The domestic energy consumption level, when expressed in terms of energy units per capita, seems quite modest but when expressed as per hectare of forest area it is quite high and probably unsustainable owing mainly to the high rural population in the subregion.

The ability of natural forests to meet domestic timber and fuelwood requirements is continuously declining. The unsatisfied requirements are often met from private plantations or from illegal ad hoc harvesting in natural forests. Uncontrolled access and excessive use of forest resources in many places is leading to forest degradation, fragmentation and deforestation.

CONCLUSION AND ISSUES

In general, the country statistics for the South Asia subregion are relatively accurate, up to date and reliable. India, the largest country in the region, has one of the most extensive national forest inventories in the world, with regular assessments and good baseline information. The main difficulties in assessing forest cover and change occurred where local definitions of forest types had changed or did not relate to FAO definitions, such as in Sri Lanka.

Forest planning and management in the subregion is increasingly guided by national forest policies that recognize the need to set aside some forests for the conservation of biodiversity and plan for the remaining forests in a manner that tries to satisfy local needs while supporting resource sustainability. The countries are emphasizing more involvement of the private sector, empowerment and participation of stakeholders, alleviation of poverty and promotion of alternative sources of renewable energy. Efforts are being made throughout the subregion to meet the growing demands of the population by increasing the stock of trees outside the forest and forest plantations to augment the production of forest products and services and to help offset reductions in supply of raw materials due to increased emphasis on sustainable management and the conservation of biodiversity.

National and international developments during the last decade have changed the way people and institutions in the subregion perceive and value their forests. Countries throughout the subregion are seeking to redefine traditional roles and to expand participation in forest management, planning, monitoring and policy. However, these new perceptions and approaches have not yet been able to make significant changes in the traditional uses of goods and services from forests. The collection of fuelwood remains the main use of the forest, and it is recognized that fundamental changes will be difficult to make without major strides in economic development and poverty reduction.

The subregion has apparently been successful in lowering the rate of deforestation in the past decade even though it suffers from a scarcity of forest land, poverty and high population levels. The major concern is human-induced degradation of forests and other natural resources that ultimately threatens the sustainability of life, livelihoods and long-term development. The countries of the subregion are working hard to lower population growth and to achieve higher rates of economic growth to provide additional employment and income. Promoting economic development while conserving the environment and natural resources is a great challenge for South Asian countries.

BIBLIOGRAPHY

Champion, H.G. & Seth, S.K. 1968. *A revised survey of the forest types of India*. Delhi, Publication Division, Government of India.

FAO. 2000a. *Forest resources of Bhutan*. FRA 2000 Working Paper No. 18. www.fao.org/forestry/fo/fra/index.jsp

FAO. 2000b. *Forest resources of Bangladesh*. FRA 2000 Working Paper No. 19. www.fao.org/forestry/fo/fra/index.jsp

FAO. 2000c. *Forest resources of Nepal*. FRA 2000 Working Paper No. 20. www.fao.org/forestry/fo/fra/index.jsp

FAO. 2000d. *Forest resources of Sri Lanka*. FRA 2000 Working Paper No. 21. www.fao.org/forestry/fo/fra/index.jsp

Chapter 24
Southeast Asia

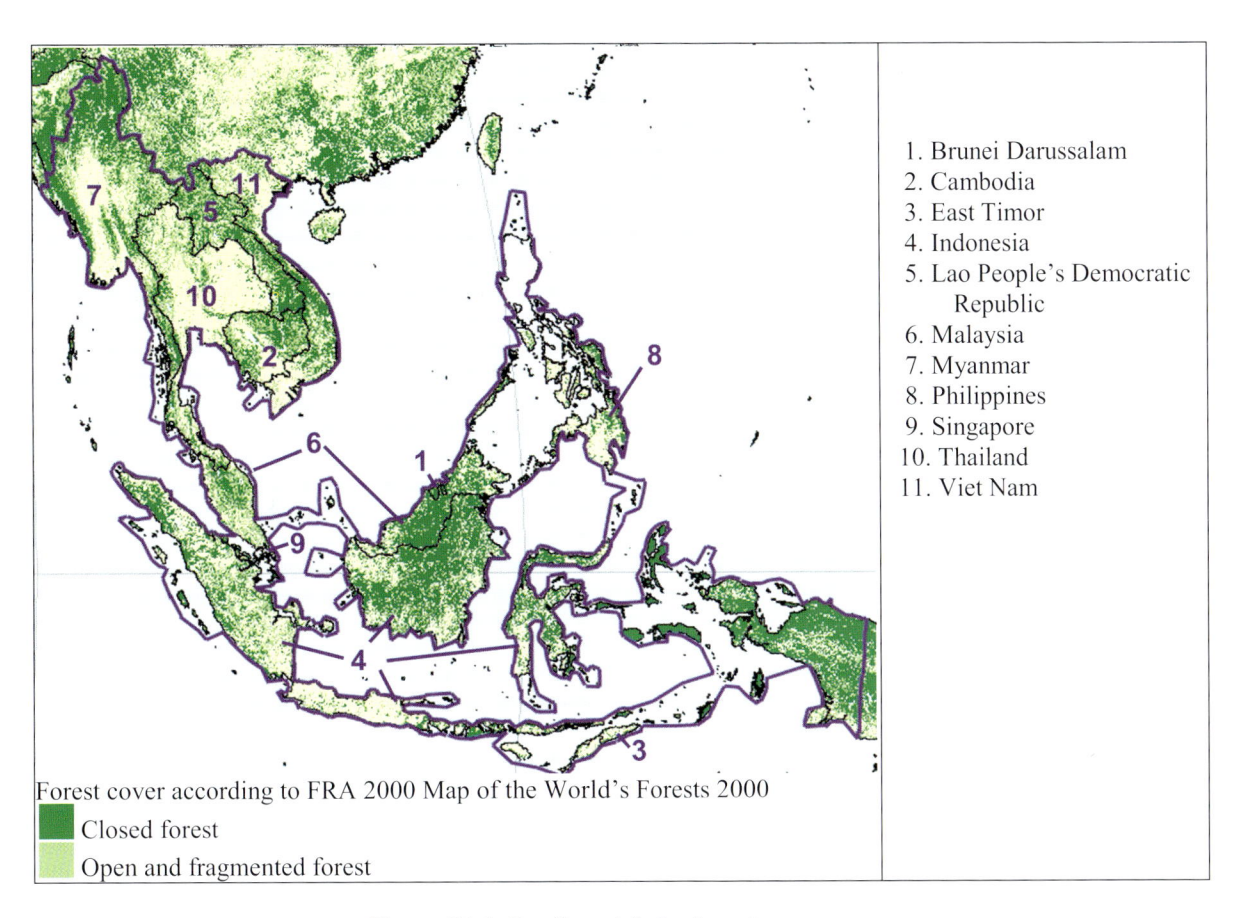

Forest cover according to FRA 2000 Map of the World's Forests 2000
- Closed forest
- Open and fragmented forest

1. Brunei Darussalam
2. Cambodia
3. East Timor
4. Indonesia
5. Lao People's Democratic Republic
6. Malaysia
7. Myanmar
8. Philippines
9. Singapore
10. Thailand
11. Viet Nam

Figure 24-1. Southeast Asia: forest cover map

The subregion consists of the countries of Brunei Darussalam, Cambodia, East Timor, Indonesia, Lao People's Democratic Republic, Malaysia, Myanmar, the Philippines, Singapore, Thailand and Viet Nam[40] (Figure 24-1).

Forests of Southeast Asia are known for their high biodiversity, arguably among the greatest in the world. They have been the subject of much international attention over the past decades.

The subregion is a major player in the tropical timber trade. Meranti timber from the dipterocarp forests and teak from Java, Myanmar and Thailand are among the better-known tropical timbers of the world. Plantation forestry is widely practised; the teak plantations of Java and the rubber plantations of Malaysia are prime examples. Special management systems for tropical natural forests have been developed in the subregion.

FOREST RESOURCES

The quality and age of data differ among countries, as do methodologies. For some countries forest cover has been estimated separately for different parts so data quality and age can differ considerably within a country. This requires adjustments to put all the data on a common basis.

Brunei Darussalam's data are based on a survey made in 1979 using aerial photos and ground surveys. This data set is kept up to date through internal reporting. Data for Cambodia, Lao People's Democratic Republic, Myanmar, the Philippines and Thailand are based on remote sensing. The Lao People's Democratic Republic data are rather old (reference year 1989). The estimate for East Timor is based on 1985 data for Indonesia and the change estimate given for that

---
[40] For more details by country, see www.fao.org/forestry

Table 24-1. Southeast Asia: forest resources and management

Country / Area	Land area	Forest area 2000			Area change 1990-2000 (total forest)		Volume and above-ground biomass (total forest)		Forest under management plan	
		Natural forest	Forest plantation	Total forest								
	000 ha	000 ha	000 ha	000 ha	%	ha/capita	000 ha/year	%	m³/ha	t/ha	000 ha	%
Brunei Darussalam	527	439	3	442	83.9	1.4	-1	-0.2	119	205	-	-
Cambodia	17 652	9 245	90	9 335	52.9	0.9	-56	-0.6	40	69	-	-
East Timor	1 479	507	-	507	34.3	0.6	-3	-0.6	79	136	-	-
Indonesia	181 157	95 116	9 871	104 986	58.0	0.5	-1 312	-1.2	79	136	72*	n.ap.
Lao People's Dem. Rep.	23 080	12 507	54	12 561	54.4	2.4	-53	-0.4	29	31	-	-
Malaysia	32 855	17 543	1 750	19 292	58.7	0.9	-237	-1.2	119	205	14 020	73
Myanmar	65 755	33 598	821	34 419	52.3	0.8	-517	-1.4	33	57	-	-
Philippines	29 817	5 036	753	5 789	19.4	0.1	-89	-1.4	66	114	6 935	120
Singapore	61	2	-	2	3.3	n.s.	n.s.	n.s.	119	205	2	100
Thailand	51 089	9 842	4 920	14 762	28.9	0.2	-112	-0.7	17	29	-	-
Viet Nam	32 550	8 108	1 711	9 819	30.2	0.1	52	0.5	38	66	-	-
Total Southeast Asia	**436 022**	**191 942**	**19 972**	**211 914**	**48.6**	**0.4**	**-2 329**	**-1.0**	**64**	**109**	**-**	**-**
Total Asia	**3 084 746**	**431 946**	**115 847**	**547 793**	**17.8**	**0.2**	**-364**	**-0.1**	**63**	**82**	**-**	**-**
TOTAL WORLD	**13 063 900**	**3 682 722**	**186 733**	**3 869 455**	**29.6**	**0.6**	**-9 391**	**-0.2**	**100**	**109**	**-**	**-**

Source: Appendix 3, Tables 3, 4, 6, 7 and 9.
*Partial result only. National figure not available.

country. Indonesian data for the Kalimantan, Maluku, Sulawesi and Sumatra provinces are based on remote sensing (1985 and 1997). Estimates for Java, Bali and Nusa Tenggara have been calculated using 1985 data and the rate of change estimates made from them. East Timor has been excluded from these analyses. For Malaysia, separate data sets for Peninsular Malaysia, Sabah and Sarawak were used to generate the estimates. The data sets are of varying age. Secondary sources were also used for Sabah and Sarawak since the original methodologies were unclear. Singapore's data were a sample survey of its forest area. For Viet Nam, secondary sources were used.

The countries of the subregion vary widely in size, population and economy. Forest cover and its annual rate of change also vary widely, typically as a function of country size. Most countries have forest cover of at least 50 percent (Table 24-1). East Timor, the Philippines, Thailand and Viet Nam have forest cover ranging between 20 and about 30 percent. Singapore has the least forest cover of the subregion with only 3 percent.

The total annual reduction of forest cover is greatest in Indonesia and Myanmar (Figure 24-2). In fact, new evidence from Indonesia indicates an annual loss of 1.8 million hectares per year (Indonesia FLB 2001), an increase of 500 000 ha over the present estimate. The only country with a positive forest cover change is Viet Nam. Brunei Darussalam and Singapore have annual change rates of zero or close to zero.

Biomass, in terms of both volume and tonnes per hectare, is somewhat lower than in tropical moist Africa and America and far lower than the international average. The reason is unclear but it should be noted that countries with the lower figures are generally countries with large areas of degraded forest.

Plantation forestry is important in the subregion. Indonesia, Thailand and Viet Nam have the largest forest plantations. Rubber (*Hevea* spp.) is the most common species. Indonesia, Malaysia and Thailand together have an area of about 7 million hectares planted to rubber. Indonesia, Myanmar and Thailand have a long tradition of raising teak (*Tectona grandis*) in plantations and these cover more than 2.5 million hectares in those countries. More recently, acacias (particularly *Acacia mangium* and *A. mearnsii*) have been planted to supply fibre for pulp mills. Some 5 million hectares are planted to miscellaneous broad-leaved species. Except for pine on Java, softwoods play a modest role.

FOREST MANAGEMENT AND USES

Three of the 11 countries and areas in Southeast Asia provided national-level information for FRA 2000 on the forest area covered by a formal, nationally approved forest management plan (Table 24-1). Malaysia reported that 14 million hectares of forest, or 73 percent of its total forest area, were covered by a formal plan. Singapore reported that all of its forest area (approximately 2 000 ha) was covered by a plan. The Philippines reported that forest management plans covered a total area of 6 935 000 ha of forestland, equivalent to 120 percent of the area classified as forest according to FRA 2000. It was confirmed that some plans included areas which were not classified as forest by FRA 2000. Indonesia, which has the largest forest area in the subregion, did not provide national-level information, but partial information was available in the form of the forest area which had obtained third party certification by the end of 2000. Information was unavailable from Myanmar where old working plans were in the process of being substituted by District Management Plans. These plans had yet to be approved at the time of reporting. Forest management practices and policies were undergoing change in Cambodia, Lao People's Democratic Republic and Viet Nam and updated information was not available at the time of reporting.

A recent ITTO study (Poore and Thang 2000) thus reported that Indonesia, Malaysia and Myanmar were among the six ITTO tropical producer countries which appeared to have established all the conditions that make it likely that they can manage their forests sustainably.

Forests are generally State-owned. Different concession systems are used, ranging from long-term leases to logging permits for specific compartments. The subregion has long been a major supplier of tropical wood, a position it still holds. Logging early last century was very selective and restricted to accessible areas. Well into the century trade was restricted to high-quality speciality timbers. Extraction levels were modest and harvesting was done using manual methods. The environmental impact was low (Walton 1954). After the Second World War, technical advances were made in wood preservation and in the use of concrete, metal and synthetic materials for construction. These advances largely eliminated the advantage of natural wood durability. However, the market for general-purpose timbers improved, fuelled by economic development. The decades following the Second World War also saw the introduction of mechanized harvesting. The natural forest resources have been increasingly depleted, while greater reliance for timber production is now placed on plantation forestry.

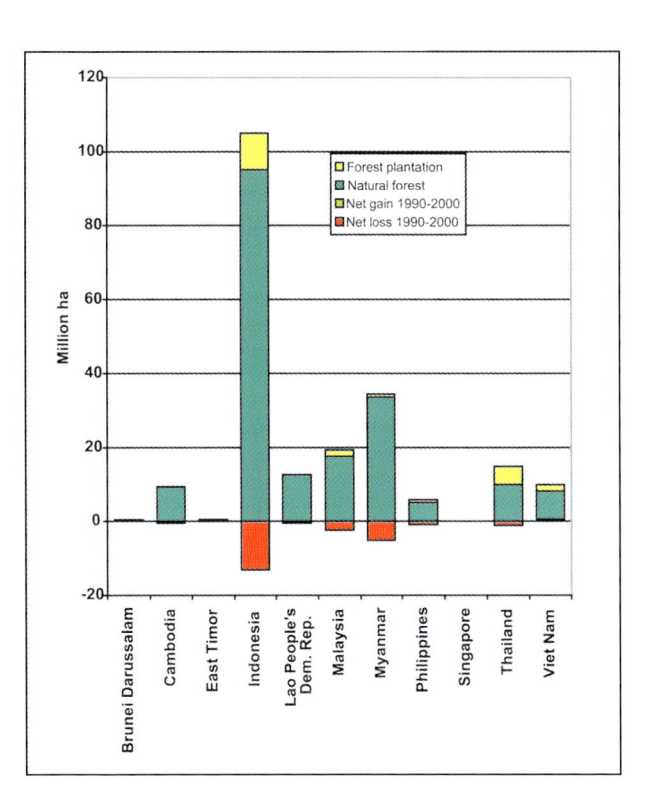

Figure 24-2. Southeast Asia: natural forest and forest plantation areas 2000 and net area changes 1990-2000

Forestry in Brunei Darussalam is strictly controlled and conservation plays a prominent role. Management and silvicultural systems have been developed. Production forests account for 65 percent of the forest estate, and the rest enjoys some form of protection.

In Cambodia, concessionaires are required to develop and follow management plans. However, no information on the size of the area actually covered by forest management plans was provided. There is a code of practice for harvesting. Protected areas make up about 18 percent of the land area. A framework for sustainable forestry practices is currently under development.

In Indonesia, colonial forest management is focused on Java. Large-scale forestry on the outer islands started upon the passage of forest legislation in 1967 when concessions were introduced. Three management systems have been developed for natural forests on the outer islands but the polycyclic TPTI (Tebang Philih Tanam Indonesia – the Indonesian Selective Cutting and Planting System) dominates. Java has teak and pine plantations with a long management history. In the early 1980s fast-growing species were introduced to provide the pulp and paper industry with raw material. Some concessions, called HTI (Hutan Tanaman Industri), have been granted where natural forest is to be replaced by plantations. Protected areas account for 44 percent of the forested area. A new Forestry Act was passed in 1999 to substitute the previous 1967 Basic Forestry Law. Recent years have seen much reorganization of the forest authorities in Indonesia, a process that has yet to be concluded. The role of local communities in forest management is receiving increasing encouragement. Forest fires are a serious concern at present, as is illegal logging.

Large-scale forestry started rather recently in Lao People's Democratic Republic. A forestry law, passed in 1996, emphasizes popular participation. Concessionaires are required to develop and operate under management plans. However, few plans exist today and there are no national guidelines for management plans. The most common form of management is selective cutting. A framework for sustainable forest management in the concessions is under development. Plans for the development of plantation forestry envisage a major community forestry component. Protected areas cover 12.5 percent of the national area. Encroachment by shifting cultivators and wild fires are constraints.

Malaysia is a federation of 13 states (including Sabah and Sarawak) and two federal territories. Forests are State-owned. Every state has its own forest department. The forest departments of Peninsular Malaysia are organized with a central department. Sabah and Sarawak have their own departments. Forest policy aims at maintaining a sustainably managed permanent forest estate while maximizing the social, economic and environmental benefits of the forest. The country has a long and impressive history of research and development in forest management. At present the Selective Management System (SMS) is used in Peninsular Malaysia. SMS prescribes a set of procedures used to determine the best silvicultural course of action for areas to be logged (Appanah and Weinland 1990). The management system in Sabah is a modification of the monocyclic Malayan Uniform System (MUS). Sarawak employs a polycyclic system based on selective logging. Considerable efforts are made to control logging damage in natural forests. Some 5.8 million hectares enjoy some form of protection. In plantation forestry, the country is best known for its rubber estates. Plantations of fast-growing species have been established in Sabah and Sarawak to supply raw material for pulp mills.

In Myanmar, a new forestry law was passed in 1992. A forest policy was formulated in 1995. The policy focuses on socio-economic, development and ecological stability. The State-owned Myanmar Timber Enterprise (MTE) is responsible for harvesting and marketing timber. MTE is engaged in a number of joint ventures with the private sector. Forest management dates back many years. The country is known for a classical selection system for management of natural forests of teak, formulated in the late 1800s, which is still in force. The taungya system for regenerating plantations was formulated in Myanmar. A programme to modernize management plans at the district level is in progress, which explains the lack of information on area of forest under approved plans. Protected forests make up 1.1 percent of the land area. Illegal logging along national borders is a problem. Illegal shifting cultivation is common.

A 25-year master plan for the forestry sector was adopted by the Philippines in 1990. The plan stipulates a mixture of management modes (community, private and State). Logging in virgin forest is not allowed, nor is logging of second-growth forests on steep terrain. The country has suffered a rapid depletion of timber stocks since the 1970s. The focus of management has now shifted from timber production to protection and rehabilitation. Popular participation is encouraged, and management plans are required. Some 2.7 million hectares of forest land are protected. Export of logs and lumber is not permitted. Fires and illegal logging occur. Policies to promote sustainable forestry have been implemented, e.g. through tax incentives.

Singapore's forests are mostly protected. Forests are managed under general environmental legislation. Forests are State-owned. Management chiefly relates to the needs of urban forestry. The major threat to the forest is their use for recreation. There are programmes to create more green corridors.

Thai forestry is regulated by the Forest Act of 1941, the National Park Act of 1961, the National Reserved Forest Act of 1964, the Wildlife Reservation and Protection Act of 1992 and the Forest Plantation Act of 1992. Current forest policy was adopted in 1997. The policy is based on a forest sector master plan. The master plan is implemented through local plans using a bottom-up approach. Earlier focus on harvesting has largely been replaced by protection. All forest is State-owned. The latter half of the last century saw a major depletion of forest area. Plantation forestry is on the increase. Popular participation in forestry is encouraged. A complete ban on logging in natural forests was introduced in 1989. Teak plays a major role in both natural and plantation forests. Eucalypt species have recently been used to rehabilitate degraded forests. It is a national goal to have 40 percent of the country in forest; today's cover is 25 percent. Protected areas (National Parks, Forest Parks and Wildlife Conservation Areas) cover 16 percent. Fire, encroachment and illegal logging are serious problems (Thailand RFD 2000).

The current forest legislation in Viet Nam was adopted in 1991. Forest land shall be allocated to organizations, households and individuals for long-term use following formal procedures and the issuing of land use certificates. The allocation of forest land shall be carried out taking into account: the availability of forest land in different localities; and the management and investment policies and projects to be approved by the competent State authorities. Popular participation is central to the national forest policy. The industrial plantations programme is another important component of forest policy. The programme aims at establishing 5 million hectares of plantations by 2010. The plantations are to meet economic demand as well as environmental concerns. Forests under State agencies are required to have management plans. Large tracts are not owned by such agencies, and may or may not have management plans. Silvicultural focus is on plantations and rehabilitation of natural forests. About 4.8 million hectares are protected in one form or another. Many protected areas are small and have been established rather recently. Shifting cultivation, encroachment and fire are problems. Timber exports are banned.

CONCLUSIONS AND ISSUES

Brunei Darussalam has a very modest annual reduction of forest cover. Forestry is strictly controlled and the country enjoys a high standard of living. Patches of forest will probably continue to be lost for infrastructure and housing development projects. The establishment of plantations may well outweigh these losses.

Information on forest cover in Cambodia is of high quality and acceptably up to date. It appears that the rate of loss of forest cover has slowed from a rather high rate during the 1980s. Forest degradation, however, remains a serious problem.

There is uncertainty about the data for Indonesia. Data published since these estimates were made (Indonesia FLB 2001) suggest a higher rate of forest cover loss. There are some questions and concerns as to how the new data were derived, but the situation in Indonesia remains serious.

Data from Lao People's Democratic Republic are probably quite reliable. The problem is their age. The most recent data are from 1989. National data suggest that forest degradation is serious.

Malaysia has separate data for Peninsular Malaysia, Sabah and Sarawak. Secondary figures had to be relied on and periods between surveys were quite long (10 years for Peninsular Malaysia, 25 years for Sabah and 20 years for Sarawak). The secondary data are probably reliable. However, extrapolation over such long periods may have caused the rate of forest cover loss to be overestimated.

Data from Myanmar are up to date and probably reliable. Myanmar has a high annual loss of forest cover. Forest degradation is also serious.

Data sets for the Philippines are rather recent and compatible. Reliability can be regarded as high. Loss of forest cover is high for the subregion, 1.4 percent per year. Innovative management initiatives to arrest this development are under way.

No major change in forest cover for Singapore should be expected. The "greening" policy and urban forest management programme are interesting examples for other large cities.

The period between the data sets of Thailand is long, 17 years, but it is unlikely that this has led to overestimation of annual forest cover loss. Interesting rehabilitation and reforestation initiatives are under way.

Viet Nam is is the only country in the subregion with a annual increase of forest cover. Data are secondary but of rather recent date. Establishment of plantations helps offset annual losses of natural forest cover in the range of 30 000 ha.

Data on forest cover for the countries of the subregion are generally of high quality and reliability. For many countries there are compatible data sets. Age of information is of concern for some countries, particularly East Timor and Lao People's Democratic Republic. Long periods have sometimes passed between inventories in some countries, particularly Brunei Darussalam, Indonesia, Malaysia and Thailand. Secondary sources have been consulted for East Timor, Malaysia and Viet Nam, and these may be less reliable.

For a number of countries forest degradation seems to be a far more serious problem than outright loss of forest cover.

Some provision has been made for protected areas in all countries but there is currently no estimate of effectiveness.

The subregion may cease to be a major exporter of large logs from natural forests, since accessible natural forests have mostly been depleted. The region has also undergone rapid economic development and there is a growing domestic demand for forest products. Forest industry has expanded during the last several decades and now includes major pulp and paper mills.

Plantation forestry is being practised on an increasingly large scale to relieve the pressure on natural forests. Large plantations exist in the subregion and many countries have major afforestation programmes. It will, however, take some time for plantations to replace natural forests as a source of raw material. In the meantime, appropriate use and management of natural forests will be crucial. Natural forests in the subregion are State-owned. There is no longer an abundance of heavily stocked natural forests to rely on.

Common issues of concern include illegal logging, forest fires and encroachment. Stakeholder participation, alternative ownership systems, resolution of land use conflicts and rehabilitation of degraded forests have started to play a more important role in forest management.

BIBLIOGRAPHY

Appanah, S. & Weinland, G. 1990. Will the management systems for hill dipterocarp forests stand up? *Journal of Tropical Forest Science,* 3(2): 140-158.

Indonesia. Forest Liaison Bureau (FLB). 2001. *Statistics on deforestation in Indonesia.* Source used by FLB: Center for Data and Mapping, Planologi Agency, Ministry of Agriculture and Forestry (2000). www.eu-flb.or.id/htm/english/references.htm.

Poore, D. & Thang, H.C. 2000. *Review of progress towards the year 2000 objective.* Report presented at the 28th Session of the International Tropical Timber Council ITTC(XXVIII)/9/Rev.2, 24-30 May 2000, Lima, Peru. Yokohama, Japan, ITTO.

Thailand. Royal Forest Department (RFD). 2000. *Forestry statistics of Thailand 1999.* Bangkok, Data Center, Information Office, Royal Forest Department.

Walton, Y.K. 1954. The regeneration of dipterocarp forest after high lead logging. *Empire Forestry Review,* 33(4): 338-344.

Chapter 25
East Asia

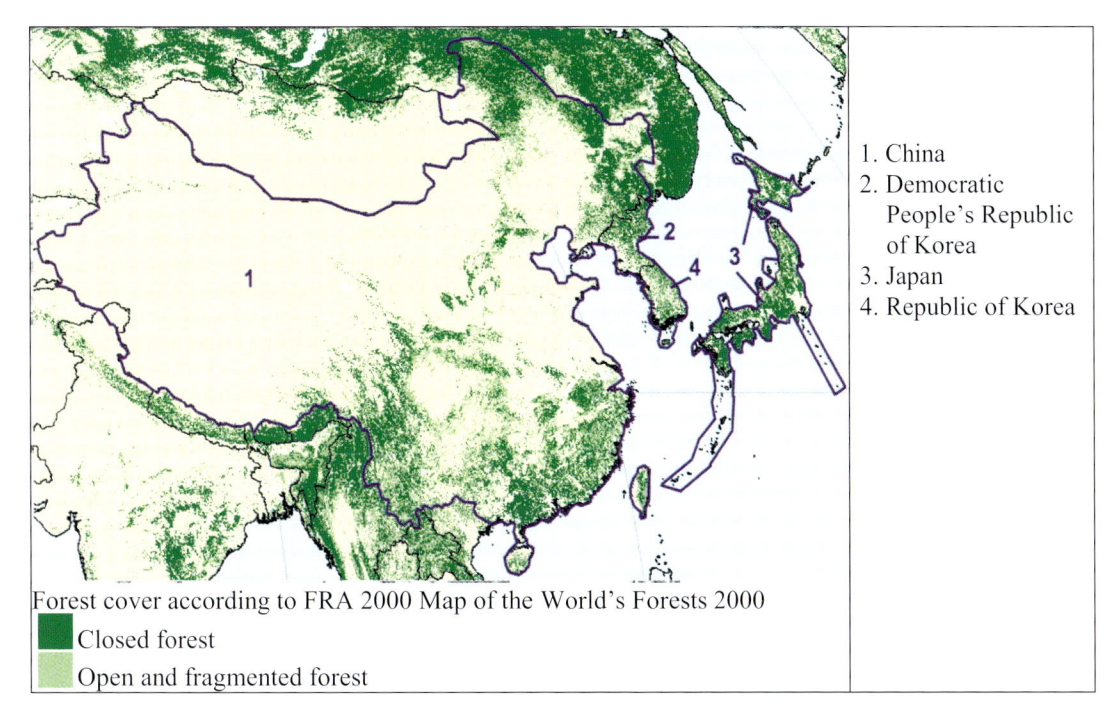

Figure 25-1. East Asia: forest cover map

The subregion of East Asia comprises China, Democratic People's Republic of Korea, Japan and Republic of Korea[41] (Figure 25-1). China is by far the largest country, with 932 million hectares, making up about 94 percent of the entire subregion.

A wide range of terrestrial ecosystems are found in the subregion, most of which occur in China alone. According to the Department of Nature Conservation 1998, China has 599 types of terrestrial ecosystems including a wide range of forests, shrublands, steppes, meadows, savannah, deserts and alpine tundra. According to preliminary statistics, there are 212 types of forest, 36 types of bamboo forest, 113 types of shrubland, 77 types of meadow (27 typical, 20 salinized, nine marshy, 21 cold), 19 types of marshland (14 herbaceous, four woody, one peaty), 18 types of mangrove, 55 types of steppe, 52 types of desert, and 17 alpine tundra, alpine-cushionlike vegetation and alpine talus vegetation. China has more than 30 000 species of higher plants and 6 347 species of vertebrates, constituting 10 percent and 14 percent of the world's total number of species, respectively. The number of freshwater and marine ecosystems have not yet been assessed (China Department of Nature Conservation 1999).

Despite its relatively small size, Japan has widely varying climatic and topographic regions, which contribute to a diverse forest vegetation. Coniferous forests or mixed coniferous and broadleaf forests are found in the boreal or alpine zones, with deciduous forests in the temperate zone, and evergreen broadleaf forests in the warm temperate or subtropical zones. Large natural forests exist only in the Hokkaido region, which has 59.5 percent of total natural forests in Japan. Natural forests also occur on the flanks of the mountains of the Northeastern and Central region of Honshu and in the Southwest Islands. In other parts of Japan, small, frequently fragmented natural forests are distributed in alpine areas or solitary islands. Wetlands occupy a very small percentage of vegetation in Japan, providing important wildlife habitats. One type of wetland is a moorland which is maintained by rainfall and composed largely of aquatic mosses. The other is

---
[41] For more details by country, see www.fao.org/forestry

a moorland which is maintained by rivers and composed of ditch reeds (Biodiversity Center of Japan 1999).

Across from the Japanese archipelago lies the Korean peninsula. Vegetation on the peninsula is associated with warm-temperate, temperate and cold-temperate climates. In the north, the forest vegetation is primarily composed of conifers, which transition into mixed conifer and broadleaf forests in the centre of the peninsula. Mixed conifer and broadleaf forests occur in the south, east and west coasts. Warm-temperate vegetation is also found in the south coast and islands. *Carpinus laxiflora* forests are found in valleys on exposed mineral soil composed of granite and granite-gneiss (Republic of Korea Ministry of the Environment 2001).

FOREST RESOURCES

The currency, accuracy and scope of data on forest resources vary considerably between countries. The National Forest Survey of China is very ambitious and surveys cover the country in five-year cycles using ground surveys and remote sensing. Data from the Democratic People's Republic of Korea were not directly available and had to be obtained from studies based on satellite remote sensing published in the Republic of Korea. Forest cover figures for the Republic of Korea are based on continuous series of reports from subnational units, revised on an annual basis. Japan's assessment data for forest and other wooded land are based on a mosaic of statistics from several different inventories, each having different dates and using different definitions (UNECE/FAO 2000).

The subregion had an annual increase of 1.8 million hectares of forest in the 1990s, which was largely due to plantation programmes in China. Small annual increases were reported in Japan, and small annual decreases were reported for the Republic of Korea. No estimates on change were calculated for the Democratic People's Republic of Korea owing to a lack of information. However, the situation there is considered to be relatively static. Forests cover about 60 percent of the land in all countries except China, which has a forest cover of about 18 percent. Plantations constitute a significant part of the forest estates of China (27 percent), Japan (44 percent) and the Republic of Korea (21 percent) (Table 25-1, Figure 25-2). Information on the extent of plantations in the Democratic People's Republic of Korea was insufficient to calculate their percentage.

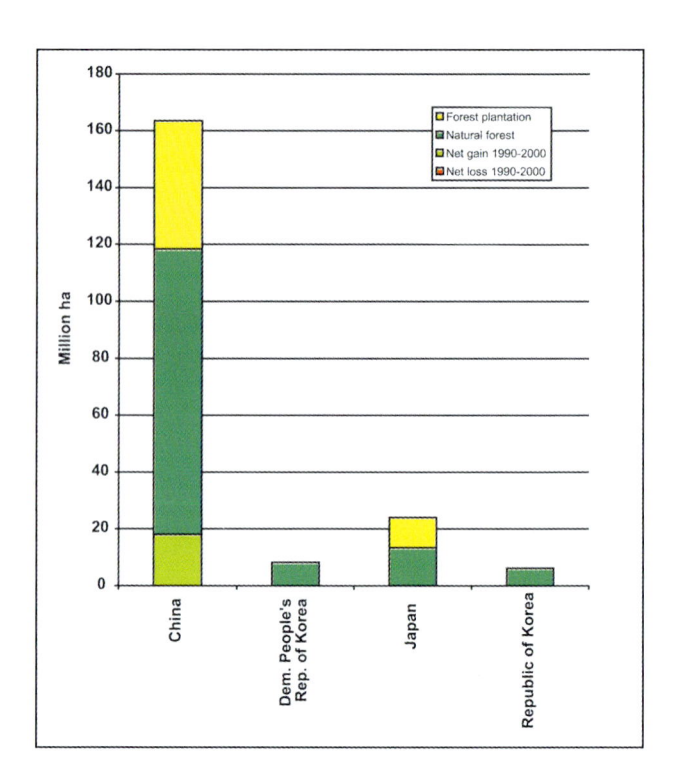

Figure 25-2. East Asia: natural forest and forest plantation areas 2000 and net area changes 1990-2000

Forest volume and biomass per hectare is much higher in Japan than in the other countries, which may be explained by their low levels of harvesting and extraction. China's low average forest volume and biomass are explained by the poor stocking of many of its forests and large areas of young plantations.

FOREST MANAGEMENT AND USES

Two definitions for forest area managed were used in East Asia. Japan reported on forest managed in accordance with a formal or informal plan with a recommendation also to include areas where a conscious decision had been taken not to undertake any management interventions. The remaining countries in the subregion were asked to report on the area of forest managed in accordance with a formal, nationally approved management plan. Japan reported that 100 percent of its forests were managed according to the first definition, whereas the Republic of Korea reported that 66 percent of its total forest area was covered by a formal plan (Table 25-1). China and the Democratic People's Republic of Korea did not provide national-level information on areas under management plans for FRA 2000.

In China, Japan and the Republic of Korea, there are several forms of ownership. In Japan most forests are privately owned. Both China and

Table 25-1. East Asia: forest resources and management

Country/area	Land area	Forest area 2000 Natural forest	Forest plan-tation	Total forest			Area change 1990-2000 (total forest)		Volume and above-ground biomass (total forest)		Forest under management plan	
	000 ha	000 ha	000 ha	000 ha	%	ha/capita	000 ha/year	%	m³/ha	t/ha	000 ha	%
China	932 743	118 397	45 083	163 480	17.5	0.1	1 806	1.2	52	61	-	-
Dem People's Rep. of Korea	12 041	8 210	-	8 210	68.2	0.3	n.s.	n.s.	41	25	-	-
Japan	37 652	13 399	10 682	24 081	64.0	0.2	3	n.s.	145	88	24 081	100
Republic of Korea	9 873	6 248	-	6 248	63.3	0.1	-5	-0.1	58	36	4 096	66
Total East Asia	992 309	146 254	55 765	202 019	20.4	0.1	1 805	0.9	62	62	-	-
Total Asia	3 084 746	431 946	115 847	547 793	17.8	0.2	-364	-0.1	63	82	-	-
TOTAL WORLD	13 063 900	3 682 722	186 733	3 869 455	29.6	0.6	-9 391	-0.2	100	109	-	-

Source: Appendix 3, Tables 3, 4, 6, 7 and 9.

the Republic of Korea have forests owned by cooperatives, although the cooperatives in the Republic of Korea are more of an umbrella organization for private forest owners. In the Democratic People's Republic of Korea, all forests are State-owned. In China, the State Forest Administration is responsible for coordinating protected areas, research and education and for controlling water resources and soil erosion.

A system to protect China's wildlife is now taking shape. Official statistics show 630 nature reserves in China at the end of 1997, which cover an area of 61.5 million hectares, or 6.4 percent of the country. Some 14 ecological zones included in the United Nation's "Man and the Biosphere" network and seven zones listed in the international list of important wetlands are protected by the system. In addition, China has established 873 forest parks across the country, covering 7.5 million hectares (China Department of Nature Conservation 1999).

Silvicultural activities in China centre on plantation establishment and management. Significant research and development has been carried out on the management of high-yielding plantation tree species. The most common genera in plantations are *Pinus*, *Larix*, *Eucalyptus* and *Populus*. The single most common species in plantations is *Cunninghamia lanceolata*, which is typically grown in rotations of 25 to 30 years and intercropped with maize or vegetables, for example. Multipurpose species, such as *Paulownia* spp. are gaining in popularity. Shelterbelt plantations are also commonly used. In natural forests, the focus is on rehabilitation and includes silvicultural practices which enhance the secondary forest growth, although work still needs to be done to rehabilitate areas of degraded forests. Clear-felling of small areas is common,

although there is a shift towards applying selective systems. Wildfires affect almost 1 million hectares annually in the subregion. Efforts to prevent and control forest fires have been carried out in some countries, such as the establishment of firebreaks in China's forests (Su Lifu 2001).

In the Democratic People's Republic of Korea, all forest land is State-owned, and frequently promoted through cooperatives. Large afforestation campaigns were carried out in the 1960s and 1970s, but have now been discontinued. Many of the country's plantations are targeted at fuelwood production, and about 1.2 million hectares of plantations have now been planted using exotic species. Management plans are required for all forests, for which the government has developed a comprehensive set of operational regulations. Coppicing is often relied on to regenerate natural forests. Clear-felling is generally used when harvesting plantations, while selective systems are more frequently used in natural forests. The extent of the country's protected areas is not clear, but is believed to be in the range of 50 000 ha.

Some 40 percent of Japan's forests are owned by the public, and the remaining 60 percent are privately owned. About 2.5 million hectares are considered formally protected. Non-wood forest products play an important role in forest use, and include the collection and use of mushrooms, bamboo shoots, chestnuts, wax and lacquer. Actual timber harvesting is well below sustainable levels. This is due in part to the high costs of extraction associated with steep terrain. Plantations, which are dominated by conifers, account for more than 10.5 million hectares. Common species include cedar, cypress and pine. Sugi (*Cryptomeria japonica*) and hinoki

(*Chamaecyparis obtusa*) are two popular and valuable local species. Many plantations in Japan are young.

Current forestry legislation in the Republic of Korea was passed in 1961 and amended in 1994. The Forestry Administration is responsible for management of the country's forests and for providing extension services to forestry cooperatives. Forest cooperatives are organized into provincial cooperatives that belong to one of four federal cooperatives. The cooperatives are an umbrella grouping for private forest owners. By 1999, 72 percent of the total forest estate was privately owned. This is composed of 57 percent in individual ownership, 8 percent in family ownership, 5 percent in cooperation ownership, 1 percent in non-cooperation ownership and 1 percent owned by temples. Twenty-eight percent of the total forest estate is in public ownership (Korea Forest Service Service 2000).

Until the 1970s, reforestation in the Republic of Korea had taken place primarily in national forests. As a result, the density of government-owned forests was about three times greater than that of private forests. Most forest owners were smallholders with inadequate financial resources to purchase and maintain seedlings. During the Saemaul Movement, however, an ambitious rural development programme launched by former President Park Chung-Hee in 1971, the performance of the village forestry associations improved significantly. Between 1972 and 1979, forestry agents and village associations planted 1.4 million hectares with 3.4 million seedlings (Korea Forest Service 2000).

Silvicultural practices are largely confined to plantations, where the clear-felling system is common. Large areas of young plantations now exist. Commercial forestry is problematic owing to high labour costs and the existence of high volumes of damaged timber, although public subsidies are now being used to support forestry enterprises. A large number of protected areas exist, many of which provide some sort of protection against soil erosion and landslides. These include 20 national parks. Wildfires are reported be a problem.

CONCLUSIONS AND ISSUES

Despite the massive undertaking of making a national inventory in so large a country, China is committed to routine surveys. Even though the periodic national inventories are not yet completely compatible with one another, they point to a large annual increase in China's forest cover through plantation establishment. In contrast, the extent and changes in the natural forest cover are less clear, as is forest degradation.

Information available from the Democratic People's Republic of Korea is weak. In fact, FRA 2000 relied on a secondary remote sensing study conducted by the Forest Research Institute of the Republic of Korea to derive the assessment results. While it is believed that drought and famine may have led to the degradation of the forests in the Democratic People's Republic of Korea, no substantive documentation is available to support this.

Information from Japan was considered to be of very high quality for all the parameters assessed.

Although information from the Republic of Korea for the assessment was considered reliable, some ambiguity existed concerning natural forest extent. According to the documents reviewed, the low-quality products from plantations and unfavourable conditions required to support commercial forestry seemed to be the most pressing challenges.

Except for China, all of the countries in the subregion are heavily forested. Even so, most forestry activities focus on meeting local needs rather than exports. Forest degradation was a major concern in all of the countries except Japan. However, standard measurement techniques for this parameter need refinement in order to assess the state of degradation.

BIBLIOGRAPHY

Biodiversity Center of Japan. 1999. *Convention on Biological Diversity. The first national report.* Japan. www.biodic.go.jp/english/biolaw/kunie/kunie_hon.html

China. Department of Nature Conservation. 1999. *A country study: the richness and uniqueness of China's biodiversity.* China Environmental Sciences Press, State Environmental Protection Administration. www.zhb.gov.cn/english/biodiv/

Korea Forest Service. 2000. *Statistical yearbook of forestry 2000.* Republic of Korea.

Korea Forest Research Institute. 1999. *Preliminary report on the state of plantation.* Republic of Korea, Forest Inventory Division.

Republic of Korea. Ministry of the Environment. 2001. *Korean biodiversity clearing-house mechanism. Convention on Biological Diversity.* www.moenv.go.kr/

Su Lifu. 2001. *The Study and planning of firebreaks in China.* Global Fire Monitoring Centre.
www.unifreiburg.de/

UNECE/FAO. 2000. *Forest resources of Europe, CIS, North America, Australia, Japan and New Zealand: contribution to the global Forest Resources Assessment 2000.* Geneva Timber and Forest Study Papers 17. New York and Geneva, United Nations. www.unece.org/trade/timber/fra/pdf/contents.htm

United States. Library of Congress. *Country studies.*
http://lcweb2.loc.gov/frd/cs/

Chapter 26
Europe

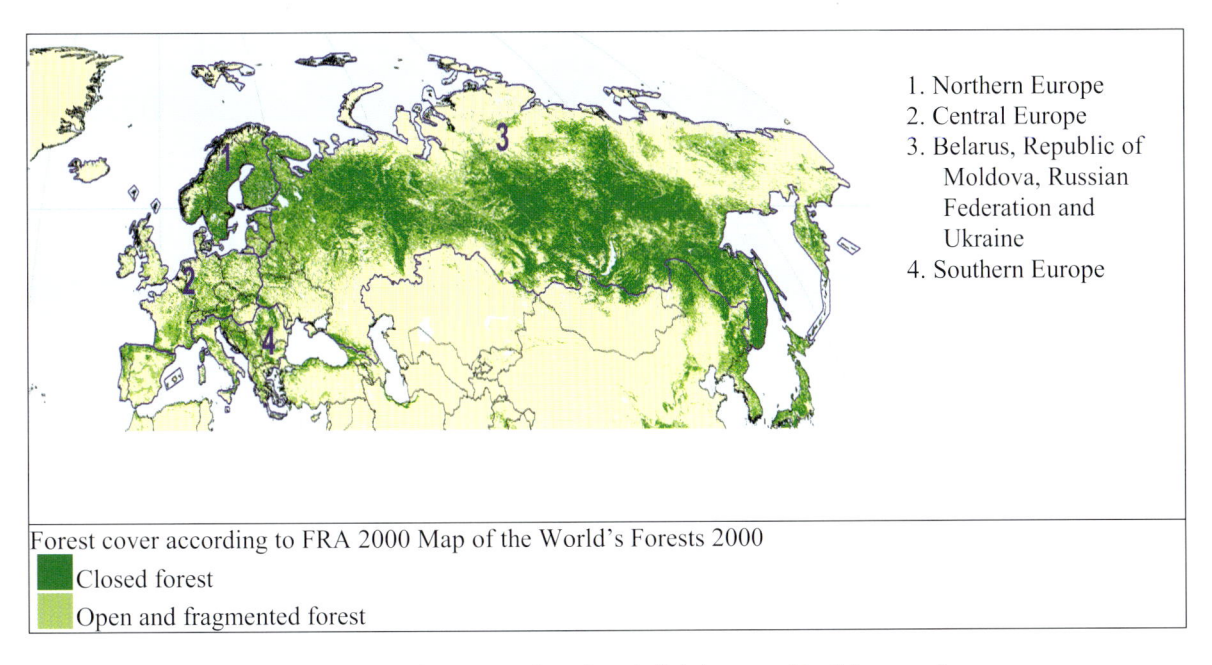

Figure 26-1. Europe: subregional division used in this report

1. Northern Europe
2. Central Europe
3. Belarus, Republic of Moldova, Russian Federation and Ukraine
4. Southern Europe

Forest cover according to FRA 2000 Map of the World's Forests 2000
- Closed forest
- Open and fragmented forest

Europe (see Figure 26-1[42] and Table 26-1) contains about 1 billion hectares of forests which corresponds to 27 percent of the world total. The Russian Federation alone accounts for 851 million hectares and Sweden and Finland for another 49 million hectares. The remaining 38 countries have together less than 15 percent of the forests in the region. Europe's forests amount to 1.4 ha per capita, which is considerably above the world average; however, the area per capita in Central and Southern Europe is much lower. Almost all forests are located in the boreal ecological domain and Europe has almost 80 percent of all boreal coniferous forest. The net change of forest area is positive at 881 000 ha per year, corresponding to 1 percent annually.

---

[42] The division into subregions was made only to facilitate the reporting at a condensed geographical level and does not reflect any opinion or political consideration in the selection of countries. The graphical presentation of country areas does not convey any opinion of FAO as to the extent of countries or status of any national boundaries.

Table 26-1. Europe: forest resources by subregion

Subregion	Land area	Forest area 2000 Natural forest	Forest plan-tation	Total forest			Area change 1990-2000 (total forest)		Volume and above-ground biomass (total forest)	
	000 ha	*000 ha*	*000 ha*	*000 ha*	*%*	*ha/capita*	*000 ha/year*	*%*	*m³/ha*	*t/ha*
Northern Europe	129 019	63 332	1 613	64 945	50.3	2.5	70	0.1	105	60
Central Europe	196 358	47 766	4 114	51 880	26.4	0.2	152	0.3	222	117
Southern Europe	163 750	47 397	4 327	51 723	31.6	0.3	233	0.5	112	60
Belarus, Republic of Moldova, Russian Federation, Ukraine	1 770 830	848 742	21 961	870 703	49.2	4.1	423	0.0	106	56
Total Europe	**2 259 957**	**1 007 236**	**32 015**	**1 039 251**	**46.0**	**1.4**	**881**	**0.1**	**112**	**59**
TOTAL WORLD	**13 063 900**	**3 682 722**	**186 733**	**3 869 455**	**29.6**	**0.6**	**-9 391**	**-0.2**	**100**	**109**

Source: Appendix 3, Tables 3, 4, 6 and 7.

Chapter 27
Europe: ecological zones

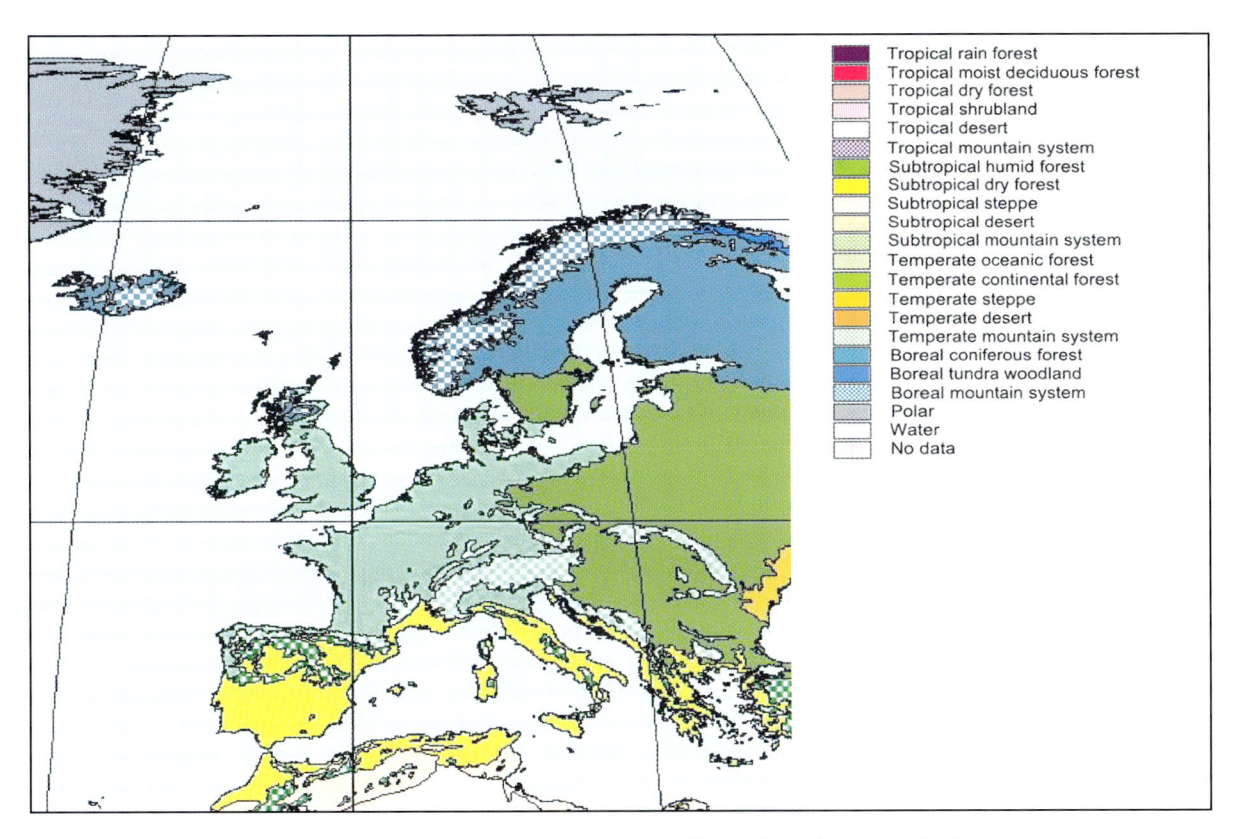

Figure 27-1. Northern, central and southern Europe: ecological zones

Figure 27-1 and Figure 27-2 show the distribution of ecological zones in Europe. Table 27-1 contains area statistics for the zones by subregion and Table 27-2 indicates the proportion of forest in each zone by subregion.

SUBTROPICAL DRY FOREST

In Europe, subtropical dry forests are found in the Mediterranean region below 800 m altitude, including the Iberian Peninsula (except the northern part), Rhone Basin, Apennines Peninsula, Dalmatia and Greece, as well as all the European islands of the Mediterranean Sea. The distribution of *Olea europaea* and *Quercus ilex* roughly defines their boundary.

The Mediterranean climate provides dry, warm summers and cool, moist winters without severe frosts. Precipitation maxima are normally in November/December and February/March. Pronounced elevational relief produces substantial local differentiation. Average annual precipitation is between 400 and 900 mm, rarely above 1 200 mm (e.g. Kerkira) or below 400 mm (southeastern Spain, southeastern Crete). The amount of precipitation decreases slightly to the east. The average temperature of the warmest month is between 25° and 28°C, that of the coldest month between 6° and 13°C.

The original vegetation was evergreen sclerophyllous forest but much of it has long been impacted by anthropogenic influences. The tree species composition is usually rather monotonous. Only one species typically dominates the canopy, often one of the evergreen oak species. *Quercus ilex* and its various subspecies compete most successfully on humid and subhumid sites. Under a 15 to 18 m tall tree layer with a closed canopy is usually a 3 to 5 m tall shrub layer.

SUBTROPICAL MOUNTAIN FOREST

This zone includes the Iberian mountains (Cordillera Cantabrica, Sistema Central, Sistema Iberico, Penibética, Pyrenees), the Apennines, the

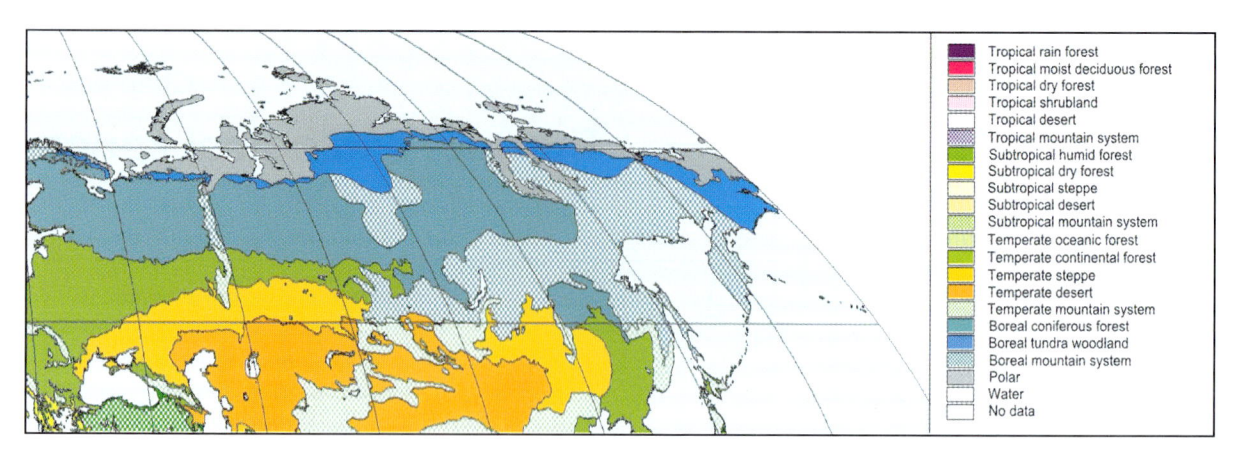

Figure 27-2. Belarus, Republic of Moldova, Russian Federation and Ukraine: ecological zones

Greek mountains (Pindus, Olympus, Peleponnesus, Crete), as well as the mountains of Corsica and Sardinia. The zone starts at about 600 to 800 m and extends up to 2 000 m, locally to 3 500 m.

The region is characterized by higher precipitation and a shorter summer drought period than the adjacent lowland region. Temperatures are lower with a greater frequency of frosts.

In contrast to the dry sclerophyllous forests, the vegetation of this zone is typically deciduous oak species. These forests are usually quite closed and shady. On the Iberian Peninsula, *Quercus pyrenaica* forests dominate on siliceous bedrock, while *Q. faginea* occupies base-rich sites. In the Pyrenees and eastwards, *Quercus pubescens* and other oak species become more important. Closed and shady *Fagus sylvatica* forests, partly with *Abies alba* or *Picea abies*, locally with *Betula pubescens,* replace the deciduous oak forests at higher elevations. In the Greek Pindus Mountains, *Abies borisii-regis* replaces *Abies alba* and is often the dominant species. At even higher altitudes the oak and beech forests are replaced by juniper and cypress woodland (*Juniperus thurifera, J. excelsa, J. foetidissima, J. polycarpos, Cupressus sempervirens*) or by pine (*Pinus nigra*), as well as fir forests (*Abies pinsapo* on the Iberian Peninsula, *A. cephalonica* in Greece).

TEMPERATE OCEANIC FOREST

The temperate oceanic forest zone combines spatially separated areas and comprises the Portugal-Spain coastline (Galicia, Asturia, Cantabrica, Euskal), the British Isles except for the Scottish Highlands and the mountainous regions, France apart from the southeastern mountainous and Mediterranean parts, Central Europe west of a rough line Danzig-Erfurt-Vienna and south of the Alps, including the Po plain. In Scandinavia, all of Denmark, southernmost Sweden and a narrow strip along the coast of Norway are included. Additionally, some climatically sheltered fjords up to 64°N belong to this zone.

The climate is influenced by the Gulf Stream and the proximity to the ocean. The influence decreases inland and is replaced in the Po plain by a different climatic parameter with similar effects. The average annual temperature ranges from 7° to 13°C and annual rainfall varies from 600 to 1 700 mm. While in coastal areas the temperature of the coldest month does not fall below 0°C, inland mean temperature is locally below 0°C.

Various types of beech forests (*Fagus sylvatica*) and mixed beech forests are the dominant vegetation. These are most extensive in Germany and neighbouring countries. Pure beech forests are relatively dense. In oceanic areas, *Ilex aquifolium* is a characteristic species of the shrub layer. On nutrient-poor, acidic soils beech is partly mixed with *Quercus robur* and *Quercus petraea* in the canopy. These stands are poor in species. Today, natural beech forests have been extensively converted into farmland or have been transformed into mixed oak-hornbeam forests. Large areas have been reforested with spruce (*Picea* spp.) and Douglas fir (*Pseudotsuga* spp.).

Outside the distribution area of beech, oak-ash forests (*Quercus robur, Fraxinus excelsior*) with *Corylus avellana* occupy base-rich, often calcareous soils. Oak-hornbeam forests (*Quercus petraea, Carpinus betulus*) dominate periodically moist soils. They often have a distinct vertical structure with a canopy and subcanopy. South of the Alps, *Quercus cerris* may occur together with oak and hornbeam. In the southwest of the zone,

Table 27-1. Europe: extent of ecological zones

Subregion	Tropical Rain forest	Tropical Moist	Tropical Dry	Tropical Shrub	Tropical Desert	Tropical Mountain	Subtropical Humid	Subtropical Dry	Subtropical Steppe	Subtropical Desert	Subtropical Mountain	Temperate Oceanic	Temperate Continental	Temperate Steppe	Temperate Desert	Temperate Mountain	Boreal Coniferous	Boreal Tundra	Boreal Mountain	Polar
Belarus, Rep. Moldova, Russian Fed. and Ukraine													247	119	9	42	553	141	476	205
Southern Europe							76				15	10	42	3		20				
Northern Europe												3	30				69		36	
Central Europe												117	51			24	1		1	
Total Europe							76				15	130	371	122	9	87	624	141	513	206
TOTAL WORLD	1 468	1 117	755	839	1 192	459	471	156	491	674	490	182	726	593	552	729	865	407	632	564

Note: Data derived from an overlay of FRA 2000 global maps of forest cover and ecological zones.

Table 27-2. Europe: proportion of forest by ecological zone

Subregion	Tropical Rain forest	Tropical Moist	Tropical Dry	Tropical Shrub	Tropical Desert	Tropical Mountain	Subtropical Humid	Subtropical Dry	Subtropical Steppe	Subtropical Desert	Subtropical Mountain	Temperate Oceanic	Temperate Continental	Temperate Steppe	Temperate Desert	Temperate Mountain	Boreal Coniferous	Boreal Tundra	Boreal Mountain	Polar
Belarus, Rep. Moldova, Russian Fed. And Ukraine													35	8		74	72	19	55	3
Southern Europe							53				38	34	29			63				
Northern Europe												48	57				70		22	
Central Europe												20	24			59				
Total Europe							53				38	22	35	8		67	71	19	53	3
TOTAL WORLD	69	31	64	7	0	26	31	45	9	2	20	25	34	4	1	26	66	26	50	2

Note: Data derived from an overlay of FRA 2000 global maps of forest cover and ecological zones.

Quercus pubescens forests occupy areas with a milder climate.

TEMPERATE CONTINENTAL FOREST

This zone has a roughly triangular shape with the corners in Oslo, Sofia and Ufa. Southern Sweden, eastern Europe south of the line Helsinki-Novgorod-Perm and north of the line Bucharest-Charkov-Ufa are included. Additionally, most of the Balkan Peninsula and the foothills of the Crimean and Caucasus Mountains are part of the zone.

Owing to less influence of the Gulf Stream, annual rainfall gradually decreases from the west (about 700 mm) to the east (about 400 mm). Summers are warm and winters are cold in most of this region. Mean annual temperature is about 6° to 13°C in the west and decreases to 3° to 9°C in the east. The temperature of the coldest month ranges from below 0°C in Scandinavia and around 0°C in the Balkans to below -10°C in the Ural Mountains. In the northern parts of the zone, more than two months of the year have a mean temperature below 0°C. Additionally, precipitation diminishes from the northwest (greater than 700 mm) to the southeast (400 mm). Locally, in the foothills of the Caucasus, rainfall is very high.

The zone has various forest types, distributed along local and regional gradients of climate and nutrient availability. In the northern parts, mixed coniferous broad-leaved forests form a belt parallel to the circle of latitude. Spruce forests (*Picea abies*) constitute most of the forest cover. On more acidic and drier soils pine forests replace spruce.

Further south, deciduous broad-leaved forests are represented by mixed oak-hornbeam and mixed lime-oak forests. The mixed oak-hornbeam forests include *Quercus robur*, *Quercus petraea*, *Carpinus betulus* and *Tilia cordata*. Associated species such as *Fraxinus excelsior* and *Acer campestre* are also important. Mixed lime-oak forests are found east of the distribution boundary of mixed oak-hornbeam forests. *Quercus robur* and *Tilia cordata* predominate in the tree layer.

Land clearing has massively decimated this type of forest.

Sessile oak, bitter oak (*Quercus petraea*) and Balkan oak (*Quercus* spp.) forests occur mainly in southeastern Europe and the Balkan countries. These species-rich, more open, mixed forests, dominated by *Quercus cerris* and *Quercus frainetto*, occupy the central part of the Balkan Peninsula. Today, these formerly dense forests have been greatly reduced and isolated after long exploitation under the coppice with standards system and for agricultural uses.

Swamp and fen woods occur in small patches across the entire zone. Extensive areas of this vegetation still exist in the lowlands of Poland and Belarus. On permanently wet sites the dominant tree species is *Alnus glutinosa* in association with *Picea abies*.

Flood-plain vegetation is prominent along the middle sections and lower courses of the large rivers Rhine, Elbe, Oder, Vistula, Pripet, Desna, Volga, Save and Danube. Owing to long-term inundation, willow and poplar alluvial forests (*Salix alba, Salix fragilis, Populus nigra* and *Populus alba*) are rather poor in species. Hardwood flood-plain vegetation is highly varied in structure with *Quercus robur, Fraxinus excelsior, Ulmus minor, Ulmus laevis* and *Fraxinus angustifolia* (in southeastern Europe). River regulation and embankment have resulted in a severe decline of near-natural habitat and nowadays only fragments of original flood-plain forests remain.

TEMPERATE MOUNTAIN SYSTEMS

This zone consists of the mountainous parts of the temperate domain, including the Cantabrican Mountains, Pyrenees, Massif Central, Jura, Alps, the highest sites of the British Isles mountains, the Central European uplands, Carpathians, Dinaric Alps, Balkan mountains, Rhodope Mountains, the High and Low Caucasus and the foothills of the Talysh Mountains as well as the southern Urals.

As the highest altitudinal belt of the temperate domain the mountain region is characterized by generally greater precipitation and lower temperature, the climate is extremely varied. Precipitation varies from less than 500 mm to more than 3 000 mm. The average annual temperature ranges from -4° to 8°C (locally 12°C) and the average January temperature at the highest altitudes fluctuates between -10° and -4°C.

Beech (*Fagus* spp.) forests, particularly mixed beech forests with *Abies alba, Picea abies, Acer pseudoplatanus, Fraxinus excelsior* and *Ulmus glabra*, comprise the vegetation of the lower belt in this region. As in the oceanic region, pure beech forests at higher altitudes are relatively dense. At higher altitudes, other tree species become more prominent. To the east *Fagus sylvatica* (subsp. *sylvatica*) is replaced by *Fagus sylvatica* subsp. *moesiaca* and further eastwards by *F. sylvatica* subsp. *orientalis*.

At even higher altitudes, fir and spruce forests (*Abies alba, A. borisii-regis, A. nordmanniana, Picea abies, P. orientalis* and *P. omorika*) replace the beech forests. Either *Abies* or *Picea* may dominate. *Pinus sylvestris, Fagus sylvatica*, some *Quercus robur*, and pioneer species such as *Sorbus aucuparia, Populus tremula* and *Betula pendula* play a minor role. Around the timberline, pine scrub (*Pinus mugo*) or *Rhododendron* spp. may occur. This scrub and krummholz grades at higher altitudes into alpine grasslands, various dwarf shrub vegetation and rock and scree vegetation of the alpine to nival belt.

In the Urals, the altitudinal zonation starts with lime-oak forests (*Quercus robur, Tilia cordata*) at the lowest level followed by herb-rich fir-spruce forests (*Abies sibirica, Picea obovata*) with broad-leaved trees such as *Ulmus glabra* and *Tilia cordata* as well as pine forests (*Pinus sylvestris*) with *Larix sibirica*.

BOREAL CONIFEROUS FOREST

This zone occurs in some parts of Norway, most of Sweden, nearly all of Finland, northern Scotland and a wide belt in the western part of the Russian Federation south of the Arctic Circle as well as the southern part of Iceland. The zone also covers major areas in the eastern parts of the Russian Federation. A small island of lowland boreal forest is in the Russian Far East, north of the Amur River.

The western part of the zone has a cool-temperate, moist climate, varying from oceanic in the west to subcontinental in the interior and the east. Mean annual temperature is generally low and ranges from 8°C in Scotland to just above 1°C in the northern parts of the Russian Federation. Precipitation ranges from more than 900 mm in the west to 400 mm in the east, with extremes of 1 200 and 300 mm. A short growing period (less than 120 days) is characteristic. Evaporation is low and prolonged periods of drought are rare. Snow generally covers the ground for several months during the winter.

The climate of boreal western Siberia is influenced by the amount of solar energy, the Atlantic Ocean to the west and the powerful east Siberian winter anticyclone from the east. The climate of the northern part is under the influence of arctic atmospheric processes. To the south, the low winter temperature contrasts a relatively high summer temperature. The maximum precipitation (on average 500 mm) is in the centre of the plain (about 60°N); to the north and south the amount of precipitation is lower. Throughout the zone, rainfall is concentrated during the growing period or warm season. Snow cover plays a significant role in western Siberia, defining the depth of frozen soils in winter and determining hydrology in summer.

In western Siberia, the average annual temperature is about -4°C (January, -22° to -24°C; July, 16° to 17°C), the growth period is about 85 days, the period with snow cover 190 to 200 days and annual precipitation 410 to 450 mm. To the south, the climate becomes significantly warmer. Between the Irtish and Yenisey Rivers, the average annual temperature increases to -0.4° to -1°C (January, -18° to -21°C; July, 16.5° to 18°C), the growth period to 100 to 115 days, snow cover 175 to 190 days and precipitation 410 to 550 mm.

In the sparse taiga of the eastern part of the middle Siberian Plateau the climate is continental, with little precipitation, dry springs and severe winters. The average annual temperature is -11° to -13°C (January, -38° to -43°C; July, 14° to 17°C), the growth period is 63 to 73 days, snow cover is 228 to 237 days and precipitation 200 to 290 mm. A major part of this area is covered by continuous permafrost, very deep (up to 600 m) and cold (-8° to -12°C) in the north, which crucially impacts the structure and functioning of forest ecosystems. The melting layer is from 0.2 to 0.5 m on wetlands and up to 0.5 to 0.8 m on drained sites.

The glaciers of northern Europe essentially wiped the land clean of most plant species. This great natural perturbation is still reflected in the species and vegetation diversity of the region. Most boreal forests are dominated by only a few conifer tree species, primarily spruce (*Picea abies*) on moister ground and pine (*Pinus sylvestris*) on drier ground. East of the White Sea, mainly closer to the Ural Mountains, Siberian conifer species such as *Pinus sibirica*, *Abies sibirica* and *Larix sibirica* may also occur. Deciduous species such as birch (*Betula* spp.), aspen (*Populus tremula*), alder (*Alnus* spp.) and willow (*Salix* spp.) are characteristic of the early successional stages (especially birch and aspen) or may form smaller stands among the conifers. Stands of deciduous trees are mainly associated with special habitats, often disturbed by fire or floods, or occupy particular soils.

Mires form characteristic landscape elements in mosaics with various forest types. In parts of northern Finland, mires cover almost 50 percent of the land area. Raised bogs, with a central raised area of peat, are found in the southern part of the zone. The most common types of mire in the boreal region are fens on level or gently sloping ground, often mixed with smaller areas of open water, raised bogs, and drier, firm ground. Many of these areas, in Fennoscandia in particular, have been ditched and partly drained for agriculture or forestry. Modern technology has the potential to restructure and transform boreal forests and the landscape on a large scale.

In the eastern Russian Federation, the zonality and continentality of the climate define the distribution of vegetation. Higher humidity in the western part promotes dark coniferous forests (dominated by spruce and fir) while increasing dryness and continentality in the eastern part of the zone favours light coniferous forests (predominantly larch, but also pine to the south).

Swamps and marshland dominate the northern taiga of the western Siberian plain. Forests are confined to well-drained river valleys. They are dominated by Siberian cedar pine (*Pinus sibirica*), with a mixture of Siberian spruce (*Picea obovata*), birch (*Betula pendula*) and Siberian larch (*Larix sibirica*) in the north and slow-growing fir (*Abies sibirica*) in the south. Secondary birch forests are extensive.

Various raised and transitional bogs are prevalent in the middle taiga. Sparse cedar forests with birch usually grow in valleys. To the south, the amount of wetlands significantly decreases. Cedar-spruce and cedar-spruce-fir forests cover the uplands in the middle and southern taiga. Birch (*Betula pendula*) and aspen (*Populus tremula*) forests increase towards the south. Pine forests with lichens grow on drained sands.

To the east of the Yenisey River, dark coniferous taiga gives way to light coniferous larch and pine forests. In the northern part, in the basin of the Podkamennaja Tunguska River, larch-pine and pine forests with mosses predominate. Spruce and cedar forests with birch and aspen occur in river valleys. Hummocky peat covers significant areas. To the south, pine predominates. The most productive Asian pine forests grow in the basin of the Angara River where growing stock volume on the best sites can reach 500 to 600 m³ per hectare.

To the east, in Central Yakutia, larch is the major dominant species. Other species, primarily pine and birch, occupy less than 10 percent of forested areas. To the north, in the northwestern part of Yakutia and partially in Evenkija and the Taimir national district, sparse northern taiga larch forests cover about 95 percent of the forested areas. Dwarf pine (*Pinus pumila*) covers about 4 to 5 percent, while birch is very rare. Sparse larch forests are common in the south with a sparse low canopy layer of Siberian spruce (*Picea obovata*).

BOREAL TUNDRA WOODLAND

In Europe, boreal tundra woodland forms a narrow belt on the Kola Peninsula and along the Arctic Circle to the Ural Mountains. Beyond the Urals, the zone is a rather wide belt stretching to the Pacific coast. Vast areas of tundra and bog vegetation alternate with sparse, low-productivity forests and shrubs. The northern part of the zone, 100 to 250-300 km wide, is a "human-induced treeless belt" where lack of forests is assumed to be a consequence of anthropogenic or natural disturbance, mostly wildfires.

The climate is cold but humid. In the European part, the average annual precipitation varies between 700 mm on the Kola Peninsula to 500 to 550 mm east of the Pechora River. The mean annual temperature on the Kola Peninsula is -1° to -2°C (average in January, -10° to -12°C; July, 9° to 12°C). Permafrost is discontinuous but widespread.

To the east, the climate is strongly impacted by continental, and partially maritime, arctic air masses, moderated only in the extreme east. The severity of winter increases from the coast inwards. All territories are under continuous deep and cold permafrost. The climate is most severe in central Siberia (between the Yenisey and Lena Rivers) where the average annual temperature decreases to -12° to -15°C (January, from -31° to -42°C; July, 11° to 14°C). The minimum temperature reaches -58° to -65°C. The growth period is very short, from 35 to 60 days. Annual precipitation amounts to 240 to 400 mm. Throughout the zone most of the precipitation falls during the warm period.

The vegetation of the European part of this zone comprises open woodlands of low-growing trees, mostly 4 to 6 m tall. The stands are predominantly composed of *Betula pubescens* subsp. *czerepanovii* and *Picea obovata*. While *Picea obovata* dominates in the north of the Russian Plain and in the Urals, *Betula pubescens* subsp. *czerepanovii* forms the woodland in the suboceanic areas of northeastern Europe. Further east, open woodlands of *Larix sibirica* occur as small isolated stands on sandy soils. Mires often occupy wet depressions while the tundra woodland covers the slopes and other well-drained sites.

East of the Urals, open woodland is usually found in lower-lying and better-drained terrain along with tundra and mires. In the southern part of the zone, sparse coniferous forests follow the river valleys in narrow belts several kilometres wide. In most cases, trees are irregular in shape, with crooked boles, one-sided flag-like crowns, and sometimes a form resembling creeping arboreal plants. Soil cryogenic processes often cause the phenomenon of "tipsy forests". In western Siberia, the predominant species in the typically sparse forests is Siberian larch (*Larix sibirica*) with an admixture of Siberian spruce (*Picea obovata*). In central Siberia, *Larix gmelinii* is dominant and spruce forms the second canopy layer. To the east, in the basins of the Indigirka and Kolyma Rivers, the principal species are *Larix gmelinii* and *L. cajanderi*. The latter replaces *L. gmelinii* to the east of the Lena River. Dwarf pine (*Pinus pumila*) and bushy willows (*Salix udensis, S. schwerin*) are abundant and exceed in area the "high" forests. Mongolian poplar (*Populus suaveolens*) and Korean willow (*Chosenia arbutifolia*) grow in river valleys. The northern tree line goes along the reaches of the Kolyma River to the north of 69°N and to about 65°N in Chukotka, characterized by poplar, Korean willow and bushy alder.

BOREAL MOUNTAIN SYSTEMS

The boreal mountain zone consists of six isolated mountainous regions – the uplands of Iceland, the Scottish Highlands, the Scandinavian mountains, the Urals, the higher northern part of the Central Siberian uplands and the vast mountain territories that occupy the south of Siberia and cover the major part of Yakutia and the Russian Far East.

In the mountains of northern Europe, the average annual temperature is nearly everywhere below 4°C. Only in coastal areas of southern Norway does the temperature reach 7°C. Annual precipitation is about 400 mm in the east and increases westwards, although orographic precipitation can locally be much higher. In the eastern Russian Federation, the climate of this zone is extremely diverse but generally severe. Snow cover is usually abundant and perseveres for a considerable time. Continuous, deep permafrost predominates. The harshest climate is found in the middle Siberian uplands and the mountains of northeastern Russian Federation. Here, mean annual temperatures range from -11° to -14°C, with January temperatures as low as -35° to -43°C, and minimum temperatures of -50° to -60°C. July temperatures are 13° to 16°C but the length of the growing period in these regions is only 60 to 80 days. Annual rainfall amounts to 200 to 300 mm, predominantly as snow. Conditions are less severe in other mountain areas, particularly those with higher minimum (January) temperatures. There is high variation in the amount of precipitation; for instance the high West Altai receives up to 2 000 mm of precipitation, which, together with rather warm conditions, favours growth of dark coniferous forest vegetation. In lower East Altai precipitation is much less, which favours development of larch forests.

Birch woodlands are widely distributed in the European part of the zone. They are composed of more or less open *Betula pubescens* subsp. *czerepanovii* forests, partly with pine (*Pinus sylvestris*) in the eastern parts. Above the timberline the forest vegetation is replaced by boreal alpine as well as subnival and nival vegetation. In Iceland, sparse mountain pioneer vegetation occupies the highest altitudes while in the Scottish Highlands blanket bogs, heaths and dwarf shrub vegetation cover the rounded hills. In the Ural Mountains, coniferous forests (*Picea obovata*, *Pinus sibirica* and *Abies sibirica*) are common.

In the eastern Russian Federation, the distribution of forest vegetation, species composition and the productivity of forests vary widely over the vast mountain territories. Altitudinal ranges of vegetation belts and forests, in particular, depend on such factors as geographical location, climate, height of the mountain system, slope orientation, etc. While temperature is a major limiting factor in the north, the amount of precipitation and air humidity determines the distribution of forest altitudinal belts in the south.

In the middle Siberian Plateau, larch (*Larix gmelinii*) forests grow up to 750 to 850 m in the southern part and up to 450 to 600 m on south-facing slopes in the north. In the central and eastern parts of the plateau, forests cover only small areas at the mouths of some rivers. Dark coniferous taiga dominates in West Altai, the Salair Range, Kusnetsky Ala-Tau and the northern part of West Sajan. Above the foothill belt of aspen forests with fir (*Abies sibirica*) lies a belt of fir taiga ("chernevaja") from 400-600 to 800-900 m, with aspen in lower parts and Siberian cedar pine (*Pinus sibirica*) in upper ones. Above that (up to 1 400 to 1 500 m) are typical dark coniferous forests dominated by cedar and fir, with a very modest admixture of spruce. Cedar forests occupy a subalpine belt from 1 500 to 1 800 m. The uppermost forest belt (1 800 to 2 400 m) is usually formed of cedar-larch forests. Eastern Altai has a well-developed belt of larch forests. Forests of the Tuva region are mostly represented by larch, which covers foothills and middle elevation mountains (up to 1 400 m). Cedar forms a narrow belt above the larch forests, usually in the eastern part of Tuva (up to 1 700 to 1 900 m).

An absolute dominance of larch is typical of the mountain country around Lake Baikal, usually in association with cedar and spruce (*Picea obovata*) with *Pinus pumila* in the understorey. Dwarf pine and alder form a subalpine belt. Pine forests grow in river valleys. Towards the east, dark coniferous species do not play a significant role, but pine and birch are common. Rather productive larch and pine forests are found in the east, particularly in the mountain ranges nearer the Pacific coast.

In the mountains of southwestern Russian Asia, forest vegetation is expressed in the northern part by larch forests on cold soils and spruce (*Picea ajsnensis*) forests on warmer soils, with an admixture of *Abies nephrolepis*, *Betula platyphylla* and *Pinus sylvestris*. Poplar (*Populus maximoviczii*), bird-cherry trees (*Padus asiatica* and *P. maackii*) and others are common in the lower belt. *Pinus pumila* is widely distributed in high mountain areas. To the southeast, spruce (*Picea ajanensis*) and fir (*Abies nephrolepis*), with some admixture of Korean cedar pine (*Pinus koraiensis*) and some broad-leaved species, constitute the zonal forest vegetation. Korean cedar pine, together with spruce, fir and broad-leaved species, including *Tilia amurensis*, different maples (*Acer* spp.) etc., constitute a common forest type. There is a significant admixture of *Fraxinus mandshurica*, *Ulmus laciniata* and *Juglans mandshurica*, forming a belt of mixed coniferous broad-leaved forests, mostly in river valleys and lower parts of the mountains. *Pinus koraiensis* forests have decreased considerably during the past decades owing to insufficient management. Lowlands in the lower reaches of the Amur River are covered with spruce and fir forests, as is a major part of the forest belt in northern and middle Sikhote-Alin.

In the northeastern Russian Federation (the Yukagir Upland), sparse larch (*Larix cajanderi*) forests, either single-species or in association with *Betula cajanderi*, cover extensive areas. Korean willow and popular grow in river valleys. Dwarf pine (*Pinus pumila*) covers only a small area owing to the severity of the climate. In the central part, dominated by the mountain systems of the Vekhojansky and Chersky Ranges, a subalpine belt with *Pinus pumila* is present at 1 400 to 1 800 m. Sparse larch forests form a belt between approximately 500 and 1 400 m on southern slopes. Relatively well-stocked larch forests cover the lower altitudinal belts and river valleys. Wildfires often decrease the productivity of larch forests. Four major altitudinal belts are observed to the east, in the coastal zone of the Okhotsk Sea. From low to high altitude they consist of a belt of stocked larch forests, on average up to 400 to 500 m; sparse larch forests with *Pinus pumila* from about 400-500 to 700-1 200 m; a subalpine belt dominated by *Pinus pumila*, usually above 700-1000 m to 900-1 400 m, and covering more than 50 percent of the area; and mountain tundra. To the west, on the Oimjakon Upland, continentality of climate increases significantly. *Pinus pumila* plays a significant role in the subalpine belt where precipitation is higher. *Larix cajanderi* is a major forest-forming species, sometimes with an admixture of birch and poplar.

Forests in the mild, cool and very humid climate of the coastal part of Kamchatka are mostly dominated by stone birch (*Betula ermanii*), which forms specific open park-like forests. Dwarf pine (*Pinus pumila*), bushy alder (*Duschekia kamtschatica*) and grassy-*Sphagnum* bogs with *Myrica tomentosa* can be found at the east coast and raised bogs with *Empetrum sibiricum*, *Myrica tomentosa* and *Carex middemdorfii* on the west coast. In mountain depressions along the Kamchatka River, bottoms and foothills of the depression are covered by larch and, in small areas, spruce forests. Further uphill is a belt of park-like birch forests, followed by *Pinus pumila* and *Duschekia kamtschatica*. Peaks are covered by mountain tundra. Larch (*Larix kurilensis*) forests dominate the northern part of Sakhalin Island while *Pinus pumila* and sparse forests of *Betula ermanii* occur along the coast and at the tree line. Rather productive spruce (*Picea ajanensis*) and fir (*Abies sachalinensis*) forests grow on the middle part of the island. Elements of nemoral flora are found in forests of the southern part with *Quercus mongolica*, *Fraxinus mandshurica* and others. Bamboo brakes (*Sasa kurilensis*) cover significant areas, in particular in the southern part of the island, as a result of intensive human-caused forest fires.

BIBLIOGRAPHY

Abaimov, A.P., Bondarev, A.I., Zyrjanova, O.A. & Shitova, S.A. 1997. *Polar forests of Krasnoyarsk region*. Novosibirsk, Russian Federation, Nauka.

Bohn, U., Neuhäusl, R. *et al*. 2000. *Map of the natural vegetation of Europe*. Ed. Bundesamt für Naturschutz. Bonn-Bad Godesberg, Germany.

Bryant, D., Nielsen, D. & Tangley, L. 1997. *The last frontier forests*. Washington, DC, World Resources Institute (WRI).

Chertovskoi, V.G., Semenov, B.A. & Zvetkov, V.F. 1987. *Pretundra forests*. Moscow, Agropromizdat.

European Topic Centre on Nature Conservation (ETC/NC). 2001. Report on Europe's biodiversity. Paris. (manuscript).

Gaidamaka, E.I., Rosov, N.N. & Shashko, D.I. 1983. *Nature-agricultural regionalization and use of lands in the USSR*. Moscow, Kolos.

Isachenko, T.I., Karamysheva, Z.V., Ladygina, G.M. & Safronova, I.N. 1990. *Map of vegetation of the USSR*. Scale 1:4 000 000. Moscow, Institute of Geography, RAS.

Kurnaev, S.F. 1973. Forest vegetation regionalization of the USSR. Moscow, Nauka.

Lavrenko, E.M. & Sochava, V.B. 1956. *Vegetation cover of the USSR* (Explanatory text to the geobotanical map of the USSR). Vol.1 and Vol. 2. Moscow, Academy of Sciences of the USSR.

Ogureeva, G.N. 1999. *Zones and altitudinal zonality types of vegetation of Russia and adjacent territories*. Scale 1:8 000 000. Explanatory text and legend of the map. Moscow, Moscow State University.

Stolbovoi, V., Fisher, G., Ovechkin, V.S. & Rozhkova (Kravets), S. 1998. *The IIASA-LUC project georeferenced database of the former USSR*. Vol. 4. *Vegetation*. IR-98-114. Laxenbur, Austria, International Institute for Applied Systems Analysis.

Tseplyaev, V.P. 1965. *The forests of the U.S.S.R.* Jerusalem, Israel Program for Scientific Translation.

Utkin, A.I. 1965. *Forests of Central Jakutija.* Moscow, Nauka.

Walter, H. & Breckle, S.W. 1991. *Ökologie der Erde,* Bd. 4: *Spezielle Ökologie der Gemäßigten und Arktischen Zonen außerhalb Euro-Nordasiens.* 2nd ed. Stuttgart, Germany, G. Fischer.

Walter, H. & Breckle, S.W. 1994. *Ökologie der Erde,* Bd. 3: *Spezielle Ökologie der Gemäßigten und Arktischen Zonen Euro-Nordasiens.* 2nd ed. Stuttgart, Germany, G. Fischer.

Zhukov, A.B. (ed). *Forests of the USSR.* 1966-1970. Volumes 1-5. Moscow, Nauka.

Chapter 28
Northern Europe

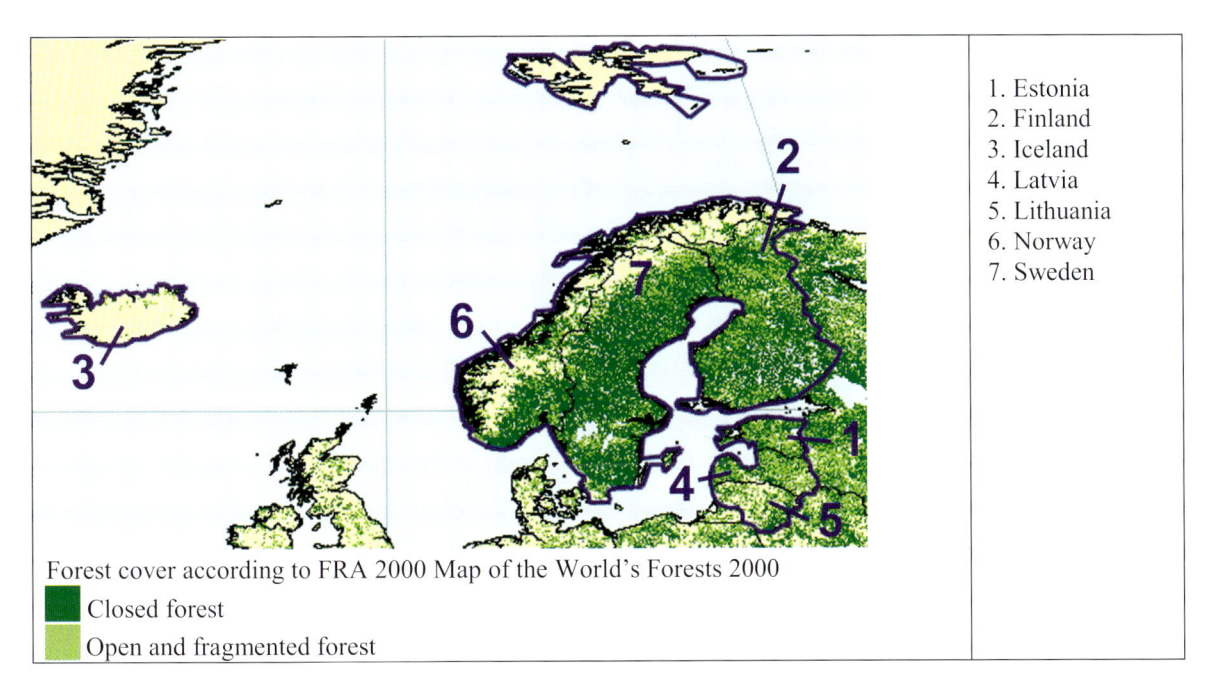

Figure 28-1. Northern Europe: forest cover map

1. Estonia
2. Finland
3. Iceland
4. Latvia
5. Lithuania
6. Norway
7. Sweden

Forest cover according to FRA 2000 Map of the World's Forests 2000
- Closed forest
- Open and fragmented forest

The subregion of Northern Europe includes the Nordic countries of Finland, Iceland, Norway and Sweden as well as the Baltic countries of Estonia, Latvia and Lithuania (Figure 28-1).[43] The total land area totals 129 million hectares, half of which is classified as forest (65 million hectares) by FRA 2000. The region extends across a wide range of climate zones, from the polar zone in the northern high-altitude areas to the moist warm-temperate zone in the southwest and the continental zone in the east. Annual precipitation varies from 300 to 3 000 mm per year depending on location.

Representative vegetation zones include the alpine, subalpine, boreal, boreal-nemoral and nemoral zones. The majority of the forests are coniferous, predominantly Scots pine (*Pinus sylvestris*) and Norway spruce (*Picea abies*), which are often mixed with broad-leaved trees such as birch (*Betula* spp.) and quaking aspen (*Populus tremula*). In the subalpine zone birch is predominant and in the nemoral zone, oak (*Quercus* spp.), beech (*Fagus sylvatica*), hornbeam (*Carpinus betulus*), ash (*Fraxinus excelsior*) and other broad-leaved trees comprise the natural tree vegetation.

Historically, forestry has played a major role in the economies of Sweden, Finland and Norway. For example, in 1999 the value of exports from the forest sector in Sweden were US$9.7 billion, and exports from Finland totalled US$10.9 billion (Sweden NBF 2001). Wood exports from the Baltic countries have increased dramatically since they gained independence in the early 1990s, much of which is imported to Finland and Sweden. At the same time, the Baltic countries have increased their industrial capacity over the last few years and are able to utilize increasing amounts of their own forest resources.

FOREST RESOURCES

The forest resources of the subregion have historically been well managed by many of the countries. For example, both forest area and timber volume have increased in Sweden, Norway and Finland since they were first inventoried early in the 1920s (Finnish Forest Research Institute 2001; Sweden. Department of Forest Resource Management and Geomatics 2001; Norwegian Institute of Land Inventory 2001). In the Baltic countries, forests increased after the Second

---
[43] For more details by country, see www.fao.org/forestry

Table 28-1. Northern Europe: forest resources and management

Country /area	Land area	Forest area 2000					Area change 1990-2000 (total forest)		Volume and above-ground biomass (total forest)		Forest under management plan	
		Natural forest	Forest plantation	Total forest								
	000 ha	*000 ha*	*000 ha*	*000 ha*	*%*	*ha/capita*	*000 ha/year*	*%*	*m³/ha*	*t/ha*	*000 ha*	*%*
Estonia	4 227	1 755	305	2 060	48.7	1.5	13	0.6	156	85	1 125	55
Finland	30 459	21 935		21 935	72.0	4.2	8	n.s.	89	50	21 900	100
Iceland	10 025	19	12	31	0.3	0.1	1	2.2	27	17	13	42
Latvia	6 205	2 780	143	2 923	47.1	1.2	13	0.4	174	93	2 923	100
Lithuania	6 258	1 710	284	1 994	31.9	0.5	5	0.2	183	99	1 938	97
Norway	30 683	8 568	300	8 868	28.9	2.0	31	0.4	89	49	7 147	81
Sweden	41 162	26 565	569	27 134	65.9	3.1	1	n.s.	107	63	27 134	100
Total Northern Europe	129 019	63 332	1 613	64 945	50.3	2.5	70	0.1	105	60	62 180	96
Total Europe	2 259 957	1 007 236	32 015	1 039 251	46.0	1.4	881	0.1	112	59	954 707	92
TOTAL WORLD	13 063 900	3 682 722	186 733	3 869 455	29.6	0.6	-9 391	-0.2	100	109	-	-

Source: Appendix 3, Tables 3, 4, 6, 7 and 9.

World War (Lithuania. Department of Forests and Protected Areas, Ministry of Environment 2001), when many farms were abandoned and reverted into forests. During the course of the last decade the overall forest area in the subregion has increased (Table 28-1). However, a point of equilibrium has been reached where afforestation and natural expansion of forests on old farmland is equal to the loss of forests due to the expansion of cities, highways and other infrastructure into once forested areas. In Sweden, preliminary figures from an evaluation of the national forest policy (Sweden NBF 2001b) show that the forest area may even have decreased somewhat during the last two decades.

The forest area per capita in Finland, Norway and Sweden is higher than for the rest of Europe and the world average. This is due to their relatively large forest areas and low populations. Many large forests are located far away from forest industries in sparsely populated areas without roads. In fact, forests are frequently the only incentive for investments in roads in these locations. In forest areas where roads do not exist, their construction adds to the costs of logging operations, making timber extraction much more expensive. There is also concern that low populations will result in an insufficient work force for forestry field operations. In Lithuania and Iceland, the forest area per capita is small in comparison.

The net annual increment is more than 220 million cubic metres per year over bark and the fellings about 150 million cubic metres per year over bark (of which approximately 6 million cubic metres per year over bark are attributed to natural losses). This accounts in part for the net annual increase of growing stocks which is close to 80 million cubic metres per year over bark from 1990 to 2000. As the forest area and stocking levels have increased since the Second World War, the total volume and biomass within the region have also increased.

The landscape in many areas has changed radically over the last 50 years. In previously open farmland, there are now often thick coniferous forests. This development has largely continued over the last two FRA reference periods (Table 28-1, Figure 28-2) (UNECE/FAO 2000). One exception is the decrease in growing stock in the late 1990s in the Baltic countries, owing to increased fellings (Lithuania. Department of Forests and Protected Areas, Ministry of Environment 2001d). However, the overall trend has helped to increase carbon sequestration in the forests of northern Europe (currently estimated at 4.7 billion tonnes, oven-dry) (UNECE/FAO 2000).

FOREST MANAGEMENT AND USES

The FAO forest plantation area estimates (Table 28-1) are fairly limited in northern Europe. Only 2 percent of the forests are considered to be in plantations. For FRA 2000, Finland reports no plantation areas and Sweden only 570 000 ha, all planted with introduced species (mostly *Pinus contorta*). However, FRA may have missed some of the plantations in the subregion. Owing to differences between national and global definitions, some of the countries were unable to supply better data on the theme within the context of their own national reporting. In fact, the major source of regeneration in the subregion following harvesting is through tree planting with one or two species. Alternatively, many countries reported these areas as "semi-natural forests" owing to their mixed composition of exotic and native species. This is commonly due to the natural seeding of native trees from the surrounding forests which continue to grow alongside planted exotics.

There are good justifications why most of the subregion's forests are considered to be semi-natural, rather than plantations. First, the geometry of the plantations is not uniform – so the resulting mature planting stock does not look like a conventional plantation. Additionally, in these same areas, regeneration is augmented by seeds supplied by the surrounding native forests, which increases their presence. The combination of these two regeneration mechanisms frequently results in the native species dominating the site until they reach about 3 to 7 m (after about 10 to 30 years). Following this period when thinnings are carried out, planted forests temporarily begin to look like conventional plantations. However, at the end of the long rotations (60 to 120 years) the forests have regained the appearance of "natural forests", except for their lack of dead wood and hollow and old trees.

About 75 percent of the forest area within the subregion is privately held (UNECE/FAO 2000). However, this figure is far from stable as there has been a rapid privatization of forest land in some countries. In the Nordic countries, the majority of the forest land has been owned privately. This increased even more in the 1990s since the majority of the State-owned Swedish forests were sold to private shareholders. Now 70 to 85 percent of the forest area is in private ownership in the four Nordic countries in the subregion. In the Baltic countries, private ownership has only recently become possible since their independence in the 1990s, although the majority of the forest land is owned by the State. Overall, State-owned lands in the subregion are less productive than the private ones. Consequently, the private forests have an even larger proportion of the production capacity than the forest land ownership figures indicate.

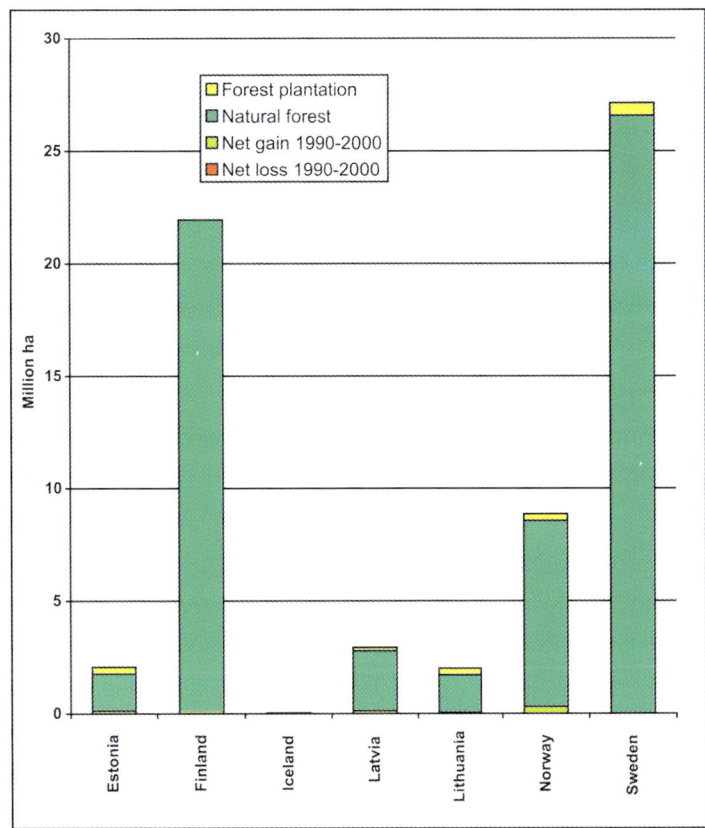

Figure 28-2. Northern Europe: natural forest and forest plantation areas 2000 and net area changes 1990-2000

The high figure of 96 percent of forest under some form of forest management plan also implies that information and knowledge about forests is abundant in the subregion. The Finnish, Swedish and Norwegian national forest surveys have all been ongoing since the 1920s. Other countries in the region also have a long history of producing national forest statistics with reliable and comparable multidate data.

Both wood and non-wood forest products (NWFP) are important in the subregion. Income from wood products represents a major portion of their national economies, with their importance at the local level being even greater. NWFP such as game-meat are important, and there is an abundance of moose, deer, game birds and other game.

Of the 65 million hectares of forests within the subregion, about 10 million hectares (16 percent) are reported as not available for wood supply (7 million hectares are under protection for conservation purposes and 3 million hectares are not available for economic reasons). In terms of percentages, this is low compared with the 30 percent not available for wood supply for all of Europe and the Russian Federation.

Protected areas may also result in forests being unavailable for wood supply, depending on their protection regime. FRA 2000 utilized the IUCN classification for protected areas with mixed results in northern Europe. For example, Sweden did not report any forests as being protected, as the IUCN scheme did not correspond well with national protected area classes. Conversely, Norway reports 26 percent of their forests as under formal protection.

CONCLUSIONS AND ISSUES

Timber resources within the subregion of northern Europe have steadily developed since the early 1900s. This is the result of long-term silvicultural work, deliberately modest levels of fellings, national forest policies and forest acts where a sustainable use of the forest has been an important objective during most of the twentieth century. In fact, a recent study by Sweden's National Board of Forestry (Sweden NBF 2000), showed that significant increases in fellings could be sustained in the country's forests. During the 1990 to 2000 period the net increase in growing stock averaged more than 1 m^3 per hectare over bark per year. This increase occurred during a period when harvesting has been greater than ever before (approximately 1 500 million cubic metres over bark for the ten-year period equivalent to 2.3 m^3 per hectare over bark per year). This has proved to be beneficial to the countries in the region and their associated industries. However, increasing public interest in the aesthetic, recreational and ecological aspects of forests have led to a rethinking of industry-oriented management practices in the forests.

During the 1970s, a large segment of the population in the countries began to recognize that timber harvesting frequently left large unsightly clearings, owing to clear-felling and mechanized regeneration. Even though these methods helped optimize the harvesting and replanting, public sentiment began to turn against them. In the 1980s, non-governmental organizations (NGOs) became stronger and showed an increasing interest in forest management, pressuring industries and, consequently, the forest owners to manage the forests in ways that would limit their impact on the biological resources and preserve their aesthetic value.

Consumers of forest products have also begun to question the ways in which their countries' forests were being managed and exploited. The result of all of these forces was that in the early 1990s industries had initiated reforms of their management practices. By this time, NGOs, the government and industries had dropped their confrontations in favour of more constructive dialogues on how to enhance both timber yield and biological diversity through forest management. For example, Sweden's forest policy of 1993 indicated that environmental goals were just as important as forest production, in contrast to its 1979 policy which was almost entirely production oriented.

In the late 1990s, the discussions between NGOs and the forest industry resulted in their cooperation on forest certification of forest estates, and by the year 2000 the majority of the forests in the Nordic countries were certified under different plans.

Forest development in the Baltic countries differs considerably from the Nordic countries for many reasons, especially those related to national politics. Since these countries regained their independence, forests have served as a ready source of badly needed capital. Much of the roundwood from the area has been exported to the Nordic countries, although the Baltic forest industries are developing fast. There is hope that the national forest industries of the Baltic countries will motivate the improved management of their forests.

The overall future for the forest sector in the subregion is good. Forest growth exceeds annual fellings and an increasing emphasis on environmental aspects of forests will help to sustain the long-term viability of the ecosystems. The prominent role of forestry in the subregion throughout recent history has helped to create strong and competent forest administrations, and stimulate research and education in various aspects of forestry. At the same time, new difficulties and challenges are emerging, such as those related to acid rain and its possible effects on forest vegetation and soils.

BIBLIOGRAPHY

Finnish Forest Research Institute. 2001. *The Finnish national forest inventory.* Statistics home page.
www.metla.fi/ohjelma/vmi/nfi-resu.htm

Lithuania. Department of Forests and Protected Areas, Ministry of Environment. 2001. *The Lithuanian Statistical Yearbook of Forestry.* Home page.
http://miskai.gamta.lt/mec/eng/index.htm

Norwegian Institute of Land Inventory. 2001. The Norwegian Institute of Land Inventory home page.
www.nijos.no/

Sweden. Department of Forest Resource Management and Geomatics. 2001. *The Swedish national forest inventory.* Statistics home page. Umeaa, Sweden, Swedish University of Agricultural and Science (SLU).
www-nfi.slu.se/

Sweden. National Board of Forestry (NBF). 2000. *Forest impact analyses 1999 (FIA 99).*
www.svo.se/ska99/newpage21.htm

Sweden. NBF. 2001a. *Statistical yearbook of forestry 2001.* Official statistics of Sweden. Jönköping, Sweden. National Board of Forestry.

Sweden. NBF. 2001b. Sweden, statistics homepage.
www.svo.se/fakta/stat/ssi/engelska/

Sweden. NBF. 2001c. National board of foresty home page. Skogsvårdsorganisationens utvärdering av skogspolitikens effekter (SUS) 2001. Sweden SUS 2001.
www.svo.se/sus

UNECE/FAO. 2000. *Forest resources of Europe, CIS, North America, Australia, Japan and New Zealand: contribution to the global Forest Resources Assessment 2000.* Geneva Timber and Forest Study Papers 17. New York and Geneva, United Nations.
www.unece.org/trade/timber/fra/pdf/contents.htm

Chapter 29
Central Europe

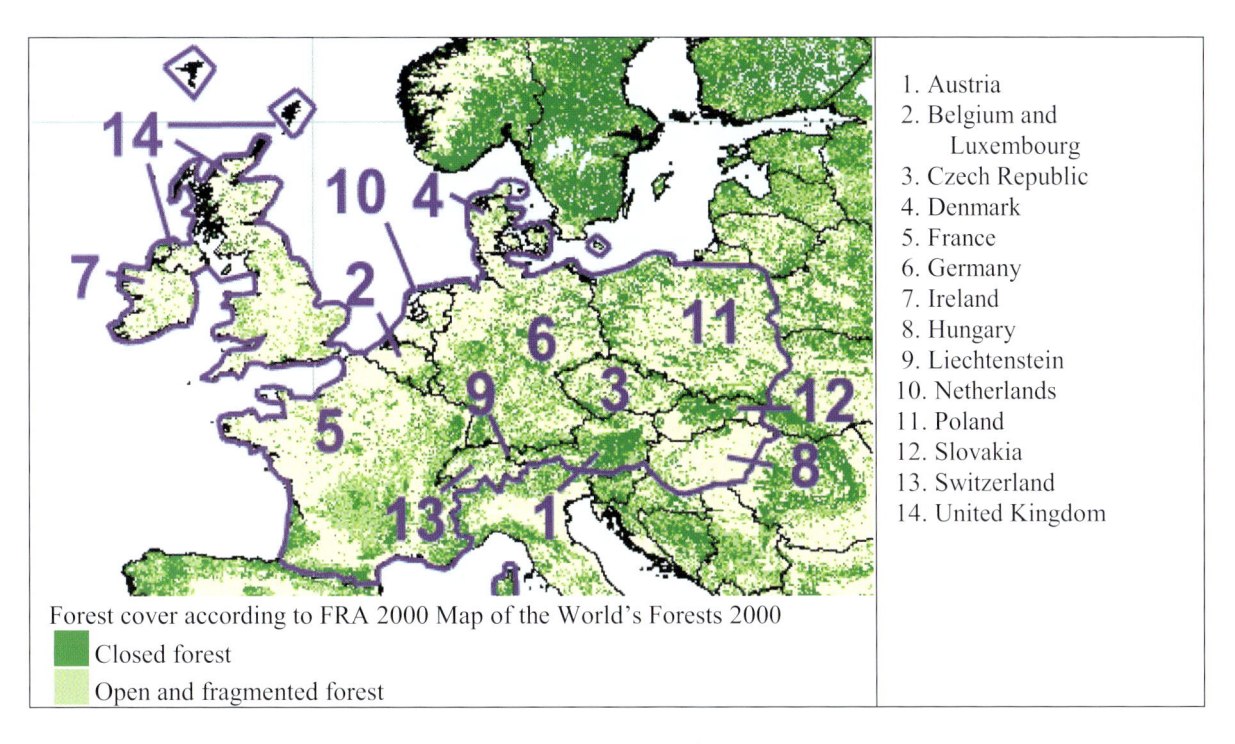

Figure 29-1. Central Europe: forest cover map

Forest cover according to FRA 2000 Map of the World's Forests 2000
- Closed forest
- Open and fragmented forest

1. Austria
2. Belgium and Luxembourg
3. Czech Republic
4. Denmark
5. France
6. Germany
7. Ireland
8. Hungary
9. Liechtenstein
10. Netherlands
11. Poland
12. Slovakia
13. Switzerland
14. United Kingdom

The 15 countries comprising this subregion are Austria, Belgium, the Czech Republic, Denmark, France, Germany, Hungary, Ireland, Liechtenstein, Luxembourg, the Netherlands, Poland, Slovakia, Switzerland and the United Kingdom[44] (Figure 29-1). All are industrially advanced countries and are important per capita users of wood products. Most are net importers of wood products, the exceptions being Austria, the Czech Republic and Poland, which are net exporters. The climate in the subregion is temperate, generally moist and cool; it is influenced by the Atlantic Ocean to the west, but becoming increasingly continental with hard winters to the east, and Mediterranean with hot dry summers in the southern part of France. France, Germany and Poland are the largest countries, accounting for three-fifths of the total land area of the group of 196 million hectares. Germany, the United Kingdom and France are the most densely populated.

FOREST RESOURCES

About one-quarter of the land area, 52 million hectares, is covered with forest, while a further 2.2 million hectares are classed as other wooded land, of which the largest part is in France. France, Germany and Poland account for two-thirds of the forest land in the subregion, France alone for 30 percent.

Until the arrival of humans, forest covered a large part of the land area, and temperate broad-leaved species made up most of the natural forest cover in the subregion. In past centuries, much of that cover has been removed to make way for agriculture and other land uses, and nearly all the remaining forest has been disturbed or modified, mostly by being brought under management. Today, there are only scattered remnants of forest undisturbed by humans; it is estimated that there are less than a quarter of a million hectares of such forest in the subregion, the largest area being in Poland (144 000 ha). Most of the forest is classed as semi-natural forest which, together with forest undisturbed by humans, makes up the total area of natural forest of 47.8 million hectares (Table 29-1). It should be noted that the term

---
[44] For more details by country, see www.fao.org/forestry

Table 29-1. Central Europe: forest resources and management

Country/area	Land area	Forest area 2000 – Natural forest	Forest area 2000 – Forest plantation	Forest area 2000 – Total forest			Area change 1990-2000 (total forest)		Volume and above-ground biomass (total forest)		Forest under management plan	
	000 ha	000 ha	000 ha	000 ha	%	ha/capita	000 ha/year	%	m³/ha	t/ha	000 ha	%
Austria	8 273	3 886		3 886	47.0	0.5	8	0.2	286	250	3 886	100
Belgium and Luxembourg	3 282	728		728	22.2	0.1	-1	-0.2	218	101	656	90
Czech Republic	7 728	2 632		2 632	34.1	0.3	1	n.s.	260	125	2 632	100
Denmark	4 243	114	341	455	10.7	0.1	1	0.2	124	58	455	100
France	55 010	14 380	961	15 341	27.9	0.3	62	0.4	191	92	15 341	100
Germany	34 927	10 740		10 740	30.7	0.1	n.s.	n.s.	268	134	10 740	100
Hungary	9 234	1 704	136	1 840	19.9	0.2	7	0.4	174	112	1 840	100
Ireland	6 889	69	590	659	9.6	0.2	17	3.0	74	25	551	84
Liechtenstein	15	7		7	46.7	0.2	n.s.	1.2	254	119	7	100
Netherlands	3 392	275	100	375	11.1	n.s.	1	0.3	160	107	375	100
Poland	30 442	9 008	39	9 047	29.7	0.2	18	0.2	213	94	9 047	100
Slovakia	4 808	2 162	15	2 177	45.3	0.4	18	0.9	253	142	1 988	91
Switzerland	3 955	1 195	4	1 199	30.3	0.2	4	0.4	337	165	1 153	96
United Kingdom	24 160	866	1 928	2 794	11.6	n.s.	17	0.6	128	76	2 319	83
Total Central Europe	196 358	47 766	4 114	51 880	26.4	0.2	152	0.3	222	117	50 990	98
Total Europe	2 259 957	1 007 236	32 015	1 039 251	46.0	1.4	881	0.1	112	59	954 707	92
TOTAL WORLD	13 063 900	3 682 722	186 733	3 869 455	29.6	0.6	-9 391	-0.2	100	109	-	-

Source: Appendix 3, Tables 3, 4, 6, 7 and 9.
Note: Belgium and Luxembourg are reported together, as they are in SOFO 2001.

"natural" forest as used in this report refers, so far as the industrialized temperate and boreal countries are concerned, to all forest that is not put under the heading of plantations. In the case of central Europe, this means that more than 90 percent of all forests are classed here as natural (Figure 29-2). There is an area of 4.1 million hectares of plantations, the largest areas being in the United Kingdom, France and Ireland. In many countries there are large areas of semi-natural forest (included in natural forest in the present document) that started out as plantations but have lost their plantation-like appearance as they have matured.

Predominantly broad-leaved and mixed broad-leaved/coniferous forests make up more than half the area of forest in the subregion. The share of predominantly coniferous forest has been increasing over the past two centuries or so as a result of management practices to encourage these species, including afforestation, primarily for wood production purposes. In recent years, the trend towards more coniferous forest has slowed down or even been reversed, as policies have evolved to encourage greater use of broad-leaved species in restocking to improve biodiversity and for other environmental and social reasons. Countries where broad-leaved species predominate include France, Hungary and Slovakia, while conifers predominate in Germany, Austria, Poland, the United Kingdom and Ireland. In the last two countries, this has been the result of active afforestation programmes since the First World War.

There is a wide diversity of forest types in the subregion, epitomized by France. Over the western and central parts of the country, broad-leaved forests predominate with beech (*Fagus sylvatica*) and oak (*Quercus* spp.) the most common species. To the east and in the mountainous areas of the Alps and the Pyrenees, conifers are the main species, notably spruce and fir, often mixed with beech, while in the southwest (Les Landes) there is the largest area of human-made coniferous forest in Europe, consisting of maritime pine (*Pinus pinaster*). To the south there is Mediterranean-type vegetation with pines and oaks as well as considerable areas of maquis and scrub. Coppice and coppice with standards are a common feature in many parts of the country and account for nearly half the total forest area. The active programme of reforestation (partly to replace coppice) and afforestation has included the use of certain exotic species, notably Douglas fir (*Pseudotsuga* spp.), as well as poplars (*Populus* spp.).

The composition of Germany's forests has been heavily influenced by management practices over the past two centuries, which emphasized the use of coniferous species in replanting. Today, more than half the forest area is predominantly coniferous, with a further fifth mixed coniferous/broad-leaved. Two thirds of the growing stock volume is coniferous. The forests are concentrated in the southern, central and eastern parts of the country, with relatively little on the northern plain. The main coniferous species are Norway spruce and Scots pine (*Pinus sylvestris*), and beech and oak the commonest broad-leaved species. The average volume of growing stock per hectare in Germany, as in other countries in central Europe, is very high, as is the net annual increment per hectare.

Austria and Switzerland share some of the same forestry features as Germany, including a preponderance of coniferous species: in Austria 88 percent of the forest area area is predominantly coniferous or mixed coniferous/broad-leaved species; in Switzerland the proportion is 77 percent. Much of their forest area is mountainous, which influences the species composition in favour of coniferous, as well as the functions of the forest. Protection against avalanches and landslides is of considerable importance in these countries. The volume of growing stock per hectare in Switzerland and Austria is the highest and second highest in Europe, exceeding 300 m^3 per hectare. Wood production is particularly important in Austria, which is the only major net exporter of wood products in the subregion (Poland and the Czech Republic are also small net exporters).

Two thirds of Poland's forest area is predominantly coniferous, with a further fifth mixed coniferous/broad-leaved. Scots pine is the main coniferous species, and oak the main broad-leaved species. Most of Poland's forests are classed as semi-natural but, as mentioned earlier,

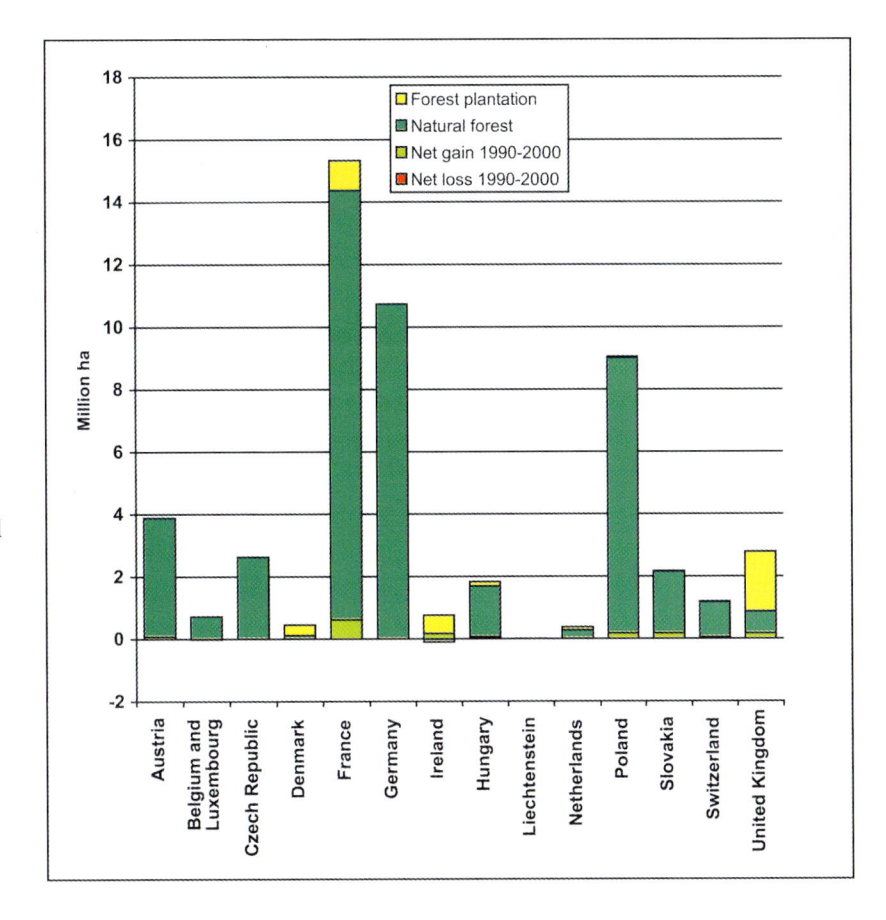

Figure 29-2. Central Europe: natural forest and forest plantation areas 2000 and net area changes 1990-2000

there are 144 000 ha of forest undisturbed by humans and an even larger area of forest not available for wood supply for conservation reasons or under some form of protection. In the Czech Republic, over half of the forest area is mixed coniferous/broad-leaved, although as much as four fifths of the growing stock volume consists of coniferous species. Norway spruce is the most important, with European larch and Scots pine, while beech is the commonest broad-leaved species. All forest is classed as semi-natural and nearly all is available for wood supply; only a small area is not available for conservation reasons.

Broad-leaved species predominate in Slovakia and Hungary; in the latter country they account for nearly nine tenths of the forest area, one of the highest proportions in the temperate and boreal regions. Beech and oak are the main broad-leaved species in Slovakia (and Norway spruce, silver fir and pines among the coniferous species), while in Hungary black locust and poplars are also very important. Young stands are overrepresented in that country as a result of reforestation and afforestation and short rotations for some species.

Belgium, Denmark, Luxembourg and the Netherlands are relatively lightly forested, apart from the Ardennes hills region of Belgium and Luxembourg. In Denmark and the Netherlands, forest cover is not much more than one-tenth. In the Netherlands, with its very dense population, the area of forest per capita, 0.02 ha, is the lowest in Europe. The forest area and growing stock volume in these countries is divided in roughly equal parts between broad-leaved and coniferous species, the main species being oak, beech and Norway spruce, with other species included in reforestation and afforestation programmes such as poplars, pines and Douglas fir.

Ireland and the United Kingdom are the two countries in the subregion, and indeed in Europe, with the highest proportion of plantations in their total forest area, with 90 percent and 69 percent, respectively. Until recently, the bulk of planting was of coniferous species, notably Sitka and Norway spruce, but also several other species such as pines, larches and Douglas fir. As a result the share of conifers (including mixed coniferous/broad-leaved species) in the total forest area has increased considerably over the past 80 years or so, reaching 86 percent in Ireland and 68 percent in the United Kingdom by the end of the twentieth century. Over the same period, afforestation raised the level of forest cover from a very low level to around one-tenth, still low by average European standards.

In the subregion as a whole, the area of forest increased during the 1990s by about 150 000 ha, or 0.3 percent a year. The largest area expansion was in France, with 62 000 ha, but there were appreciable increases also in Germany, Ireland, Poland, Slovakia and the United Kingdom. The fastest growth was in Ireland with 3 percent per annum. The area increases shown in Table 29-1 are net changes after allowing for loss of forest to other land uses, notably urbanization and communications infrastructure. Increases in the area of forest were mainly a result of afforestation (planting) and the conversion of other wooded land to forest, although in a few countries, notably France, natural colonization of non-forest land, mostly abandoned agricultural land, occurred.

While the environmental and social functions of the forest in all countries of the subregion have increased in absolute and relative importance in recent decades, wood production remains, and is likely to remain in the foreseeable future in most areas, the single most important function. Exceptions to this generalization include the Netherlands, parts of Denmark and the United Kingdom, areas with high population density where the recreation and nature conservancy uses of the forest are particularly important. Wood removals in the late 1990s from forest available for wood supply, which accounted for the bulk of the total, amounted to 156 million cubic metres under bark, with France, Germany, Poland and Austria the largest producers. After including the volume of bark on the felled wood and the volume of unrecovered fellings (harvesting losses), the equivalent volume of fellings was about 217 million cubic metres over bark. That volume may be compared with the annual volume of growth as measured by net annual increment (NAI), in order to obtain an idea of the net change in growing stock. The volume of NAI on forest available for wood supply, as reported by the countries in the subregion, was 366 million cubic metres over bark in the late 1990s. Accordingly, fellings were only about 59 percent of NAI, resulting in an appreciable expansion in the volume of growing stock. This is a phenomenon that has been occurring over several decades and is common to all the constituent countries. The fellings per NAI percentage or ratio is not, by itself, a reliable indication of the potential to increase fellings or of the sustainability of the forest resource, especially where the age-class structure of the forest is oriented towards less mature stands, as in Hungary, Ireland and the United Kingdom. There could also be environmental and practical reasons why it is unlikely that fellings could be raised to the level of NAI on a sustainable basis. Nevertheless, there could be scope in most of the countries to expand wood production without any risk of straining the sustainability of the forest resource.

FOREST MANAGEMENT AND USES

All the countries in central Europe provided national-level information on the forest area managed (Table 29-1), applying the definition used by industrialized countries – i.e. forests managed in accordance with a formal or an informal plan applied regularly over a sufficiently long period (five years or more) and including areas where a decision had been taken not to undertake any management interventions. The reported figures ranged from 83 percent of the total forest area in 2000 (United Kingdom) to 100 percent in most countries. In total, approximately 51 million hectares, or 98 percent of the total forest area in central Europe, were reported as being managed in accordance with a formal or informal plan.

Among the countries in the subregion there are three main types of forest ownership: by the State, by other public bodies such as communes and municipalities and by private individuals. Other types exist, for example by private institutions and corporations and by forest industries, but these are less important. The pattern of ownership varies from country to country as a result of historical, political and social influences, but on average in the subregion the ownership distribution is: by the State 36 percent; by other public bodies 13 percent; by private individuals 43 percent; by others 8 percent. These proportions refer to the ownership of forest available for wood supply, which in the countries of the subregion accounts for most of the forest area. Countries where ownership by the State accounts for the major share include Poland, the Czech Republic and Hungary (81, 71 and 63 percent, respectively), which with Slovakia (43 percent) were formerly centrally planned economies but are in transition to forms of market economy and part of whose forest estates are in the process of privatization or restitution. The shares of State forest are likely to continue to fall in these countries. Of the other countries, only Ireland has the larger part of its forests in State ownership (66 percent), while in the United Kingdom it is also important (42 percent). In both these countries, the State acquired land for its afforestation programmes, although during the 1990s some State forests were sold back to the private sector. In Germany, the State owns 33 percent of forest available for wood supply, the relatively high figure being explained by total State ownership in the eastern Länder dating from the time before reunification. In other countries, the State generally owns a modest share of the forest estate.

In several countries, communes, municipalities and other public bodies other than the State are important forest owners, notably in Belgium, France, Germany, Liechtenstein, Luxembourg and Switzerland. In Switzerland, the share of total ownership is as high as 65 percent. The biggest areas of forest available for wood supply owned by public bodies other than the State are in France and Germany, with 2.3 million and 2.0 million hectares, respectively.

The highest proportions of forest available for wood supply owned by private individuals are found in Austria and France, with 69 and 62 percent, respectively, but this type of ownership is also important in Belgium, Denmark, Germany, Ireland and the United Kingdom. In many cases, forest ownership is linked with farming and the owners live near to the forest and derive part of their income from it. There has been a trend in some countries, however, partly associated with the population drift from the countryside to towns, towards an increasing share of absentee ownership, sometimes resulting in neglect of their forest properties. The number of private forest owners in the subregion runs into the millions – more than three and a half million in France alone – and the average size of privately owned forests is small, probably less than 5 ha. This complicates the problems of organizing efficient management and achieving profitability of forest operations, although on many of the smaller properties commercial wood production is not the most important function.

There are substantial areas of forest by private institutions or corporations in France, as well as in Hungary, the Netherlands, Slovakia and the United Kingdom. In the case of the Netherlands and the United Kingdom, a feature in recent years has been the acquisition of forest (and other land) by nature conservancy organizations to be managed as wildlife habitats, nature reserves, etc. In Hungary, cooperatives have been formed to manage forest on behalf of private owners, while there and in Slovakia areas still in the process of restitution have been included in the "other private ownership" category. The only country in the subregion with forest owned by forest industries is the United Kingdom where, however, the area is small.

The type of ownership, and more particularly the size of holdings, provide an indication of the intensity of forest management. The data for "forest under management plan" in Table 29-1 are based on information provided by countries on areas managed in accordance with a formal or an informal plan applied regularly over a sufficiently long period. Furthermore, a decision not to manage an area at all, for example to preserve it as a wilderness area or nature reserve, also qualified it as being managed. Although the Table shows for several of the countries that most or all of their forest is managed, it does not give an indication of how much of the area is being satisfactorily managed. In the past, management was usually directed primarily towards wood production, but this has been shifting towards a multifunction approach, with increasing emphasis on non-wood goods and services. As a generalization, it may be said that the quality of management is good or at least adequate on practically all publicly owned forest and most of the larger private forests, whether owned by individuals or institutions or corporations. As mentioned above, providing good management on smaller properties is more problematic, except where they can be grouped into some kind of cooperative.

CONCLUSIONS AND ISSUES

The foregoing has pointed to a number of issues with important policy implications. One concerns the need to adapt management and silviculture to the changing pattern of demands by society on the forest, notably the increasing emphasis on the environmental and social functions, and the actual or potential impact on the "traditional" wood supply function. The countries of the subregion are mostly densely populated and largely urbanized and have high standards of living; their needs for benefits from the countryside are increasingly diverse, both material and other. Many of them have well-developed forest industries which will continue to depend on roundwood supplies from the forest. However, industry will increasingly use other raw materials, such as waste paper and industrial residues. It will be a major challenge to maintain the economic viability of the forest sector, while at the same time ensuring that it provides the non-market goods and services that are increasingly in demand.

One type of pressure faced by all countries is to remove a greater part of their forest resource from wood production for environmental protection reasons, especially for the preservation of biodiversity – the protection of rare species of fauna and flora. Targets have been set in some countries of the proportion of forest to be classed in this way. Another development has been to adapt silvicultural practices to enhance biodiversity and sustainability, for example to transform coniferous monocultures to stands with a range of species, notably by the introduction of broadleaves. These and other measures, for instance the lengthening of rotations, will take a long time to achieve but may eventually have an impact on the quantity of wood harvested, although it is very difficult to assess how great an impact.

Since UNCED in 1992 and the Second Ministerial Conference on the Protection of Forests in Europe, held in Helsinki in 1993, all European countries, not least those in the subregion, have been giving increased attention to ensuring that their forests are managed according to the principles of sustainable forest management. Several international and national schemes have been introduced to allow owners, both public and private, to certify that their forests are managed in accordance with the principles of sustainable forest management and that the products passing along the chain of custody through the forest industries, trade and commerce to the final consumer come from such forests. The extent to which certification has been embraced by forest owners in the subregion has varied from country to country and among the different classes of owner. The likely cost of certification and doubts about its benefits have often inhibited smaller private owners from adopting it.

The countries in the eastern part of the subregion, in transition towards market economies, inherited generally well-managed forests from the previous regimes, but often dilapidated forest industries and infrastructure. They face enormous tasks of modernizing industry and institutions, as well as of privatization and restitution. Among the problems they face are that of raising the living standards of their populations towards the European average while carrying out the necessary measures to improve environmental quality, which had often been neglected by the previous regimes. Privatization of forests, while politically and socially justified, has brought certain problems in its wake, such as how to ensure the continuation of acceptable levels of sustainable management, of environmental protection and access by the public.

Grave concerns arose during the 1980s and 1990s about the health condition of the forests in the subregion, especially the impact of air pollution. An increasing proportion of both conifers and broadleaves were observed to be suffering loss of foliage, while the number of dead and dying trees rose. Further research seemed to show, however, that air pollution by itself was the cause of tree mortality only in extreme cases and that probably it was a combination of causes, including climatic conditions and past silvicultural practices, for example the establishment of stands of species outside their natural range, that was resulting in loss of vitality, with air pollution an important contributory factor. Other health concerns were the decimation of the elm (*Ulmus* spp.) population in the subregion as a result of the accidental importation of a particularly virulent virus from

North America, and a widespread decline in the health of oaks. Several heavy storms within a relatively short period, the latest at the end of 1999, caused severe damage to forests. This raised the question of whether there might be a link with possible changes in climate and the accumulation of greenhouse gases in the atmosphere. Whether or not such a link exists, the possible role of forests as a sink for carbon dioxide has entered policy debates. For example, the extensive establishment of new plantations has come under consideration, although the relative shortage of suitable land in the subregion would probably limit the possibilities for countries to contribute in more than a minor way to such an initiative.

BIBLIOGRAPHY

UNECE/FAO. 2000. *Forest resources of Europe, CIS, North America, Australia, Japan and New Zealand: contribution to the global Forest Resources Assessment 2000.* Geneva Timber and Forest Study Papers 17. New York and Geneva, United Nations. www.unece.org/trade/timber/fra/pdf/contents.htm

Chapter 30
Southern Europe

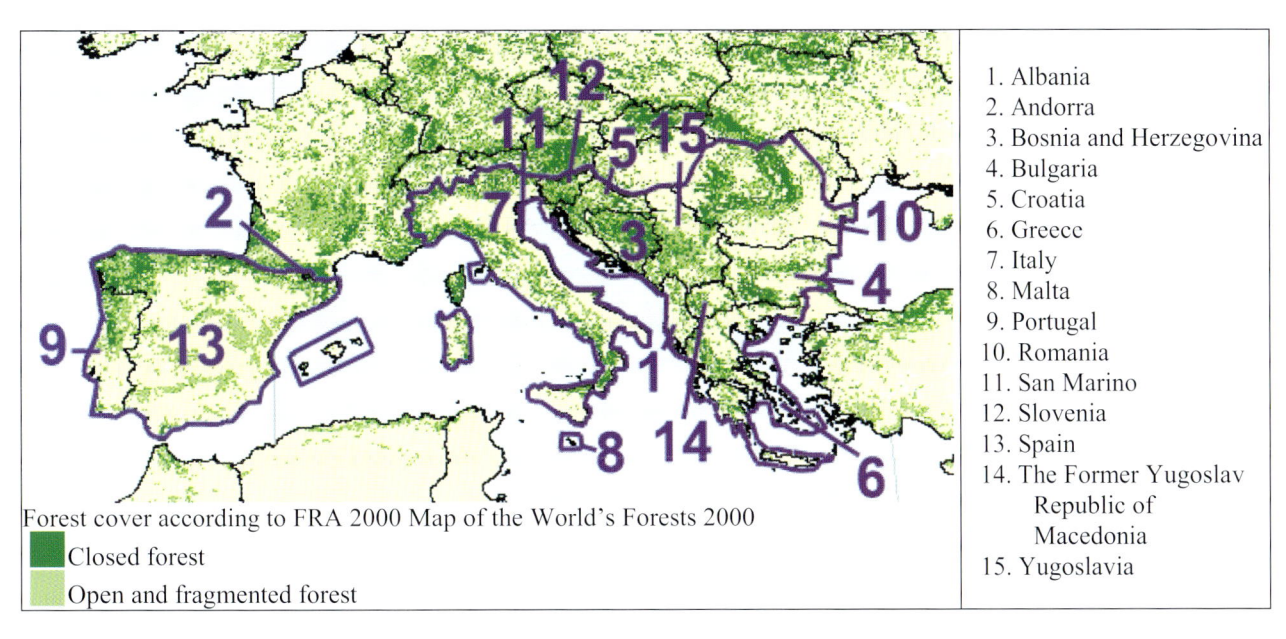

Figure 30-1. Southern Europe: forest cover map

1. Albania
2. Andorra
3. Bosnia and Herzegovina
4. Bulgaria
5. Croatia
6. Greece
7. Italy
8. Malta
9. Portugal
10. Romania
11. San Marino
12. Slovenia
13. Spain
14. The Former Yugoslav Republic of Macedonia
15. Yugoslavia

Forest cover according to FRA 2000 Map of the World's Forests 2000
- Closed forest
- Open and fragmented forest

The 15 countries[45] which make up this subregion are Albania, Andorra, Bosnia and Herzegovina, Bulgaria, Croatia, Greece, Italy, Malta, Portugal, Romania, San Marino, Slovenia, Spain, The Former Yugoslav Republic of Macedonia and Yugoslavia (Figure 30-1). Most of them border the northern and eastern rim of the Mediterranean Sea; two of them, Bulgaria and Romania, lie on the western coast of the Black Sea; and Portugal has a coastline along the Atlantic Ocean. Over most of the subregion a Mediterranean-type climate prevails, characterized by hot, dry summers, although in some areas, for example northern parts of Spain, Italy, Romania and Slovenia, there is considerable precipitation providing good conditions for forest growth. There are very marked differences between countries in the stage of economic development and living standards, those that are members of the European Union (Greece, Italy, Portugal and Spain) being more advanced than those that are in transition from centrally planned to market economies (Albania, Bosnia and Herzegovina, Bulgaria, Croatia, Romania, Slovenia, The Former Yugoslav Republic of Macedonia and Yugoslavia). Three countries, Spain, Italy and Romania, account for nearly two-thirds of the total land area of 168 million hectares in the subregion; they are also the most heavily populated. Three countries, Andorra, Malta and San Marino, are very small and have little significance for the forest economy of the subregion.

FOREST RESOURCES

Forests covered 52 million hectares in the subregion in 2000 (Table 30-1), with other wooded land covering a further 19 million hectares. Forest and other wooded land thus accounted for two-fifths of the total land area, forest alone for 30 percent. On average, there was 0.3 ha of forest per inhabitant in the subregion, but there were quite large variations among countries, ranging from 0.6 ha per capita in Bosnia and Herzegovina and Slovenia to 0.2 ha per capita in Italy.

Since ancient times there has been a history of forest destruction to make way for agriculture and other land uses, as well as degradation from excessive utilization and grazing, notably by goats. A further hazard, especially because of the climatic conditions, has been fire, largely caused by humans through negligence or arson. Many of the remaining forests are therefore in poor

---

[45] For more details by country, see www.fao.org/forestry

Table 30-1. Southern Europe: forest resources and management

Country/area	Land area	Forest area 2000			Area change 1990-2000 (total forest)		Volume and above-ground biomass (total forest)		Forest under management plan	
		Natural forest	Forest plantation	Total forest							
	000 ha	000 ha	000 ha	000 ha	%	ha/capita	000 ha/year	%	m³/ha	t/ha	000 ha	%
Albania	2 740	889	102	991	36.2	0.3	-8	-0.8	81	58	406	41
Andorra	45	-	-	-	-	-	-	-	0	0	n.a.	n.a.
Bosnia and Herzegovina	5 100	2 216	57	2 273	44.6	0.6	n.s.	n.s.	110	-	2 007	88
Bulgaria	11 055	2 722	969	3 690	33.4	0.4	20	0.6	130	76	3 690	100
Croatia	5 592	1 736	47	1 783	31.9	0.4	2	0.1	201	107	1 531	86
Greece	12 890	3 479	120	3 599	27.9	0.3	30	0.9	45	25	2 009	56
Italy	29 406	9 870	133	10 003	34.0	0.2	30	0.3	145	74	1 117	11
Malta	32	n.s.	0	n.s.	n.s.	-	n.s.	n.s.	232	-	n.s.	100
Portugal	9 150	2 832	834	3 666	40.1	0.4	57	1.7	82	33	1 201	33
Romania	23 034	6 357	91	6 448	28.0	0.3	15	0.2	213	124	6 448	100
San Marino	6	-	-	-	-	-	-	-	0	0	n.a.	n.a.
Slovenia	2 012	1 106	1	1 107	55.0	0.6	2	0.2	283	178	1 107	100
Spain	49 945	12 466	1 904	14 370	28.8	0.4	86	0.6	44	24	11 694	81
The Former Yugoslav Republic of Macedonia	2 543	876	30	906	35.6	0.5	n.s.	n.s.	70	-	906	100
Yugoslavia	10 200	2 848	39	2 887	28.3	0.3	-1	-0.1	111	23	2 723	94
Total Southern Europe	163 750	47 397	4 327	51 723	31.6	0.3	233	0.5	112	60	34 839	67
Total Europe	2 259 957	1 007 236	32 015	1 039 251	46.0	1.4	881	0.1	112	59	954 707	92
TOTAL WORLD	13 063 900	3 682 722	186 733	3 869 455	29.6	0.6	-9 391	-0.2	100	109	-	-

Source: Appendix 3, Tables 3, 4, 6, 7 and 9.

condition or reduced to scrub and brushland with scattered trees (other wooded land). Less than 700 000 ha, or only about 1.5 percent of the forest area, are classed as undisturbed by humans, the largest areas being reported in Bulgaria and Romania. By 2000, plantations covered about 4.3 million hectares, or more than 8 percent of the forest area, the largest areas being reported in Spain, Bulgaria and Portugal. A distinction should be made between those plantations established primarily for wood production purposes and those whose function is soil stabilization and environmental protection. Although data are not available to separate these categories, much of the planting in Bulgaria and central and southern Spain fall into the latter, and in some cases the rate of growth is quite modest. Plantations in Portugal and along the northern coastal area of Spain are mainly for wood production, many of them with high growth rates, for example of maritime pine (*Pinus pinaster*) and radiata pine (*Pinus radiata*) and *Eucalyptus globulus*. The natural forest area of more than 47 million hectares (Table 30-1) includes the small area of forest undisturbed by humans mentioned above, but consists mostly of what was defined as "semi-natural" forest in the Temperate and Boreal Forest Resources Assessment 2000 (TBFRA) (UNECE/FAO 2000). This is forest which has been used by humans in the past or is being used, with or without being brought under management.

During the 1990s, the area of forest in southern Europe increased at an average annual rate of about 230 000 ha. In order of importance, the increases occurred in Spain, Portugal, Italy, Greece and Bulgaria. Part of the expansion was the result of regeneration of forest on other wooded land. The rest was due to recolonization by artificial (planting) or natural means on non-forest land, mainly abandoned agricultural land. It is not possible to determine how much of the afforestation and reforestation was through planting and how much took place naturally, but it is probable that most was by planting. As shown in Figure 30-2, there were net gains in all countries in the subregion with the exception of Albania, one of the few countries in Europe where the area of forest declined.

In terms of forest cover, broad-leaved species predominate in southern Europe, particularly in the eastern part of the subregion. Overall, predominantly broad-leaved forest makes up over three-fifths of the forest area, with a further 10 percent mixed broad-leaved/coniferous. Because in some countries the average growing stock volume per hectare in coniferous stands is greater than that in broad-leaved stands, conifers account for a half or more of total growing stock in Bosnia and Herzegovina, Greece, Portugal and Spain, and almost half in Slovenia. Oaks (*Quercus* spp.), both deciduous and evergreen species, are the most common broad-leaved species throughout the subregion, with beech (*Fagus* spp.) also common at higher altitudes, while other species include chestnut, poplar and eucalyptus, the last two mostly in plantations. There are extensive areas of broad-leaved coppice and coppice with standards, notably in Italy, Greece, Spain, Bulgaria and Yugoslavia. Altogether about one-quarter of the forest area in the subregion consists of coppice and coppice with standards. Among coniferous species, pines, notably Aleppo (*Pinus halepensis*), Scots (*Pinus sylvestris*), maritime and radiata, are the most common, with spruce (*Picea* spp.), fir (*Abies* spp.) and larch (*Larix* spp.) also found in certain localities.

Spain, which occupies the major part of the Iberian Peninsula, has the largest forest area in the subregion, with 14.4 million hectares or more than one-quarter of the total. It has a further 12.5 million hectares of other wooded land, so that forest and other wooded land accounts for about half the country's land area. The area of forest has been expanding strongly as a result of planting and conversion of other wooded land to forest, despite setbacks due to forest fires. More than two-fifths of the forest area is predominantly coniferous, with a further one-fifth mixed coniferous/broad-leaved. The strongest growth occurs in the maritime and radiata pine and eucalyptus stands in the northern part of Spain, where a large part of wood production also takes place. Elsewhere, forest serves an important soil protection function. About one-quarter of the forest area is not available for wood supply, mainly for reasons of conservation and protection. Portugal, occupying the western part of the Iberian Peninsula, has extensive areas of cork oak and is the world's leading producer and exporter of cork products. It is also an important net exporter of wood products based on its forests of maritime pine and eucalyptus.

Because of its long north-south extension and range of altitudes, Italy possesses a large variety of forest types and of flora and fauna. With 10 million hectares, it has the largest area of forest in the subregion after Spain, and nearly

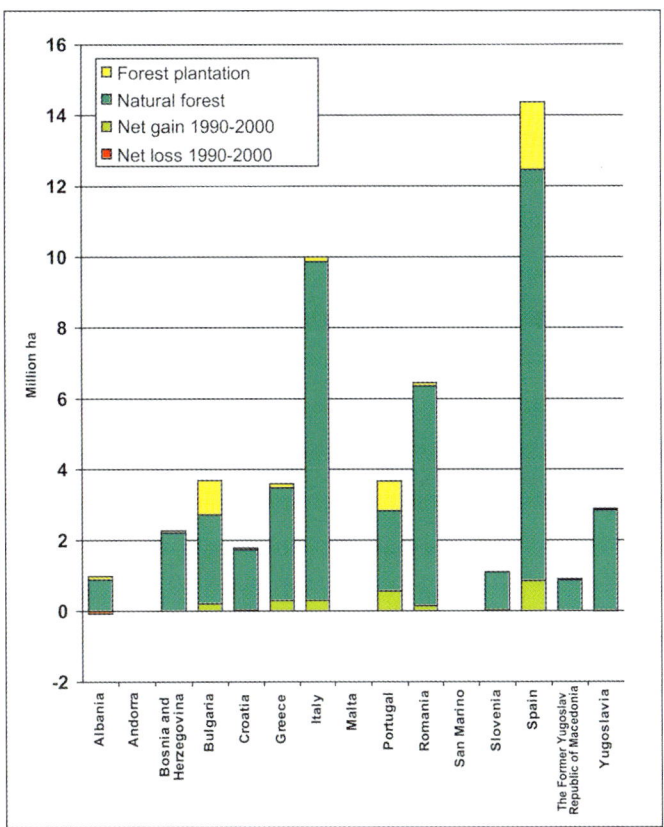

Figure 30-2. Southern Europe: natural forest and forest plantation areas 2000 and net area changes 1990-2000

1 million hectares of other wooded land. Predominantly broad-leaved stands make up more than 70 percent of the forest area, of which about half is coppice and coppice with standards. As elsewhere in the Mediterranean area, forest fires are an annual hazard. With 0.2 ha per capita, Italy has the lowest forest area per inhabitant among the countries of the subregion, and is a major net importer of primary wood products (although a large exporter of furniture).

The countries that were formerly part of Yugoslavia, namely Bosnia and Herzegovina, Croatia, Slovenia, The Former Yugoslav Republic of Macedonia and Yugoslavia, between them have about 9 million hectares of forest and 1.5 million hectares of other wooded land. In most of this area, mainly broad-leaved stands predominate, as is also the case in Greece and Albania, also on the Balkan Peninsula. Growing conditions in these countries, with the exception of Slovenia, are poor in many places, with degraded soils, and forest fires are frequent.

Bulgaria and Romania are the two most easterly countries of the subregion. Between them they possess over 10 million hectares of forest, in which predominantly broad-leaved stands are the major part and beech and oak are the most common species. The structure of Romania's forests is oriented towards the medium-age classes, and increment, which exceeds fellings by a considerable margin, is above the European average on a per hectare basis. This is also the case for its per hectare volume of growing stock. In Bulgaria, there is an active programme of afforestation and forest improvement, more for soil protection reasons than for wood production; growth is appreciably higher than present cutting levels.

Because of the large areas of hills and mountains, fragile soils, difficult climatic conditions and the risk of forest fires, the protection role of forests is important in many parts of southern Europe. About one-quarter of the forest area is not available for wood supply, mainly for conservation and protection reasons, but in some localities also for economic reasons, that is to say inaccessibility. The largest areas of forest not available for wood supply are found in Italy, Spain, Portugal and Bosnia and Herzegovina. Roundwood removals amounted to about 54 million cubic metres under bark per year in the late 1990s on forest available for wood supply, where the major part of total removals occurred. After adding the on-bark percentage on the felled trees and unrecovered volumes, fellings (removals plus unrecovered harvesting losses) on forest available for wood supply amounted to more than 65 million cubic metres over bark, the largest volumes being recorded by Romania, Portugal, Spain and Italy. This volume is barely half of the volume of net annual increment, meaning that the volume of growing stock is rising quite strongly in the subregion. The extent to which the net annual increment (NAI) exceeds fellings varies considerably from country to country. Fellings to NAI ratios are particularly low in Bosnia and Herzegovina, Slovenia and Spain. Fellings are almost equal to NAI in Portugal, where there has been a marked expansion in wood-processing capacity in recent decades, and in The Former Yugoslav Republic of Macedonia. In several countries, the low fellings to NAI ratio is explained by the orientation of the age-class structure towards young and medium-aged stands.

There is also a wide range of NAI volumes per hectare between countries, as well as within countries, reflecting differences in growing conditions. In Romania and Slovenia, average NAI is more than 5 m^3 per hectare, which is higher than the European average. In Portugal and Spain the average is 6.8 and 2.7 m^3 per hectare respectively, but some pine and eucalyptus plantations in these countries are among the fastest-growing in Europe. Fast-growing plantations are also found in other countries, for example poplars in Italy. In Albania, Greece and The Former Yugoslav Republic of Macedonia, NAI is reported as being only around 1 m^3 per hectare.

FOREST MANAGEMENT AND USES

In southern Europe as a whole, more than half the forest area is publicly owned and less than half is in private hands. This conceals, however, wide differences in ownership patterns between countries. In Albania and Bulgaria, all forest is State property, and the proportion is about three-quarters in Bosnia and Herzegovina, Croatia and The Former Yugoslav Republic of Macedonia and more than 90 percent in Romania. In Italy, Portugal and Spain, the State owns relatively small areas of forest, while other forms of public ownership, mainly by municipalities and communes, are more important. In these three countries, ownership by private individuals is by far the most important ownership category, with more than three-quarters of all forest in Portugal and Spain and two-thirds in Italy. Particularly in Spain and Italy, the average size of holdings is very small and the number of private owners correspondingly large. Ownership by forest industries is quite important in Portugal, with 10 percent of the total, with smaller areas in this category also in Italy and Spain. Other forms of private ownership, for example by cooperatives, are found in Spain and Greece.

In some countries that formerly had planned economies but are now changing to forms of market economy, privatization or restitution of some forest areas has been taking place. This has progressed particularly rapidly in Slovenia, where three-quarters of the forest is now owned by private individuals, and to a lesser extent in Bosnia and Herzegovina, Croatia, Romania and The Former Yugoslav Republic of Macedonia. In Slovenia, private forests were not nationalized but only managed by the State.

With the exception of Andorra and San Marino, all the countries in southern Europe provided national-level information on the forest area managed (Table 30-1), applying the definition used by industrialized countries of forests managed in accordance with a formal or an informal plan applied regularly over a sufficiently long period (five years or more) and including areas where a decision had been taken not to undertake any management interventions. The extent to which countries' forests are managed varies considerably. Bulgaria, Romania and The Former Yugoslav Republic of Macedonia report that all their forests are managed. In contrast, relatively low proportions are reported to be so by Albania, Greece and Portugal, and especially Italy. In Italy only 11 percent is shown as managed, as only formal management plans were included, which mostly applied to forest in public ownership. The wide differences reported by countries in the extent of management of their forests may partly reflect the difficulties in applying the definition of "managed forest" used for TBFRA to national conditions or in obtaining comprehensive data. The definition includes forest under both formal and informal management plans; it may be difficult to determine the areas under informal plans, especially in the private sector. In total, approximately 35 million hectares, or 67 percent of the total forest area in southern Europe, were reported as being managed in accordance with a formal or informal plan.

Wildlife and hunting activities have been recognized as relevant practices in this subregion.

CONCLUSIONS AND ISSUES

There is a wide variation in the types and quality of the forests growing in southern Europe. Because of the long-term historical pressures on the forest from overexploitation and the often difficult growing conditions, it would be true to say that many of the "natural" forests are of relatively poor quality. This represents a major challenge to managers, who often lack the resources, human and financial, to apply more intensive silviculture to these forests. At the other extreme, there are important areas of plantations managed for wood production, where levels of management and silviculture are very high. As elsewhere in Europe, there is a growing awareness on the part of the public and the policy-makers of the importance of forests in providing environmental and social benefits, as well as wood, and this is reflected in the growing interest in the way in which forests of all types, both natural and plantation, are being managed. Certain practices, such as the use of exotic species like eucalyptus in plantations and the replacement of existing broad-leaved scrub by pines, are sometimes being called into question.

The perennial problem of forest fires remains acute in most of the countries of southern Europe. A large proportion of all fires each year in Europe occur in these countries, and the principal cause is human. Many are started accidentally or through negligence, for example, through fires started on agricultural land spreading into the forest, but a large number are also set deliberately for a variety of social, economic or political reasons. The degradation of ecosystems through overgrazing has also made them much more vulnerable to human-caused fires. Local population density has declined with urbanization, so fires are not detected and extinguished as fast as they would be if the rural areas were more densely populated.

The increasing number of visitors to forests also poses an additional fire threat. Demographic changes have often led to reduced silvicultural and harvesting activities and fewer grazing animals entering the forest, resulting in a build-up in the amount of fuel and greater risk of any fire that breaks out being of greater intensity. Foresters have a major task not only in fighting forest fires but also in educating the public on their prevention.

The climatic conditions in the subregion, especially the hot dry summers, are a major factor in raising the risk of fire. Linked to possible changes in global climate, there have also been growing concerns in some parts of the subregion, such as the south of the Iberian Peninsula and southern Italy, about the threat of desertification. This has emphasized the need for forest protection measures and, where appropriate, afforestation for soil conservation.

With increasing industrialization and the gradual shift of populations from the countryside to towns, the problem of maintaining viable rural communities has been becoming increasingly acute in most of the countries of the subregion where in the past agriculture was a major activity. Considerable areas of marginal agricultural land are being abandoned. While forestry might in many cases seem to be a solution, the economic and social problems of this as a suitable land use alternative are considerable, not least the funding for afforestation and reforestation projects and the organizational arrangements needed to ensure the proper maintenance and eventual use of new forest and rejuvenated areas. It remains uncertain, therefore, how fast and how far the expansion of the area of forests that occurred in recent decades will continue in the subregion.

The volume of annual fellings in southern Europe is only about half the net annual increment, even less in some countries. This is leading to a rise in the volume of growing stock and contributing to carbon sequestration, and in most of the countries is a consequence of the relatively young age-class structure of the forests. It does raise the question, however, whether sooner or later greater use should be made of the wood production potential of the subregion, not only to enable the forests to make a greater contribution to the economic welfare of society but also for ecological and fire prevention reasons. This is not to suggest that fellings could be doubled to equal the present NAI, but there is undoubtedly scope for some expansion without jeopardizing in any way the sustainability of the forest resource. It would involve the development of the wood-processing capacity in some countries, the main exception being Portugal where it has already expanded to the point where virtually all available wood supplies are fully committed. Italy and Spain are major net importers of wood products, and there could be scope for import substitution in these and some of the other countries. Modernization of its industries could also enable Romania to regain its position as a sizeable exporter of wood products.

Romania, Bulgaria, Albania and the countries of former Yugoslavia are all in the process of transforming themselves towards forms of market economy and of raising the living standards of their people towards the European average. In all of them, consumption of wood products is still fairly low and would respond to the stimulation of economic activity, notably in the construction sector; provision of more and better housing is likely to be one of their priorities. This is a long-term process and involves the replacement or modernization of their industries, including the wood-processing industries, which are generally obsolete or undercapitalized, and development of their infrastructures. It also involves the privatization of parts of the economy and attracting foreign assistance, including direct foreign investment. Given the long-term potential to increase the wood supply of most of these countries, Albania being the exception, wood and wood products could play an increasing role in strengthening their economies.

The countries of former Yugoslavia have a special problem arising from the conflicts and disturbances that have occurred during the break-up into the present five separate countries. Considerable damage has been done to the industries and infrastructure of these countries, as well as to the forests in some areas. Only Slovenia appears to have been relatively spared. The possibilities to restore and strengthen the forest and forest industries sector in these countries are largely dependent on the stabilization of the political situation, which will contribute to raising their attractiveness to foreign investors.

A major obstacle to achieving fuller use of the forest resources in the subregion, whether for wood supply or for environmental or social objectives, is the ownership of a substantial part of the forest area by a very large number of small private owners. Small-scale forestry operations are seldom as profitable as larger ones, and it is more difficult to achieve effective management and planning. For political reasons, consolidation of properties into larger units, either privately or publicly owned, is generally not acceptable, and consequently other solutions need to be found. Among these could be the grouping of properties into management or marketing cooperatives, as has been done in a number of instances. Before getting involved in the cost and effort of such initiatives, however, it would be necessary to determine what is the actual and potential contribution that small forest holdings make to the overall forest economy, notably through the production and sale of roundwood. At present, it is probable that it is much less than their share of forest area, and it would need to be determined whether their contribution could be significantly improved by joint action. If it should prove difficult to justify action on economic grounds, it would still be necessary to consider whether, by leaving things as they are, there would be long-term negative consequences regarding the environmental and social functions, increased risk of forest fires being one example.

Given the fragile environmental situation and evolving social conditions in several parts of southern Europe, there may indeed be a strong case for supporting a greater role for forestry. That still leaves open the difficult question of how such support should be organized and who should pay for it.

BIBLIOGRAPHY

UNECE/FAO. 2000. *Forest resources of Europe, CIS, North America, Australia, Japan and New Zealand: contribution to the global Forest Resources Assessment 2000.* Geneva Timber and Forest Study Papers 17. New York and Geneva, United Nations. www.unece.org/trade/timber/fra/pdf/contents.htm

Chapter 31
Belarus, Republic of Moldova, Russian Federation and Ukraine

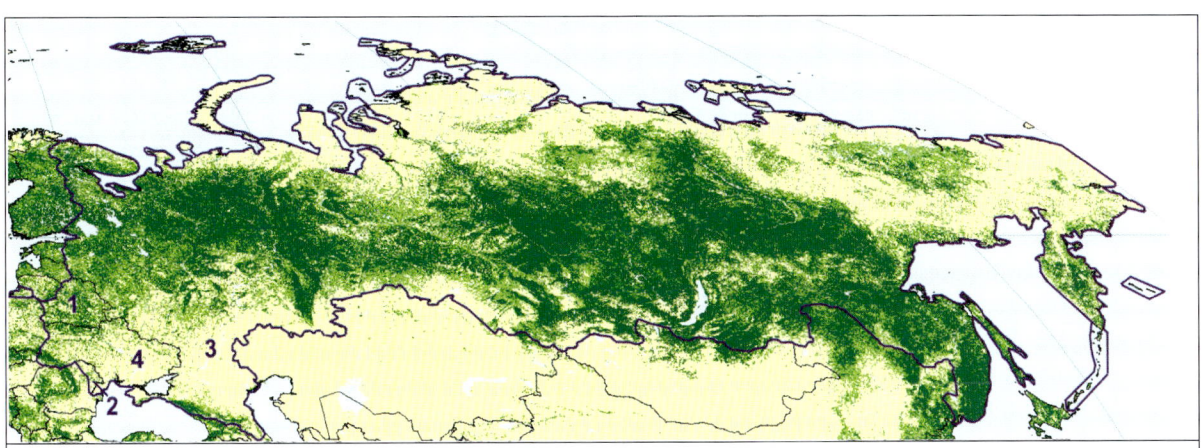

1. Belarus
2. Republic of Moldova
3. Russian Federation
4. Ukraine

Forest cover according to FRA 2000 Map of the World's Forests 2000
- Closed forest
- Open and fragmented forest

Figure 31-1. Belarus, Republic of Moldova, Russian Federation and Ukraine: forest cover map

The four countries of this subregion – Belarus, Republic of Moldova, the Russian Federation and Ukraine (Figure 31-1) – were part of the former Union of Socialist Soviet Republics (USSR) until its break-up into 15 separate countries in the early 1990s.[46] Belarus, the Republic of Moldova and Ukraine occupy the southwestern part of the subregion, while the Russian Federation spreads across two continents (Europe and Asia), eight time zones and 7 000 km from east to west. The land area of the Russian Federation of 1.69 billion hectares is three times the area of all other European countries put together, and with a population of 147 million it is one of the most sparsely populated: 11 ha of land per capita compared with about 1 ha per capita for the rest of Europe. Ukraine, with 58 million hectares, has the second largest land area in Europe after the Russian Federation. The climate of the subregion is boreal to the north, with part of the Russian Federation lying within the Arctic Circle, and temperate to the south, with a correspondingly wide range of ecosystems. The living standards of the populations in the four countries is currently low by European standards, having fallen considerably after the economic collapse which followed the dissolution of the USSR.

FOREST RESOURCES

The forest area in the four countries of the subregion amounts to 871 million hectares, to which can be added 71 million hectares of other wooded land. It is one of the subregions of the world with the most forest and accounts for well over one-fifth of the global forest area. In terms of forest cover, it is considerably more heavily forested than the world average: nearly 50 percent compared with 30 percent; and in terms of forest area per inhabitant it is also very well endowed, with more than 4 ha per capita compared with a world average of 0.6 ha per capita (Table 31-1).

These impressive figures are due very largely to the situation in one country, the Russian Federation, which alone has 851 million hectares of forest, the largest of any country in the world, and a further 70 million hectares of other wooded

---
[46] For more details by country, see www.fao.org/forestry

Table 31-1. Belarus, Republic of Moldova, Russian Federation and Ukraine: forest resources and management

| Country/area | Land area | Forest area 2000 ||| Area change 1990-2000 (total forest) || Volume and above-ground biomass (total forest) || Forest under management plan ||
| | | Natural forest | Forest plantation | Total forest |||||||||
	000 ha	000 ha	000 ha	000 ha	%	ha/capita	000 ha/year	%	m³/ha	t/ha	000 ha	%
Belarus	20 748	9 207	195	9 402	45.3	0.9	256	3.2	153	80	7 577	81
Republic of Moldova	3 296	324	1	325	9.9	0.1	1	0.2	128	64	325	100
Russian Federation	1 688 851	834 052	17 340	851 392	50.4	5.8	135	n.s.	105	56	851 392	100
Ukraine	57 935	5 159	4 425	9 584	16.5	0.2	31	0.3	179	-	9 584	100
Total subregion	1 770 830	848 742	21 961	870 703	49.2	4.1	423	0.0	106	56	868 878	100
Total Europe	2 259 957	1 007 236	32 015	1 039 251	46.0	1.4	881	0.1	112	59	954 707	92
TOTAL WORLD	13 063 900	3 682 722	186 733	3 869 455	29.6	0.6	-9 391	-0.2	100	109	-	-

Source: Appendix 3, Tables 3, 4, 6, 7 and 9.

land, and accounts for nearly 98 percent of the forest area of the subregion and 22 percent of the world total. The data for the Russian Federation tend to overshadow those for the other three countries of the subregion, but it should not be forgotten that Belarus and Ukraine both have well over 9 million hectares of forest, which puts them among the European countries with large forest areas. Nevertheless, owing to its sheer size, particular attention is given to the Russian forest resource in this assessment.

According to the classification used in this report, 98 percent of the Russian Federation's forests are "natural", the remainder (17 million hectares) being plantations (Table 31-1, Figure 31-2). In contrast to other European countries, where very little really natural (old growth) forest remains, the area in the Russian Federation of forest undisturbed by humans [according to the definition of the Temperate and Boreal Forest Resources Assessment 2000 (TBFRA) (UNECE/FAO 2000)] is very extensive, amounting to 749 million hectares, with only 50 million hectares semi-natural. About two-fifths of the forest undisturbed by humans is classified as not available for wood supply, and most of that for economic reasons, that is to say inaccessibility, although there are 24 million hectares not available for wood supply for conservation and protection reasons. The area being withdrawn from actual or potential harvesting is increasing, as more emphasis is given to nature conservation and protection. The forest undisturbed by humans is mostly mature or overmature and is at risk of damage from natural causes, notably fire, pests and diseases. Fires in the more remote and inaccessible areas have to be left to burn themselves out and consequently may be very extensive in area. Although, in more inhabited areas, human error is the most common cause of fires, lightning is a frequent cause in more remote areas, which may be considered in some locations as a positive ecological element by inducing the rejuvenation of overmature stands. However, about half the Russian forest grows on permafrost, where the ecosystems are fragile and regeneration slow and difficult.

Most of the semi-natural forest is located in the European part of the country or within exploitation distance of the trans-Siberian railway. Many of these areas have suffered overexploitation in the past, and their present state is degraded or unsatisfactorily restocked, for example by alder, aspen and birch on areas that were formerly coniferous stands. Over half the forest area in the Russian Federation is occupied by predominantly coniferous stands, with a further two-fifths mixed coniferous/broad-leaved, leaving less than 10 percent predominantly broad-leaved. The last group is mainly in the southern, more temperate parts of the country, consisting of such species as beech, oak, lime and hornbeam. In the more northerly, boreal areas, the main species in the western parts of the Russian Federation are Norway spruce (*Picea abies*) and Scots pine (*Pinus sylvestris*). In Siberia and the Far East, larch (*Larix* spp.) is the most common species, with Siberian stone pine (*Pinus sibirica*), dwarf pine (*Pinus pumila*), spruces (*Picea* spp.) and firs (*Abies* spp.) also present, and birch (*Betula* spp.) and aspen (*Populus tremula*) among the broad-leaved species. In terms of growing stock volume, coniferous species make up about four-fifths of the total, larch being the most important.

Only about one-fifth of the forest area lies in the European part of the Russian Federation, where the major part of the population and of the wood-processing capacity and wood products consumption occurs, and four-fifths lies in the lightly populated parts to the east of the Ural mountains.

Only part of the area of 525 million hectares, which is reported as available for wood supply, is or has been under exploitation or accessible for harvesting. Much of it, although not under any harvesting restriction, is currently and likely to remain inaccessible for the foreseeable future, being located in parts of Siberia and the Far East without road or rail infrastructure as well as some areas in the northern part of the European Russian Federation. Growing stock on forest available for wood supply amounts to 61 billion cubic metres over bark, or about 70 percent of the total volume on the forest area, while the net annual increment (NAI) is estimated at 742 million cubic metres over bark or 1.4 m^3 over bark per hectare. NAI is difficult to calculate where much of the forest is undisturbed by humans (old growth) and where natural losses are likely to more or less offset gross increment. The NAI per hectare is only about one-third of the level achieved in northern Europe, where growing conditions are roughly similar and most of the forest is under management, giving an indication of the potential for improvement if and when management is extended over a wider area in the Russian Federation. Moreover, despite the low NAI per hectare figure, it is still many times higher than the level of fellings on forest available for wood supply in the late 1990s of about 125 million cubic metres over bark. The barriers to higher fellings are economic and infrastructural, including the capacity of the domestic wood-processing industries and access to export markets. At least since 1990, the difference between NAI and fellings has led to a rising trend in the volume of growing stock and a tendency for the proportion of mature and overmature forest to increase.

Total removals in the Russian Federation in the late 1990s were running at about 104 million cubic metres under bark a year, which compare with volumes of between 300 and 400 million cubic metres under bark a year in the 1970s and 1980s. In former periods the quantities of unrecovered harvesting losses were very large, but improvements in logging methods in more recent times have reduced the proportion of such losses in the total volume of fellings. According to the data provided for TBFRA, under bark removals on forest available for wood supply were 69 percent of the over bark volume of fellings, so that after allowing for the bark percentage on removals, the proportion of unrecovered harvesting losses was not significantly higher than in some other temperate and boreal countries.

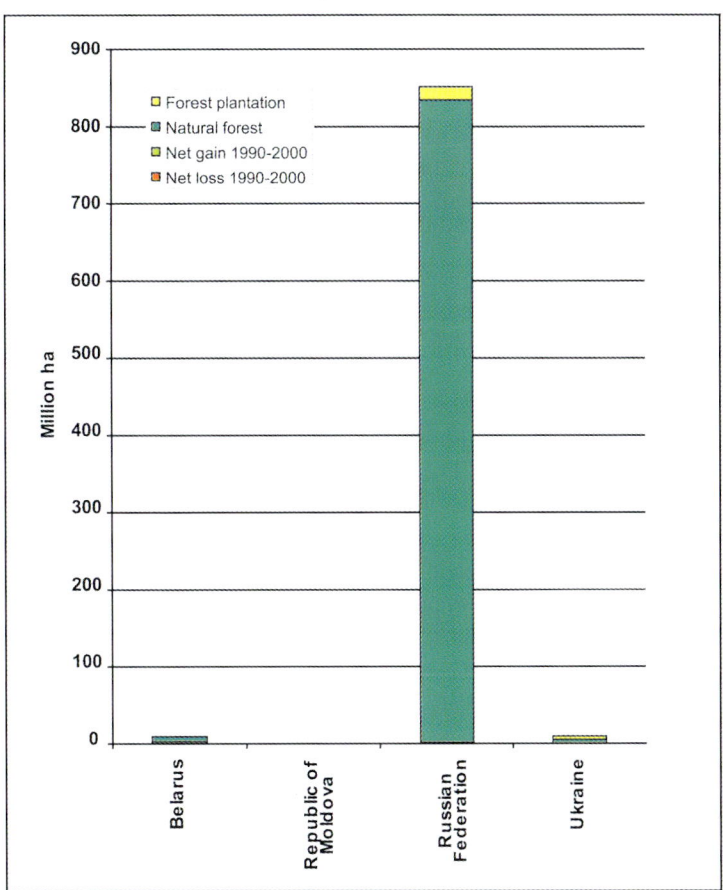

Figure 31-2. Belarus, Republic of Moldova, Russian Federation and Ukraine: natural forest and forest plantation areas 2000 and net changes 1990-2000

Natural losses in Russian forests, which are defined in TBFRA as mortality from causes other than cutting by humans, such as natural mortality, diseases, insect attacks, fire, windthrow and other physical damage, were reported as amounting to 359 million cubic metres over bark per year in the late 1990s, equivalent to 37 percent of gross annual increment and between two and three times the volume of fellings. The high level of natural losses is linked to the large proportion of old growth, mature and overmature forests.

Belarus, lying to the west of the Russian Federation, has some similar features so far as its forest resource is concerned, but also some differences. Nearly two-fifths of the forest is classified as predominantly coniferous and a similar area as mixed coniferous/broad-leaved; forest covers about 45 percent of the land area. However, three-quarters of its forest is available for wood supply, and it has very little forest undisturbed by humans, the bulk of which is in the semi-natural category, similar to other European countries. NAI in Belarus, averaging more than 4 m^3 over bark per hectare is comparable to that of northern Europe, but it is still more than twice the volume of fellings in the late 1990s. This is probably associated with the fact that most of its forests are in the younger age classes – less than 80 years.

The forests of Ukraine and the Republic of Moldova, lying to the southwest of the Russian Federation, have certain features which resemble those of southern European countries more than those of the Russian Federation and Belarus. In the first place, forest cover is fairly low: 16 and 10 percent, respectively, of the land area, while forest area per inhabitant is very low: 0.2 and 0.1 ha per capita, respectively. Second, broad-leaved forests are in the majority: in Ukraine half the forest area is predominantly broad-leaved with a further 10 percent in the mixed broad-leaved/coniferous category. In Moldova, all but a very small area is classed as predominantly broad-leaved. About two-thirds of both countries' forest is available for wood supply; most of the area not available is in that category for conservation and protection reasons. Ukraine has only a small area of forest undisturbed by humans and the Republic of Moldova has none, but whereas most of the latter's is classed as semi-natural, as much as 45 percent of Ukraine's is reported to be plantations. Considerable afforestation has been carried out for the protection of soils against wind and water erosion.

The annual average change in forest area between 1990 and 2000 in the subregion was an increase of 423 000 ha (Table 31-1, Figure 31-2), of which 60 percent was in Belarus and most of the rest in the Russian Federation. Information is not available on how these data were calculated. What they should show is the net change after deducting losses in forest to other use from additions as the result of afforestation and natural colonization of non-forest land. Problems may arise if the classification and definition of land categories changes between one reference period and another. In the case of the Russian Federation, it is interesting to note that its original response to the TBFRA enquiry showed an average annual decrease in forest area between 1988 and 1993 of 1.1 million hectares, which was more than offset by an increase of 1.6 million hectares in the area of other wooded land. In recalculating the data for the present report to cover the period 1990 to 2000, the result was an average annual increase in forest area of 135 000 ha (no information is available on the change in other wooded land). For Belarus, the average annual increase in forest area of 256 000 ha or 3.2 percent, a remarkably strong rate of expansion, is the same in both reports. The conclusion to be drawn seems to be that care should be taken in accepting the change data for these countries until more is known about how they were derived. With regard to the Russian Federation, a long-term programme is being carried out for the creation of shelterbelts to protect agricultural land from wind and water erosion, which would presumably count as afforestation. On the other hand, clear-felling of forests has not always been followed by regeneration, artificial or natural, which might change the classification of some areas from forest to other wooded land, if some sort of scrub vegetation appeared, or to non-forest land, either agriculture or built-up or waste land. Further examination of the changes taking place in the Russian Federation and the other countries of the subregion would clearly be useful.

FOREST MANAGEMENT AND USES

Ownership of forest and other wooded land in all four countries of the subregion is entirely by the State. In the Russian Federation, the Federal Forest Service controlled 94 percent of the country's forests until the recent reorganization of the administration, which involved the absorption of the service into a new ministry. Other departments, such as the Committee on Environmental Protection, the Ministries of Agriculture, Education and Defence, and some municipalities, were also responsible for some forest areas. Unlike other European countries in transition towards a market economy, there has been no move towards the privatization or restitution of forest in the four countries of the subregion. In the Russian Federation, for example, the policy has been to retain all land in public hands, although some plots, including forest land, may be made available to citizens and legal entities on the basis of leases, rights of use or concessions.

In Belarus, all but 19 percent of the forest is reported to be under management plans. In the other countries of the subregion the coverage is reported to be 100 percent. Given that in the Russian Federation, a large part of the forest area remains inaccessible and has not been intensively ground surveyed, the nature and extent of management in such areas has been simplified compared with that in the more accessible ones.

In the Russian Federation, all forests have been subdivided into three management groups, in relation to their protective functions and to the degree they can be exploited for wood. Group I, protection forests, includes forests with mainly water- and soil-protection, sanitary and health restoration functions. These are belts along the banks of rivers, lakes, reservoirs, etc., forests preventing erosion, including those on steep slopes, shelterbelts, urban forests, forest parks, green belts, natural and national parks and so on. In 1998 they made up about 21 percent of the forest area but 36 percent in the European part of the Russian Federation (Pisarenko *et al.* 2001). Strict felling regimes are maintained within this Group. Group II, multipurpose forests, includes forests in areas where the population density is high and the road network is good. The forests have protective and limited exploitation importance, and the group also includes forests without sufficient wood reserves. Wood harvesting is restricted to amounts equal to annual growth. This group accounts for about 6 percent of the forest area. Group III, forests for commercial use, accounts for the remaining 73 percent of the forest and includes forests in richly forested areas, predominantly exploitable and designed to provide a continuous wood supply without damaging their protective functions. Clear-cutting is allowed in these forests. The proportion of forests in Group III diminished between 1966 and 1988, while that in Group I and Group II increased.

CONCLUSIONS AND ISSUES

In terms of its forest resources, the Russian Federation is a giant on the world scene. In terms of wood production and trade, on the other hand, its share of the world total is relatively modest. A major question is, therefore, whether, when, to what extent and how it might raise its production and trade towards the potential of its forest resources. The answers lie to a large extent beyond the control of those responsible for the resource. In the first place, it depends on the way and pace at which the country is transformed into a modern economy, with an efficient long-distance transportation system and other infrastructural developments, as well as reconstructed and expanded industries. Latent demand for wood products is considerably higher than present levels, with a large potential for recovery and growth in the use of sawnwood and wood-based panels in construction and for all categories of paper and paperboard. The domestic market is likely to remain the principal outlet for wood products, but exports have been, and are likely to remain, an important source of foreign currency and will continue to be supported. The largest markets are Europe and the Near East (fed mainly from the European part of the Russian Federation) and Japan and other Pacific Rim countries (fed from the Russian Far East). The importance of China as an outlet for wood from the Siberian forests is increasing strongly. Up to now a large part of Russian exports has been in the form of raw material and semi-processed products, such as sawnwood, with relatively low unit values. The expansion of production and export of wood products in the Russian Federation is dependent on the possibilities to shift harvesting to hitherto underexploited forest areas in the central and northern parts of the country and to develop economic systems for transporting them to domestic and overseas markets, as well as the establishment of new wood-processing capacity. At best, this will be a gradual process.

With the liberalization of the economy, there has been increasing awareness within the country of the environmental damage that has occurred in many sectors, including forestry and forest industries. Logging practices, including large-scale clear-felling, had caused serious forest degradation and had sometimes not been followed up by proper regeneration. Biodiversity had been compromised, well-publicized examples being the threats to the survival of the Siberian tiger by logging in the Russian Far East and to the purity of the water and the unique ecosystem of Lake Baikal by pulp and paper mill activities in the vicinity. Measures have been strengthened to protect the environment by increasing the areas under nature conservation and other protection measures, e.g. by increasing the extent of Group I and Group II forests. It is not clear, however, to what extent the good intentions are being implemented in practice, given the inadequacy of resources to monitor the forests and logging activities and the difficulties of ensuring that regulations are followed.

The fallout from the explosion at the Chernobyl nuclear power station in Ukraine affected about 1 million hectares of forests in the Russian Federation, as well as large areas in Belarus and Ukraine. These areas will remain out of bounds to the population for the foreseeable future and will be excluded from any kind of utilization, either for wood or other forest products, apart from some research activity into the effects of this major environmental disaster.

Unless a change in policy occurs, privatization of forests in the countries of the subregion will not be an issue in the coming years. On the other hand, there do seem to be possibilities for extending the private ownership and management of wood-processing industries; in the Russian Federation most of these industries have already been converted into joint stock companies. With regard to the management of forests, there appears to be need for clarification about the intensity of management, especially in the more remote areas of the Russian Federation. Although reportedly all forests are under management, in practice this does not appear feasible according to the internationally used definition of the term.

There also seems to be some ambiguity about the figures of change in forest area over time in Belarus and the Russian Federation: are the reported figures net changes in the real area or do they arise from differences in definitions or land classifications between one period and another? Reliable information on change is considered to be important for policy discussions in international fora and, given the importance of the Russian Federation in the world forest total, it would be highly desirable to have as accurate an indication as possible of the extent and type of changes that are taking place in its resource. No doubt measurements from remote sensing at different periods would provide such information.

The area of forest per inhabitant is very low in Ukraine and the Republic of Moldova, and their resources are insufficient to meet their populations' need for wood or to ensure an adequate supply of other goods and services. They do not have the reserves to be able to import wood products to cover the latent demand. Consequently, the possibilities need to be considered of extending their forest resources, both to increase wood supply in the medium to long term and to provide other essential services in the shorter term, notably soil protection and nature conservation. Finding the means to do this could be a major challenge.

BIBLIOGRAPHY

Pisarenko, A.I., Strakov, V., Päivinen, R., Kuusela, K., Dyakun, F.A., & Sdobnova, V.V. 2001. *Development of forest resources in the European Part of the Russian Federation.* European Forest Institute Research Report 11. Leiden, the Netherlands, Koninklijke Brill NV.

UNECE/FAO. 2000. *Forest resources of Europe, CIS, North America, Australia, Japan and New Zealand: contribution to the global Forest Resources Assessment 2000.* Geneva Timber and Forest Study Papers 17. New York and Geneva, United Nations. www.unece.org/trade/timber/fra/pdf/contents.htm

UNECE/FAO. 2001. *Forest and forest products country profile: Russian Federation.* ECE/TIM/SP/18, United Nations Publication.

Chapter 32

North and Central America and the Caribbean

Subregions: 1. North America (excluding Mexico); 2. Mexico and Central America; 3. Caribbean
Forest cover according to FRA 2000 Map of the World's Forests 2000
■ Closed forest
▨ Open and fragmented forest

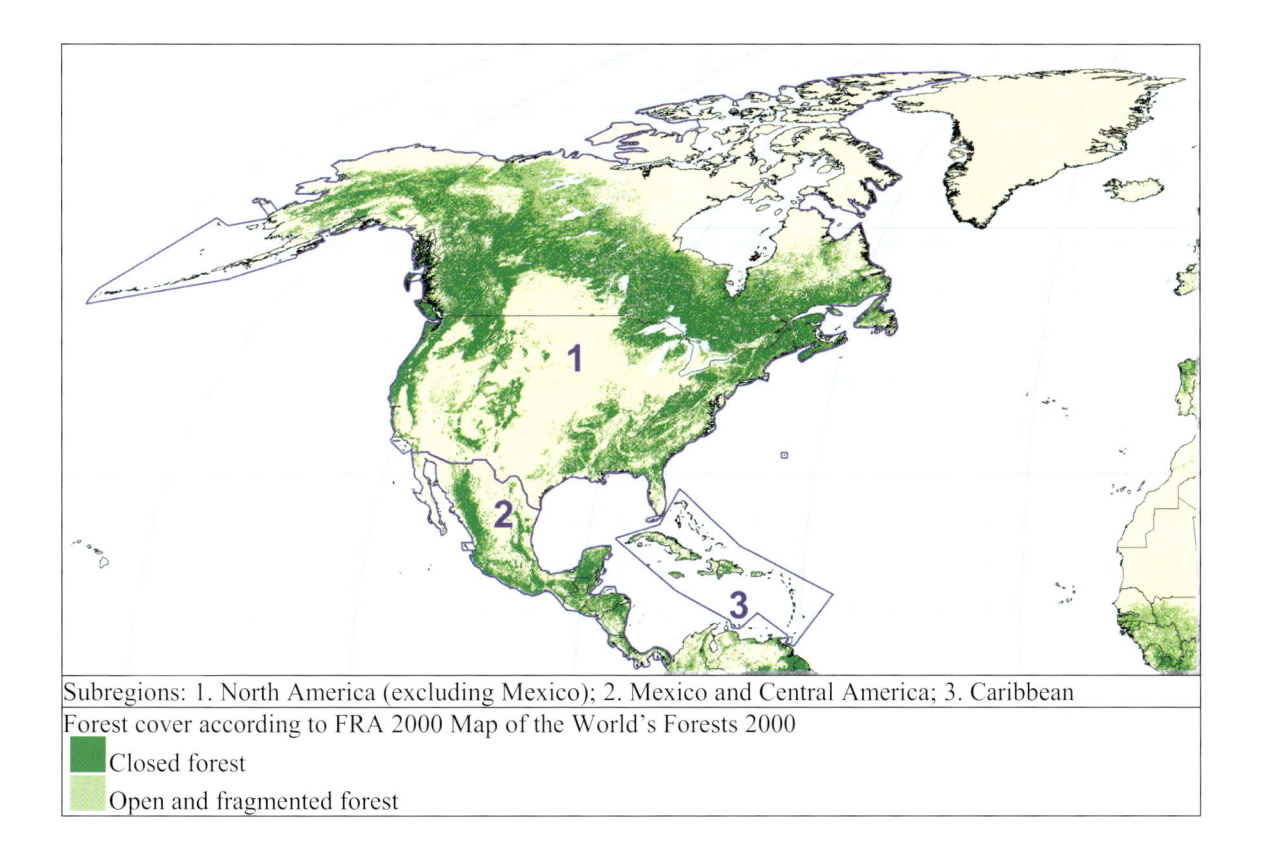

Figure 32-1. North and Central America: subregional division used in this report

North America, Central America and the Caribbean (see Figure 32-1[47] and Table 32-1) together contain about 549 million hectares of forests, corresponding to 14 percent of the world total. The forests of North America, Central America and the Caribbean amount to 1.1 ha per capita, which is above the world average. The forest areas of Central America and the Caribbean are located mainly in the subtropical ecological domain. Forest cover in North America is distributed between the temperate and boreal ecological zones. Some 86 percent of the region's forests is in two large countries – Canada and the United States. The forests in the region do not constitute a major proportion of any ecological zone; however, this is the most diversified region with all but two ecological zones represented. The net change of forest area is -570 000 ha per year, which represents the sum of a high net loss in Central America and a considerable increase in the United States.

---

[47] The division into subregions was made only to facilitate the reporting at a condensed geographical level and does not reflect any opinion or political consideration in the selection of countries. The graphical presentation of country areas does not convey any opinion of FAO as to the extent of countries or status of any national boundaries. The reporting units Greenland and Saint Pierre and Miquelon were not included in any subregion.

Table 32-1. North and Central America: forest resources by subregion

Subregion	Land area	Forest area 2000 – Natural forest	Forest plantation	Total forest 000 ha	Total forest %	Total forest ha/capita	Area change 1990-2000 (total forest) 000 ha/year	Area change 1990-2000 (total forest) %	Volume (total forest) M³/ha	Above-ground biomass (total forest) t/ha
	000 ha	000 ha	000 ha	000 ha	%	ha/capita	000 ha/year	%	M³/ha	t/ha
Central America	241 942	72 300	729	73 029	30.2	0.5	-971	-1.2	86	93
Caribbean	22 839	5 145	566	5 711	25.0	0.2	13	0.2	57	98
North America	1 837 992	454 326	16 238	470 564	25.6	1.5	388	0.1	128	95
Other North and Central America	34 193	-	-	-	-	-	-	-	-	-
Total North and Central America	2 136 966	531 771	17 533	549 304	25.7	1.1	-570	-0.1	123	95
TOTAL WORLD	13 063 900	3 682 722	186 733	3 869 455	29.6	0.6	-9 391	-0.2	100	109

Source: Appendix 3, Tables 3, 4, 6 and 7.

Chapter 33
North and Central America: ecological zones

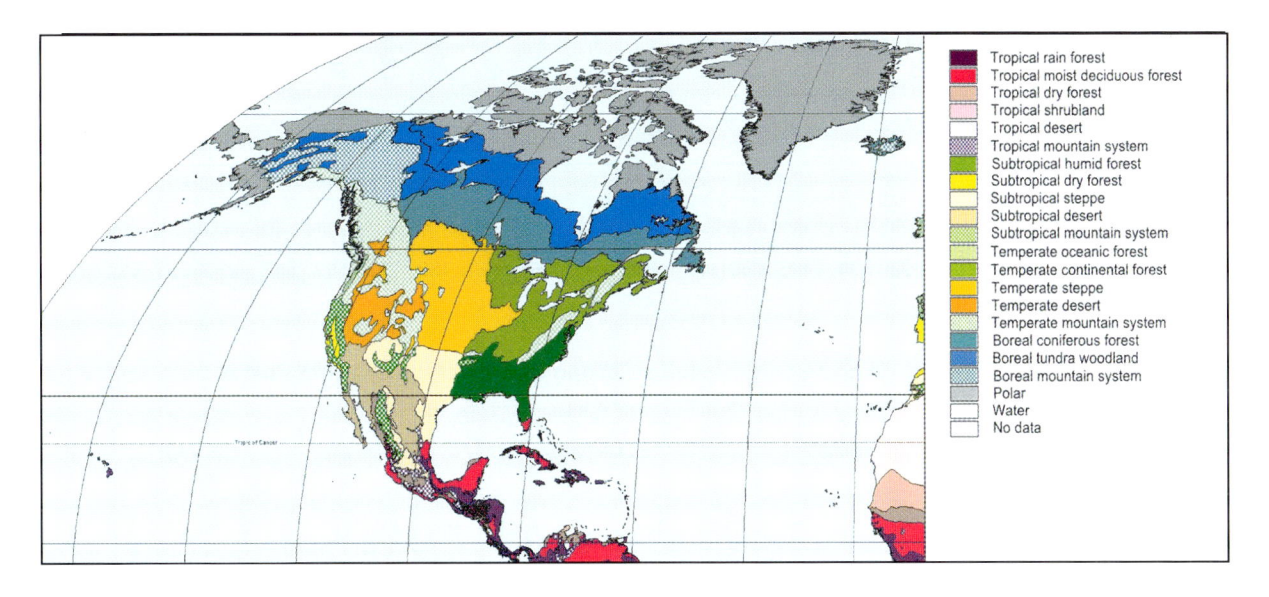

Figure 33-1. North and Central America: ecological zones

Figure 33-1 shows the distribution of ecological zones in North and Central America, as identifed and mapped by FRA 2000. Table 33-1 contains area statistics for the zones by subregion, and Table 33-2 indicates the proportion of forest in each zone by subregion.

BOREAL TUNDRA WOODLAND

The Hudson Plain occupies a major area of this zone, while the western portion consists of subdued broad lowlands and plateaus incised by major rivers. The climate, influenced by cold arctic air, is characterized by short, cool summers and long, cold winters. Mean annual temperature ranges from -10° to 0°C, with mean temperature in summer from 6° to 14°C and in winter from -26° to -16°C. Snow and ice persist for six to eight months of the year. The mean annual precipitation is low in the west, ranging from 200 to 500 mm, but reaches 500 to 800 mm in the east, with portions of Labrador reaching 1 000 mm.

Vegetation associations of the Hudson Bay lowlands consist of arctic tundra and some boreal forest transition types. The better-drained sites support open woodlands of black spruce (*Picea mariana*), tamarack (*Larix laricina*) and some white spruce (*Picea glauca*). Balsam poplar (*Populus balsamifera*), white spruce and white or paper birch (*Betula papyrifera*) are common along rivers.

East of these lowlands are large open stands of black spruce woodland as well as stunted black spruce and tamarack on the windswept plateaus. White spruce is also present. Alder (*Alnus incana*) thickets are common along riverbanks and other drainage areas. Other species include quaking aspen (*Populus tremuloides*) and balsam fir (*Abies balsamea*). Limited tree vegetation occurs along the exposed headlands of the Atlantic Coast and within the interiormost windswept barrens.

West of Hudson Bay, open stands of black and white spruce and tamarack dominate. Sometimes these open forests include jack pine (*Pinus banksiana*) as well. The western limits of the zone are characterized by open, generally slow-growing black spruce. Upland and foothill areas and southerly locales tend to be better drained and are somewhat warmer. Here, mixed-wood forests of white and black spruce, lodgepole pine (*Pinus contorta*), tamarack, white birch, trembling aspen and balsam poplar are common. Along nutrient-rich alluvial flats, white spruce and balsam poplar grow to sizes comparable to the largest in the boreal forest to the south.

Table 33-1. North and Central America: extent of ecological zones

Subregion	Tropical						Subtropical					Temperate					Boreal			
	Rain forest	Moist	Dry	Shrub	Desert	Mountain	Humid	Dry	Steppe	Desert	Mountain	Oceanic	Continental	Steppe	Desert	Mountain	Coniferous	Tundra	Mountain	Polar
Central America	33	55	22			24			30	63	20									
Caribbean	10	11	1			2														
North America		2					106	9	86	45	39	4	225	211	75	197	225	266	118	324
Total North and Central America	43	68	23			26	106	9	116	108	59	4	225	211	75	197	225	266	118	358
TOTAL WORLD	1468	1117	755	839	1192	459	471	156	491	674	490	182	726	593	552	729	865	407	632	564

Note: Data derived from an overlay of FRA 2000 global maps of forest cover and ecological zones.

Table 33-2. North and Central America: proportion of forest by ecological zone

Subregion	Tropical						Subtropical					Temperate					Boreal			
	Rain forest	Moist	Dry	Shrub	Desert	Mountain	Humid	Dry	Steppe	Desert	Mountain	Oceanic	Continental	Steppe	Desert	Mountain	Coniferous	Tundra	Mountain	Polar
Central America	69	59	44			65			11	6	74									
Caribbean	46	28				55														
North America							46	23	8	3	54	39	34	2	8	47	52	31	36	2
Total North and Central America																				
TOTAL WORLD	64	53	44			65	46	23	8	5	61	39	34	2	8	47	52	31	36	2

Note: Data derived from an overlay of FRA 2000 global maps of forest cover and ecological zones.

BOREAL CONIFEROUS FOREST

A broadly rolling mosaic of uplands and associated wetlands dominates this zone. The climate is generally continental with long, cold winters and short, warm summers, modified in the east by the Atlantic Ocean. The mean annual temperature ranges from -4°C in central Canada to 5.5°C in the boreal regions of Newfoundland. Mean summer temperature varies between 11° and 15°C, with mean winter temperature from -20.5°C in the west to -1°C in the east. Mean annual precipitation varies between 100 and 625 mm with the exception of boreal Newfoundland, where average precipitation is higher, from 900 to 1 600 mm.

Much of the zone is distinguished by closed stands of conifers, largely white spruce (*Picea glauca*), black spruce (*P. mariana*), balsam fir (*Abies balsamea*) and tamarack (*Larix laricina*). Common deciduous species include white birch (*Betula papyrifera*), trembling aspen (*P. tremuloides*) and balsam poplar (*P. balsamifera*). In the south, conifers such as eastern white pine (*Pinus strobus*), red pine (*P. resinosa*) and jack pine (*P. banksiana*) are evident. At the transition with forests to the south, species such as sugar maple (*Acer saccharum*), black ash (*Fraxinus nigra*) and eastern white cedar (*Thuja occidentalis*) are found.

Towards the western boundary of the zone the vegetation is medium to tall closed stands of trembling aspen, balsam poplar and jack pine with white and black spruce occurring in late successional stages. Lodgepole pine (*Pinus contorta*) may dominate in some of the upland areas along with white spruce and balsam fir. Black spruce tends to be concentrated in the poorly drained valleys. Trembling aspen and balsam poplar characterize the transition to the south. White spruce and balsam fir are the climax species but are not widespread because of the frequent occurrence of fire.

Both open and closed black spruce and balsam fir forests are characteristic in the east. White birch and trembling aspen are typical of disturbed sites. White spruce is generally more tolerant of ocean spray and is more prevalent near the ocean.

Wetlands are extensive, with a cover of stunted black spruce, tamarack and shrubs.

The northern part of the zone is transitional to the boreal tundra. Pure stands of jack pine or mixed stands of jack pine, white birch and trembling aspen are typical of the drier sites, while black spruce and balsam fir dominate wet sites.

BOREAL MOUNTAIN SYSTEMS

Mountain ranges with numerous high peaks and extensive plateaus separated by wide valleys and lowlands characterize this zone. The climate ranges from cold, subhumid to semi-arid with long, cold winters and short, warm summers. Mean annual temperatures range from -10°C in the north to 5°C in the south. Mean summer ranges are 6.5° to 11.5°C and mean winter temperatures range between -13° and -25°C. Annual precipitation is lowest in valleys in the rain shadow of the Coast Range (less than 300 mm) and increases up to 1 500 mm at higher elevations of the interior mountains.

Vegetation at higher elevations ranges from arctic to alpine tundra. At lower elevations in the north, open woodlands of white spruce (*Picea glauca*) and white birch (*Betula papyrifera*) are mixed with dwarf birches and willows. The unglaciated Old Crow Basin has stunted stands of black spruce and tamarack with some white spruce. To the south, vegetative cover ranges from closed to open forest of white and black spruce, subalpine fir (*Abies lasiocarpa*), lodgepole pine (*Pinus contorta*), trembling aspen (*Populus tremuloides*), balsam poplar (*P. balsamifera*) and white birch. Lodgepole pine and subalpine fir tend to disappear rapidly towards the north.

TEMPERATE OCEANIC FOREST

This relatively small ecological zone occupies a north-south depression between the Pacific Coast Range and the Cascade Mountains. The nearness of the ocean profoundly moderates the climate, and annual temperatures average 9° to 13°C. Average rainfall ranges from around 400 to 1 500 mm, but more typically is from 750 to 1 150 mm. Fog partially compensates for the summer drought.

These forests are composed of mixtures of western red cedar (*Thuya plicata*), western hemlock (*Tsuga heterophylla*) and Douglas fir (*Pseudotsuga menziesii*). In the interior valleys, the forest is less dense than along the coast and often contains such deciduous trees as big-leaf maple (*Acer macrophyllum*), black cottonwood (*Populus trichocarpa*) and, to the south, Oregon ash (*Fraxinus latifolia*). There are woodlands that support open stands of oaks or are broken by groves of Douglas fir and other trees such as Oregon white oak (*Quercus garryana*) and Pacific madrone (*Arbutus menziesii*). Clearing for cultivation has greatly reduced the area of these forests.

TEMPERATE CONTINENTAL FOREST

Warm summers and cool winters are typical of this zone. The weather is highly changeable. Mean annual temperatures range from 2° to 10°C. The mean summer temperature ranges from 16° to 18°C, with the winter mean ranging from -2.5° to -7°C. Annual precipitation over much of the zone ranges from 720 to 1 000 mm, reaching 1 500 mm near the Atlantic Coast. The proximity of the Atlantic Ocean moderates the climate of the eastern portion of the zone.

At one time the entire zone was heavily forested, but most of the forests around the Great Lakes and in the northeastern United States have succumbed to urbanization and conversion to agriculture. Forest cover varies from mixed coniferous/deciduous stands of white and red pine (*Pinus strobus* and *P. resinosa*), eastern hemlock (*Tsuga canadensis*), red oak (*Quercus rubra*), sugar maple (*Acer saccharum*) and white birch (*Betula papyrifera*) in the northern portions to the rich diversity of the deciduous Carolinian forest in the southwest.

The mixed mesophytic association, the deciduous forest with the greatest diversity, occupies well-drained sites. Widespread dominants include sugar maple, American beech (*Fagus grandifolia*), white elm (*Ulmus americana*), basswood (*Tilia americana*), red and white oak (*Quercus rubra, Q. alba*), walnut (*Juglans nigra, J. cinerea*), hickory (*Carya ovata, C. cordiformis*), buckeye (*Aesculus* spp.) and eastern hemlock (*Tsuga canadensis*) in addition to 20 to 25 other species. An oak association, with white oak and northern red oak as dominant species, occurs east of the Appalachian Mountains.

Further inland, where precipitation is lower, the drought-resistant oak-hickory association is dominant, with white oak, red oak, black oak (*Quercus velutina*), bitternut hickory (*Carya cordiformis*) and shagbark hickory (*Carya ovata*). Wetter sites typically feature American or white elm, tulip-tree (*Liriodendron tulipifera*) and sweetgum (*Liquidambar styraciflua*). Northern reaches of this association contain maple, beech and basswood (*Tilia americana*).

Forests in the northeastern portion of the zone are generally mixed stands of conifers and deciduous species characterized by red spruce (*Picea rubens*), balsam fir (*Abies balsamea*), yellow birch (*Betula alleghaniensis*) and sugar maple. Red and white pine and eastern hemlock occur to a lesser but significant degree. Some boreal species are present, including black spruce (*Picea mariana*), white spruce (*P. glauca*), balsam poplar (*Populus balsamifera*) and white birch. Jack pine (*Pinus banksiana*) is prominent on sandy soils. Pine-oak forest occupies dry sandy soils along the northern coastal plain of the United States and is frequently exposed to naturally occurring fires. Eastern white cedar (*Thuja occidentalis*) occurs on mesic sites.

TEMPERATE STEPPE

The climate of this zone is greatly influenced by its location in the heart of the continent. The zone has a continental climate that is subhumid to semi-arid with short, hot summers and long, cold winters. Generally, precipitation is low and evaporation is high. Mean annual temperature ranges from 1.5° to 3.5°C. Mean winter temperature ranges from -12.5° to -8° C, with summer means from 14° to 16°C. Annual precipitation is variable, from 250 mm in the arid grasslands to near 700 mm in the higher-elevation wooded portions.

Park-like stands of trembling aspen (*Populus tremuloides*) and balsam poplar (*P. balsamifera*) lie at the northern edge of this zone, a transition to the boreal forest to the north. The aspen parkland has expanded considerably southwards since prairie fires were effectively eliminated. Patches of scrubby aspen and cottonwood (*Populus* spp.), willow (*Salix* spp.) and box-elder (*Acer negundo*) occur on shaded slopes of valleys and river terraces. To the east, the zone consists of a mosaic of trembling aspen, bur oak (*Quercus macrocarpa*) and grasslands. Further south, oak and hickory become the dominant tree species in the transition zone with the eastern broadleaf forests.

TEMPERATE DESERT

This zone covers the Great Basin, the northern Colorado Plateau in Utah and the plains and tablelands of the Columbia-Snake River Plateaus and the Wyoming Basin. The aridity of this zone is the result of the rain shadow of the Sierra Nevada and Cascade Mountains. Summers are hot and winters are cold, with stronger seasonal temperature extremes on the higher plateaus. The average annual temperature ranges from 4° to 13°C. Annual precipitation averages about 130 to 400 mm. Almost no rain falls during the summer months. Part of the winter precipitation falls as snow.

The main vegetation, sometimes called sagebrush steppe, is made up of sagebrush and other shrub species mixed with short grasses. Above the sagebrush belt lies a woodland zone dominated by pinyon pine (*Pinus edulis*) and juniper (*Juniperus* spp.).

TEMPERATE MOUNTAIN SYSTEMS

This zone includes the Coast Range, the Rocky Mountains and the Appalachian Mountains. The climate is extremely varied, from a relatively humid maritime climate at low elevations along the Pacific Coast to cold, arctic conditions above the tree line in the Rocky Mountains. Along the coast the mean annual temperature ranges from 4.5°C in the north to 9°C in the south. Average annual precipitation is extremely variable, from 600 mm in the Gulf Islands to 4 000 mm to the north. The interior portion of the zone is similarly variable. The climate of the Appalachian Highlands is more temperate, with a distinct summer and winter. Average annual temperatures range from below 10°C in the north to about 18°C at the southern end. Average annual precipitation varies from 900 mm in the valleys to 2 000 mm on the highest peaks.

The temperate rain forests of the Pacific Coast Mountains are among the most productive in North America and contain some of the world's largest and longest-lived trees. This vegetation association is dominated by western hemlock (*Tsuga heterophylla*) and amabilis or Pacific silver fir *(Abies amabilis)* as climax species, although several other species are common. Big-leaf maple (*Acer macrophyllum*) reaches its northern extension in the southern portion of this zone. It is generally found along creek beds and in other alluvial areas along with red alder (*Alnus rubra*) and black cottonwood (*Populus trichocarpa*). At high elevations, up to 2 000 m, mountain hemlock (*Tsuga mertensiana*), subalpine fir (*Abies lasiocarpa*) and amabilis fir assume prominence along with yellow or Pacific cedar (*Chamaecyparis nootkatensis*), becoming open and stunted at higher elevations.

Amabilis fir, lodgepole pine (*Pinus contorta*) and Sitka spruce (*Picea sitchensis*) are common in the north. At lower elevations in the north, western hemlock and western red cedar (*Thuya plicata*) dominate with red alder pioneering on disturbed sites. The coastal Douglas fir association is found in the lee of the coastal mountains. Douglas fir (*Pseudotsuga menziesii*) dominates. Western red cedar (*Thuja plicata*) is typical on wetter sites, and Garry oak (*Quercus garryana*) and arbutus (*Arbutus menziesii*) are abundant on drier sites.

Interior Douglas fir associations dominate in the rain shadow of the Coast Range and other mountain ranges. Fires have resulted in even-aged lodgepole pine stands at higher elevations, while ponderosa pine (*Pinus ponderosa*) is the common seral tree at the warmer and drier lower elevations.

At mid elevations of the interior plateau regions, closed stands of Englemann spruce (*Picea engelmannii*) and subalpine fir are common. Lodgepole pine, western white pine (*Pinus monticola*), Douglas fir and trembling aspen reflect past fire history. At higher elevations the Englemann spruce-subalpine fir association begins to dominate. The forest often has an open parkland appearance. Under drier conditions, extensive stands of lodgepole pine and whitebark pine (*Pinus albicaulis*) are common. Wetter areas may be dominated by mountain hemlock (*Tsuga mertensiana*).

A western red cedar-western hemlock forest with a wide variety of conifer trees is characteristic of the interior wet belt of this zone. In addition to the two dominant species, other common trees include white spruce (*Picea glauca*), Englemann spruce and subalpine fir. Douglas fir and lodgepole pine occur in drier areas. Englemann spruce, white spruce and subalpine fir are the dominant trees in subboreal plateau areas. Even-aged lodgepole pine and trembling aspen (*Populus tremuloides*) cover extensive areas of previously burned sites.

In the Appalachian Highlands, a vertical zonation prevails, with the lower limits of each forest belt rising towards the south. The valleys of the southern parts support a mixed oak-pine forest. Above this zone lies the Appalachian oak forest, dominated by a dozen species of white and black oaks. At higher elevations is hardwood forest composed of birch, American beech (*Fagus americana*), maple (*Acer* spp.), elm (*Ulmus* spp.), red oak (*Quercus rubra*) and basswood (*Tilia americana*), with an admixture of eastern hemlock (*Tsuga canadensis*) and white pine. Spruce-fir forest and meadows are found on the highest peaks. Mixed mesophytic forest extends into narrow valleys of the southern Appalachians, where oak vegetation predominates. The northern reaches are located in the transition zone between the boreal spruce-fir forest to the north and the deciduous forest to the south. Growth form and species are very similar to those found to the north, but red spruce (*Picea rubens*) tends to replace white spruce. Here the valleys contain a hardwood forest dominated by sugar maple, yellow birch (*Betula alleghaniensis*) and beech. Low mountain slopes are covered with a mixed forest of spruce, fir, maple, beech and birch. Above the mixed-forest zone lie pure stands of balsam fir (*Abies balsamea*) and red spruce.

SUBTROPICAL HUMID FOREST

This zone comprises the Atlantic and Gulf coastal plains and piedmont. Winters are mild and summers hot and humid. Average annual temperature is 15° to 21°C. Annual precipitation ranges from around 1 000 to 1 500 mm, relatively evenly distributed throughout the year.

On the coastal plains temperate evergreen rain forest is the dominant natural vegetation. Subtropical rain forest has fewer tree species than its tropical counterpart; trees are not as tall, leaves are usually smaller and more leathery and the leaf canopy is less dense. Common species include evergreen oaks (*Quercus myrtifolia, Q. virginiana, Q. laurifolia*) and species of laurel (*Cordia alliodora, C. bicolor*) and magnolia (*Magnolia grandiflora, M. virginiana*). Further inland, the climax vegetation is medium-tall to tall forests of broadleaf deciduous and needleleaf evergreen trees. Loblolly pine (*Pinus taeda*), shortleaf pine (*Pinus echinata*) and other southern yellow pine species dominate the stands, singly or in combination. Common associates include oak (*Quercus* spp.), hickory (*Carya* spp.), sweetgum (*Liquidambar styraciflua*), blackgum (*Nyssa sylvatica*), red maple (*Acer rubrum*) and winged elm (*Ulmus alata*). Gum and cypress dominate the extensive coastal marshes and interior swamps along the Atlantic and Gulf Coasts.

Along the Mississippi River, small patches of riverine deciduous forests still occur, with an abundance of green ash (*Fraxinus pennsylvanica*), Carolina poplar (*Populus deltoides*), elm, cottonwood, sugarberry (*Celtis laevigata*), sweetgum and water tupelo (*Nyssa aquatica*), as well as oak and baldcypress (*Taxodium distichum*). Pecan (*Carya illinoiensis*) is also present, associated with American sycamore (*Platanus occidentalis*), American elm (*Ulmus americana*) and roughleaf dogwood (*Cornus drummondii*).

Today, extensive forests of loblolly and slash pine (*Pinus taeda, P. elliottii*) are widespread in this zone, predominantly as plantations or second-growth forest following fire.

SUBTROPICAL DRY FOREST

This ecological zone is situated on the Pacific Coast between approximately 30° and 45°N latitude. The climate is typically Mediterranean, characterized by hot, dry summers and mild winters, with precipitation associated with winter storms. Annual temperatures average about 10° to 18°C, with average summer temperature above 18°C and average winter temperatures above 0°C. Annual rainfall ranges from 200 to 1 000 mm depending on latitude and altitude, always with a pronounced summer drought. Extreme droughts are not uncommon. Coastal fog is typical, particularly from May through July.

Redwood (*Sequoia sempervirens*) is characteristic of the fog belt on seaward slopes in coastal northern California. Associated with it are Douglas fir (*Pseudotsuga menziesii*) and other conifers such as western hemlock (*Tsuga heterophylla*) and western red cedar (*Thuya plicata*). Along the coast in a narrow, patchy belt lies pine-cypress forest. Inland, the south-facing mountain slopes are covered by mixed forest, including tanoak (*Lithocarpus densiflorus*), live oak (*Quercus* spp.), madrone (*Arbutus menziesii*) and Douglas fir.

The central and southern coastal areas are covered by chaparral, a mostly evergreen shrub vegetation. Several tree species are endemic to this region, including Monterey cypress (*Cupressus macrocarpa*), Torrey pine (*Pinus torreyana*), Monterey pine (*P. radiata*) and Bishop pine (*P. muricata*). Patches of live oak (*Quercus* spp., mostly *Q. agrifolia*) or valley oak (*Quercus lobata*) woodland are found on the hills and lower mountains.

A blue oak (*Quercus douglasii*)-foothill pine (*Pinus sabiniana*) woodland community forms a ring around the Central Valley of California. Most of the coastal plains and interior valleys have been converted to urban use or irrigated agriculture.

SUBTROPICAL STEPPE

This zone is dominated by flat to rolling plains and plateaus. The climate is semi-arid subtropical. Summers are long and hot and winters are generally short and mild. Annual temperatures average 14° to 21°C. Annual precipitation varies considerably, from about 250 mm in the drier (mostly western) regions, to about 1 000 mm in the northeastern Prairie Parkland region. The zone is also subject to periodic intense droughts and frosts.

A variety of natural vegetation is found in this zone. Grasslands in which shrubs and trees grow singly or in bunches are predominant. Locally, oak and juniper are mixed with grasses and mesquite (*Prosopis* spp.). Because of the low rainfall they rarely grow higher than 5 to 7 m. The most characteristic tree is Ashe juniper (*Juniperus ashei*). Live oak (*Quercus virginiana*) forest is found along the Gulf Coast. In the northeastern part of the zone, oak savannah, dominated by post oak (*Quercus stellata*) and blackjack oak (*Quercus marilandica*), forms a transition with the more humid subtropical forest zone.

The generally higher Colorado Plateau has distinct vegetation. Woodland is the most extensive vegetation type, dominated by open stands of pinyon pine and several species of juniper (*Juniperus* spp.). Cottonwoods and other trees grow along some of the permanent streams.

SUBTROPICAL MOUNTAIN SYSTEMS

This zone comprises the southernmost portion of the Cascade Mountains and the Rocky Mountains, the Sierra Nevada, the Coast Range and the Western Sierra Madre. The climate is extremely diverse, with variation related to latitude, altitude and exposure. The prevailing west winds influence climatic conditions; the eastern slopes are much drier than the western slopes. Winter and annual precipitation increases with elevation; at high altitude precipitation is mostly snow.

Vegetation zones are well differentiated, generally in altitudinal belts. In the Sierra Nevada, southern Cascades and northern Coast Range, the slopes, from about 500 to 1 200 m, are covered by coniferous and shrub associations. On higher slopes, foothill pine (*Pinus sabiniana*) and blue oak (*Quercus douglasii*) dominate, forming typical open or woodland stands. Above this belt, between 600 and 1 800 m in the Cascades and between 1 500 and 2 400 m or higher in the south, the most important trees are ponderosa pine (*Pinus ponderosa*), Jeffrey pine (*P. jeffreyi*), Douglas fir (*Pseudotsuga menziesii*), sugar pine (*P. lambertiana*), white fir (*Abies concolor*), California red fir (*Abies magnifica*) and incense cedar (*Calocedrus decurrens*), but several other conifers are also present. The spectacular giant sequoia (*Sequoiadendron giganteum*) grows in a few groves on the western slopes. On the dry eastern slopes, Jeffrey pine replaces ponderosa pine. The subalpine zone begins at 1 800 to 2 500 m and extends upslope for about 300 m. Mountain hemlock (*Tsuga mertensiana*), California red fir, lodgepole pine (*Pinus contorta*), western white pine (*P. monticola*) and whitebark pine (*P. albicaulis*) are important. The timberline ranges from about 2 100 m in the north to 3000 m in the south.

Further south in the drier California Coastal Range, the vegetation consists of sclerophyll forest and chaparral. Chaparral is found on south-facing slopes and drier sites, while forest appears on northfacing slopes and wetter sites. The most important evergreen trees are California live oak (*Quercus agrifolia*), canyon live oak (*Q. chrysolepis*), interior live oak (*Q. wislizeni*), tanoak (*Lithocarpus densiflorus*), California laurel (*Umbellularia californica*), Pacific madrone (*Arbutus menziesii*), golden chinkapin (*Castanopsis chrysophylla*) and Pacific bayberry (*Myrica californica*). At higher elevations and near the ocean, chaparral is often interspersed with coniferous forest.

Vegetation zones in the southern Rocky Mountains resemble those further north but occur at higher elevations. The foothill zone, reaching as high as 2 000 m, is characterized by mixed grasses, chaparral brush, oak-juniper woodland and pinyon-juniper woodland. At about 2 000 m, open forests of ponderosa pine are found, although pinyon and juniper occupy south-facing slopes. In Arizona, the pine forests are strongly infused with Chihuahuan pine (*Pinus leiophylla* var. *chihuahuana*) and Apache pine (*Pinus engelmannii*). Pine forest is replaced at about 2 400 m by Douglas fir. Aspen (*Populus tremuloides*) is common in this zone and limber pine (*Pinus flexilis*) grows in places that are rockier and drier. At about 2 700 m the Douglas fir zone merges into a belt of Englemann spruce (*Picea engelmannii*) and corkbark fir (*Abies lasiocarpa* var. *arizonica*). Limber pine and bristlecone pine (*Pinus longaeva*) grow in the rockier places. The alpine zone starts around 3 400 m.

The vegetation of the western Sierra Madre in Mexico includes both evergreen and deciduous forest, primarily composed of conifers and oaks. These grow usually from 10 to 30 m, sometimes reaching 50 m. Mountain cloud forest also occurs. Mexico has about 40 species of pine and more than 150 species of oak.

TROPICAL RAIN FOREST

This zone encompasses parts of the Gulf coastal plain and the lowlands of the Chiapas Sierra Madre in Mexico as well as lowlands along the Caribbean Coast and small areas along the Pacific Coast in Central America. Parts of the Caribbean islands are also included. Year-round temperatures average between 20° and 26°C with little seasonal variation. The average annual precipitation range is 1 500 to 3 000 mm and in some areas may total more than 4 000 mm. The dry season lasts less than three months, occurring in winter. North of about 12°S latitude, hurricanes (tropical cyclones) bring very heavy regional rains from August to October.

The evergreen to semi-evergreen forest along the Atlantic Coast is tall and dense. The forest has a complex and diverse flora with approximately 5 000 vascular plant species. Canopy trees reach 30 to 40 m high, with emergent trees up to 50 m. The subcanopy layer is dense, with trees from 5 to 25 m tall. The understorey layers present a great variety of palms and tree ferns. Common tree species include paque or paleto (*Dialium guianense*), allspice tree (*Pimenta dioica*), breadnut (*Brosimum alicastrum*), manteco (*Ampelocera hottlei*), masica (*Brosimum alicastrum*), masaquilla (*Pseudolmedia* cf. *spurea*), laurel (*Cordia alliodora, C. bicolor*), maria (*Calophyllum brasiliense),* hule (*Castilla elastica, C. tunu),* cuajada (*Dendropanax arboreus),* caobina (*Mauria sessiliflora*), seliyon (*Pouteria izabalensis),* sangre de pozo (*Pterocarpus officinalis*), varillo (*Symphonia globulifera*), caoba (*Swietenia macrophylla*), cumbillo or sombrerete (*Terminalia amazonia*), sangre real (*Virola koschnyi*) and San Juan or copai-yé wood (*Vochysia hondurensis*). There are also well-developed rain forests in specific places on the Pacific side of Central America. Pine grows in infertile locations, alone or in association with oak.

An evergreen forest, intermediate in height, with two or three strata, grows between 400 and 1 300 m altitude on the wetter (Atlantic) side of the Central American ranges. Canopy trees are mostly 30 to 40 m tall. The subcanopy is very dense with trees 15 to 25 m tall.

TROPICAL MOIST DECIDUOUS FOREST

This zone consists of the lower Pacific part of the central mountain ranges in Central America, the plains and hills of the Yucatan Peninsula, humid parts of the Gulf of Mexico plains and the Everglades in the United States. The climate is drier than in the rain forest zone and the dry season is more pronounced (three to five months). Average annual precipitation is around 1 300 mm in El Salvador. It falls to less than 1 000 mm in Honduras and increases again from Nicaragua to Costa Rica. Most of the Yucatan Peninsula in Mexico receives 1 000 to 1 500 mm.

The predominant vegetation is deciduous high forest with three or four storeys and approximately 100 tree species in association on fertile soils. The most typical tree species are *Cordia alliodora, Carapa guianensis, Guarea* spp., *Vitex* spp., *Virola* spp., *Calophyllum brasiliense, Terminalia chiriquensis, Dialium guianense, Tabebuia pentaphylla, Ochroma lagopus* and *Manilkara* spp. From Nicaragua southwards the associations are enriched by many South American species such as *Anacardium excelsum, Dipteryx panamensis, Eschweilera calyculata, Lecythis* spp. and *Prioria copaifera*. Certain distinct associations include pure stands of cativo (*Prioria copaifera*) on riparian flood lands, palm swamps and mangrove swamps on tidal estuaries.

A two-layer semideciduous, seasonal forest of medium height grows in the drier parts of the zone, from 600 to about 1 600 m. The canopy is are mostly dry-season deciduous trees about 25 m tall. Understorey trees are 10 to 20 m tall.

TROPICAL DRY FOREST

This zone comprises flat narrow lowlands or low hilly areas up to 1 000 m altitude, located mainly along the Pacific Coast but also including interior depressions of the Sierra Madre and the northwestern plain of the Yucatan Peninsula in Mexico. The tropical climate of the zone is characterized by short intense episodes of rainfall, especially during the summer. Overall, average annual precipitation is between 600 and 1 600 mm. The dry season varies from five to eight months.

The dominant vegetation formation is dry deciduous forest. A diverse flora is present and low deciduous and semideciduous forests predominate. The forests are from 4 to 15 m tall and have three distinct strata. Southern floristic elements are prominent along with numerous endemic genera on the Pacific side. Legumes dominate the tree flora. On very infertile soils, *Curatella americana* and *Byrsonima crassifolia* form a distinctive association. Since these two species are extremely fire resistant, they are often found on soils seriously degraded by excessive cropping and burning. In northwestern Costa Rica, in the Guanacaste region, a similar association occurs on pumice soils. This association differs in that *Quercus oleoides* accompanies the other two species.

The two vegetation associations covering the major part of the zone on the Pacific Coast differ little in tree species but are quite distinct in terms of dominant species. Characteristic species include *Cedrela mexicana, Swietenia humilis, Enterolobium cyclocarpum, Pithecellobium saman, Hymenaea courbaril, Andira inermis, Platymiscium* spp., *Chlorophora tinctoria, Astronium graveolens, Dalbergia* spp., *Sweetia panamensis, Achras zapota* and *Tabebuia chrysantha*. From Mexico to Honduras *Cybistax donnell-smithii* is another important species, while from Nicaragua southwards *Bombacopsis quinata* is a conspicuous tree. In Mexico, the low deciduous forests contain about 6 000 vascular plant species, of which 40 percent are endemic.

Where the water table is high in fertile soils, as on river flats, a taller and more luxuriant forest occurs; *Brosimum* spp. and *Anacardium excelsum* are common species.

TROPICAL MOUNTAIN SYSTEMS

The climate in the mountain areas varies enormously. Wind-exposed areas are normally wet, while interior valleys are usually moist or dry. Monthly mean temperature shows little seasonal variation but ranges from 12°C at about 1 500 m to less than 6°C at 3 800 m on mountain summits.

Broadleaf forests prevail in highland areas of Mexico, Guatemala, Honduras and Nicaragua, but pine forests are also very common. In the mountain areas of Guatemala where annual rainfall is less than 1 000 mm, the most notable trees are *Pinus pseudostrobus* and several species of *Quercus*. Other genera from the temperate zone such as *Salix, Sambucus, Ostrya* and *Acer* are also represented. On sites where annual precipitation exceeds 1 000 mm, the climax forest consists of mixed broadleaf forest, including species of *Prunus* and *Cornus*, members of the Lauraceae and Ericaceae families and several other species. The forest here is tall and very dense, with canopy trees generally reaching 30 m in height, and it has a dense shrub layer.

The high area of Costa Rica and Panama includes several altitudinal belts. The so-called coffee belt, between 600 and 1 600 m, is an important zone in Central America since most of the population lives there. This belt is part of the previously described tropical lowland zone. From 1 600 m to approximately 2 800 m, the vegetation can is either very tall oak forest or mixed Lauraceae-rich forest. The tall oak forest is a high, comparatively open stand, characterized by emergent, large-crowned oaks, *Quercus copeyensis* and *Q. seemanni*, reaching up to 50 m, and a lower stratum of relatively small to medium-sized trees. The Lauraceae-rich forest is not as tall as the oak forest but still reaches 30 m in height. The forest is very dense, with multiple strata. Genera represented from the Lauraceae family include *Ocotea, Phoebe, Nectandra* and *Persea*. From 2 800 to 3 500 m there are many shrubs and a bamboo species. In the primary forest, evergreen oaks, including *Quercus costaricensis,* dominate the tree canopy, which reaches a height of some 25 to 30 m.

BIBLIOGRAPHY

Bailey, R.G. 1995. *Description of the ecoregions of the United States.* USDA FS Publication No 1391, Washington DC, USA.

Commission for Environmental Cooperation (CEC). 1997. *Ecological regions of North America -toward a common perspective.* Montreal, Quebec.

Dulin, P. 1982. *Distribución de la estación seca en los países Centroamericanos.* Proyecto Leña y Fuentes Alternas de Energia (CATIE-ROCAP No. 596-0089. Turrialba, Costa Rica.

Ecological Stratification Working Group. 1995. *A National ecological framework for Canada.* Agriculture and Agri-Food Canada, Research Branch, Centre for Land and rBiological Resources Research and Environment Canada, State of the Environment Directorate, Ecozone Analysis Branch, Ottawa/Hull. Report and map at 1:7.500 000 scale.

González, L., Ramirez, M., Peralta, R. & Hartshorn, G. 1983. *Estudio ecológico y dendrológico, zonas de vida y vegetación del proyecto Plan de uso de la tierra unidad de manejo bonito oriental.* Programa Forestal ACDI – COHDEFOR. Tegucigalpa, Honduras.

Hartshorn, G., Hartshorn, L., Atmella, A., Gomez, L.D., Mata, A., Morales, R., Ocampo, R., Pool, D., Quesada, C., Solera, C., Solorzano, R., Stiles, G., Tosi, J.A., Umaña, A., Villalobos, C. & Wells, R. 1982. *Costa Rica country environmental profile: A field Study.* USAID Contract NO. 000-C-00-1004-00. Tropical Science Center. San Jose, Costa Rica.

Hirvonen, H.E. 1984. *The Atlantic region – An ecological perspective.* Lands and Integrated Programs Directorate, Environment Canada, Halifax, Nova Scotia.

Holdridge, L.R. 1957. *The Vegetation of ainland Middle America.* Reprinted from the Proceedings of the Eighth Pacific Science Congress, Volume IV, pp. 148-161. National Research Council of the Philippines. University of the Philippines. Diliman, Quezon City.

Holdridge, L.R., Lamb, B. & Masson, B. 1950. *Los bosques de Guatemala: Informe general de silvicultura, manejo y posibilidades industriales de los recursos forestales de Guatemala.* Instituto Interamericano de Ciencias Agricolas y el Instituto de Fomento de la Producción de Guatemala. Turrialba, Costa Rica.

Lopoukhine, N., Prout, N. & Hirvonen, H. 1979. *The ecological land classification of Labrador – reconnaissance.* Fisheries and Environment Canada, Halifax, Nova Scotia.

Oswald, E.T. & Senyk, J.P. 1977. *Ecoregions of Yukon Territory.* Publication Number BC-X-164. Canadian Forestry Service, Environment Canada, Victoria, British Columbia.

Ricketts, T., Olson, D., Loucks, C. *et al*. 1999. *Terrestrial ecoregions of North America –* A Conservation Assessment. World Wildlife Fund – United States and Canada. Island Press, Washington, D.C.

Chapter 34
North America, excluding Mexico

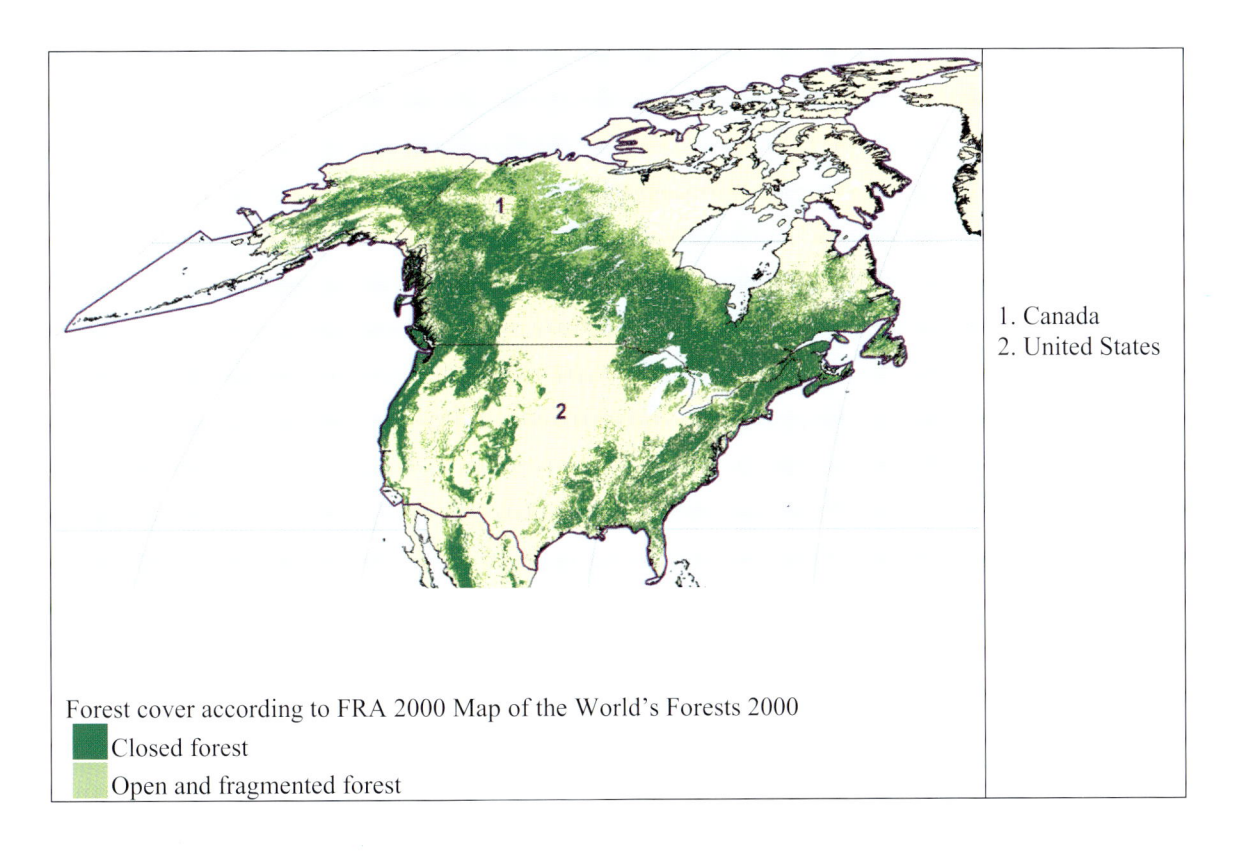

Figure 34-1. North America excluding Mexico: forest cover map

For purposes of discussion, Canada and the United States are considered together in this chapter (Figure 34-1) Mexico has historically been considered part of either North or Central America, depending on the context. In terms of forest ecosystems, Mexico tends to have more in common with countries in Central America, and it is therefore discussed in the next chapter along with those countries.

The forests of Canada and the United States are among the largest, most diverse and most intensively utilized in the world. The combined forests of Canada and the United States account for 14 percent of the world's land area, 12 percent of the global forest area and 28 percent of the world's temperate and boreal forests. North America is about 26 percent forested, slightly below the global average of 30 percent (Table 34-1). An additional 11 percent of the region is "other wooded land" (between 5 and 10 percent canopy cover). In Canada, forest and other wooded land together comprise 45 percent of the land area, when inland water areas are not considered; in the United States, the respective figure is 31 percent.

Canada is the world's second largest country in terms of total land area (behind the Russian Federation), and Canada ranks third in total forest area behind the Russian Federation and Brazil. The United States follows close behind, ranking third in land area and fourth in forest area.

FOREST RESOURCES

Canada's Forest Inventory 1991 is the authoritative national database on the distribution and structure of Canada's forest resource. The inventory is aggregated from many sources, including existing data available in the provincial and territorial forest services. Over the years the specifications of the modern source inventories have become more complete, and most provinces and territories have programmes of periodic inventory renewal for the active areas of forest management. The oldest source inventories, with

Table 34-1: North America excluding Mexico: forest resources and management

Country/area	Land area	Forest area 2000					Area change 1990-2000 (total forest)		Volume and above-ground biomass (total forest)		Forest under management plan	
		Natural forest	Forest plantation	Total forest								
	000 ha	*000 ha*	*000 ha*	*000 ha*	%	ha/capita	*000 ha/year*	%	m³/ha	t/ha	*000 ha*	%
Canada	922 097	244 571		244 571	26.5	7.9	n.s.	n.s.	120	83	173 400	71
United States	915 895	209 755	16 238	225 993	24.7	0.8	388	0.2	136	108	125 707	56
Total North America	1 837 992	454 326	16 238	470 564	25.6	1.5	388	0.1	128	95	299 107	66
Total North and Central America	2 136 966	531 771	17 533	549 304	25.7	1.1	-570	-0.1	123	95	-	-
TOTAL WORLD	13 063 900	3 682 722	186 733	3 869 455	29.6	0.6	-9 391	-0.2	100	109	-	-

Source: Appendix 3, Tables 3, 4, 6, 7 and 9.

the most missing values in the data, tend to be those from more remote areas.

The United States Forest Service has conducted periodic forest inventories of all forested lands in the United States for more than 70 years, providing scientifically reliable data on the status, condition, trends and health of the nation's forests. The national forest inventory utilizes a systematic random grid sample design with remote sensing samples (30 m to 1 km resolution) and field samples (every 5 km) distributed uniformly across the landscape. Field crews collect a variety of ecosystem data. Samples are permanent, remeasured on a five- to ten-year cycle, and designed to an accuracy of ±1 percent per million hectares for forest area estimates and ±3 percent per billion cubic metres or volume estimates. Additional resource data are derived from surveys that monitor private forest landowner objectives, inputs to primary wood processing facilities, residential fuelwood use, participation in outdoor recreation and wildlife activity.

The forest cover in the two countries is 96.5 percent natural forest. Natural forests showed a net increase of 0.1 percent during the period 1990-2000. Canada reported zero plantation forests in 2000, while forest plantations accounted for 7 percent of total forest area in the United States (Table 34-1, Figure 34-2).

The volume of wood above ground averages 128 m³ per hectare in North America, considerably higher than the global average of 100 m³. In contrast, the average woody biomass in the region was 95 tonnes per hectare, considerably below the world average of 109 tonnes per hectare. Relative to tropical forests, the typical temperate or boreal forest has larger trees but lower tree density, especially in boreal forests. Hence it is not surprising that the woody volume per hectare in this region is higher than the global average, while the biomass is below the world average.

In comparison with other developed regions of the world, particularly Western Europe, North America still has relatively large areas of natural forests, especially in Canada and in the western United States.

Canada has a broad belt of coniferous forest, much of it boreal, across the country, with tundra to the north. In the temperate southern and eastern parts of the country (Ontario, Quebec and the maritime provinces), broadleaf species including maples (*Acer* spp.) and oaks (*Quercus* spp.) predominate – hence the famous maple leaf on the national flag. Species of birch (*Betula* spp.), alder (*Alnus* spp.) and willow (*Salix* spp.) occur widely throughout the country. British Columbia in the west has specific forest types determined by the montane and coastal nature of the province. Coniferous species make up the major part of the growing stock, the main species being spruces (*Picea* spp.), pines (*Pinus* spp.), firs (*Abies* spp.) and larches (*Larix* spp.). Along the western coast of British Columbia other species, which grow to very large sizes, include Douglas fir (*Pseudotsuga* spp.), western hemlock (*Tsuga heterophylla*) and western red cedar (*Thuja* spp.). There are about 180 species of forest trees in Canada and a very wide range of forest types in 15 different major ecological zones.

Forests in the United States are among the most diverse in the world, ranging from boreal forests in Alaska to tropical forests in Hawaii. In the "lower 48" states, forests in the east reside in a temperate humid climate zone to the north and a subtropical humid climate zone to the south and comprise both broadleaf deciduous and coniferous evergreen trees. The eastern temperate zone is heavily forested with second- and third-growth forests dominated by spruce-fir with northern pine forests (*Pinus strobus*) interspersed. Oaks,

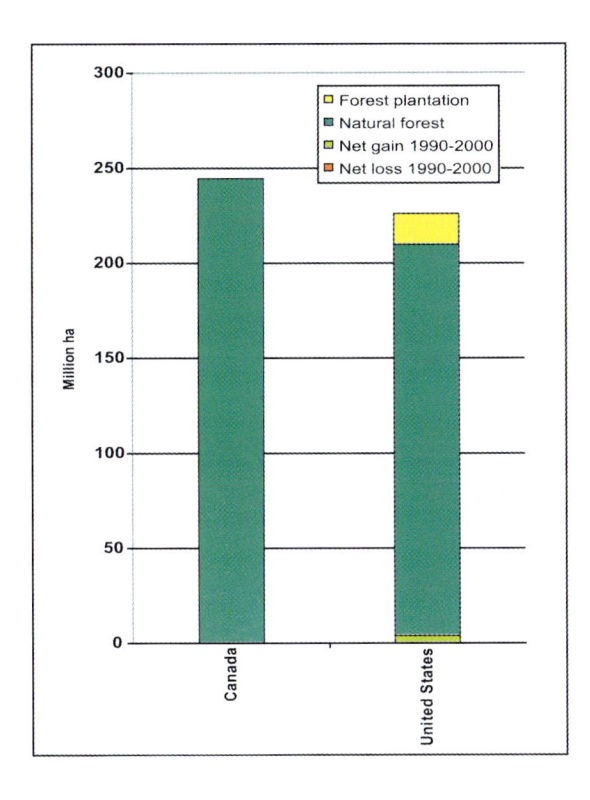

Figure 34-2. North America excluding Mexico: natural forest and forest plantation areas in 2000 and net changes 1990-2000

hickories (*Carya ovata*), yellow poplar (*Liriodendron tulipifera*), maples and beech (*Fagus* spp.) on the uplands and elm (*Ulmus* spp.), ash (*Fraxinus* spp.) and maple in the lowlands dominate the forests of the central and southern reaches of the eastern temperate zone. The temperate zone gives way to the subtropical zone in the middle and lower latitudes of the east, with extensive southern oak and hickory forests on the uplands with mixed oak and southern pine (*Pinus elliottii*) on the drier sites. Oak, gum (*Eucalyptus* spp.) and cypress (*Cupressus* spp.) dominate lowland forests throughout the subtropical zone of the east.

Forests in the western United States reside in arid and semi-arid conditions in the interior and in temperate oceanic and Mediterranean climates along the West Coast. Conifers, including spruce, pines, firs, cedars and hemlock, dominate western forests. In Alaska, boreal forests generally consist of closed stands of conifers (spruce, tamarack and fir) interspersed with birch and aspen. Hemlock, cedar and spruce dominate Alaska's southeastern coast. A small area of tropical humid climate is also found at low latitudes. Hawaii and extreme southern Florida support this regime. While southern Florida is dominated by wet savannah, Hawaii has evergreen and semideciduous forests of great diversity.

FOREST MANAGEMENT AND USES

Canada and the United States are both developed countries with major forest resources. Both countries face similar challenges as they enter the new millennium, with increasing pressure to conserve or sustainably manage their large areas of natural forests. Both countries are among the world leaders in the production and export of forest products, and the United States is also the world's largest importer of forest products (including imports from Canada). Canada produces large quantities of all forest products and is particularly important as a producer of sawn timber and wood pulp (Natural Resources Canada/Canadian Forest Service 2000).

The United States produces around 30 percent of global industrial roundwood, and its share of global production and consumption of sawn timber, wood-based panels, pulp and paper is of a similar magnitude. Private forest lands provided 89 percent of the timber harvest as of 1996 (USDA Forest Service 2001).

The two North American countries differ significantly in the ownership of their forest resources. This has a major influence on approaches to forest management and political positions on international forest policy issues, seen most notably in their opposite positions on the merits of a global forest convention.

Over 93 percent of Canada's forests are publicly owned; provincial governments have jurisdiction over more than 70 percent of Canada's forest and other wooded land, and 23 percent is under federal and territorial government jurisdiction. Although privately owned forests constitute less than 7 percent of the forest area, there are more than 425 000 private landowners.

In the United States over 60 percent of forests are privately owned, with over 10 million private forest owners. Public forest ownership is concentrated in the west, while most private forests are in the east, with the result that forest politics tend to be influenced by geography. Vast tracts of private forests are owned by large companies, amounting to about 10 percent of the total forest area and the greatest part of the forest plantations. Historically, much of the timber production in the United States came from public lands, but in the past decade this was reduced to less than 10 percent. A major shift in public policy has greatly reduced timber harvesting in National Forests, which are increasingly used for recreation and environmental conservation. National Forests account for 17 percent of forest land and 19 percent of theoretically available timber supply; however, as of 1996 only 5 percent of the United States timber harvest came from National Forests.

Canada reported that 71 percent of its forest land is under management. Silvicultural systems used in managing even-aged forests for timber production include clear-cutting, seed-tree and shelterwood harvesting. Clear-cutting remains the most widely used silvicultural system in Canada, but harvesting techniques are changing. Canadian forestry officials have reported widespread use of advanced and appropriate regeneration techniques to ensure that most harvested areas will regenerate naturally, supplemented by planting or seeding on sites where regeneration fails to meet stocking standards (Natural Resources Canada/Canadian Forest Service 1999). More than 16 million hectares of Canadian forest land are certified under one of the three systems used in Canada: those of the Canadian Standards Association (CSA), the International Organization for Standardization (ISO) and the Forest Stewardship Council (FSC).

In the United States, a reported 55 percent of the total forest area is under management. While 100 percent of public forests can be considered covered by management plans, an estimated 70 percent of public forest is managed for multiple objectives and the remaining 30 percent is in protected areas. Only about 5 percent of the private landowners have written management plans, but these cover 39 percent of the private forest area because most large forest owners have management plans (USDA Forest Service 2001). Private lands are regulated by the states, and all states have forest management laws. Forest policies and legislation are heavily influenced by the constitutional and customary rights of private property owners. For both public and private lands, forest management decisions are usually decentralized to the local level.

Forests throughout the North American region are vulnerable to forest fires and forest pests. For the past half century, Canada and the United States, together with Mexico (the third member of the North American Forestry Commission), have collaborated on research and management approaches to protect their forest resources from fire and pests. The extent to which the three countries cooperate in fighting forest fires could serve as a model for other countries.

Several native North American insect species – spruce budworm, forest tent caterpillar, hemlock looper and jack pine budworm – annually defoliate areas of Canada's forest. The spruce budworm (*Choristoneura fumiferana*) is considered the most destructive pest of fir and spruce forests in eastern Canada (Natural Resources Canada/Canadian Forest Service 1999). In the United States, native pest outbreaks tend to be triggered by conditions such as weather or timber stands that are overmature, overcrowded or otherwise under stress. In addition, severe impacts have resulted from introduced pests, including the gypsy moth, chestnut blight, Dutch elm disease, white pine blister rust and the Asian longhorn beetle.

Forest fires are among the most critical forest management challenges in the region. Increasingly, attention has been turned to the positive as well as the negative effects of fire. One of the great ironies of the history of forest management and protection in North America is that successful efforts to prevent and control forest fires have contributed to increasing the overall threats that fire and pests pose to the health and productivity of forests.

Wildfire performs many valuable ecological services in Canada's forests. Several species are adapted to and may even require fire for reproduction. Other species, however, are very averse to fire and may disappear entirely from an area if fire becomes too frequent or severe. It is therefore important to track not only the national area burned, but also the location of fire activity in different ecological zones and forest types. Furthermore, while fire suppression may allow an increase in the mean stand age in an area, it may also allow greater accumulation of organic material which may fuel more severe fires in the future. For this reason, minimizing fire is not always desirable. In fact, long-term forest sustainability includes a role for naturally variable fire activity (Canadian Council of Forest Ministers, annual).

Similarly, in the United States during the 1990s fire came to be increasingly viewed as a management tool as well as a deadly enemy. The worst damage from forest fires occurred in areas where fires had been successfully prevented and suppressed in the previous decades, which had resulted in an increase in fuel.

On average, 91.5 percent of all fires in Canada in the 1990s were found to burn less than 10 ha;

these fires accounted for 0.4 percent of the total area burned. Conversely, the 1.4 percent of all fires that exceeded 1 000 ha accounted for 93.1 percent of the total area burned. Some 58 percent of all fires in Canada started as a result of human carelessness, but these fires burned only 15 percent of the total area burned. Lightning, on the other hand, started 42 percent of all fires, accounting for 85 percent of the total area burned. Six of the ten most severe years of recorded forest fires were in the 1990s.

Forest fires are a serious problem in the United States, where on average 108 597 fires occurred per year during the 1990s. In the interior west, heavy fuel loads, exacerbated until recently by a strict fire control policy, combined with a ready source of ignition (lightning) and rugged terrain, have resulted in serious and difficult to control fires in some locations almost every year. In addition, some forest pest outbreaks appear to have been related to ill health of forest stands which was partly a result of their successful protection from fire – including natural lightning fire – over a long period of time.

Canada and the United States both have large populations of indigenous people. In Canada, aboriginal communities are often more dependent on products from the forest than non-aboriginal communities. The income in-kind represented by subsistence products (i.e. the replacement cost of purchasing similar products from a store) accounts for a significant proportion of total household income. Equally important is the role of subsistence forest products in maintaining the social fabric of the community and in preserving aboriginal culture. A significantly high percentage of food that is gathered by individuals in the community is shared or bartered with other members of the community. Moreover, members of aboriginal communities consider living from the land as an important aspect of traditional culture.

In the United States, 555 federally recognized Native American tribes own about 6.9 million hectares of forest and other wooded land. In addition, Native Americans have rights of harvest and collection on an estimated 70 million hectares of federal lands. Many of the tribes or their members own forest products businesses, commercial fishing operations or guiding and outfitting operations for hunting and other types of recreation. Products harvested for tribal use include fish, fur-bearing animals, game for meat and hides to make clothing and other goods, fuelwood, plants for food and medicinal uses and materials for crafts such as basketry. The forest also has important symbolic and cultural value, with certain sites having particular spiritual or cultural significance (Birch 1996).

Throughout North America there is increasing commercial demand for non-wood forest products such as mushrooms, honey, various species of nuts, medicinal and herbal plants and decorative foliage. Hunting is a source of significant income for both private landowners and public management agencies. Fishing is also frequently associated with forests. Recreation and tourism have become increasingly important to national and local economies.

CONCLUSIONS AND ISSUES

The 1990s saw major changes in approaches to forest management in both the United States and Canada. Perhaps the most dramatic change was the large increase in local consultation and conflict resolution processes in decision-making about forests. Some observers noted that these two major developed countries were learning to use social and community forestry techniques that had been pioneered in developing countries over the previous several decades.

In Canada the 1990s witnessed increasing consultation with stakeholders (forest owners, industries, aboriginal groups, policy-makers, etc.) to identify appropriate forest strategies, legislation, and management plans. Strategies varied and included buyer regulations, land use planning, regulations for forestry practices on forest land and agricultural land, and licences to reduce clear-cutting, among others.

In the United States, much attention was focused on the management of National Forests. Legal decisions and public opinion at the national level continued to shift towards an increased emphasis on recreation, amenity values and biodiversity protection on public lands, setting off confrontations with local communities who have relied on timber harvesting and other development activities on National Forest lands for jobs and income. In some cases, local approaches to conflict resolution helped to resolve conflicts, but in other cases disagreements over forest management contributed to community polarization.

As the 1990s drew to a close, the brown spruce longhorn beetle, native to central Europe and Asia, was found in Nova Scotia, a grim reminder of the vulnerability of North America's forests to exotic pests. In its home environment the beetle eats only dead and dying trees, but in

Canada it appears to feed on living red spruce trees which are native to North America. Dutch elm disease is again spreading fast across eastern Ontario; this may be a result of the 1998 ice storm, as beetles that carry the disease have moved into breaks in branches and are spreading the infection.

In 1999 the Canadian Senate released a report on Canada's progress in achieving the national goals of sustainable forest management and the protection of biodiversity in boreal forests. According to the report, Canadians must find better ways to manage the boreal forest to meet the competing needs of preserving the resource, maintaining the lifestyle and values of boreal communities, extracting economic wealth and preserving ecological values. Portions of Canada's remaining undisturbed boreal forest and its areas of old growth are now at risk from both climate change and overcutting. The report concluded that the demands being placed on Canada's forests could no longer be met under the current system of management. It was recommended that the boreal forest be divided into three categories: 20 percent to be intensively managed for timber production; roughly 60 percent to be reserved for multiple use which would include some less intensive timber production; and up to 20 percent to be protected.

The sustainable management of forests in Canada continued to gain momentum as a goal for all stakeholders. There was progress towards a network of protected areas. On the economic front, in 2000 the Canadian pulp and the paper industry experienced record exports (Natural Resources Canada/Canadian Forest Service 2000).

In the United States, the 2000 Assessment of Forest and Range Lands carried out by the Forest Service (USDA Forest Service 2001) found that the area of forest land has remained relatively stable at about one-third of the total land area. Prior to European settlement about 50 percent of the land in the United States was forested. The United States was the first country to set aside forests in protected areas, and by 2000 the protected forest area accounted for 40 percent of total forests (essentially all public forests) – by far the largest protected forest area in the world. Deforestation has not been a national problem in the United States since about 1920, although at the local level valuable forests were sometimes victims of urban expansion. Forest losses have been more than offset by reversion of pasture and cropland to forests, both naturally and through afforestation and reforestation. Over the coming decades, the area of privately owned forest land is expected to decline and more outputs will need to be produced from a stable or perhaps slowly declining land base. Fragmentation of privately owned forest land will make management of these lands for timber production an increasing challenge. Forest health and productivity are major concerns in the United States.

In the United States, population and income are projected to continue to grow. While per capita consumption of timber products is projected to remain relatively stable, total consumption is expected to increase because of increased population, including increases due to immigration. As a result of major policy changes, there has been a substantial decline in the volume harvested from the National Forests in the western United States (from 57 million cubic metres in 1986 to 23 million cubic metres per year in 1996). As a consequence timber harvesting has shifted towards private lands, especially on softwood plantations in the south. As in the past, it is anticipated that technology will continue to lead to increased output per unit of roundwood input.

A lack of information about the supply of non-wood forest products in North America makes it difficult to assess the sustainability of current use or appropriate management techniques. Growth in demand, both domestic and international, for many of these products has led to the potential threat of overuse, destructive production techniques and possible harm to the productivity of the resources. A coherent monitoring and management policy for non-wood products is needed. At a minimum, future assessments should determine what products are important to report at the international scale, provide clearer definitions for the products and require better information about the source and coverage of the data. Finally, existing and potential conflicts between users, combined with the increasing demand, are creating immediate challenges for managers.

In both the United States and Canada, it has recently been recognized that in order to prevent catastrophic loss from fire, insects and diseases, forests must be maintained in a healthy condition. Forests that were formerly maintained in seral stages by frequent fires have experienced a change in character with fire exclusion. Fuel loads have increased and understory trees and shrubs have become established which result in catastrophic, stand-replacing fires. Senescent stands and climax species are also more vulnerable to attack by insects and diseases. It is difficult to duplicate the natural conditions that formerly existed. Prescribed burning, especially with the terrain and weather conditions prevalent in the west, is difficult, costly and risky. Numerous groups oppose harvesting, especially on publicly owned lands. Yet the alternative is an increasing frequency and magnitude of catastrophic fires and insect and disease attacks.

Forest policy-makers in Canada and the United States will continue to be confronted with difficult choices in the face of greatly divergent opinions about priorities for managing forest resources.

BIBLIOGRAPHY

Birch, T.W. 1996. *Private forest-land owners of the United States, 1984.* Resour. Bull. NE-RB-134. Radnor, Pennsylvania, USA, USDA Forest Service, Northeastern Forest Experiment Station.

Canadian Council of Forest Ministers. (annual). *Compendium of Canadian Forestry Statistics.* http://nfdp.ccfm.org

Canadian Council of Forest Ministers. 2000. *Criteria and indicators of sustainable forest management in Canada – national status 2000.* Ottawa. www.nrcan.gc.ca/cfs/proj/ppiab/ci/

Howard, J.L. 1997. *United States timber production, trade, consumption, and price statistics 1965-1994.*

Natural Resources Canada/Canadian Forest Service. 1999. *Forest health in Canada: an overview 1998.* Ottawa.

Natural Resources Canada/Canadian Forest Service. 2000. *The state of Canada's forests 1999-2000 – forests in the new millennium.* Ottawa. www.nrcanrncan.gc.ca/cfs/proj/ppiab/sof/common

Smith, W.B., Faulkner, J.L. & Powell, D.S. 1994. *Forest statistics of the United States 1992 metric units.* Gen. Tech. Rep. NC-168. St Paul, Minnesota, USA, USDA Forest Service.

Smith, W.B. & Sheffield, R.M. 2000. *A brief overview of the forest resources of the United States.* Washington, DC, USA, USDA Forest Service.

Statistics Canada. (annual). *Canada Year Book.* Ottawa.

United States Department of Agriculture (USDA) Forest Service. 1997. Forest Inventory and Analysis National Program Office. http://fia.fs.fed.us./

USDA Forest Service. 2001a. *United States forest facts and historical trends.* FS-696. Washington, DC.

USDA Forest Service. 2001b. *2000 RPA Assessment of Forest and Range Lands.* FS-687. Washington, DC. www.fs.fed.us/pl/rpa/

United States Department of Commerce, Bureau of the Census. 1990. *Decennial Census Statistics.*

Wiken, E.B. (undated). *An introduction to ecozones.* www.cprc.uregina.ca/ccea/ecozones/intro.html

Chapter 35
Central America and Mexico

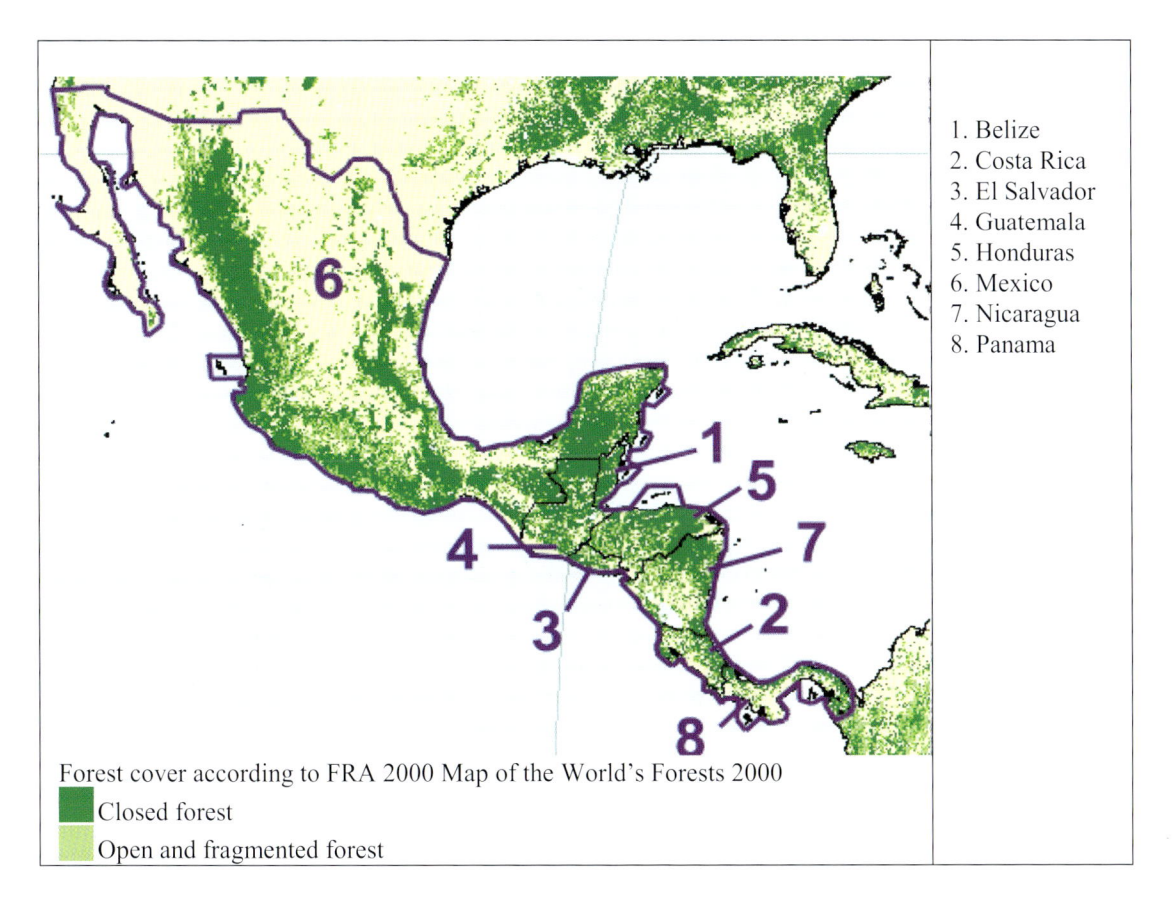

Figure 35-1. Central America and Mexico: forest cover map

This subregion includes the countries of Belize, Guatemala, Costa Rica, El Salvador, Honduras, Mexico, Nicaragua and Panama (Figure 35-1). Mexico has the largest land area, more than 190 million hectares, with 29 percent under forest cover. The remaining countries together have a land area of 51 million hectares, with 34 percent covered by forest.

The natural vegetation of Mexico can be divided into three approximately equal areas. The tropical/subtropical region includes tropical rain forests originally covering 6 percent of the country. The temperate region occupies the main cordilleras, about 15 percent of the country, with forests consisting of a wide variety of pines (*Pinus* spp.) and oaks (*Quercus* spp.). About 80 percent of the plant species found in the pine forests are endemic. In the higher parts of the cordilleras, up to 3 300 m, forests of silver fir (*Abies* spp.) occur. The semi-arid/arid zone, found mainly in the north and centre (Sonoran and Chihuahuan deserts and central altiplano), includes mostly open shrubland (matorral), cacti and xerophytic monocots (Mexico Comisión Nacional para el Conocimiento y Uso de la Biodiversidad 1998).

The countries to the south of Mexico from Guatemala to Panama are recognized as a biological corridor between North and South America. In addition, the influences of the Atlantic and Pacific Oceans and the mountains create environmental conditions also conducive to high biodiversity. Holdridge identified 14 life zones in Guatemala, 13 in Costa Rica, 12 in Panama and 9 in Belize (Godoy 1997).

FOREST RESOURCES

A workshop, with participation from all countries, was organized in Costa Rica in 1999 for the collection of FRA 2000 information for this

Table 35-1. Central America and Mexico: forest resources and management

Country/area	Land area	Forest area 2000 – Natural forest	Forest area 2000 – Forest plantation	Forest area 2000 – Total forest			Area change 1990-2000 (total forest)		Volume and above-ground biomass (total forest)		Forest under management plan	
	000 ha	000 ha	000 ha	000 ha	%	ha/ capita	000 ha/ year	%	m^3/ha	t/ha	000 ha	%
Belize	2 280	1 345	3	1 348	59.1	5.7	-36	-2.3	202	211	1 000	74
Costa Rica	5 106	1 790	178	1 968	38.5	0.5	-16	-0.8	211	220	116*	n.ap.
El Salvador	2 072	107	14	121	5.8	n.s.	-7	-4.6	-	-	-	-
Guatemala	10 843	2 717	133	2 850	26.3	0.3	-54	-1.7	355	371	54	2
Honduras	11 189	5 335	48	5 383	48.1	0.9	-59	-1.0	58	105	821	15
Mexico	190 869	54 938	267	55 205	28.9	0.6	-631	-1.1	52	54	7 100	13
Nicaragua	12 140	3 232	46	3 278	27.0	0.7	-117	-3.0	154	161	236	7
Panama	7 443	2 836	40	2 876	38.6	1.0	-52	-1.6	308	322	20*	n.ap.
Total Central America	241 942	72 300	729	73 029	30.2	0.5	-971	-1.2	86	93	-	-
Total North and Central America	2 136 966	531 771	17 533	549 304	25.7	1.1	-570	-0.1	123	95	-	-
TOTAL WORLD	13 063 900	3 682 722	186 733	3 869 455	29.6	0.6	-9 391	-0.2	100	109	-	-

Source: Appendix 3, Tables 3, 4, 6, 7 and 9.
*Partial result only. National figure not available.

subregion. In addition, FAO assembled historical data to estimate forest cover as of 2000 and change from 1990 to 2000.

The quality of information varies from country to country. Mexico has carried out systematic forestry inventories since the 1960s. The latest one was published in 1994 based on remote sensing images from 1993. Good, updated information on forest cover is available for Costa Rica, Guatemala and Honduras from detailed forest maps and forest inventories. Forest information for Belize and El Salvador was extracted from general land use maps. The estimates for Nicaragua and Panama were based on secondary sources.

This subregion has one of the highest negative rates of forest area change in the world. In terms of gross area, Mexico and Nicaragua have the highest negative change in the subregion. In relation to the amount of forest cover, however, the highest rates of negative change are found in El Salvador, Nicaragua and Belize, whereas Costa Rica and Honduras have the lowest rates. The countries with the highest proportion of forest cover are Belize and Honduras, with 59 percent and 48 percent, respectively (Figure 35-2 and Table 35-1). Even though broadleaf forest covers the largest area in the region, coniferous formations are economically more important in some countries; for example, in Honduras *Pinus oocarpa* and *Pinus caribaea* are very important, as are *Pinus montezumae* and *Pinus ayacahuite* in Mexico. It was also noted that in all countries where coniferous formations exist, deforestation rates for these formations are higher than for other forest types.

Mexico and Costa Rica have the largest area of plantations in the subregion, promoted by forest incentive policies in both countries. These plantations are both for industrial purposes and fuelwood production. Belize has the lowest plantation area in the subregion. The contribution of the region to the world's plantation area is less than 1 percent.

The volume and biomass estimates for Central America are based on a regional project that estimated carbon sequestration potential in these seven countries. The estimate of biomass for Mexico is based on the commercial volume of the different forest types of the country, adjusted to arrive at the total volume. It is noteworthy that the forests of this region have the highest level of biomass per unit area in the world.

According to Calvo (2000), this relatively small subregion contains 7 percent of the world's biological diversity. It has approximately 4 million hectares of natural tropical pine forests, ranging from Mexico to southern Nicaragua and Panama, and approximately 7 million hectares of tropical hardwood forests as well as mangrove areas along both the Atlantic and Pacific Coasts (Calvo 2000).

On about 4 million hectares of the area under hardwood forest cover, valuable species such as mahogany (*Swietenia* spp., mainly *S. macrophylla*, *S. humilis* and *S. mahagoni*) and "cedro" (*Cedrela odorata*) are present at the rate of about one commercial tree for every 6 to

7 ha of forest, or 4 percent of the total commercial volume per hectare. There are up to 100 commercial hardwood species. In their natural habitat these species grow relatively slowly, usually less than 1 m³ per hectare per year; however, this rate can double or triple in forests under sustainable forest management. In well-managed plantations, the volume growth per year of both pine and hardwood species can be exceptionally high, in some cases reaching 30 m³ per hectare per year (Calvo 2000).

FOREST MANAGEMENT AND USES

Five of the eight countries in Central America and Mexico provided national-level information on the area of natural forest under management for the biennial meeting of the Latin American and Caribbean Forestry Commission in 2000 (Table 35-1). The figures provided ranged from 2 to 74 percent of the total forest area in 2000. Partial data were available from Costa Rica[48] and Panama.[49] No information was provided by El Salvador.

In each country the prerequisites for authorizing forest management activities are different, but in general the State and the users agree on the implementation of certain forest practices on a specific area for a specified period of time. All the administrative, technical and legal requirements are to be met by the parties that assume responsibility and obtain the benefits from the use of the area. The technologies used are supposed to be compatible with conservation of the environment and to guarantee the future services and functions of the forest. While this is the goal of the management plans, the results cannot always be quantified with regard to appropriate practices or sustainability.

According to the *Revista Forestal Centroamericana* (1999), during the past decade there has been increased interest in improving the monitoring and management of natural forests. The area of forest plantations has increased. The forestry industry, the "campesinos" (local people who work the land and may also depend on the forests) and the national forestry institutions have begun to work together to improve forest management of areas under communal and industrial concessions. The use of external evaluators to support "green certification" has increased. Traditional use of communal forests by indigenous peoples (such as the Mizquitos in Nicaragua and Honduras, the Cunas in Panama, the Garifunas along the Atlantic Coast of the subregion and the Mayas in Guatemala and Mexico) has not been quantified (*Revista Forestal Centroamericana* 1999).

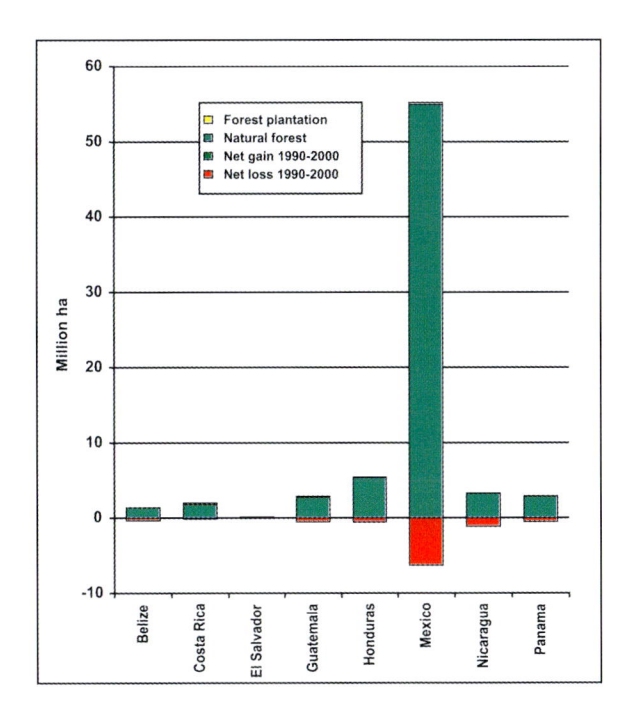

Figure 35-2. Central America and Mexico: natural forest and forest plantation areas in 2000 and net area changes 1990-2000

Surveys of forest resource use in Mexico have mainly related to commercial uses. National demand for wood has not been met since 1997; in 2000 the deficit was 43 percent. The forest sector contributed US$369 million to the national economy in 1997, but forest products imports, mainly paper and cellulose, amounted to US$1 169 million. Communal management in the rural areas has focused mainly on resin extraction and the use of fuelwood (Mexico SEMARNAP 2000).

In the other seven countries the contribution of the forestry sector to the national economy, mainly timber production, is largely unknown since the statistics are combined with those for agricultural production. In addition, information on the extraction of non-wood forest products is limited to those that are regulated for export (e.g. resins, rubber, fruits, spices). There is a need to quantify the economic contribution of timber and non-wood forest products consumed locally, as these are significant uses (FAO 1998).

---

[48] The figure provided was the area of forest taken under management for the period 1998-1999.
[49] Production forests only.

Export values for some hardwoods (especially *Swietenia* spp. and *Cedrela* spp.) can exceed US$300 per cubic metre, and they are therefore highly desired and often exploited. However, pine forests may be more frequently exploited since they are more accessible, the trees are generally smaller (thus requiring simpler technology) and the demand is high, especially for construction. (Calvo 2000).

Studies carried out by FRA 2000 indicate that, while there is increased interest in sustainable forest management in the region, the percentage of forest that is under management is still low. One of the reasons is that the development and approval of forest management plans largely depends on external financial support, especially from international non-governmental organizations (NGOs), mainly because the national forest authorities do not have enough personnel and funds to respond to users' needs. In Guatemala, instability and the ambiguity of the laws make forestry an unattractive sector for investment. While forest cover may stabilize through protected areas policies, the forest industry will not be an area of major investment as long as the laws and regulations change so often (see FRA Working Papers 13, 34, 35, 36, 37, 40, 41 and 44).

As discussed in the FRA Working Papers, fuelwood is still a major source of energy in the region and fuelwood extraction is a major forest use. In Mexico, around 3 million families in rural areas depend on forests for fuelwood. In Guatemala and El Salvador, more than 80 percent of the population uses fuelwood for domestic and industrial energy needs. In Costa Rica, there has been a decline in fuelwood consumption in homes but an increase in fuelwood consumption by industry.

Only general information on forest fires is available for most of the countries of the subregion. Fires have seriously affected the forests of Mexico. The total forest area burned in 1995 was around 300 000 ha (Mexico SEMARNAP 1995). In the Central American countries 450 000 ha were burned in 1998 (Comisión Centroamericana de Ambiente y Desarrollo 1998).

Protected areas in Mexico are facing management problems from irregularities in land tenure and pressure from settlements in and around protected areas. Most protected areas have been established on communal land or *ejidales*. This has led to conflicts between nature conservation and local utilization.

There are 411 declared and 391 proposed protected areas in Central America based on the IUCN classification scheme. Only 83 have management plans, and only 171 have patrols and controls in place. Private reserves exist in Costa Rica (85) and Guatemala (10), but most of the protected areas are national property (Godoy 1997).

CONCLUSIONS AND ISSUES

Based on the country reports sent to FRA 2000 there is reasonably reliable and accurate forest cover information for six of the eight countries in this subregion. The main difficulties in estimating forest cover and change were for those countries where local definitions of forest types had changed or did not correspond to FAO definitions. Information is most accurate for those countries with baseline forest information: Costa Rica, Guatemala and Mexico. For these countries, forest area and change estimates have been produced at the national level, with good accuracy and easy integration into the global database.

All the countries have policies in place promoting sustainable management of forest resources. They recognize the sector's role as a source of rural employment as well as the valuable environmental functions of forests, and they are making efforts to evaluate the contribution of the sector to the national economy. Nevertheless, the forestry authorities do not have enough funds or personnel to be able to provide management advice to the various forest owners and users. Forest management plans exist mainly in those areas that are part of industrial and communal concessions, most of them supported by external grants or international aid funds. The impact of fires has been reduced in these areas (Rodríguez 1999).

FRA 2000 conducted an analysis of the historic causes of deforestation for the eight countries of the subregion, and there is no general agreement on the causes of forest cover change. However, agricultural demand for forest land and the conflict and competition that exist between the agriculture and forestry sectors are suspected to have had a significant impact. The causes of forest cover change also appear to have changed over time. In the 1950s, during a period of agrarian reform, forested areas were considered "useless". Property rights were often established by converting forested areas into agricultural land or cattle ranches. Cattle ranching was identified as a cause of deforestation during the 1980s. Internal political conflicts also impacted the region's forests. During conflicts, mainly in Guatemala, El Salvador and Nicaragua, agricultural land was abandoned and regenerated to forest. However, after the conflict, repatriation was mostly to forested areas. Neighbouring countries (Belize, Mexico, Honduras and Costa Rica) received refugees with negative consequences to their forested areas (see FRA Working Papers 13, 34, 35, 36, 37, 40, 41 and 44).

In El Salvador and Belize the main cause of deforestation is the expansion of urban areas. In El Salvador, economic development policies since the war have moved from support of agriculture and the rural economy to support of manufacturing industry. The drop in coffee prices has also resulted in a reduction of forest cover used for coffee shade.

In Nicaragua, cotton and sugar-cane production along the Pacific Coast has displaced traditional subsistence farmers. These people have migrated to urban areas and to the north to what is called the "agricultural frontier". A credit programme for agricultural production has also promoted the conversion of forest land to agriculture. In Guatemala, alternative employment outside the agricultural sector is difficult to find in rural areas, especially in areas that are not connected by road to markets or government services. When people cannot support themselves with their own food production, they often encroach into adjacent forest areas.

Local forestry experts felt that environmental movements during the 1990s had a positive impact on forest cover, particularly with the declaration of protected areas and the development of environmental policies, especially in Costa Rica. However, some experts maintain that preservation is not an appropriate strategy because of the dependency on and traditional use of forest resources by local communities. Monitoring and control of encroachment and other illegal uses in protected areas is difficult and expensive.

In general, popular participation in forest management has increased – both through communal and industrial concessions and through the participation of NGOs, universities and local people in the elaboration of forest management and protected area plans – but it needs to be increased further. Certification is also increasing, together with research programmes to determine indicators of sustainable forest management (Galloway 1999). Other important issues include giving an appropriate economic value to forests and their products in the national economy, quantifying the environmental services provided by forests, diversifying the species used in forest industries, documenting local knowledge for appropriate forest management, and research on biodiversity and forest plantations (FAO 1998).

BIBLIOGRAPHY

Calvo, J. 2000. *El estado de la caoba en Mesoamerica: memorias del taller.* Costa Rica, PROARCAS-CAPAS, Centro Científico Tropical.

Comisión Centroamericana de Ambiente y Desarrollo. 1998. *Atlas Centroamericano de incendios. Las quemas e incendios de la temporada 1998 en la región Centro Americana.* Panama.

FAO. 1998. *Los programas forestales nacionales y el desarrollo forestal sostenible en América Latina,* by J. Gamboni & C. Carneiro. Proyecto GCP/RLA/127/NET. Santiago, Chile, FAO Regional Office for Latin America and the Caribbean.

Galloway, G. 1999. Avances en Centro América. *Revista Forestal Centroamericana,* 25.

Godoy, J.C. 1997. *Hacia el consenso del sistema centroamericano de áreas protegidas (SICAP).* Guatemala, PROARCAS-CAPAS.

Mexico. Comisión Nacional para el Conocimiento y Uso de la Biodiversidad. 1998. *La diversidad biológica de México. Estudio del país. 1998.* www.conabio.gob.mx/biodiversidad/territo.htm

Mexico. Secretaría de Medio Ambiente, Recursos Naturales y Pesca (SEMARNAP). 1995. *Incendios forestales. Resultados 1995.* Mexico.

Mexico. SEMARNAP. 2000. *Jornadas nacionales de consulta.* www.ecouncil.ac.cr/rio/national/reports/americ a/mexico.htm

Revista Forestal Centroamericana. 1999. *Control y monitoreo en aprovechamientos forestales. Deforestación y pobreza.*

Rodríguez, J. 1999. *Plan estratégico del CCAB-AP. Borrador para consulta.* Guatemala, PROARCAS-CAPAS.

Chapter 36
Caribbean

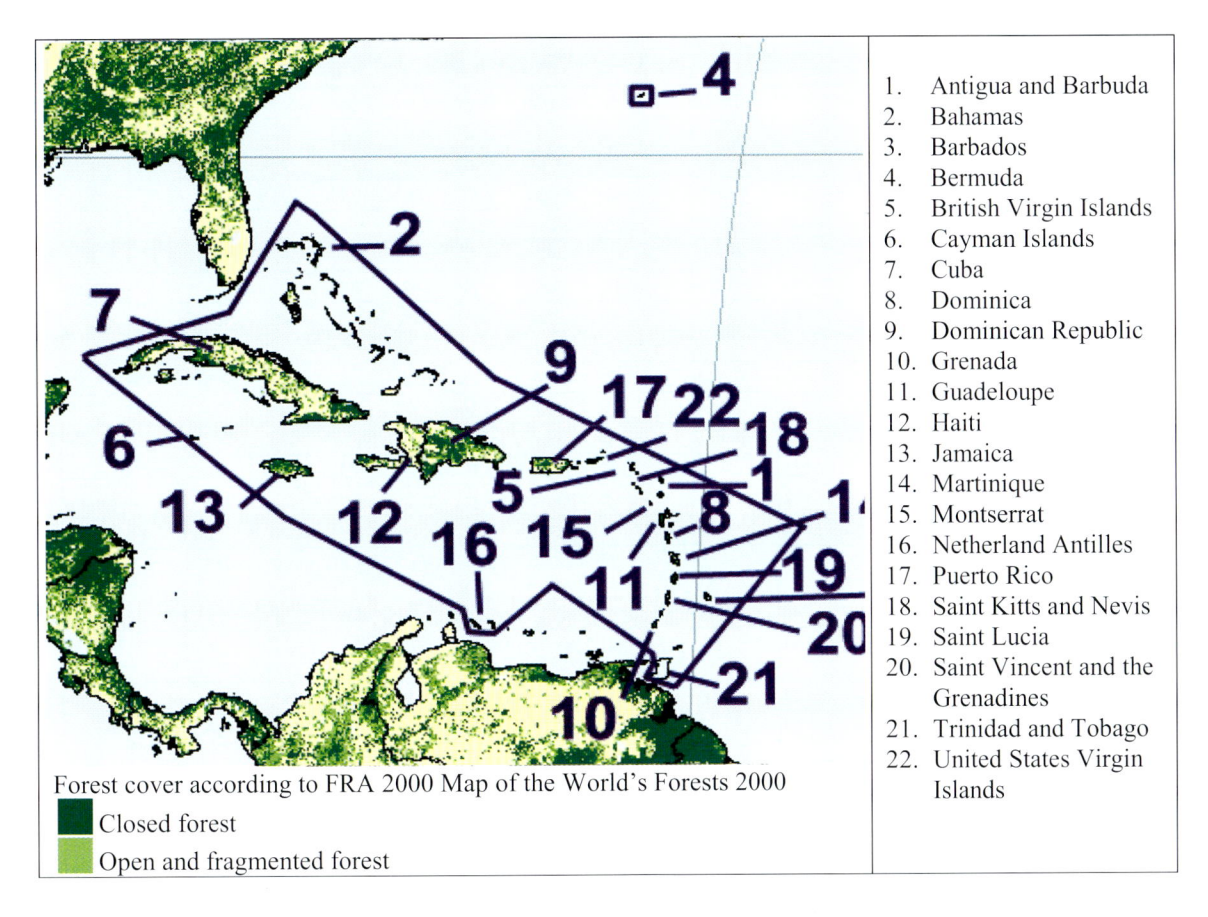

Figure 36-1. Caribbean subregion: Forest cover map

1. Antigua and Barbuda
2. Bahamas
3. Barbados
4. Bermuda
5. British Virgin Islands
6. Cayman Islands
7. Cuba
8. Dominica
9. Dominican Republic
10. Grenada
11. Guadeloupe
12. Haiti
13. Jamaica
14. Martinique
15. Montserrat
16. Netherland Antilles
17. Puerto Rico
18. Saint Kitts and Nevis
19. Saint Lucia
20. Saint Vincent and the Grenadines
21. Trinidad and Tobago
22. United States Virgin Islands

Forest cover according to FRA 2000 Map of the World's Forests 2000
- Closed forest
- Open and fragmented forest

This subregion includes the three major physiographic divisions of the West Indies: the Greater Antilles, comprising the islands of Cuba, Jamaica, Hispaniola (Haiti and the Dominican Republic) and Puerto Rico; the Lesser Antilles, including the Virgin Islands, Saint Kitts and Nevis, Antigua and Barbuda, Montserrat, Guadeloupe, Dominica, Martinique, Saint Lucia, Saint Vincent and the Grenadines, Barbados, and Grenada; and the isolated island groups of the North American continental shelf, the Bahamas, and the islands of the South American shelf, including Trinidad and Tobago and the Netherlands Antilles (Aruba, Curaçao, and Bonaire). Bermuda and the Cayman Islands are also reported in this subregion (Figure 36-1).

The Greater Antilles, which are continental remnants, have a total land area of around 21 million hectares with more than 4 million hectares (22 percent) in forest cover. The Lesser Antilles and all the isolated islands of this subregion have a total land area of around 2 million hectares, of which more than 1 million hectares (59 percent) is covered by forest.

The natural vegetation of the Lesser Antilles was well studied by Beard (1949). The islands are mainly of volcanic origin. According to Beard the climax vegetation formations of these islands can be divided into the rain forest formation, montane formations (lower montane rain forest, montane thicket and elfin woodland), seasonal formations (evergreen, semi-evergreen and deciduous) and dry formations (bushland and littoral woodlands). Swamp formations and mangroves are present in all islands with the exception of Dominica (Beard 1949).

The Caribbean subregion contains a rich variety of complex ecosystems with a great

Table 36-1. Caribbean: forest resources and management

Country/area	Land area	Forest area 2000 Natural forest	Forest plantation	Total forest			Area change 1990-2000 (total forest)		Volume and above-ground biomass (total forest)		Forest under management plan	
	000 ha	000 ha	000 ha	000 ha	%	ha/capita	000 ha/year	%	m³/ha	t/ha	000 ha	%
Antigua and Barbuda	44	9		9	20.5	0.1	n.s.	n.s.	116	210	-	-
Bahamas	1 001	842	-	842	84.1	2.8	n.s.	n.s.			-	-
Barbados	43	2	0	2	4.7	n.s.	n.s.	n.s.			-	-
Bermuda	5			-	-	-	-	-			-	-
British Virgin Islands	15	3	-	3	20.0	0.1	n.s.	n.s.			-	-
Cayman Islands	26	13	-	13		0.4	n.s.	n.s.			-	-
Cuba	10 982	1 867	482	2 348	21.4	0.2	28	1.3	71	114	730	31
Dominica	75	46	0	46	61.3	0.6	n.s.	-0.7	91	166	-	-
Dominican Republic	4 838	1 346	30	1 376	28.4	0.2	n.s.	n.s.	29	53	152	11
Grenada	34	5	0	5	14.7	0.1	n.s.	0.9	83	150	-	-
Guadeloupe	169	78	4	82	48.5	0.2	2	2.1			28*	n.ap.
Haiti	2 756	68	20	88	3.2	n.s.	-7	-5.7	28	101	-	-
Jamaica	1 083	317	9	325	30.0	0.1	-5	-1.5	82	171	44	14
Martinique	107	45	2	47	43.9	0.1	n.s.	n.s.	5	5	10	21
Montserrat	11	3	-	3	27.3	0.3	n.s.	n.s.			-	-
Netherlands Antilles	80	1	-	1	n.s.	n.s.	n.s.	n.s.			-	-
Puerto Rico	887	225	4	229	25.8	0.1	-1	-0.2			57	25
Saint Kitts and Nevis	36	4		4	11.1	0.1	n.s.	-0.6			-	-
Saint Lucia	61	8	1	9	14.8	0.1	-1	-4.9	190	198	-	-
Saint Vincent and the Grenadines	39	6	0	6	15.4	0.1	n.s.	-1.4	166	173	-	-
Trinidad and Tobago	513	244	15	259	50.5	0.2	-2	-0.8	71	129	120	46
United States Virgin Islands	34	14	-	14	41.2	0.1	n.s.	n.s.			-	-
Total Caribbean	22 839	5 145	566	5 711	25.0	0.2	13	0.2	57	98	-	-
Total North and Central America	2 136 966	531 771	17 533	549 304	25.7	1.1	-570	-0.1	123	95	-	-
TOTAL WORLD	13 063 900	3 682 722	186 733	3 869 455	29.6	0.6	-9 391	-0.2	100	109	-	

Source: Appendix 3, Tables 3, 4, 6, 7 and 9.
*Partial result only. National figure not available.

abundance of plant and animal species and a variety of coastal and marine habitats. The subregion, taken together with the coastal regions of North, Central and South America, represents the greatest concentration of biodiversity in the Atlantic Ocean basin (UNEP 2000). Cuba's species richness is of major regional importance. Cuba has the highest species diversity and highest degree of endemism in the West Indies. Over 50 percent of the flora and 32 percent of the vertebrate fauna are endemic (WCMC 2001). Because the nations in this region depend heavily on the health and beauty of the natural world to generate tourism income, the conservation of the region's biological diversity is not only linked to social, cultural and political conditions but also to the economic realities of the region. Coral reefs, sea grass meadows and mangroves are among the best known marine and coastal ecosystems in the region and are large contributors to its biodiversity (UNEP 2000). The main economic trend in the Caribbean countries during the 1990s has been the rapid transition from agriculture-based economies to service-based ones, mainly centred on tourism. Relative to the first half of the 1990s, the prospects for GDP growth have improved somewhat, owing mainly to the expansion of the tourism sector and other services to substitute the banana industry as the main foreign exchange earner. The islands' relative proximity to the markets of North America and Europe and their attractive nature offer them the opportunity to further develop their tourism sector, as well as to further diversify their economies. For example, in Saint Lucia agriculture represented approximately 16 percent of the total GDP value added in 1977, while tourism represented 21 percent. In 1997, the agriculture and tourism contributions to GDP were 8 percent and 33 percent respectively (World Bank 2000).

There are two major implications for forestry. The forest acts as a lure for ecotourism, while protecting the surrounding environment.

However, with tourism comes the pressure for land and infrastructure development, which can impinge on the forest. A careful balance is required. The loss of the forest may well lead to the loss of tourism. This balance is one of the major problems facing the Caribbean forest sector and tourism industry today (Fripp 2000).

FOREST RESOURCES

Forest data for the Caribbean countries were collected with the support of the FAO Subregional Office for the Caribbean and through a Workshop on Data Collection and Outlook Study for Forestry in the Caribbean, held in Trinidad in February 2000. The most accurate and up-to-date forestry information in the Caribbean is found in Cuba (1998) and Jamaica (1997), provided by detailed forest mapping. The Dominican Republic has recent information but it was not possible to compare it with previous years because of differences in definitions and methodologies. Fifteen of the 22 reporting units provided data from land use maps, secondary sources or forest estimates. In most of them the exact method used to make these estimates was not clear.

The greatest conversion of forest cover is in Haiti and Saint Lucia, while Cuba and Grenada showed increase in forest cover change. In Cuba this was due to an intensive forest plantation programme which now totals almost 500 000 ha. The report from Grenada mentions that it was not possible to differentiate between shrub and open forest on the satellite images used to prepare the map. Thus, the area of forest could be overestimated. Countries with a high proportion of forest cover are the Bahamas, Dominica and Trinidad and Tobago. Those with less forest cover are Barbados and Haiti (Figure 36-2, Table 36-1).

The forest cover of the Caribbean region represents only about 0.1 percent of the total forest cover of the world. Nevertheless the high endemism of the plants of the region, the particular characteristics of Caribbean wetlands and the importance of green cover in local economies, especially for tourism, make the forest cover important in this subregion (UNEP 2000).

Volume and biomass were calculated for 10 of the 22 islands, based on commercial volume with appropriate adjustments. The highest per hectare levels are found in Saint Lucia and Saint Vincent and the Grenadines, while the level in Martinique is low.

FOREST MANAGEMENT AND USES

Information on areas covered by forest management plans was lacking from most countries in the Caribbean. Three countries or areas (Jamaica, Martinique and Puerto Rico) provided national-level information on the area of forest subject to a formal, nationally approved forest management plan for FRA 2000. An additional three countries (Cuba, the Dominican Republic and Trinidad and Tobago) provided information on the area of natural forest under management for the 2000 meeting of the Latin American and Caribbean Forestry Commission (Table 36-1). The figures provided ranged from 11 to 46 percent of the total forest area in 2000. Partial data was available from Grenada.

It is important to keep in mind that the total area reported in forest management plans is not necessary equivalent to the total forest area that is under sustainable forest management. The information reported did not indicate if the plans are appropriate, implemented as planned or having the intended effect.

Most of the Caribbean islands do not have any wood processing industry and import their forest products, mainly from Belize and Guyana. Many of the major islands are trying to increase timber production through plantation programmes. Cuba plans to increase its timber production for industrial purposes by 2.5 times in the period 1998-2015; 78 percent will be provided by plantations (FAO 2000).

The islands face environmental challenges such as hurricanes, soil erosion, flooding, forest fires and drought. Therefore, watershed management to reduce the negative impact of natural disasters is identified as a priority in all the islands. Throughout the region, forestry institutions are promoting programmes for soil and water conservation, to support tourism and recreation activities to conserve biodiversity and to increase protected areas (FAO 2000).

UNEP's Caribbean Environmental Programme has undertaken the development of a regional framework for integrated coastal planning and management in the wider Caribbean region. The governments of the region, with the assistance of the UNEP, have developed innovative approaches for the protection of coastal and marine ecosystems through the Special Protected Areas and Wildlife Protocol (SPAW) of the Cartagena Convention. The

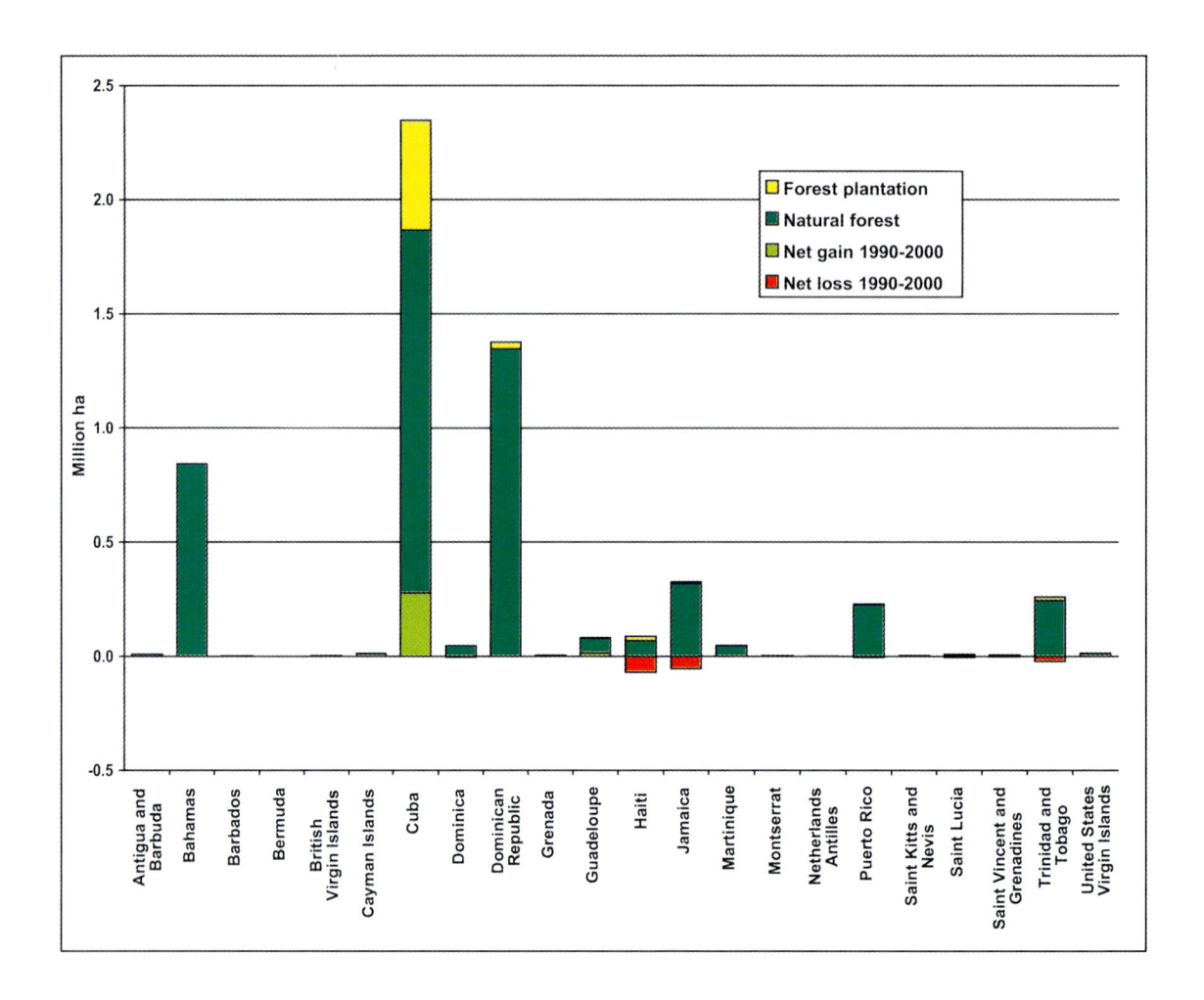

Figure 36-2. Caribbean: natural forest and forest plantation areas 2000 and net area changes 1990-2000

parties agreed to protect key ecosystem components such as coral reefs, sea grasses and mangroves. They also agreed that coastal and marine ecosystems should be regionally managed and monitored in order to maintain the integrity of coastal ecosystems; ensure the propagation of ecologically important and commercially harvestable marine and estuarine species; restore ecosystems and populations of depleted and endangered species; and further develop the region's ecotourism industry. The last item points out that there is a clear economic benefit to maintaining a healthy, systematically managed regional system of parks and protected areas (UNEP 2000).

Trees outside the forest and urban forests are of increasing importance. Almost all countries of the region recognized the need to increase the number of trees and there are programmes to encourage planting in both urban and rural areas. There has been no assessment of the amount and value of trees outside the forest. They serve mostly social purposes such as production of fruit, landscaping and recreation. Trees are planted for fodder, fuelwood, windbreaks and other purposes (FAO 2000).

CONCLUSIONS AND ISSUES

The current status of data and information about forest products and services in the Caribbean is variable, but all countries mentioned the need to obtain higher-quality, more accurate and more relevant data than are currently collected and to better monitor forest resources and services. Data are needed on the use of fuelwood and non-wood forest products, forest recreation and tourism, forestry's role in watershed management, forest employment, the contribution of forests to the national economy and the social and community benefits of forestry (FAO 2000).

With the exception of Cuba and the Dominican Republic, the larger islands have experienced a reduction in forest cover. In Jamaica and Haiti this situation is due to the increased need for agricultural land, while in Puerto Rico the main reason is urban development. In some of the small islands, where commercial agriculture is expected to decrease, there is the potential that agricultural land may revert to forest. On the other hand, tourist development and increased urbanization may further reduce forest cover around cities and villages (FAO 2000).

Growth of the tourism sector often encroaches on forest resources for development, diminishing natural beauty, watershed protection and other values. This then decreases the attraction of the forest for tourism.

Natural phenomena such as hurricanes and drought affect forests in this subregion. Declining profitability of trade in agricultural products such as bananas may lead to a decline in foreign exchange earnings and employment, fostering greater interest in forestry or ecotourism as a substitute (Fripp 2000).

For the Caribbean islands, the classification and establishment of protected areas, land use policy and institutional constraints are important issues (FAO 2000). The role of forests in watershed protection is extremely important. Most countries are moving towards sustainable management of their forests through the formulation and implementation of sustainable forest management plans (FAO 2000).

BIBLIOGRAPHY

Beard, J.S. 1949. *The natural vegetation of the Windward & Leeward Islands.* Oxford, UK, Clarendon Press.

Caribbean Tourism Organization. 1999. *Ecotourism statistical fact sheet.*

FAO. 2000. *Caribbean Workshop on Data Collection and Analysis for Sustainable Forestry Management.* Report of EC-FAO workshop, February 2000.

Fripp, E. 2000. *Socio-economic trends and outlook: implications for the Caribbean forestry sector to 2020.* Baseline study carried out in the framework of the Data Collection and Outlook Effort for Forestry in the Caribbean. London.

United Nations Environment Programme (UNEP). 2000. *Maintenance of biological diversity.* Caribbean Environmental Programme. www.cep.unep.org/issues/biodiversity.html

World Bank. 2000. *Latin America and Caribbean region report.* http://wbln0018.worldbank.org/external/lac/lac.nsf

World Conservation Monitoring Centre (WCMC). 2001. *National biodiversity profiles. Cuba.* www.wcmc.org.uk /nbp/index.html

Chapter 37
Oceania

1. Australia and New Zealand
2. Other Oceania

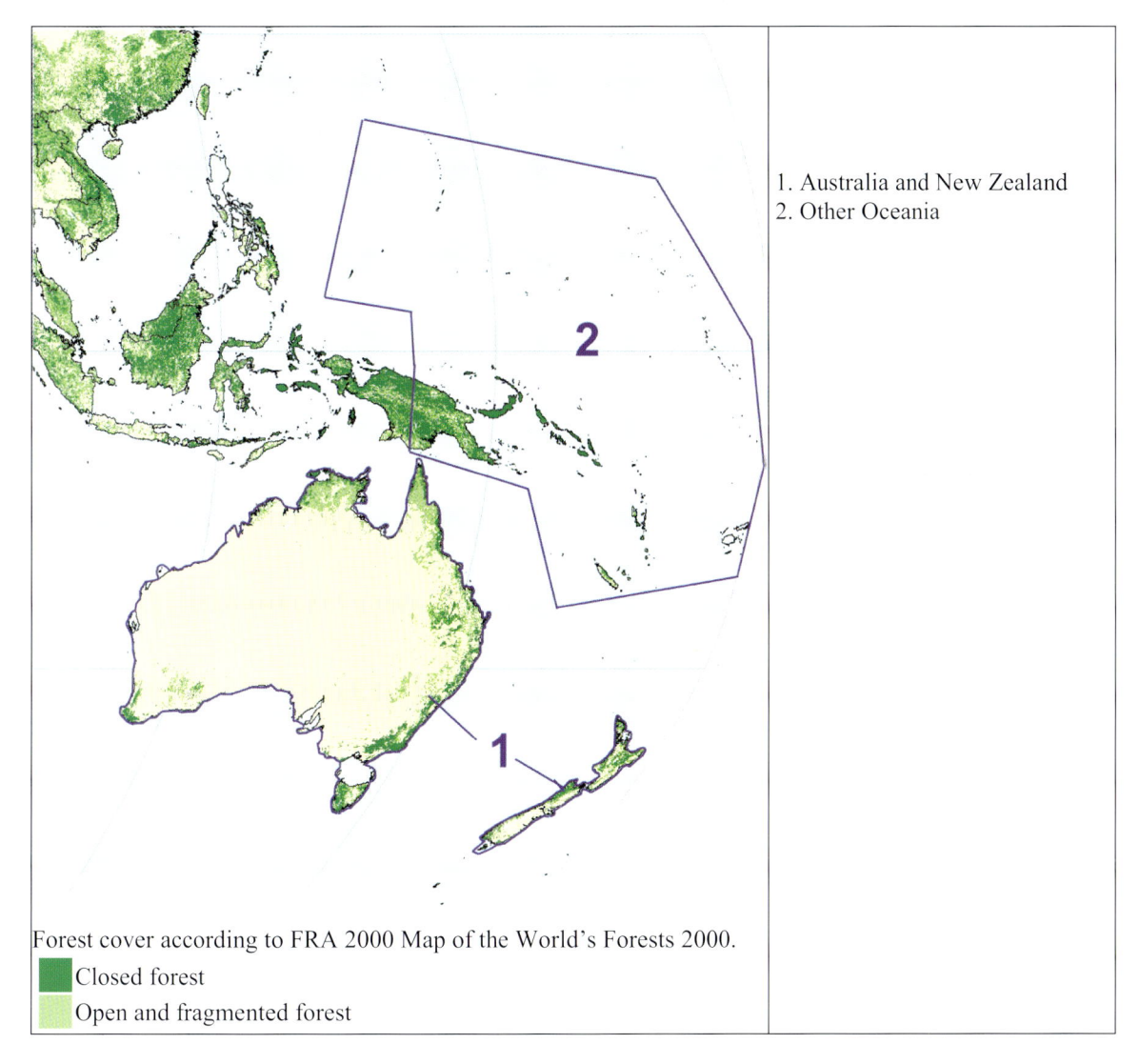

Forest cover according to FRA 2000 Map of the World's Forests 2000.
- Closed forest
- Open and fragmented forest

Figure 37-1. Oceania: Subregional division used in this report

Oceania (see Figure 37-1[50] and Table 37-1) as a whole contains less than 200 million hectares of forests corresponding to 5 percent of the world total. Oceania's forests amount to 6.6 ha per capita, which is the highest at world level. Almost all forests are located in the tropical ecological domain. The dry forest types in Australia dominate the region's forest area. Forest plantation areas are located mainly in Australia and New Zealand and represent 1.4 percent of the total forest area. The annual net loss, based on country reports, is estimated at 365 000 ha, corresponding to 0.2 percent annually.

---

[50] The division into subregions was made only to facilitate the reporting at a condensed geographical level and does not reflect any opinion or political consideration in the selection of countries. The graphical presentation of country areas does not convey any opinion of FAO as to the extent of countries or status of any national boundaries.

Table 37-1. Oceania: forest resources by subregion

Subregion	Land area	Forest area 2000					Area change 1990-2000 (total forest)		Volume and above-ground biomass (total forest)	
		Natural forest	Forest plantation	Total forest						
	000 ha	000 ha	000 ha	000 ha	%	ha/capita	000 ha/year	%	M³/ha	t/ha
Australia and New Zealand	795 029	159 547	2 938	162 485	20.4	7.2	-243	-0.1	58	65
Other Oceania	54 067	34 875	263	35 138	65.0	4.7	-122	-0.3	34	58
Total Oceania	**849 096**	**194 775**	**2 848**	**197 623**	**23.3**	**6.6**	**-365**	**-0.2**	**55**	**64**
TOTAL WORLD	13 063 900	3 682 722	186 733	3 869 455	29.6	0.6	-9 391	-0.2	100	109

Source: Appendix 3, Tables 3, 4, 6 and 7.

Chapter 38
Oceania: ecological zones

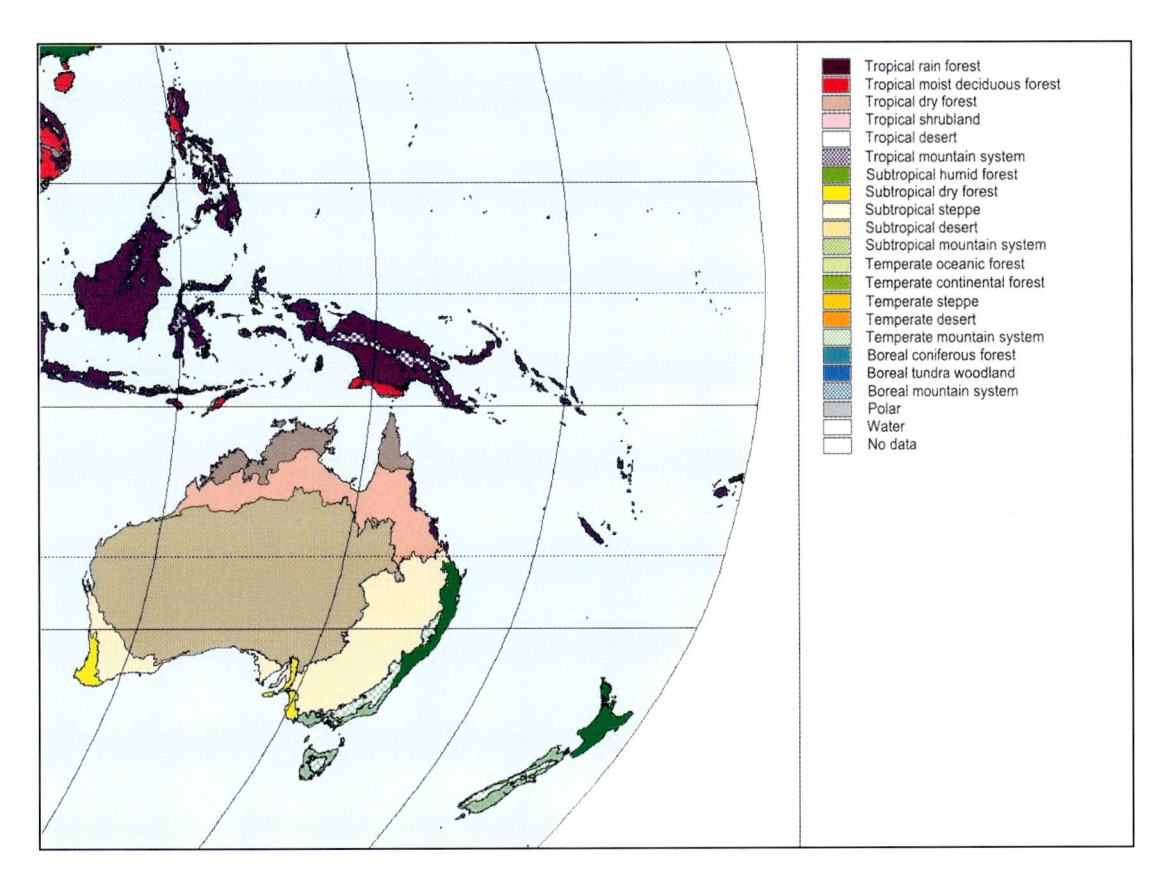

Figure 38-1. Oceania: ecological zones

Figure 38-1 shows distribution of ecological zones in Oceania. Table 38-1 contains area statistics for the zones by subregion and Table 38-2 indicates the proportion of forest in each zone by subregion.

Oceania comprises of Australia, New Zealand, Papua New Guinea and the Pacific Islands (Micronesian, Melanesian and Polynesian archipelagos). The descriptions of the ecological zones of Papua New Guinea are dealt with under Asia. This country forms an ecological entity with the western half of the island of New Guinea, Irian Jaya, a province of Indonesia.

TROPICAL RAIN FOREST

Oceania, the Pacific Islands and small patches in northeastern Australia (Queensland) constitute this zone, in addition to a large portion of Papua New Guinea.

The climate of the Pacific Islands is dominated by the trade winds and most of islands have ample precipitation. The average annual precipitation generally varies between 1 500 and 4 000 mm and the dry season is seldom severe. Locally, rainfall depends on the relief and the leeward side may be fairly dry. Mean temperature at sea level is about 23°C near the Tropics and 27°C at the equator, with little difference between the hottest and coolest months. Cyclonic disturbances mainly affect the western Pacific archipelagos (Melanesia and western Micronesia).

The coastal area of northeastern Australia has a tropical wet climate and receives the highest annual rainfall in Australia. It has a mean annual precipitation of 1 500 to 2 500 mm with some areas exceeding 4 500 mm. There is a marked summer maximum (January to March). The mean annual temperature is around 23°C.

Table 38-1. Oceania: extent of ecological zones

Subregion	Total area of ecological zone (million ha)																		
	Tropical						Subtropical					Temperate					Boreal		
	Rain forest	Moist	Dry	Shrub	Desert	Mountain	Humid	Dry	Steppe	Desert	Mountain	Oceanic	Continental	Steppe	Desert	Mountain	Coniferous	Tundra	Mountain	Polar
Australia and New Zealand	3		46	107			28	12	147	416		22				20				
Other Oceania	42	3	1			7														
Total Oceania	46	3	47	107		7	28	12	147	416		22				20				
TOTAL WORLD	1 468	1 117	755	839	1 192	459	471	156	491	674	490	182	726	593	552	729	865	407	632	564

Note: Data derived from an overlay of FRA 2000 global maps of forest cover and ecological zones.

Table 38-2. Oceania: proportion of forest by ecological zone

Subregion	Forest area as proportion of ecological zone area (percentage)																			
	Tropical						Subtropical					Temperate					Boreal			
	Rain forest	Moist	Dry	Shrub	Desert	Mountain	Humid	Dry	Steppe	Desert	Mountain	Oceanic	Continental	Steppe	Desert	Mountain	Coniferous	Tundra	Mountain	Polar
Australia and New Zealand	73		*	17			40	63	22	1		36				36				
Other Oceania	79	56	*			55														
Total Oceania	78	56	*	17		55	40	63	22	1		36				36				
TOTAL WORLD	69	31	64	7	0	26	31	45	9	2	20	25	34	4	1	26	66	26	50	2

Note: Data derived from an overlay of FRA 2000 global maps of forest cover and ecological zones.
* Estimate uncertain because of discrepancies in global forest cover map.

The rain forests of the tropical Pacific Islands are generally evergreen. Their structure is comparable to that of the Indo-Malayan forests but the flora of the dominant strata is often relatively poor. The tallest hardwood forests, with heights ranging from 30 to 45 m, are found on deep volcanic soils. About a dozen species (in the genera *Calophyllum, Campnosperma, Dillenia, Elaeocarpus, Endospermum, Gmelina, Maranthes, Parinari, Schizomeria* and *Terminalia*) are the main constituents of the canopy, overtopped occasionally by banyan figs (*Ficus* spp.) and *Terminalia calamansanai*. In Vanuatu, Fiji and Samoa this forest type is somewhat lower (about 30 m) and floristically slightly different. New Caledonian flora is totally different from that of the forests in other parts of Melanesia. Clusiaceae (*Calophyllum* spp. and *Montrouziera* spp.), Cunoniaceae, Myrtaceae, Myrtoideae, Proteaceae and Sapotaceae predominate in the upper stratum. A poorer forest grows on the limestone atolls. In certain special environments a single species dominates the upper stratum. Examples are the *Nothofagus* spp. forests in New Caledonia and the *Metrosideros collina* forest that is found throughout the tropical Pacific. Coniferous forests belonging to the Araucariaceae, Cupressaceae, Podocarpaceae and Taxaceae families have a limited distribution throughout the Pacific.

Mangroves cover rather large areas in the Melanesian archipelagos and in the Caroline Islands. They can reach a height of 25 m and the main constituents are Rhizophoraceae together with species of the genera *Avicennia, Lumnitzera, Sonneratia* and *Xylocarpus*.

Tropical rain forests constitute around one million hectares of Australia's forests. The forest canopy ranges from around 30 to 40 m high with emergent trees up to 50 m. They resemble the rain forests of Indo-Malaya in floristic composition except for the complete absence of Dipterocarpaceae. Australian endemics of the emergent tree strata include species of *Flindersia, Cardwellia, Musgravea, Placospermum, Buckinghamia, Darlingia, Backhousia, Blepharocarya, Castanospermum, Ceratopetalum* and *Doryphora*. The presence of several primitive and restricted angiosperm genera – *Idiospermum, Austrobaileya, Sphenostemon, Bubbia, Ostrearia, Neostrearia, Eupomatia* and *Galbulimima* – add a further distinctive character to the rain forests. In swamp forests, limited to the coastal zone, *Melaleuca viridiflora* paperbark forest often constitutes the main canopy species along with numerous palms. In the well-drained lowlands, woodlands and forests include *Eucalyptus tereticornis, E. tessellaris, E. intermedia* and *E. pellita*.

TROPICAL MOIST DECIDUOUS FOREST

Papua New Guinea is the only location in Oceania where this ecological zone is found. The zone is described under Asia. Small areas of this type of forest may be found in northern Australia and the Pacific Islands; however, they are too small to map and are thus included in other ecological zones.

TROPICAL DRY FOREST

This zone is confined to the northern parts of Australia. These northern tropics have a marked seasonal alternation in moisture conditions, with an intense drought lasting six to eight months throughout the winter, followed by monsoon rainfall. The zone receives an average annual precipitation of 1 000 to 1 400 mm with around 75 percent falling in the monsoon period. The mean annual temperature is around 27°C with a mean summer maximum of 33°C. Average minimum temperatures during the monsoon period are around 23°C.

The main natural vegetation is eucalypt forest and woodland. Various types occur, characterized by different dominant *Eucalyptus* species. The *Eucalyptus tetrodonta-E. miniata* suballiance occurs mainly west of the Carpentaria Gulf. It forms open to closed forests to 30 m high in the wettest areas or, in drier areas, woodlands 10 to 30 m high. In the Kimberly region this alliance often gives way to a *Eucalyptus tectifica* and *E. grandifolia* alliance. *Callitris intratroopica*, now mostly removed for timber, once formed local associations or with *E. miniata*. *Melaleuca* forests occur throughout the zone on damp or wet sites. Often these forests are narrow strips of dense pure stands along streams and swamps. The dominant canopy species include *Melaleuca dealbata, M. leucadendra, M. minutifolia* and *M. viridiflora*.

Small patches of so-called semi-evergreen vine forests or monsoon forests occur along watercourses, around lagoons and on patches of soil fed by springs or runoff water from the uplands. The dominants are chiefly deciduous.

Along the northern Australian coasts, which have tides of up to 10 m, are mangrove forests. Typically, there is a pioneer outer zone of *Sonneratia caseolaris* or *Avicennia marina*.

Inland is a *Rhizophora* forest dominated by *R. stylosa*, a *Bruguiera gymnorhiza* dominated zone and a *Ceriops tagal* community.

TROPICAL SHRUBLAND

This zone is located in the northern part of Australia immediately inland of the more humid coastal zones.

The semi-arid tropics of northern Australia have a marked seasonal variation in moisture conditions with a pronounced winter drought lasting six to eight months followed by substantial monsoonal rainfall. The zone receives an average annual precipitation of 700 mm, ranging from around 350 mm to 1 000 mm. Most of the precipitation occurs during December to March, with drought conditions for the remainder of the year. The mean annual temperature is around 26°C.

The natural vegetation is largely eucalypt forests and woodlands. *Eucalyptus tetrodonta* and *E. miniata* forests and woodlands dominate the northern Kimberly while low and open woodlands of *Eucalyptus brevifolia* and *E. setosa* dominate the southern Kimberly. The vegetation of the centre of the zone is mainly eucalypt woodlands and acacia forests and woodlands. Common dominants are *Eucalyptus terminalis* and *E. brevifolia. E. brevifolia* often forms mosaics with other species, for instance *E. tetradonta, E. dichromophloia* or *E. pruinosa*. Lance wood (*Acacia shirleyi*) is the most widespread of the central northern acacia woodlands. *A. shirleyi* is a species up to 18 m tall that forms low woodland in the drier parts of its range, often intermingled with eucalypt woodlands.

In the eastern part, *Eucalyptus drepanophylla* is the most common species. Another characteristic vegetation are the "boxes", medium-height eucalypt woodlands in drier areas. The main species are *Eucalyptus leptophleba, E. microneuro* and *E. normantonensis*. *Callitris glauca*, a common associate in some of the woodlands, is an important timber species. At the southern end of the zone, silverleaf ironbark (*Eucalyptus melanophloia*) becomes dominant as does *Callitris glauca* and brigalow (*Acacia harpophylla*), which has now largely been cleared.

SUBTROPICAL HUMID FOREST

The subtropical humid forest zone comprises the east coast of Australia, roughly between 23° and 35°S, and the North Island of New Zealand. The coastal areas of southern Queensland and northern

New South Wales have a subtropical humid climate with mild winters and hot summers. Mean annual precipitation across the region is 1 100 mm, with areas on the Queensland/New South Wales border receiving in excess of 2 200 mm and rain-shadow areas receiving as little as 700 mm annually. Precipitation is reasonably well distributed. The mean annual temperature of the region is around 18°C with the northern extent 3° hotter and the southern extent 2° colder. The climate of the North Island of New Zealand is strongly influenced by the ocean. Extremes of heat and cold are absent. The mean summer temperature is 16° to 18°C with mean winter temperature around 10°C. Rainfall is high, rather regular over the island and ranges from around 1 000 mm to more than 1 500 mm (on the central plateau), with the maximum during winter.

The dominant vegetation in Australia is open eucalypt forest that generally exceeds 30 m tall and can often reach 50 m, while in the moist valley bottoms, warm temperate rain forests are the dominant life form. The vegetation in the centre of this region is extremely diverse. In the north the inland medium-open eucalypt forests are dominated by *Eucalyptus tereticornis* and *Corymbia maculata* (formally *E. maculata*) while the coastal forests are dominated by bloodwoods such as *E. intermedia* and *E. acmenioides*. Further to the west numerous rain shadows occur that are dominated by dry ironbark forests and woodlands with *Eucalyptus crebra, E. fibrosa, E. tessellaris* and *E. melanophloia*.

At the centre of the region, on the Queensland/New South Wales border, warm temperate rain forest is the dominant forest type. Outside this area it mainly occurs as narrow strips in the valley bottoms of eucalypt forest. Coachwood (*Ceratopetalum apetalum*) characterizes the rain forests between latitudes 37° and 28°S. Codominants include *Doryphora sassafras, Schizomeria ovata, Acmena smithii, Tristania laurina* and *Argyrodendron* spp. The forests have three tree layers and in this respect resemble the richest rain forests in the tropics. *Argyrodendron actinophyllum* and *A. trifoliolatum* are present in the stands, and other tree species belong to the Lauraceae, Simaroubaceae, Rutaceae, Meliaceae and succulent-fruited Myrtaceae, especially *Syzygium*. In areas with lower rainfall, a drier type of rain forest appears, characterized by *Drypetes australasica, Araucaria* spp., *Brachychiton discolor* and *Flindersia* spp.

To the south of the Queensland border, medium to tall open eucalypt forests dominate the landscape, with dozens of distinct floristic communities. The main medium-open forest types include *Eucalyptus pilularis, E. saligna* and *E. maculata*, while *E. acmenioides* and *E. microcorys* dominate the tall forests.

Conifer-broadleaf forest represents the "subtropical" or warm-temperate evergreen forests of the North Island of New Zealand. Conifers, where present, form the tallest storey, usually as well-spaced, large-crowned trees, but they can also form continuous canopies. Most of the tree species are podocarps of the genera *Podocarpus, Dacrycarpus, Dacrydium* and *Phyllocladus*, the tallest species reaching heights of over 40 m, exceptionally 60 m. There are also two species of *Libocedrus* (Cupressaceae) and, north of 38°S, the massive kauri (*Agathis australis*). Hardwoods and some of the less-tall podocarps form the next storey, which is usually the main canopy. Species include *Beilschmiedia, Knightia, Laurelia, Litsea* and *Nestegis*. A host of small trees form a subcanopy and fill gaps. Small patches of beech forest (*Nothofagus* spp.) occur on poor soils and at higher altitudes.

SUBTROPICAL DRY FOREST

This climatically very distinct zone is found in two locations in southern Australia: the southwestern tip around Perth and the central east around Adelaide. The climate occurs in two slightly different Mediterranean forms and has a significant rainfall gradient that has a major impact on the type of vegetation. The area approximately 200 km south and east to 500 km north of Perth in Western Australia has hot, dry summers. Mean annual precipitation within is around 750 mm to 1 000 mm, mostly falling between May and August. The annual average temperature is around 16°C. The southern tip of Western Australia and areas to the south of Adelaide in South Australia have slightly cooler summers and are subject to a significant rainfall gradient. The region receives 400 to 800 mm of annual precipitation in Victoria and South Australia and between 1 000 mm and 1 300 mm on the southern coast of Western Australia, with approximately 60 percent falling between May and September. The annual average temperature is 15°C. The south coast of Western Australia is generally around two degrees warmer than the rest of the zone.

The vegetation in the southwest is floristically distinct from the rest of Australia. On fertile soils derived from granite, two tall forests occur: karri (*Eucalyptus diversicolor*), where rainfall exceeds 1 000 mm in the south, and red tingle (*E. jacksonii*). On laterite and lateritic strew, jarrah (*E. marginata*) and marri (*E. calophylla*) are dominant and on the coastal limestones, tuart (*E. gomphocephala*). Karri is one of the tallest eucalypts in Australia and can reach a height of about 85 m and a diameter of about 7 m. The *Eucalyptus marginata-E. calophylla* association is most widely distributed in this zone, between the 600 and 1 300 mm isohyets. Forests up to 40 m high, with an almost closed canopy, occur in the wetter areas while in drier areas the forests reach a height of 12 to 24 m and are more open.

The original vegetation covering the Lofty Block and the Naracoorte Coastal Plain was significantly different from the agricultural lands and low open eucalypt woodlands that occur there today. The region was originally dominated by low to medium eucalypt woodlands in the lower rainfall areas with gum and peppermint species such as *Eucalyptus leucoxylon* and *E. odorata* and shrubby understoreys. Medium-open stringybark forests comprising *Eucalyptus baxteri, E. obliqua* and *E. viminalis* and shrubby understoreys dominated the higher rainfall areas. Vegetation of the Naracoorte Coastal Plain was similar in many areas to that of the Lofty Block, with the addition of heaths in the poorly drained lowlands and inter-dune swales and eucalypt mallee formations.

SUBTROPICAL STEPPE

This zone is confined to Australia and separated into two distinct units, a northeastern part with typical subtropical characteristics and a southern part with "warm temperate" influences.

The northeastern area has a significant climatic gradient that has a major impact on vegetation. Southwestern Queensland and northwestern New South Wales have a subtropical semi-arid climate with mild winters and hot summers. The mean annual precipitation of 350 mm is fairly evenly distributed throughout the year, with a slight increase from December to February. The mean annual temperature of the region is around 2°C. The region is commonly known as the Mulga Lands.

Southern central Queensland and northern central New South Wales have a subtropical semi-arid climate with mild winters and hot summers. The mean annual precipitation is 560 mm, decreasing to 350 mm towards the interior and increasing to 700 mm on the western slopes of the Great Dividing Range. Precipitation is evenly distributed throughout the year, with a slight increase from December to February. The mean annual temperature of the region is around 19°C. This zone covers regions commonly known as the Southern Brigalow Belt, the Darling Riverine Plain, the South Western Slopes of New South Wales and the Cobar Peneplain.

The southern part has a semi-arid climate with a marked winter increase in precipitation. It has average annual precipitation of 375 mm with as little as 250 mm in inland areas and up to 600 mm at higher altitudes (300 m) towards the coast. Precipitation is markedly winter dominant, increasing from east to west. The mean annual temperature is around 17°C.

Low *Acacia aneura* woodlands and shrublands commonly known as "mulga" dominate the Mulga Lands. This species occurs as small trees in the higher rainfall eastern margins and as low shrub towards the interior.

Five primary vegetation types occur within the Southern Brigalow Belt. These are: ironbark woodlands on the eastern margins (*Eucalyptus crebra, E. alba*); ironbark and *Callitris* forests (*E. crebra, E. fibrosa* and *Callitris glauca*); brigalow forests and woodlands (*Acacia harpophylla*) and poplar box woodlands (*E. populnea*) in the central and interior regions. All also occur as mixed forest and mosaics of relatively pure stands. *Callitris glauca* is a very important commercial species that can form very pure stands over extensive areas.

River redgum (*E. camaldulensis*) and blackbox (*E. largiflorens*) dominate the Darling Riverine Plain. The Cobar Peneplain is dominated by mulga (*Acacia aneura*) shrublands. Other species include myall (*A. pendula*), nelia (*A. loderi*) and gidgee (*A. cambagei*). Box woodlands dominate the South Western Slopes: *Eucalyptus albens, E. melliodora* and *E. blakelyi* on the slopes and greybox (*E. microcarpa*) and ironbark (*E. sideroxylon*) woodlands in the lower rainfall regions.

All the above vegetation communities have considerable economic importance. They all provide grazing for domestic stock and large tracts have been cleared for cultivation.

Mallee is the dominant natural vegetation over large areas of the Murray-Darling, Riverina, Eyre and York Block and Mallee regions of Western Australia. The term "mallee", an aboriginal word, describes eucalypts with many stems arising at ground level from a large, bulbous woody structure called a lignotuber or "mallee" root. There are more than 100 mallee species and many species that occur as both mallee and tree forms. Common species include: white mallee (*Eucalyptus diversifolia*), which dominates the wetter communities in South Australia; lerp mallee (*E. incrassata*) and narrow-leaved red mallee (*E. foecunda*), occurring on deep sands; giant mallee (*E. socialis*), congoo mallee (*E. dumosa*), yorell (*E. gracilis*) and redwood (*E. oleosa*) characterizing the main mallee alliance in the east; and tall sand mallee (*E. eremophila*), confined to Western Australia found over a wide range of soil types. In more arid areas mallee is usually replaced by acacias and at the upper rainfall limit (circa 400 mm per year) by single-stemmed eucalypts, often of the same species.

The Wheatbelt region of Western Australia has been highly modified for agriculture and today only remnants of the original vegetation exist. Medium-height eucalypt woodlands 10 to 30 m high with low understoreys were once dominant with *Eucalyptus marginata* (jarrah) forests in the higher rainfall areas to the west giving way to *E. wandoo* (wandoo) and then *E. salmonophloia* (salmon gum) with decreased rainfall.

TEMPERATE OCEANIC FOREST

This zone covers the southeastern coast of Australia, Tasmania and the lowlands of South Island, New Zealand.

The southeastern coast of mainland Australia and Tasmania has a humid, mild winter climate. Annual precipitation varies from around 600 mm in the Gippsland region in Victoria to in excess of 2 000 mm in western Tasmania. Precipitation is distributed throughout the year with a slight winter dominance, more pronounced in western Tasmania. The annual average temperature varies from around 9°C in western Tasmania to 13°C in southern Victoria and eastern Tasmania.

The western, coastal part of South Island of New Zealand has a humid climate. Annual rainfall ranges from around 1 800 mm to locally more than 4 000 mm, rather evenly distributed throughout the year. To the east of the Southern Alps, the climate is distinctly drier, with annual rainfall from 400 to 800 mm, locally below 400 mm. Also, temperatures become more extreme here, as the region is sheltered from the moderating western ocean winds. The mean annual temperature ranges from 13°C in the north to 9° in the south.

Cool temperate rain forests are found in the wetter parts of western Tasmania. These forests are often dominated by myrtle (*Nothofagus cunninghamii*) with conifers such as huon pine (*Lagarostrobos franklinii*), celery top pine (*Phyllocladus aspleniifolius*) and King Billy pine (*Athrotaxis selaginoides*). In lowland areas, the rain forests are dominated by *Anodopetalum biglandulosum*. In Victoria, cool temperate rain forests occur in restricted areas in the coastal ranges. Dominant canopy species include southern sassafras (*Atherosperma moschatum*), *Acacia melanoxylon* and mountain quandong (*Elaeocarpus holopetalus*).

Dry ash, stringybark and peppermint forests (*Eucalyptus sieberi, E. gummifera, E. botryoides, E. radiata* and *E. dives*) dominate areas of moderate rainfall to the east of this zone on the mainland and Tasmania. Tall forests dominated by *Eucalyptus viminalis, E. fastigata, E. obliqua* and *E. cypellocarpa* replace these forests in higher rainfall and protected areas. Many of the wetter areas of this zone in Tasmania are dominated by tall messmate/stringybark forest (*Eucalyptus obliqua* and *E. nitida*). The basalt plains of western Victoria were once dominated by wet *E. obliqua* and *E. cypellocarpa* forest but most of these have since been cleared.

Beech and conifer-beech-broadleaf forests dominate the western lowlands and lower hills of New Zealand's South Island. *Nothofagus fusca* is characteristic of conifer-beech-broadleaf forests in the northwest. In these forests, conifers form a scattered overstorey with *Dacrydium cupressinum* and *Podocarpus ferruginea* as the main species. Beeches form the main canopy, with *Nothofagus fusca* predominating on the deeper, more freely drained sites, but usually mixed with *N. truncata, N. menziesii* and *N. solandri*. On optimal sites, *Weinmannia racemosa* and, in places, *Quintinia acutifolia* form a tall subcanopy. In the extremely humid fjord country in the southwest, where rainfall exceeds 6 000 mm, the *Nothofagus* forests are similar in nature to those of southern Chile. *Nothofagus menziesii* is the dominant species.

The east of South Island has little forest vegetation owing to much lower rainfall. Patches of beech-conifer-broadleaf forest occur, adjoining a wide variety of mostly anthropogenic vegetation. There is evidence that, prior to human intervention, a zone of microphyllous woodland, consisting of species such as *Coprosma virescens, Discaria toumatou, Leptospermum ericoides, Olearia lineata* and *Sophora microphylla*, grew under moisture regimes intermediate between those supporting forest and semi-arid grasslands.

TEMPERATE MOUNTAIN SYSTEMS

In Australia, this zone consists of the Tasmanian Highlands, the Southeastern Highlands, the Australian Alps and the New England Tablelands. New Zealand's Southern Alps on South Island are also part of the zone.

The highlands and tablelands of southeastern Australia have a cool temperate climate with annual precipitation ranging from around 600 mm at lower elevations to 1 200 mm at higher elevations. Precipitation is evenly distributed throughout the year, with most months receiving 70 to 80 mm. The annual mean temperature is around 12°C with mainland areas around 2° hotter and Tasmania 4° cooler. The Alps region of southeastern Australia receives average annual precipitation of 1 300 mm, with higher elevation areas receiving in excess of 2 000 mm, much of it as snow. Precipitation is fairly evenly distributed throughout the year. The annual average temperature for the region is around 9°C.

The climate of the Southern Alps in New Zealand is cold temperate, characterized by high annual rainfall, particularly on the western slopes. Frost and snow are abundant in winter and to some extent at all seasons.

The lower-elevation rolling hills of the southeast highlands and the elevated plateaus and hills of the New England Tablelands were originally covered with eucalypt forests and woodlands of stringy bark/peppermint/box species, including *Eucalyptus caliginosa, E. laevopinea, E. nova-anglica, E. melliodora, E. albens* and *E. blakelyi*. Today, these communities mainly occur as open woodlands used for grazing.

In sheltered areas receiving more than 1 000 mm annual rainfall, tall wet eucalypt forests dominate with species such as alpine ash (*Eucalyptus delegatensis*), mountain white gum (*E. dalrympleana*) and manna gum (*E. viminalis*) forming open forests where the canopy exceeds 40 m. The outstanding example of these forests occurs in the southern ranges of southern Victoria and Tasmania where mountain ash (*E. regnans*) trees commonly exceed 70 m in height and can reach over 90 m on the best sites. In Tasmania, cool temperate rain forests are dominated by myrtle (*Nothofagus cunninghamii*) while blackwood (*Acacia melanoxylon*) often forms an understorey 10 to 30 m tall.

The lower- and medium-altitude zones of the mountains of South Island, New Zealand are mostly covered by beech forest. *Nothofagus solandri* var. *cliffortioides* or *N. menziessi* constitute most of the subalpine forests. The timberline is at around 1 200 m in the north and decreases to around 850 m in the south. Locally, beech forest is absent and depauperate conifer-broadleaf forest extends into the subalpine belt. Its conifer storey consists of *Podocarpus halii*, often accompanied by *Libocedrus bidwillii*, while the main canopy consists of *Weinmannia racemosa, Metrosideros umbellata* or, in certain circumstances, small trees such as *Dracophyllum traversi, Griselinia litoralis* and *Olearia ilicifolia*.

BIBLIOGRAPHY

Australian Surveying and Land Information Group (AUSLIG). 1990. *Atlas of Australian resources.* Volume 3. Vegetation. Commonwealth of Australia.

Australia. National Forest Inventory. 1998. *Australia's state of the forests report 1998.* Canberra, Bureau of Rural Sciences.

Beadle, N.C.W. 1981. *The vegetation of Australia.* Cambridge University Press.

Cockayne, L. 1921. *The vegetation of New Zealand.* Die Vegetation der Erde, Volume XIV. Leipzig, Germany, Engelmann.

Groves, R.H. 1981. *Australian vegetation.* Cambridge University Press.

Schmid, M. 1989. The forests in the tropical Pacific Archipelagos. In *Tropical rain forest ecosystems: biogeographical and ecological studies.* Eds H. Lieth & M.J.A. Werger. Ecosystems of the world 14b. Amsterdam, The Netherlands, Elsevier.

Stocker, G.C. & Unwin, G.L. 1989. The rain forests of Northeastern Australia – their environment, evolutionary history and dynamics. In *Tropical rain forest ecosystems: biogeographical and ecological studies.* Eds H. Lieth & M.J.A. Werger. Ecosystems of the world 14b. Amsterdam, The Netherlands, Elsevier.

Thackway, R. & Cresswell, I.D. (eds). 1995. *An interim biogeographic regionalisation for Australia: a framework for setting priorities in the national reserves system cooperative programme.* Version 4.0. Canberra, Australian Nature Conservation Agency.

Wardle, P., Bulfin, M.J.A. & Dugdale, J. 1983. Temperate broad-leaved evergreen forests of New Zealand. In *Temperate broad-leaved evergreen forests.* Ecosystems of the world 10. Ed. J.D. Ovington. Amsterdam, The Netherlands, Elsevier.

Chapter 39
Australia and New Zealand

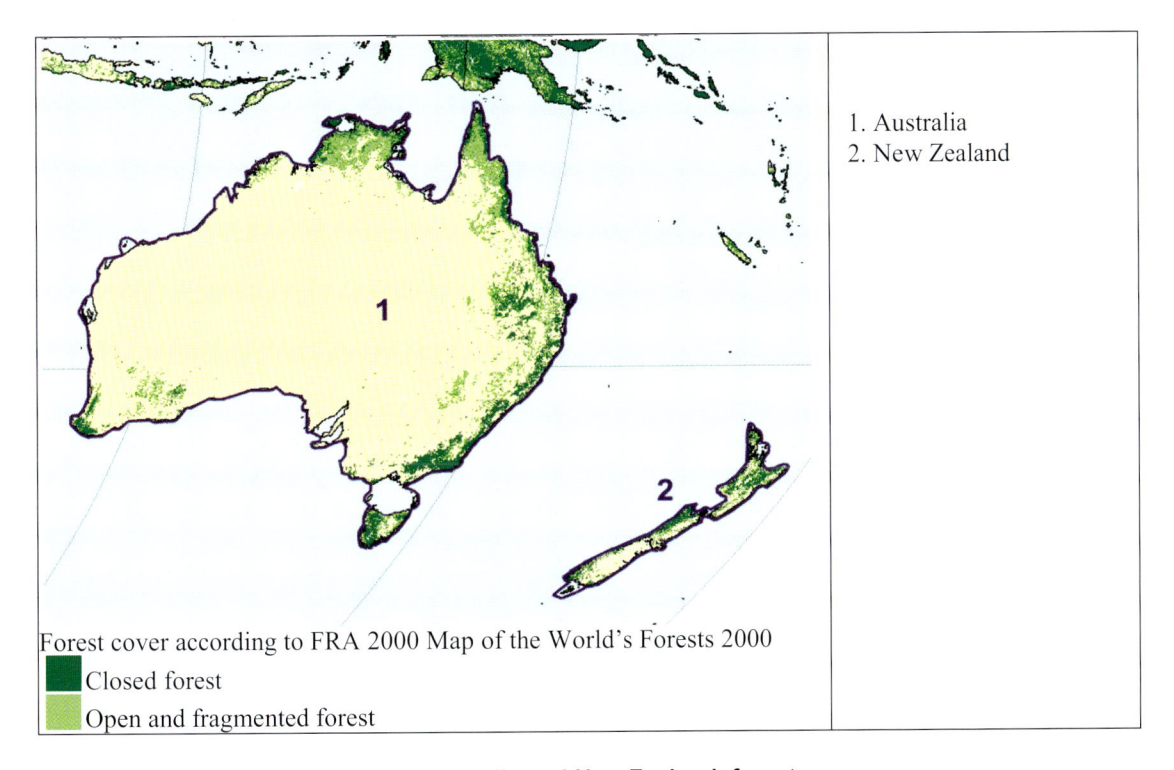

Figure 39-1. Australia and New Zealand: forest cover map

Australia and New Zealand (Figure 39-1)[51] are among the world's least densely populated countries and this absence of population pressure is among the defining characteristics of forestry in this subregion. Australia, the world's sixth largest country, has 154.5 million hectares of forests covering 20.1 percent of the country's land area. Forest cover in New Zealand amounts to 7.9 million hectares or 29.7 percent of land area.

Natural forests in New Zealand comprise mainly cool temperate rain forests extending along much of the western side of South Island and through the mountainous axes of North Island. In northernmost areas there is a gradual transition to warm temperate rain forests. Plantation forests have been established throughout the country, with the largest concentration (around one-third of the total area) planted on the volcanic plateau of central North Island (FAO 1997a).

In general, Australia's forests and woodlands form a broad crescent around coastal Australia extending from the Kimberley Plateau in the north, to Perth in the southwest, and as much as 700 km inland. Closed canopy forests mainly occur in relatively narrow coastal zones, primarily in tracts along the eastern and southeastern coasts (including Tasmania), and in the far southwest of Western Australia. These tracts of closed forest are generally encircled by larger areas of open forests (primarily eucalypt forest). Further inland, where average annual rainfall begins to decline below 900 mm, open forests give way to eucalypt woodlands, which in turn are supplanted by acacia shrubland in areas where annual rainfall is below 400 mm (Bureau of Rural Sciences 2000).

FOREST RESOURCES

Australia and New Zealand participated in the temperate and boreal component process of FRA 2000. In Australia, on-the-ground forestry data are collected by the individual states and territories, and compiled at national level by the National Forest Inventory group in the Bureau of Rural Sciences. Inventory data compilation is a continuous process culminating in the periodic

---

[51] For more details by country, see www.fao.org/forestry

Table 39-1. Australia and New Zealand: forest resources and management

Country/area	Land area	Forest area 2000 – Natural forest	Forest plantation	Total forest			Area change 1990-2000 (total forest)		Volume and above-ground biomass (total forest)		Forest under management plan	
	000 ha	000 ha	000 ha	000 ha	%	ha/capita	000 ha/year	%	m³/ha	t/ha	000 ha	%
Australia	768 230	153 143	1 396	154 539	20.1	8.3	-282	-0.2	55	57	154 539	100
New Zealand	26 799	6 404	1 542	7 946	29.7	2.1	39	0.5	125	217	6 912	87
Total Australia and New Zealand	795 029	159 547	2 938	162 485	20.4	7.2	-243	-0.1	58	65	-	-
Total Oceania	849 096	194 775	2 848	197 623	23.3	6.6	-365	-0.2	55	64	-	-
TOTAL WORLD	13 063 900	3 682 722	186 733	3 869 455	29.6	0.6	-9 391	-0.2	100	109	-	-

Source: Appendix 3, Tables 3, 4, 6, 7 and 9.

publication of National Forest Inventory datasets and components such as the National Plantation Inventory and the National Forest Cover Database. In New Zealand, a National Exotic Forest Description is published annually to provide the latest plantation inventory data. A GIS land cover database, based on satellite imagery and differentiating areas of natural and plantation forests, was published in 2000. A comprehensive forest inventory for the natural forests has not, however, been carried out since the early 1950s. Work is currently under way to implement a carbon monitoring system for natural forests, scrublands and soils. This system, when fully operational, will provide updated and comprehensive statistics on many of the traditional forest inventory parameters for New Zealand's natural forests.

New Zealand's plantation forests cover more than 1.5 million hectares (Table 39-1), with *Pinus radiata* constituting about 90 percent of the plantation estate, and *Pseudotsuga menziesii* and *Eucalyptus* spp. accounting for the bulk of the remainder. On favourable and well-managed sites, *Pinus radiata* attains exceptional growth rates, with mean annual increments (MAIs) commonly approaching 24 m³ per hectare per year. Natural forests cover 6.4 million hectares and can be broadly divided into two main types: beech forests, dominated by four species of *Nothofagus* spp. ("false beech"); and conifer-hardwood forests comprising a complex association of species with typical canopy species including *Podocarpus totara*, *Dacrydium cupressinum* and *Agathis australis*. In general, New Zealand's lowland rain forests have a high forest canopy (20 to 35 m) and dense understory, while at higher altitudes the canopy trees become progressively lower (5 to 15 m) and more dense (with *Nothofagus solandri* var. *cliffortiodes* the primary species). Thus, New Zealand's forests have a relatively high average per hectare biomass (217 tonnes per hectare) (Crowe 1992).

Australia's plantation estate covers around 1.4 million hectares with more than 70 percent of the estate planted in softwood species. *Pinus radiata* is the most extensively planted softwood, while *Eucalyptus* spp. comprise almost the entire hardwood plantation estate. Australian-grown *Pinus radiata* typically achieves MAIs of around 20 m³ per hectare per year, while *Eucalyptus* spp. MAIs are generally in the range of 12 to 19 m³ per hectare per year.

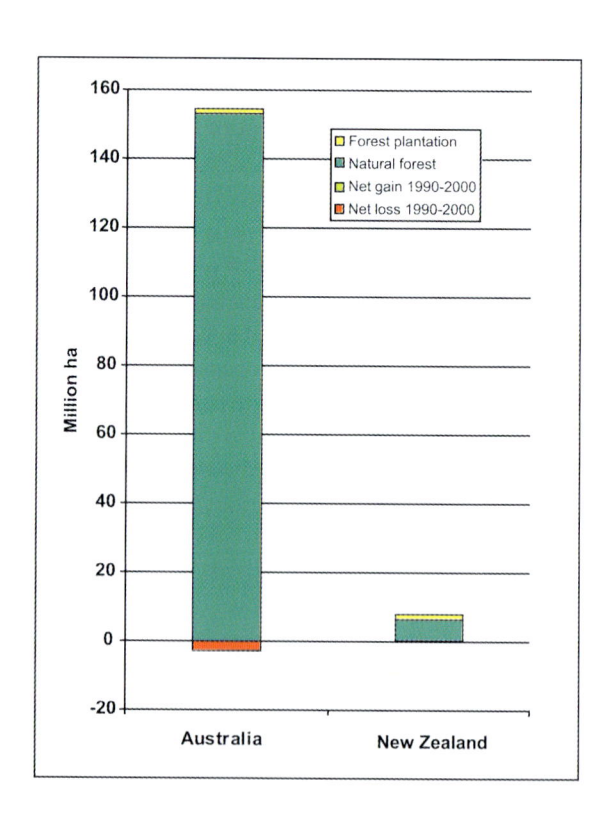

Figure 39-2. Australia and New Zealand: natural forest and plantation areas 2000 and net area changes 1990-2000

The predominant natural forest types in Australia are characterized as eucalypt forests and acacia forests. Eucalypt forests are easily the most widespread forest type, comprising around 80 percent of Australia's forest cover. The bulk of eucalypt forest has a relatively open "woodland type" canopy (20 to 50 percent crown cover), with most of the remainder classified as wet or dry sclerophyll forests. (Florence 1996). Acacia forests are widespread throughout the country, covering around 12 million hectares and generally predominate in areas with annual rainfall below 500 mm. At the more arid extent of their range the density of trees declines and low acacia woodlands are formed. The low density of these prevailing forest types is reflected in the relatively low estimates of per hectare forest biomass for Australia (Table 39-1). Tropical rain forests in Australia extend along the coasts of Arnhem Land and the Cape York Peninsula, and along the eastern seaboard of northern Queensland. In southern Queensland the seaboard forests are best characterized as warm temperate rain forests, while further south, cool temperate rain forests occur across coastal New South Wales, Victoria and much of Tasmania. Other major forest types include those characterized by a respective dominance of *Melaleuca* spp., *Casuarina* spp. and *Callitris* spp. while mangroves occur in many coastal areas (Bureau of Rural Sciences 2000).

Changes in forest area cover in Australia and New Zealand in the period 1990 to 2000 are relatively small in a global context. During this period Australia reported deforestation of 282 000 ha per year, while New Zealand reported an average net gain in forest area of 39 000 ha per year (Table 39-1, Figure 39-2). The net forest loss of 243 000 ha per year in the subregion constitutes only 2.6 percent of global deforestation. Australia's reported decline in forest area is, in part, the result of improved forest assessment methods. Australia's generally dry climate means large areas of the country are susceptible to wildfires, and significant areas of forest and woodland are burnt each year.

In 1994, for example, severe bushfires in New South Wales burned across 800 000 ha of forests and woodlands. In New Zealand, significant increases in the national plantation estate more than offset a modest decline in the area of natural forests (much of which is the result of more accurate assessment, rather than the physical clearance of forests) (Emergency Management Australia 2000).

FOREST MANAGEMENT AND USES

Both Australia and New Zealand provided national-level information on the forest area managed, applying the definition used by industrialized countries of forests managed in accordance with a formal or an informal plan applied regularly over a sufficiently long period (five years or more) and including areas where a decision had been taken not to undertake any management interventions. For Australia, all forests were reported as being managed, whereas for New Zealand, where the figure provided was limited to forests managed primarily for wood supply, not for conservation or protection purposes, the area equalled 87 percent of the total forest area. For the subregion as a whole, approximately 161 million hectares, or 99 percent of the total forest area, was reported as being managed in accordance with a formal or informal plan.

The basis for forest management planning in Australia is a system of Regional Forest Agreements negotiated between the Commonwealth and State governments to provide a blueprint for long-term management and use of forests in a particular region. Regional Forest Agreements have a 20-year lifespan and aim to: establish a world class forest reserve system in Australia; provide planning certainty for industries and regional communities; and ensure ecologically sustainable management of the national forest estate. Regional Forest Agreements apply to Australia's predominant commercially productive forest areas. Other forest areas are subject to a variety of State government legislation and management planning requirements (Commonwealth of Australia 2000).

More than 90 percent of New Zealand's plantation forests are under private ownership and virtually all the plantations are managed for commercial wood production. There is no strict legislative requirement for plantations to be managed under formal forest management plans, although the vast majority are subject to detailed plans. All forests are subject to the requirements of the Resource Management Act 1991, which regulates land use activities and under which many forestry operations (particularly harvesting and planting) require that a local government Resource Consent be obtained. A large majority (77 percent) of natural forests in New Zealand are government owned and managed as protected areas by the Department of Conservation. All these forests are subject to conservation management plans. In 1992, the Forests Act 1949 was amended to require that privately owned natural forest areas be managed under government-approved Sustainable Forest Management Plans if they are to be subject to commercial harvesting (Environment Australia 1997).

Both Australia and New Zealand are strongly committed to principles of sustainable forest management. Australia's commitment to sustainable forest management is formally expressed through its National Forest Policy Statement 1992, which aims for sustainable management of all its forests for future generations, whether the forest is within reserves or in production forests or plantations, and on public and private land. The development of Regional Forest Agreements is a key initiative in realizing this commitment. Other initiatives include: the development of an Australian Forestry Standard as a means of certifying forest management practices in Australia; and the development of a framework of subnational criteria and indicators of sustainable forest management in Australia. In New Zealand, the commitment to sustainable forest management and sustainable resource use is enshrined in the Resource Management Act 1991 and amendments to the Forests Act 1949. Voluntary measures that enhance the protection and sustainable management of New Zealand's forest resources include the New Zealand Forest Code of Practice and the New Zealand Forest Accord 1991. Several New Zealand forests have obtained Forest Stewardship Council certification, and a process to develop a national certification process consistent with international standards is under way. Both countries are active participants in the major international fora and processes aimed at the achievement of sustainable forest management.

Production of industrial roundwood in New Zealand is centred on plantation forests, which provide more than 99 percent of the country's annual harvest. New Zealand produces a significant volume of wood, surplus to its own requirements, with around 60 percent of current production exported in some form. Large areas of plantations are approaching maturity and New Zealand's annual harvest is projected to increase markedly from the current 18 million cubic metres, to more than 30 million cubic metres by 2010. Australia also has significant areas of maturing plantation forests and is expected to become a net exporter of forest products during the next decade. At present, Australia's annual wood harvest totals around 21 million cubic metres, which is evenly divided between coniferous and non-coniferous wood.

Household use of woodfuels in Australia and New Zealand is significant, but is not regularly monitored in either country. One estimate (FAO 1997) suggests that the current consumption of woodfuel in the subregion is around 3.5 million cubic metres. Woodfuel does not constitute a major source of electricity production in either country, although there are examples of wood by-products being used to generate electricity in particular industrial plants.

Both Australia and New Zealand have developed comprehensive protected area networks. Australia's terrestrial protected area network currently covers around 8 percent of the country's land area. Most recently, the signing of Regional Forest Agreements has led to a significant boost in the area of forests in protected areas. At present, around 42 percent of land covered by Regional Forest Agreements is in conservation reserves. In New Zealand, the Department of Conservation (DOC) administers State-owned protected areas; the DOC estate encompasses almost 5 million hectares (77 percent) of natural forests. An additional 70 000 ha of privately owned natural forests have protected-area status through a variety of covenant arrangements (Environment Australia 1997; Commonwealth Forests Taskforce 2000).

CONCLUSIONS AND ISSUES

Data reported to FRA 2000 by Australia and New Zealand are both reliable and indicative of the countries having well-developed forest monitoring and inventory systems in place. Systems for collecting information and data pertaining to plantation forests in each country are among the most comprehensive in the world. Natural forest inventory systems are less well developed, but both countries are making substantive efforts to upgrade data and the next several years will see systems in place comparable with those of leading forestry countries.

Both countries are well placed to deliver on commitments to sustainable forest management. As economically developed countries, with very low population and land use pressures, both have the physical and financial capacity, as well as the apparent political will, to achieve very high standards of forest management. These capacities are reflected in relatively high proportions of forests in formally protected areas, the establishment of large plantation forest estates as a means of reducing industrial pressures on natural forests and significant progress in implementing mechanisms to support sustainable forest management.

The key forestry issues requiring attention in Australia mainly relate to achieving an acceptable balance between the economic, social and environmental dimensions of forestry. Regional Forest Agreements provide a mechanism for attaining this balance, at least in terms of achieving agreement between Commonwealth and State governments; however, there remains considerable disparity in broader stakeholder perceptions of the appropriate emphasis that should be placed on nature conservation objectives compared with economic development objectives. A separate dimension relates to social aspects of forestry and, in particular, how the rights and aspirations of Aboriginal peoples and Torres Strait Islanders in respect to their forest interests can be reconciled within national and regional frameworks for sustainable forest management.

The principal challenges faced by the New Zealand forestry sector lie in two distinct spheres. In the natural forests there remains a distinct tension between preservationist and multiple-use management philosophies. In recent years, there has been a marked shift towards further reducing the already modest industrial forestry activities in natural forests. At the same time, this has removed a significant component of the natural forests' ability to generate funds for improved management. Natural forests managers have consequently become increasingly reliant on direct government funding for effective management, and in some areas this has fallen short in providing adequate protection from degradation by introduced pests, most notably by red deer and the Australian brush-tailed opossum. In New Zealand's plantation forests the principal issues relate to challenges in effectively marketing rapidly increasing wood supplies. This incorporates the development of significant value-added processing capacity and the opening of new export markets, but also ensuring that plantation-grown wood meets more environmentally conscious market expectations through, for example, the development of an internationally accepted certification system and continuous improvement in plantation forest management.

BIBLIOGRAPHY

Bureau of Rural Sciences. 2000. *Forest information.* Agriculture, Fisheries and Forestry – Australia.
www.affa.gov.au/docs/rural_science/nfi/forestinfo/info.html

Commonwealth Forests Taskforce. 2000. The CAR reserve system. *RFA Forest news – August 2000.* Commonwealth of Australia.

Commonwealth of Australia. 2000. *Regional Forest Agreements.*
www.rfa.gov.au/

Crowe, A. 1992. *Which native tree?* Auckland, New Zealand, Penguin Books.

Emergency Management Australia. 2000. *Wildfire prevention in Australia.* United Nations 2000 World Disaster Reduction Campaign.
www.unisdr.org/unisdr/infokitaustra.htm

Environment Australia. 1997. *National reserve system – terrestrial and marine protected areas in Australia.* Canberra, Department of Environment and Heritage.
www.ea.gov.au/parks/nrs/protarea/paaust/index.html

FAO. 1997a. *In-depth country study – New Zealand,* by C. Brown. Asia Pacific Forestry Sector Outlook Study. Working Paper No. 5. Rome.
www.fao.org/forestry

FAO. 1997b. *Country report Australia,* by Commonwealth Department of Primary Industries and Energy. Asia Pacific Forestry Sector Outlook Study. Working Paper No. 13. Rome.
www.fao.org/forestry/

Florence, R.G. 1996. *Ecology and silviculture of eucalypt forests.* Commonwealth Scientific and Industrial Research Organisation (CSIRO).

New Zealand. Ministry of Agriculture and Forestry (MAF). 1997. *Indigenous forestry sustainable management – a guide to plans and permits.* Wellington, New Zealand.
www.maf.govt.nz/MAFnet/sectors/forestry/indig/httoc.htm

New Zealand. MAF. 2000. *National exotic forest description: national and regional wood supply forecasts 2000.* Wellington, New Zealand.

Chapter 40
Other Oceania

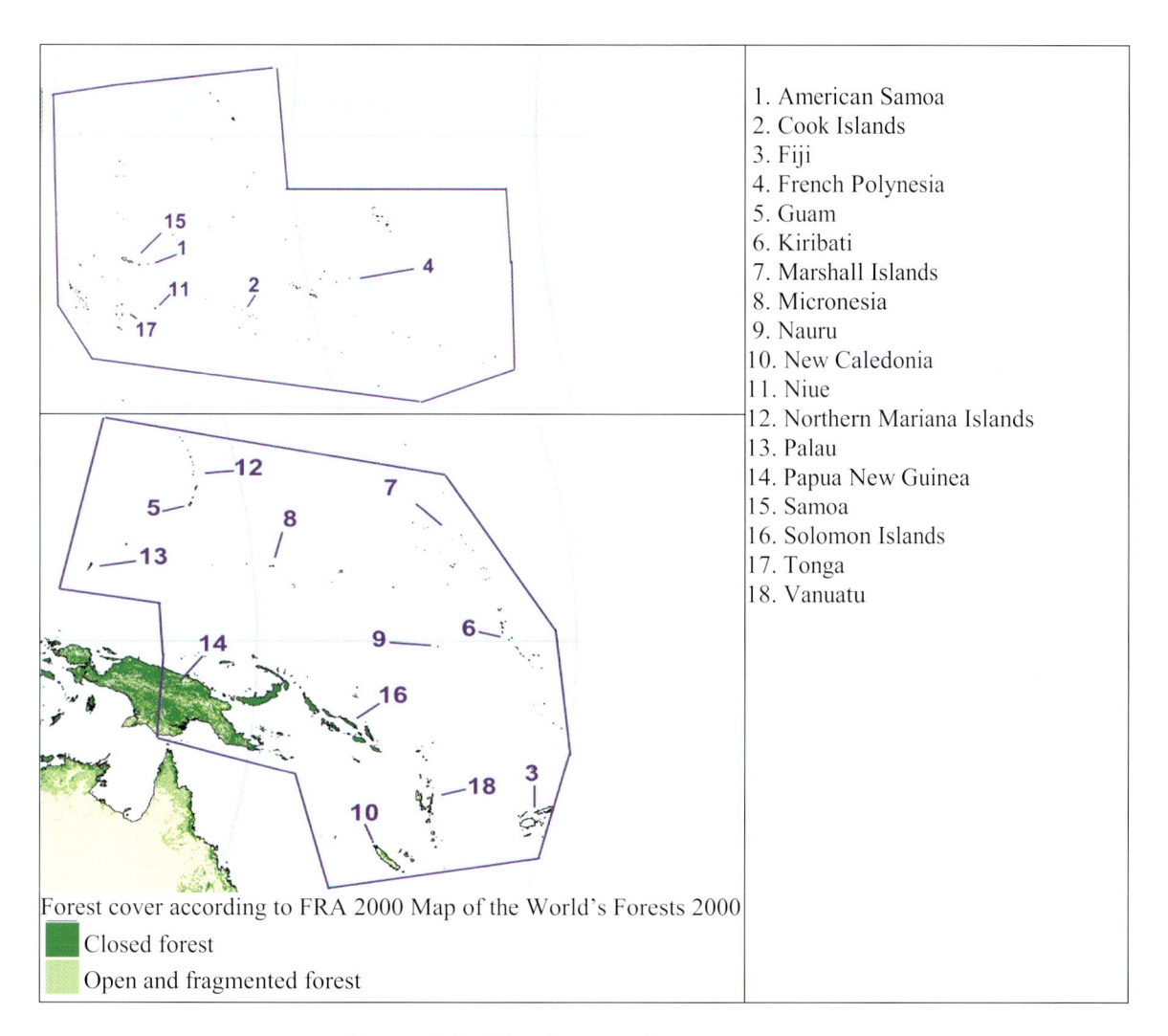

Figure 40-1. Other Oceania: forest cover map

	1. American Samoa
	2. Cook Islands
	3. Fiji
	4. French Polynesia
	5. Guam
	6. Kiribati
	7. Marshall Islands
	8. Micronesia
	9. Nauru
	10. New Caledonia
	11. Niue
	12. Northern Mariana Islands
	13. Palau
	14. Papua New Guinea
	15. Samoa
	16. Solomon Islands
	17. Tonga
	18. Vanuatu

Forest cover according to FRA 2000 Map of the World's Forests 2000
- Closed forest
- Open and fragmented forest

The vast area of the Pacific Ocean is dotted with a number of island countries and territories,[52] including American Samoa, Cook Islands, Fiji, French Polynesia, Guam, Kiribati, Marshall Islands, Federated States of Micronesia, Nauru, New Caledonia, Niue, Northern Mariana Islands, Palau, Samoa, Solomon Islands, Tonga and Vanuatu. American Samoa and Guam are United States Territories, Northern Mariana Islands is a United States Commonwealth and French Polynesia and New Caledonia are French Territories; the others are independent States. In addition, the subregion includes Papua New Guinea, which shares the island of New Guinea with Irian Jaya, a province of Indonesia (reported under Asia) (Figure 40-1).

Landforms vary from low atolls, barely above sea level, to the mountains of Papua New Guinea, reaching 4 500 m. Most of the islands are of volcanic origin, although New Guinea and some western Pacific islands are continental.

The subregion can be divided into three geographic areas, Polynesia, Micronesia and Melanesia, located roughly in the central, northwest and southwest Pacific, respectively. The flora of Melanesia is the richest, given its proximity to the Indo-Malaysia region and Australia. Similarly, the vegetation of Micronesia is largely of Indo-Malaysian origin, becoming less

---
[52] For more details by country, see www.fao.org/forestry

Table 40-1. Other Oceania: forest resources and management

Country/area	Land area	Forest area 2000					Area change 1990-2000 (total forest)		Volume and above-ground biomass (total forest)		Forest under management plan	
		Natural forest	Forest plantation	Total forest								
	000 ha	000 ha	000 ha	000 ha	%	ha/capita	000 ha/year	%	m³/ha	t/ha	000 ha	%
American Samoa	20	12	0	12	60.1	0.2	n.s.	n.s.	-	-	-	-
Cook Islands	23	21	1	22	95.7	1.2	n.s.	n.s.	-	-	-	-
Fiji	1 827	718	97	815	44.6	1.0	-2	-0.2	-	-	-	-
French Polynesia	366	100	5	105	28.7	0.5	n.s.	n.s.	-	-	-	-
Guam	55	21	0	21	38.2	0.1	n.s.	n.s.	-	-	-	-
Kiribati	73	28		28	38.4	0.3	n.s.	n.s.	-	-	-	-
Marshall Islands	18	n.s.	-	n.s.		-	n.s.	n.s.	-	-	-	-
Micronesia	69	15	0	15	21.7	0.1	-1	-4.5	-	-	-	-
Nauru	2	n.s.	-	n.s.		-	n.s.	n.s.	-	-	-	-
New Caledonia	1 828	362	10	372	20.4	1.8	n.s.	n.s.	-	-	-	-
Niue	26	6	0	6	-	3.0	n.s.	n.s.	-	-	-	-
Northern Mariana Islands	46	14	-	14	30.4	0.2	n.s.	n.s.	-	-	-	-
Palau	46	35	0	35	76.1	1.8	n.s.	n.s.	-	-	-	-
Papua New Guinea	45 239	30 511	90	30 601	67.6	6.5	-113	-0.4	34	58	5 341	17
Samoa	282	100	5	105	37.2	0.6	-3	-2.1	-	-	-	-
Solomon Islands	2 856	2 486	50	2 536	88.8	5.9	-4	-0.2	-	-	43*	n.ap.
Tonga	73	3	1	4	5.5	n.s.	n.s.	n.s.	-	-	-	-
Vanuatu	1 218	444	3	447	36.7	2.4	1	0.1	-	-	-	-
Total Other Oceania	**54 067**	**34 875**	**263**	**35 138**	**65.0**	**4.7**	**-122**	**-0.3**	**34**	**58**	**-**	**-**
Total Oceania	**849 096**	**194 775**	**2 848**	**197 623**	**23.3**	**6.6**	**-365**	**-0.2**	**55**	**64**	**-**	**-**
TOTAL WORLD	**13 063 900**	**3 682 722**	**186 733**	**3 869 455**	**29.6**	**0.6**	**-9 391**	**-0.2**	**100**	**109**	**-**	**-**

Source: Appendix 3, Tables 3, 4, 6, 7 and 9.
*Partial results only. National figure not available.

rich from west to east. Polynesia, while still having many species of Indo-Malaysian origin, also includes species from America and New Zealand. The many coral atolls, with difficult growing conditions, have limited vegetation of strand plants (Mueller-Dombois and Fosenberg 1998).

The entire area is tropical. In addition to origin, temperature and rainfall largely influence the vegetation. Temperature varies by elevation and latitude. Overall rainfall is determined by location in relation to the intertropical convergence zone, with rainfall up to 5 000 mm in the centre of the zone. Rainfall patterns are also influenced by the trade winds, which promote orographic precipitation on the windward side of the mountains of the high islands and drier areas on the lee side. Tropical cyclones affect some of the islands in the western Pacific. El Niño and the Southern Oscillation have long-term, multiseasonal effects.

Halophytic strand vegetation forms a narrow belt around high islands and covers most atolls. Tidal zones, particularly those protected from the open ocean by coral reefs or river estuaries, often host mangrove swamps, with vegetation varying from scrub to forest. Lowland tropical rain forest, multistoried and with epiphytes and shrubs when undisturbed, was originally the most extensive vegetation type but has been eliminated or highly altered by human activity in most areas. At higher elevations, on moist hilltops and slopes, this forest grades into a lower, epiphyte- and shrub-rich montane rain forest. Cloud forests are often found at the highest elevations. On leeward slopes are more mesophytic forests or even seasonally dry evergreen forests, grading into xerophytic types on the more severe sites (Mueller-Dombois and Fosenberg 1998).

FOREST RESOURCES

A workshop, with participation from many of the Pacific countries, was organized in Apia, Samoa in 2000 for the collection of information for this subregion. In addition, FAO assembled historical data to estimate forest cover as of 2000 and change from 1990 to 2000.

The quality of information varies greatly from country to country, but much of it is incomplete or out of date. Surveys were conducted in

American Samoa, the Federated States of Micronesia, Northern Mariana Islands and Palau in the 1980s. New inventories are beginning for these locations, but will not be ready for several years. An inventory was completed for Solomon Islands in 1995, for Tonga in 1999 and for Vanuatu in 1992. An inventory, mainly focused on merchantable volume, was conducted in Samoa in 1977. Fiji has had national forest inventories in 1966-1969 and 1991-1993. Papua New Guinea's recent inventories are mostly project-specific surveys and maps of areas being considered for exploitation.

Papua New Guinea dominates the statistics for this region, with by far the bulk of the forest land (Table 40-1, Figure 40-2). Its proportion of forest cover remains high, even though it has experienced a continuing loss of forest cover. Although much smaller in area, Solomon Islands likewise has a high proportion of land in forest cover and a high area per capita. Fiji, New Caledonia and Vanuatu all have considerable forest land. The amount of forest land in the remaining island countries and territories is small.

Fiji, Papua New Guinea and Solomon Islands have done the most in establishing plantations, but a number of those in Papua New Guinea are reported to be neglected or abandoned (Papua New Guinea Forest Authority undated). Small areas of plantations have been established in the other countries, both for protection and production.

Volume and biomass are estimated only for Papua New Guinea.

FOREST MANAGEMENT AND USES

Only one of the 18 countries and areas reported on in this subregion provided national data on the area of forest managed according to a formal, nationally approved forest management plan (Table 40-1). However, this country alone (Papua New Guinea) accounted for 87 percent of the total forest area in the subregion. A reported 5 million hectares, or 17 percent of the total forest area in Papua New Guinea, is subject to a formal management plan. Partial information was available from Solomon Islands in the form of the forest area that had obtained third party certification by the end of 2000.

Forests contribute significantly to the economy and foreign exchange earnings of Papua New Guinea. Exports in 1999 reportedly totalled US$151 952 000, although this was down from the mid-1980s when export earnings were approximately US$500 million. At present a total of 10.98 million hectares is under logging concessions, with a further 3.0 million hectares that have yet to be allocated. The annual log harvest from these concessions in 1999 was 2 097 000 m^3, excluding the volume that was harvested using small-scale portable sawmills and removals as a result of land clearing for agriculture or other land uses. Most of this was

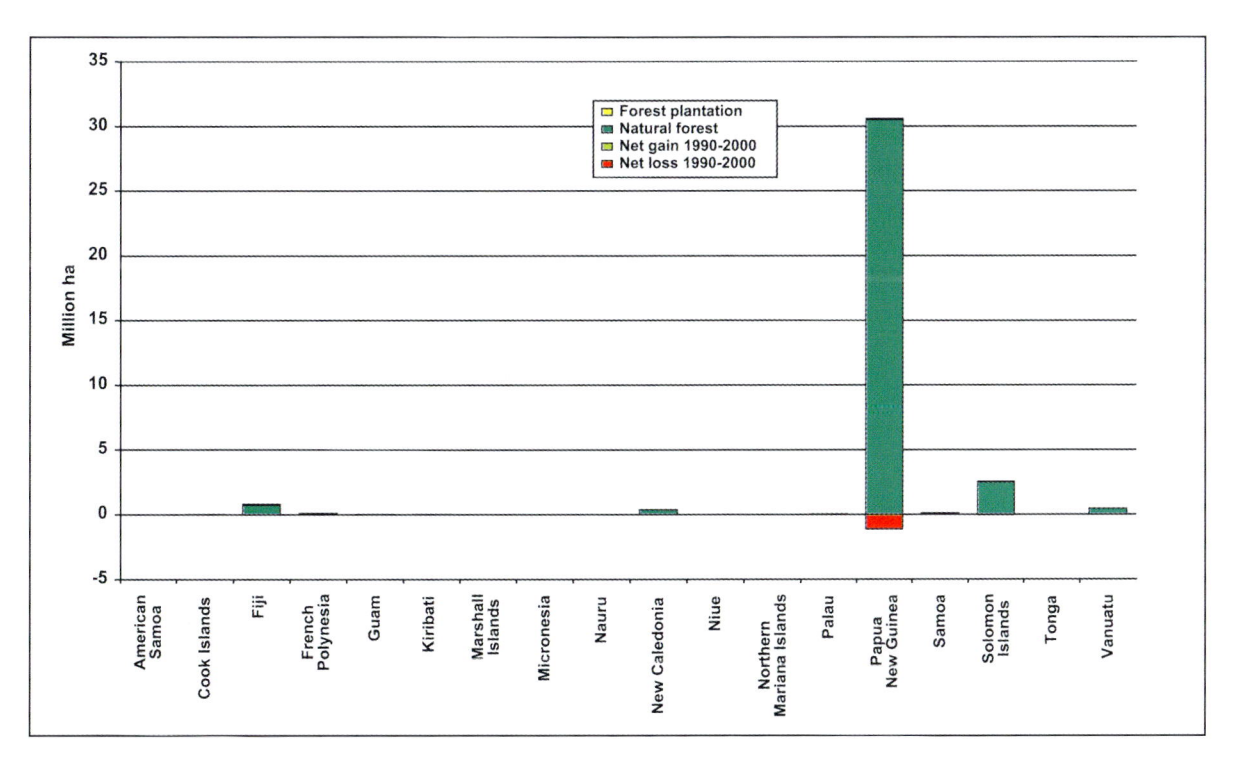

Figure 40-2. Other Oceania: natural forest and forest plantation areas 2000 and net area changes 1990-2000

exported in the form of round logs to China (including Taiwan Province), Japan, the Republic of Korea and the Philippines. Some wood chips are exported to Japan as well as some sawn timber, plywood and veneer, mostly to Australia and New Zealand (FAO in press).

Solomon Islands is heavily forested; slightly more than 88 percent of the country is in forest, mostly tropical rain forest. However, like Papua New Guinea, a large proportion of the forests are inaccessible owing to steep terrain. Approximately 1 000 ha per annum of plantations have been established, mostly on Government-owned land. As with Papua New Guinea, log exports in the past have provided significant revenue. Harvest levels in the recent past appear to have been above sustainable levels (FAO 1997).

Fiji has experienced continuing loss of natural forests, particularly in the lowlands. Approximately 40 percent of the remaining natural forest has already been logged over and an additional 30 percent is in protected reserves. The remaining 30 percent is being cut over at a rate of about 8 000 ha per year (FAO in press). Fiji has a substantial plantation resource, both softwood and hardwood. An aggressive plantation programme is continuing. Fiji has a substantial forest-based industry and is a producer and exporter of woodchips, sawn timber and plywood/veneer (FAO 1997).

Vanuatu has a substantial forest resource, but much of it is on steep, inaccessible sites, and from a commercial standpoint many species are of limited commercial use. At the present time, harvesting is well within sustainable levels and the loss of forest land is modest. Efforts to establish plantations have been small and largely unsuccessful. A notable export is sandalwood (FAO in press).

In Samoa, during a 15-year period from 1978 on, it is estimated that 50 percent of the merchantable forest and 30 percent of the non-merchantable forest was cleared (FAO 1997). Natural forests remain on 36 percent of the land area, with another 1 percent in plantations. The bulk of the remaining commercial forest is on the larger island of Savai'i, with small areas on Upolu. Deforestation, primarily conversion to agricultural use, remains a significant factor. Timber exports once provided significant employment and export earnings, but have decreased significantly since cyclones in 1990 and 1991 resulted in severe damage to the forests (FAO in press).

The Marshall Islands and Kiribati, as well as parts of other Pacific states, consist of low atolls and the natural forest is mostly strand vegetation. The forests are important for protection and local use. The high islands of American Samoa, the Cook Islands, French Polynesia and the Federated States of Micronesia have significant areas of forest land, but the terrain is rugged. These forests are highly valuable for watershed protection and local use. Little undisturbed forest remains on Guam or on the major Northern Mariana Islands. Palau has a significant amount of forest for its size. There is little natural forest left on Nauru and Tonga. Niue's forests are mostly second growth (FAO 1997).

Mangroves are a significant resource in the Federated States of Micronesia, Kiribati, New Caledonia, Palau, Papua New Guinea, Solomon Islands and Vanuatu (FAO in press).

In addition to providing a very important source of fuelwood and local forest products, they provide highly valuable shore protection and habitat. Most countries recognize these valuable benefits and are working to try to prevent the loss of mangrove cover.

Very little forest land in the Pacific is under any sort of a formal management plan. Where there is such a plan, the provisions of the plan are not always strictly enforced.

Fire is of little concern on many of the islands owing to a generally wet climate with few dry periods. The exceptions usually occur on the lee side of mountains, such as in Papua New Guinea, Fiji and Solomon Islands, or during the occasional drought periods. Guam, however, has a severe fire problem. A tradition of deliberately set fires has reduced most of the vegetation on the southern half of the island to grass. On other islands, moreover, introduced grasses, such as *Melinis minutiflora,* carry fire and prevent other species from regenerating, perpetuating the fire-grass regime (D'Antonio and Vitousek 1992). Fires tend to be especially destructive to plantations.

Occasional insect and disease problems have occurred with introduced plantation species, such as an attack on *Cordia alliodora* by the fungus *Phellinus noxious* reported from Vanuatu (FAO in press) and a shoot borer that attacks *Swietenia macrophylla* (Oliver 1999). Cyclones, particularly in the western Pacific, can cause serious damage. They have often been particularly damaging to plantations of introduced species, which may be more poorly adapted to these conditions than native species growing in mixed forests. Introduced plants are a serious problem throughout the subregion, in some cases completely changing the character of the forests. Some species were introduced for forestry purposes, but most were introduced for other purposes and have escaped (Meyer and Malet in press; Space in press). An introduced small tree, *Miconia calvescens,* has become a huge problem in Tahiti and some of the other islands of French Polynesia, choking out the natural forests.

Traditional agriculture in the Pacific subregion is closely tied to forests. These tree-gardens include trees producing edible fruits, fuelwood and other products. The undergrowth and small openings produce bananas, cassava and root crops. In wetter areas, taro is grown. This traditional agroforestry provided sustenance and useful non-timber products to the people while maintaining protective forest cover (Thaman and Whistler 1995). Archaeological and historical evidence shows that many of these Pacific islands were able to sustain very high population levels using traditional systems. There is renewed interest in the use of these systems from both the standpoint of improving land management and providing a better diet than imported food.

A ubiquitous resource throughout the Pacific is the coconut palm (*Cocos nucifera*). In the past, copra was a major component of the economy and source of foreign exchange. Now, use is mainly local. Coconut wood can, however, be sawn into building material and some countries are utilizing senescent coconut trees for this purpose (FAO in press).

The land tenure system often makes the establishment of protected areas difficult. However, many of the countries have managed to formally declare parks and reserves. Papua New Guinea has the largest area, but some smaller countries have made significant efforts in relation to their size, including American Samoa, Fiji, French Polynesia, Niue, Samoa and Vanuatu. New Caledonia has done an exceptional job of creating parks and reserves. The most successful efforts have focused on the strong involvement of local people (FAO in press).

CONCLUSIONS AND ISSUES

In the past, timber harvesting in Papua New Guinea was largely exploitive. Prior to the amendment of the Forestry Act in 1991, "there was a headlong rush throughout the country by landowners ... to allow their forest resources to be logged at totally exploitive rates. This has resulted in the almost complete exhaustion of commercial forest resources in the New Ireland and West New Britain Provinces." (Papua New Guinea Forest Authority undated). Most of the readily accessible production forest areas have already been logged out, nearly logged out or have been committed for exploitation. What are left are forests that are in the hinterland and rugged terrain areas (FAO in press). At the present rate of cutting and with a moratorium on new concessions currently in place, Papua New Guinea appears to be operating within sustainable forest management when the country is viewed as a whole. However, none of the existing forest concessions are being managed on a sustainable basis. While these are being renegotiated as they come up for renewal, a long-term strategy for the assessment, development and use of forest resources is still needed. Thus, while a substantial area is reported as being under management plans, the effectiveness of many of these may be minimal.

While the inaccessible upland areas of Solomon Islands will undoubtedly remain in forest cover, Oliver (1992) estimated that approximately half the commercially exploitable natural forest was already logged over by 1990. Maturing plantations may be able to alleviate the situation for a while, but harvest levels will have to decrease significantly to be sustainable. Brown in FAO (1997) states: "The basic story at present appears to be that of a resource being harvested too quickly and unsustainably with few of the benefits being reinvested to ensure the long-term viability of the forest industry".

New Caledonia's forest situation appears to be stable, with reasonable harvest levels, progressive forest management and a modest plantation programme. Substantial areas have been set aside as parks and reserves.

Fiji has policies and organizations in place that should permit future sustainable management of its forest resources. In addition to trying to achieve full acceptance of sustainable forest management, it is working to strengthen the involvement and active participation of landowners in forestry and the wood-based industries as well as to further strengthen value-added processing and promote alternative and non-destructive uses of natural forests (e.g. ecotourism) (FAO in press). Continuing loss of natural forest is of concern, but is offset to some degree (from a production standpoint) by a strong plantation programme.

Vanuatu has been working hard to put appropriate policies and procedures in place for sustainable management of its forests. A National

Forest Policy was promulgated in 1997. The Forestry Act is currently being revised. Harvesting licences have been reduced to a level well within sustainable forest management. A logging practice code is in place and reduced impact logging is being promoted by the Forestry Department. There is the potential for a substantial increase in forest plantations.

The National Forest Policy of Samoa calls for the sustainable utilization and management of the remaining merchantable indigenous forests. Neither a code of logging practices nor reduced impact logging guidelines has yet been established, although a practices code has been drafted. A sustainable forest management project, carried out with donor assistance, is currently under way that aims to develop a model sustainable harvesting system. Most of the previous plantations were damaged by cyclones, and the current rate of planting is only between 50 and 100 ha per year (FAO in press). Given the current situation, it appears that Samoa will increasingly depend on timber imports (FAO 1997).

Forest management throughout the Pacific is complicated by the customary land tenure system, which is a combination of individual and communal rights. Although land itself usually cannot be sold, the forest resources on the land can be. This has often led to exploitive practices that turn these resources into cash or other land uses with little thought to the future or to the benefit or loss to society as a whole. In the past, claim to a piece of land was usually established by clearing it. Only recently have most countries begun efforts to work within the customary system to promote responsible and sustainable forest practices. Likewise, the land tenure system makes it difficult to obtain commitment for enough land for a sufficiently long period to establish protected areas or a viable plantation system (FAO 1997; FAO in press).

The demand for living space and agricultural land continues to have a significant effect on the reduction of forest land, particularly in American Samoa, the Federated States of Micronesia, Samoa and Solomon Islands. The Second World War and subsequent urbanization had a significant effect on the forests of Guam and some of the Northern Mariana Islands. Forestry organizations are underfinanced (often dependent on donor funding) and poorly staffed. Even when funding is available, it is very difficult to find trained foresters and technicians. Enforcement of forestry laws and regulation is often lax. There is frequently a lack of awareness of the importance of forest resources and the inability of local people to make informed decisions on their use (FAO in press).

Nevertheless, most countries have a forest management organization in place and operating, at least to some degree. Most countries have put in place forestry policies, usually recognizing the need for sustainable forest management. Some countries have forest practice codes in place and are implementing reduced impact logging. Awareness on the part of the people and the political leadership seems to be steadily increasing. Most Pacific nations are signatories of the international conventions and treaties dealing with conservation. Regional institutions are in place that focus assistance and promote intercountry cooperation. Much remains to be done, but on balance significant progress is being made.

BIBLIOGRAPHY

D'Antonio, M. & Vitousek, P.M. 1992. Biological invasions by exotic grasses, the grass-fire cycle, and global change. *Annual Review of Ecology and Systematics,* 23: 63-87.

FAO. 1997. *Regional study – the South Pacific,* by C. Brown. Asia-Pacific Forestry Sector Outlook Study Working Paper APFSOS/WP/01. Rome.

FAO. In press. *Proceedings of the workshop on Data Collection for the Pacific Region.* FRA Working Paper. Rome.

Meyer, J.Y. & Malet, J.P. In press. *Forestry and agroforestry alien trees as invasive plants in the Pacific Islands.* In FAO Workshop Data Collection for the Pacific Region. FRA Working Paper. Rome, FAO.

Mueller-Dombois, D. & Fosenberg, F.R. 1998. *Vegetation of the tropical Pacific islands.* New York, Springer-Verlag.

Oliver, W.W. 1992. *Plantation forestry in the South Pacific: a compilation and assessment of practices.* Project RAS/86/036. FAO/UNDP.

Oliver, W.W. 1999. *An update of plantation forestry in the South Pacific.* SPC/UNDP/AusAID/FAO. Pacific Islands Forests and Trees Support Programme.

Papua New Guinea Forest Authority. Undated. *Country Report – Papua New Guinea.* Asia-Pacific Forestry Sector Outlook Study Working Paper APFSOS/WP/47. Rome, FAO.

Space, J.C. In press. *Invasive plants threatening Pacific ecosystems: the Pacific Islands Ecosystems at Risk Project.* In FAO Workshop on Data Collection for the Pacific Region. FRA Working Paper. Rome, FAO.

Thaman, R.R. & Whistler, W.A. 1995. *Samoa, Tonga, Kiribati and Tuvalu: a review of uses and status of trees and forests in land use systems with recommendations for future actions.* Project RAS/92/T04. Rome, FAO.

Whistler, W.A. 1992. Vegetation of Samoa and Tonga. *Pacific Science,* 46(2): 159-178.

Chapter 41
South America

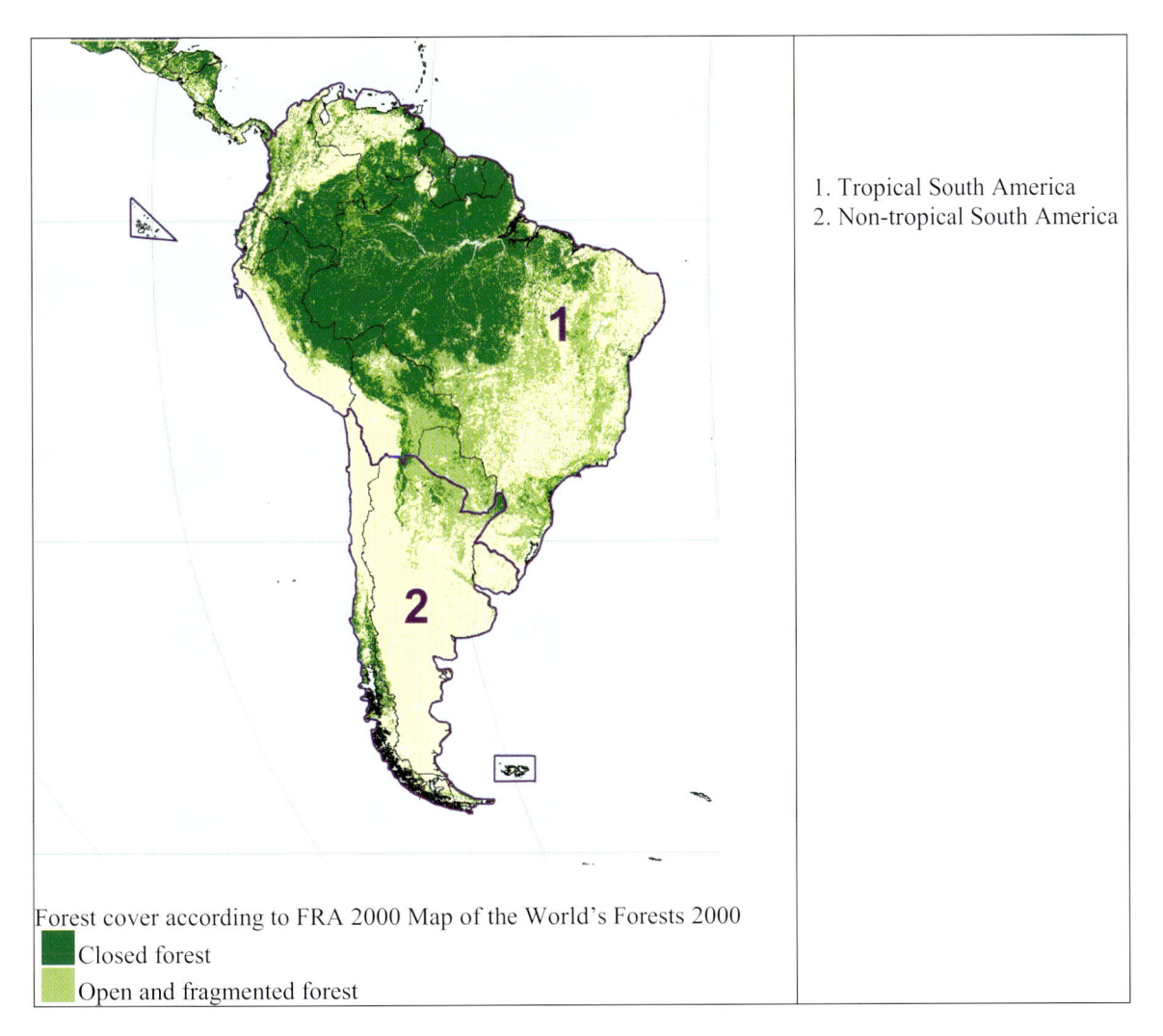

Figure 41-1. South America: subregional division used in this report

South America (see Figure 41-1[53] and Table 41-1) contains about 885 million hectares of forests which corresponds to 23 percent of the world total. South American forests amount to 2.6 ha per capita, which is considerably above the world average. Almost all forests are located in the tropical ecological domain and South America has about 54 percent of all tropical rain forests and the proportion of forest cover in the tropical rain forest zone is 82 percent. Forest plantations represent just 1 percent of the total forest cover. The annual net loss, based on country reports, is high at 3.7 million hectares annually, corresponding to 0.4 percent annually.

---

[53] The division into subregions was made only to facilitate the reporting at a condensed geographical level and does not reflect any opinion or political consideration in the selection of countries. The graphical presentation of country areas does not convey any opinion of FAO as to the extent of countries or status of any national boundaries.

Table 41-1. South America: forest resources by subregion

Subregion	Land area	Forest area 2000 – Natural forest	Forest plantation	Total forest			Area change 1990-2000 (total forest)		Volume and above-ground biomass (total forest)	
	000 ha	000 ha	000 ha	000 ha	%	ha/capita	000 ha/year	%	m^3/ha	t/ha
Non-tropical South America	367 248	47 911	3 565	51 476	14.0	0.9	-255	-0.5	67	130
Tropical South America	1 387 493	827 252	6 890	834 142	60.1	2.9	-3 456	-0.4	129	208
Total South America	**1 754 741**	**875 163**	**10 455**	**885 618**	**50.5**	**2.6**	**-3 711**	**-0.4**	**125**	**203**
TOTAL WORLD	**13 063 900**	**3 682 722**	**186 733**	**3 869 455**	**29.6**	**0.6**	**-9 391**	**-0.2**	**100**	**109**

Source: Appendix 3, Tables 3, 4, 6 and 7.

Chapter 42

South America: ecological zones

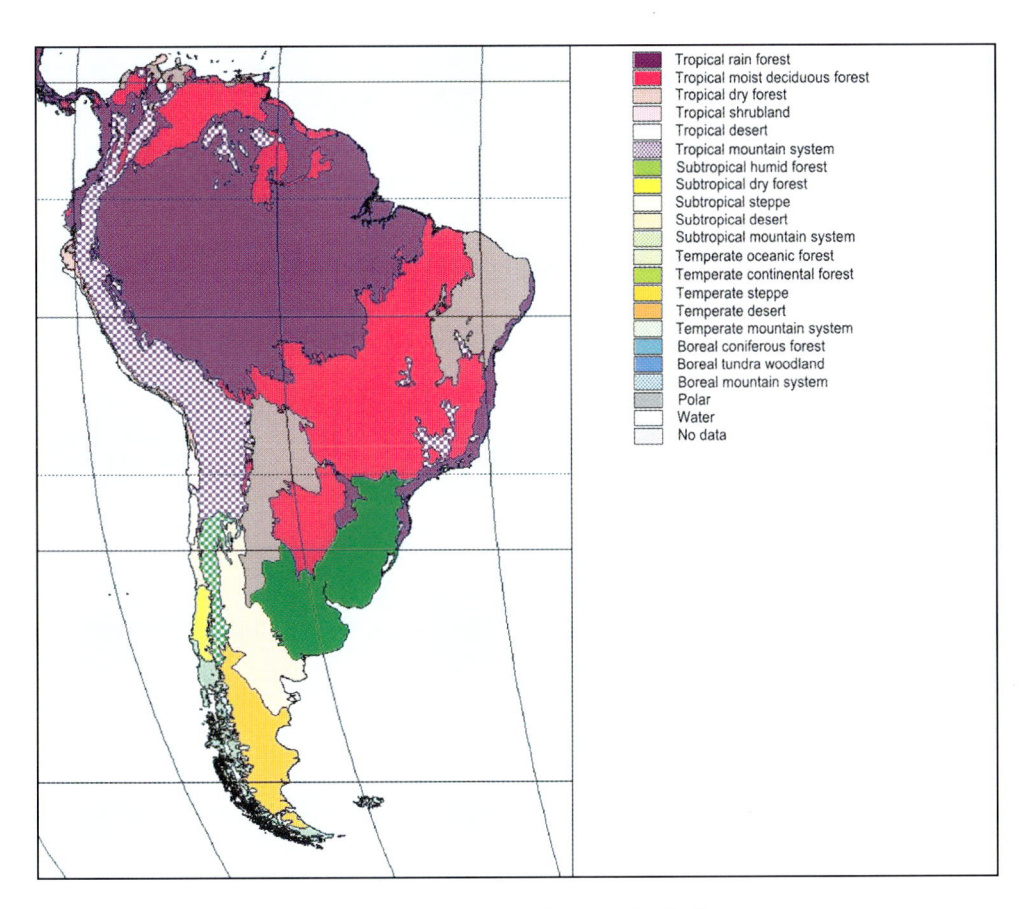

Figure 42-1. South America: ecological zones

Figure 42-1 shows the distribution of ecological zones in South America. Table 42-1 contains area statistics for the zones by subregion, and Table 42-2 indicates the proportion of forest in each zone by subregion.

TROPICAL RAIN FOREST

The tropical rain forests of South America extend over the whole Amazonian Basin, the Pacific coast of Colombia and Ecuador and the Atlantic coast and Iguaçu and Parana River valleys of Brazil. Huge amounts of rain fall in the heart of the Amazon Basin and along the western coast (more than 3 000 mm, even up to 8 000 mm). Elsewhere, rainfall is between 1 000 and 3 000 mm, often with a short dry period in winter. Temperatures are high, especially in the Amazonian region, where the mean temperature of the coldest month is always above 20°C. On the Atlantic coast, mean temperatures decrease as latitude increases (15° to 20°C).

The Amazon Basin contains the world's largest area of tropical rain forest. In this vast extent at least 10 to 20 different vegetation types might be distinguished. The wettest type is found in the upper basin of the Amazon River, the State of Amapà in Brazil and the west coast of Colombia. The vegetation is luxuriant, multilayered evergreen forest, up to 50 m tall, with emergent trees. The most important tree families are Annonaceae, Bombacaceae, Burseraceae, Clusiaceae, Euphorbiaceae, Leguminosae, Moraceae and Sterculiaceae.

The most extensive rain forest is somewhat drier and occurs in the Amazon Basin and on the eastern foothills of the central Andes. It is a multilayered forest up to 40 m tall, with or without emergent trees, mainly evergreen but with

Table 42-1. South America: extent of ecological zones

Subregion	Total area of ecological zone (million ha)																	
	Tropical						Subtropical				Temperate						Boreal		
	Rain forest	Moist	Dry	Shrub	Desert	Mountain	Humid	Dry	Steppe	Desert	Mountain	Oceanic	Continental	Steppe	Desert	Mountain	Coniferous	Tundra	Mountain	Polar
Non-tropical South America	3	36	36	1	9	32	74	10	64		24	26		50		8				
Tropical South America	665	397	133	9	5	158	46													
Total South America	**668**	**433**	**169**	**10**	**14**	**190**	**120**	**10**	**64**		**24**	**26**		**50**		**8**				
TOTAL WORLD	**1 468**	**1 117**	**755**	**839**	**1 192**	**459**	**471**	**156**	**491**	**674**	**490**	**182**	**726**	**593**	**552**	**729**	**865**	**407**	**632**	**564**

Note: Data derived from an overlay of FRA 2000 global maps of forest cover and ecological zones.

Table 42-2. South America: proportion of forest by ecological zone

Subregion	Forest area as proportion of ecological zone area (percentage)																	
	Tropical						Subtropical				Temperate						Boreal		
	Rain forest	Moist	Dry	Shrub	Desert	Mountain	Humid	Dry	Steppe	Desert	Mountain	Oceanic	Continental	Steppe	Desert	Mountain	Coniferous	Tundra	Mountain	Polar
Non-tropical South America	72	27	75			7	2	89	1		4	29		1		20				
Tropical South America	82	27	89	15		26	19													
Total South America	**82**	**27**	**86**	**13**		**23**	**9**	**89**	**1**		**4**	**29**		**1**		**20**				
TOTAL WORLD	**69**	**31**	**64**	**7**	**0**	**26**	**31**	**45**	**9**	**2**	**20**	**25**	**34**	**4**	**1**	**26**	**66**	**26**	**50**	**2**

Note: Data derived from an overlay of FRA 2000 global maps of forest cover and ecological zones.

marked leaf reduction during the short dry season. The chief families are Bignoniaceae, Bombacaceae, Euphorbiaceae, Moraceae and Sterculiaceae. In Brazil, Leguminosae (*Parkia* spp., *Tachiglia* spp., *Hymenolobium* spp., *Swartzia* spp. and others) are particularly important. In Peru, the most common species include *Bombax munguba, Calycophyllum spruceanum, Castilla ulei* and *Cedrela odorata* while in Venezuela *Calophyllum brasiliense, Carapa guianensis, Cedrela fissilis* and *Ceiba pentandra* are among the dominant species.

Evergreen swamp forest covers large areas in the Amazon region, particularly in the delta. Characteristic species are *Bombax aquaticum, Calophyllum brasiliense, Macrolobium acaciaefolium, Triplaris surinamensis* and many palms, including *Euterpe oleracea, Manicaria saccifera, Mauritiella pacifica* and *Raphia taedigera.*

Mangrove forests are well established in the larger estuaries along the Atlantic and, to a lesser extent, Pacific coasts. The largest mangroves are found in Brazil. From the sea inland is first a belt of *Rhizophora mangle*, then *Avicennia tomentosa* and *A. nitida* and, finally, on higher ground vegetation dominated by *Laguncularia racemosa*, often edged on its landward side by a fringe of palms. Other common trees and shrubs include *Ardisia granatensis, Avicennia tomentosa, Conocarpus erectus, Conostegia polyandra, Rhizophora brevistyla* and *Rustia occidentalis.*

TROPICAL MOIST DECIDUOUS FOREST

This zone roughly corresponds with the Brazilian and Guiana Shields of eastern South America. A wide area with rather high rainfall but a pronounced dry season extends around the wet Amazonian Basin.

This large zone is mainly covered by cerrado in Brazil, a mosaic of grasslands, tree savannahs and woodlands with patches of semi-deciduous forest. The flora is rich, with Leguminosae and Myrtaceae very prevalent in the tree and shrub canopies. The most common species are *Caryocar brasiliense, Curatella americana, Kielmeyera coriacea* and *Qualea* spp. In some areas a real forest occurs, the cerradao – a short semi-deciduous forest, 10 to 15 m tall, of medium density. The flora includes such forest species as *Bowdichia, Hymenaea, Piptadenia inaequalis* and *Machaerium* and also cerrado species. In northern Argentina, around Salta, a similar forest grows on the foothills of the Andes. The higher trees are

Aspidosperma peroba, Astronium spp., *Cedrela fissilis* and *Gallesia gorazema* (guararema).

An evergreen seasonal or semi-deciduous forest grows on the edge of the Amazonian Basin and in the Andean foothills. In Argentina and Paraguay, this fairly dense forest includes three tree canopies, the tallest reaching 30 m. Characteristic trees include *Apuleia leiocarpa, Aspidosperma polyneuron, Balfourodendron riedlianum, Cabralea* spp. and *Cedrela* spp. In Bolivia, *Astronium urundeuva, Ateleia guaraya, Ficus* spp. and *Hura crepitans* are dominant species.

In Venezuela, the flora and physiognomy of the llanos have some similarity with Brazilian cerrados. These are tall grasslands with evergreen broad-leaved trees including *Acacia caven, Celtis spinosa, Prosopis alba* and *P. nigra*. A deciduous thorn forest occurs in some places with *Caesalpinia coriaria, Capparis coccolobifolia, Cercidium praecox, Mimosa* spp., *Piptadenia flava* and other species in addition to the main llanos species.

The zone also includes the grasslands of the Pantanal, those around the junction of the Paraguay and Parana Rivers in Argentina and the residual forest on the low plain of the Cauca River in Columbia.

TROPICAL DRY FOREST

In areas sheltered from the humid trade winds, the climate is drier. These regions may be close to the sea, as in northeastern Brazil and the Caribbean coast, or inland, such as the Argentine chaco. Rainfall varies between 500 and 1 000 mm or less with a dry season of five to eight months. Temperatures are always high near the Equator (mean temperature of the coldest month greater than 20°C) but lower in the chaco, which extends to 34°S.

In Brazil, the typical vegetation is the caatinga, xerophytic vegetation types varying from dense to very open. The trees are more or less deciduous, thin-stemmed and with a low canopy (5 to 10 m). The flora is rich, with fairly numerous Leguminosae, especially *Amburana, Caesalpinia* and *Mimosa* species, and often includes Cactaceae. The palms *Cocos comosa* and *Copernicia cerifera* (carnauba) assume considerable importance in flood plains.

In Argentina, the chaco is a wooded region of relative ecological homogeneity between the tropical and subtropical zones. The prevailing vegetation is deciduous dry forest with many climatic and, above all, edaphic variations. All these types are characterized by "quebrachos" (*Schinopsis* spp. and *Aspidosperma* spp.). The most humid forests occur in the east, a drier forest in the west and xerophilous forest on the lower Andean foothills.

In the coastal region of the Caribbean, deciduous forests and woodlands rich in Leguminosae once occupied a large part of the plain. Agriculture and thickets have largely replaced these forests. Similar woodlands with Cactaceae grow along the Gulf of Guayaquil in Peru and Ecuador.

TROPICAL SHRUBLAND

In addition to the drier parts of the Caribbean coast this zone extends along the Pacific coast of South America from south of the Gulf of Guayaquil to the Tropic of Capricorn, forming a narrow belt between the lower slopes of the Andes and the coastal desert. Rainfall is less than 500 mm, with a long dry season of eight to nine months and high temperatures (always more than 20°C). To the south, in Peru, rainfall is even less than 100 mm, but a light drizzle maintains high humidity and allows some plants to live.

Xeromorphic woodlands are represented by algarrobo, found on the southern coast of the Gulf of Guayaquil, a perennial-leaved woodland dominated by *Prosopis chilensis*. In western Venezuela, a deciduous thorn woodland grows under the same conditions. It is a multilayered woodland 8 to 15 m high with the canopy dominated by *Bulnesia arborea, Capparis* spp., *Pithecellobium unguis-cati, P. saman, Prosopis* spp. and *Pterocarpus* spp.

TROPICAL MOUNTAIN SYSTEMS

The tropical mountains are mainly the Andean Range, extending from northern Colombia and Venezuela to 28° to 29°S. However, some areas in Venezuela and Brazil have similar climatic conditions. The mountain regions experience lower temperatures, leading to specific vegetation types above 1 000 to 1 500 m. Precipitation varies greatly but the region is everywhere tropical, with a low annual range of temperature. Ecofloristic zones can generally be differentiated by altitude.

In the northern Andes (Colombia and Venezuela), both the eastern and western faces of the mountains are well watered. Precipitation ranges from 1 500 to 5 000 mm. The mean temperature of the coldest month is often close to 15°C, but drops down to 10°C or lower with increasing elevation. There is generally no dry season, or a very short one. In some places there is heavy cloud cover and very frequent fog. Frost occurs above 2 000 m.

South of Ecuador there is a contrast between the very wet eastern side of the Andes and the drier Andean valleys and western side. On the eastern face, the climate is similar to that of the northern Andes. In the inter-Andean valleys, even in Colombia and Venezuela, precipitation is 1 000 to 1 500 mm (sometimes less) and the dry season is two to five months. On the western face, in Peru, precipitation is lower (less than 500 mm) and the climate is very dry or semi-arid. In Venezuela, the southern part of the Guiana Shield reaches 1 000 to 3 000 m with a fairly even annual distribution.

Between 1 000 and 1 800 to 2 400 m in the northern Andes many of the lowland taxa still persist, such as species of *Licania* and *Eschweilera*, but a number of distinctly highland elements also enter the lower montane forest. For example, in the Colombian Andes *Alchornea bogotensis, Brunellia comocladifolia* and *Cinchona cuatrecasasii* are present. The montane or upper montane forest, starting at 1 800 to 2 400 m, may extend in places up to 3 400 m. An increasing number of typical montane species enter the flora, for example, *Brunellia occidentalis, Symplocos pichindensis* and *Weinmannia balbisiana*. In the drier parts, montane forests are evergreen seasonal. Above this zone, subalpine forests may extend up to 3 800 m in some places. The characteristic highland flora includes many species of *Befaria, Brunellia, Clusia, Gynoxys, Miconia, Rhamnus* and *Weinmannia*. On the high ridges exposed to wet winds there is montane cloud forest with an "elfin woodland" of low gnarled trees with abundant mosses and lichens.

A unique submontane formation is *Podocarpus* spp. forest, today existing mainly in the lower montane region in northern Peru. The conifer *Podocarpus oleifolius* dominates this forest, where *Drimys winteri, Ocotea architectorum* and *Weinmannia* spp. are also common trees.

In Peru and Bolivia, the wet eastern face of the Andes bears submontane and montane forests similar to those of the northern Andes. In the drier inter-Andean valleys the forest often becomes deciduous, even xerophilous, but often very degraded and transformed into thicket or scrub. On the western slopes of the Andes, under a very dry climate, scrub woodland replaces forest.

In the non-Andean highlands, the submontane level is rather similar to the lowland forest but of lower stature and with a slightly different flora.

SUBTROPICAL HUMID FOREST

This zone includes plateaus and lowlands on the Atlantic side of the continent in southern Brazil, Uruguay and Argentina. The two main climatic characteristics are lower temperatures in winter (mean temperature of the coldest month less than 15°C) and rainfall evenly distributed throughout the year. However, rainfall decreases from the north (1 000 to 2 500 mm) to the south (600 to 1 000 mm).

The natural vegetation of the wetter northern parts of the zone is evergreen coniferous forest dominated by *Araucaria angustifolia*. The *Araucaria* forest, some 25 m tall, may be almost pure, but more often it dominates a dense forest with a profusion of *Cedrela fissilis, Phoebe porosa, Tabebuia* spp., *Parapiptadenia* spp. and the shrub *Ilex paraguariensis*. Today, only residual areas remain, as this forest has been much exploited for timber production.

Grasslands are the main vegetation in Rio Grande do Sul as well as the lowlands of Uruguay and eastern Argentina. Riparian forests fringe the main rivers.

SUBTROPICAL DRY FOREST

This zone of lowlands, less than 200 km wide, lies between the Andes foothills and the Pacific Ocean. The rainfall regime is of the Mediterranean type, with summer drought (two to seven months) and winter rains. Annual precipitation varies from 500 mm in the northern coastal region to 2 000 mm on the Andean foothills. Winter temperatures are cool (10° to 15°C).

The climax is sclerophyllous evergreen forest or woodland with xerophytic species such as *Lithraea caustica, Quillaja saponaria, Peumus boldus* and species of the genera *Cryptocarya* and *Beilschmiedia*. The endemic palm *Jubaea chilensis* grows in a narrow area northeast of Valparaiso. Much of the forest has been degraded and replaced by secondary thorny thicket with *Acacia caven* or replaced by agriculture.

Towards the south or in the Andean foothills, where precipitation is higher, the sclerophyllous forest gives way to open deciduous "mesophytic" forest dominated by various *Nothofagus* species (*N. obliqua, N. dombeyi, N. procera*) associated with *Aextoxicon punctatum, Araucaria araucana, Drimys winteri, Laurelia serrata* and others.

SUBTROPICAL STEPPE

Two regions belong to this ecological zone. One is located to the west of the Andes, covering most of the Chilean Norte Chico and forming a transitional area between the previous zone and the Atacama Desert. The other is to the east of the Andes, an extensive region in central Argentina of transition between the tropical chaco, subtropical pampa and temperate steppes to the south. Rainfall ranges from 100 to 800 mm and the dry period is very long, up to nine months. The mean temperature of the coldest month may be less than 10°C. In Chile, rainfall is even lower, from less than 100 to 400 mm. Temperatures are warmer than in Argentina, with mean temperature of the coldest month between 13° and 15°C.

In this zone the densest vegetation type is a deciduous thicket with various species of *Prosopis*, turning into large areas of thorn woodland. In the drier inland plain is subdesert shrubland with *Bougainvillea* spp., *Cercidium* spp. and various Rhamnaceae. In Chile, *Acacia caven* and *Puya* spp. dominate the subdesert thorn scrub of the Norte Chico.

SUBTROPICAL MOUNTAIN SYSTEMS

The subtropical Andes lie roughly from 26° to 40°S. From 1 000 m to nearly 7 000 m altitude, the climate is cold everywhere. The area is bordered to the west by the highest peaks, forming a barrier against the winds blowing from the Pacific Ocean. As a result, precipitation is low, generally less than 300 mm. The dry season mainly occurs in spring and summer (October-December). Strong winds make the effects of aridity and cold more pronounced.

In the lower reaches of the Andes, between 1 000 m and 1 800 to 2 400 m, we find submontane beech forest on the wetter slopes. It is a deciduous low forest or woodland containing species such as *Nothofagus dombeyi, N. obliqua, N. procera, Aetoxicon punctatum, Araucaria araucana, Drimys winteri, Laurelia serrata* and *Persea lingue*. Drier slopes are covered with evergreen sclerophyllous shrubs or xerophytic deciduous woodland. Higher up, the vegetation changes gradually into a steppe.

TEMPERATE OCEANIC FOREST

South of 38°S the western side of the Andes is well watered owing to oceanic influences. The dryness decreases from north to south, together with decreasing temperatures. Rainfall ranges from 1 000 to 3 500 mm, evenly distributed throughout the year. The mean temperature of the coldest month is lower than 10°C in the north and decreases to about 0°C in the south. In eastern Patagonia, rainfall is less than 1 000 mm with mean monthly temperatures always lower than 10°C.

The northern part of the region harbours a broad-leaved, very dense evergreen forest up to 40 to 45 m tall, with equally dense undergrowth. Species of *Nothofagus* dominate the tree canopy, including *Nothofagus obliqua, N. dombeyi* and *N. procera* in association with *Aextoxicon punctatum, Drimys winteri* and *Eucryphia cordifolia*. A slight lowering of temperature at higher altitude or latitude gives rise to a less species-rich, mixed broad-leaved/coniferous forest with *Nothofagus antarctica, N. dombeyi, N. nitida, Fitzroya cupressoides, Pilgerodendron uvifera* and *Podocarpus nubigena*.

TEMPERATE MOUNTAIN SYSTEMS

The central part of the Patagonian Andes, up to 52°S, reaches 2 000 to 3 000 m elevation. The western upper slopes are wet, whereas the eastern side is drier. The most striking climatic features are cold, snow and winds.

Subalpine beech forest, dominated by *Nothofagus betuloides*, lies below the timberline on the wettest slopes. This elfin type has low multistemmed trees, greatly deformed by the weight of snow. These forests are transitional to scrub and grasslands at higher altitudes. On the drier slopes and towards the eastern drier zone a beech forest of *Nothofagus betuloides* and *N. pumilio* occurs. It is transitional between the purely evergreen lowland forests and the deciduous *N. pumilio* forests that lie below the timberline on the drier sites.

BIBLIOGRAPHY

Cabrera, A. 1971. *Fitogeografia de la República Argentina*. Boletín de la Sociedad Argentina de Botánica, vol. XIV.

Cuatrecasas, J. 1934. *Observaciones geobotánicas en Colombia.* Trabajos del Museo Nacional de Ciencias Naturales, Serie botánica n° 27. Madrid.

Cuatrecasas, J. 1958. Aspectos de la vegetación natural de Colombia. *Revista de la Academia Colombiana de Ciencias Físicas y Naturales*, Vol. 10.

Ecological Laboratory of Toulouse (LET). 2000. *Ecofloristic zones and global ecological zoning of Africa, South America and tropical Asia.* Prepared for FAO-FRA 2000 by M.F. Bellan. Toulouse, France.

Espinal, L.S. & Montenegro, E.M. 1963. *Formaciones vegetales de Colombia. Memoria explicativa sobre el mapa ecológico.* Departamento Agrológico, Instituto Geográfico "Agustin Codazzi". Map 1:1 000 000. Bogotá.

Ferreyra, H.R. 1960. *Algunos aspectos fitogeográficos del Perú.* Publicaciones del Instituto de Geografía, Facultad de Letras, Universidad Nacional Mayor de San Marcos, Serie I: Monografías y ensayos geográficos, N° 3. Lima.

Ferreyra, H.R. 1972. *Protección del medio ambiente y de los recursos naturales en Perú.* Simposio internacional sobre la protección del medio ambiente y de los recursos naturales. Instituto italo-latinoamericano en colaboración con el Consejo Nacional de Ciencias y Tecnología, IILA. Rome.

Koecklin, J. 1968. *Végétation et mise en valeur dans le sud du Mato Grosso.* Travaux du Centre d'études de géographie tropicale (CEGET), Bordeaux-Talence, France.

Prance, G.T. 1989. American tropical forests. In *Tropical rain forest ecosystems: biogeographical and ecological studies.* Eds. H. Lieth & M.J.A. Werger. Ecosystems of the World 14b. Amsterdam, the Netherlands, Elsevier.

Rizzini, C.T. 1963. Nota brevia sobre a divisao fitogeografica (floristico-sociologica) do Brasil. *Revista Brasileira de Geografia*, 25(1).

Tosi, J.A. 1960. *Zonas de vida natural en el Perú.* Memoria explicativa sobre el mapa ecológico del Perú. Boletín técnico n° 5. Lima, Instituto Interamericano de Ciencias Agrícolas, zona Andina.

UNESCO. 1981. *Vegetation map of South America – Explanatory notes.* Natural Resources Research, XVII. Paris, Les presses de l'UNESCO.

Walter, H. 1973. *Vegetation of the earth – in relation to climate and the eco-physiological conditions.* Heidelberg Science Library. New York, Springer-Verlag.

Weberbauer, A. 1945. *El mundo vegetal de los Andes peruanos; estudio fitogeográfico.* Estación experimental agrícola de La Molina. Lima.

Chapter 43
Tropical South America

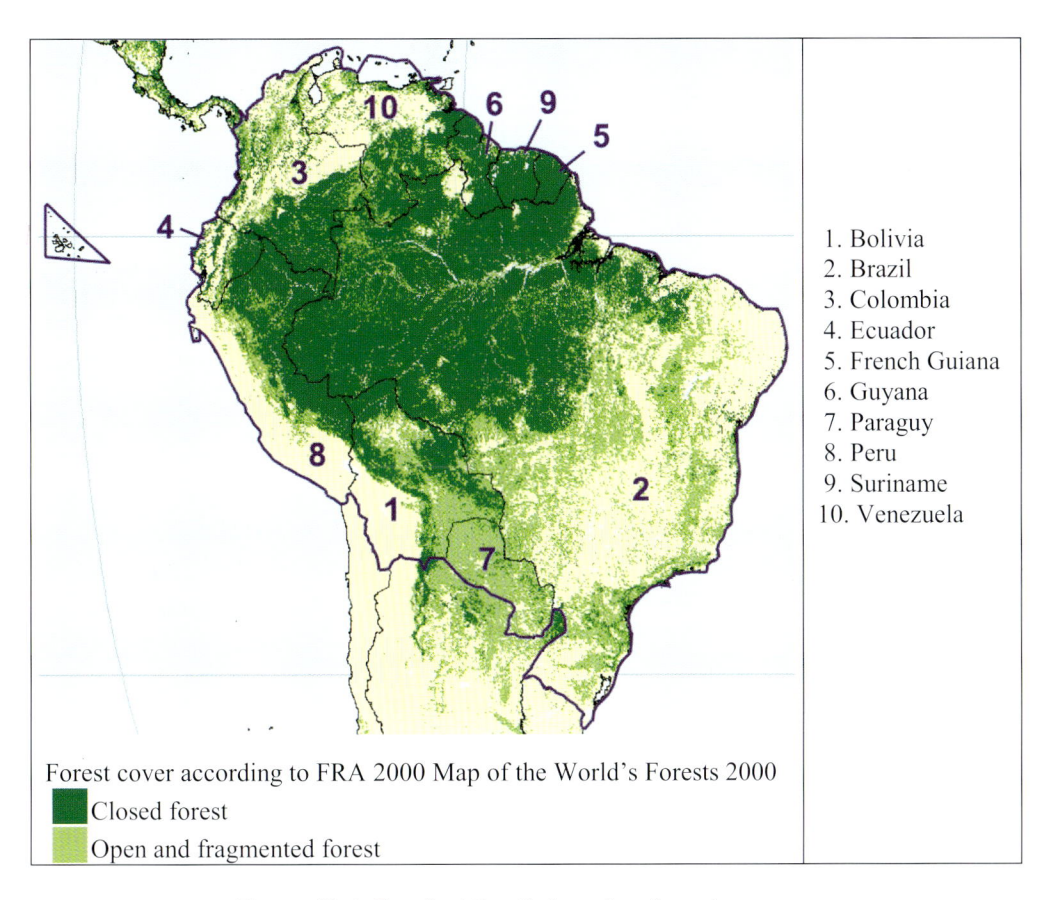

Figure 43-1. Tropical South America: forest cover map

Forest cover according to FRA 2000 Map of the World's Forests 2000
- Closed forest
- Open and fragmented forest

1. Bolivia
2. Brazil
3. Colombia
4. Ecuador
5. French Guiana
6. Guyana
7. Paraguy
8. Peru
9. Suriname
10. Venezuela

The tropical South America subregion,[54] comprising Colombia, French Guiana, Suriname, Guyana, Venezuela, Ecuador, Peru, Bolivia, Paraguay and Brazil, represents the greatest concentration of tropical rain forest in the world, with approximately 885 million hectares in the Amazon Basin and another 85 million hectares in the Orinoco and Paraná watershed complex. The total land area of tropical South America is 1 387 million hectares (Figure 43-1, Table 43-1).

The Amazonian tropical rain forest is considered to be the world's richest ecosystem in terms of biodiversity. By country, Brazil ranks first, Colombia fourth and Peru seventh. This ecozone accounts for 85 percent of the total forest cover and approximately 60 percent of the total land cover of the subregion, playing a very important role in the economic as well as the environmental context of these countries. However, climates and associated forest types vary from arid and semi-arid to pluvial. The dominant ecological zone is the tropical rain forest, representing 36 percent of the total area, followed by tropical moist deciduous forest with 24 percent, tropical mountain forest with 10 percent and tropical dry forest with 9.5 percent. In the northern part of the subregion the llanos in Venezuela and Colombia are typical open subhumid forests, as is the cerrado in the central-west part of Brazil. The certão or caatinga in the Brazilian northeast is a typical semi-arid ecosystem, as are the Paraguayan chaco and the dry forest formations along the Peruvian Pacific littoral.

The tropical rain forest of the Amazon Basin starts in the Andes chain in Bolivia, Peru, Ecuador, Colombia and Venezuela at more than 3 000 m elevation. It borders the immense

---

[54] For more details by country, see www.fao.org/forestry

Table 43-1. Tropical South America: forest resources and management

Country/area	Land area	Forest area 2000					Area change 1990-2000 (total forest)		Volume and above-ground biomass (total forest)		Forest under management plan	
		Natural forest	Forest plantation	Total forest								
	000 ha	000 ha	000 ha	000 ha	%	ha/capita	000 ha/year	%	m³/ha	t/ha	000 ha	%
Bolivia	108 438	53 022	46	53 068	48.9	6.5	-161	-0.3	114	183	6 900	13
Brazil	845 651	538 924	4 982	543 905	64.3	3.2	-2 309	-0.4	131	209	4 000	1
Colombia	103 871	49 460	141	49 601	47.8	1.2	-190	-0.4	108	196	85	0
Ecuador	27 684	10 390	167	10 557	38.1	0.9	-137	-1.2	121	151	14	0
French Guiana	8 815	7 925	1	7 926	89.9	45.6	n.s.	n.s.	145	253	400	5
Guyana	21 498	16 867	12	16 879	78.5	19.7	-49	-0.3	145	253	4 200	25
Paraguay	39 730	23 345	27	23 372	58.8	4.4	-123	-0.5	34	59	3 000	13
Peru	128 000	64 575	640	65 215	50.9	2.6	-269	-0.4	158	245	1 573	2
Suriname	15 600	14 100	13	14 113	90.5	34.0	n.s.	n.s.	145	253	1 568	11
Venezuela	88 206	48 643	863	49 506	56.1	2.1	-218	-0.4	134	233	3 970	8
Total Tropical South America	1 387 493	827 252	6 890	834 142	60.1	2.9	-3 456	-0.4	129	208	-	-
Total South America	1 754 741	875 163	10 455	885 618	50.5	2.6	-3 711	-0.4	125	203	-	-
TOTAL WORLD	13 063 900	3 682 722	186 733	3 869 455	29.6	0.6	-9 391	-0.2	100	109	-	-

Source: Appendix 3, Tables 3, 4, 6, 7 and 9.

Amazonian plain, mostly inside Brazil, and has a strong ecological and socio-economic relationship with the low parts of the basin. The contribution of the forest resource to the national economies of the subregion is still very low, providing less than 2 percent of GNP, except for Brazil where it is estimated to be 5 percent. Nevertheless, in the informal economy, particularly in rural and native settlements, forests play a crucial role, furnishing the main source of the population's livelihood, including food, water, housing materials and other forest products (FAO 1989).

Forest resources have experienced serious deforestation and degradation during the last four or five decades. Deforestation started in the highest part of the Amazon Basin in Peru, Bolivia and Colombia and spread to the lower part. In Brazil, it started on the border of the Amazonian region in the northeast and southeast and rapidly progressed to the north and northwest, following the Trans-Amazon Highway and main river courses. The occupation of the tropical rain forest by immigrant populations began with rubber exploitation at the beginning of the twentieth century, then progressed to coffee, cacao and oil palm plantations, oil exploration and exploitation and large cattle ranches, especially in the Brazilian cerrado, Venezuelan llanos and Paraguayan chaco. Spontaneous or government-sponsored colonization by landless people during the 1950s and 1960s continued the deforestation process.

FOREST RESOURCES

Tropical South America has 79 percent of the total land, 95 percent of the population, 94 percent of the natural forest and 65 percent of the plantations of South America. *Vis-à-vis* the world it has 10 percent of the total land, 5 percent of the population, 21.5 percent of the natural forest and 3 percent of the plantations. The smallest country in terms of forest cover is French Guiana and the largest is Brazil, accounting for 0.9 percent and 65 percent of the subregion, respectively. The largest forest area per capita belongs to French Guiana, Guyana and Suriname, with 45, 34 and 19 ha per capita, respectively. The lowest are Ecuador, Colombia and Venezuela with 0.8, 1.19 and 2.0 ha per person, respectively. Bolivia, Paraguay, Brazil and Peru are in between, with 6.5, 4.4, 3.2 and 2.6 ha per person, respectively (FAO 2000) (Figure 43-2, Table 43-1).

Peru has the second largest area of tropical rain forest cover in the subregion, after Brazil, but a significant percentage of this area is located in the foothills of the Andes where the Amazon Basin begins. Ecuador, Bolivia and Colombia also have a similar pattern. Most of the population is in the Andes region but there is constant and increasing migration to the low plains in search of new land for cultivation and grazing. The other forest types previously mentioned have been subject to pressure for a long time and their area has already been significantly reduced.

The average annual deforestation rate in the subregion is approximately 0.4 percent, ranging from 0.3 percent in Guyana and Bolivia to 1.2 percent in Ecuador. Brazil, Peru, Colombia and Venezuela have 0.4 percent. While this probably results in large areas of secondary forest (mostly short fallow), this is not reflected in the forest cover or vegetation cover statistics reported by the countries. General estimates show that there are more than 60 million hectares of secondary forest in the Amazon Basin (FAO 1989).

Although deforestation rates are high in the dense tropical forest, they are still higher in the tropical moist deciduous formations, such as in the Brazilian northeastern and central-eastern regions, the Venezuelan and Colombian llanos and the Bolivian and Peruvian tropical mountain systems. Logging or exhaustive exploitation of some high-value species contributes to degradation and loss of value and biodiversity when concentrated on a few high-value species. However, it cannot be blamed for the entire deforestation process, as small farmers commonly follow in the tracks of logging, establishing new agricultural and grazing settlements.

Wood volume per hectare is high compared to other forest regions, but commercial volume (high-value species) is, in general, less than 10 percent of the total volume, which averages about 120 m^3 per hectare (trees larger than 30 cm DBH). Although forest inventories usually include all species, with subsamples for natural regeneration purposes (>10 cm DBH), the tables for standing volume only report volume outside bark for trees above 25 or 30 cm DBH as commercial volume. Thus, biomass estimates for the Amazon are calculated using expansion models, in the majority of the cases resulting in estimates of more than 200 tonnes per hectare (FAO 1997).

The total area of forest plantations in South America is approximately 10.6 million hectares, with about 7.0 million hectares in this subregion, of which 70 percent belongs to Brazil. The main species planted in these countries for pulp and paper, timber and fuelwood are *Eucalyptus* spp. and *Pinus* spp. Scarcity of fuelwood in the highlands has led to increased interest in reforestation but plantations are mostly located far away from the ecozones where deforestation occurs and most plantations use exotic species. For example, Peru, Ecuador, Colombia and Venezuela are reforesting or afforesting the highlands or semi-dry plains and Brazil the subtropical and temperate areas of the southern regions. Brazilian industries seem to be much more interested in plantations to supply their pulp and paper, plywood and furniture plants than logging natural forest. However, medium and large sawmills are interested in high-value species coming from the natural forest. In the other countries, where large pulp and paper industries are not established, selective exploitation of the natural forest for high- and medium-value species will continue to be the main activity in the medium and long term, but with more value added through secondary manufacturing processes.

Forest fire is a very important issue in the subregion. Even though extensive forest fires do not affect large areas, they are a problem in the dry, semi-dry and open forest formations in northeastern Brazil, northern Colombia and Venezuela, the chaco formations in Bolivia and Paraguay, the dry forest in the northern part of Peru and the mountain deciduous forest formations. Unfortunately, very little is known about these fires in terms of numbers and affected areas in these countries. Slash-and-burn practices, used to clear the forest to establish agriculture and grazing, are the main problem in all of the countries. Extensive areas, equal to the area deforested annually (3.5 million hectares), are burned every year in the subregion, producing huge emissions of carbon to the atmosphere, estimated to be 80 to 100 tonnes of carbon emitted per hectare in the form of CO_2 (Fernside 1997).

Brazil has implemented an early warning system for forest fires in the Amazon region, differentiating forest fires from *queimadas* (slash and burn). Other countries, such as Peru, Venezuela, Colombia, Ecuador and Bolivia, are implementing statistical databases on forest fires. French Guiana, Guyana and Suriname are much less affected by forest fires owing to the predominance of tropical rain forest (IBAMA 2001).

FOREST MANAGEMENT AND USES

Management of tropical rain forests has always been considered an extremely difficult task, owing to the complex ecological ecosystems of the tropics and the lack of control and consistent action plans implemented by the governments. However, with more and more international markets for tropical wood demanding that it should come from forests under management, the governments and the private sector are being pushed to implement sustainable forest management as extensively as possible. In the subregion, Bolivia, Paraguay and Brazil have started intensive programmes to establish management plans for timber production.

All countries in tropical South America have information on the size of the forest area subject to a formal management plan (Table 43-1). Most countries included only natural forests in their reporting to the meeting of the FAO Forestry Commission for Latin America and the Caribbean in 2000, and Guyana, Suriname and Venezuela only included production forests or areas under concession agreements. The area subject to a forest management plan varied between 0.1 and 25 percent of the total forest area in each country. For the subregion as a whole, approximately 26 million hectares, or 3 percent, of the total forest area was reportedly subject to a formal management plan. This figure may seem low. However, it should be kept in mind that many countries in this subregion have large expanses of forests which are located in remote areas with lack of access or with very limited human intervention and which may not require a management plan. It is also uncertain whether all countries included protected forest areas in their reporting on areas covered by forest management plans. A recent ITTO study (Poore and Thang 2000) thus reported that Guyana is one of only six ITTO tropical producer countries which appeared to have established all the conditions that make it likely that they can manage their forest management units sustainably.

Protected areas have significantly increased during the last decade. In 1990, less than 10 percent of the forest cover was estimated to be protected, while in 2000 this area is estimated to have increased to approximately 14 percent of the subregion. Bolivia has the greatest proportion of protected forest area, 31 percent, while Guyana has 25 percent, Colombia 24 percent, Ecuador 20 percent, Brazil 17 percent, Suriname 11 percent, Peru 10 percent and the other countries 5 percent or less (FAO 2000).

Although all countries' legislation obliges forest owners or concessionaires to implement management plans, forestry administrations do not have enough resources and efficient organizations to control the hundreds or thousands of properties and concessions, spread over immense areas, often with poor accessibility (FAO 2000).

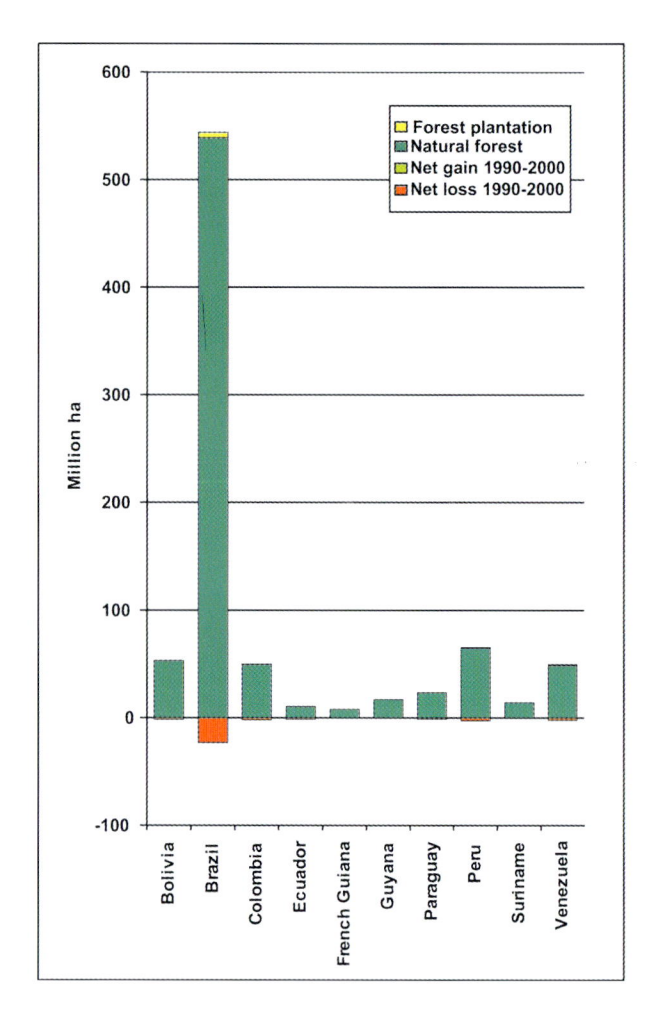

Figure 43-2. South America: natural forest and forest plantation areas 2000 and net area changes 1990-2000

Ownership is one of the main issues related to forest management. The majority of the countries do not recognize private property rights on forest land (i.e. Bolivia, Peru, Venezuela and Colombia) and logging concessions under management plans are the normal way to accede forest resources. Brazil, on the other hand, allows the private ownership of forest land; approximately 80 percent of forest land is already in private hands, regulated by a Forestry Code that established the norms for forest management and land use change. In the latter case, the owner can be authorized to clear a maximum of 20 percent of the forest cover for conversion to agricultural land (FAO 2000).

Informal use of the forest, either for logging or resulting in a change of land use is, without doubt, the main problem that all governments in the subregion have to face. Almost open access to the forestry domain facilitates encroachment on the natural forest. These practices are extremely difficult to control or stop owing to a severe lack of resources and weak institutional capacity.

The lack of current and reliable information about forest resources makes consistent planning and efficient use of natural resources difficult. Only a few countries have adequate systems for data collection and analysis. Field forest inventories are increasingly rare or limited to small areas in which the private sector is interested. Forestry or vegetation maps are not prepared following standardized methods, classification systems, scales, etc. Only Brazil has systematically monitored deforestation in the Amazon region and provides relevant and reliable information on a yearly basis.

CONCLUSIONS AND ISSUES

Forest cover in tropical South America is still very important in terms of percentage of the total land area. Suriname, French Guiana and Guyana have the highest percentage of forest cover, with 80 percent or more of their total land in forests. Brazil's forests cover 64 percent of its land area, but in the Amazon region the percentage is much higher, approximately 85 percent. Other countries are below 60 or 50 percent. However, forest change is often concentrated in some particular ecological zones, which can be subject to severe deforestation or degradation.

In all countries, deforestation is the main problem facing the forestry sector. Although deforestation rates seem to have slowed down, it is not yet possible to establish a constant or clear trend over time. Cultural and socio-economic problems in these countries will have a very strong influence in increasing or reducing deforestation rates. More stable and consistent policies and administration of natural resources can contribute to a positive trend, but lack of alternative income sources and extreme poverty will continue to provide the incentive to clear forests for agricultural purposes.

Special ecological zones such as wetlands, coastal forest formations, highland forests and dry or semi-dry forest are under much higher pressure from deforestation and are disappearing more rapidly than humid forests. National and subregional plans and strategies must take this issue into account.

Data and information systems related to forest resources are, in general, very poor. Countries need strong support in the short and medium term to improve data collection and analysis to provide information for decision-makers, stakeholders, researchers and teachers to help achieve sustainable forest management.

BIBLIOGRAPHY

FAO. 1989. *La deforestación en Latino América: orígenes, causas y efectos,* by J. Malleux. Monitoreo de los procesos de deforestación en bosques húmedos tropicales. Proyecto manejo de recursos forestales naturales en América Latina. Lima.

FAO. 1993. *Forest Resources Assessment 1990. Tropical countries.* FAO Forestry Paper No. 112. Rome.

FAO. 1997. *State of the World's Forests 1997.* Rome.

FAO. 2000. *Informes nacionales de los países.* Latin American and Caribbean Forestry Commission (LACFC). Bogota.

Fernside, O.M. 1997. Greenhouse gases from deforestation in Brazilian Amazonia: net committed emissions. *Climatic Change,* 35: 321-360.

Instituto Brasileiro do Meio Ambiente e dos Recursos Naturais Renováveis (IBAMA). 2001. Home page. www.ibama.gov.br

Instituto Geográfico "Agustin Codazzi" (IGAC). 1974. *La colonización de la selva pluvial en el piedemonte Amazónico de Colombia.* Bogota.

Poore, D. & Thang, H.C. 2000. *Review of progress towards the year 2000 objective.* Report presented at the 28th Session of the International Tropical Timber Council ITTC(XXVIII)/9/Rev. 2, 24-30 May 2000, Lima. Yokohama, Japan, ITTO.

Chapter 44
Non-tropical South America

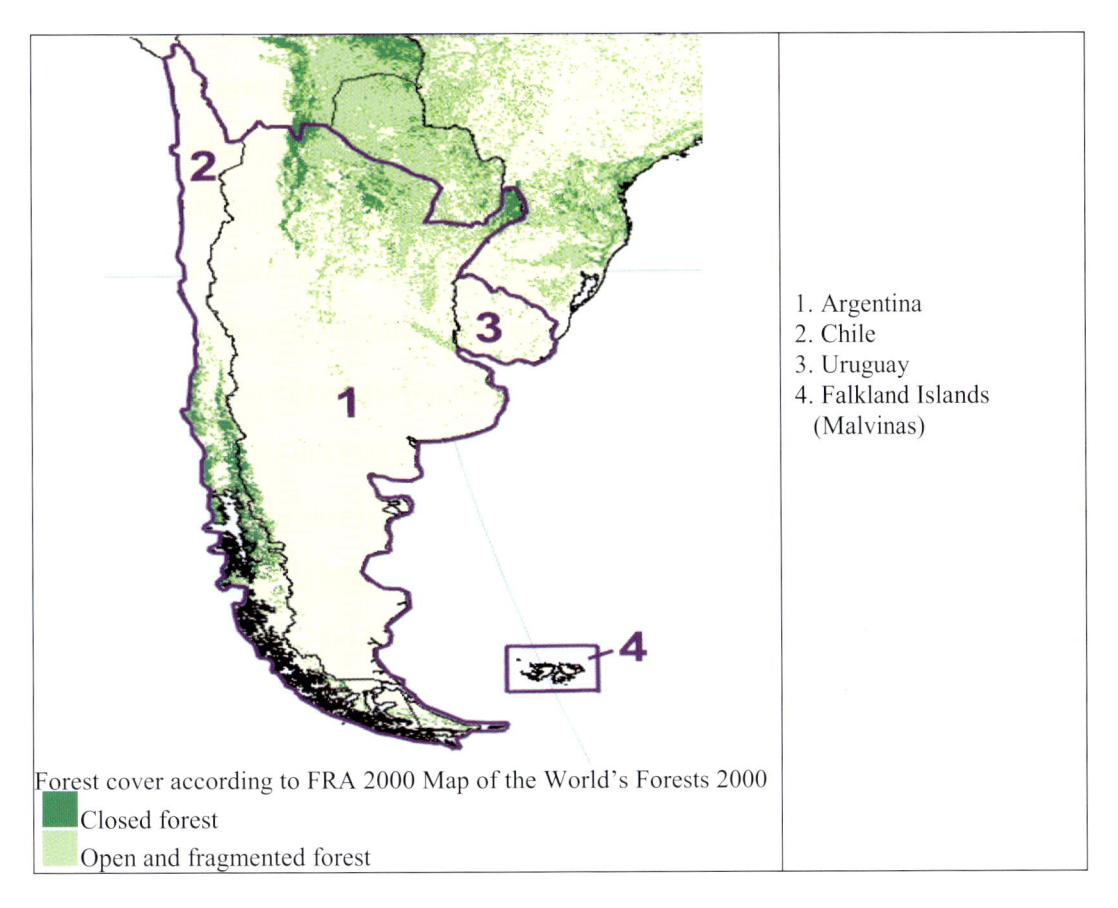

Figure 44-1. Non-tropical South America: forest cover map

Only three countries, Argentina, Chile and Uruguay,[55] are part of the temperate so-called "southern cone" of South America. Falkland Islands (Malvinas) also fall within the subregion. Argentina is the largest country with a total land area of 2.73 million square kilometres, followed by Chile with 0.74 million square kilometres and Uruguay with 0.17 million square kilometres. The total land area is 3.66 million square kilometres. This region of South America, located on and south of the Tropic of Capricorn, is considered to have subtropical and temperate ecosystems. The climate of most of the subregion has four well-defined seasons, with the cool winters and hot summers typical of the temperate zone. The southern part of Chile and Argentina, across the Straight of Magellan, belongs to the Antarctic zone with large areas covered by permanent snow or ice (Figure 44-1).

Several coniferous and broadleaf tree species typical of temperate zones are part of the native flora, but intensively managed forest plantations established in the last three decades are replacing the natural forest formations. Human activity since colonial settlement, including huge ranches and farms established since the beginning of the twentieth century, has also changed the landscape of large areas originally covered by natural forest and shrub formations. Temperate steppe, mainly located in the southern part of Argentina, is the dominant ecozone of that country. Temperate oceanic forest is very important in the central-southern part of Chile. Some of the southern areas in Chile and Argentina are part of the polar forest ecozone. The northern parts of Argentina, Chile and Uruguay are covered by subtropical forest and shrub formations.

---

[55] For more details by country, see www.fao.org/forestry

In Argentina, four different climates and associated forest formations can be identified. The first is the temperate mountains of the southern Andes, very mountainous and cold. The second is the chaco formation on the border with Paraguay and Bolivia (a typically semi-arid subtropical zone). The so-called pampas is a very flat and treeless zone in the central part of the country where most of the big cattle ranches are located. The Patagonia zone has desolate steppes and poor soils.

Chile has three contrasting ecozones: in the northern part the Atacama Desert; in the central part the temperate southern Andes; and, in the southern part, boreal climes covered by boreal forest.

Grass prairies dominate the central and southern parts of Uruguay while the northern part is predominately covered by the so-called *serranias* with low mountains or hills. Almost 90 percent of the area of the country is occupied by agriculture or cattle farms.

FOREST RESOURCES

The estimated total forest cover (forest and other wooded land) in 1990 was 68 453 000 ha, of which 43 283 000 ha were in forest (dense and open), representing approximately 12 percent of the total land. In 2000, the total forest cover was estimated at 51 476 000 ha. The subregion's natural forest cover represents approximately 5 percent of the total forest of South America and 0.85 percent of the world. It comprises 10 percent of the world's temperate forests, but natural forest is no longer the main source for timber in this part of the world where, as mentioned above, forest plantations are rapidly replacing the native forests. Nevertheless, in the poor rural areas of Chile and Argentina fuelwood from the natural woody vegetation still supplies about 35 to 50 percent of the energy consumed as fuelwood. The estimated forest plantations for the subregion total 3 575 000 ha. Chile has the largest plantation area, more than 2 million hectares (Figure 44-2, Table 44-1).

The forestry sector in Chile contibutes more than 10 percent of the GNP. Exports have totalled about US$2 billion per year during the last three years, mainly pulp, paper and sawn wood (pine). The national forestry policy in Chile aims to integrate forest plantations and management of native forests into productive systems as part of the natural patrimony of the country (FAO 2000).

Argentina is seriously concerned about the situation of its natural forest cover, which has been reduced to less than 13 percent of the total land area. A national forest inventory is being carried out to evaluate the situation fully and final results will be available by mid-2001. Forest plantations are receiving very strong support from the government through the recently approved Law No. 25.080 (FAO 2000) as well as the desire to reverse the current trade deficit in forest products (US$1 billion in 1999).

Uruguay has the smallest amount of forest cover in South America, only 5.72 percent of its total land area. The current forestry policy in Uruguay is similar to that of Argentina, i.e. to preserve natural forests while enlarging the reforested area of the country. In both cases, the government is providing some economic incentives or subsidies, especially for reforestation (FAO 2000).

Plantations are definitely the most important forestry activity in these three countries, supplying more than 90 percent of the wood for local consumption and export. Chile almost balances annual deforestation of natural forests with plantations, but plantations are established using exotic species such as *Eucalyptus* spp. and

Table 44-1. Non-tropical South America: forest resources and management

| Country/area | Land area | Forest area 2000 ||||| Area change 1990-2000 (total forest) || Volume and above-ground biomass (total forest) || Forest under management plan ||
| | | Natural forest | Forest plan-tation | Total forest ||||||||
	000 ha	000 ha	000 ha	000 ha	%	ha/capita	000 ha/year	%	m³/ha	t/ha	000 ha	%
Argentina	273 669	33 722	926	34 648	12.7	0.9	-285	-0.8	25	68	-	-
Chile	74 881	13 519	2 017	15 536	20.7	1.0	-20	-0.1	160	268	-	-
Uruguay	17 481	670	622	1 292	7.4	0.4	50	5.0	-	-	99	8
Falkland Islands (Malvinas)	1 217	-	-	-	-	-	-	-	-	-	-	-
Total non-tropical South America	367 248	47 911	3 565	51 476	14.0	0.9	-255	-0.5	67	130	-	-
Total South America	1 754 741	875 163	10 455	885 618	50.5	2.6	-3 711	-0.4	125	203	-	-
TOTAL WORLD	13 063 900	3 682 722	186 733	3 869 455	29.6	0.6	-9 391	-0.2	100	109	-	-

Source: Appendix 3, Tables 3, 4, 6, 7 and 9.

Pinus radiata, comprising 17 and 83 percent of the total planted area, respectively. Approximately the same percentage applies to Uruguay. Regarding Argentina, 50 percent of the plantations are conifers, 30 percent *Eucalyptus* spp., 16 percent *Salix* spp. and *Populus* spp. and 4 percent others. These three countries all have economic or fiscal incentives for plantations, subsidizing part of the cost of reforestation, pruning and thinning. Between 1992 and 1998, 220 000 ha of forest were planted, providing 35 000 new jobs and representing an investment of more than US$1.2 billion in the industrial forestry sector. When high-value and native species are considered in the reforestation plan, subsidies are granted with an additional 20 percent.

Forest biomass per hectare is significantly lower than plantations, averaging 60 tonnes per hectare, while plantations are above 120 tonnes per hectare.

FOREST MANAGEMENT AND USES

Uruguay was the only country in non-tropical South America to provide national-level information in the form of the area of natural forest under management, as reported at the year 2000 Meeting of the FAO Forestry Commission for Latin America and the Caribbean (Table 44-1). Uruguay reported that 99 000 ha, equivalent to 8 percent of its total forest area in 2000, were under management. Significant efforts have, however, begun to establish the framework for field-level implementation of sustainable forest management practices in the subregion and almost all the planted areas are considered to be under management monitored by the Forest Services.

Chile is initiating a very interesting programme for the management of natural forests. The plan foresees the establishment of 35 000 ha of natural forests under sustainable management within five years, starting in 1998. However, by the end of 1999, more than 30 000 ha and 640 management plans had been approved, representing 86 percent of the target. A special project called Bosque Modelo Chiloe was initiated in 1998 as a pilot project to promote forest conservation and sustainable use of natural forests and associated ecosystems. After finishing the national forest inventory, financed by the World Bank, Argentina is seriously considering the implementation of a programme for sustainable management of natural forests.

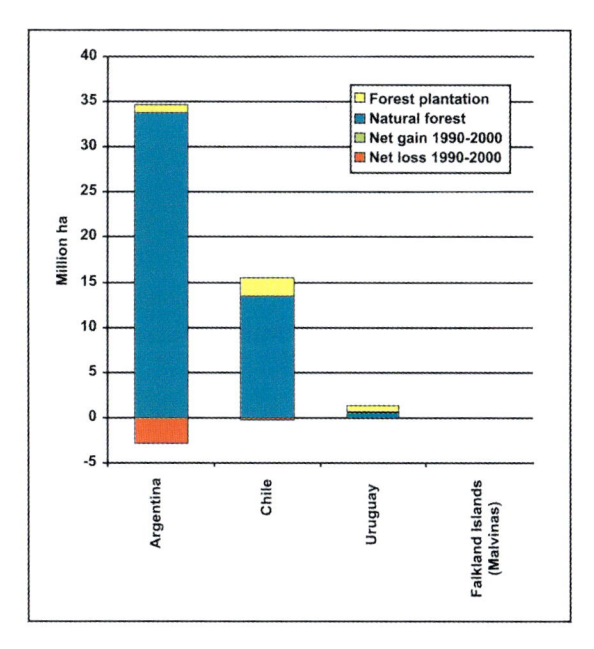

Figure 44-2. Non-tropical South America: natural forest and forest plantation areas 2000 and net area changes 1990-2000

Uruguay is planning to incorporate 20 percent of the total natural forest area under sustainable management in the medium-term plan. During 1998-1999, more than 16 000 ha were placed under sustainable forest management. The second phase of forestry mapping using TM satellite imagery is nearly finished (INFOR 1992).

Another common characteristic of these countries is the ownership framework. Practically all the land is privately owned, including natural forests. Only national reserves and equivalent units belong to the State or are in the public domain.

Argentina will finish its first national inventory of forest plantations, together with a national inventory of natural forests formations, in 2001. Chile recently completed the National Forestry Cadastro, which gives very detailed information about natural and planted forests for the whole country by province, region, forest type, species, size, etc. (Universidad Austral de Chile *et al.* 1999).

Forest fire is a problem. In Argentina, more than 2 000 fires burned 171 277 ha during the period 1997 to 1998. The most seriously affected areas were natural formations (57 percent shrub-grass areas, 41 percent native forest and 2 percent plantations). Plantations of *Populus* and *Salix* species were also strongly attacked by insect pests. In Chile, more than 84 000 ha of natural forests were affected by fire in 1998 and more than 56 000 ha in 1999. There are no reports about forest fires in plantations. There have been no reported problems with forest fires or pests in Uruguay during the last few years (FAO 2000).

When soil conservation activities are not the main objective, conversion of natural or native forest to agricultural land, in these three countries, requires a special authorization from the forestry authority. In Chile, more than 10 000 ha were authorized for land use change from natural forest cover to agriculture during the last two years (1998 and 1999).

Total removals per year are more than 50 million cubic metres, mostly from plantations. Chile is cutting more than 15 million cubic metres and producing 4.5 million cubic metres of sawnwood, 2.2 million cubic metres of pulp and 0.64 million cubic metres of paper and cartons. Argentina is cutting approximately 10 million cubic metres of wood from forest plantations and producing 0.87 million cubic metres of sawn wood, 0.75 million cubic metres of plywood, 0.72 million cubic metres of pulp and, 0.98 million cubic metres of cartons. The total annual harvest in Uruguay is 1.1 million cubic metres of industrial wood and 1.7 million cubic metres of fuelwood.

CONCLUSIONS AND ISSUES

The three countries in the temperate zone of South America have been very active in updating their forestry policy framework, especially in regard to the promotion of reforestation through economic incentives or subsidies to the private sector. This has resulted in the planting of more than 150 000 ha of new forests per year during the last decade. More than 90 percent of the annual removals and wood supply comes from planted forests, except fuelwood, which mostly comes from natural formations (FAO 2000).

Natural forest formations were drastically reduced during and after the colonial period, especially since the beginning of the twentieth century. Large areas of forested land, particularly in Argentina and Uruguay, were converted to agriculture and pasture. In Chile, conversion of native forest to agricultural land was prevalent and, in some cases, natural forest was partially replaced by forestry plantations. According to the newly adopted policies and legal frameworks in these countries, deforestation of natural forest is to be stopped or drastically reduced. High priority is to be given to the preservation of natural areas and remaining forest cover. Detailed forest inventories are proposed at the national level, both for natural forests and plantations. Sustainable management plans are to be implemented together with the private sector and with the active participation of local people and farmers. According to this new policy, forestry activities should be incorporated into agricultural systems (FAO 2000).

In terms of percentages, the deforestation rate is very low (Chile) or reforestation is higher than deforestation (Uruguay). This, however, is due to the fact that natural forests have been reduced to a minimal amount and forest plantations have become the primary activity in the forestry sector during the last three or four decades (FAO 2001).

The recent efforts by Argentina, Chile and Uruguay to evaluate their forestry resources at the national level and the development of sustainable management plans demonstrate the serious concern of these countries for the state of their forests. The adoption of criteria and indicators following the Montreal Process is also of high priority, together with strengthening the conservation of natural forests and ecosystems.

BIBLIOGRAPHY

FAO. 2000. *National reports of Argentina, Chile and Uruguay.* Latin American and Caribbean Forestry Commission (LACFC), 21st session, September 2000, Bogota.

FAO. 2001. *Forest cover assessment in the Argentinean regions of monte and espinal,* by D. Altrell. Rome.

FAO. In press. *Causas y tendencias de la deforestación en América Latina,* by J. Malleux. Rome.

Instituto Forestal de Chile. INFOR. 1992. *El sector forestal en Chile.* Santiago.

Universidad Austral de Chile & Universidad Católica de Temuco. 1999. *Catastro y evaluación de recursos vegetacionales nativos de Chile.* Santiago, Corporación Nacional Forestal (CONAF)-Comisión Nacional del Medio Ambiente (CONAMA).

Part III: Processes and methodologies

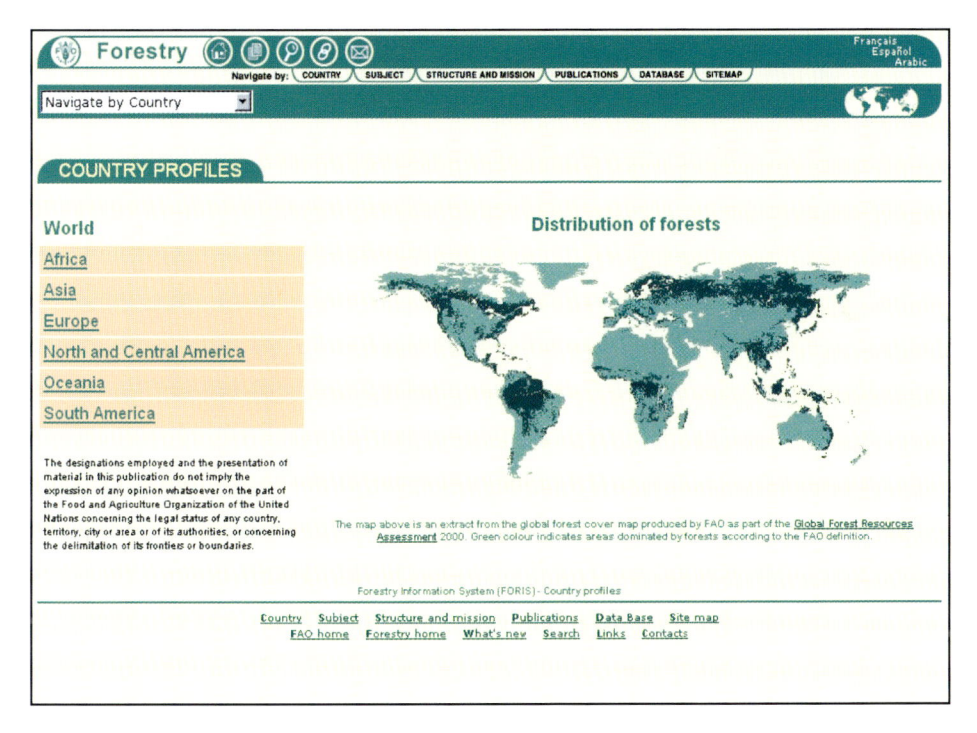

Navigation page to FAO Forestry country profiles
www.fao.org/forestry/fo/country/nav_world.jsp

Chapter 45
Framework for implementation and country participation

ABSTRACT

FRA 2000 was developed according to the guidance of major United Nations policy fora, including the United Nations Conference on the Environment and Development (UNCED), held in 1992, and especially its Agenda 21. In 1997, the FAO Committee on Forestry (COFO) approved the plan for the Global Forest Resources Assessment 2000 according to the recommendations of a formal FAO Expert Consultation held in Kotka, Finland in 1996. The Intergovernmental Panel on Forests (IPF) also reviewed and endorsed results of the Kotka meeting and provided important feedback to FAO on conducting the assessment.

The Kotka meeting developed an agenda, outlined major issues and defined ways of compiling the information needed for the assessment. Later, a great amount of operational work and fundraising was required to execute FRA 2000. In practice, the assessment required the active participation of countries and areas throughout the world. Of the 212 countries represented in the assessment, 160 participated actively in workshops or worked with FAO staff in their own countries. Countries provided specific technical information used as baseline data for the assessment and worked with FAO in adjusting national data to global standards.

Countries were involved in the review of the results of the assessment as well as in its planning and implementation. In late 2000, all countries were given the opportunity to preview and check the results of the assessment before their publication. Through this process, 56 countries provided additional material and feedback to FAO. The preliminary results were also reviewed during the COFO meeting in 2001, which provided formal and positive feedback on the implementation and findings of FRA 2000.

INTRODUCTION

The foundation for FAO's global assessments lies in its Constitution, which states that "the Organization shall collect, analyse, interpret and disseminate information relating to nutrition, food and agriculture", where agriculture is defined to include fisheries, marine products, forestry and primary forest products (FAO 1992). After reviewing the results of FAO's first world survey of forests in 1947, the sixth session of the FAO Conference in 1951 recommended that the Organization "maintain a permanent capability to provide information on the state of forest resources worldwide on a continuing basis" (FAO 1951). Since that time, FAO has conducted global or regional assessments about every five to ten years.

The importance of forest resources assessments was highlighted at the United Nations Conference on Environment and Development (UNCED) in Rio de Janeiro, Brazil, in 1992. In fact, much of the information that led to the environmental concerns highlighted at UNCED came from previous global forest resources assessments, particularly change information from the 1990 assessment. UNCED devoted a full chapter of Agenda 21, "Combating deforestation", to the issues of forest conservation and development and adopted the Non-Legally Binding Authoritative Statement of Principles for a Global Consensus on the Management, Conservation and Sustainable Development of all Types of Forests (known as the "Forest Principles").

Chapter 11 has as a key element Programme D, "Establishing and/or strengthening capacities for the planning, assessment and systematic observations of forests and related programmes, projects and activities, including commercial trade and processes". Programme D contains a series of relevant recommendations for periodic assessments, which are underscored in the "Basis for Action" as follows:

> Assessment and periodical evaluations are essential components of long term planning, for evaluating effects, quantitatively and qualitatively, and for rectifying inadequacies. This mechanism, however, is one of the often neglected aspects of forest resources, management, conservation and development. In many cases, even the basic information related to the area and type of forests, existing potential and volume of harvest, etc. is lacking (UNCED 1992).

Other principles contained in Chapter 11 provide additional guidance to international organizations and countries regarding the importance and conduct of periodic assessments. The Rio+10 conference in 2002 will provide the opportunity to review progress on these proposals since they were elaborated in 1992.

In all policy fora of relevance to FRA 2000, the need for involving countries and their professionals in the global assessment was stressed. This is one of the guiding principles of the FRA Programme. Along these lines, the fourth session of the Intergovernmental Panel on Forests (IPF IV) noted that:

> FRA 2000 should be a partnership exercise facilitated by FAO but also involving United Nations organizations, national institutions and other interested parties, including relevant major groups. Cooperation at the national level should involve all interested parties, both within and outside the forest sector (UN 1997).

FAO fulfilled this obective and surpassed all past assessments to ensure that countries were involved in the assessment, that their information was utilized and that their perspectives were included in the final analyses.

INSTITUTIONAL FRAMEWORK FOR IMPLEMENTATION

Committee on Forestry

FAO serves as a steward of the data that are proffered by countries regarding their forests, which are often sensitive – especially those concerning tropical deforestation. Misuse of such data has the potential to impair a country's economy and therefore the welfare of its citizens. For an assessment to succeed, countries must be convinced that it is in their best overall interest to share such information with the rest of the world and to participate actively in the assessment. Therefore, a formal endorsement by FAO member countries for each periodic assessment is sought prior to initiation of the work. This is conducted through the FAO Forestry Department's highest policy forum, the Committee on Forestry (COFO).

During biennial COFO meetings each member country has the right to endorse or veto an assessment, as well as to make specific requests concerning its execution. As almost all FAO member countries attend COFO, approval to move forward with an assessment signifies that these countries are aware of the work that will be required of them and politically committed to the task. In 1997, COFO approved the agenda for FRA 2000 according to the recommendations of a formal FAO Expert Consultation held in Kotka, Finland in June 1996.

Expert consultations

Expert consultations are key to developing an agenda, outlining major issues and defining ways of compiling the information needed for an assessment.

The June 1996 Expert Consultation on Global Forest Resources Assessment 2000 (known as Kotka III) was attended by 45 forestry and environment experts from 32 countries and representatives from five non-governmental agencies and three international organizations. The meeting provided a unique opportunity for some of the world's foremost experts on forest assessments to discuss the scope and execution of FRA 2000. Participants provided valuable technical advice on the parameters needed for the assessment, as well as insight into its political and operational complexities. FRA 2000 also benefited from the review and endorsement of the Kotka recommendations by IPF IV.

As expected, Kotka III underscored the need for FRA 2000 to provide basic information on worldwide forest area, volume and biomass – its state in the year 2000 and changes since 1990 and 1980. In addition, the Kotka participants emphasized the need to include in the assessment a number of non-traditional parameters to provide a more holistic vision of forests. These include non-wood products and services, protected forest areas, trees outside the forest and others. The Kotka participants recommended a multifaceted approach for amassing FRA 2000 information, to include information provided by countries based on remote sensing surveys of forest cover change, low-resolution mapping and a number of special studies. Important decisions were also reached on a core set of comparable global definitions for all countries and on a division of labour between FAO headquarters in Rome (developing countries) and the United Nations Economic Commission for Europe (UNECE) (industrialized countries) (Finnish Forest Research Institute 1996).

Because of the difficulty and importance of providing the best possible information on forest change, FAO convened a second official meeting, the Expert Consultation on Forest Change, in March 2000. At this meeting leading specialists in forest inventory from around the world reviewed FAO's past methods of estimating forest change and submitted proposals for estimating forest change in developing countries for FRA 2000. During the meeting, methods for conducting change assessment were tested and analysed using representative data sets from FRA 2000.

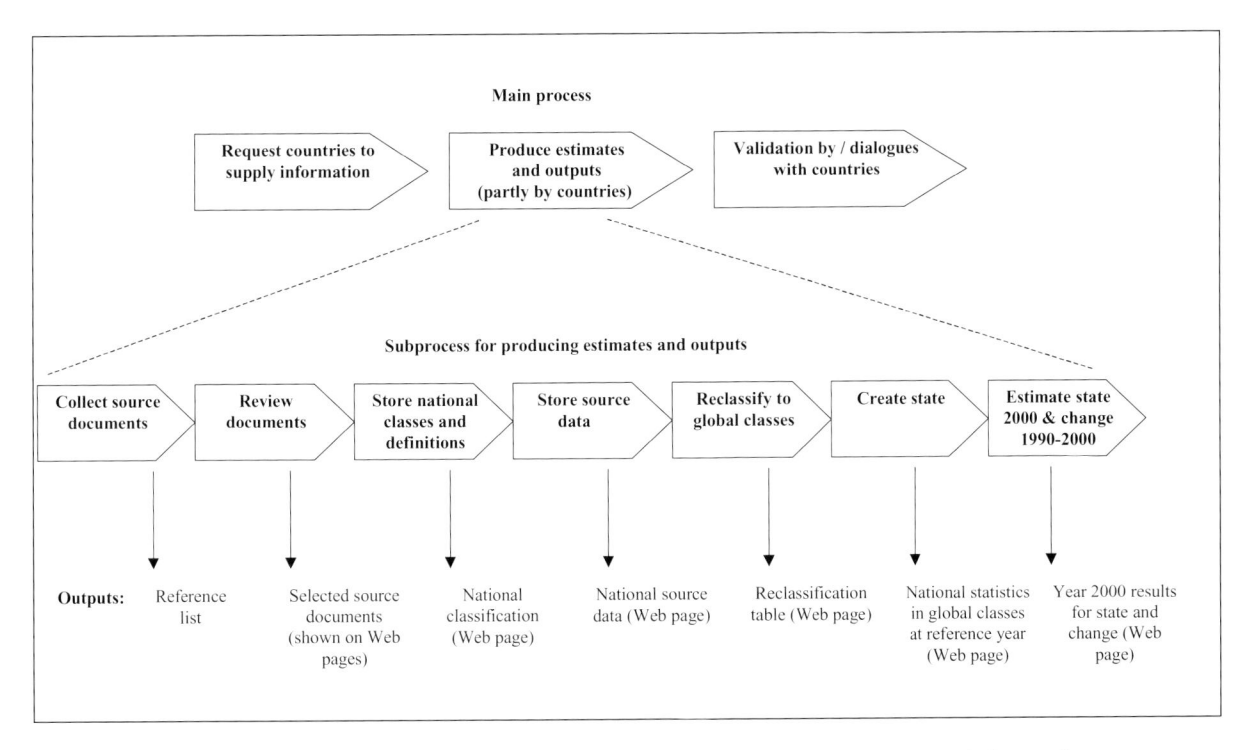

Figure 45-1. Process of developing forest cover estimates using country information

The 1990 population-deforestation regression model was tested against new data coming from countries which could be compared against the FRA 1990 predicted values – with only sketchy results. The tests showed the tendency of the model to overestimate deforestation, particularly in countries lacking comparable multi-date inventories. Eventually, the meeting confirmed that "deforestation is such a complex process, involving physical, climatic, political, and socio-economic forces which are themselves very complex, that simple generalized models of forest change have so far not been developed. Current models are oversimplified and yield similar predictions of forest cover change rates for countries which are known to be very different" (Päivinen and Gillespie 2000). Based on these findings, FRA 2000 discontinued using the FRA 1990 model for predicting or extrapolating forest loss based on population dynamics.

The expert panel also tested and finally recommended a variation of the "convergence of evidence" method for estimating forest change, as it could be tailored to the available information from a particular country. This method was adopted for FRA 2000, as countries with greater amounts of ancillary information could use it to generate more precise estimates of forest change. Guidelines for the use of the method were developed in the meeting, and computer modules were written for the Forest Resources Information System (FORIS) to aid analysts in extracting, graphing and analysing multiple data sets needed to generate national estimates of change for FRA 2000.

Intergovernmental Panel on Forests

IPF IV specifically reviewed and commented on the Kotka agenda for FRA 2000 and made recommendations to FAO on its global assessments in general. The panel noted the importance of the assessment and commended the broadening of the scope to include non-traditional roles of forests and trees in the survey. However, it also noted that as of February 1997 the assessment was still unfunded and FAO had yet to produce a relevant working plan for the operation. In conclusion, the panel endorsed the Kotka agenda:

> The Panel expressed strong support for FRA 2000 and the arrangements being made following the recommendations of the FAO Expert Consultation on Global Forest Assessment in Finland in June 1996 (Kotka III) ... [and requested] FAO to implement the Global Forest Resources Assessment 2000 in collaboration with international organizations, countries and other organizations with competence in assessments, and to share the results of the assessment effectively with the international community (UN 1997).

Use of country information

FRA 2000 relied on information from countries as a source for national-level statistics (see Figure 45-1). To collect the country data, the FRA Programme made formal requests to developing country representatives in 1996 and 1998 for their latest forest inventory reports and initiated a dialogue with them to ensure understanding of the information contained within the reports. The 1998 request was accompanied by specific guidelines (FAO 1998b) for all FRA 2000 assessment parameters to ensure that the information collection was well structured, along with a publication on FRA 2000 terms and definitions (FAO 1998a). UNECE/Geneva sent an enquiry, guidelines and terms and definitions to industrialized countries to initiate collection in those countries.

FAO relies primarily on the statistics from technical reports from the countries, rather than quoted or secondary sources. The use of data published in primary source documents ensures that FAO has the most objective, scientific and statistically valid information – and the necessary background to understand how it may best be used. In the few countries that have no applicable national forest inventories, FAO has had to compile information from various partial inventories or to use subjective estimates. This complex work could only be carried out with the direct collaboration of professionals from the various countries.

Once information from developing countries was compiled in FAO and its utility for FRA 2000 assessed, it was archived in the Forestry Information System (FORIS). Each entry included the original statistics from the source, terms and definitions, a description of the utility of the information and complete bibliographic references. Even information not directly relevant to the assessment was entered into the system, in the event that it might eventually be useful for other purposes. The detailed information on sources was also archived in order to achieve the maximum amount of transparency in the generation of FAO estimates.

To make the highly variable country information useful for global reporting, FAO employed a set of standards for its harmonization. First, all country information was classified according to a common set of terms and definitions. This was a difficult task because of the sheer magnitude and variability of the information produced by countries and the wide range of forest formations, ecological conditions and cover types that exist on a global scale. For example, in FRA 2000 more than 650 definitions of forest were assembled from 132 developing countries (from 110 independent surveys). FORIS was used to compute and archive the relationships established between national and global definitions.

FAO experts visited over 100 countries to work with national professionals on the use of their national data for FRA 2000. FAO also conducted numerous workshops for training related to data collection, analysis of country statistics and adjustment of information to global reporting standards required by FRA 2000 (see Box). UNECE/Geneva held a series of workshops and meetings to guide the implementation for the assessment in industrialized countries.

Review of results

Prior to publication of the results, FAO asked that countries review the country results and provide their comments. A formal letter was sent to each country requesting its cooperation, along with a Country Validation Profile containing the results and source information. Some countries requested changes in the FAO estimates.

The results of FRA 2000 were presented at COFO 2001 and were reviewed and commented on by the member countries. Member countries were asked to give any final comments to FAO by the end of March 2001. By the end of May 2001, all comments from countries had been taken into account. To modify the statistics, countries submitted primary technical material (inventory reports) which improved the results already compiled by FAO. Revised estimates were then calculated as a collaborative exercise between the countries and FAO.

The report of the fifteenth session of COFO summarizes the countries' final official position on the results of the assessment:

> The Committee commended FAO for carrying out the Forest Resources Assessment 2000 (FRA 2000) and for presenting the findings in a comprehensive and transparent way. It acknowledged the difficulties posed by the limited availability of timely and accurate national inventory reports, and by the lack of adequate financial resources to ensure the elaboration of these inventories. It recognized the considerable efforts involved in harmonizing national inventory information in a global synthesis. While recognizing that the rate of global deforestation may have slowed in the 1990-2000 period, the Committee nonetheless noted with concern the continued high level of deforestation. It urged countries to consider FRA 2000 findings when carrying out policy development and planning (FAO 2001).

FUNDING AND CONTRIBUTIONS

Beginning in 1997, the FAO Regular Programme provided three staff positions to FRA, equivalent to US$423 000 per year, and US$404 000 per year in non-staff funding. This represented 5.4 percent of the Forestry Department's annual allocation, and

FRA 2000 workshops

Africa

Data Collection and Analysis for Sustainable Forest Management in ACP Countries: Linking National and International Efforts (Nakuru, Kenya, 12-16 October 1998)
Countries attending: Eritrea, Ethiopia, Kenya, Somalia, the Sudan, United Republic of Tanzania, Uganda

Data Collection and Analysis for Sustainable Forest Management in ACP Countries: Linking National and International Efforts (Mutare, Zimbabwe, 30 November-4 December 1998)
Countries attending: Angola, Botswana, Lesotho, Malawi, Mozambique, Namibia, South Africa, Swaziland, Zambia, Zimbabwe

Data Collection and Analysis for Sustainable Forest Management: Linking National and International Efforts (Lambarene, Gabon, 27 September-1 October 1999)
Countries attending: Cameroon, Chad, Congo, Equatorial Guinea, Madagascar, Central African Republic, Rwanda

Data Collection and Analysis for Sustainable Forest Management: Linking National and International Efforts (Yamaussoukro, Côte d'Ivoire, 13-18 December 1999)
Countries attending: Benin, Cape Verde, Côte d'Ivoire, Guinea, Mali, the Niger, Nigeria

Regional Workshop on Forestry Information Services (Stellenbosch, South Africa, 12-17 February 2001)
Countries attending: Angola, Botswana, Lesotho, Malawi, Mozambique, Namibia, South Africa, Swaziland, Zambia, Zimbabwe

Latin America and the Caribbean

Workshop on the Forest Resources Assessment 2000 (Turrialba, Costa Rica, 17-21 May 1999)
Countries attending: Belize, Costa Rica, Colombia, Ecuador, El Salvador, Guatemala, Honduras, Mexico, Nicaragua, Panama, Venezuela

Subregional Workshop on Data Collection and Oulook Effort for Forestry in the Caribbean (Port of Spain, Trinidad and Tobago, 21-25 February 2000)
Countries attending: Bahamas, Barbados, Belize, Cuba, Dominica, the Dominican Republic, Grenada, Guyana, Haiti, Jamaica, St Lucia, St Kitts and Nevis, St Vincent and the Grenadines, Suriname, Trinidad and Tobago

Asia and Oceania

South Asian Regional Workshop on Planning, Database and Networking for Sustainable Forest Management (Thimpu, Bhutan, 23-26 May 2000)
Countries attending: Bangladesh, Bhutan, India, Myanmar, Nepal, Pakistan, Sri Lanka

Data Collection for Pacific Region (Apia, Samoa, 4-8 September 2000)
Countries attending: American Samoa, Cook Islands, Fiji, French Polinesia, Kiribati, Micronesia, Niue, Papua New Guinea, Samoa, Tonga, Vanuatu

Industrialized countries

Temperate and Boreal Country Forest Resources Assessment Team of Specialists Meetings and Ad Hoc meetings on FRA 2000 in Industrialized Countries (Geneva, Switzerland, April 1996; Birmensdorf/Zurich, Switzerland, March 1997; Geneva, November 1997; Ispra, Italy, March 1998; Geneva, March 1999; Joensuu, Finland, May 2000; Victoria, British Columbia, Canada, June 2001)
Countries attending: Australia, Austria, Canada, Czech Republic, Denmark, Finland, France, Germany, Hungary, New Zealand, Poland, Portugal, Russian Federation, Slovakia, Sweden, Switzerland, United States
Other organizations represented: FAO, European Commission, Joint Research Centre (EU), UNEP, World Wildlife Fund, European Forest Institute, World Conservation Monitoring Centre and UNECE/FAO Secretariat

TBFRA Meeting for Countries in Transition (Gmunden, Austria, 1-4 October 1997)
Countries attending: Albania, Armenia, Austria, Bulgaria, Croatia, Czech Republic, Estonia, Hungary, Latvia, Lithuania, Moldova, Poland, Romania, Russian Federation, Slovak Republic, Slovenia, Ukraine, Yugoslavia

Table 45-1. Trust Funds

Donor	Project code	Amount US$
Finland	GCP/INT/723/FIN	1 046 000
Japan	GCP/INT/162/JPN	650 000
Sweden	GCP/INT/702/SWE	1 596 924
Switzerland	GCP/INT/692/SWI	355 950
UNEP	EP/RAF/652/UEP	30 000
United Kingdom	TEMP/INT/928/UK	550 582

0.25 percent of the Organization's annual budget. A number of Trust Funds to support the assessment, totalling about US$4.2 million, were established through donations (Table 45-1).

The following countries provided Associate Professional Officers (APOs) to work on the assessment: Austria (duty station Rome), Denmark (duty station Bangkok), Finland (duty station Geneva), France (duty station Rome), Italy (duty station Cairo), Japan (duty station Rome) and Sweden (duty stations in Rome and Santiago).

MAJOR PARTNERS

Additional goods and services were rendered by a number of institutions in the form of in-kind contributions. These contributors include the Brazilian Institute of Environment and Natural Resources (for country data collection and remote sensing interpretations), the Canadian Forest Service (for global ecological zoning), the EROS Data Center in the United States (for cost sharing and implementation of global mapping), the United States National Aeronautics and Space Administration (for remote sensing imagery), the Forest Survey of India (for remote sensing interpretations), the International Institute of Applied Systems Analysis (for global ecological zoning), the Swedish National Board of Forestry (for information systems and data collection in Africa) and the Forest Service of the United States Department of Agriculture (for global mapping, information collection in the Caribbean and technique development for estimating global change). Many other individuals contributed their time and labour to the assessment either in agreement with their organizations or as non-affiliated experts.

BIBLIOGRAPHY

FAO. 1951. *Sixth session of the FAO Conference.* Rome.

FAO. 1992. *Basic texts.* Rome.

FAO. 1997. *Report of the thirteenth session of the Committee on Forestry.* Rome.

FAO. 1998a. *FRA 2000 terms and definitions.* FRA Working Paper No. 1. Rome.

FAO. 1998b. *FRA 2000 guidelines for assessments in tropical and subtropical countries.* FRA Working Paper No. 2. Rome.

FAO. 2001. *Report of the fifteenth session of the Committee on Forestry.* Rome.

Finnish Forest Research Institute. 1996. *Proceedings of FAO Expert Consultation on Global Forest Resources Assessment 2000 in cooperation with ECE and UNEP with the support of the Government of Finland (Kotka III)*, Kotka, Finland, 10-14 June 1996, ed. A. Nyyssönen & A. Ahti. Research Papers No. 620. Helsinki, Finland.

Päivinen, R. & Gillespie, A.J.R. 2000. *Estimating global forest change 1980-1990-2000.* Background document prepared for an international panel of experts convened to review methods to be used in completing the FAO Global Forest Resource Assessment (FRA) 2000. Rome, March 2000.

Space, J. 1997. *Strategic plan, Global Forest Resources Assessment 2000.* Unpublished paper. Rome, FAO.

UN. 1997. *Report of the Ad Hoc Intergovernmental Panel on Forests on its fourth session.* New York, 11-27 February 1997. E/CN.17/1997/12. New York.

UNCED. 1992. Combating deforestation. *Agenda 21*, Chapter 11. Rio de Janeiro, Brazil.

Chapter 46

Pan-tropical survey of forest cover changes 1980-2000

ABSTRACT

The FRA 2000 pan-tropical remote sensing survey complemented the assessment based on country information and focused on change processes in tropical forests during the 1980s and 1990s. Stratified random sampling (10 percent) of the world's tropical forests was employed through 117 sample units representing 87 percent of the tropical forests. For each of the sample units, three Landsat satellite images from different dates provided the raw material for producing statistics on forest and other land cover changes from the period 1980 to 1990 and from 1990 to 2000.

Important products generated through the survey include change matrixes for the tropics as a whole (developing country areas) and for Africa, Asia and Latin America separately. The matrixes show the various forest and land cover classes and how they have changed over the past two decades. The study is the first to provide a consistent methodology for assessing forest cover change between two assessment periods (1980 to 1990 and 1990 to 2000). Correlations between the remote sensing survey results and the country statistical data for the tropics summed at regional levels were good, although the remote sensing survey indicated a lower level of deforestation than the aggregate national findings for Africa.

Results of the study at the pan-tropical level indicate that the world's tropical forests within the surveyed area were lost at the rate of about 8.6 million hectares annually in the 1990s, compared to a loss of around 9.2 million hectares during the previous decade. While this change fell within the margin of error for the tropics as a whole, statistically significant decreases in deforestation were detected in tropical moist deciduous forests. In contrast, smaller increases in deforestation (not statistically significant) were detected in both tropical rain forests and dry forests. Across the tropics, most of the deforestation was due to the direct conversion of forests to permanent agriculture or pastures and, to a lesser degree, to the gradual intensification of shifting agriculture.

INTRODUCTION

The FRA 2000 estimates of forest area and change are largely based on national statistics and inventory reports, which contain detailed information on the forests of individual countries. However, differences among data sets from the various countries can be great owing to the methods applied, the terms and definitions employed and the currency of the information in the individual inventories. Despite adjustments made to accommodate these differences, uncertainties can still arise when statistics from different countries are compared, especially those relating to forest change.

To bolster FAO's understanding on land-cover change processes in the tropics, especially deforestation, and to complement the country-specific statistics, FAO carried out an independent survey of land cover changes in the tropics. The survey, which emphasized quantifying forest cover change, was based on 117 sampling units covering 10 percent of the survey area. Each sample unit was composed of three multitemporal Landsat satellite images acquired from about 1980 through 2000.

The results of the survey complement the estimates of forest area based on country data and provide unique information on trends in forest change since the 1980s at ten-year intervals. The survey is the first to generate a consistent overview of forest change processes at the pan-tropical, regional and ecological zone levels between two assessment periods. Principal products of the study include change matrixes which quantify changes in forests and other land cover classes. From these, several forms of change were identified – deforestation, degradation, fragmentation and shifting cultivation, among others. Analysis in the shifts between classes attributed to these change processes has helped to identify cause-and-effect

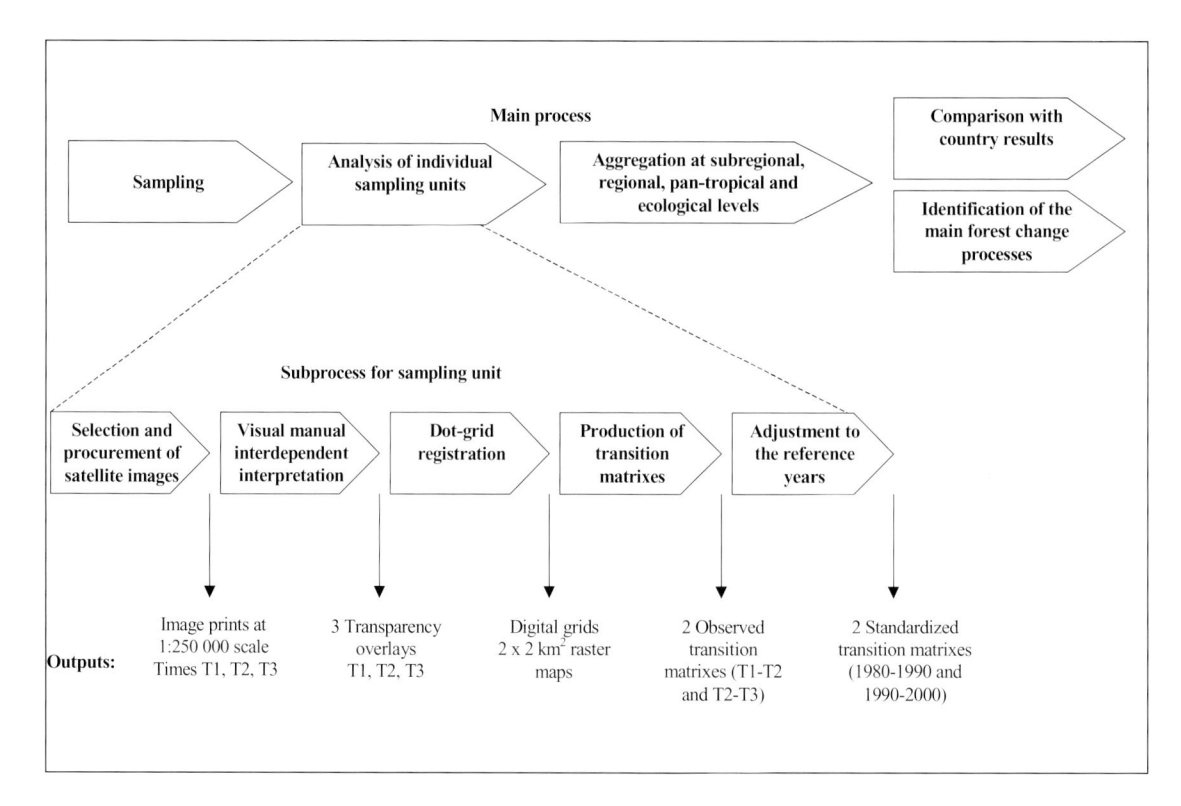

Figure 46-1. Remote sensing survey processes

relationships useful in understanding the complex processes of deforestation. In contrast, the country-based studies were only able to generate single estimates of forest change, without showing how or why the forest area had changed.

The most recent acquisitions of satellite imagery were used in conjunction with the same sample units established for the FRA 1990 pan-tropical survey. Archives from the 1990s containing two multi-date images for the sample units were available for most areas and were complemented by the later image acquisition. The same methodologies and definitions were applied for FRA 2000. The three dates of imagery for each sample unit made it possible to conduct the study over 20 years and to produce statistics at ten-year intervals.

The objectives of the FRA 2000 remote sensing survey were to:
- monitor tropical forest cover state and change for the past 20 years at regional and pan-tropical levels;
- analyse trends in forest cover change between 1980-1990 and 1990-2000;
- study the dynamics of change in forest cover and identify causal mechanisms of deforestation;
- complement existing country information by providing spatially and temporally consistent data on forest state and change.

METHODS

Figure 46-1 illustrates the different steps of the survey processes, further explained below.

The time-series analysis for the survey was developed to ensure a high level of consistency by the use of uniform data sources and interpretation techniques. Data used for each sample unit were composed of three images acquired as close as possible to the reference years 1980, 1990 and 2000.

The three dates of imagery made it possible to analyse and calculate changes in land cover over two sequential time periods, and to assess differences in the land cover changes between the two periods. The use of a third date of imagery in the time series introduced some complexity in the calculation of the estimates for the reporting periods and reference years.

The main features of the survey's methodology were:
- the statistical sampling design;
- a standard classification oriented towards forest assessment;
- an interdependent interpretation procedure;
- standardization of results to reference years;
- calculation of aggregated estimates.

For further details refer to Chapter 1 in FAO (1996) and FAO (2001).

Table 46-1. Land cover classification used for the survey

Land cover categories	Land cover classes (main classes)	Brief description
Natural forest		
Continuous forest cover	Closed canopy	Canopy cover > 40 %
	Open canopy	Canopy cover 10-40%
	Long fallow	Forest affected by shifting cultivation
Fragmented forest	Fragmented forest	Mosaic of forest/non-forest
Non forest		
Other wooded land	Shrubs	
	Short fallow	Agricultural areas with short fallow period
Non woody areas	Other land cover	Includes urban and agricultural area, area with less than 10% woody vegetation cover
	Water	
Human-made woody vegetation	Plantations	Forest and agricultural plantations
Non-visible	Non-interpreted	Clouds, burnt woodland, shadow, outside study area

Note: Classes are grouped as forest/non-forest according to the f3 forest definition.

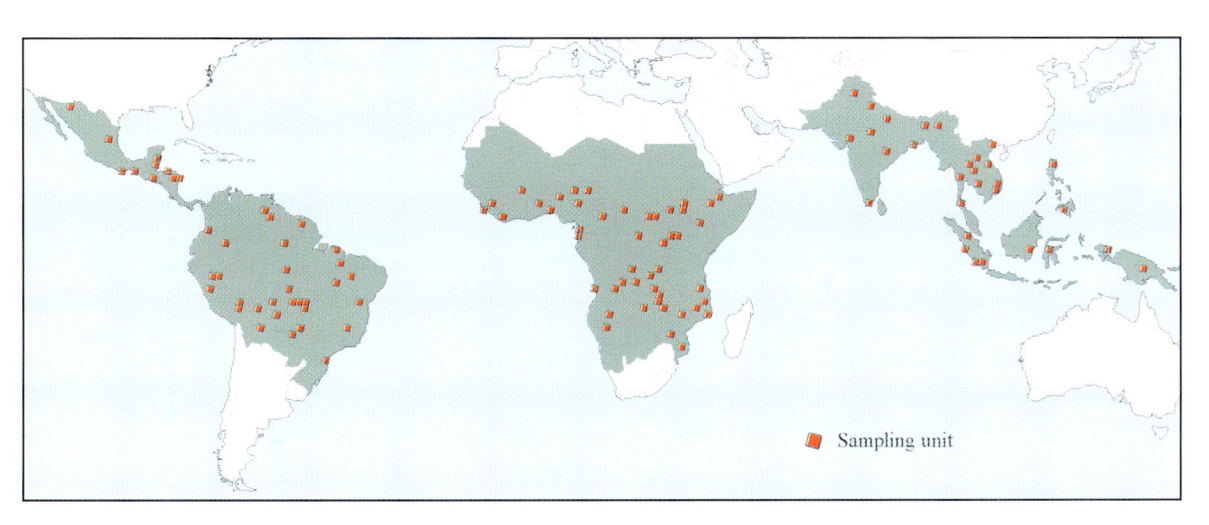

Figure 46-2. Distribution of sampling units in the pan-tropical remote sensing survey

Statistical design

FRA 2000 employed the same sampling design as FRA 1990 (FAO 1996; Czaplewski 1994). Its main characteristics are the following:

- Two-stage stratified random sampling was used. The surveyed area was divided into regions and subregions. Each subregion was further stratified into a maximum of three strata corresponding to forest cover (Latin America and Asia) or forest dominance (Africa).
- The sample population consisted of 1 203 Landsat frames representing all frames where the forest cover is above 10 percent and the land area is above 1 million hectares, i.e. 51 percent of the total number of frames in the pan-tropical area. Within the population, the sample covered all tropical forests in wet, moist and dry conditions. According to FRA 1990 country data statistics, 87 percent of the tropical forests are located in the sampled area.

Following this method, 117 sampling units were selected: one sampling unit corresponds to a Landsat frame. Figure 46-2 shows the distribution of the sampling units.

Land cover classification

A uniform land cover classification was used for all the sampling units to map, gather statistics and describe the vegetation (particularly woody vegetation). It included ten cover classes, of which nine were visible classes (Table 46-1).

Because the land cover classification contains many different classes of woody vegetation, they may be aggregated relative to various reporting and analysis needs. In the case of forests, three distinct definitions were derived by grouping different classes of woody vegetation. The first and most exclusive definition, referred to as forest 1 (f1) includes only the closed forest class. The second, forest 2 (f2), was constructed to match the forest definition used in the country reporting, and comprises the closed and open forest classes, and a fraction (two-ninths) of the fragmented forest class. The third definition, forest 3 (f3), is the broadest and includes the classes of long fallow and a higher fraction (one-third) of the fragmented forest class than the f2 definition (see also FAO 1996). The last definition allows the most detailed differentiation among changes.

Interpretation of the sampling units and data compilation

The interpretation of the sampling units was carried out for FRA 2000 by experts in photo interpretation of satellite imagery. Many regional and national organizations contributed to the work including the Tropical Agricultural Research and Higher Education Center (CATIE) in Costa Rica, the École Nationale du Génie Rural des Eaux et des Forêts (ENGREF) in Montpellier, France, the Forest Survey of India (FSI) and the Instituto Brasileiro do Meio Ambiente (IBAMA) in Brazil. FAO conducted training for staff at the cooperating institutions. Each interpretation was carefully checked and reviewed at FAO headquarters in Rome to ensure that the interpretations were consistent from sample to sample. The interpretative work was aided by the national experts' knowledge of the conditions and vegetation in the areas surveyed. In some locations, including many in Brazil, substantial fieldwork was conducted (aerial and ground survey).

Interpretations were done using conventional manual methods for interpretation of temporal series of satellite images. Landsat data (MSS and TM) were used for most sample units, but a few SPOT and IRS images were also needed where Landsat data were not available or were of low quality. Images were processed in three bands as standard false-colour infrared prints. Interpretations were carried out on hard copies at 1:250 000 scale.

Each sampling unit was interpreted at three points in time with imagery acquired as close to the reference years 1980, 1990 and 2000 as possible. The average dates of the imagery for the three image sets were 1977, 1989 and 1998. The designations T1, T2 and T3 were assigned to imagery corresponding to the data sets 1980, 1990 and 2000 respectively.

The T1 and T2 images had already been studied in FRA 1990 through interdependent interpretation. The same technique was employed to interpret the T3 images acquired for FRA 2000. This method required that the change analysis be conducted using continuous image-to-image comparison with other images from the time series. Moreover, all of the images were geometrically registered to the T2 image through local registration techniques, as the interpreter progressed in the interpretation. Although more time consuming than independent interpretation, this method has been shown to eliminate classification errors in both state and change estimates. It substantially reduced errors that would have been caused by geometric offsets in the images, as well as those from differences in satellite scenes due to varying contrast enhancements or seasonal differences in vegetation.

The image acquired for the third date added substantially more information to the analysis. Most of the T3 imagery was acquired digitally by FRA. New ancillary information such as vegetation maps had become available since FRA 1990 and was used to improve the interpretations of the entire time series. The T1 and T2 interpretations were consequently revised when necessary. This contributed to very slight differences in statistics for the 1980-1990 period, compared to those generated for FRA 1990 for the same period.

Interpretations were made on transparent overlays. Over 900 million hectares were interpreted, with a common visible area over all time series covering about 250 million hectares. Data capture was achieved by using a 2 x 2 km^2 dot grid. The interpreted class was registered at each dot for three points in time, and the resulting data grids were entered into the Forestry Information System (FORIS). All data were archived in FORIS, which can also display the interpretations and aggregate the results for the various reporting levels. The data grids were also geo-referenced and integrated in a Geographic Information System.

The information set for each sample unit consists of three states (1980, 1990 and 2000) and two area transition matrixes (1980-1990 and 1990-2000). Data grids were used to determine the state (i.e. the areas for the various land cover classes at each of the three points in time) and to estimate the class-to-class changes during the two time periods. The changes within a sampling unit between two dates were compressed into a single area transition matrix, which quantifies the various shifts between the classes.

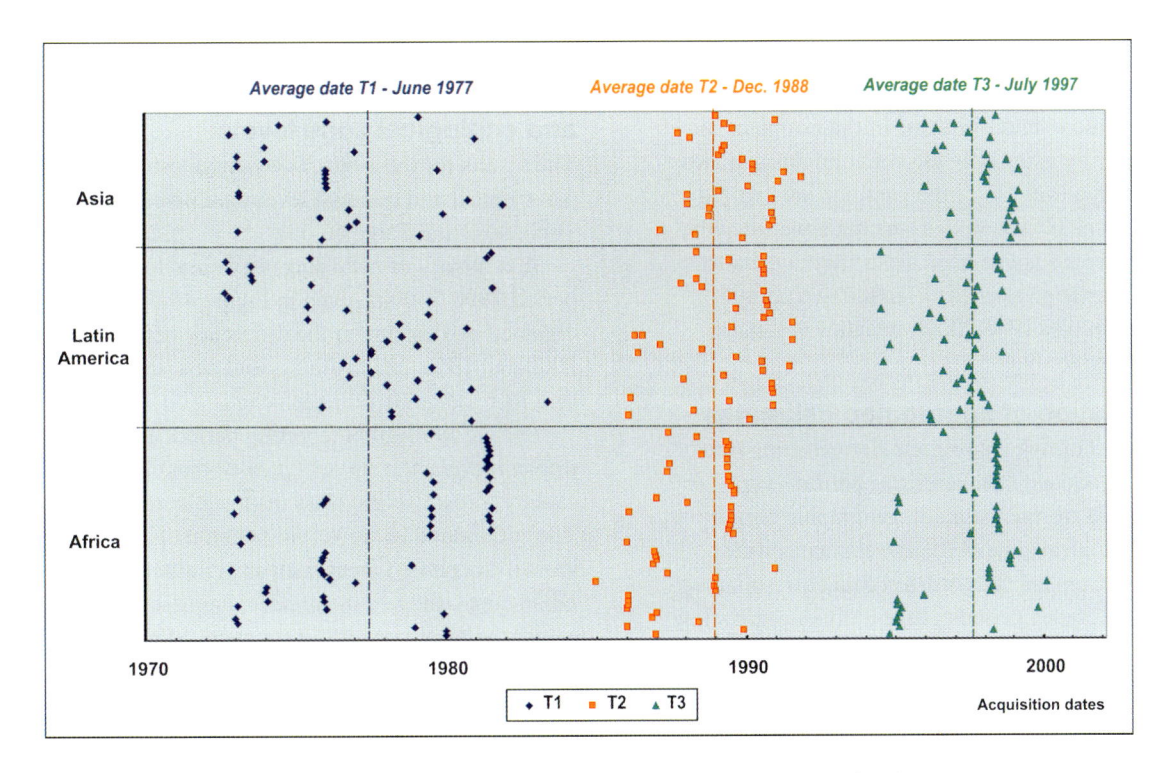

Figure 46-3. Temporal distribution of satellite images used for the survey

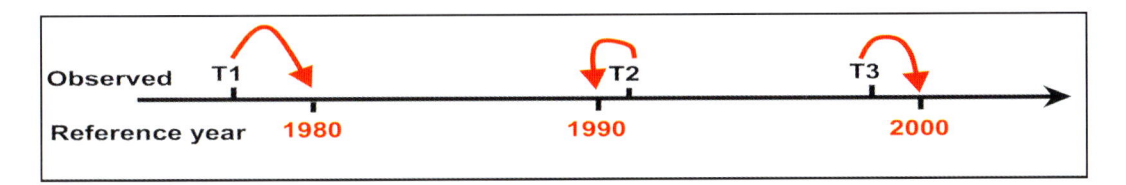

Figure 46-4. Illustration of standardization to reference years

Standardization of data to the reference years 1980, 1990 and 2000

While the images were selected to be as close as possible to the reference years 1980, 1990 and 2000, the date rarely corresponded exactly to the reference year and varied among sampling units (Figure 46-3).

Before estimates could be made at the various aggregate levels, data had to be standardized to the reference years 1980, 1990 and 2000. The statistics had to be either extrapolated or interpolated from the sample units from the original date of acquisition of the imagery to the various reference years (Figure 46-4).

Data for each sample unit were first organized into a series of computerized matrixes and then processed using one of two algorithms. The algorithms were developed by FAO to project the interpreted information from the satellite images to the standard reference years. Two methods were used – the constant method and the linear method. In the constant method the annual changes in land cover were assumed to be constant and unchanging during the period. They are calculated using only one date of reference information. Conversely, the linear method assumes that annual changes in land cover occur gradually and linearly and requires the use of two sequential data sets (T1-T2 or T2-T3).

The linear method was computationally more complex and was considered the preferred method since it did not provoke abrupt class-to-class transitions at the second point in time, as did the constant method. The linear method also had the advantage of giving the same results for the 1990 reference year whether they were interpolated or extrapolated from the T1-T2 or T2-T3 data sets. However, sometimes the results produced by the linear method were not supported by the original data. In these cases (23 percent) the more robust constant method was used. The constant method was also used when extrapolation outside the observed time series was necessary for the adjustment to the years 1980 or 2000.

Calculations for the estimates were based on the common area of all three images. About

90 percent of the area common to T1 and T2 was also found in the T3 image. It would theoretically be possible to improve some of the estimates in the study by using only the common areas of two images at a time (since the common area would be greater). However, this was not done since the method could not be used for estimating changes in deforestation rates between the two periods (1980-1990 or 1990-2000), which was a major aim of the study.

Calculation of aggregated estimates

Estimates of forest cover and deforestation rates over the two ten-year reference periods were calculated for each stratum, geographic region and ecological zone. All these estimates were derived from the standardized data corresponding to the reference periods of the individual sampling units.

The sample of scenes within each stratum of the survey was considered a cluster sample. The estimators were generally ratio estimators or combinations thereof. As the number of samples was relatively low per stratum, the method of the combined estimators over strata was used to limit the bias.

Standard errors (SE) were calculated as a measure of precision of the estimates, for constructing 95 percent confidence intervals and for testing hypotheses. The standard errors of the basic estimators were calculated according to common theory (ratio estimators in stratified sampling) (Raj 1968). For more complicated estimators, Taylor expansions (Raj 1968) were used to derive the standard error formulas. FRA Working Paper No. 49 (FAO 2001) contains detailed explanations of the statistical methods employed in the survey.

RESULTS AND FINDINGS

The results cover most of pan-tropical forests under a wide range of ecological conditions, from tropical rain forests to dry forests. The survey is the first assessment tool to provide consistent and comparable information over two reporting periods (1980-1990 and 1990-2000), allowing the calculation of both changes and the change in changes between the two periods. Past assessments have not been able to provide such information owing to various inconsistencies in information between subsequent FRA reports.

An example of results from the interpretation is given for a sample located in Zimbabwe in Figure 46-5.

States and changes for the period 1990-2000 at pan-tropical, regional and ecological zone levels

The results for the 1990-2000 period, estimated at pan-tropical and regional levels, are presented in Table 46-2 and Table 46-3.

A summary of net changes by class for the period 1990-2000 is given in Figure 46-6. It was obtained by calculating the difference between the 2000 and 1990 area estimates and describes the area lost and gained for each class.

For the 1990-2000 reporting period, at pan-tropical levels, the survey revealed that closed canopy forest was the class most subject to loss. The "other land cover" class, which includes sparsely vegetated areas such as agriculture and urban areas, showed the greatest increase in area across the tropics. Most forests were converted to other land cover at the pan-tropical level. The implication of this finding is that most tropical closed canopy forests were lost as a result of their conversion to agriculture (an insignificant portion went to urban areas). At the regional level the results varied somewhat.

In Africa, during the 1990s, the amount of closed canopy forest converted into other land cover was relatively low in comparison with other regions. Large portions of both closed and open canopy forests were converted into fragmented forest and short-fallow classes in the region. Significant areas of fragmented forest were also converted into other land cover. The open canopy forest in Africa sustained greater losses than in the other regions.

Forest change in Latin America was characterized by a marked large transition from closed canopy forests into other land cover (which was about twice as great as in the other regions). While the findings were similar in Asia, that region also had large areas of closed canopy forest that were transformed into both long and short fallow. Substantial areas of shrubs were also converted into other land cover in Latin America, but not in Asia or Africa. Changes from other land cover and closed forests to plantations (human-made woody vegetation) were also notably observed in Asia.

Positive transitions are those in which the woody content of the area increased. While they were not common during the 1990s, some positive changes were observed when other land cover recuperated into short fallow and shrubs in Latin America. Shifts from other land cover to fragmented forest were more uniformly

Table 46-2. Area transition matrixes for the period 1990-2000 at pan-tropical level (million ha)

1990-2000 area transition matrix Pan-tropical

(Million ha) Land cover classes in 1990	Closed Forest	Open Forest	Long Fallow	Fragmented Forest	Shrubs	Short Fallow	Other Land Cover	Water	Plantations	Total 1990	% of total land area
Closed Forest	1131.6	1.2	5.7	9.4	1.3	9.8	43.1	1.1	1.9	1205.1	39.3
Open Forest	0.2	287.3	0.5	6.8	0.7	2.2	6.6	0.1	0.0	304.5	9.9
Long Fallow	1.1	0.1	63.2	0.2	0.0	4.8	4.7	0.0	0.2	74.4	2.4
Fragmented Forest	0.5	0.4	0.2	202.1	0.5	2.2	11.2	0.1	0.2	217.5	7.1
Shrubs	0.1	0.1	0.0	0.1	143.5	0.6	9.7	1.8	0.1	155.9	5.1
Short Fallow	1.0	0.3	1.2	1.5	0.2	122.7	11.6	0.2	0.4	139.0	4.5
Other Land Cover	0.6	0.5	0.5	2.3	3.7	4.9	928.4	1.3	2.3	944.4	30.8
Water	0.2	0.0	0.0	0.0	0.8	0.0	1.2	5.6	0.0	7.8	0.3
Plantations	0.0	0.0	0.0	0.0	0.0	0.0	1.1	0.0	18.0	19.3	0.6
Total 2000 ---->	1135.2	290.0	71.5	222.5	150.6	147.3	1017.6	10.2	23.2	3068.0	
% of total land area	37.0	9.5	2.3	7.3	4.9	4.8	33.2	0.3	0.8		

Notes: Classes are ordered according to decreasing indicative woody biomass content, with the exception of the plantation class, so negative changes (from higher to lower biomass) correspond to the values above the diagonal while positive changes are below. The diagonal values represent the areas that did not change during the period.

Table 46-3. Area transition matrixes for the period 1990-2000 by region (million ha)

1990-2000 area transition matrix Africa

(Million ha) Land cover classes in 1990	Closed Forest	Open Forest	Long Fallow	Fragmented Forest	Shrubs	Short Fallow	Other Land Cover	Water	Plantations	Total 1990	% of total land area
Closed Forest	261.4	0.6	0.7	5.7	0.0	5.0	2.0	0.0	0.1	275.6	22.5
Open Forest	0.0	186.1	0.2	5.8	0.1	1.8	2.9	0.0	0.0	197.0	16.1
Long Fallow	0.0	0.1	16.3	0.1	0.0	0.5	0.2	0.0	0.0	17.2	1.4
Fragmented Forest	0.2	0.2	0.0	139.8	0.1	1.8	6.0	0.0	0.0	148.2	12.1
Shrubs	0.0	0.0	0.0	0.0	42.9	0.3	1.5	0.0	0.0	44.9	3.7
Short Fallow	0.8	0.2	0.3	0.9	0.2	65.3	2.9	0.1	0.0	70.7	5.8
Other Land Cover	0.1	0.2	0.0	0.4	0.2	0.4	467.0	0.3	0.1	468.7	38.3
Water	0.0	0.0	0.0	0.0	0.0	0.0	0.2	0.2	0.0	0.4	0.0
Plantations	0.0	0.0	0.0	0.0	0.0	0.0	0.1	0.0	1.2	1.3	0.1
Total 2000 ---->	262.6	187.4	17.6	152.8	43.5	75.1	483.0	0.6	1.4	1224.0	
% of total land area	21.5	15.3	1.4	12.5	3.6	6.1	39.5	0.0	0.1		

1990-2000 area transition matrix Asia

(Million ha) Land cover classes in 1990	Closed Forest	Open Forest	Long Fallow	Fragmented Forest	Shrubs	Short Fallow	Other Land Cover	Water	Plantations	Total 1990	% of total land area
Closed Forest	193.0	0.4	3.6	0.7	0.1	3.2	9.4	0.7	1.8	213.0	34.9
Open Forest	0.2	24.7	0.2	0.2	0.2	0.3	1.1	0.0	0.0	26.8	4.4
Long Fallow	0.5	0.0	41.7	0.0	0.0	3.9	2.8	0.0	0.2	49.3	8.1
Fragmented Forest	0.1	0.0	0.2	16.8	0.2	0.2	1.3	0.0	0.1	19.0	3.1
Shrubs	0.0	0.0	0.0	0.0	8.6	0.2	0.8	0.0	0.0	9.7	1.6
Short Fallow	0.1	0.0	0.7	0.0	0.0	41.3	6.3	0.1	0.4	49.0	8.0
Other Land Cover	0.2	0.2	0.2	0.4	0.2	0.3	222.7	0.3	2.0	226.3	37.1
Water	0.0	0.0	0.0	0.0	0.0	0.0	0.1	1.2	0.0	1.4	0.2
Plantations	0.0	0.0	0.0	0.0	0.0	0.0	0.7	0.0	15.3	16.0	2.6
Total 2000 ---->	194.2	25.3	46.6	18.3	9.3	49.5	245.1	2.5	19.8	610.5	
% of total land area	31.8	4.2	7.6	3.0	1.5	8.1	40.1	0.4	3.2		

1990-2000 area transition matrix Latin America

(Million ha) Land cover classes in 1990	Closed Forest	Open Forest	Long Fallow	Fragmented Forest	Shrubs	Short Fallow	Other Land Cover	Water	Plantations	Total 1990	% of total land area
Closed Forest	677.1	0.2	1.4	3.0	1.1	1.7	31.7	0.3	0.0	716.6	58.1
Open Forest	0.0	76.6	0.1	0.8	0.3	0.1	2.7	0.0	0.0	80.7	6.5
Long Fallow	0.5	0.1	5.3	0.1	0.0	0.3	1.6	0.0	0.0	7.9	0.6
Fragmented Forest	0.2	0.2	0.0	45.5	0.2	0.1	3.9	0.1	0.1	50.3	4.1
Shrubs	0.0	0.1	0.0	0.0	92.0	0.0	7.4	1.7	0.0	101.3	8.2
Short Fallow	0.1	0.1	0.2	0.5	0.0	16.1	2.4	0.0	0.0	19.4	1.6
Other Land Cover	0.4	0.1	0.3	1.5	3.3	4.2	238.7	0.7	0.2	249.3	20.2
Water	0.1	0.0	0.0	0.0	0.8	0.0	0.9	4.1	0.0	6.0	0.5
Plantations	0.0	0.0	0.0	0.0	0.0	0.0	0.3	0.0	1.6	1.9	0.2
Total 2000 ---->	678.5	77.2	7.4	51.4	97.7	22.7	289.6	7.0	2.0	1233.5	
% of total land area	55.0	6.3	0.6	4.2	7.9	1.8	23.5	0.6	0.2		

Notes: See Table 46-2.

distributed throughout the tropics, while changes from short fallow to long fallow were observed mostly in Asia.

Table 46-4 reports the forest area estimates for the f3 definition of forest. The forest area for the surveyed area in 2000 was estimated at 1.6 billion hectares, or about 50 percent of the surveyed area. Half of this area was in Latin America.

Deforestation was defined as the sum of all area transition from forest to non-forest classes. The net area change was estimated as the difference of the transitions resulting from non-forest into forest classes minus deforestation. The deforestation rate was estimated at 0.52 percent per year, or 9.2 million hectares per year, for the pan-tropical zone for the time period 1990-2000. It corresponds to a net area change of -8.6 million hectares per year during the period (Table 46-5). Standard errors at the regional level were relatively high, and differences of deforestation rates among geographical regions were not statistically significant at the 5 percent level.

Reporting on forests through the remote sensing survey was classified according to ecological zones by grouping classes from the FRA 2000 global ecological zone map to obtain three aggregate zones (see Chapter 47, Figure 46-7 and Table 46-6):

- tropical rain forest (no or short dry season);
- tropical moist deciduous forest (three to seven dry months);
- tropical dry forest (more than six dry months).

To aggregate the statistics for the ecological zone of interest, the sampling units were classified according to their location relative to the ecological zone covering most of the sampling unit area, since zones transected some of the sample units.

The distribution of forests by ecological zones showed that the surveyed forests were concentrated mainly in the tropical rain forests. Deforestation estimates by ecological zones show that the forest loss is also concentrated in the rain forests.

Comparison of the forest changes, 1980-1990 and 1990-2000

Statistical tests showed no significant difference in the estimates of deforestation at the 5 percent level of significance for the two study periods (1980-1990 and 1990-2000) at either regional or pan-tropical level (Figure 46-8).

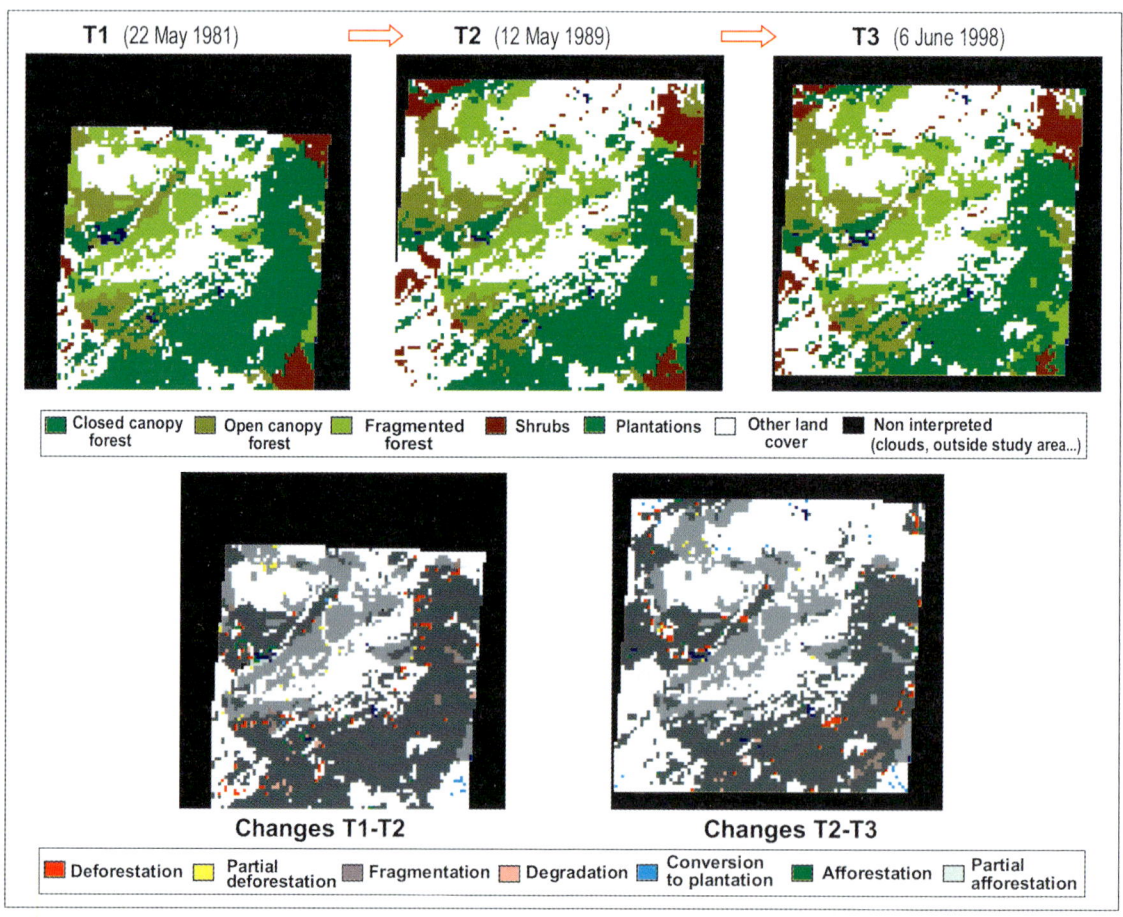

Note: Pixel size is 2 x 2 km².

Figure 46-5. Results for a sampling unit in Zimbabwe: raster maps based on dot-grid registrations

At the ecological zone level, deforestation in the tropical moist deciduous forest zone was found to be significantly different between the two study periods (1980-1990 and 1990-2000) (Figure 46-9). In this zone, both the net forest area change and the deforestation rate decreased significantly at the 5 percent level of significance. For the other ecological zones, differences in the net forest area change and annual deforestation rate was not significant.

Main forest change processes by region

Standardized transition matrixes were used to depict major forest change processes and to quantify their relative importance at the pan-tropical and regional levels. Change processes were classified according to the extent of forest degradation, the size of the activity contributing to the deforestation, the main driving forces involved in the change and the types of land use involved. According to these criteria four deforestation processes were differentiated:

- ***Expansion of shifting cultivation into undisturbed forests.*** This process occurred in forests where shifting cultivation or degradation began after 1980. The impact on the forests was moderate and gradual, as the shifting cultivation incrementally expanded into them. This process was denoted by transitions from closed and open forest classes to the long fallow class, and from closed forest to open forest.

Table 46-4. Estimates of forest area by region and at pan-tropical level in 2000

Region	Forest area			
	Million ha		%	
	Estimate	SE	Estimate	SE
Africa	519	37	42	3
Latin America	780	49	63	4
Asia	272	23	45	4
Pan-tropical	1 571	66	51	2

Notes: SE = Standard error of the mean. The figures are related to the surveyed area, representing about 90 percent of the total forest land in the pan-tropical region. The estimates refer to the f3 definition of forest.

Table 46-5. Annual deforestation and net forest area change during the period 1990-2000 by region and at pan-tropical level

Region	Annual deforestation million ha/year	Annual net forest area change million ha/year		Deforestation rate %/year	
	Estimate	Estimate	SE	Estimate	SE
Africa	2.3	-2.1	0.4	0.34	0.06
Asia	2.5	-2.3	0.6	0.79	0.20
Latin America	4.4	-4.2	1.1	0.51	0.15
Pan-tropical	9.2	-8.6	1.3	0.52	0.08

Notes: SE = Standard error of the mean. The f3 definition of forest was used.

Table 46-6. Annual deforestation and net forest area change during the period 1990-2000 by ecological zone

Ecological zone	Annual deforestation million ha/year	Annual net forest area change million ha/year		Annual deforestation rate million ha/year	
	Estimate	Estimate	SE	Estimate	SE
Tropical rain forest	6.0	-5.7	1.2	0.59	0.14
Tropical moist deciduous forest	2.4	-2.2	0.4	0.43	0.07
Tropical dry forest	0.8	-0.7	0.3	0.38	0.13

Notes: SE = Standard error of the mean. The f3 forest definition was used.

- ***Intensification of agriculture in shifting cultivation areas.*** This process occurred in forests already impacted by shifting agriculture practices in 1980. It also occurred in areas where shifting cultivation had become more intense (where the fallow period

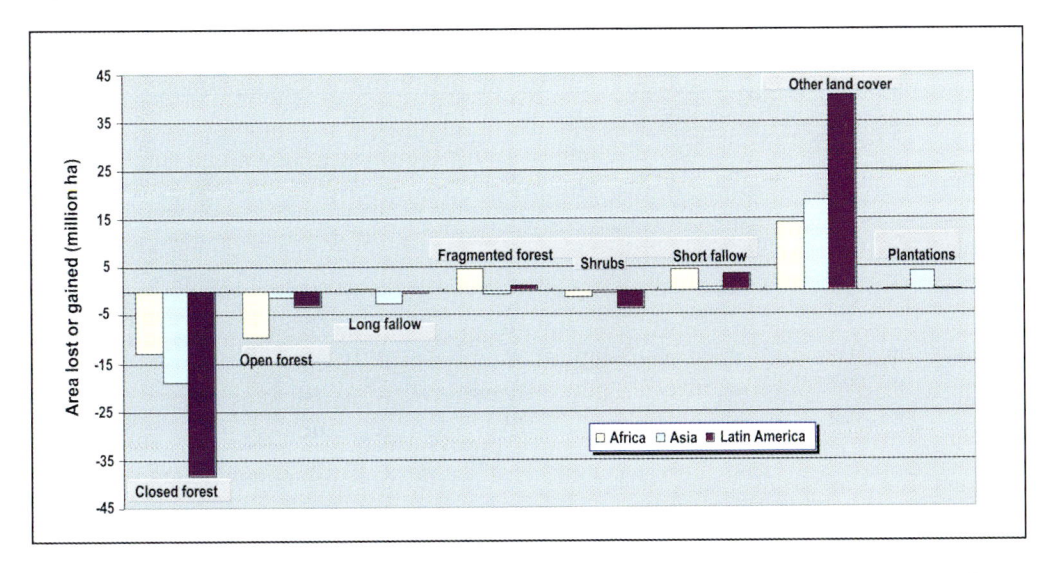

Figure 46-6. Summary of net changes during the period 1990-2000 by land cover classes by region

decreased) or where a complete transition from shifting to permanent agriculture had occurred from the 1980s to the 1990s. For this study, it included the transitions from the long fallow class to fragmented forest and short fallow; and from the short fallow class to other land cover.

- **Direct conversion of forests to small-scale permanent agriculture.** In this process, small areas of forest (less than 25 ha) were converted to agriculture. For this study, the transitions were represented in changes from closed and open forest to fragmented forest and short fallow, and from fragmented forest to either short fallow or other land cover.
- **Direct conversion of forest area to large-scale agriculture.** In this process, large areas (greater than 25 ha) of closed forest, open forest and long fallow were converted to other land cover. (This could also be represented by the more or less simultaneous conversion of smaller adjoining areas which, when aggregated, occupied an area of more than 25 ha. Such areas were indistinguishable in satellite imagery from large uniformly converted areas of forests.)

At the pan-tropical level, deforestation in undisturbed forests was prevalent and evenly distributed between large- and small-scale conversions to agriculture. Regional variations (Figure 46-10) in change processes are summarized as follows.

- **Africa.** The major process of deforestation was due to the conversion of forest for the establishment of small-scale permanent agriculture.
- **Latin America.** Deforestation due to conversion to large-scale permanent agriculture was the predominant process.
- **Asia.** The major process was the direct conversion of forest to large-scale agriculture, with other processes contributing substantially to deforestation as well.

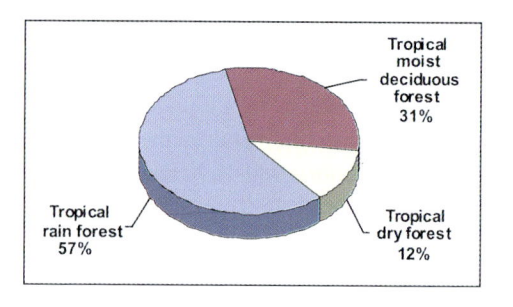

Figure 46-7. Distribution of the forest by ecological zone in 2000 (f3 definition)

Comparison with FRA 2000 statistics from countries

FRA 2000 included a separate assessment of forest state and change using existing information from countries. The results of the two studies were compared to analyse the relationships between the two and to find ways of using the two data sets together to obtain an integrated estimate at the worldwide level.

It was observed that the two assessment components differed in the following respects.

- **Resolution.** The country statistics provided estimates at the national level, while the remote sensing survey was designed to provide information at the pan-tropical and regional levels.
- **Definitions.** The forest definitions used were close but did not correspond exactly between the two approaches. Country statistics were adjusted to a FRA 2000 global forest definition based on both use and cover, while the remote sensing survey used a uniform land cover definition based on photo-interpretation criteria.
- **Geographic coverage.** The areas surveyed were different. While the assessment based on country information was conducted worldwide, the remote sensing survey covered only 63 percent of the land area in the tropics,

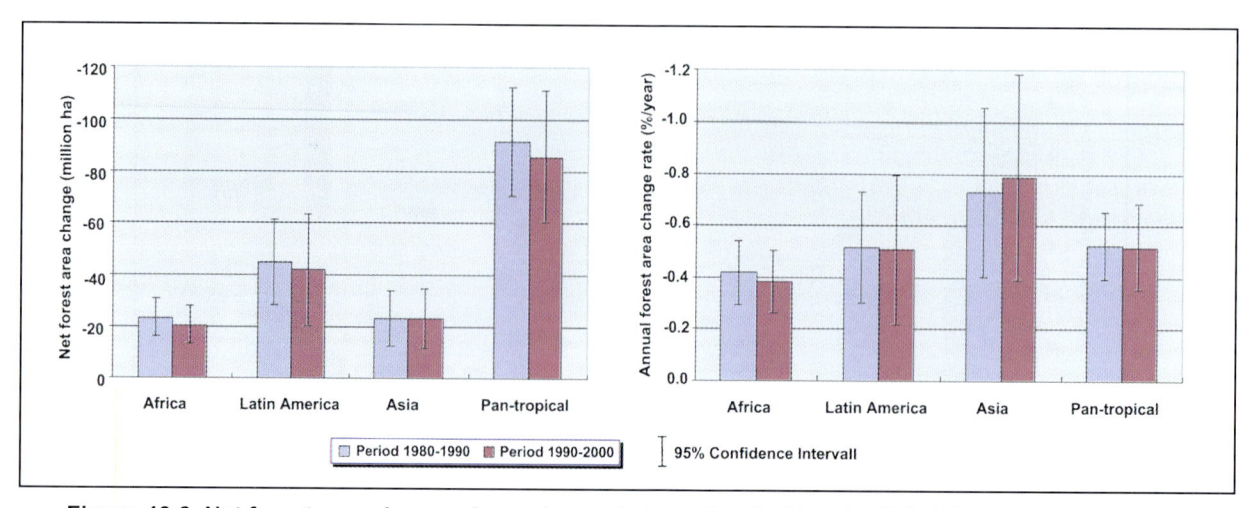

Figure 46-8. Net forest area changes by region and at pan-tropical level, 1980-1990 and 1990-2000 (left); annual deforestation rate by region and at pan-tropical level, 1980-1990 and 1990-2000 (right)

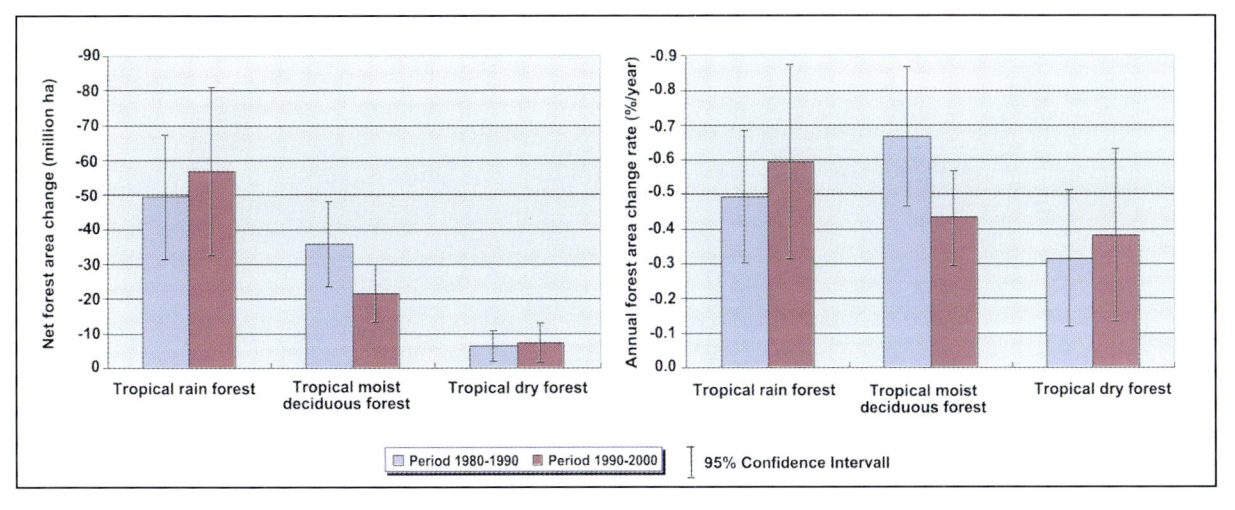

Figure 46-9. Net forest area change by ecological zone, 1980-1990 and 1990-2000 (left); annual forest area change by ecological zone, 1980-1990 and 1990-2000 (right)

representing about 87 percent of the world's tropical forests. Within the land area of the survey, Landsat scenes with less than 10 percent forest were placed into a stratum that was not sampled. Landsat frames with land area of less than 1 million hectares were also not included, whereas information from countries theoretically covered the entire land area.

- **Measurement techniques.** Country statistics were based on a wide range of reference data derived from a number of methods (expert opinion, maps based on satellite imagery, field surveys and sampling), while the remote sensing survey relied on interpreted satellite imagery and objective statistical sampling.
- **Currency of information.** The remote sensing survey was based on imagery acquired near the reference years 1980, 1990 and 2000 (with some variations), while the average date of the country information from developing countries was 1994, although some of the country data were older or more recent.

Variations between the two information sets could contribute to differences in the respective estimates; consequently a direct comparison between the two was impossible. However, because the remote sensing survey was conducted under relatively controlled conditions and employed the application of statistical sampling, it was used as a calibration tool at the regional level to improve some of the overall findings for the tropics.

Comparisons between the country-based findings and the remote sensing survey estimates were limited to the 73 countries that were covered by the remote sensing survey (Table 46-7). Sixty of these countries were covered by at least a part of one sampling unit. Only results at the subregional, regional and pan-tropical levels were examined (as the remote sensing survey was not used for generating national level results) using the f2 definition of forests (since it corresponds most closely to the definition used for the country statistical data).

Forest area estimates from the remote sensing survey were in general lower than estimates from the country data in the tropics, throughout the regions, and in most subregions. Nevertheless, there is a good correlation between the country data and the remote sensing estimates, observable at the subregional and regional levels (Figure 46-11).

The forest area change estimates from the two information sets were comparable for Asia and Latin America. However, the data for Africa were not comparable and consequently the correlation at the pan-tropical level was also low. The subregions contributing most to the disparity of the two data sets were East Africa and southern Africa. The disparity could be attributed primarily to two causes.

- **Seasonality and ecological conditions.** In dry areas, difficulties are commonly encountered in the use of satellite imagery to classify and interpret vegetation and to detect change. Leaf cover in such forests is low, exception during the short rainy season. When leaves are green the forests show up well in the imagery, but when they are absent it is difficult to detect and interpret the vegetation.
- **Inconsistencies in specific countries.** Country data from a few countries – the People's Democratic Republic of the Congo, the Sudan and Zambia – contributed to the high deforestation rate in Africa. Deforestation

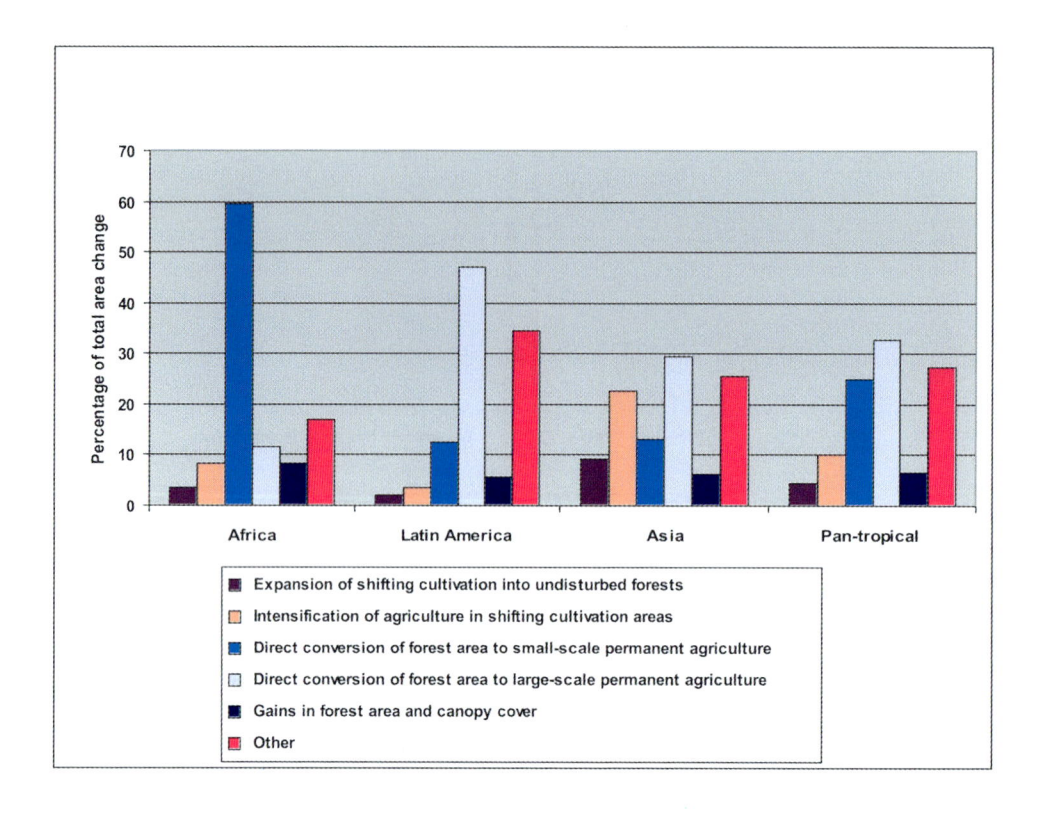

Figure 46-10. Percentage of total area change by individual change processes at regional and pan-tropical level for the period 1990-2000

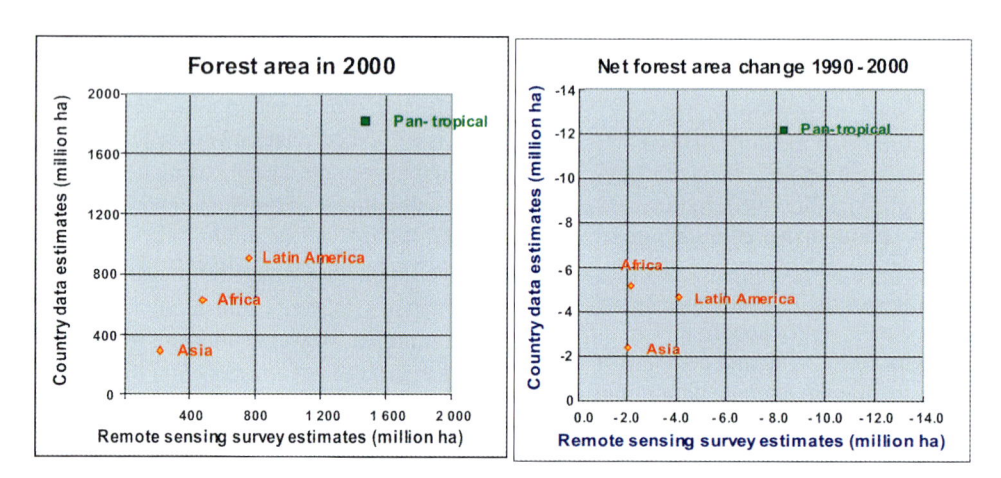

Figure 46-11. Forest area in 2000 (left) and net forest area change (right) - comparison between country data and remote sensing survey estimates

rates for the sampling units in the Sudan and Zambia were much lower than those calculated from the country data. This is not unexpected, as sampling units were not designed to provide representative national statistics and may have been located in areas that had lower deforestation rates within the countries. It is also possible that the country data from the Sudan and Zambia overestimated deforestation. For example, the baseline data for Zambia were from 1978, and the data for the Sudan from 1990 covered only one-third (the gum belt) of the country.

Moreover, the change estimates were based on expert opinion or on estimates from surrounding countries owing to the absence of comparable time series of information for both countries.

SOURCES OF ERROR AND IMPACT ON RESULTS

Statistical errors

Statistical errors identified in the survey were sampling errors, measurement errors, missing

Table 46-7. Comparison of forest area and forest area change estimates from the remote sensing survey with those from country data (using the f2 forest definition)

Region	Forest area 2000 million ha			Annual net forest area change million ha/year			Annual deforestation rate %/year		
	Country data	Remote sensing survey	Significant difference	Country data	Remote sensing survey	Significant difference	Country data	Remote sensing Survey	Significant difference
Africa	622	484	**	-5.2	-2.2	***	0.77	0.43	***
Asia	289	224	**	-2.4	-2.0	n.s.	0.78	0.84	n.s.
Latin America	892	767	**	-4.4	-4.1	n.s.	0.45	0.51	n.s.
Pan-tropical	1 803	1 475	***	-12.0	-8.3	**	0.62	0.54	n.s.

Notes: Only the results from the countries included in the remote sensing survey were compiled to obtain the country data given in the table. The hypothesis tested in the table is that the country data value is the true value of the sampled population of the remote sensing survey. The level of significance of the difference between country data and remote sensing estimates: *** = 0.01 percent level of significance, ** = 1 percent level of significance, * = 5 percent level of significance, n.s. = not significant at the 5 percent level.

values and discrepancies between the target population and sampled population.

Sampling errors. The sampling error depends on the sample design and the variation within the population and is quantified by the standard error. For each estimate calculated in the survey the corresponding standard error (or more precisely, the root mean square error, since estimates are ratios) was calculated. Some of these error estimates were covered in the previous sections.

Both estimated values and standard errors for relative forest area were close to those reported in FRA 1990 (FAO 1996). The estimates of the relative forest cover for 1980 and 1990 deviated slightly from the FRA 1990 report and the standard errors were slightly higher. One explanation for this deviation is the restriction of statistical calculations to the common area of the images for all three dates.

The estimated values of the deforestation rate for 1980-1990 were somewhat lower for FRA 2000 than those generated in FRA 1990. One explanation for this difference is the use of different standardization methods for adjustment of the information to standard reference years.

The standard errors of the estimator of the deforestation rate 1980-1990 were of the same magnitude as (or smaller than) those reported in FRA 1990. The standard errors were somewhat larger for the 1990-2000 period. The differences in the errors for the two reporting periods may be due to chance (a consequence of the sampling error of the standard deviation) or may indicate that the variance of the deforestation had increased in the surveyed area. A third reason could be that the year 2000 statistics were almost entirely extrapolated, which could potentially magnify observed variations.

The calculations show that relatively few sampling units contributed a great deal to the standard error, indicating a true large variation among the units with respect to the variables studied. The stratification and allocation of the number of units per stratum were neutral with respect to some characteristics to be estimated and guaranteed an approximately area-proportional coverage of the surveyed area, but they were not the most efficient for estimating, for example, the changes in deforestation rates.

For a few strata the sample sizes were lower than planned because of the lack of suitable data available in some locations (owing to high cloud cover). Four sampling units, of which three belonged to the same stratum, could not be interpreted.

The estimate of the forest area 2000 could be slightly improved by considering only the 1990-2000 common area instead of the common area for all three observation dates.

Measurement errors. The direct influence of moderate measurement errors on the results have been by numerical and theoretical studies not presented here, shown to be of minor importance.

Missing values. There are missing values in the FRA 2000 remote sensing survey, since parts of scenes were covered by clouds. The presence of clouds might very well be correlated to the proportions of different land cover classes, which would explain why the missing values can cause bias in certain estimates.

Discrepancy between target population and sampled population. Discrepancies between target and sampled populations occur because the entire population cannot be sampled. In the present study, scenes with small land area (e.g. coastal regions) were excluded for reasons of cost efficiency. The sampled population covered about 87 percent of the tropical forest land. The excluded scenes are likely to be different from those that were sampled; thus the results cannot be considered valid for all tropical forest land.

Interpretation accuracy

The accuracy of the interpretative work is difficult to estimate without further study and quality control and accuracy assessment. Several aspects can be considered as sources of error:
- classification accuracy (discrimination between classes);
- change detection (identification of the changes between two observations);
- consistency of interpretation and homogeneous use of the classification among photo interpreters;
- data registration errors.

For further details see FAO 1996.

Modelling effects (effect of the standardization)

The standardization process was motivated by the necessity to adjust the observed transition matrixes to the reference years 1980, 1990 and 2000. There is no general theory for describing effects of modelling errors. Empirical studies could help in evaluating the effects of the model but could not be carried out within the scope of FRA 2000. However, examples can elucidate the impact of these errors.

For both methods, constant and linear, there is a risk that errors were induced in the standardized matrixes, which do not reflect actual changes during the periods between the data acquisition. This risk is greatest when the first observation date of the imagery (T1) was well before 1980. For example, if the date of the T1 image was 1974 and that of T2 was 1991, forest cover at 1980 and deforestation for the period 1980-1990 will be underestimated if most deforestation actually took place after 1980. Conversely, considering the same theoretical acquisition dates, if large amounts of deforestation actually took place between 1974 and 1980, the deforestation rate for 1980-1990 would be overestimated. (The average acquisition dates for T1 and T2 are 1977 and 1989 respectively.) The phenomenon does not apply only to deforestation but to any class-to-class transition. The same is also the case for the 1990-2000 period, since most T3 images are from before 2000. If there were significant changes in about the last two years these would not be reflected in the standardized time-adjusted information sets (1998 is the average date for T3 images).

In many cases, when the observation dates are close to the reference years, the two adjustment methods generate similar results. However, when the two observed transition matrixes are very different, the 1990 state statistics and the standardized matrixes can differ. This is explained by the predicting property of the linear method. By analogy with ordinary first and second degree interpolation, it can be assumed that the linear method will result in a smaller difference (in absolute value) between the two consecutive deforestation rates than the constant method. This is also derived intuitively, since the constant method places all the difference between the two rates at the time T2, close to 1990.

The effects of moderate "random" interpretation errors on the resulting standardized matrixes were studied to the extent possible. The effects seem small and no great risk for error propagation should exist. The initial error is in principle first "transformed" into an annual error and then multiplied by the number of years needed to adjust the observed transition matrix. This indicates that the final error is often smaller than the initial error and seldom larger than doubled.

TECHNICAL RECOMMENDATIONS

Statistical and design improvements

The FRA 2000 estimates of forest area could be improved by eliminating the restriction to only the common area of all three observation dates. Also forest area change estimates for the latest period could be improved by using the common area of the last two dates.

In the present survey, general information in the form of vegetation maps was used for stratification within each subregion. The allocation is roughly area-proportional and is thus not likely to be optimal for estimating changes in deforestation rate.

The precision of the estimates of the survey could be improved in essentially two ways – either by increasing the sampling efficiency (through a larger sample area or through better distribution of the sampled area, for example by using smaller sampling units) or by using external information for a more efficient sample or a more efficient estimator. The following are some possible improvements.
- The estimators could be improved by using collateral information for a two-phase (or perhaps multiphase) sampling design. This

- could be first tested on a small scale (e.g. for a subregion) and with data already available.
- The precision questions and sample sizes could be reconsidered. It might be worth taking some extra samples for improving the precision for some important characteristic.
- The stratification could be reconsidered, especially the allocation of the sample. There are problems with changing the stratification, but the allocation can be changed without very large complications.

Standardization improvement

Much work has been done in the FRA 2000 remote sensing survey to overcome the problems of standardizing statistics to the reference dates. This will continue to be a challenge for the remote sensing survey in the next global forest resources, if a four-date change analysis is carried out.

The relative reliability of the results obtained using any standardization method is difficult to assess since real and sometimes dramatic transitions can occur between the two consecutive observation dates. Problems in reliability are expected to be greatest when the observation dates deviate greatly from the reference years. Therefore, an obvious recommendation is to acquire imagery as close as possible to the reference dates.

CONCLUSION

The remote sensing fulfilled its objectives by providing a detailed set of information describing the state and change of tropical forest at different aggregation levels for the periods 1980-1990 and 1990-2000. One major accomplishment of the survey was to produce a comparable set of information on forest change in the tropics spanning two decades.

In precision the results are consistent with the FRA 1990 findings, and correspond with expected levels. Improvements in future designs could increase the precision of the forest area change estimates and the comparison between two periods.

The major results of the current survey are as follows.

- The net forest area change was estimated at -8.6 million hectares annually for the 1990-2000 period.
- No significant difference in deforestation could be identified between the two periods at the pan-tropical or regional levels, although the decrease in the rate of deforestation in the 1990s in tropical moist deciduous forests was significantly less than in the period 1980-1990.
- The main deforestation process was the direct conversion of forests to permanent agriculture.
- Comparisons with FRA 2000 country data showed a high and statistically significant difference in the forest area change estimates for Africa for the period 1990-2000.

BIBLIOGRAPHY

Czaplewski, R. 1994. Statistical evaluation of FRA 90 Results. FAO.

FAO. 1996. *Forest Resources Assessment 1990. Survey of tropical forest cover and study of change process.* Forestry Paper No. 130. Rome.

FAO. 2001. *FRA 2000: Pan-tropical survey of forest cover changes 1980-2000.* FRA Working Paper No. 49. www.fao.org/forestry/fo/fra/index.jsp.

Raj, D. 1968. *Sampling theory.* New York, McGraw-Hill.

Chapter 47
Global mapping

ABSTRACT

FRA 2000 developed new global forest and ecological maps which give spatial definition to area statistics of the survey findings from individual countries and regions, providing a synoptic view of worldwide forest cover. The global ecological zoning map provides an important means of aggregating global information on forests or other natural resources according to their ecological character. Together the maps are useful for the analysis and depiction of worldwide forest cover according to the forests' ecological character.

The forest cover map was developed using coarse-resolution satellite imagery. In previous global assessments, the means and technology did not exist to produce a global map based on satellite imagery. Thus the technical map based on state-of-the-art technology replaces mere illustrations of global forests.

The ecological zoning map, based on a standard global classification, was produced using existing national and regional potential vegetation maps, climate data and satellite imagery.

A third map of protected forests was also developed and used in estimating the area of forest under formal protection worldwide. Inputs were collected from countries around the world.

Each map is generated from a corresponding computerized geographic information system (GIS) database, which makes it possible to combine the maps with different spatial and statistical data, permitting new perspectives on the world's forests. Computerized maps and databases are more easily updated than conventional maps, and they set the groundwork for future assessments as well. Digital versions of the maps are available to researchers and the general public through the FRA Web site (www.fao.org/forestry/fo/fra/index.jsp).

INTRODUCTION

FRA 2000 produced three global maps: a forest map, an ecological map and a protected areas map. Each map is generated from a corresponding computerized geographic information system (GIS) database. This makes it possible to combine the maps with spatial and statistical data from other sources for computation of statistics at the global, regional and ecological zone levels, permitting new perspectives on the world's forests. Computerized maps and databases are more easily updated than conventional maps and lay the groundwork for future assessments.

The global forest map shows the extent and location of major forest formations throughout the world (Figure 47-1). The ecological map can be combined with other maps or data to help quantify or depict global forests according to their ecological character (Figure 47-2). The protected area map depicts the location, extent and type of protected area for each country of the world.

The forest and ecological zone maps are useful to a scale of 1:40 000 000, although enlargements are possible up to 1:10 000 000.

Digital versions of the maps are available to researchers and the general public through the FRA Web site, with the exception of the protected areas map, which is managed exclusively by the UNEP World Conservation Monitoring Centre (UNEP-WCMC).

FAO worked with several cooperators in the development of the various maps, including the EROS Data Center (EDC), United States; UNEP-WCMC, United Kingdom; the International Institute for Applied Systems Analysis (IIASA), Austria; the Ecological Laboratory of Toulouse (LET), France; the Tropical Science Center, Costa Rica; the Autonomous University of Mexico (UNAM); the Canadian Forest Service (CFS); the Forest Service of the United States Department of Agriculture (USDA); Damascus University, Syrian Arab Republic; the Institute of Remote Sensing Applications (IRSA), China; the Chinese Academy of Sciences; and the Australian Bureau of Rural Sciences (BRS). EDC conducted all the image processing for the forest map and the

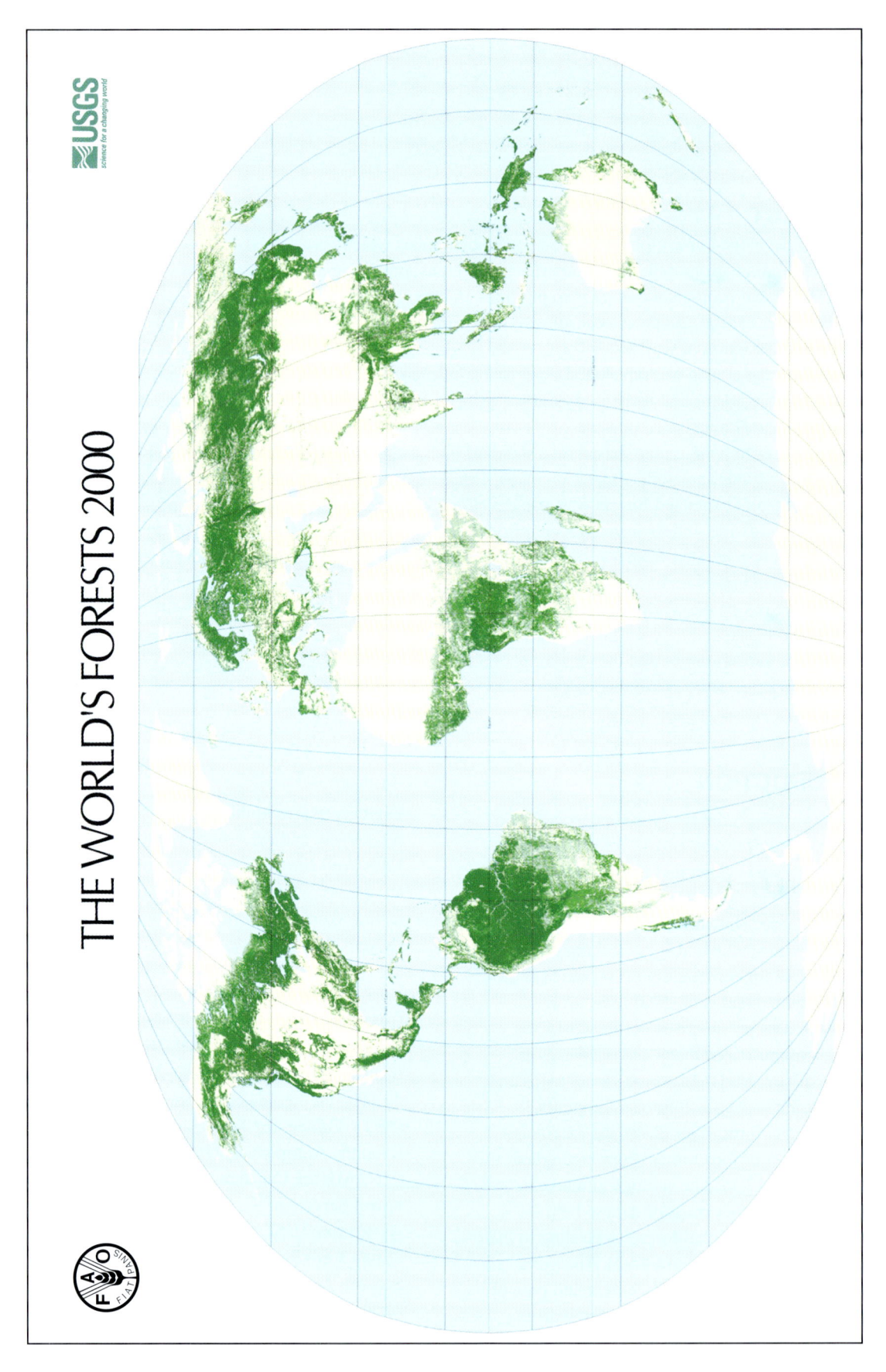

Figure 47-1. FRA 2000 global forest cover map

Global mapping 323

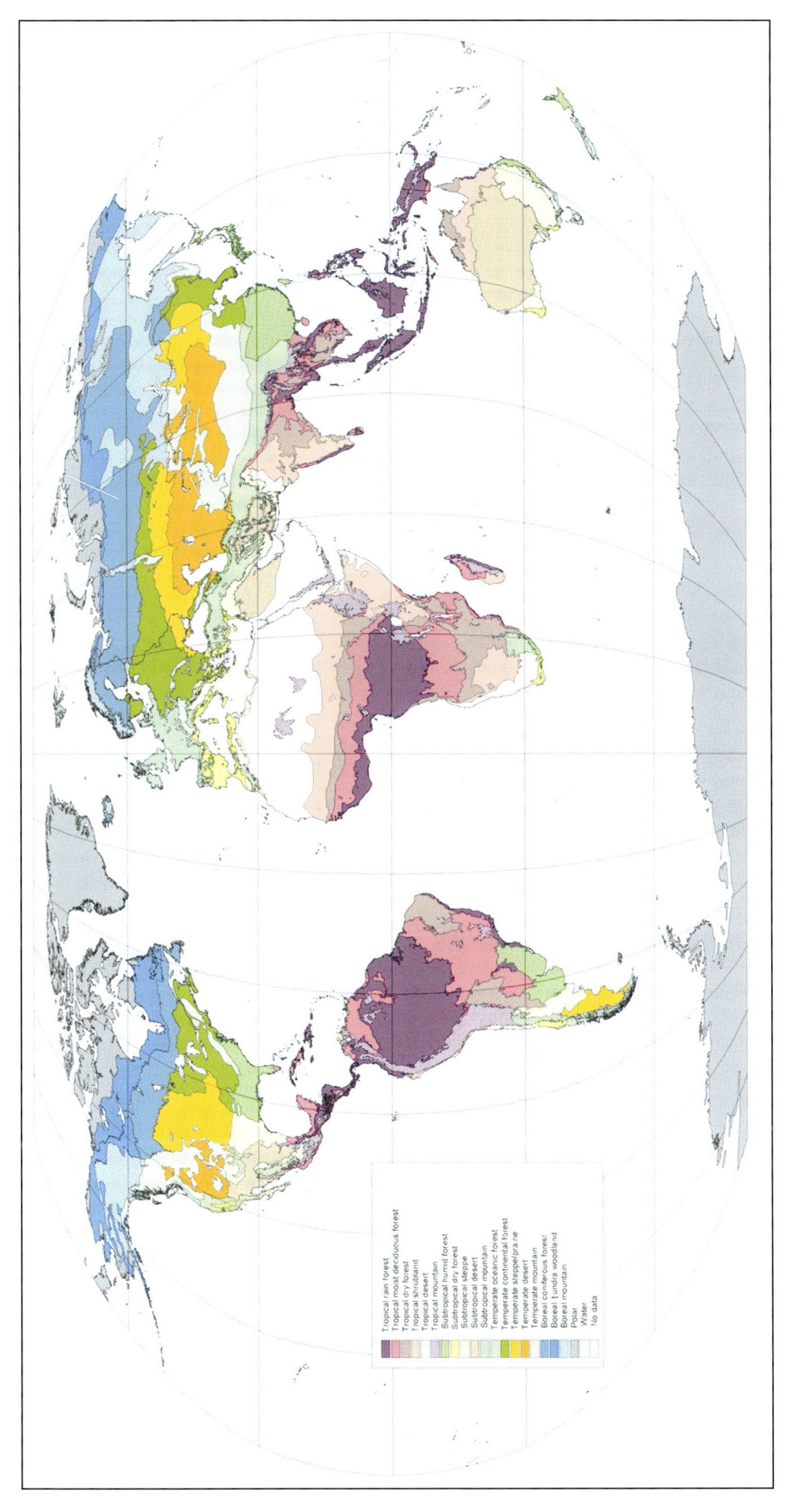

Figure 47-2. FRA 2000 global ecological zone map

Table 47-1. FRA 2000 global land cover map legend, definitions and representative land cover types

FRA 2000 class	FAO definition	Representative land cover
Closed forest	Land covered by trees with a canopy cover of more than 40 percent and height exceeding 5 m. Includes natural forests and forest plantations.	Tropical/subtropical moist forest Temperate broadleaf mixed forest Subtropical/temperate conifer plantation Boreal conifer forest
Open or fragmented forest	Land covered by trees with a canopy cover between 10 and 40 percent and height exceeding 5 m (open forest), or mosaics of forest and non-forest land (fragmented forest). Includes natural forests and forest plantations.	Northern boreal/taiga open conifer or mixed forest Southern Africa woodland Tropical fragmented/degraded forest
Other wooded land	Land either with a 5 to 10 percent canopy cover of trees exceeding 5 m height, or with a shrub or bush cover of more than 10 percent and height less than 5 m.	Mediterranean closed shrubland Tropical woody savannah
Other land cover	All other land, including grassland, agricultural land, barren land, urban areas.	Grassland, cropland, non-woody wetland, desert, urban
Water	Inland water.	Inland water

global mosaic for the ecological zoning map. UNEP-WCMC compiled the global protected areas map. FAO organized and coordinated the work and carried out final quality control and edge-matching for all the maps. Other partners contributed valuable assistance to the technical construction and thematic content of the maps.

The development of the three global maps represented a major technical challenge for FRA 2000. Each was produced using the best available information for the purpose. The forest cover map was developed using coarse-resolution satellite imagery, the ecological map from national and regional potential vegetation maps and climate data, and the protected areas map from independent maps and point information supplied by countries.

The global maps provide a synoptic overview of the worldwide situation regarding forests, ecological zones and protected areas. They were used in conjunction with statistical data for the FRA 2000 reports on forest area by ecological zone, forest area under protection, protected areas within ecological zones and other parameters.

FOREST COVER MAP

The FRA 2000 forest map took three years to complete and shows the location and distribution of forests according to FRA 2000 classification criteria.

Overall consistency at the global level was viewed as an important objective for all the maps. For forest cover mapping, this could only be achieved by using a common input source such as satellite imagery and applying similar classification criteria for all areas. The classification scheme for the map was developed using the same criteria that were used in FAO's global assessment based on country statistical data and in its high-resolution remote sensing sampling programme (Table 47-1). Consequently, the global map could be integrated into the overall FRA framework and can be used in conjunction with other data sets. The map can also be used as a simple visual aid to show the location and extent of forests around the world, according to FAO's terminology.

One of the difficult and expensive tasks in producing a global map from satellite data, including imagery from the Advanced Very High Resolution Radiometer (AVHRR), is piecing together a large amount of data to produce a single cloud-free data set. Because clouds obscure forests, as well as other land cover, they must be eliminated before mapping can begin. Therefore, the forest map relied to a large extent on the Global Land Cover Characteristics Database (GLCCD) produced by EDC. This database proved invaluable to the mapping, since many of the problems of cloud cover and reflectance anomalies had been resolved through preprocessing and the use of a multi-date composite containing only the best image data. Nevertheless, the lack of good imagery prevented the mapping of several Pacific Islands. Source data for the forest map were drawn from the 1995-1996 data set, which was the latest imagery available in the GLCCD archive. This imagery consisted of five calibrated AVHRR bands and a Normalized Difference Vegetation Index (NDVI) band (Zhu and Waller 2001).

Although correspondence between the GLCCD and FRA classification schemes was considered generally good, the entire range of

GLCCD classes could not be recoded easily into FRA 2000 classes. For example, in South America 34 of the original 167 classes required further processing for FAO's map (Zhu *et al.* 1999). Consequently, EDC adopted a methodology based on a combination mixture model with scaling of NDVI values and the visible band based on pixel positions along the infrared band. Regional variations in forest cover and associated reflectance required stratification of the processing into geographic divisions and adjustment of the models according to their respective conditions (Zhu and Waller 2001).

Once an advanced draft forest map was developed in 1999, EDC worked with FAO and UNEP-WCMC on validation and quality control. UNEP-WCMC used its extensive map archive to identify areas that needed further processing, and FAO sent copies to experts and FAO field offices around the world for feedback. This input was used over the final year to refine the map before assessment of its accuracy.

Determining the map's accuracy was viewed as an important step in the mapping exercise, since the final map was going to be used for technical work in conjunction with other data sets. EDC employed the use of an existing set of validation points from the International Geosphere and Biosphere Programme (IGBP) and full land-cover data sets available from the United States and Chinese Governments (Zhu and Waller 2001). FAO also conducted an evaluation of the map using 117 interpreted thematic mapper scenes from the tropics. The results of all these evaluations showed that the average accuracy of the map for all forest classes is about 80 percent. The closed forests are more accurately mapped than the average, and the open and fragmented forests are somewhat less accurately mapped. Other wooded lands are the least accurately mapped of the three classes.

The global forest cover map provides spatial definition for the area statistics and survey findings from the individual countries and regions. In previous global assessments, the means and technology did not exist to produce a global map based on satellite imagery. Thus, FRA 2000 has replaced artists' depictions of global forests with a technically correct map based on state-of-the-art technology.

For illustrative purposes the map was reproduced in Robinson Projection. However, because it exists in a GIS format, it is possible to transform the map (or portions of the map) into other projections according to specific requirements.

Forest area estimates were not derived from the map. However, the map served as a spatial framework for the integration of country statistics which were then used in conjunction with the ecological zoning and protected areas maps to estimate the fraction of forests under protection and in the various ecological zones.

ECOLOGICAL ZONE MAP

The underlying strategy for the FRA ecological zoning reflected both the thematic and technical needs of the map as well as the many operational constraints that were expected in its development. In terms of ecosystem principles, the map requirements were such that zones or classes were defined and mapped using a holistic approach. That is, both biotic and abiotic components of ecosystems were considered in the zoning scheme. Beyond the thematic content and zoning, practical aspects of digital cartographic production, such as data availability, currency, scale and the associated reliability of the map inputs, were also taken into account (Simons 2001).

FAO conducted two preliminary studies to identify specific alternatives and constraints in the development of a global ecological zone (GEZ) map appropriate for FRA 2000 purposes (Preto 1998; Zhu 1997). Findings from these studies, experience in the development of the tropical ecological zone map for FRA 1990, and recommendations from other parties consulted in the process indicated that FAO could not complete an entirely new global ecological zoning map by 2000 because of the large amount of scientific, organizational and financial resources and time required. FAO therefore focused on identifying an existing scheme that might be used or adapted to the programme's needs. A Workshop on Global Ecological Zones Mapping, held in Cambridge, United Kingdom in July 1999, and attended by experts from 15 countries, helped set the framework.

Because of the enormity of conducting the work on a global scale, a classification scheme had to be chosen that would meet FAO's thematic requirements, be practical to construct with available resources and meet the scrutiny of diverse users from all parts of the world. Existing schemes were each developed for specific purposes according to various environmental criteria. Macroclimate (temperature and precipitation) was an element used by most (Preto

Table 47-2. Ecological zone breakdown used in FRA 2000

EZ Level 1 – Domain		EZ Level 2 – Global Ecological Zone		
Name	Criteria (equivalent to Köppen-Trewartha climatic groups)	Name (reflecting dominant zonal[a] vegetation)	Code	Criteria (approximate equivalent of Köppen–Trewartha climatic types, in combination with vegetation physiognomy, and one orographic zone within each domain)
Tropical	All months without frost: in marine areas over 18°C	Tropical rain forest	TAr	Wet: 0-3 months dry,[b] during winter
		Tropical moist deciduous forest	TAwa	Wet/dry: 3-5 months dry, during winter
		Tropical dry forest	TAwb	Dry/wet: 5-8 months dry, during winter
		Tropical shrubland	TBSh	Semi-arid: evaporation > precipitation
		Tropical desert	TBWh	Arid: all months dry
		Tropical mountain systems	TM	Approximately > 1 000 m altitude (local variations)
Subtropical	Eight months or more over 10°C	Subtropical humid forest	SCf	Humid: no dry season
		Subtropical dry forest	SCs	Seasonally dry: winter rains, dry summer
		Subtropical steppe	SBSh	Semi-arid: evaporation > precipitation
		Subtropical desert	SBWh	Arid: all months dry
		Subtropical mountain systems	SM	Approximately > 800-1000 m altitude
Temperate	Four to eight months over 10°C	Temperate oceanic forest	TeDo	Oceanic climate: coldest month over 0°C
		Temperate continental forest	TeDc	Continental climate: coldest month under 0°C
		Temperate steppe	TeBSk	Semi-arid: evaporation > precipitation
		Temperate desert	TeBWk	Arid: All months dry
		Temperate mountain systems	TM	Approximately > 800 m altitude
Boreal	Up to three months over 10°C	Boreal coniferous forest	Ba	Vegetation physiognomy: coniferous dense forest dominant
		Boreal tundra woodland	Bb	Vegetation physiognomy: woodland and sparse forest dominant
		Boreal mountain systems	BM	Approximately > 600 m altitude
Polar	All months below 10°C	Polar	P	Same as domain level

[a] Zonal vegetation: resulting from the variation in environmental, i.e. climatic, conditions in a north-south direction.
[b] A dry month is defined as the month in which the total precipitation expressed in millimetres is equal to or less than twice the mean temperature in degrees Centigrade.

1998; WCMC 1992). Since macroclimate correlates well with the potential vegetation associated with a particular locale, it was considered a logical basis for the FRA ecological zoning as well.

However, a climatic map showing such key features as temperature and precipitation is not necessarily an ecological map until the boundaries are shown to correspond to significant biological boundaries. Likewise, maps of landform types (derived from digital elevation data) are not necessarily ecological maps until it has been shown that the types co-vary with other components of the ecosystem, such as vegetation (Bailey 1998).

For the choice of climatic parameters to be used in the FRA 2000 map a number of global systems were surveyed (Köppen 1931; Trewartha 1968; Thornthwaite 1933; Holdridge 1947). Köppen modified by Trewartha was selected as the best candidate because of the number of classes that corresponded well to FRA 2000 needs. Moreover, while Köppen-Trewartha is based on climate, there is a demonstrated good correspondence between its subzones or climatic types and the natural climax vegetation types and soils within them (Bailey 1996).[56]

FAO, in cooperation with EDC and UNEP-WCMC, thus developed a prototype zoning scheme for FRA 2000 based on Köppen-Trewartha. The zoning was made hierarchical using Köppen-Trewartha's climatic groups and climatic types as FAO ecological zone levels 1 and 2, respectively (Table 47-2). A third level was also tested during the pilot project, representing the differentiation within the first two levels according to landform – distinguishing mountains with altitudinal zonation from lowland plains. This third level was ultimately not used.

At level 1, the broadest level, equivalent to Köppen-Trewartha's climatic groups, five domains are distinguished based on temperature: tropical, subtropical, temperate, boreal, polar.

At the second level, 20 classes or ecological zones are distinguished, which indicate broad zones of relatively homogeneous vegetation, such as tropical rain forest, tropical dry forest and boreal coniferous forest. The names of the global ecological zones reflect the dominant zonal vegetation. Typical azonal vegetation types, for instance mangroves, heath and swamps, are not separately classified and mapped.

Level 2 is the reference or working level for the GEZ mapping. The ecological zones were delineated by using both macroclimate data and existing climax or potential vegetation maps. Use of vegetation maps ensured a more precise delineation of the ecological zones. If generalized climate maps had been used alone, the zones of the final map would probably have corresponded poorly to boundaries of homogeneous vegetation transitions.

Within each domain (level 1) a zone of mountain systems is distinguished at level 2. Mountain systems usually contain a variety of vegetation types and include forests, alpine shrubs, meadows and bare rock. The current global framework cannot address the high, mostly small-scale diversity of mountain habitats. The polar domain is not further subdivided, as it is treeless, and only very sparse shrub or grass vegetation occurs locally. Here the second level is equivalent to the first.

A main principle in delineating the global ecological zones involves aggregating or matching regional ecological or potential vegetation maps into the global framework. The following steps can be distinguished:

- identification of Köppen-Trewartha climatic types and mountains occurring in a region to approximate the level 2 ecological zone class of the FAO scheme;
- establishment of correspondence between regional/national potential vegetation types and the global ecological zones;
- final definition and delineation of the global ecological zones, using the maps and source data consulted in the first two steps;
- edge-matching between adjacent maps;
- validation.

To ensure the best use of regional knowledge and information, existing regional/national maps on vegetation, biogeography, ecology and climate were used to generate the GEZ map. In some countries, such as the United States, classification is based on the Köppen-Trewartha climate system and translation to the FAO scheme was straightforward. In other cases, a more thorough study of mapping criteria, including physiognomy, phenology, floristics and dynamics of vegetation types, was needed to establish the correspondence. A benefit of using the existing country/regional maps is that they could form the basis or provide supporting information for more detailed regional ecological zoning beyond FRA 2000 (see Table 47-3).

The country/regional vegetation maps also helped in harmonization of ecological zone boundaries across countries or regions. The experts who attended the Cambridge workshop contributed in a major way to definition of the ecological zones of their respective regions as well as to edge-matching between adjoining geographic regions.

Both the existing FRA 1990 ecofloristic zone map and several existing regional maps were produced using the ESRI Arc/Info GIS software. Thus, it was convenient for the rest of the work to be conducted using Arc/Info, or at least to be Arc/Info importable. After study of the digital map in the Arc/Info coverage environment and confirmation that the digital version had appropriate attributes for the ecological zones (represented in the map by polygons), the coverage was edited and attributes for FAO ecological zone levels 1 and 2 were added.

Two problems occurred in polygon edge-matching along country and regional boundaries. One was mismatch of polygon definition translations between polygons in adjacent maps. This problem was generally easy to solve by

---

[56] This is largely because Köppen derived his climate classes from observations on the distribution of natural vegetation types on various continents (Köppen 1931).

Table 47-3. Source maps used for the delineation of FAO global ecological zones

Region	Name of map	Scale	Projection	Thematic information / classification criteria
Canada and Mexico	Ecological regions of North America (CEC 1997)	1:10 million	Lambert Azimuthal Equal Area	Holistic classification system based on climate, soils, landform, vegetation and also land use. Hierarchical system: 15 Level I ecological regions and 52 Level II regions.
United States	Ecoregions of the United States (Bailey 1995)	1:7.5 million	Lambert Azimuthal Equal Area	Classification based on Köppen climate system: broad domains equivalent to climate groups, subdivided into divisions approximately equivalent to climate types.
Central America	National Holdridge life zone maps, transformed to a regional base map (Bolanos & Watson 1991; De la Cruz 1976; Hartshorn 1984; Holdridge 1962; Holdridge & Tosi 1971; Tosi 1970; Tosi & Hartshorn 1978)	Various scales Base map at 1:1.5 million	x	Holdrige life zones are defined using the parameters (bio)temperature, rainfall and evapotranspiration.
South America, Africa, Tropical Asia	Ecofloristic zones maps (LET 2000)	1:5 million	Lat-Long	28 groups of ecofloristic zones are defined, based on climate, vegetation physiognomy and physiography, i.e. altitude. The EFZ identifies the most detailed ecological units, based on the additional criteria of flora and geographic location.
Near East	Vegetation map of the Mediterranean zone (UNESCO/FAO 1970)	1:5 million	x	Distribution of potential vegetation formations in relation to climate. The various formations are distinguished mainly on the basis of physiognomy.
Europe	General map of the natural vegetation of Europe (Bohn et al. 2000)	1:10 million	Equidistant_Conic	Distribution of potential natural plant communities corresponding to the actual climate and edaphic conditions. At broadest level 19 vegetation formations defined, of which 14 zonal and 5 azonal formations.
Former USSR	Vegetation map of the USSR (Isachenko et al. 1990)	1:4 million	Lambert Azimuthal Equal Area	Distribution of broad vegetation formations related to climate, altitude and also current land use. 133 vegetation classes are aggregated into 13 categories of vegetation.
China	Geographic distribution of China's main forests (Zhu 1992)	x	x	Main aim to identify and map China's forest vegetation. A hierarchical classification is used based on climate and distribution of forest types and tree species. 27 forest divisions are mapped.
Australia	Interim biogeographic regionalisation for Australia (Thackway & Cresswell 1995)	1:15 million	Albers Equal Area	Major attributes to define biogeographic regions are: climate, lithology/geology, landform, vegetation, flora and fauna and land use. A total of 80 IBRA regions have been mapped.
Caribbean, Mongolia, Korean Peninsula, Japan, New Zealand, Pacific Islands	Terrestrial ecoregions of the world (WWF 2000)	x	Lat-Long	Ecoregions are defined by shared ecological features, climate and plant and animal communities. Main use is for biodiversity conservation.

going back to the original maps, checking the translation and modifying as needed. The other problem was the misalignment of lines of the polygons on both sides, even though they may have had the same labels. To resolve this problem, FAO manually edited the coverage and changed the locations of the boundaries. This sometimes required verification using ancillary data and maps such as composites of United States National Oceanic and Atmospheric Administration (NOAA) AVHRR spectral bands, classified continental-scale land cover data (such as the United States Geological Survey [USGS] global land cover database) and digital elevation model (DEM) data.

Following the classification and guidelines outlined above, the global map was compiled in a region-by-region approach. Case studies on North America and South America provided useful experiences and guidelines for GEZ mapping in other regions. In the course of the work regional experts actively participated or were consulted. EDC was responsible for producing the ecological zone maps for the temperate and boreal regions and jointly with FAO compiled the global map and database, while LET, Toulouse produced the ecological zone maps for the tropical regions, i.e. South America, Africa and Asia. FAO provided overall technical and conceptual guidance. After the Cambridge meeting in July 1999, it took one year to produce a draft global map. The draft map was reviewed at a meeting in Salt Lake City, Utah, United States (5-7 July 2000), and the final

Table 47-4. Distribution of forests by ecological zone, 2000

Ecological zone	Total forest %	Africa %	Asia %	Oceania %	Europe %	North and Central America %	South America %
Tropical rain forest	28	24	17	-	-	1	58
Tropical moist deciduous	11	40	14	6	-	9	31
Tropical dry	5	39	23	-	-	6	33
Tropical mountain	4	11	29	-	-	30	30
Total tropical forests	**47**	**28**	**18**	**1**	**-**	**5**	**47**
Subtropical humid forest	4		52	8	-	34	6
Subtropical dry forest	1	16	11	22	30	6	14
Subtropical mountain	3	1	47	-	13	38	1
Total subtropical forests	**9**	**2**	**42**	**7**	**7**	**37**	**5**
Temperate oceanic forest	1	-	-	33	33	9	25
Temperate continental forest	7	-	13	-	40	46	-
Temperate mountains	3	-	26	5	40	29	-
Total temperate forests	**11**	**-**	**17**	**4**	**39**	**39**	**2**
Boreal coniferous forest	19	-	2	-	74	24	-
Boreal tundra woodland	3	-	-	-	19	81	-
Boreal mountain	11	-	1	-	63	36	-
Total boreal forests	**33**	**-**	**2**	**-**	**65**	**34**	**-**
Total forests	**100**	**17**	**14**	**5**	**27**	**14**	**23**

map and database were completed by October 2000.

After production of the regional GEZ maps the global GEZ map was composed from all the regional tiles. Edge-matching was an issue, particularly for the vast area of Europe and Asia, where a number of different tiles had to be brought together with large bordering areas. The delineation of ecological zones between bordering areas of Europe and the former Union of Socialist Soviet Republics (USSR) matched well, with only small adjustments needed. The same applies to the ecological zone boundaries between Europe and the Near East. More work was needed to match the tiles for tropical Asia, China and the former USSR; the task was complicated by the presence of extensive mountain systems on the border areas. After the edge-matching problems were resolved, the regional tiles were registered to a global base map, ESRI's *Digital chart of the world*, 1st edition, December 1994 (base scale 1:1 000 000). The GEZ map, together with other global maps produced by FRA 2000, is presented on the FAO Forestry Web site (www.fao.org/forestry/fo/fra/index.jsp) under "Global maps".

The GEZ map can be used to aggregate information on forest resources by ecological zone. Consequently, it is now possible to produce reports according to the natural characteristics of the vegetation rather than by national boundaries, which frequently cut across natural ecosystems. This is particularly important today, with the growing awareness that many environmental problems are not national in character. For example, analysis of global change in climate and forest resources and of change in regional biological corridors requires information with a broad geographical context. Through ecological zone mapping, valuable insight is being obtained about the characteristics of forest resources which may serve to identify and resolve issues of importance to many countries, entire regions or the planet as a whole. For FRA 2000 reporting purposes, an overlay of the forest cover map with the ecological zoning map was used to derive area statistics on forests according to ecological zones (Table 47-4).

PROTECTED AREAS MAP

UNEP-WCMC served as the lead collaborator in mapping protected areas and was wholly responsible for compiling the information. UNEP-WCMC maintains a database for protected areas around the world and worked in cooperation with FAO to update this information for FRA 2000 under a formal Letter of Agreement.

A draft protected areas map for each country was circulated to over 200 countries in 1997 and 1998. The maps depicted the location and boundaries of previously registered protected areas. The maps were accompanied by a survey form to facilitate the information collection. About 25 percent of the countries responded to the survey and provided new information to UNEP-WCMC. After determining its suitability, UNEP-WCMC digitized the data and entered them into the geographic information system.

Table 47-5. International and national data for protected areas

Region	Polygons			Points		
	National	International	Total	National	International	Total
Africa	1 926	293	2 219	2 088	74	2 162
Asia	3 907	288	4 195	2 384	107	2 491
Europe	2 1468	1 587	23 055	19 478	1 915	21 393
North and Central America	10 119	352	10 471	4 722	92	4 814
Oceania	816	427	1 243	2 739	53	2 792
South America	2 436	158	2 594	1 413	48	1 461
Antarctica			0	28		28
Other	25	12	37	3 156	517	3 673
Total	40 697	3 117	43 814	36 008	2 806	38 814

Some follow-up with the countries – with little resulting additions to the information base – was carried out through May 1999, when the activity was formally closed (UNEP-WCMC 2000).

Information entered into the database was classified according to two aggregations of the six IUCN categories (categories Ia to II and III to VI) and digitized as either points or polygons. The attribute data included the IUCN designation and various metadata needed to understand the source and currency of the information.

The protected areas map contains the latest and best overall compilation of spatial information on the world's protected areas. The database consists of over 43 000 polygons and 38 000 points representing over 55 000 national and international protected areas (Table 47-5). UNEP-WCMC will continue to update the information as a core part of its programme.

The protected areas map was used in FRA 2000 to estimate the status of the protection of forests around the world.

CONCLUSIONS

The global forest map is a useful visual aid for perceiving the location and extent of the major forest areas of the world (although FAO country statistics are still derived through other means). In addition, the map can be used as an overlay to combine it with the protected areas map to show areas of protected forest or with the GEZ map to show forest distribution by ecological zone. Produced using advanced image processing techniques and satellite imagery, it is the first map of its kind for a global assessment.

The GEZ map provides an important means of aggregating global information on forests and other natural resources according to their ecological characteristics. It is the only global tool of its nature, in the sense that it has been compiled and reviewed by a body of experts through an international process, is based on technical input from around the world and is digital, geometrically corrected and registered to a map base. The map provides an important tool for all users conducting global studies with ecological parameters. This is especially important, as the use of global ecological zoning is expected to grow in importance with the increasing need for information relating to climate change (Kyoto Protocol), desertification and biological diversity conservation. The map will also continue to be important for FAO's periodic global assessments.

The protected areas map shows the worldwide location and distribution of protected areas according to UNEP-WCMC data. In conjunction with statistical and spatial data on forests, the map can be used to estimate the amount of forest at present under some sort of protection. UNEP-WCMC plans to update the map regularly. Illustrative examples of protected areas are available on the FAO Web site.

Thanks to FRA 2000, future efforts in mapping of global forest cover, ecological zoning and protected areas now have a sound basis on which to build. Because the information is digital and geometrically corrected to a geographic map base, new data can be relatively easily integrated with existing information. The forest cover and ecological zoning maps are available to users around the world at no charge on the Internet. FAO hopes that the newly available information will be useful to other global change projects and scientific endeavours.

BIBLIOGRAPHY

Bailey, R.G. 1989. Explanatory supplement to ecoregions of the continents. *Environmental Conservation*, 16(4).

Bailey, R.G. 1995. *Description of ecoregions of United States*. USDA Forest Service Publication No. 1391, Washington, DC.

Bailey, R.G. 1996. *Ecosystem geography.* New York, Springer Verlag.

Bailey, R.G. 1998. *Ecoregion map of North America.* USDA Forest Service Publication No. 1548, Washington, DC.

Bohn, U., Gollub, G. & Hettwer, C. 2000. *General map of the natural vegetation of Europe.* Scale 1:10 million. Bonn, Germany, Federal Agency for Nature Conservation.

Bolanos, R. & Watson, V. 1991. *Mapa ecológico de Costa Rica.* Scale 1:200 000. San Jose, Costa Rica, Tropical Science Center.

Commission for Environmental Cooperation (CEC). 1997. *Ecological regions of North America.* Montreal, Canada.

De la Cruz, R. 1976. *Mapa de zonas de vida de Guatemala.* Scale 1:500 000. Instituto Nacional Forestal (INAFOR), Ministerio de Agricultura, Guatemala.

Ecological Laboratory of Toulouse (LET). 2000. *Ecofloristic zones and global ecological zoning of Africa, South America and tropical Asia.* Prepared for FAO-FRA 2000 by M.F. Bellan. Toulouse, France.

Hartshorn, G. 1984. Ecological life zones of Belize. Scale 1:1 400 000. In *Belize country environmental profile: a field study.* San Jose, Costa Rica, Trejos Hnos. Suc.

Holdridge, L.R. 1947. Determination of world plant formations from simple climatic data. *Science*, 105: 367-368.

Holdridge, L.R. 1962. *Mapa ecológico de Honduras.* Scale 1:1 000 000. Organization of American States.

Holdridge, L.R. & Tosi, J.A. 1971. *Mapa ecológico de la República de Nicaragua.* Scale 1:500 000.

Isachenko, T.I., Karamysheva, Z.V., Ladygina, G.M. & Safronova, I.N. 1990. *Map of vegetation of the USSR.* Scale 1:4 million. Moscow, Institute of Geography, RAS. (in Russian)

Köppen. 1931. *Grundrisse der Klimakunde.* Berlin, Walter de Gruyter Co.

Preto, G. 1998. *A proposal for the preparation of the global eco-floristic map for FRA 2000.* Rome, FAO. (unpublished)

Simons, H. 2001. *Global ecological zones mapping.* FRA Working Paper No. 56. Rome, FAO.

Thackway, R. & Cresswell, I.D. (eds). 1995. *An interim biogeographic regionalisation for Australia: a framework for setting priorities in the National Reserves system cooperative program.* Version 4.0. Canberra, Australia, Australian Nature Conservation Agency.

Thornthwaite, C.W. 1933. The climates of Earth. *Geographic Review*, 23.

Tosi, J.A. 1970. *Mapa ecológico de Panamá.* Scale 1:500 000. Proyecto de Inventario y Demostraciones Forestales. Panama/UNDP/FAO.

Tosi, J.A. & Hartshorn, G.S. 1978. *Mapa ecológico de El Salvador: sistema de zonas de vida del Dr. L. R. Holdridge.* Scale 1:300 000. Ministerio de Agricultura y Ganadería de El Salvador/Centro Agronómico Tropical de Investigación y Enseñanza, Subprograma de Suelos Análogos de Centro América.

Trewartha, G.T. 1968. *An introduction to weather and climate.* New York, McGraw-Hill.

UNEP-WCMC. 2000. *Global FRA 2000 final report.* United Kingdom. (unpublished)

UNESCO/FAO. 1970. *Vegetation map of the Mediterranean zone.* Explanatory notes. Arid Zone Research Series No. 30.

WCMC. 1992. *Global biodiversity: status of the earth's living resources.* London, Chapman & Hall.

WWF. 2000. *Terrestrial ecoregions of the world.* Washington, DC.

Zhu, Z. 1992. *Geographic distribution of China's main forests.* Nanjing, China, Nanjing Forestry University.

Zhu, Z. 1997. *Develop a new global ecological zone map for GFRA 2000.* Rome, FAO.

Zhu, Z. & Waller, E. 2001. *Global forest cover mapping for the United Nations Food and Agriculture Organization Forest Resources Assessment 2000 Program.* Project Report to FAO. Sioux Falls, South Dakota, USA, EROS Data Center.

Zhu, Z., Waller, D., Davis, R. & Lorenzini, M. 1999. *Global forest cover map.* Interim Progress Report. FRA Working Paper No. 19. Rome, FAO.

Chapter 48

Forestry information system development

ABSTRACT

This chapter documents the development of an electronic Forestry Information System (FORIS) which was carried out within FRA 2000. The chapter discusses central concepts and principles, basic technology aspects and prospects for future development. It also describes the procedures used for generating country information for FRA 2000 with the help of FORIS functionality.

INTRODUCTION

The development of an electronic, Web-based Forestry Information System (FORIS) was a significant component of FRA 2000. In 1998 the FAO Forestry Department made an overall initiative to upgrade the communication of global forestry information, and the timing coincided with FRA 2000 requirements to organize, generate and disseminate large volumes of forest and forestry information. Combined efforts by the FRA team and a departmental core group working to enhance the Forestry Department Web site proved fruitful and led to the development of an integrated system, FORIS. Responsibility for FORIS has now shifted away from FRA 2000, and the system is being maintained and further developed for the Forestry Department as a whole. The main user interface for FORIS is the FAO Forestry Department Web site (FAO 2001a; Figure 48-1).

To cater to a wide range of communication needs, the information in FORIS is organized by various criteria, such as country, subject, species, publication and organizational entity. The system architecture allows for presentation of all information items in all of FAO's five official languages (Arabic, Chinese, English, French and Spanish). The system is integrated with FAO corporate-level data and systems, e.g. for correct presentation of country names in all languages and for the use of officially reported land area, to conform to corporate standards and reduce the maintenance needs of standard data sets.

This chapter focuses on the information-by-country aspect, as this is the most relevant to FRA 2000. When navigating the FAO Forestry Web site by country (FAO 2001b, Figure 48-2), users will find profiles for all countries which aim at a comprehensive presentation of the forest sector in each country. These country profiles have a standardized structure covering forestry-related subjects, currently under the three general headings Resources, Management, and Products & Trade (Table 48-1). FRA 2000 initiated and developed contents for a number of the subjects under these headings, including geography, forest cover, volume and biomass, forest plantations, trees outside the forest, forest management, protected areas, removals and non-wood forest products – following roughly the thematic studies in Part I of this report. Each of these subjects is further subdivided into several Web pages in the country profiles.

The standardized structure (i.e. the composition of subjects and pages) of the country profiles is continuously improved and expanded. The structure is displayed as a table of contents to the country profile (Figure 48-3) which is dynamically loaded and indicates which sections

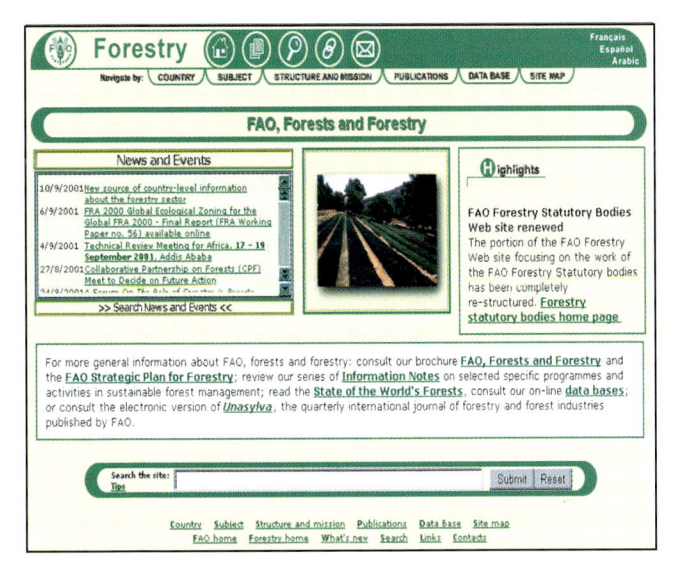

Figure 48-1. FAO Forestry homepage, from which FORIS contents are reached

Table 48-1. Currently identified categories and subjects in the FAO Forestry country profiles

Category	Subjects
Resources	Geography
	Forest cover
	Plantations
	Trees outside the forest
	Volume and biomass
	Fires
Management	Legislation
	Policy
	Forest management
	Protected areas
	Forest services
Products and trade	Industrial wood products
	Industrial products trade
	Removals
	Non-wood forest products
Further information	Contacts
	Institutions
	Photographs
	Publications

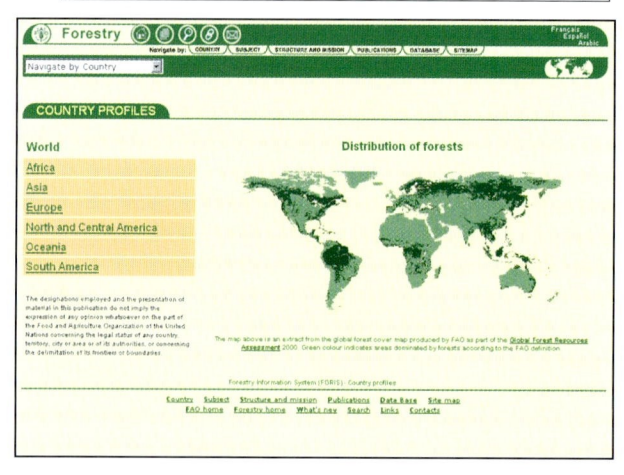

Figure 48-2. Navigation to FAO Forestry country profiles, www.fao.org/forestry/fo/country/nav_world.jsp

of the profile are available for the given country in the chosen language.

SYSTEM DEVELOPMENT DURING FRA 2000

The need for a comprehensive information system to support FRA 2000 work was indisputable. Several thousands of source documents have been consulted by a large number of FAO staff members and associates based throughout the world. Data have been extracted from the references for further processing, involving analytical steps requiring thorough documentation. In addition to tabular data, other types of information, e.g. texts, maps and lists of references, have been developed for each country. Large amounts of text have been translated. To enter and maintain these data sets, it was necessary to develop functionality that allowed many users to input data simultaneously. Administrative functions for managing data ownerships and editing privileges were therefore required. Finally, the aspiration to provide full transparency and availability to users and to embark on a continuous improvement of global forestry information beyond FRA 2000 made it necessary to begin an ambitious information system development (FAO 1999b).

It should be noted that UNECE/FAO in Geneva developed a separate database to meet the needs of the FRA 2000 work related to industrialized countries, the Temperate and Boreal Forest Resources Assessment 2000 (TBFRA). This was necessary because the assessment process progressed faster for these countries, and the report contributing to the global assessment was released before FORIS had been established and could support this work. The TBFRA 2000 database has been released on a CD-ROM (UNECE 2001), but data are not currently presented on the Web. Since some country data related to forest area were adjusted after the release of TBFRA 2000, there are some discrepancies between the databases. It is planned that in future assessments UNECE/FAO in Geneva would maintain data directly within FORIS to ensure conformity.

Previous global assessments had also identified the need for information system development, and FRA 1990 established a predecessor to the current FORIS, named the Forest Resources Information System (FAO 1995). As general information technology at the time was less advanced and the user requirements were perhaps less pronounced, the information system work focused more on the internal processing of data and less on broad-scale electronic dissemination to users. The earlier system did not include support for multiple users for data entry. Obviously some tools, especially the World Wide Web, were not available to earlier assessments, and the expectations for providing public access to data were lower. Some features of information management have, however, not changed, including the need to document source data used and processing steps to derive final results. In many cases the current assessment was hindered by difficulties in tracing the background information to estimates in previous assessments. Despite the strides made in

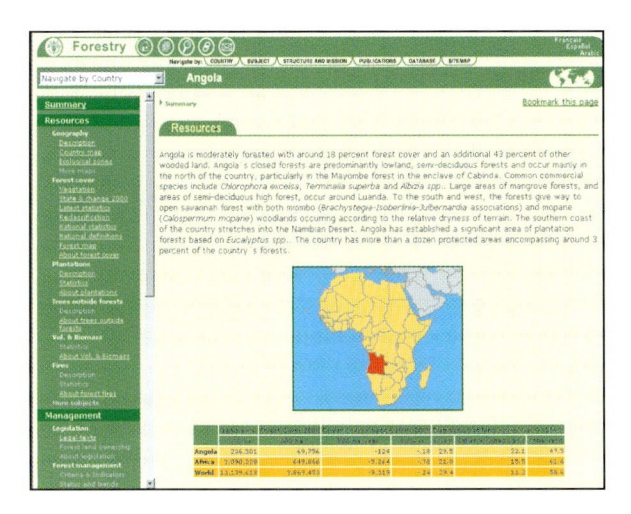

Figure 48-3. Example of an FAO Forestry country profile: summary page for Angola

making data and information publicly available, a main objective and desired feature of the current information system remains to document the work well for the benefit of future global assessments.

Since its initiation in the second half of 1998, the system development process has largely followed the work progress of FRA 2000 in what can be characterized as an interactive application development where user needs were identified and prioritized according to the overall FRA 2000 objectives. Early system modules included functions for entering source references and source data, followed by modules for reclassifying national data into global classification schemes and for creating country-wide forest area estimates at reported reference years. Following development of the data processing support, functions for maintenance of texts in several languages were developed. During the second half of 1999, the development of a dynamic Web application was initiated. In January 2000, the FRA 2000 Web application was joined with the Forestry Department Web application in the current forestry country profile approach, which was launched in early 2000.

The next step was to make editing functions available through Web browsers to allow for a more distributed maintenance of the contents. Over the past year the overall system performance has been further enhanced and a number of functions have been added. In January 2001, overall responsibility for FORIS was shifted away from the FRA team and FORIS became in the formal sense a departmental system. Current main objectives include:

- to provide functions for direct maintenance of statistics through Web browsers, thus setting the stage for decentralization of responsibilities for core data maintenance to officers at headquarters, units at regional offices or member countries directly;
- to expand the scope of FORIS usage to other areas besides global country profiles, for example to other subject areas or to country-run national information services.

Currently, the Web country profiles consist of more than 20 000 published Web pages covering more than 200 countries and in four languages. The country pages are accessed by users outside FAO at a rate of about 1 000 pages per day.

CONCEPTS

FORIS provides dynamic access to forestry data by country to Web users. Dynamic access means that all contents of the page, including the table of contents, are drawn from database tables and not from static html files. Dynamic presentation brings some major advantages. For example, it makes it possible to develop effective maintenance functions for all types of contents; to use the same original data items for different presentations, thus reducing the risk of unsynchronized reporting; and to restructure presentations by changing the virtual structure rather than having to replace large numbers of static files. Over the long term, perhaps the most significant advantage of a well-structured dynamic system is that the scope of the contents is expandable; new subjects or entire new Web sites, for example, can be included with a relatively small effort.

Data ownership and partnerships

One important principle that FORIS supports is decentralized data ownership for core forest and forestry data. Obviously, forestry expertise and/or local knowledge resides with a wide range of FAO staff and partners depending on the subject and geographic location. From an organizational point of view, these persons (it is noted that a data owner should always be one person) also have technical responsibilities as focal points for a particular subject or a particular region. It seems that the most feasible way to maintain a wide range of forestry information for all countries is to have the system support the existing distribution of subject responsibilities and to provide tools to the respective officers (and to those to whom these officers delegate the hands-on work) for maintaining their information segment. For each page in the country profiles, the currently identified data owner is indicated in the footer,

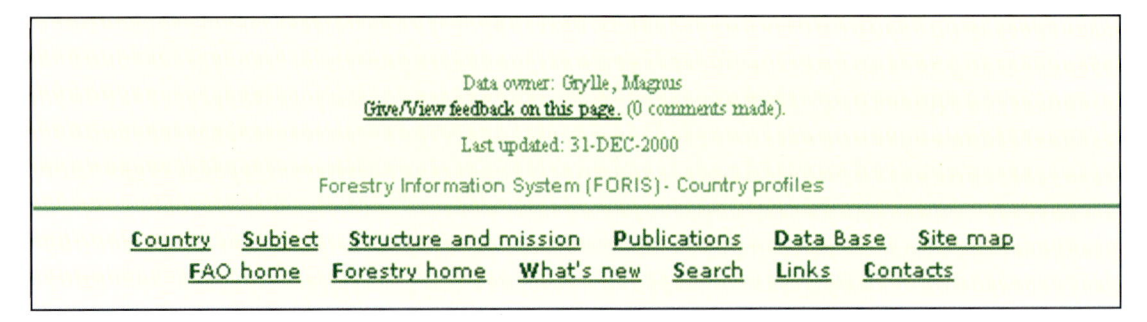

Figure 48-4. Example of footer that appears on all FORIS country profile pages; indicating the data owner and providing a link for giving feedback to this person

and readers can give feedback to these persons directly (Figure 48-4).

Given that a potentially large number of data owners will maintain their information from various parts of the world, the contemporary solution of choice is to carry out information maintenance through a standard web browser (e.g. Internet Explorer or Netscape Navigator). With log-in and editing functions provided on Web pages, no extra installations are required on the user's computer (Figure 48-5). This keeps down the cost of software maintenance and makes it possible to decentralize maintenance tasks to anyone with an Internet connection and a standard computer configuration.

While the responsibility and mandate for maintaining a core set of global forestry variables lies with the FAO Forestry Department, there are related and relevant data sets for which it may be efficient to establish a partnership for data sharing, rather than to acquire copies of the data sets and potentially face a duplication of effort in keeping the copies up to date. Current information technology and the Internet facilitates such partnership arrangements.

Two different types of partnerships can be identified:

- Some partner data sets are maintained within FORIS, i.e. the information structure and editing privileges are set up in FORIS and the data are maintained with the same functions as internal data sets. This can be a suitable solution for partners that have limited capacity to develop their own information system and/or to invest in secure information platforms, for example forestry agencies in developing countries.
- Other partner data sets are maintained outside FORIS, but identified in the country profile table of contents and actively linked into the country presentation. This is a suitable solution where partners have an established information system and maintenance routines. Examples include the current presentation of legal texts, maintained by the FAO Legal Office, and the ongoing development of a direct link to UNEP-WCMC data on protected areas.

System implementation

From the above, it becomes clear that the "system" concept is wider than computer hardware and software. The "system" also includes the various information processes involved, which in turn build on the organizational structures, including formal delegation of authority. It is clear that the system owner, in this case FAO through the Forestry Department, has a role to invest discriminately in further system development as well as content development, and to put these investments in the context of the overall activities and role of the department.

Conformity to corporate FAO standards – regarding hardware, software, corporate-level data and procedural aspects – is central to a cost-effective and successful implementation. For hardware, the corporate database and Web servers must be used. Regarding software, all data reside

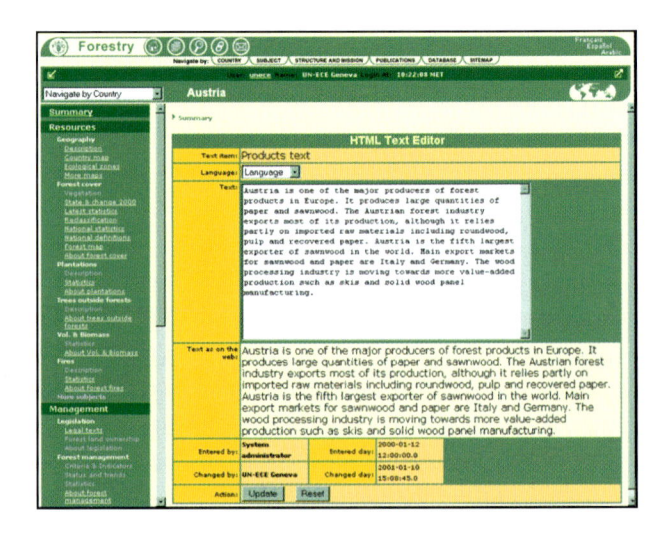

Figure 48-5. Example of user input screen to FORIS, using a standard Web browser; the selected text (from the summary page for Austria) can be directly modified by the logged-in user, in this case UNECE in Geneva

in the corporate Oracle database. The engine to generate dynamic Web pages is based on Java and Java Server Page (jsp) techniques, identified as a corporate standard. Corporate-level data, for example official country names and official areas of countries, are directly accessed and incorporated. By conforming with corporate standards, FORIS benefits from general security procedures, including backup, and also over the long term achieves cost-efficiency.

Assembly of country information and generation of estimates

This section explains how country-specific information for forest area was assembled in FRA 2000, based on existing and available reports and using the functionality in FORIS. FRA 2000 reported on 213 countries and areas. These reporting units are all identified as official geographic units by FAO, which takes into account considerations of a political and practical nature. Several units are not constitutionally independent countries, but their remoteness from the mainland territory of the country motivates an independent presentation. For simplicity, the word "country" refers in the following to any of the 213 reporting units, i.e. it includes also areas that are not officially considered countries.

The assembly of country information followed and implemented the recommendations made by an FAO Expert Consultation on FRA 2000 methodologies, held in March 2000 (FAO 2000). The three most important principles of the approach were:

- traceability – to make it possible to trace the FAO estimates to the source documents;
- transparency – to publish and make available the details of each processing step, including source data;
- continuity – to provide the possibility for continuous upgrading of the estimates when new information becomes available.

The main process in assembling country information involves interaction with countries, which is elaborated in Chapter 45. This section describes the subprocess of producing estimates and outputs (Figure 48-6). Each step in the subprocess is supported by functions in FORIS. Note that only estimations for developing countries strictly followed the process. For industrialized countries, the estimates were prepared by the countries themselves, and FAO did not therefore document all steps.

Step 1. Through requests to countries, copies of source documents containing primary data from inventories or surveys were obtained. In this step no distinction as to the quality and relevance of the primary data was made. Instead the goal was to include all known inventories that could be used for country-wide estimates. In several cases, partial inventories were recorded as these were the best data available and as partial data could be combined into country-wide estimates later in the process. The reference citations were entered into FORIS.

Step 2. The collected documents were reviewed with respect to the subject considered – in this case forest area estimates for the country. In this step the the quality of the information contained in each document was evaluated and decisions were made about whether to continue to work with the document. Documents containing secondary data were rejected in favour of primary sources, for example; and some documents were rejected for their use of methodology that does not generate reliable data. The reviews, including comments, were entered into FORIS. In all, more than 1 500 documents were reviewed with respect to forest area estimates for developing countries. Citations for reviewed documents are shown on the FORIS country profile Web pages.

Step 3. Most countries apply their own forest classification, adapted to local conditions and uses but seldom corresponding to the global classifications applied in FRA 2000. As one ambition was to make it possible to trace FRA 2000 estimates back to the source, the national classes and corresponding definitions were typed into FORIS from the source document. The national classifications are shown on the FORIS country profile Web pages.

Step 4. The next step was to enter the source data as given by the source document for each of the national classes. When data were at the subnational unit level, the names of the subnational units were also entered and data for each unit incorporated. These data are displayed on the FORIS country profile Web pages. By including data for subnational units, a higher spatial resolution is reported for national data for many countries in comparison with the global tables where only national totals are reported.

Step 5. To provide results that are comparable among countries, the national classifications had to be reclassified into the global classification scheme developed over past decades for the global assessments. For FRA 2000 reporting purposes, national data were reclassified into

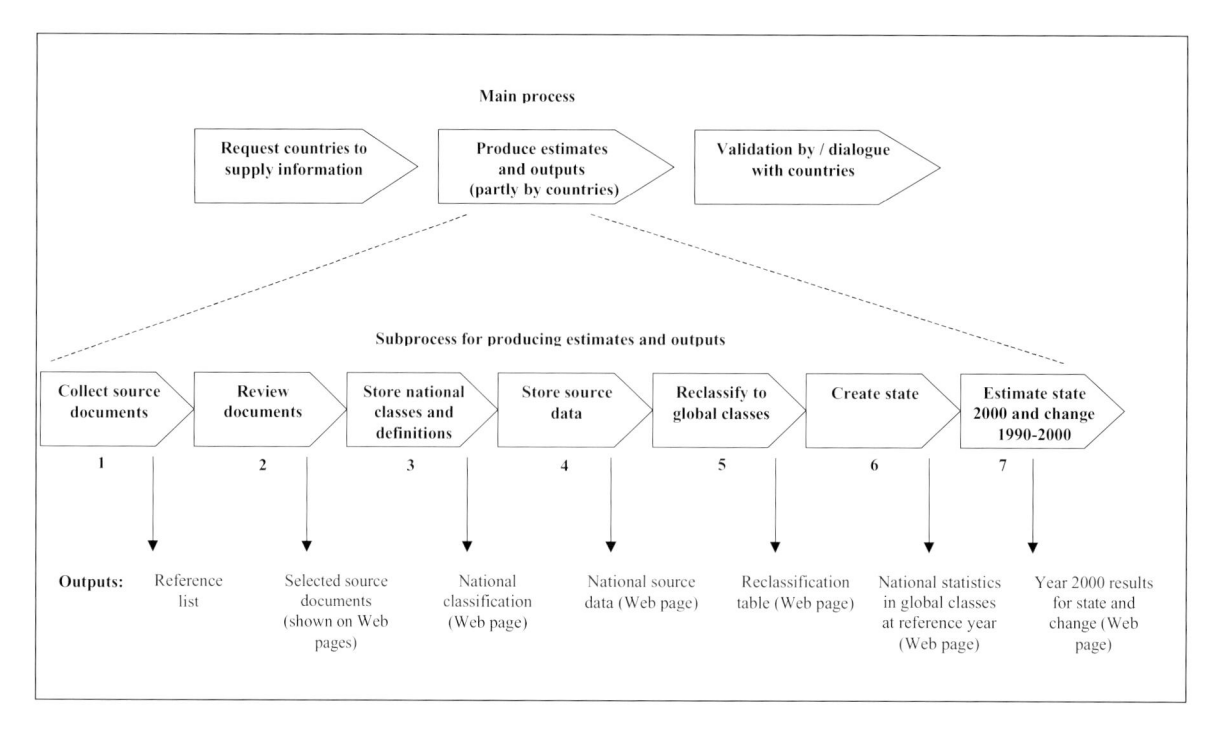

Figure 48-6. Main process for assembly of country information and subprocess for producing estimates and outputs

global land use classes. Definitions for these classes are given in Appendix 2. In most cases, the reclassification was simply a remapping to a corresponding global class, but sometimes national definitions overlapped with several global classes and the national class had to be split between two or more global classes.

The reclassifications made for FRA 2000 are displayed on the FORIS country profile Web pages.

Step 6. Using national source data that had been reclassified into the global classes, country-specific area estimates ("states") were created. These states were created for all reference years required for change estimates (see Step 7). For some countries only one state could be created as survey information only existed for one point in time. The most recent and reliable state for each country is shown on the FORIS country profile Web pages, and in the global FRA 2000 tables (Appendix 3, Table 5). A state has the following properties.

- It is a set of statistics using the global classification scheme that represents exactly the land area of the country, i.e. no areas are missing and no areas are counted twice. Establishing a state can be simple (when data from one reference provide full coverage) or complex (when several references provide data from different parts of the country and need to be patched together).

- It has a reference year, which is the area-weighted average year when the source data were registered. For a remote sensing survey, this is the average year of image acquisition. For field-based surveys it is the average year of field data collection.

- In the states, the total area of a country equals the area recorded by FAOSTAT, which is the official total area as provided to FAO by national survey agencies. As forest inventories for different reasons often report slightly different total area, the statistics needed to be calibrated to match with officially reported areas for the countries. To the extent possible, the land area (i.e. the total area minus the area of inland water) was also taken directly from FAOSTAT, although for some countries new forest-related data sets appeared more reliable and were therefore used.

Step 7. The final step was to extrapolate the observed states from the reference years to the year 2000 and in the same process to estimate the area change between 1990 and 2000. As quality and availability of information varied greatly among countries, a unique analysis had to be made for each country. Area state 2000 and area change 1990-2000 were estimated for the total forest area, i.e. including closed natural forest, open natural forest and forest plantations together. The standard model was to use the two most recent states and make a linear extrapolation of the area to the year 2000. The slope of the line would then represent the rate of change between 1990 and 2000. This approach worked well when two comparable states were available, with reference years that approximately fit with the 1990s (Figure 48-7).

In some cases, when more than two states were available, a regression was made to determine the rate of area change. The regression was deemed more suitable in cases where it was difficult to select two states for the change estimate.

In cases where only one reliable area state was available, the area change estimate had to rely on ancillary information such as expert judgement, partial inventory data that could be studied over time, and results from samples of the FRA 2000 remote sensing survey falling inside the country.

Each area state 2000 and area change 1990-2000 is shown on the FORIS country profile Web pages, as well as in the FRA 2000 global tables (Appendix 3, Table 4). The country Web page also includes a note on how the extrapolation was made for the country.

CONCLUSION

Some major steps towards a broad Forestry Information System have been taken with the help of FRA 2000, made possible through extrabudgetary support to the FRA 2000 project. The system has been fully integrated into the Forestry Department Web site, and its development continues on a departmental level.

The information system efforts are closely linked to the overall ambitions set out in the FAO Strategic Framework (FAO 1999a), particularly strategy E which relates to knowledge management, including integrated information systems and assessments. The efforts also reflect the overall strategies of the FAO Forestry Department (FAO 2000).

The functions developed within FORIS provided essential support to the assembly of country information and generation of estimates in FRA 2000 and to the presentation of the results, at www.fao.org/forestry/fo/country/nav_world.jsp.

Future development and expansion of FORIS will include many possibilities and challenges for forestry knowledge management and communication. Possibilities include, for example, expanded information partnerships – particularly with member countries, but also with other international initiatives such as the Global Forest Information Service (GFIS); increased

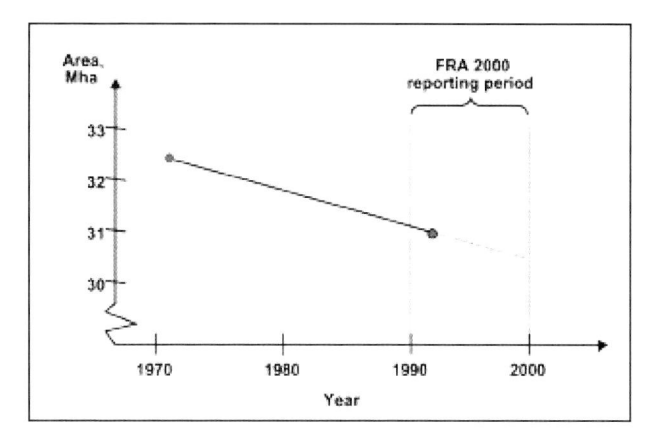

Figure 48-7. Example (Mozambique) of extrapolation of forest area to 2000 based on two area states with reference years 1972 and 1991

support to international processes such as the United Nations Forum on Forests (UNFF); and involvement of the wider public in knowledge sharing. Challenges include mobilization of resources; committment to maintaining vital global data sets over the long term; and continued efforts to find cost-reducing synergies with other knowledge sharing efforts, inside FAO as well as outside.

BIBLIOGRAPHY

FAO. 1995. *Forest resources assessment 1990.* FAO Forestry Paper No. 124. Rome.

FAO. 1999a. *The Strategic Framework for FAO 2000-2015.* Rome. www.fao.org/docrep/X3550E/X3550E00.htm

FAO. 1999b. *Forest Resources Information System – concepts and status report.* FRA Working Paper No. 7. Rome.

FAO. 2000a. *FAO Strategic Plan for Forestry.* Rome. www.fao.org/FORESTRY/FO/STRATEGY/vision-e.stm

FAO. 2000b. *Proceedings of the FAO Expert Consultation to Review FRA 2000 Methodology for Regional and Global Forest Change Assessment.* FRA Working Paper No. 42. Rome. www.fao.org/forestry/fo/fra/index.jsp

FAO. 2001a. FAO Forestry homepage. www.fao.org/forestry.

FAO. 2001b. Forestry country profiles navigation page. www.fao.org/forestry/fo/country/nav_world.jsp

UNECE. 2001. *TBFRA 2000 database.* Compact disc distributed by the Timber Section of the Trade Division, UNECE, Geneva.

Part IV: Conclusions and recommendations

Field sample inventory of forests and forest benefits, Thailand

Chapter 49
Conclusions

FRA 2000 is the latest of the global forest assessments that FAO has carried out at approximately ten-year intervals since 1948. FRA 2000 improved on previous assessments in several ways. It covered more countries and parameters, and it used for the first time a single global definition of forest. The average national inventory year for information used in the assessment was closer to the global reporting year than in previous assessments. More support was given than in the past to country capacity building; and new technologies, such as remote sensing, were extensively used. The reliability of the results is thus believed to be greatly enhanced. Nevertheless there are many gaps in information, and reliability still needs to be improved for future assessments – see the Process review and the Recommendations (Chapters 50 and 51).

In FRA 2000 a uniform definition of forest – 10 percent canopy cover – was used for all regions of the world.[57] This will make comparisons among future assessments more reliable. For this assessment, however, it was necessary to revise the estimates made for the area of temperate and boreal forests in 1990 using the definition and methodology adopted in 2000, since the 1990 estimates were based on a definition of 20 percent forest cover. Details will be documented in a forthcoming FRA Working Paper.

The total estimated global forest area in 2000 was nearly 3.9 billion hectares, of which 95 percent was natural forest and 5 percent was forest plantations.

About 47 percent of the world's forests occur in the tropical zone, 9 percent in the subtropics, 11 percent in the temperate zone and 33 percent in the boreal zone.

The world's natural forests continued to be lost or converted to other land uses at a very high rate. During the 1990s, the total loss of existing natural forests was 16.1 million hectares per year, of which 15.2 million hectares occurred in the tropics (Table 49-1). This means that 4.2 percent of the natural forest area that existed in 1990 was lost by 2000. For the tropics, the loss of existing natural forest was 7.8 percent.

Not all loss of natural forests was deforestation, as 1.5 million hectares of natural forests were converted to forest plantations. Global deforestation thus amounted to 14.6 million hectares per year during the 1990s (Table 49-1), or 3.6 percent for the ten-year period as a whole.

The overall area of forest plantations increased by an average of 3.1 million hectares per year during the 1990s, including the 1.5 million hectares converted from natural forest and 1.6 million hectares of afforestation on land previously under non-forest land use.

Expansion of natural forests, mainly in areas previously under agriculture, occurred at a rate of 3.6 million hectares per year worldwide, including 1 million hectares per year in the tropics. At the global level, natural forests and forest plantations together expanded by 5.2 million hectares per year. The net change in forest area was therefore -14.6 + 5.2 = -9.4 million hectares per year (Table 49-1). The net reduction in forest area was 2.4 percent for the 1990s as a whole.

Although the global rate of change figures for the 1980s and 1990s are not directly comparable because of changes in definitions and methodologies and updated inventory information, it appears that the estimated net loss of forest (i.e. the balance of the loss of forest area by deforestation and the gain through afforestation and natural expansion of forests) was lower in the 1990s than in the 1980s. One major reason is that secondary natural forests have expanded more rapidly in recent years. This expansion may be underestimated, as it is not always captured by national reports and is accounted for by a relatively small number of countries. The general process seems to be that forests return to areas where agriculture is being discontinued (notwithstanding that deforestation in tropical forests remains a serious problem; see below).

The implication is that forest products and services may in the future be provided from

---

[57] The full definitions of forest and the other parameters measured in FRA 2000 are given in the text of the appropriate chapters.

Table 49-1. Forest area changes 1990-2000 in tropical and non-tropical areas (million ha/year)

Domain	Natural forest					Forest plantations			Total forest
	Losses			Gains	Net change	Gains		Net change	Net change
	Deforestation (to other land use)	Conversion to forest plantations	Total loss	Natural expansion		Conversion from natural forest (reforestation)	Afforestation		
Tropical	-14.2	-1	-15.2	+1	-14.2	+1	+0.9	+1.9	**-12.3**
Non-tropical	-0.4	-0.5	-0.9	+2.6	+1.7	+0.5	+0.7	+1.2	**+2.9**
Global	**-14.6**	**-1.5**	**-16.1**	**+3.6**	**-12.5**	**+1.5**	**+1.6**	**+3.1**	**-9.4**

secondary forests, perhaps reducing the pressure on primary forests. Further, the biological impact of losses of primary forest formations over time may be alleviated as secondary forests develop into more diverse systems over time.

LOSSES OF NATURAL FOREST

The loss of natural forest area remains at about the same level as reported in previous global assessments (a slight reduction was observed but may not be significant because it is within the margin of error of the estimate). The processes of natural forest loss were studied in the FRA 2000 remote sensing survey (Chapter 46). The survey showed different patterns among regions within the tropics, which may reflect general land use patterns and land use policies. In Latin America, large-scale direct conversion of forests dominates. Direct conversions also dominate in Africa, but on a smaller scale. In Asia, the area of gradual conversions (intensification of shifting agriculture) is equal to the direct conversions from forests to other land uses. At the global level, direct conversions dominate the picture, accounting for about three-quarters of the converted area. Most tropical deforestation is thus a result of rapid, planned or large-scale conversion to other land uses, mainly agriculture. Policies to address deforestation may therefore have more impact if they address the causes and mechanisms of direct and permanent conversion of forests to other land uses.

The influence of population pressure on forest change was emphasized in FRA 1990, partly because it relied on a population-driven model to estimate deforestation. In FRA 2000 the use of this model was abandoned in favour of transparency and to preserve the integrity and representativity of source data in the final results. New studies indicate that links between population density/growth and land conversion are weak and an oversimplification of the situation. Other factors such as the development of the overall economy, urbanization, policies, legislation, culture and tradition may explain a relatively large proportion of the variation in the rate of forest area change among countries.

Further cross-sectoral studies are thus needed to shed light on land use and land use change processes. Studies might include the rights to use forest land under different conditions and the effects of varying levels of capital investments and subsidies in agriculture.

WOOD VOLUME AND WOODY BIOMASS

Wood volume, defined as stem volume outside bark excluding branches, was included in FRA 2000 as an indicator of the capability of forests to meet demands for wood-based products. The total wood volume in 2000 was estimated at 386 billion cubic metres, or about 2 percent higher than in 1990, since increases in volume in temperate and boreal forests offset declines in tropical regions.

Above-ground woody biomass, defined as the above-ground woody parts of trees, shrubs and bushes, alive or dead, was estimated as an indicator of stored carbon and of the contribution of forests to climatic stability. It was estimated at 422 billion tonnes (dry), of which 27 percent was in Brazil alone. This measure was about 1.5 percent lower than in 1990 owing to the loss of tropical forests with high biomass content.

Information on volume and biomass was limited, particularly for tropical forests. The need for reliable and comparable measurements of both of these parameters, and in particular of their change over time, will continue to grow. Estimates of wood volume, not only for natural forests but also increasingly for forest plantations and trees outside the forest, will be required at country and regional levels for trend studies, policy development and planning. The need for estimates of woody biomass is related to the possibility of carbon offset payments under the Kyoto Protocol of the United Nations Framework Convention on Climate Change (UNFCCC), which may be of great future importance as a payment for environmental services provided by the forest sector in many countries.

PLANTATIONS

Forest plantations were estimated to cover 187 million hectares in 2000, of which 62 percent were in Asia. This is a significant increase over the 1995 estimate of 124 million hectares. Globally, the estimated annual rate of successful new planting is about 3 million hectares, with Asia and South America accounting for 89 percent. Globally, half the forest plantation estate is for industrial end-use.

The reported expansion of forest plantations is impressive, but about half of the reported area appeared to be planted on land previously under natural forest. Furthermore, the verified success rate of plantations is low relative to national reports from some countries, and a small number of countries account for most of the expansion. However, taking into account also the expansion of trees outside the forest in many countries, a major and increasing part of wood and fibre supply is likely to come from planted tree resources in the future. For example, although forest plantations accounted for only 5 percent of global forest cover in 2000, it is estimated that they supplied about 35 percent of global roundwood. This figure is expected to increase to 44 percent by 2020. In some countries forest plantation production already contributes most of the industrial wood supply.

In developing countries about one-third of the total plantation estate was primarily grown for woodfuel in 1995. Yet it should be noted that the often underestimated contribution of planted trees on farmland, in villages and homesteads and along roads and waterways, together with other fuelwood sources such as twigs and shrubs, played a large part in explaining why the woodfuel crisis feared in the 1980s in developing countries did not occur.

The results of FRA 2000 tend to support the prediction that plantations will provide an increasingly large part of future wood supply. The need to use natural forests to provide wood should decrease, at least in relative terms, in areas where investments have been and are being made in planted tree resources.

There is increasing interest in development of forest plantations as carbon sinks; however, failure to resolve international debates on legal instruments, mechanisms and monitoring remain constraints.

TREES OUTSIDE THE FOREST

Trees outside the forest (TOF) represent an important resource which is not included in FRA 2000 definitions of "forest" and "other wooded land". They are often, but not always, planted trees and they include trees in cities, on farms, along roads, and in many other locations which are by definition not part of a forest. TOF make major contribution to the environment and to the social and economic well-being of humankind, including contributions to food security.

FRA 2000 did not attempt a comprehensive global assessment of TOF, nor has such an assessment ever been carried out although many studies have been made of TOF for specific countries or land areas. Given the scale of the goods and services provided by TOF and their almost complete exclusion from policy-making and planning at present, future forest resources assessments should assist countries to assess TOF, thus supporting moves towards a more comprehensive global assessment.

BIOLOGICAL DIVERSITY

FRA 2000 addressed a number of important indicators of biological diversity such as information on forests by ecological zones, protection status, naturalness, endangered species and aspects related to fragmentation. It is hoped that the information provided in this report will contribute towards a better understanding of the status and trends in forest biological diversity. Two studies carried out within the framework of FRA 2000 were also presented, one addressing the number of forest-occurring ferns, palms, trees, amphibians, reptiles, birds and mammals by country, and the other examining spatial attributes of forests that define one aspect of "naturalness", applicable at the global level.

The assessment of biological diversity in forests at the global level presents a number of conceptual difficulties which must be resolved for the success of future assessments.

FOREST MANAGEMENT

An assessment of trends in management of forest resources highlights the slowly increasing appreciation of the concept of, and need for, sustainable forest management. For example, as of 2000, 149 countries were involved in one or more of the nine ecoregional initiatives to develop and implement criteria and indicators for sustainable forest management, although the degree of implementation varies considerably. The area of forests worldwide under formal or informal management plans has apparently increased – another indicator of efforts to improve forest. It was reported that 89 percent of forests in industrialized countries were being managed according to a "formal or informal management plan". Figures for developing countries were far from complete. Nevertheless, preliminary results showed that at least 123 million hectares, or about 6 percent of the total forest area in these countries, were covered by a "formal, nationally approved forest management plan covering a period of at least five years". The difference in definitions of "management plan" make it difficult to compare the two groups.

It must also be kept in mind that the existence of a formal or informal forest management plan does not necessarily signify that forest is sustainably managed. The study did not indicate whether plans were appropriate, being implemented as planned or having the intended effects; thus some areas reported as being covered by a management plan may not be sustainably managed, while other areas, not currently under a formal management plan, may be.

While primarily a marketing tool, certification may also contribute to the promotion of sustainable management of forests. The global area of certified forests had grown to about 80 million hectares by 2000.

The practice of sustainable forest management and the quantity and quality of information on the subject should continue to improve with the growing implementation of criteria and indicators in many countries. If significant improvement is to be made, however, there will have to be continued growth in political awareness of the challenges (possibly catalysed through the high-level sessions of the United Nations Forum on Forests [UNFF]), better sharing of information and experiences, greater capacity building and increased support to effective field programmes in forest management, especially (but not only) in developing countries.

PROTECTED AREAS

At the global level, the FAO/UNEP-WCMC mapping project indicated that 12 percent of the world's forest area was in one of the IUCN protected area management categories. However, discrepancies between results from the global map analysis and the areas reported by national FRA 2000 correspondents indicated differences in interpretation of the IUCN classification and its implementation in the national context. Continuous improvement of definitions and assessment approaches is highly desirable.

At the global level, the proportion of forests in protected areas estimated in FRA 2000 exceeds 10 percent, a figure that has been suggested as a minimum target for protected forest areas. However, it should be noted that statistics at the global level may not be representative of the protection afforded to forests in different ecological zones or in different countries. It should also be noted that varying levels of protection are included in the six IUCN categories, and that not all legally protected forests are effectively managed.

FOREST FIRES

The widespread physical damage, economic disruption and threat to public health caused by outbreaks of forest wildfires during the past decade have attracted public attention. There has been greater awareness that the cause of these fires frequently lay in the unforeseen effects of public policies developed for land use applications in other sectors than forestry. There has also been greater appreciation of the beneficial biological effects of wildland fires under certain circumstances. Despite increased public attention and unprecedented intersectoral and international cooperation, there is a lack of reliable global data on the extent and impacts of forest wildfires and on the use of fire as a land clearing and vegetation management tool.

The development of integrated land use policies affecting fires and appreciation of the need to use fire as a tool are expected to continue to improve. It is hoped that these advances may have an effect on forest fire outbreaks, but more information must be collected before a reliable evaluation of trends can be made.

WOOD SUPPLY

From a study on forest area accessible for wood supply or other uses it was estimated that 51 percent of the world's forests are within 10 km of major transportation infrastructure and are potentially accessible for wood supply. This increased to 75 percent for forests within 40 km from transportation infrastructure. The highest accessibility was found in subtropical forests (73 percent within 10 km of transport) and the lowest accessibility was found in boreal forests (34 percent within 10 km of transport).

Since harvesting is one of the most important management interventions in forests, information regarding wood removals and harvesting was analysed for all major industrialized countries. However, very few tropical countries reported this information. Accordingly, a study was carried out for 43 tropical countries which account for approximately 90 percent of the world's tropical forest resources. The study showed that about 11 million hectares of tropical forests were harvested annually in the 1990s, with harvesting intenstity varying widely, from 1 to 34 m^3 per hectare.

NON-WOOD FOREST PRODUCTS

Non-wood forest products (NWFP) make a major contribution to food security and sustainable livelihoods. Few countries assess the resource supplying NWFP or monitor their contribution to the national economy, so an accurate global assessment was difficult. FRA 2000 summarized NWFP for which data had been collected and described the most important NWFP in each region, with estimates of economic values where available. Some of the major problems associated with collecting and analysing data on NWFP have been identified; these should be overcome in order to improve future assessments.

NWFP have an important socio-economic role in many countries, both developing and developed, but because of the paucity of information on NWFP they are at present not effectively included in policy dialogue, formulation and implementation. As with trees outside the forest, future forest resources assessments should assist countries in assessing their NWFP resources, thus furthering the move towards a more comprehensive global assessment.

Chapter 50
Process review of FRA 2000

AVAILABILITY OF INFORMATION

FRA 2000 was launched with the ambition to cover a broad set of variables relevant to the forest sector at the international level. The intention was to broaden the global assessment approach from the previous strong focus on forest area statistics and address more complex forestry issues. The new subjects included qualitative aspects of forests such as biological diversity, biomass and availability for wood production as well as management parameters such as the status of forest management planning and protected areas.

FRA 2000 used the best available and most relevant country information on forest resources. Although some countries notably improved their inventories, and although the number of reports on forest resources increased in the 1990s, many countries still lack the basic data needed to accurately assess the state and changes of their forests. Most countries updated their forest cover estimates during the 1990s, often through remote sensing mapping, but in many cases the methodology was not directly compatible with that of previous surveys, making change estimates difficult. There is a scarcity of comparable multiple-date inventories and a need to improve both the accuracy and depth of information provided in forest inventories.

In the course of the study on forest area and area change there were found to be many documents and publications related to forest area, but some data were not representative or were derived from secondary sources. Information on forest area change could be derived with some precision, but data on qualitative changes such as forest degradation were generally missing, even in developed countries with relatively advanced forest inventory methodology.

Systematic field inventories that measure volume, biomass and productivity of the forests were carried out in many countries, but often within limited areas. As a result, national estimates for volume and biomass had to be extrapolated from local studies.

Although biological diversity has been widely studied, most studies focus on a specific ecosystem or species, and little quantitative and systematic information has been generated at the country level. Basic concepts for larger-scale assessment methodologies are still being developed. Simplified indicators such as the number of species under threat or spatial analyses of the degree of forest disturbance were proposed for FRA 2000 but no major advances were made in assessing this important aspect of forests.

Information on areas under forest management plans has generally not improved over the past decade, although the focus on certification has brought a higher level of information quality for the areas included in certification schemes. FRA 2000 collected estimates of forest areas that are covered by management plans and that are certified, but more work is needed for assessing the effectiveness of forest management for large areas. The increasing national commitments to implementing criteria and indicators for sustainable forest management offer hope that the next decade will see major strides in this important area. But a parallel important prerequisite to achieving sustainable forest management is an increase in efforts (including funding) to put more forest area under effective management.

Significant improvements have been made in obtaining estimates of protected forest areas. The data problem in this area is the apparent lack of consistency in countries' interpretation of the IUCN protected area management categories. Until the recent development of new methodology by the IUCN World Commission on Protected Areas (WCPA 2000) and the WWF/World Bank Alliance Rapid Assessment Methodology based on the WCPA framework, no generally applicable methodology for assessing management effectiveness in protected areas was available. It is still too early to see if the new methods will facilitate standardization of the assessment of management effectiveness in protected areas.

Several assessment parameters, such as forest fires, removals and non-wood forest products, would be relatively straightforward to assess if countries were willing and able to adopt a common approach to monitoring and reporting. FAO proposes to facilitate the development of definitions and reporting standards and to work with countries to implement them.

In summary, the availability of global and country information was not satisfactory for many subjects considered important for forest policy development. Furthermore, until baseline information is improved for important forest parameters, including diversity, degradation and productivity and their change over time, there is a danger that international policies and agreements such as the UN Framework Convention on Climate Change (UNFCCC) and the Convention on Biological Diversity (CBD) will be guided by general assumptions or extrapolations from partial, incomplete or invalid scientific studies.

ESTIMATING FOREST AREA AND AREA CHANGE

As in previous global assessments, the estimates of forest area and area change were the core of FRA 2000. Whether overemphasized or not, these parameters continue to be the most sought after from the global forest resources assessments. It is inevitable that the estimates will arouse controversy because of the political sensitivity of the subject material, especially as regards tropical deforestation (Matthews 2001; Stokstad 2001).

FRA 2000 provides transparent estimates that are traceable back to primary data and source documents. This approach will contribute to the establishment of a primary international data set and will help to counteract the recycling of inaccurate statistics in many reports. By indicating where information quality is low, FAO hopes to create an incentive to improve the baseline data. FRA 2000 reporting was done with the intention to establish a continuous improvement process where new information can be incorporated when it becomes available.

This approach does not make the results more reliable – only more accessible. Reliability can only stem from improvements in the quality and timeliness of national surveys and submissions to the global assessment in addition to the regional-level information generated directly by the assessment team, including remote sensing analysis. The collaboration with countries and national experts was very productive in FRA 2000 and the final findings are built on all available and relevant data on forest area and area change at the country level.

As concluded in Chapter 1, the precision of global-level estimates of forest area and area change is statistically valid, even though some source material may contain deviations from the true values. It is likely that an objective analysis would reveal some cases where forest area statistics reported by countries are too high, and others where they are too low. For example, one recent study in a moist tropical region suggested that regrowth of secondary forests is grossly underestimated in the forest area statistics reported by countries in the region. Therefore, it is important to continue to improve the science, methodology and consistency of forest area assessment and to try to ensure that the results of this work are processed as objectively as possible.

FRA 2000 conducted independent studies to support and complement information obtained from countries. Global mapping of forest cover, using 100 percent coverage by coarse-resolution remote sensing, provided an overview of the distribution of forests as well as a tool for global spatial analyses. However, the low spatial resolution of the remote sensing data meant that the results could not be used to enhance the findings on forest area or area change. By contrast, the FRA 2000 remote sensing survey of forest cover changes in tropical forests using higher-resolution images was very useful. The 10 percent sample, comparing the same areas that were sampled in 1990, provided statistically valid information on the area change dynamics at the regional and pan-tropical levels. The findings were not used to estimate forest area at the national level, but they provided an important validation of country data when aggregated at the regional level. This was especially useful for tropical Africa, where it was otherwise difficult to calibrate the area change estimate.

The documentation of the basis for estimating forest area and area change exceeds that of previous global forest assessments. A combination of well-developed inputs was used to draw the conclusions on the changes in forest area from 1990 to 2000. These include FRA 2000 national estimates on forests and plantations and the remote sensing survey of the tropics.

The result for net area change at the global level (-9.4 million hectares per year) indicates a lower net loss of forests compared with the results of earlier global assessments, mainly attributed to a greater expansion rate of new forests. While this lower rate of change is based on the most ambitious and accurate global assessment to date, the meaning and significance of the apparent trend is not yet known. While it may be a real trend, it could also be due to temporary conditions in the 1990s.

Confirming a shift in the rate of change is always a challenge, requiring long time series and careful analyses. In studying changes based on country reports, it is necessary to keep in mind that the average area-weighted reference year was 1994 for the world, and some years before that in many areas (Africa, for example). As a consequence the trends estimated for African countries relate mainly to the 1980s, with the change rates extrapolated to the 1990s.

In FRA 1990, the lack of updated information in many developing countries was addressed by developing a model that predicted the rate of change by country based mainly on population density and growth and climatic zone. This approach may be valid for a one-time estimate, but it has limited value in the study of trends over time, as the model will generate a time series based on its driving parameters rather than on real-world observations. Because of this potential bias and the desire to produce transparent estimates, the model approach used in FRA 1990 was not used in FRA 2000 to estimate area changes.

To support the country-level estimates, the FRA 2000 remote sensing survey was designed to study changes at the latest possible date, and to compare the 1980-1990 and 1990-2000 trends for the tropics. The statistical design made it possible to establish confidence intervals for the estimates. The area change estimate for the 1990s was lower than that for the 1980s, but the difference was not statistically significant. Thus, no reduced area loss could be confirmed for the tropics based solely on the remote sensing survey.

For the non-tropics as a whole, a considerable increase of forests in the 1990s was reported. Since the reference year was considerably later for these countries, it was valid to draw the conclusion that forests were expanding more in the non-tropics in the 1990s than in the previous decade.

Finally, it is useful to examine the possible reasons for this change in a larger social and land use context. The change of forest area depends to a large degree on the demand for land for other purposes. Two trends support the conclusion that the rate of deforestation may be decreasing. First, reduced demand for land by the agricultural sector and active afforestation programmes are leading to expansion of forests in temperate and boreal countries. Second, urbanization processes resulting from the development of national economies are significant in most parts of the world, which may reduce the demand for agricultural land in rural areas. In this case, agricultural land that is no longer needed often reverts to or is converted to forest, while land converted to urban uses may or may not be forest land.

In summary, FRA 2000 concluded that the global net change in forest area was lower in the 1990s than in the 1980s but that the rate of loss of natural forests remained at approximately the same level.

REMOTE SENSING

Remote sensing technologies are an area where there is promising potential to improve future assessments. Remote sensing technologies can provide images of physical or biological characteristics of the same land area at different points in time. FRA 2000 used remote sensing technology to create new global forest maps and to validate forest area change estimates. In theory, remote sensing is cheaper than traditional ground inventories when applied to large land areas (although the FRA 2000 remote sensing survey was paradoxically limited by financial constraints).

However, remote sensing can only address parameters that are well correlated with the information visible in images from above; it is thus excluded for many essential parameters and provides only limited precision and accuracy for basic biophysical variables such as wood volume and biomass. Furthermore, technology-intensive approaches generally preclude the participation of local people, thus limiting the ownership and local utilization of the information. Thus, a combination of remote sensing and ground-based assessment methodologies will continue to be needed for the foreseeable future.

STRENGTHS AND WEAKNESSES OF THE FRA 2000 APPROACH

FRA 2000 was carried out in response to global demand, represented notably by the recommendations of the FAO Committee on Forestry (COFO). The fourth session of the Intergovernmental Panel on Forests (IPF) endorsed the plan for the assessment elaborated during the Expert Consultation on FRA 2000 (Kotka III). The foundation for FRA 2000 was information provided by the countries of the world. It was recognized from the start that there would be considerable variation in the quality and completeness of this information, just as there is considerable variation in the institutional capabilities of nations in all other areas of endeavour. It was understood that major efforts would be required to validate and extrapolate from information provided by the countries. The problems related to inconsistent data quality and incomplete country information are the major weakness of FRA 2000. But the fact that FRA 2000 is built on country information is also its greatest strength. It is human nature that those who own and supply the information and are responsible for it are the most likely to be committed to use it and to improve it to influence policy decisions.

FRA 2000 also involved the active participation of international specialists and organizations in all phases of the work, starting with the expert consultation (Kotka III) that laid the framework for the assessment. For example, the global maps were produced in collaboration with the EROS Data Center in the United States. The protected forest estimates were made in collaboration with the UNEP World Conservation Monitoring Centre (UNEP-WCMC) in the United Kingdom. A transparent and forward-looking presentation of results was adopted, which encourages the continuous improvement of the baseline information. For the future, any individual, organization or country that develops more reliable or current information is encouraged to contribute it as soon as it is available so that it can be used to strengthen the next global assessment.

THE EXPANSION OF KNOWLEDGE IN FOREST ASSESSMENTS

FRA 2000 shared a major problem with most other forestry processes and programmes. Most of the process interaction was within the traditional forest sector. National forestry agencies provided data, and forestry experts in international organizations, universities and NGOs provided professional expertise. But most negative impacts on forests originate in other sectors. Furthermore, most countries that have succeeded in stabilizing their forest area are developed countries whose citizens do not need to exploit forest resources to try to escape from poverty. FRA 2000 would have benefited from greater cross-sectoral interaction. A major challenge for future global forest assessments will be to involve other sectors of society, to better integrate the assessment with other disciplines and to find new ways to use knowledge about forests to improve the lives of the world's citizens.

Interaction between the agricultural and forest sectors is central to how land is used and thus to the dynamics of forests. Economic development often leads to more capital-intensive agriculture and to a decrease in the area needed to produce agricultural outputs; in such cases (for example, throughout rural areas in the eastern United States) forests often expand on to former agricultural lands. Policies related to the development of infrastructure (e.g. roads and energy supply) often influence the use and size of forests. General development of economies may create employment opportunities in urban areas and less reliance on forest resources for basic needs such as fuelwood. Most of these topics fall outside the framework of a global forest resources assessment at the moment, but it would be important to incorporate cross-sectoral studies in the design of future global (and national) forest assessments.

At the local level, knowledge about forest resources is often relevant or essential to forest management. Such knowledge may also be relevant when aggregated to national or global levels. For example, the distribution of soil types and productivity (affecting carbon storage and fluxes), the location and dynamics of rare species (affecting biological diversity) and the impacts and benefits of uses of the land (affecting sustainable forest management) are important issues that are frequently studied and are discussed in this report. People who live and work in forests have huge amounts of knowledge about these subjects. This knowledge is usually reported in the form of local studies or case studies.

It is difficult to envision ways in which a global forest assessment can effectively address controversial issues such as illegal logging, but such areas should not be ruled out as impossibilities when plans are made for future assessments.

The present era is an exciting time of knowledge expansion, with new tools making it possible to share knowledge in unprecedented ways. One of the great challenges of forest assessments in the future will be to expand the participation of local people and of experts from other sectors and disciplines and to share and use knowledge in new ways.

BIBLIOGRAPHY

IUCN World Commission on Protected Areas (WCPA). 2000. *Evaluating effectiveness: a framework for assessing management of protected areas*, by M. Hockings with S. Stolton and N. Dudley. Best Practice

Protected Area Guidelines Series No. 6. Cambridge, UK, WWF/IUCN Forest.

Matthews, E. 2001. *Understanding FRA 2000.* World Resources Institute Forest Briefing No. 1. Washington, DC, WRI. www.wri.org/pdf/fra2000.pdf.

Stokstad, E. 2001. UN report suggests slowed forest losses. *Science*, 291(5512): 2294.

Chapter 51
Recommendations for future assessments

At its fifteenth session in March 2001, the FAO Committee on Forestry (COFO)[58] was informed about the main findings of FRA 2000 as well as proposals for future assessments. COFO made a number of recommendations (FAO 2001) which formed the starting point for the recommendations below. In particular, COFO recommended that the global FRA programme continue to be a priority for the FAO Forestry Department. FAO was requested to provide continued technical and financial assistance to build national capacities for carrying out forest assessments.

SCOPE OF ASSESSMENTS

FRA 2000 expanded the scope of previous assessments, and COFO recommended that FAO continue its efforts to carry out broad assessments, including various aspects of forest resources such as biological diversity, forest health and resource use. Future global forest resource assessments should continue to expand the number of parameters that are assessed.

National and international information requirements should guide the design and implementation of inventories and assessments so that the results will be useful for scenario development, planning processes and policy formulation. Information requirements should be holistic, multisectoral and multidisciplinary (Figure 51-1).

INFORMATION REQUIREMENTS AT THE NATIONAL LEVEL

Forest information requirements are determined by overall policy objectives, and the parameters are chosen to indicate, evaluate or predict to what extent these objectives are being fulfilled. The requirements must by necessity address not only biophysical status and development of the forests, but also parameters that tell how the forest is used, as well as types and quantities of various benefits that are derived from the forest.

Each country needs to identify the information that is required to develop and implement effective forest policies and programmes, including the need to monitor criteria and indicators for sustainable forest management (in accordance with the criteria and indicators process with which the country is affiliated).

National forest information must have a number of key characteristics to be trusted and useful for the complex analyses of the forest sector and impacts on forest ecosystems. More specifically, the information must be objectively collected, be representative of all forests or lands, have both a high precision and good accuracy, and capture relevant variations of key parameters. FRA 2000 has clearly shown that such information is lacking in most countries. This lack of information impedes the provision of qualified input to national policy processes and makes international assessments and reporting of key indicators difficult.

Investments in information directed at national needs should be proportional and relevant to the national issues. For example, although detailed mapping of forests and other land uses to achieve better area estimates is feasible, the costs involved must be weighed against the cost and importance of assessing other variables such as productivity, values of products and services and other indicators adopted to monitor sustainability of forest management.

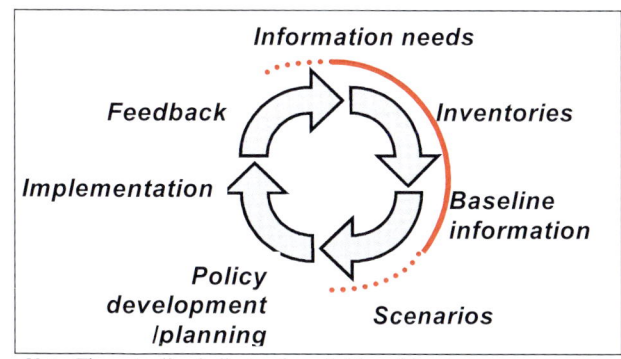

Note: The outer line indicates the activities of forest resources assessments.

Figure 51-1. Forestry knowledge management at the local, national or international level

---
[58] The Committee on Forestry (COFO) is the most important of the FAO Forestry Statutory Bodies. The biennial sessions of COFO (held at FAO headquarters in Rome) bring together heads of forest services and other senior government officials, usually representing more than 100 countries, to identify emerging policy and technical issues, to seek solutions and to advise FAO and others on appropriate action. Other international organizations and, increasingly, non-governmental groups participate in COFO.

As many forestry parameters are inherently local by nature and also affected by local management decisions, they must be inventoried and monitored at a local scale. Further, such local-scale observations must be aggregable to the national level. The consequence is that national-level assessments should be based on systematic field sampling where direct measurement and observations of relevant parameters can be made. This approach ensures not only that representative estimates can be achieved, but also that the variation of important parameters can be described. The variables to be assessed should be relevant for national-level policies as well as local needs.

To the extent feasible, international standards and reporting requirements should be taken into consideration when developing national information requirements and forest assessment processes. The use of internationally agreed standards and definitions is fundamental. This will greatly improve the consistency and comparability of data among countries as well as greatly simplifying the compilation of global assessments.

The fifteenth session of COFO (FAO 2001) recommended that FAO should lead the development and implementation of capacity building initiatives for developing countries and countries in transition, with an emphasis on improving national capacity for routine forest surveys related both to resources and to uses of forests. Such initiatives should be integrated with efforts to foster national information and knowledge management capabilities (FAO 2000a). COFO asked FAO to further develop this concept and to discuss it in the Regional Forestry Commissions. The goal is to initiate systematic knowledge collection and management at the national level and at the same time to develop internationally consistent data.

Capacity building for technical work should start at the field level and work up. At the same time, decision-makers should be increasingly brought into the picture, to ensure that information is relevant and available for their needs.

PERIODICITY OF GLOBAL ASSESSMENTS

Previous global assessments have been carried out at approximately ten-year intervals. While this historically reflected a need to balance cost, reporting requirements and the availability of new information, the demands for more complex and timely information have increased. At the same time, requests for increasingly detailed documentation of forest resources are proliferating with the needs of international fora, global treaties and other opportunities for forest-related discussion. With this in mind, it is relevant to reconsider the format and cycle of global forest resources assessments.

The fifteenth session of COFO in 2001 recommended that FAO should begin staging the next global assessment and should present a plan to the sixteenth session of COFO in 2003. The fourth session of the Intergovernmental Panel on Forests (IPF) recommended that FAO perform global assessments every five years instead of every ten, or carry out "rolling regional assessments" in the fifth year of the ten-year cycle. The twenty-third session of the Joint FAO/ECE Working Party on Forest Economics and Statistics (May 2001), representing the industrialized countries, recommended a ten-year cycle for the full global assessment.

To the extent possible, the next assessment should be coordinated with national criteria and indicator processes and with the United Nations Forum on Forests, with the assistance of the participating members of the Collaborative Partnership on Forests (CPF).

National statistics, updated in tables and in-depth reports on selected assessment issues, will be published every two years in FAO's *State of the World's Forests*.

As countries update their national inventories or undertake other related national assessments, this information should be reported concurrently to FAO for updating of databases and for posting on the country pages of the FAO Forestry Web site, www.fao.org/forestry.

DISSEMINATION OF RESULTS

Future global assessments should continue the approach introduced by FRA 2000 to present findings in a fully transparent manner. Steps should be taken to continue to improve the distribution and hence the impact of FRA results, both electronically and in print, and to improve access to Internet technology in developing countries.

PROCESSES

As recommended by an expert consultation in March 2000 (FAO 2000b), a global multidisciplinary team of specialists should be established to provide guidance on content, methodologies, definitions and other relevant issues to improve the quality of future assessments. The team should interact on a regular basis through meetings or electronic networking. Team members would also participate in capacity building and knowledge management processes.

To facilitate future information gathering, data ownership for global core variables should be further distributed among countries, the CPF and UNECE/FAO. In addition, further partnerships should be developed to support core information on subjects such as protected areas, threatened species, forest fires and illegal logging, taking advantage of the comparative advantages of partners.

The FAO Forestry Information System (FORIS) should continue to be developed as a platform for effective and transparent monitoring and reporting of forestry knowledge.

PROPOSED IMMEDIATE ACTIVITIES

Upon completion of FRA 2000, FAO will undertake:

- to begin an information needs assessment that balances information needs against the availability of data, funding, human resources, partnerships, etc., preparatory to the design of the next global assessment, interim assessments and updates;
- to convene a global expert consultation in 2002 to evaluate FRA 2000 and the information needs assessment and to make recommendations for future assessments, particularly the next global assessment;
- to involve all six Regional Forestry Commissions in the review of FRA 2000 and in planning for the next assessment and for national capacity building during 2002;
- to develop background information and recommendations for future assessments for consideration by COFO in 2003;
- to continue to develop a framework for capacity building and to implement it in pilot countries.

BIBLIOGRAPHY

FAO. 2000a. *Global Forest Survey – concept paper.* FRA Working No. Paper 28. Rome. www.fao.org/forestry/fo/fra/index.jsp

FAO. 2000b. *Proceedings of the FAO Expert Consultation to Review the FRA 2000 Methodology for Regional and Global Forest Change Assessment.* Rome, 6-10 March 2000. FRA Working Paper No. 42. Rome.

FAO. 2001. *Report of the fifteenth session of the Committee on Forestry.* Rome, 12-16 March 2001. Rome. www.fao.org/forestry/fo/statbod/Cofo/cofo_sessions-e.stm

Appendices

Appendix 1. FRA 2000 contributors

FRA 2000 was made possible through the committment and contributions of a large number of dedicated professionals worldwide.

First of all, many forestry professionals working for their respective national institutions provided country data, analyses and information – often after considerable efforts to meet the detailed requests from FAO. Their submissions ensured that FRA 2000 was based on the most reliable information in every country. National experts also reviewed and validated the findings related to their respective countries and many participated in FRA 2000 related meetings or workshops around the world.

Secondly, a large number of persons were assigned to specific tasks in the assessment, ranging from literature surveys to remote sensing and mapping and covering a broad set of forestry subjects. The following 238 persons representing 72 nationalities are hereby acknowledged for their efforts in various assignments within FRA 2000:
M. Ikermoud (Algeria); J. Beltran, S. Maradei (Argentina); R. Penny, D. Race, R. Lesslie, P. Tickle (Australia); A. Knieling, M. Noebauer, N. Winkler (Austria); L. De Pontbriand, T. Jacobs (Belgium); L. Akpado (Benin); C. Lisboa, M. Shuchvoski, C. Foelkel, S. Kengen, P. Moura Costa, J. Siqueira, I. Soressini, G. Stacioli, I. Tomasini, D. Campos Jansen, L. Cirineu, E. da Sila Fagundes Filho, A. Harausz, T. Krug, M. Marcos Andrey Hermogenes, M. Maristela Félix de Lima, S. Velásquez, E. Veras Micheletti (Brazil); J. Compaore, C. Kabore (Burkina Faso); A. Bararwandika (Burundi); R. Eba' Atyi (Cameroon); F. Banoun, M. Gillis, W. Glen, J. Innes, J. Williams, H. Hirvonen, D. Leckie, Z. Kalensky (Canada); P. Mangala, A. Zanga (Central African Republic); B. Djekourbian (Chad); Z. Du, X. Li, (China); M. Andrade (Colombia); P. Boundzanga, G. Nkéoua (Congo); K. Mbizi (Democratic Republic of the Congo); L. Corrales, F. Franklin, D. Morales, X. Soto, L. Ugalde, K. Oliveira Barvosa, G. Robles Valle, A. Salas, C. Scholz, R. Solano Alvarado, J. Tosi, R. Villalobos Soto, V. Watson (Costa Rica); O. Koffi (Côte d'Ivoire); V. Henzlik, J. Pavlis (Czech Republic); H. Olesen (Denmark); T. Omran (Egypt); P. Bololo (Equatorial Guinea); M. Bekele (Ethiopia); S. Jaakkola, P. Kauppi, K. Korhonen, A. Lehtonen, J. Liski, P. Tiusanen, E. Tomppo, P. Vourinen, R. Päivinen (Finland); I. Amsallem, C. Bernard, P. Couteron, A. Le Magadoux, J. Valdenaire, M. Pain-Orcet, S. Petit, M. Bellan, O. Ostermann, D. Richard (France); J. Bile Allogho, P. Koumba Zaou (Gabon); J. Sillah (Gambia); J. Goldammer, M. Köhl, F. Schmitz, U. Bohn, C. Kleinn (Germany); F. Odoom (Ghana); P. Baeza (Guatemala); K. Diombera (Guinea Bissau); P. Csoka (Hungary); K. Govil, D. Pandey, J. Singh, A. Wahal, V. Bhalla, D. Bist, T. Bist, R. Chaturvedi, B. Chauhan, P. Joshi, S. Nautiyal, S. Pipara, H. Singh (India); M. Mirsadeghi (Iran); A. Shakir (Iraq); S. Iremonger (Ireland); I. Ambrosini, A. Baccini, P. De Salvo, A. Dell'Agnello, C. Dibari, R. Drigo, M. Garzuglia, U. Leonardi, M. Maggi, R. Messina, O. Pfister, V. Robiglio , R. Scotti, F. Urbani (Italy); A. Al Wadi (Jordan); C. Kahuki, H. Kojwang, J. Legilisho-Kiyiapi, J. Ndambiri, G. Ochieng (Kenya); F. Asmar (Liban); S. Ramakavelo, B. Ramamonjisoa, H. Randrianjafy (Madagascar); W. Wan Mohd (Malaysia); A. Kouyaté, A. Maiga, H. Yossi (Mali); N. Ould Taleb (Mauritania); L. Morales (Mexico); M. Et-Tobi, O. M'Hirit (Morocco); M. Monjane (Mozambique); F. Deodatus (Netherlands); J. Barton, D. Mead, C. Perley (New Zealand); Mahamane (Niger); C. Ameh, O. Olaleye (Nigeria); A. Ennals, S. Folving (Norway); R. Michalak, M. Suchorowski (Poland); M. Duarte (Portugal); A. Filipchuk, I. Lysenko, G. Ogureeva, A. Shvidenko, B. Zhimbiev (Russian Federation); S. Murererehe (Rwanda); K. Al Mousa (Saudi Arabia); S. Gueye (Sénégal); J. Ilavsky (Slovakia); K. Tjon, J. Wirjosentono (Surinam); E. Andersson, J. Cedergren, Z. Fazakas, S. Holm, A. Nylén, E. Sollander, S. Svensson, R. Persson, T. Thuresson (Sweden); P. Brassel, C. Elliott, S. Jaakkola (Switzerland); Y. Barkoudah, Z. Jebawi (Syrian Arab Republic); K. Akakpo, K. Kokou (Togo); K. Selmi (Tunisia); P. Drichi (Uganda); S. Blyth, N. Cox, N. Dudley, C. Elliott, H. Gillett, C. Lusty, C. Ravilous, G. D'Souza, J. Evans, N. Jewell, T. Peck, L. Wearne, J. Wong (United Kingdom); W. Ciesla, P. Kennedy, W. Libby, R. Mutch, A. Rice, J. Space, T. Waggener, A. Gillespie, J. Hutchison, D. Johnson, V. Kapos, T. Loveland, D. Waller, Z. Zhu, B. Smith, R. Bailey, N. Burgess, L. Langner, J. Morrison (United States);

A. Al Attas (Yemen); M. Chihambakwe, D. Duwa, C. Gumbie, D. Gwaze, D. Kwesha, C. Marunda, C. Phiri (Zimbabwe).

Finally, FAO staff formed an institutional backbone to the work. FRA 2000 included a range of forestry subjects and therefore involved staff members from several units and decentralized offices. The following 60 staff members, representing 25 nationalities, are hereby acknowledged for their efforts: G. Allard, D. Altrell, A. Al-Fares, M. Amatiste, J. Ball, J. Bourke, A. Branthomme, S. Braatz, C. Brown, J. Carle, C. Carneiro, F. Castaneda, J. Clement, S. Dalsgaard, R. Davis, A. Del Lungo, P. Durst, C. Eckelman, H. El-Lakany, I. Eriksson, T. Etherington, T. Frisk, M. Gauthier, M. Grylle, S. Hirai, P. Holmgren, M. Kashio, W. Killmann, D. Kneeland, A. Korotkov, M. Laverdière, J. Lejeune, J. Lorbach, M. Lorenzini, J. Malleux, A. Mathias, M. Mancurti, A. Marzoli, F. Ndeckere-Ziangba, J. Nichols, H. Ortiz-Chour, C. Palmberg, A. Perlis, G. Preto, C. Prins, P. Pugliese, L. Ransom, L. Russo, M. Saket, E. Sène, H. Simons, K. Singh, T. So, K. Thelen, P. Van Laake, P. Vantomme, J. Varjo, S. Walter and M.L. Wilkie.

Appendix 2. Terms and definitions

The terms and definitions used in FRA 2000 were based on the consensus agreement of the participants of the Expert Consultation on Global Forest Resources Assessment 2000 in 1996 (Kotka III) (Finnish Forest Research Institute 1996). In 1997, the Intergovernmental Panel on Forests (IPF) endorsed the findings of the consultation and, in particular, supported the single global definition of forest developed in the meeting (UN 1997). The following year, FAO released FRA Working Paper No. 1 (1998) which contained the terms and definitions advocated at Kotka III. Some of the definitions were clarified in 2000 in FRA Working Paper No. 33 (FAO 2000) and in the *State of the World's Forests 2001* (FAO 2001) in such a way as to enhance their understanding, without changing their meaning.

during FRA 2000. One driving factor behind this re-examination was the request for input to the Kyoto Protocol process and the elaborations on carbon sequestration in forests. A clear and complete outline of forest change processes was sought. Some of the original formulations for FRA 2000 were therefore modified for clarity and completeness, without changing the meaning of the definitions. The process of further refining forest terms and definitions will continue under the facilitation of FAO, with the clear objective of keeping the base FRA definitions used and ensuring that the time series of forest area data can be continued.

The following terms and definitions, illustrated below, represent the comprehension at the end of the FRA 2000 project.

DEFINITIONS OF FOREST AND RELATED LAND USE CLASSIFICATIONS AND FOREST CHANGE PROCESSES

Background
In 2000, the basic forest and forest change terms were revisited in light of the experiences gained

Forest and related land use classifications

Forest	Forest includes natural forests and forest plantations. It is used to refer to land with a tree canopy cover of more than 10 percent and area of more than 0.5 ha. Forests are determined both by the presence of trees and the absence of other predominant land uses. The trees should be able to reach a minimum height of 5 m. Young stands that have not yet but are expected to reach a crown density of 10 percent and tree height of 5 m are included under forest, as are temporarily unstocked areas. The term includes forests used for purposes of production, protection, multiple-use or conservation (i.e. forest in national parks, nature reserves and other protected areas), as well as forest stands on agricultural lands (e.g. windbreaks and shelterbelts of trees with a width of more than 20 m), and rubberwood plantations and cork oak stands. The term specifically excludes stands of trees established primarily for agricultural production, for example fruit tree plantations. It also excludes trees planted in agroforestry systems.
Natural forest	A forest composed of indigenous trees and not classified as forest plantation.
Forest plantation	A forest established by planting or/and seeding in the process of afforestation or reforestation. It consists of introduced species or, in some cases, indigenous species.
Other wooded land	Land that has either a crown cover (or equivalent stocking level) of 5 to 10 percent of trees able to reach a height of 5 m at maturity; or a crown cover (or equivalent stocking level) of more than 10 percent of trees not able to reach a height of 5 m at maturity; or with shrub or bush cover of more than 10 percent.

Forest change processes

Afforestation	Establishment of forest plantations on land that, until then, was not classified as forest. Implies a transformation from non-forest to forest.
Natural expansion of forest	Expansion of forests through natural succession on land that, until then, was under another land use (e.g. forest succession on land previously used for agriculture). Implies a transformation from non-forest to forest.
Reforestation	Establishment of forest plantations on temporarily unstocked lands that are considered as forest.
Natural regeneration on forest lands	Natural succession of forest on temporarily unstocked lands that are considered as forest.
Deforestation	The conversion of forest to another land use *or* the long-term reduction of the tree canopy cover below the minimum 10 percent threshold (see definition of forest and the following explanatory note). Explanatory note: Deforestation implies the long-term or permanent loss of forest cover and implies transformation into another land use. Such a loss can only be caused and maintained by a continued human-induced or natural perturbation. Deforestation includes areas of forest converted to agriculture, pasture, water reservoirs and urban areas. The term specifically excludes areas where the trees have been removed as a result of harvesting or logging, and where the forest is expected to regenerate naturally or with the aid of silvicultural measures. Unless logging is followed by the clearing of the remaining logged-over forest for the introduction of alternative land uses, or the maintenance of the clearings through continued disturbance, forests commonly regenerate, although often to a different, secondary condition. In areas of shifting agriculture, forest, forest fallow and agricultural lands appear in a dynamic pattern where deforestation and the return of forest occur frequently in small patches. To simplify reporting of such areas, the net change over a larger area is typically used. Deforestation also includes areas where, for example, the impact of disturbance, overutilization or changing environmental conditions affects the forest to an extent that it cannot sustain a tree cover above the 10 percent threshold.
Forest degradation	Changes within the forest which negatively affect the structure or function of the stand or site, and thereby lower the capacity to supply products and/or services.
Forest improvement	Changes within the forest which positively affect the structure or function of the stand or site, and thereby increase the capacity to supply products and/or services.

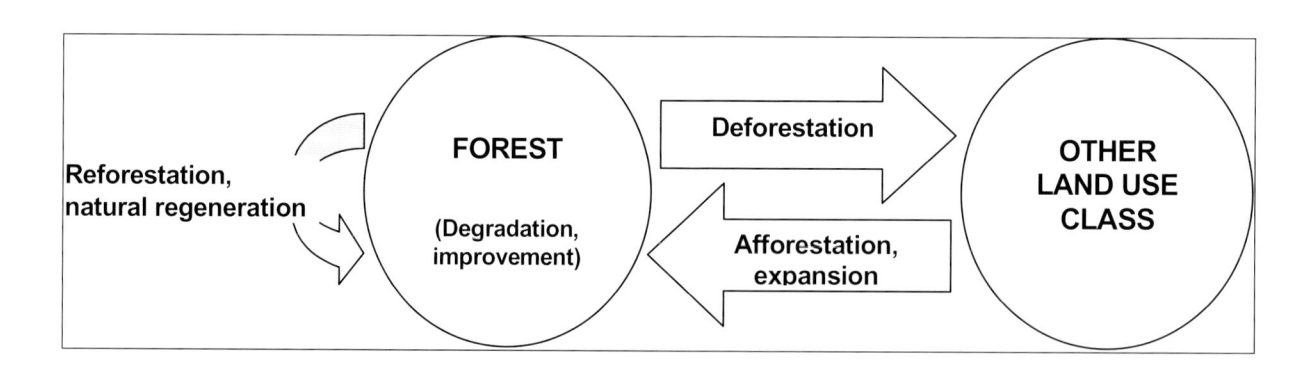

TERMS AND DEFINITIONS AS PRESENTED IN FRA WORKING PAPER NO. 1

Background

The terms and definitions below are taken directly from FRA Working Paper No. 1 (FAO 1998). They represent the formulations used at the beginning of the FRA 2000 process. Although no changes in definitions have occurred during FRA 2000, it is important to note that some slight adjustments and clarifications have been adopted. For example, the general classification of land was called "land cover classification" in 1998, whereas in this report the division into forest, other wooded land and other land is a "land use classification", in that "forest" is defined both by the presence of trees and by the absence of other land uses.

Land classifications

Land cover, general classification	
Total area[59]	Total area (of country), including area under inland water bodies, but excluding offshore territorial waters.
Forest	Land with tree crown cover (or equivalent stocking level) of more than 10 percent and area of more than 0.5 ha. The trees should be able to reach a minimum height of 5 m at maturity *in situ*. May consist either of closed forest formations where trees of various storeys and undergrowth cover a high proportion of the ground; or open forest formations with a continuous vegetation cover in which tree crown cover exceeds 10 percent. Young natural stands and all plantations established for forestry purposes which have yet to reach a crown density of 10 percent or tree height of 5 m are included under forest, as are areas normally forming part of the forest area which are temporarily unstocked as a result of human intervention or natural causes but which are expected to revert to forest. Includes: forest nurseries and seed orchards that constitute an integral part of the forest; forest roads, cleared tracts, firebreaks and other small open areas; forest in national parks, nature reserves and other protected areas such as those of specific scientific, historical, cultural or spiritual interest; windbreaks and shelterbelts of trees with an area of more than 0.5 ha and width of more than 20 m; plantations primarily used for forestry purposes, including rubberwood plantations and cork oak stands. Excludes: Land predominantly used for agricultural practices
Other wooded land	Land either with a crown cover (or equivalent stocking level) of 5-10 percent of trees able to reach a height of 5 m at maturity *in situ*; or a crown cover (or equivalent stocking level) of more than 10 percent of trees not able to reach a height of 5 m at maturity *in situ* (e.g. dwarf or stunted trees); or with shrub or bush cover of more than 10 percent.
Other land	Land not classified as forest or other wooded land as defined above. Includes agricultural land, meadows and pastures, built-on areas, barren land, etc.
Inland water	Area occupied by major rivers, lakes and reservoirs.
Forest classification	
Plantation	Forest stands established by planting or/and seeding in the process of afforestation or reforestation. They are either: • of introduced species (all planted stands), or • intensively managed stands of indigenous species, which meet all the following criteria: one or two species at plantation, even age class, regular spacing. See also *afforestation* and *reforestation*. Note: Area statistics on forest plantations provided by countries should reflect the actual forest plantations resource, excluding replanting. **Replanting** is the re-establishment of planted trees, either because afforestation or reforestation failed, or tree crop was felled and regenerated. It is not an addition to the total plantation area.

---

[59] The "Total land area" is defined as the total area, but excluding inland water.

Natural forest	Natural forests are forests composed of indigenous trees, not planted by humans. Or in other words forests excluding plantations. Natural forests are further classified using the following criteria: • forest formation (or type): closed/open, • degree of human disturbance or modification, • species composition.
Subdivisions of natural forests: forest formation	
Closed forest	Formations where trees in the various storeys and the undergrowth cover a high proportion (> 40 percent) of the ground and do not have a continuous dense grass layer (cf. the following definition). They are either managed or unmanaged forests, primary or in advanced state of reconstitution and may have been logged-over one or more times, having kept their characteristics of forest stands, possibly with modified structure and composition. Typical examples of tropical closed forest formations include tropical rain forest and mangrove forest.
Open forest	Formations with discontinuous tree layer but with a coverage of at least 10 percent and less than 40 percent. Generally there is a continuous grass layer allowing grazing and spreading of fires. (Examples are various forms of cerrado, and chaco in Latin America, wooded savannahs and woodlands in Africa).
Subdivisions of natural forests: degree of human disturbance or modification	
Natural forest undisturbed by humans	Forest which shows natural forest dynamics such as natural species composition, occurrence of dead wood, natural age structure and natural regeneration processes, the area of which is large enough to maintain its natural characteristics and where there has been no known human intervention or where the last significant human intervention was long enough ago to have allowed the natural species composition and processes to have become re-established.
Natural forest disturbed by humans	Includes: • logged over forests associated with various intensity of logging, • various forms of secondary forests, resulting from logging or abandoned cultivation.
Semi-natural forest	Managed forests modified by man through sylviculture and assisted regeneration.
Subdivisions of natural forests: forest composition by species groups	
Broad-leaved forest	Forest with a predominance (more than 75 percent of tree crown cover) of trees of broad-leaved species.
Coniferous forest	Forest with a predominance (more than 75 percent of tree crown cover) of trees of coniferous species.
Bamboo/palm formations	Forest on which more than 75 percent of the crown cover consists of tree species other than coniferous or broad-leaved species (e.g. tree-form species of the bamboo, palm and fern families).
Mixed forest	Forest in which neither coniferous nor broad-leaved species nor palms nor bamboos account for more than 75 percent of the tree crown cover.
Subdivision of other wooded land	
Shrubs	Refer to vegetation types where the dominant woody elements are shrubs i.e. woody perennial plants, generally of more than 0.5 m and less than 5 m in height on maturity and without a definite crown. The height limits for trees and shrubs should be interpreted with flexibility, particularly the minimum tree and maximum shrub height, which may vary between 5 and 7 m approximately.

Forest fallow system	Refers to all complexes of woody vegetation deriving from the clearing of natural forest for shifting agriculture. It consists of a mosaic of various reconstitution phases and includes patches of uncleared forests and agriculture fields, which cannot be realistically segregated and accounted for area-wise, especially from satellite imagery. Forest fallow system is an intermediate class between forest and non-forest land uses. Part of the area may have the appearance of a secondary forest. Even the part currently under cultivation sometimes has appearance of forest, due to presence of tree cover. Accurate separation between forest and forest fallow may not always be possible.
Protected areas – IUCN classification for nature protection	
I - Strict nature reserve/ wilderness area	**Protected area managed mainly for science or wilderness protection.** These areas possess some outstanding ecosystems, features and/or species of flora and fauna of national scientific importance, or they are representative of particular natural areas. They often contain fragile ecosystems or life forms, areas of important biological or geological diversity, or areas of particular importance to the conservation of genetic resources. Public access is generally not permitted. Natural processes are allowed to take place in the absence of any direct human interference, tourism and recreation. Ecological processes may include natural acts that alter the ecological system or physiographic features, such as naturally occurring fires, natural succession, insect or disease outbreaks, storms, earthquakes and the like, but necessarily excluding man-induced disturbances.
II – National park	**Protected area managed mainly for ecosystem protection and recreation.** National parks are relatively large areas, which contain representative samples of major natural regions, features or scenery, where plant and animal species, geomorphological sites, and habitats are of special scientific, educational and recreational interest. The area is managed and developed so as to sustain recreation and educational activities on a controlled basis. The area and visitors' use are managed at a level which maintains the area in a natural or semi-natural state.
III - Natural monument	**Protected area managed mainly for conservation of specific natural features.** This category normally contains one or more natural features of outstanding national interest being protected because of their uniqueness or rarity. Size is not of great importance. The areas should be managed to remain relatively free of human disturbance, although they may have recreational and touristic value.
IV – Habitat/ species management area	**Protected area managed mainly for conservation through management intervention.** The areas covered may consist of nesting areas of colonial bird species, marshes or lakes, estuaries, forest or grassland habitats, or fish spawning or seagrass feeding beds for marine animals. The production of harvestable renewable resources may play a secondary role in the management of the area. The area may require habitat manipulation (mowing, sheep or cattle grazing, etc.).
V - Protected landscape/ seascape	**Protected areas managed mainly for landscape/seascape conservation and recreation.** The diversity of areas falling into this category is very large. They include those whose landscapes possess special aesthetic qualities which are a result of the interaction of man and land or water, traditional practices associated with agriculture, grazing and fishing being dominant; and those that are primarily natural areas, such as coastline, lake or river shores, hilly or mountainous terrains, managed intensively by humans for recreation and tourism.
VI - Managed resource protection area	**Protected area managed for the sustainable use of natural ecosystems.** Normally covers extensive and relatively isolated and uninhabited areas having difficult access, or regions that are relatively sparsely populated but are under considerable pressure for colonization or greater utilization.

Forest area available for wood supply	
Forest available for wood supply	Forest where any legal, economic, or specific environmental restrictions do not have a significant impact on the supply of wood.
	Includes: Areas where, although there are no such restrictions, harvesting is not taking place, for example areas included in long-term utilization plans or intentions.
Forest not available for wood supply	Forest where legal, economic or specific environmental restrictions prevent any significant supply of wood.
	Includes:
	• Forest with legal restrictions or restrictions resulting from other political decisions, which totally exclude or severely limit wood supply, *inter alia* for reasons of environmental or biological diversity conservation, e.g. protection forest, national parks, nature reserves and other protected areas, such as those of special environmental, scientific, historical, cultural or spiritual interest;
	• Forest where physical productivity or wood quality is too low or harvesting and transport costs are too high to warrant wood harvesting, apart from occasional cuttings for autoconsumption.

Forest parameters

Volume and biomass	
Growing stock	Stem volume of all living trees more than 10 cm diameter at breast height (or above buttresses if these are higher), over bark measured from stump to top of bole.
	Excludes: all branches
Commercial growing stock	Part of the growing stock, that consists of species considered as actually or potentially commercial under current local and international market conditions, at the reported reference diameter (DBH).
	Includes: species which are currently not utilized, but potentially commercial having appropriate technological properties.
	Note: When most species are merchantable, i.e. in the temperate and boreal zone, the commercial growing stock, in a given area or for a country, can be close to the total growing stock. In the tropics however, where only a fraction of all species are merchantable, it may be much smaller.
Woody biomass	The mass of the woody part (stem, bark, branches, twigs) of trees, alive and dead, shrubs and bushes.
	Includes: Above ground woody biomass, stumps and roots.
	Excludes: foliage, flowers and seeds.
Above-ground woody biomass	The above ground mass of the woody part (stem, bark, branches, twigs) of trees, alive or dead, shrubs and bushes.
	Excludes: stumps and roots, foliage, flowers and seeds.

Fellings and removals	
Fellings	Average volume of all trees, living or dead, measured over bark to a minimum diameter of 10 cm (DBH), that are felled during a given period (e.g. annually), whether or not they are removed from the forest or other wooded land.
	Includes: silvicultural and pre-commercial thinnings and cleanings of trees more than 10 cm (DBH) left in the forest, and natural losses of trees above 10 cm (DBH).

Removals	(Annual) removals that generate revenue for the owner of the forest or other wooded land or trees outside the forest. They refer to "volume actually commercialized" (VAC), i.e. volume under bark actually cut and removed from the forest. This volume may include wood for industrial purposes (e.g. sawlogs, veneer logs, etc.) and for local domestic use (e.g. rural uses for construction).
	Includes: removals during the given reference period of trees felled during an earlier period and removal of trees killed or damaged by natural causes (natural losses), e.g. fire, wind, insects and diseases.
	Excludes: removals for fuelwood.
	Note: Removals as defined above refer to commercial removals, i.e. harvested timber, both for industrial and local domestic uses. In many developing countries, removals for fuelwood make up a considerable part of the total harvested wood. However, data on fuelwood removals are generally scarce and/or unreliable, and need to be reported separately when national or local data are available.

Non-wood forest products and forest services

Non-wood forest products	Products for human consumption: food, beverages, medicinal plants, and extracts (e.g. fruits, berries, nuts, honey, game meats, mushrooms, etc.).
	Fodder and forage (grazing, range).
	Other non-wood products (e.g. cork, resin, tannins, industrial extracts, wool and skins, hunting trophies, Christmas trees, decorative foliage, mosses and ferns, essential and cosmetic oils, etc.).
Forest services	Protection (against soil erosion by air or water, avalanches, mud and rock slides, flooding, air pollution, noise, etc.).
	Social and economic values (e.g. hunting and fishing, other leisure activities, including recreation, sport and tourism).
	Aesthetic, cultural, historical, spiritual and scientific values (including landscape and amenity).

Changes

Forest cover changes

Deforestation	Refers to change of land cover with depletion of tree crown cover to less than 10 percent. Changes within the forest class (e.g. from closed to open forest) which negatively affect the stand or site and, in particular, lower the production capacity, are termed forest degradation.
Forest degradation	Takes different forms, particularly in open forest formations, deriving mainly from human activities such as overgrazing, overexploitation (for fuelwood or timber), repeated fires, or due to attacks by insects, diseases, plant parasites or other natural sources such as cyclones. In most cases, degradation does not show as a decrease in the area of woody vegetation but rather as a gradual reduction of biomass, changes in species composition and soil degradation. Unsustainable logging practices can contribute to degradation if the extraction of mature trees is not accompanied with their regeneration or if the use of heavy machinery causes soil compaction or loss of productive forest area.
New plantations: afforestation	Artificial establishment of forest on lands which previously did not carry forest within living memory.
New plantations: reforestation	Artificial establishment of forest on lands which carried forest before.

Additional terms

Broad-leaved tree	All trees classified botanically as Angiospermae. They are sometimes referred to as "non-coniferous" or "hardwoods".
Coniferous tree	All trees classified botanically as Gymnospermae. They are sometimes referred to as "softwoods".
Endangered species	Species classified by an objective process (e.g. national "Red Book") as being in IUCN categories "critically endangered" and "endangered". A species is considered to be "critically endangered" when it is facing an extremely high risk of extinction in the wild in the immediate future. It is considered "endangered" when it is not critically endangered but is still facing a very high risk of extinction in the wild in the near future.
Endemic species	Species is endemic when found only in a certain strictly limited geographical region, i.e. restricted to a specified region or locality.
Indigenous tree species	Tree species which have evolved in the same area, region or biotope where the forest stand is growing and are adapted to the specific ecological conditions predominant at the time of the establishment of the stand. May also be termed native species or autochthonous species.
Introduced tree species	Tree species occurring outside their natural vegetation zone, area or region. May also be termed non-indigenous species.
Managed forest/other wooded land	Forest and other wooded land that is managed in accordance with a formal or an informal plan applied regularly over a sufficiently long period (five years or more).
Protection	The function of forest/other wooded land in providing protection of soil against erosion by water or wind, prevention of desertification, the reduction of risk of avalanches and rock or mud slides; and in conserving, protecting and regulating the quantity and quality of water supply, including the prevention of flooding. Includes: Protection against air and noise pollution.
Tree	A woody perennial with a single main stem, or in the case of coppice with several stems, having a more or less definite crown. Includes: bamboos, palms and other woody plants meeting the above criterion.

BIBLIOGRAPHY

FAO. 1998. *FRA 2000: Terms and definitions.* FRA Working Paper No. 1. Rome. www.fao.org/forestry/fo/fra/index.jsp

FAO. 2000. *On definitions of forest and forest change.* FRA Working Paper No. 33. Rome.

Finnish Forest Research Institute. 1996. *Expert consultation on Global Forest Resource Assessment 2000. Kotka III.* Proceedings of FAO Expert Consultation on Global Forest Resources Assessment 2000 in cooperation with ECE and UNEP with the Support of the Government of Finland (KOTKA III). Kotka, Finland, 10-14 June 1996. Eds. Nyyssonen, A. & Ahti, A. Research Papers No. 620. Helsinki. Finland.

UN. 1997. *Report of the Ad Hoc Intergovernmental Panel on Forests on its Fourth Session.* New York, 11-27 February 1997. E/CN.17/1997/12. New York.

Appendix 3. Global tables

OVERVIEW OF GLOBAL TABLES AND PAGE REFERENCES:

Table	Title	Notes	Table
1	Basic country data	372	379
2	Forest cover – information status	372	383
3	Forest cover 2000	372	387
4	Change in forest cover 1990-2000	372	391
5	Forest cover – latest national statistics	373	395
6	Forest plantations 2000	373	399
7	Volume and biomass in forest	373	403
8	Forest fires 1990-2000	373	407
9	Status and trends in forest management	373	411
10	Removals	376	415
11	Comparison of forest management areas	376	419
12	Non-wood forest products – major product groups	376	423
13	Endangered, endemic species for seven species groups	376	427
14	Distribution of total forest area by ecological zone	376	431
15	Forest in protected areas / available for wood supply	376	435
16	FRA 2000 country interaction	377	439

NOTES TO GLOBAL TABLES

General notes
The 16 tables included represent a summary of FRA 2000 findings. The tables are available on the FAO Forestry Web site (www.fao.org/forestry/fo/fra/index.jsp)

Country nomenclature and regional groups used in the tables
The country names and order used in these tables follow standard UN practice regarding nomenclature and alphabetical listing of countries. Data for China incorporate values for China (including Hong Kong and Macao) and for Taiwan Province of China, as consistent with UN practice. The regional groups used in these tables represent FAO's standardized regional breakdown of the world according to geographical (note: not economic or political) criteria.

The designations employed and the presentation of material in this publication do not imply the expression of any opinion whatsoever on the part of the Food and Agriculture Organization of the United Nations concerning the legal status of any country, territory, city or area or of its authorities, or concerning the delimitation of its frontiers or boundaries.

Totals
Numbers may not tally because of rounding.

Abbreviations
n.s. = not significant, indicating a very small value
-     = not available (n.a.)
n.ap. = not applicable
000   = thousands
      = empty data cells indicate a zero value

Further information
For many of the country estimates, further background and explanation of the numbers are available in the FAO Forestry country profiles at: www.fao.org/forestry/fo/country/nav_world.jsp

Table 1: Basic country data (page 379)
The "Land area" figure refers to total area of the country, excluding areas under inland water bodies. These data have been derived from the total area of the country (including inland water) as maintained by FAOSTAT (http://apps.fao.org), minus the area of inland water as reported by FRA 2000 (see Table 5 below). Population statistics on total population, population density, and annual rate of change are taken from *World population prospects – the 1998 revision* (UN 1999). The source of "percentage rural population" data is *World urbanization prospects – the 1996 revision* (UN 1997).

The source of the economic data is *World development indicators 1999* (World Bank 2000). The gross national product (GNP) per capita figure represents the GNP divided by the mid-year population. The data are in constant 1995 US dollars. The annual percentage growth rate of gross domestic product (GDP) is based on constant local currency.

Table 2: Forest cover – information status (page 383)
The table indicates the availability of forest cover information and its compatibility with definitions used in FRA 2000. "Latest available statistics" refers to the latest available and relevant source data covering the entire country (see also Table 5). "Reference year" is the average year of the field survey or the remote sensing material used. "Method" specifies how the information was obtained (FS = nationwide field sampling, DM = detailed mapping, GM = general mapping, ES = expert estimate). "Compatibility" indicates how well the national forest classifications could be transformed to FRA 2000 global classes (H = High, M = Medium, L = Low). The "Time series" columns indicate whether a country time series for forest cover could be constructed and used in FRA 2000, and how compatible the observations along the time series were (H = High, M = Medium, L = Low). The number of references is the number of publications used in FRA 2000 to estimate forest cover and plantation extent for the country.

Table 3: Forest cover 2000
Table 4: Change in forest cover 1990-2000 (page 387)
The tables show the forest cover in 2000 and the estimated change between 1990 and 2000. FAO made adjustments to the standard reference years 1990 and 2000 based on available national statistics (see also Tables 2 and 5). "Total forest" is the sum of natural forest and plantations. "Forest cover change" is the net change in forests and includes expansion of plantations, and losses and gains in the area of natural forests. Changes to FRA 2000 estimates have been included up to 19 January 2001. The statistics will be updated as new information becomes available; latest updates are posted on the FAO Forestry Web site (www.fao.org/forestry/fo/country/nav_world.jsp).

Appendix 3. Global tables

Table 5: Forest cover – latest national statistics (page 395)

The table presents the latest forest cover statistics with national coverage, reclassified from national classifications into the global classification system used in FRA 2000 (see also Table 2). In some cases the national coverage was obtained by combining several surveys into a national scenario. "Reference year" is the average year of the field survey or the remote sensing material used. Refer to FRA Working Paper No. 1 (FAO 1998) for an explanation of definitions used (www.fao.org/forestry/fo/fra/docs/FRA_WP1eng.pdf). The statistics have been obtained from analysis of national documentation which is fully referenced in the country profiles on the FAO Forestry Web site (www.fao.org/forestry/fo/country/nav_world.jsp).

Table 6: Forest plantations 2000 (page 399)

Forest plantations are forest stands of introduced species, or intensively managed stands of indigenous species of even age class and regular spacing (see also Annex 1). The table shows the total plantation area in 2000 and the currently reported annual expansion of plantation forest. The total area is broken down by major species groups. The statistics have been obtained from analysis of national documentation which is fully referenced in the country profiles on the FAO Forestry Web site (www.fao.org/forestry/fo/country/nav_world.jsp), where a further breakdown of plantation areas according to ownership and purpose is presented.

Table 7: Volume and biomass in forest (page 403)

The table shows estimates of volume (total volume over bark of living trees above 10 cm diameter at breast height) and biomass (above-ground mass of the woody part (stem, bark, branches, twigs) of trees, alive or dead, shrubs and bushes. "Information source" refers to the type of source data used for the estimate (NI = national inventory, PI = partial inventory, ES = expert estimate, EX = data extrapolated from other countries).

For industrialized countries (Europe, the countries of the Commonwealth of Independent States, Japan, Australia, New Zealand, Canada and the United States) the stem volume for all living trees has been used for the volume figure. Some variation as to the minimum diameter applied are reported in UNECE/FAO (2000).

Table 8: Forest fires 1990-2000 (page 407)

Forest wildfire statistics for the 1990s are presented by number of fires and area affected. The average, smallest (min) and largest (max) values are shown. Note that the figures are derived on partial time series when data from some years are missing.

Table 9: Status and trends in forest management (page 411)

Criteria and indicators for sustainable forest management

Membership of ecoregional processes on criteria and indicators are listed using the following acronyms:

ATO = African Timber Organization
DZAf = Dry-Zone Africa Process on Criteria and Indicators for Sustainable Forest Management
DFAs = Regional Initiative for the Development and Implementation of National Level Criteria and Indicators for the Sustainable Management of Dry Forests in Asia
EUR = Pan-European Forest Process on Criteria and Indicators for Sustainable Forest Management
ITTO = International Tropical Timber Organization
LEP = Lepaterique Process of Central America on Criteria and Indicators for Sustainable Forest Management
MON = Montreal Process on Criteria and Indicators for the Conservation and Sustainable Management of Temperate and Boreal Forests
NE = Near East Process on Criteria and Indicators for Sustainable Forest Management
TARA = Tarapoto Proposal of Criteria and Indicators for Sustainability of the Amazon Forest

Four countries that were invited to join the Pan-European Forest Process (Bosnia and Herzegovina, Georgia, San Marino and Yugoslavia) as of December 2000 have been included in the table.

Area under forest management plans in 2000

For industrialized countries (Europe, CIS countries, Cyprus, Israel, Turkey, Japan, Australia, New Zealand, Canada and the United States), the areas listed include all forest areas managed irrespective of whether a formal plan exists or not. See further explanation under the geographical regions below. For additional information on these countries refer to UNECE/FAO (2000).

Some countries (including all the industrialized countries and many in South America) provided information both on the total area of forest managed or under a management plan (in hectares) and on the area in percentage of the latest figure available for the total area of forest.

In this study, the area figures provided have been used and the percentage figures (which are in percentage of the estimated forest area in 2000) may therefore differ from national statistics and should be treated with some caution. One notable exception concerns countries that reported that all forest areas were under management. In these cases the percentage figure (100 percent) was used and the actual area figure was recalculated to correspond to the 2000 forest area figure. The figure of the area under management may, therefore, differ from national statistics for these countries.

All national-level information was provided as part of FRA 2000 reporting or as national reports presented to Regional Forestry Commission meetings. Partial data were obtained from a variety of sources.

Africa. The definition used for area under forest management plans in Africa is: "The area of forest which is managed for various purposes (conservation, production, other) in accordance with a formal, nationally approved management plan over a sufficiently long period (five years or more)".

Asia. Two definitions for area under forest management plans were used in Asia. Industrialized countries (CIS countries, Cyprus, Israel, Japan and Turkey) reported on "Forest [and other wooded land] which is managed in accordance with a formal or an informal plan applied regularly over a sufficiently long period (five years or more). The management operations include the tasks to be accomplished in individual forest stands (e.g. compartments) during the given period". It was also recommended that any areas where a decision had been made not to manage the area at all should be included. The figures used are those pertaining to forests only, excluding other wooded lands.

The remaining countries reported on "The area of forest which is managed for various purposes (conservation, production, other) in accordance with a formal, nationally approved management plan over a sufficiently long period (five years or more)".

For Georgia, forests classified as "undisturbed" were listed as not managed.

For the Philippines, the area under forest management plans included forest land with less than 20 percent crown cover.

Oceania. With two exceptions (Australia and New Zealand), the definition used for area under forest management plans in Oceania was: "The area of forest which is managed for various purposes (conservation, production, other) in accordance with a formal, nationally approved management plan over a sufficiently long period (five years or more)".

For Australia and New Zealand, the definition included informal management plans and areas where a decision had been made not to manage the area at all.

For Australia, only the forests managed for wood supply were included in the figure provided.

Europe. The definition used for area under forest management plans in all the European countries was: "Forest [and other wooded land] which is managed in accordance with a formal or an informal plan applied regularly over a sufficiently long period (five years or more). The management operations include the tasks to be accomplished in individual forest stands (e.g. compartments) during the given period". It was also recommended that areas where a decision had been made not to manage the area at all should be included. The figures used are those pertaining to forests only, excluding other wooded lands.

For Italy, only forests with specific management plans were included in the figure given for forests under management. All other forests in the country are submitted to general silvicultural prescriptions.

For Finland, the original figure provided on the area of forest managed was 18 609 000 ha. However, as of December 2000 a total of 21.9 million hectares had been certified. Since this implies the existence of a management regime, this latter, more recent figure has been used.

North and Central America. With two exceptions (Canada and the United States), the definition used for area under forest management plans in North and Central America was: "The area of forest which is managed for various purposes (conservation, production, other) in accordance

Appendix 3. Global tables 375

with a formal, nationally approved management plan over a sufficiently long period (five years or more)".

For Canada and the United States, the definition included informal management plans and areas where a decision had been made not to manage the area at all.

South America. The definition used for area under forest management plans in South America was: "The area of forest which is managed for various purposes (conservation, production, other) in accordance with a formal, nationally approved management plan over a sufficiently long period (five years or more)".

For Guyana, the figure provided on area under management equals the area under concession agreements, as all concessionaires must prepare a long-term forest management plan to be approved by the government.

Areas under forest management in 1990 and 1980

Figures for areas under forest management in 1990 and 1980 are taken from FAO (1988), FAO/UNEP (1982), UNECE/FAO (1985) and UNECE/FAO (1992). The percentages represent the percentage of the respective forest areas in 1980 and 1990 as provided in these references.

The definitions of forest under management were as follows:

- For tropical countries in 1980, "Area of forest under intensive management" was defined as follows: "The concept of intensive management is used here in a restricted way and implies not only the strict and controlled application of harvesting regulations but also silvicultural treatments and protection against fires and diseases".
- For UNECE countries in 1980, the definition was "Area of forest being managed according to a forest management plan".
- For UNECE countries in 1990, "Forest under active management" was defined as "Forest and other wooded land that is managed according to a professionally prepared plan or is otherwise under a recognized form of management applied regularly over a long period (five years or more)".

Note that the definition of forest changed for industrialized countries between 1990 and 2000 (from crown cover of 20 percent to crown cover of 10 percent), so the figures are not directly comparable in some cases.

Europe. For Bulgaria, the area under forest management plans (1980) included other wooded land and the percentage is thus above 100.

For Yugoslavia, the figures from 1980 and 1990 correspond to the former Yugoslavia, hence the sharp decrease in area under management plans for the year 2000.

For further details, please refer to the references cited.

Certified forest areas

The cumulative area of forests certified under the following schemes is listed:

ATFP = American Tree Farm Program (as of December 2000)

CSA = Canada's National Sustainable Forest Management System Standard (as of 21 December 2000)

FSC = Forest Stewardship Council – Accredited Certification Bodies (as of 31 December 2000)

GT = Green Tag (United States) (as of 31 December 2000)

PEFC = Pan-European Forest Certification (National schemes endorsed by the PRFC Council) (as of December 2000)

SFI = Sustainable Forest Initiative Program, American Forest and Paper Association (for Canada as of 21 December 2000, for the United States as of October 2000)

Although about 29 million hectares of land are enrolled in the SFI program in the United States and Canada, and plans are to have 56 million hectares under third-party certification by the end of 2001, only those areas which had already been independently certified by the end of 2000 have been included (12 million hectares in the United States and 1.04 million hectares in Canada).

Areas certified under the ISO 14001 Environmental Management System Standard scheme have only been included if also certified under specific forest certification schemes.

In Canada, a total of 30 980 046 ha of forest has been certified under the ISO 14001 Environmental Management System Standard scheme (as of 21 December 2000). However, only those area which have also been certified under CSA, FSC or SFI – equivalent to 3 615 000 ha – have been included in this table.

In New Zealand, more than 300 000 ha have been certified under the ISO 14001 Environmental Management System Standard scheme (as of May 2000). These areas have not been included in the table.

Ghana, Malaysia and Indonesia, among others, are developing national certification schemes and additional areas may soon be certified under these. A total of 2 325 356 ha of forests in three states of Malaysia (Pahang, Terengganu and Selangor) have, as a first step, been assessed to the requirements of a mutually agreed standard and were awarded audit statements by an independent third-party assessor (the Keurhout Foundation) under the Malaysia/Netherlands cooperation programme (H. Singh, National Timber Certification Council, Malaysia, personal communication, 2001).

Table 10: Removals (page 415)
For tropical countries, removals are reported as total area under harvesting scheme (short- to very long-term), area actually harvested annually and the harvesting intensity range in volume per hectare. For industrialized countries, the total volume annually harvested is given.

Table 11: Comparison of forest management areas (page 419)
Data from Tables 9, 10 and 15 are combined to compare the areas under forest management from different aspects. As the assessment procedure has been different for the different management categories, the numbers may not always match. Note that the reported areas overlap to some extent (e.g. area under management plan and protected areas). Notes for Tables 9 and 10 apply.

Table 12: Non-wood forest products – major product groups (page 423)
Major product groups are identified with an "x" by country, using the standard product groups developed by FAO for NWFP country profiles. Detailed descriptions of the products and data by country can be found in the FAO Forestry Web site country profiles (www.fao.org/forestry/fo/country/nav_world.jsp).

Table 13: Number of endangered, endemic species for seven species groups (page 427)
The table is based on a study carried out by the World Conservation Monitoring Centre of the United Nations Environment Programme (UNEP-WCMC) for FRA 2000 (FAO 2001). The study examined the presence and threatened status (whether the species were considered globally endangered) for seven species groups: amphibians, birds, ferns, mammals, palms, reptiles and trees. The threatened status was derived from the World Conservation Union (IUCN) lists of threatened species.

The total number of species (sum of the seven groups) present in each country is given in the first column, and the number thereof that are considered endangered in the second.

Of the total number of threatened species, the number of country-endemic species are then reported in the third column.

Finally, the last seven columns list how many of the endangered and country-endemic species occur in forests, for each species group.

Table 14: Distribution of total forest area by ecological zone (page 431)
The table is derived from FRA 2000 global maps of forest cover and ecological zones. The distribution of forest area over ecological zones was produced by overlaying these maps and a country mask. The distribution is reported in percentage of total forest area.

The distributon of ecological zones has been analysed for each country individually based on the FRA 2000 global maps. Totals for regions and the world are not given, as they would not tally exactly with the global distribution of ecological zones given in Chapter 47.

Table 15: Forest in protected areas / available for wood supply (page 435)
Forest protected areas refer to areas within IUCN categories I to VI for nature protection. "Country report" refers to the country submissions to FRA 2000 from industrialized countries, in which the term "protection" was interpreted broadly, particularly for IUCN categories V and VI, and may include areas under general forest management. "Global maps" refers to an overlay (implemented by UNEP-WCMC) of FRA 2000 global maps of forest cover and the FRA 2000 global map of protected areas with legal protection status. Percentages refer to total forest area.

"Forest available for wood supply" refers to a study based on FRA 2000 global maps (see Chapter 9). It was assumed that forests inside protected areas are not available for wood supply and that forests above an altitude threshold (tropical domain, 3 000 m; subtropical, 2 500 m; temperate, 2 000 m; boreal, 1 000 m) were economically inaccessible. The remaining forest area was measured within different distances to existing infrastructure (roads and railroads, but not rivers). Results for distances of 10, 20, 30 and

50 km are reported, as well as for an unlimited distance.

Table 16: FRA 2000 country interaction (page 439)

This table lists for each country the FRA 2000 country correspondent – the official contact point for information requests and validation of results.

The table also indicates the countries in which FRA 2000 assignments were carried out to support the national assessments, the countries that participated in workshops and meetings organized within the framework of FRA 2000, and additional FRA 2000 documents that are available for the country.

The document codes are as follows:

WP*x* = FRA Working Paper No. *x*, www.fao.org/forestry/fo/fra/index.jsp

WP*x, y* = two different FRA Working Papers, Nos. *x* and *y*

UNECE = UNECE/FAO (2000)

EC-FAO = proceedings from workshops carried out within the EC-FAO projects in support of outlook studies and FRA 2000. Documents available on line at: www.fao.org/forestry/FON/FONS/outlook/Africa/ACP/acp-proc.stm

BIBLIOGRAPHY

FAO. 1988. *An interim report on the state of forest resources in the developing countries.* Miscellaneous paper FO:MISC/88/7. Rome.

FAO. 1998. *FRA 2000 terms and definitions.* FRA Working Paper No. 1. Rome. (www.fao.org/forestry/fo/fra/docs/FRA_WP1eng.pdf)

FAO. 2001. *Forest occurring species of conservation concern: review of status of information for FRA 2000.* FRA Working Paper No. 53. Rome. www.fao.org/forestry/fo/fra/index.jsp

FAO/UNEP. 1982. *Tropical forest resources.* FAO Forestry Paper No. 30. Rome.

UNECE/FAO. 1985. *The forest resources of the ECE Region (Europe, the USSR, North America).* Geneva.

UNECE/FAO. 1992. *The forest resources of the temperate zones. The UNECE/FAO 1990 Forest Resources Assessment.* Vol. 1. *General forest resource information.* New York, UN.

UNECE/FAO. 2000. *Forest resources of Europe, CIS, North America, Australia, Japan and New Zealand: contribution to the global Forest Resources Assessment 2000.* Geneva Timber and Forest Study Papers No. 17. New York and Geneva, UN. www.unece.org/trade/timber/fra/pdf/contents.htm

United Nations. 1997. *World urbanization prospects – the 1996 revision.* New York.

UN. 1999. *World population prospects – the 1998 revision.* New York.

World Bank. 2000. *World development indicators 1999.* Washington, DC.

Appendix 3. Global tables

Table 1. Basic country data

Country/area	Land area Total 1998 000 ha	Population Total 1999 000	Population Density 1999 n/km²	Population Annual change 1995-2000 %	Population Rural 1999 %	Economic indicators GNP per capita 1997 USD	Economic indicators Annual GDP change 1997 %
Algeria	238 174	30 774	12.9	2.3	41.5	1 409	1.3
Angola	124 670	12 479	10.0	3.3	66.5	159	7.6
Benin	11 063	5 937	53.3	2.7	58.5	381	5.6
Botswana	56 673	1 597	2.8	1.9	29.4	3 307	6.9
Burkina Faso	27 360	11 616	42.5	2.8	82.1	250	5.5
Burundi	2 568	6 565	255.6	1.7	91.3	141	0.4
Cameroon	46 540	14 693	31.6	2.7	51.9	587	5.1
Cape Verde	403	418	103.7	2.4	39.5	1 108	3.0
Central African Republic	62 297	3 550	5.7	1.9	59.2	341	5.1
Chad	125 920	7 458	5.9	2.7	76.5	218	6.5
Comoros	186	676	303.1	2.8	67.3	413	0.0
Congo	34 150	2 864	8.4	2.8	38.3	633	-1.9
Côte d'Ivoire	31 800	14 526	45.7	1.8	54.1	727	6.0
Dem. Rep. of the Congo	226 705	50 335	22.2	2.6	70.0	114	-5.7
Djibouti	2 317	629	27.1	1.2	17.0	-	0.5
Egypt	99 545	67 226	67.5	1.9	54.3	1 097	5.5
Equatorial Guinea	2 805	442	15.8	2.5	52.9	892	76.1
Eritrea	11 759	3 719	36.8	3.9	81.6	222	7.9
Ethiopia	110 430	61 095	61.1	2.5	82.8	112	5.6
Gabon	25 767	1 197	4.6	2.6	45.9	3 985	4.1
Gambia	1 000	1 268	126.8	3.3	68.2	342	5.4
Ghana	22 754	19 678	86.5	2.7	62.2	384	4.2
Guinea	24 572	7 360	30.0	0.8	68.0	552	4.8
Guinea-Bissau	3 612	1 187	42.2	2.2	76.7	232	5.0
Kenya	56 915	29 549	51.9	2.0	67.9	330	2.1
Lesotho	3 035	2 108	69.5	2.2	72.9	734	8.0
Liberia	11 137	2 930	30.4	8.6	52.7	-	-
Libyan Arab Jamahiriya	175 954	5 471	3.1	2.4	12.8	-	-
Madagascar	58 154	15 497	26.6	3.0	71.1	229	3.6
Malawi	9 409	10 640	113.1	2.5	85.1	163	5.1
Mali	122 019	10 960	9.0	2.5	70.6	259	6.7
Mauritania	102 522	2 598	2.5	2.8	43.6	452	4.5
Mauritius	202	1 150	566.5	0.8	58.9	3 796	5.0
Morocco	44 630	27 867	62.4	1.8	45.4	1 281	-2.0
Mozambique	78 409	19 286	24.6	2.5	61.1	131	12.4
Namibia	82 329	1 695	2.1	2.3	60.2	2 196	1.8
Niger	126 670	10 400	8.2	3.2	79.9	202	3.4
Nigeria	91 077	108 945	119.6	2.4	56.9	239	3.9
Réunion	250	691	276.4	1.3	29.8	-	-
Rwanda	2 466	7 235	293.3	8.0	93.9	207	10.9
Saint Helena	31	6	19.4	0.8	33.3	-	-
Sao Tome and Principe	95	144	150.0	2.1	54.2	297	1.0
Senegal	19 252	9 240	48.0	2.6	53.7	554	5.2
Seychelles	45	77	171.1	1.1	41.6	7 031	4.3
Sierra Leone	7 162	4 717	65.9	3.0	64.1	150	-20.2
Somalia	62 734	9 672	15.4	4.2	72.9	-	-
South Africa	121 758	39 900	32.7	1.5	49.9	3 377	1.7
Sudan	237 600	28 883	12.2	2.1	64.9	255	4.6
Swaziland	1 721	980	57.0	2.9	65.3	1 555	3.7
Togo	5 439	4 512	83.0	2.7	67.3	337	4.7
Tunisia	16 362	9 460	60.9	1.4	35.2	2 092	5.4
Uganda	19 964	21 143	105.9	2.8	86.2	326	5.4
United Republic of Tanzania	88 359	32 793	37.1	2.3	72.9	183	4.1
Western Sahara	26 600	284	1.1	3.4	4.9	-	-
Zambia	74 339	8 976	12.1	2.3	55.8	387	3.5
Zimbabwe	38 685	11 529	29.8	1.4	65.4	656	3.2
Total Africa	**2 978 394**	**766 627**	**25.9**	**2.4**	**63.0**	**...**	**...**

Table 1. Basic country data (cont.)

Country/area	Land area Total 1998 000 ha	Population Total 1999 000	Population Density 1999 n/km²	Population Annual change 1995-2000 %	Population Rural 1999 %	Economic indicators GNP per capita 1997 USD	Economic indicators Annual GDP change 1997 %
Afghanistan	64 958	21 923	33.6	2.9	78.5	-	-
Armenia	2 820	3 525	125.0	-0.3	30.3	896	3.1
Azerbaijan	8 359	7 697	88.9	0.4	43.1	472	3.2
Bahrain	69	606	878.3	2.1	8.1	-	-
Bangladesh	13 017	126 947	975.2	1.7	79.4	352	5.9
Bhutan	4 701	2 064	43.9	2.8	93.1	406	-
Brunei Darussalam	527	322	61.1	2.2	28.6	-	4.0
Cambodia	17 652	10 945	62.0	2.3	77.2	303	1.0
China	932 743	1274 106	136.6	0.9	66.2	668	8.8
Cyprus	925	778	84.2	1.1	43.8	-	-
Dem People's Rep. of Korea	12 041	23 702	196.8	1.6	37.5	-	-
East Timor	1 479	871	58.6	1.7	92.5	-	-
Gaza Strip	38	1 077	2 834.2	4.4	5.5	-	-
Georgia	6 831	5 006	71.8	-1.1	39.8	689	11.0
India	297 319	998 056	335.7	1.7	71.9	392	5.2
Indonesia	181 157	209 255	115.5	1.4	60.8	1 096	4.9
Iran, Islamic Rep.	162 201	66 796	41.2	1.7	38.9	1 581	-
Iraq	43 737	22 450	51.3	2.8	23.6	-	-
Israel	2 062	6 101	295.9	2.2	8.9	15 456	2.2
Japan	37 652	126 505	336.0	0.2	21.3	43 574	0.8
Jordan	8 893	4 823	54.2	3.1	26.4	1 479	1.7
Kazakhstan	267 074	16 269	6.1	-0.3	38.7	1 277	1.7
Kuwait	1 782	1 897	106.5	3.1	2.5	-	-
Kyrgyzstan	19 180	4 669	24.3	0.6	60.2	817	9.9
Lao People's Dem. Rep	23 080	5 297	23.0	2.6	77.1	414	6.5
Lebanon	1 024	3 236	316.3	1.8	10.7	-	-
Malaysia	32 855	21 830	66.4	2.0	43.5	4 469	7.8
Maldives	30	278	926.7	2.8	71.9	1 107	6.2
Mongolia	156 650	2 621	1.7	1.7	37.0	391	3.3
Myanmar	65 755	45 059	68.5	1.2	72.7	-	-
Nepal	14 300	23 385	163.5	2.4	88.4	216	4.0
Oman	21 246	2 460	11.6	3.4	17.8	-	-
Pakistan	77 087	152 331	197.6	2.8	63.5	502	-0.4
Philippines	29 817	74 454	249.7	2.1	42.3	1 170	5.2
Qatar	1 100	589	53.5	1.8	7.8	-	-
Republic of Korea	9 873	46 480	470.8	0.8	14.8	11 028	5.5
Saudi Arabia	214 969	20 899	9.7	3.4	14.9	6 739	1.9
Singapore	61	3 522	5 773.8	1.4	0.0	32 486	7.8
Sri Lanka	6 463	18 639	288.4	1.0	76.7	770	6.4
Syrian Arab Republic	18 377	15 725	85.6	2.6	46.0	1 138	4.0
Tajikistan	14 087	6 104	43.4	1.5	67.3	319	-
Thailand	51 089	60 856	119.1	0.9	78.8	2 821	-0.4
Turkey	76 963	65 546	85.2	1.7	25.9	3 119	7.7
Turkmenistan	46 992	4 384	9.3	1.8	54.6	642	-
United Arab Emirates	8 360	2 398	28.7	2.0	14.5	-	-
Uzbekistan	41 424	23 942	57.8	1.6	57.9	-	5.4
Viet Nam	32 550	78 705	241.8	1.6	80.3	299	8.8
West Bank	580	1 660	286.2	-	-	-	-
Yemen	52 797	17 488	33.1	3.8	62.9	223	5.4
Total Asia	**3 084 746**	**3 634 278**	**117.8**	**1.4**	**63.0**	**...**	**...**

Appendix 3. Global tables 381

Table 1. Basic country data (cont.)

Country/area	Land area Total	Population Total	Population Density	Population Annual change	Rural	Economic indicators GNP per capita	Economic indicators Annual GDP change
	1998	1999	1999	1995-2000	1999	1997	1997
	000 ha	000	n/km^2	%	%	USD	%
American Samoa	20	66	330.0	3.7	48.5	-	-
Australia	768 230	18 701	2.4	1.0	15.3	19 689	1.7
Cook Islands	23	19	82.6	0.6	36.8	-	-
Fiji	1 827	806	44.1	1.2	58.1	2 340	-1.8
French Polynesia	366	231	63.1	1.8	43.3	-	-
Guam	55	164	298.2	2.1	61.0	-	-
Kiribati	73	82	112.3	1.4	63.4	839	3.0
Marshall Islands	18	62	344.4	3.3	29.0	1 473	-5.2
Micronesia	69	116	165.7	2.0	70.7	1 886	-4.0
Nauru	2	11	550.0	1.9	-	-	-
New Caledonia	1 828	210	11.5	2.1	36.2	-	-
New Zealand	26 799	3 828	14.3	1.0	13.3	15 233	2.4
Niue	26	2	7.7	-1.9	50.0	-	-
Northern Mariana Isl.	46	74	160.9	5.9	45.9	-	-
Palau	46	19	41.3	2.4	26.3	-	-
Papua New Guinea	45 239	4 702	10.4	2.2	82.9	931	-6.5
Samoa	282	177	62.5	1.4	78.5	1 239	4.0
Solomon Islands	2 856	430	15.4	3.2	80.9	797	-0.5
Tonga	73	98	136.1	0.3	55.1	1 635	-1.7
Vanuatu	1 218	186	15.3	2.4	80.1	1 315	2.7
Total Oceania	**849 096**	**30 014**	**3.5**	**1.3**	**29.8**	**...**	**...**
Albania	2 740	3 113	113.6	-0.4	61.3	757	-7.0
Andorra	45	75	166.7	4.0	5.3	-	-
Austria	8 273	8 177	98.8	0.5	35.4	29 309	4.0
Belarus	20 748	10 274	49.5	-0.3	26.3	2 047	10.4
Belgium & Luxembourg	3 282	10 579	322.3	0.1	3.0	28 284	2.9
Bosnia & Herzegovina	5 100	3 839	75.3	3.1	57.3	-	-
Bulgaria	11 055	8 279	74.9	-0.7	30.3	1 273	-6.9
Croatia	5 592	4 477	80.1	-0.1	42.7	4 092	-
Czech Republic	7 728	10 262	132.8	-0.2	33.9	5 111	1.0
Denmark	4 243	5 282	124.5	0.3	14.4	36 418	3.3
Estonia	4 227	1 412	33.4	-1.2	26.0	3 689	11.4
Finland	30 459	5 165	17.0	0.3	35.4	26 020	6.3
France	55 010	58 886	107.0	0.4	24.6	27 437	2.4
Germany	34 927	82 178	235.3	0.1	12.7	30 133	1.7
Greece	12 890	10 626	82.4	0.3	40.1	11 343	-
Hungary	9 234	10 076	109.1	-0.4	33.5	4 517	4.6
Iceland	10 025	279	2.8	0.9	7.9	-	-
Ireland	6 889	3 705	53.8	0.7	41.7	17 739	10.0
Italy	29 406	57 343	195.0	0.0	33.1	19 104	1.5
Latvia	6 205	2 389	38.5	-1.5	26.0	2 815	6.6
Liechtenstein	15	32	200.0	1.3	81.3	-	-
Lithuania	6 258	3 682	56.8	-0.3	25.9	2 015	5.7
Malta	32	386	1 206.25	0.7	9.8	9 368	2.9
Netherlands	3 392	15 735	463.9	0.4	10.7	27 402	3.4
Norway	30 683	4 442	14.5	0.5	26.0	35 947	3.4
Poland	30 442	38 740	127.3	0.1	34.8	3 472	6.9
Portugal	9 150	9 873	107.9	0.0	62.5	11 243	4.0
Republic of Moldova	3 296	4 380	132.8	0.0	45.5	641	1.3
Romania	23 034	22 402	97.3	-0.4	42.3	1 399	-6.6
Russian Federation	1 688 851	147 196	8.7	-0.2	22.7	2 235	0.8
San Marino	6	26	433.3	1.3	3.8	-	-
Slovakia	4 808	5 382	111.9	0.1	39.4	3 645	6.5
Slovenia	2 012	1 989	98.9	0.0	47.7	10 163	3.8
Spain	49 945	39 634	79.4	0.0	22.6	14 800	3.7
Sweden	41 162	8 892	21.6	0.2	16.7	25 685	1.2
Switzerland	3 955	7 344	185.7	0.7	37.7	46 448	1.7
The FYR of Macedonia	2 543	2 011	79.1	0.6	38.4	1 053	1.5

Table 1. Basic country data (cont.)

Country/area	Land area Total	Population Total	Population Density	Population Annual change	Population Rural	Economic indicators GNP per capita	Economic indicators Annual GDP change
	1998	1999	1999	1995-2000	1999	1997	1997
	000 ha	000	n/km^2	%	%	USD	%
Ukraine	57 935	50 658	87.4	-0.4	28.0	1 452	-3.2
United Kingdom	24 160	58 974	244.1	0.2	10.7	19 946	3.5
Yugoslavia	10 200	10 637	104.3	0.1	40.8	-	-
Total Europe	**2 259 957**	**728 932**	**32.2**	**0.0**	**25.4**	**...**	**...**
Antigua and Barbuda	44	67	152.3	0.5	64.2	7 331	-
Bahamas	1 001	301	30.1	1.8	12.0	-	-
Barbados	43	269	625.6	0.5	50.6	-	-
Belize	2 280	235	10.3	2.4	53.6	2 547	2.6
Bermuda	5	64	1280.0	0.8	n.a	-	-
British Virgin Islands	15	21	140.0	2.7	38.1	-	-
Canada	922 097	30 857	3.3	1.0	23.0	19 267	5.4
Cayman Islands	26	37	142.3	3.7	-	-	-
Costa Rica	5 106	3 933	77.0	2.5	48.7	2 626	3.2
Cuba	10 982	11 160	101.6	0.4	22.5	-	-
Dominica	75	71	94.7	-0.1	29.6	2 940	1.9
Dominican Republic	4 838	8 364	172.9	1.7	35.5	1 659	8.2
El Salvador	2 072	6 154	297.0	2.1	53.7	1 684	4.0
Greenland	34 170	56	0.2	0.1	17.9	-	-
Grenada	34	93	273.5	0.3	62.4	3 052	-
Guadeloupe	169	450	266.3	1.4	0.2	-	-
Guatemala	10 843	11 090	102.3	2.7	59.9	1 481	4.3
Haiti	2 756	8 087	293.4	1.7	65.8	364	1.1
Honduras	11 189	6 316	56.4	2.8	53.7	723	4.5
Jamaica	1 083	2 560	236.4	0.9	44.4	1 525	-2.4
Martinique	107	392	369.8	0.9	5.4	-	-
Mexico	190 869	97 365	51.0	1.6	25.8	3 304	7.0
Montserrat	11	11	110.0	-0.3	81.8	-	-
Netherlands Antilles	80	215	268.8	1.1	30.2	-	-
Nicaragua	12 140	4 938	40.7	2.8	35.8	408	-
Panama	7 443	2 812	37.8	1.7	42.7	2 993	-
Puerto Rico	887	3 839	432.8	0.8	25.1	-	-
Saint Kitts and Nevis	36	39	108.3	-0.8	66.7	6 032	-
Saint Lucia	61	152	249.2	1.4	62.5	3 454	-
Saint Pierre & Miquelon	23	7	30.4	0.3	14.3	-	-
Saint Vincent and Grenadines	39	113	289.7	0.7	46.9	2 335	-
Trinidad and Tobago	513	1 289	251.3	0.5	26.5	4 119	3.2
United States	915 895	276 218	30.2	0.8	23.0	28 310	6.9
US Virgin Islands	34	94	276.5	-0.8	54.3	-	-
Total North and Central America	**2 136 966**	**477 791**	**22.4**	**1.6**	**26.8**	**...**	**...**
Argentina	273 669	36 577	13.4	1.3	10.9	8 755	8.6
Bolivia	108 438	8 142	7.5	2.4	36.0	912	4.2
Brazil	845 651	167 988	19.9	1.3	19.3	4 514	3.2
Chile	74 881	15 019	20.1	1.4	15.5	4 478	7.1
Colombia	103 871	41 564	40.0	1.9	25.5	2 039	3.1
Ecuador	27 684	12 411	44.8	2.0	38.3	1 531	3.4
Falkland Islands	1 217	2	0.2	0.5	-	-	-
French Guiana	8 815	174	2.0	4.3	22.4	-	-
Guyana	21 498	855	4.3	0.7	62.3	766	-
Paraguay	39 730	5 358	13.5	2.6	44.8	1 946	3.5
Peru	128 000	25 230	19.7	1.7	27.6	2 580	7.2
Suriname	15 600	415	2.7	0.4	48.4	940	-
Uruguay	17 481	3 313	19.0	0.7	8.9	6 076	5.1
Venezuela	88 206	23 706	26.9	2.0	13.0	3 499	5.1
Total South America	**1 754 741**	**340 754**	**19.4**	**1.5**	**20.7**	**...**	**...**
TOTAL WORLD	**13 063 900**	**5 978 396**	**45.8**	**1.3**	**53.0**	**...**	**...**

Table 2. Forest cover – information status

Country/area	Reference year	Method FS/DM/GM/ES	Compatibility H / M / L	Used	Compatibility H / M / L	Number of references consulted
Algeria	2000	ES	H	yes	H	8
Angola	1983	DM/FS	M	yes	M	20
Benin	1996	GM	H	yes	M	11
Botswana	1990	GM	L	no	M	22
Burkina Faso	1991	GM	M	yes	M	6
Burundi	1998	ES	L	yes	L	6
Cameroon	1999	ES	L	yes	L	11
Cape Verde	1999	ES	H	yes	H	9
Central African Republic	1994	FS/ES	M	no	n.ap.	11
Chad	1988	ES	H	no	n.ap.	11
Comoros	1984	GM	M	yes	H	9
Congo	1999	ES	M	yes	M	15
Côte d'Ivoire	1987	GM	L	no	n.ap.	22
Dem. Rep. of the Congo	1989	DM	M	yes	M	8
Djibouti	1985	GM	H	no	n.ap.	1
Egypt	1996	ES	H	yes	L	8
Equatorial Guinea	1998	ES	L	yes	L	7
Eritrea	1997	FS	M	no	n.ap.	3
Ethiopia	1997	ES	H	yes	L	9
Gabon	1999	ES	L	yes	L	5
Gambia	1993	FS	M	yes	H	13
Ghana	1996	ES	L	no	n.ap.	10
Guinea	1988	DM/ES	L	no	n.ap.	16
Guinea-Bissau	1990	DM	H	yes	H	5
Kenya	1993	ES	H	no	n.ap.	18
Lesotho	1994	DM/FS	H	yes	H	10
Liberia	1990	ES	H	yes	L	9
Libyan Arab Jamahiriya	1980	ES	H	no	n.ap.	3
Madagascar	1996	FS	H	yes	L	8
Malawi	1991	DM/FS	M	yes	H	23
Mali	1991	FS	H	no	n.ap.	7
Mauritania	1991	ES	L	yes	L	6
Mauritius	1979	ES	L	yes	M	7
Morocco	1995	FS	H	no	n.ap.	9
Mozambique	1995	DM/FS	H	yes	H	40
Namibia	1992	FS	M	yes	L	9
Niger	1992	ES	H	yes	M	9
Nigeria	1994	FS/DM	M	yes	H	17
Réunion	1997	ES	M	yes	M	5
Rwanda	1999	ES	L	yes	L	7
Saint Helena	1980	ES	H	no	n.ap.	3
Sao Tome and Principe	1989	FS	L	no	n.ap.	3
Senegal	1985	FS	M	yes	M	8
Seychelles	1993	ES	M	no	n.ap.	2
Sierra Leone	1986	ES	M	yes	H	6
Somalia	1980	ES	H	no	n.ap.	2
South Africa	1994	DM	M	no	n.ap.	14
Sudan	1990	DM	M	no	n.ap.	11
Swaziland	1999	DM/FS	M	yes	H	7
Togo	1991	ES	M	yes	M	8
Tunisia	1994	FS	M	no	n.ap.	12
Uganda	1992	DM/DM	M	yes	L	8
United Republic of Tanzania	1995	DM/FS	H	yes	M	10
Western Sahara	1995	FS	M	no	n.ap.	1
Zambia	1978	FS/ES	M	yes	L	7
Zimbabwe	1992	DM	M	no	n.ap.	12
Total Africa	**1991**					**547**

Table 2. Forest cover – information status (cont.)

Country/area	Reference year	Method FS/DM/GM/ES	Compatibility H / M / L	Used	Compatibility H / M / L	Number of references consulted
Afghanistan	1993	GM	H	no	n.ap.	1
Armenia	1996	-	-	yes	M	1
Azerbaijan	1988	-	-	yes	M	1
Bahrain	1999	n.ap.	n.ap.	n.ap.	n.ap.	1
Bangladesh	1996	DM/ES	H	yes	H	12
Bhutan	1990	DM	H	yes	M	7
Brunei Darussalam	1998	DM	H	yes	H	6
Cambodia	1997	DM	H	yes	H	11
China	1996	ES	H	yes	H	23
Cyprus	1999	-	-	yes	M	1
Dem People's Rep. of Korea	1993	ES	H	no	n.ap.	2
East Timor	1985	GM	H	no	n.ap.	1
Gaza Strip		-	-	-	-	0
Georgia	1995	-	-	yes	M	1
India	1997	DM	H	yes	H	12
Indonesia	1997	GM	H	yes	H	32
Iran, Islamic Rep.	1999	ES	H	no	n.ap.	2
Iraq	1990	ES	H	no	n.ap.	4
Israel	1997	-	-	yes	M	1
Japan	1995	-	-	yes	H	1
Jordan	2000	ES	H	no	n.ap.	5
Kazakhstan	1993	-	-	yes	M	1
Kuwait	2000	n.ap.	n.ap.	n.ap.	n.ap.	2
Kyrgyzstan	1993	-	-	yes	M	1
Lao People's Dem. Rep	1989	GM	M	yes	H	7
Lebanon	1996	ES	H	no	n.ap.	1
Malaysia	1995	ES	M	yes	M	31
Maldives	1999	ES	L	no	n.ap.	1
Mongolia	1987	ES	M	no	n.ap.	2
Myanmar	1997	DM/ES	H	yes	L	17
Nepal	1994	GM	L	yes	L	16
Oman	2000	n.ap.	n.ap.	n.ap.	n.ap.	1
Pakistan	1990	DM	M	yes	M	13
Philippines	1997	ES	H	yes	H	12
Qatar	2000	n.ap.	n.ap.	n.ap.	n.ap.	1
Republic of Korea	1999	DM	H	no	n.ap.	3
Saudi Arabia	1994	ES	H	no	n.ap.	2
Singapore	1990	FS	H	no	n.ap.	2
Sri Lanka	1992	GM	M	yes	L	14
Syrian Arab Republic	1992	ES	H	no	n.ap.	4
Tajikistan	1995	-	-	yes	M	1
Thailand	1998	DM	H	yes	H	10
Turkey	1985	-	-	yes	M	2
Turkmenistan	1995	-	-	yes	M	1
United Arab Emirates	2000	n.ap.	n.ap.	n.ap.	n.ap.	1
Uzbekistan	1995	-	-	yes	M	1
Viet Nam	1995	DM	H	yes	H	11
West Bank		-	-	-	-	0
Yemen	1993	ES	H	no	n.ap.	1
Total Asia	**1995**					**284**

Appendix 3. Global tables

Table 2. Forest cover – information status (cont.)

Country/area	Latest available statistics			Time series		Number of references consulted
	Reference year	Method FS/DM/GM/ES	Compatibility H / M / L	Used	Compatibility H / M / L	
American Samoa	1999	ES	L	no	n.ap.	3
Australia	1992	-	-	no	n.ap.	9
Cook Islands	1998	ES	H	no	n.ap.	2
Fiji	1995	ES	H	yes	H	9
French Polynesia	1991	ES	H	no	n.ap.	2
Guam	1999	ES	H	no	n.ap.	1
Kiribati	1997	ES	H	no	n.ap.	3
Marshall Islands	1999	ES	H	no	n.ap.	2
Micronesia	1983	DM	H	yes	H	6
Nauru	1993	ES	H	no	n.ap.	2
New Caledonia	1993	ES	H	no	n.ap.	2
New Zealand	1996	-	-	yes	H	10
Niue	1981	-	-	-	n.ap.	2
Northern Mariana Isl.	1984	DM	H	no	n.ap.	3
Palau	1985	DM	H	no	n.ap.	5
Papua New Guinea	1996	DM	L	yes	H	7
Samoa	1992	ES	H	yes	M	5
Solomon Islands	1993	DM	M	yes	H	3
Tonga	1990	ES	H	no	n.ap.	3
Vanuatu	1993	ES	M	no	n.ap.	6
Total Oceania	**1992**					**85**
Albania	1995	-	-	yes	M	1
Andorra		-	-	-	n.ap.	0
Austria	1994	FS	H	yes	H	1
Belarus	1994	-	-	yes	M	1
Belgium & Luxembourg	1997	-	-	yes	H	1
Bosnia & Herzegovina	1995	-	-	yes	H	1
Bulgaria	1995	-	-	yes	H	1
Croatia	1996	-	-	yes	H	1
Czech Republic	1995	-	-	yes	H	1
Denmark	1990	-	-	yes	H	1
Estonia	1996	-	-	yes	H	1
Finland	1994	FS	H	yes	H	1
France	1997	-	-	yes	H	1
Germany	1987	-	-	yes	M	1
Greece	1992	-	-	yes	H	1
Hungary	1996	-	-	yes	H	1
Iceland	1998	-	-	yes	H	1
Ireland	1996	-	-	yes	H	1
Italy	1995	-	-	yes	H	1
Latvia	1997	-	-	yes	H	1
Liechtenstein	1995	-	-	yes	H	1
Lithuania	1996	-	-	yes	H	1
Malta	1996	-	-	yes	H	1
Netherlands	1994	-	-	yes	H	1
Norway	1995	-	-	yes	H	1
Poland	2000	-	-	yes	H	1
Portugal	1995	-	-	yes	H	3
Republic of Moldova	1997	-	-	yes	M	1
Romania	1990	-	-	yes	H	1
Russian Federation	1998	-	-	yes	H	2
San Marino		-	-	-	n.ap.	0
Slovakia	2000	-	-	yes	H	1
Slovenia	1996	-	-	yes	H	1
Spain	1990	-	-	yes	H	4
Sweden	1994	FS	H	yes	H	1
Switzerland	1994	-	-	yes	H	1
The FYR of Macedonia	1995	-	-	yes	M	1

Table 2. Forest cover – information status (cont.)

Country/area	Latest available statistics			Time series		Number of references consulted
	Reference year	Method FS/DM/GM/ES	Compatibility H / M / L	Used	Compatibility H / M / L	
Ukraine	1996	-	-	yes	M	1
United Kingdom	2000	-	-	yes	H	1
Yugoslavia	1995	-	-	yes	H	1
Total Europe	**1997**					**44**
Antigua and Barbuda	1983	ES	H	no	n.ap.	7
Bahamas	1986	ES	M	yes	H	5
Barbados	1998	ES	H	no	n.ap.	4
Belize	1993	GN	H	yes	H	19
Bermuda		-	-	-	n.ap.	0
British Virgin Islands	1980	ES	H	no	n.ap.	4
Canada	1994	-	-	no	n.ap.	2
Cayman Islands	1998	ES	H	no	n.ap.	2
Costa Rica	1997	DM	H	yes	H	39
Cuba	1998	DM	H	yes	H	13
Dominica	1984	DM	H	yes	H	12
Dominican Republic	1998	DM	H	no	n.ap.	7
El Salvador	1990	GM	H	yes	H	10
Greenland		-	-	-	n.ap.	0
Grenada	1992	GM	H	yes	H	5
Guadeloupe	1991	ES	H	yes	H	5
Guatemala	1999	DM	H	yes	H	16
Haiti	1995	ES	H	yes	H	8
Honduras	1995	DM	M	yes	H	17
Jamaica	1997	DM	H	yes	H	12
Martinique	1998	DM	H	no	n.ap.	4
Mexico	1993	DM	H	yes	M	36
Montserrat	1983	ES	H	yes	H	4
Netherlands Antilles	1991	ES	H	no	n.ap.	1
Nicaragua	1999	ES	H	yes	H	14
Panama	1998	ES	H	yes	H	15
Puerto Rico	1990	DM	H	yes	H	3
Saint Kitts and Nevis	1992	ES	H	yes	H	6
Saint Lucia	1992	ES	H	yes	H	7
Saint Pierre & Miquelon		-	-	-	n.ap.	0
Saint Vincent and Grenadines	1993	DM	H	yes	H	10
Trinidad and Tobago	1997	ES	H	yes	H	12
United States	1997	-	-	yes	H	5
US Virgin Islands	1976	ES	H	no	n.ap.	0
Total North and Central America	**1995**					**304**
Argentina	1993	DM	H	yes	H	32
Bolivia	1993	DM	H	yes	M	19
Brazil	1989	DM	H	yes	M	48
Chile	1995	DM	H	yes	H	42
Colombia	1996	DM	H	yes	L	32
Ecuador	1992	DM	H	yes	H	20
Falkland Islands	2000	ES	H	no	n.ap.	1
French Guiana	1990	ES	H	no	n.ap.	2
Guyana	1999	ES	H	yes	H	4
Paraguay	1991	DM	M	yes	H	12
Peru	1990	DM	H	yes	H	16
Suriname	1995	DM	H	no	n.ap.	8
Uruguay	1998	ES	H	yes	H	17
Venezuela	1995	DM	H	yes	H	27
Total South America	**1991**					**280**
TOTAL WORLD	**1994**					**1544**

Table 3. Forest cover 2000

Country/area	Land area	Total forest 2000 Area	Percentage of land area	Area per capita
	000 ha	*000 ha*	*%*	*ha*
Algeria	238 174	2 145	0.9	0.1
Angola	124 670	69 756	56.0	5.6
Benin	11 063	2 650	24.0	0.4
Botswana	56 673	12 427	21.9	7.8
Burkina Faso	27 360	7 089	25.9	0.6
Burundi	2 568	94	3.7	n.s.
Cameroon	46 540	23 858	51.3	1.6
Cape Verde	403	85	21.1	0.2
Central African Republic	62 297	22 907	36.8	6.5
Chad	125 920	12 692	10.1	1.7
Comoros	186	8	4.3	n.s.
Congo	34 150	22 060	64.6	7.7
Côte d'Ivoire	31 800	7 117	22.4	0.5
Dem. Rep. of the Congo	226 705	135 207	59.6	2.7
Djibouti	2 317	6	0.3	n.s.
Egypt	99 545	72	0.1	n.s.
Equatorial Guinea	2 805	1 752	62.5	4.0
Eritrea	11 759	1 585	13.5	0.4
Ethiopia	110 430	4 593	4.2	0.1
Gabon	25 767	21 826	84.7	18.2
Gambia	1 000	481	48.1	0.4
Ghana	22 754	6 335	27.8	0.3
Guinea	24 572	6 929	28.2	0.9
Guinea-Bissau	3 612	2 187	60.5	1.8
Kenya	56 915	17 096	30.0	0.6
Lesotho	3 035	14	0.5	n.s.
Liberia	11 137	3 481	31.3	1.2
Libyan Arab Jamahiriya	175 954	358	0.2	0.1
Madagascar	58 154	11 727	20.2	0.8
Malawi	9 409	2 562	27.2	0.2
Mali	122 019	13 186	10.8	1.2
Mauritania	102 522	317	0.3	0.1
Mauritius	202	16	7.9	n.s.
Morocco	44 630	3 025	6.8	0.1
Mozambique	78 409	30 601	39.0	1.6
Namibia	82 329	8 040	9.8	4.7
Niger	126 670	1 328	1.0	0.1
Nigeria	91 077	13 517	14.8	0.1
Réunion	250	71	28.4	0.1
Rwanda	2 466	307	12.4	n.s.
Saint Helena	31	2	6.5	0.3
Sao Tome and Principe	95	27	28.3	0.2
Senegal	19 252	6 205	32.2	0.7
Seychelles	45	30	66.7	0.4
Sierra Leone	7 162	1 055	14.7	0.2
Somalia	62 734	7 515	12.0	0.8
South Africa	121 758	8 917	7.3	0.2
Sudan	237 600	61 627	25.9	2.1
Swaziland	1 721	522	30.3	0.5
Togo	5 439	510	9.4	0.1
Tunisia	16 362	510	3.1	0.1
Uganda	19 964	4 190	21.0	0.2
United Republic of Tanzania	88 359	38 811	43.9	1.2
Western Sahara	26 600	152	0.6	0.5
Zambia	74 339	31 246	42.0	3.5
Zimbabwe	38 685	19 040	49.2	1.7
Total Africa	**2 978 394**	**649 866**	**21.8**	**0.85**

Table 3. Forest cover 2000 (cont.)

Country/area	Land area	Total forest 2000 Area	Total forest 2000 Percentage of land area	Total forest 2000 Area per capita
	000 ha	*000 ha*	*%*	*ha*
Afghanistan	64 958	1 351	2.1	0.1
Armenia	2 820	351	12.4	0.1
Azerbaijan	8 359	1 094	13.1	0.1
Bahrain	69	n.s.	n.s.	-
Bangladesh	13 017	1 334	10.2	n.s.
Bhutan	4 701	3 016	64.2	1.5
Brunei Darussalam	527	442	83.9	1.4
Cambodia	17 652	9 335	52.9	0.9
China	932 743	163 480	17.5	0.1
Cyprus	925	172	18.6	0.2
Dem People's Rep. of Korea	12 041	8 210	68.2	0.3
East Timor	1 479	507	34.3	0.6
Gaza Strip	38	-	-	-
Georgia	6 831	2 988	43.7	0.6
India	297 319	64 113	21.6	0.1
Indonesia	181 157	104 986	58.0	0.5
Iran, Islamic Rep.	162 201	7 299	4.5	0.1
Iraq	43 737	799	1.8	n.s.
Israel	2 062	132	6.4	n.s.
Japan	37 652	24 081	64.0	0.2
Jordan	8 893	86	1.0	n.s.
Kazakhstan	267 074	12 148	4.5	0.7
Kuwait	1 782	5	0.3	n.s.
Kyrgyzstan	19 180	1 003	5.2	0.2
Lao People's Dem. Rep	23 080	12 561	54.4	2.4
Lebanon	1 024	36	3.5	n.s.
Malaysia	32 855	19 292	58.7	0.9
Maldives	30	1	3.3	n.s.
Mongolia	156 650	10 645	6.8	4.1
Myanmar	65 755	34 419	52.3	0.8
Nepal	14 300	3 900	27.3	0.2
Oman	21 246	1	0.0	n.s.
Pakistan	77 087	2 361	3.1	n.s.
Philippines	29 817	5 789	19.4	0.1
Qatar	1 100	1	0.1	n.s.
Republic of Korea	9 873	6 248	63.3	0.1
Saudi Arabia	214 969	1 504	0.7	0.1
Singapore	61	2	3.3	n.s.
Sri Lanka	6 463	1 940	30.0	0.1
Syrian Arab Republic	18 377	461	2.5	n.s.
Tajikistan	14 087	400	2.8	0.1
Thailand	51 089	14 762	28.9	0.2
Turkey	76 963	10 225	13.3	0.2
Turkmenistan	46 992	3 755	8.0	0.9
United Arab Emirates	8 360	321	3.8	0.1
Uzbekistan	41 424	1 969	4.8	0.1
Viet Nam	32 550	9 819	30.2	0.1
West Bank	580	-	-	-
Yemen	52 797	449	0.9	n.s.
Total Asia	**3 084 746**	**547 793**	**17.8**	**0.15**

Table 3. Forest cover 2000 (cont.)

Country/area	Land area	Total forest 2000		
		Area	Percentage of land area	Area per capita
	000 ha	*000 ha*	*%*	*ha*
American Samoa	20	12	60.1	0.2
Australia	768 230	154 539	20.1	8.3
Cook Islands	23	22	95.7	1.2
Fiji	1 827	815	44.6	1.0
French Polynesia	366	105	28.7	0.5
Guam	55	21	38.2	0.1
Kiribati	73	28	38.4	0.3
Marshall Islands	18	n.s.	-	-
Micronesia	69	15	21.7	0.1
Nauru	2	n.s.	-	-
New Caledonia	1 828	372	20.4	1.8
New Zealand	26 799	7 946	29.7	2.1
Niue	26	6	-	3.0
Northern Mariana Isl.	46	14	30.4	0.2
Palau	46	35	76.1	1.8
Papua New Guinea	45 239	30 601	67.6	6.5
Samoa	282	105	37.2	0.6
Solomon Islands	2 856	2 536	88.8	5.9
Tonga	73	4	5.5	n.s.
Vanuatu	1 218	447	36.7	2.4
Total Oceania	**849 096**	**197 623**	**23.3**	**6.58**
Albania	2 740	991	36.2	0.3
Andorra	45	-	-	-
Austria	8 273	3 886	47.0	0.5
Belarus	20 748	9 402	45.3	0.9
Belgium & Luxembourg	3 282	728	22.2	0.1
Bosnia & Herzegovina	5 100	2 273	44.6	0.6
Bulgaria	11 055	3 690	33.4	0.4
Croatia	5 592	1 783	31.9	0.4
Czech Republic	7 728	2 632	34.1	0.3
Denmark	4 243	455	10.7	0.1
Estonia	4 227	2 060	48.7	1.5
Finland	30 459	21 935	72.0	4.2
France	55 010	15 341	27.9	0.3
Germany	34 927	10 740	30.7	0.1
Greece	12 890	3 599	27.9	0.3
Hungary	9 234	1 840	19.9	0.2
Iceland	10 025	31	0.3	0.1
Ireland	6 889	659	9.6	0.2
Italy	29 406	10 003	34.0	0.2
Latvia	6 205	2 923	47.1	1.2
Liechtenstein	15	7	46.7	0.2
Lithuania	6 258	1 994	31.9	0.5
Malta	32	n.s.	n.s.	-
Netherlands	3 392	375	11.1	n.s.
Norway	30 683	8 868	28.9	2.0
Poland	30 442	9 047	29.7	0.2
Portugal	9 150	3 666	40.1	0.4
Republic of Moldova	3 296	325	9.9	0.1
Romania	23 034	6 448	28.0	0.3
Russian Federation	1 688 851	851 392	50.4	5.8
San Marino	6	-	-	-
Slovakia	4 808	2 177	45.3	0.4
Slovenia	2 012	1 107	55.0	0.6
Spain	49 945	14 370	28.8	0.4
Sweden	41 162	27 134	65.9	3.1
Switzerland	3 955	1 199	30.3	0.2
The FYR of Macedonia	2 543	906	35.6	0.5

Table 3. Forest cover 2000 (cont.)

Country/area	Land area	Total forest 2000		
		Area	Percentage of land area	Area per capita
	000 ha	*000 ha*	*%*	*ha*
Ukraine	57 935	9 584	16.5	0.2
United Kingdom	24 160	2 794	11.6	n.s.
Yugoslavia	10 200	2 887	28.3	0.3
Total Europe	**2 259 957**	**1 039 251**	**46.0**	**1.43**
Antigua and Barbuda	44	9	20.5	0.1
Bahamas	1 001	842	84.1	2.8
Barbados	43	2	4.7	n.s.
Belize	2 280	1 348	59.1	5.7
Bermuda	5	-	-	-
British Virgin Islands	15	3	20.0	0.1
Canada	922 097	244 571	26.5	7.9
Cayman Islands	26	13	-	0.4
Costa Rica	5 106	1 968	38.5	0.5
Cuba	10 982	2 348	21.4	0.2
Dominica	75	46	61.3	0.6
Dominican Republic	4 838	1 376	28.4	0.2
El Salvador	2 072	121	5.8	n.s.
Greenland	34 170	-	-	-
Grenada	34	5	14.7	0.1
Guadeloupe	169	82	48.5	0.2
Guatemala	10 843	2 850	26.3	0.3
Haiti	2 756	88	3.2	n.s.
Honduras	11 189	5 383	48.1	0.9
Jamaica	1 083	325	30.0	0.1
Martinique	107	47	43.9	0.1
Mexico	190 869	55 205	28.9	0.6
Montserrat	11	3	27.3	0.3
Netherlands Antilles	80	1	n.s.	n.s.
Nicaragua	12 140	3 278	27.0	0.7
Panama	7 443	2 876	38.6	1.0
Puerto Rico	887	229	25.8	0.1
Saint Kitts and Nevis	36	4	11.1	0.1
Saint Lucia	61	9	14.8	0.1
Saint Pierre & Miquelon	23	-	-	-
Saint Vincent and Grenadines	39	6	15.4	0.1
Trinidad and Tobago	513	259	50.5	0.2
United States	915 895	225 993	24.7	0.8
US Virgin Islands	34	14	41.2	0.1
Total North and Central America	**2 136 966**	**549 304**	**25.7**	**1.15**
Argentina	273 669	34 648	12.7	0.9
Bolivia	108 438	53 068	48.9	6.5
Brazil	845 651	543 905	64.3	3.2
Chile	74 881	15 536	20.7	1.0
Colombia	103 871	49 601	47.8	1.2
Ecuador	27 684	10 557	38.1	0.9
Falkland Islands	1 217	-	-	-
French Guiana	8 815	7 926	89.9	45.6
Guyana	21 498	16 879	78.5	19.7
Paraguay	39 730	23 372	58.8	4.4
Peru	128 000	65 215	50.9	2.6
Suriname	15 600	14 113	90.5	34.0
Uruguay	17 481	1 292	7.4	0.4
Venezuela	88 206	49 506	56.1	2.1
Total South America	**1 754 741**	**885 618**	**50.5**	**2.60**
TOTAL WORLD	**13 063 900**	**3 869 455**	**29.6**	**0.65**

Table 4. Change in forest cover 1990-2000

Country/area	Total forest 1990	Total forest 2000	Annual change	Annual change rate
	000 ha	*000 ha*	*000 ha*	*%*
Algeria	1 879	2 145	27	1.3
Angola	70 998	69 756	-124	-0.2
Benin	3 349	2 650	-70	-2.3
Botswana	13 611	12 427	-118	-0.9
Burkina Faso	7 241	7 089	-15	-0.2
Burundi	241	94	-15	-9.0
Cameroon	26 076	23 858	-222	-0.9
Cape Verde	35	85	5	9.3
Central African Republic	23 207	22 907	-30	-0.1
Chad	13 509	12 692	-82	-0.6
Comoros	12	8	n.s.	-4.3
Congo	22 235	22 060	-17	-0.1
Côte d'Ivoire	9 766	7 117	-265	-3.1
Dem. Rep. of the Congo	140 531	135 207	-532	-0.4
Djibouti	6	6	n.s.	n.s.
Egypt	52	72	2	3.3
Equatorial Guinea	1 858	1 752	-11	-0.6
Eritrea	1 639	1 585	-5	-0.3
Ethiopia	4 996	4 593	-40	-0.8
Gabon	21 927	21 826	-10	n.s.
Gambia	436	481	4	1.0
Ghana	7 535	6 335	-120	-1.7
Guinea	7 276	6 929	-35	-0.5
Guinea-Bissau	2 403	2 187	-22	-0.9
Kenya	18 027	17 096	-93	-0.5
Lesotho	14	14	n.s.	n.s.
Liberia	4 241	3 481	-76	-2.0
Libyan Arab Jamahiriya	311	358	5	1.4
Madagascar	12 901	11 727	-117	-0.9
Malawi	3 269	2 562	-71	-2.4
Mali	14 179	13 186	-99	-0.7
Mauritania	415	317	-10	-2.7
Mauritius	17	16	n.s.	-0.6
Morocco	3 037	3 025	-1	n.s.
Mozambique	31 238	30 601	-64	-0.2
Namibia	8 774	8 040	-73	-0.9
Niger	1 945	1 328	-62	-3.7
Nigeria	17 501	13 517	-398	-2.6
Réunion	76	71	-1	-0.8
Rwanda	457	307	-15	-3.9
Saint Helena	2	2	n.s.	n.s.
Sao Tome and Principe	27	27	n.s.	n.s.
Senegal	6 655	6 205	-45	-0.7
Seychelles	30	30	n.s.	n.s.
Sierra Leone	1 416	1 055	-36	-2.9
Somalia	8 284	7 515	-77	-1.0
South Africa	8 997	8 917	-8	-0.1
Sudan	71 216	61 627	-959	-1.4
Swaziland	464	522	6	1.2
Togo	719	510	-21	-3.4
Tunisia	499	510	1	0.2
Uganda	5 103	4 190	-91	-2.0
United Republic of Tanzania	39 724	38 811	-91	-0.2
Western Sahara	152	152	n.s.	n.s.
Zambia	39 755	31 246	-851	-2.4
Zimbabwe	22 239	19 040	-320	-1.5
Total Africa	**702 502**	**649 866**	**-5 262**	**-0.78**

Table 4. Change in forest cover 1990-2000 (cont.)

Country/area	Total forest 1990	Total forest 2000	Forest cover change 1990-2000 Annual change	Forest cover change 1990-2000 Annual change rate
	000 ha	*000 ha*	*000 ha*	*%*
Afghanistan	1 351	1 351	n.s.	n.s.
Armenia	309	351	4	1.3
Azerbaijan	964	1 094	13	1.3
Bahrain	n.s.	n.s.	n.s.	14.9
Bangladesh	1 169	1 334	17	1.3
Bhutan	3 016	3 016	n.s.	n.s.
Brunei Darussalam	452	442	-1	-0.2
Cambodia	9 896	9 335	-56	-0.6
China	145 417	163 480	1 806	1.2
Cyprus	119	172	5	3.7
Dem People's Rep. of Korea	8 210	8 210	n.s.	n.s.
East Timor	541	507	-3	-0.6
Gaza Strip	-	-	-	-
Georgia	2 988	2 988	n.s.	n.s.
India	63 732	64 113	38	0.1
Indonesia	118 110	104 986	-1 312	-1.2
Iran, Islamic Rep.	7 299	7 299	n.s.	n.s.
Iraq	799	799	n.s.	n.s.
Israel	82	132	5	4.9
Japan	24 047	24 081	3	n.s.
Jordan	86	86	n.s.	n.s.
Kazakhstan	9 758	12 148	239	2.2
Kuwait	3	5	n.s.	3.5
Kyrgyzstan	775	1 003	23	2.6
Lao People's Dem. Rep	13 088	12 561	-53	-0.4
Lebanon	37	36	n.s.	-0.4
Malaysia	21 661	19 292	-237	-1.2
Maldives	1	1	n.s.	n.s.
Mongolia	11 245	10 645	-60	-0.5
Myanmar	39 588	34 419	-517	-1.4
Nepal	4 683	3 900	-78	-1.8
Oman	1	1	n.s.	5.3
Pakistan	2 755	2 361	-39	-1.5
Philippines	6 676	5 789	-89	-1.4
Qatar	n.s.	1	n.s.	9.6
Republic of Korea	6 299	6 248	-5	-0.1
Saudi Arabia	1 504	1 504	n.s.	n.s.
Singapore	2	2	n.s.	n.s.
Sri Lanka	2 288	1 940	-35	-1.6
Syrian Arab Republic	461	461	n.s.	n.s.
Tajikistan	380	400	2	0.5
Thailand	15 886	14 762	-112	-0.7
Turkey	10 005	10 225	22	0.2
Turkmenistan	3 755	3 755	n.s.	n.s.
United Arab Emirates	243	321	8	2.8
Uzbekistan	1 923	1 969	5	0.2
Viet Nam	9 303	9 819	52	0.5
West Bank	-	-	-	-
Yemen	541	449	-9	-1.9
Total Asia	**551 448**	**547 793**	**-364**	**-0.07**

Table 4. Change in forest cover 1990-2000 (cont.)

Country/area	Total forest 1990	Total forest 2000	Annual change	Annual change rate
	000 ha	*000 ha*	*000 ha*	*%*
American Samoa	12	12	n.s.	n.s.
Australia	157 359	154 539	-282	-0.2
Cook Islands	22	22	n.s.	n.s.
Fiji	832	815	-2	-0.2
French Polynesia	105	105	n.s.	n.s.
Guam	21	21	n.s.	n.s.
Kiribati	28	28	n.s.	n.s.
Marshall Islands	n.s.	n.s.	n.s.	n.s.
Micronesia	24	15	-1	-4.5
Nauru	n.s.	n.s.	n.s.	n.s.
New Caledonia	372	372	n.s.	n.s.
New Zealand	7 556	7 946	39	0.5
Niue	6	6	n.s.	n.s.
Northern Mariana Isl.	14	14	n.s.	n.s.
Palau	35	35	n.s.	n.s.
Papua New Guinea	31 730	30 601	-113	-0.4
Samoa	130	105	-3	-2.1
Solomon Islands	2 580	2 536	-4	-0.2
Tonga	4	4	n.s.	n.s.
Vanuatu	441	447	1	0.1
Total Oceania	**201 271**	**197 623**	**-365**	**-0.18**
Albania	1 069	991	-8	-0.8
Andorra	-	-	-	-
Austria	3 809	3 886	8	0.2
Belarus	6 840	9 402	256	3.2
Belgium & Luxembourg	741	728	-1	-0.2
Bosnia & Herzegovina	2 273	2 273	n.s.	n.s.
Bulgaria	3 486	3 690	20	0.6
Croatia	1 763	1 783	2	0.1
Czech Republic	2 627	2 632	1	n.s.
Denmark	445	455	1	0.2
Estonia	1 935	2 060	13	0.6
Finland	21 855	21 935	8	n.s.
France	14 725	15 341	62	0.4
Germany	10 740	10 740	n.s.	n.s.
Greece	3 299	3 599	30	0.9
Hungary	1 768	1 840	7	0.4
Iceland	25	31	1	2.2
Ireland	489	659	17	3.0
Italy	9 708	10 003	30	0.3
Latvia	2 796	2 923	13	0.4
Liechtenstein	6	7	n.s.	1.2
Lithuania	1 946	1 994	5	0.2
Malta	n.s.	n.s.	n.s.	n.s.
Netherlands	365	375	1	0.3
Norway	8 558	8 868	31	0.4
Poland	8 872	9 047	18	0.2
Portugal	3 096	3 666	57	1.7
Republic of Moldova	318	325	1	0.2
Romania	6 301	6 448	15	0.2
Russian Federation	850 039	851 392	135	n.s.
San Marino	-	-	-	-
Slovakia	1 997	2 177	18	0.9
Slovenia	1 085	1 107	2	0.2
Spain	13 510	14 370	86	0.6
Sweden	27 128	27 134	1	n.s.
Switzerland	1 156	1 199	4	0.4
The FYR of Macedonia	906	906	n.s.	n.s.

Table 4. Change in forest cover 1990-2000 (cont.)

Country/area	Total forest 1990 000 ha	Total forest 2000 000 ha	Annual change 000 ha	Annual change rate %
Ukraine	9 274	9 584	31	0.3
United Kingdom	2 624	2 794	17	0.6
Yugoslavia	2 901	2 887	-1	-0.1
Total Europe	**1 030 475**	**1 039 251**	**881**	**0.08**
Antigua and Barbuda	9	9	n.s.	n.s.
Bahamas	842	842	n.s.	n.s.
Barbados	2	2	n.s.	n.s.
Belize	1 704	1 348	-36	-2.3
Bermuda	-	-	-	-
British Virgin Islands	3	3	n.s.	n.s.
Canada	244 571	244 571	n.s.	n.s.
Cayman Islands	13	13	n.s.	n.s.
Costa Rica	2 126	1 968	-16	-0.8
Cuba	2 071	2 348	28	1.3
Dominica	50	46	n.s.	-0.7
Dominican Republic	1 376	1 376	n.s.	n.s.
El Salvador	193	121	-7	-4.6
Greenland	-	-	-	-
Grenada	5	5	n.s.	0.9
Guadeloupe	67	82	2	2.1
Guatemala	3 387	2 850	-54	-1.7
Haiti	158	88	-7	-5.7
Honduras	5 972	5 383	-59	-1.0
Jamaica	379	325	-5	-1.5
Martinique	47	47	n.s.	n.s.
Mexico	61 511	55 205	-631	-1.1
Montserrat	3	3	n.s.	n.s.
Netherlands Antilles	1	1	n.s.	n.s.
Nicaragua	4 450	3 278	-117	-3.0
Panama	3 395	2 876	-52	-1.6
Puerto Rico	234	229	-1	-0.2
Saint Kitts and Nevis	4	4	n.s.	-0.6
Saint Lucia	14	9	-1	-4.9
Saint Pierre & Miquelon	-	-	-	-
Saint Vincent and Grenadines	7	6	n.s.	-1.4
Trinidad and Tobago	281	259	-2	-0.8
United States	222 113	225 993	388	0.2
US Virgin Islands	14	14	n.s.	n.s.
Total North and Central America	**555 002**	**549 304**	**-570**	**-0.10**
Argentina	37 499	34 648	-285	-0.8
Bolivia	54 679	53 068	-161	-0.3
Brazil	566 998	543 905	-2 309	-0.4
Chile	15 739	15 536	-20	-0.1
Colombia	51 506	49 601	-190	-0.4
Ecuador	11 929	10 557	-137	-1.2
Falkland Islands	-	-	-	-
French Guiana	7 926	7 926	n.s.	n.s.
Guyana	17 365	16 879	-49	-0.3
Paraguay	24 602	23 372	-123	-0.5
Peru	67 903	65 215	-269	-0.4
Suriname	14 113	14 113	n.s.	n.s.
Uruguay	791	1 292	50	5.0
Venezuela	51 681	49 506	-218	-0.4
Total South America	**922 731**	**885 618**	**-3 711**	**-0.41**
TOTAL WORLD	**3 963 429**	**3 869 455**	**-9 391**	**-0.22**

Table 5. Forest cover – latest national statistics

Country/area	Ref. year	Total area	Forest Closed	Forest Open	Plantation	Other wooded land Shrubs/trees	Other wooded land Forest fallow	Other land	Inland water
	year	000 ha	000 ha	000 ha	000 ha	000 ha	000 ha	000 ha	000 ha
Algeria	2000	238 174	673	754	717	1 662		234 368	
Angola	1983	124 671	859	6 350		5 095	123	112 243	1
Benin	1996	11 263	546	2 277	106	3 731		4 403	200
Botswana	1990	58 174	675	12 936		34 517		8 545	1 501
Burkina Faso	1991	27 400	2 233	4 993		7 430	238	12 466	40
Burundi	1998	2 783	55		69			2 444	215
Cameroon	1999	47 544	19 985	4 015		2 000		20 540	1 004
Cape Verde	1999	403			80			323	
Central African Rep.	1994	62 297	4 826	18 347		10 021		29 103	
Chad	1988	128 400	211	13 451		9 698		102 560	2 480
Comoros	1984	186	14					172	
Congo	1999	34 200	22 000			3 000		9 150	50
Côte d'Ivoire	1987	32 246	3 248	8 475		6 536	84	13 457	446
Dem. Rep. of the Congo	1989	234 486	126 236	14 440		14 877	10 282	60 870	7 781
Djibouti	1985	2 319	2	3		220		2 092	2
Egypt	1996	100 145						99 545	600
Equatorial Guinea	1998	2 805	1 774			19		1 012	
Eritrea	1997	11 759	659	930		5 032		5 138	
Ethiopia	1997	110 430	1 680	4 075		31 554		73 121	
Gabon	1999	26 767	21 800		30			3 937	1 000
Gambia	1993	1 130	80	368		72	89	391	130
Ghana	1996	23 854	1 634	5 001				16 119	1 100
Guinea	1988	24 586	1 750	5 586		5 850		11 386	14
Guinea-Bissau	1990	3 612	742	1 659				1 211	
Kenya	1993	58 038	5 038	13 099	148	20 584		18 046	1 123
Lesotho	1994	3 035			2	822		2 211	
Liberia	1990	11 137	4 124					7 013	
Libyan Arab Jamahiriya	1980	175 954	78	112		446		175 318	
Madagascar	1996	58 704	11 550		319	1 472		44 813	550
Malawi	1991	11 849	338	2 763	169	48	3 010	3 081	2 440
Mali	1991	124 019	7 500	6 580		17 020		90 919	2 000
Mauritania	1991	102 552		410		3 110		99 002	30
Mauritius	1979	203	6		10	10		176	1
Morocco	1995	44 655	1 455	1 091	490	1 265		40 329	25
Mozambique	1995	80 159	7 710	23 163	39	22 688	19 461	5 348	1 750
Namibia	1992	82 429	3 965	4 662		8 950	1 012	63 740	100
Niger	1992	126 700	1 101	669		334		124 566	30
Nigeria	1994	92 377	4 456	10 897	277	9 645		65 802	1 300
Réunion	1997	251	70		1	23		156	1
Rwanda	1999	2 633	58	4		77		2 327	167
Saint Helena	1980	31	2			8		21	
Sao Tome and Principe	1989	96	27				37	31	1
Senegal	1985	19 671	324	6 457		8 086	3 957	428	419
Seychelles	1993	45	25		5	7		8	
Sierra Leone	1986	7 174	725	829	2	499	3 879	1 228	12
Somalia	1980	63 766	1 540	7 510				53 684	1 032
South Africa	1994	122 102	5 013	2 469	1 614	63 679		48 983	344
Sudan	1990	250 581	17 622	52 300		52 088		115 590	12 981
Swaziland	1999	1 737	71	285	123	311		931	16
Togo	1991	5 679	272	393		348		4 426	240
Tunisia	1994	16 362	275	127		328		15 632	
Uganda	1992	24 103	1 443	3 445	35	1 419		13 622	4 139
United Republic of Tanzania	1995	94 509	8 305	30 842	135	12 964	9 232	26 881	6 150
Western Sahara	1995	26 600		152		859		25 589	
Zambia	1978	75 261	43 984	9 452	44	4 090	679	16 090	922
Zimbabwe	1992	39 076	13 941	7 535	156	5 502		11 551	391
Total Africa	**1991**	**3 031 122**	**352 700**	**288 906**	**4 571**	**377 996**	**52 083**	**1 902 138**	**52 728**

Table 5. Forest cover – latest national statistics (cont.)

Country/area	Ref. year	Total area	Land area - Forest Closed	Forest Open	Forest Plantation	Other wooded land Shrubs/trees	Other wooded land Forest fallow	Other land	Inland water
	year	000 ha	000 ha	000 ha	000 ha	000 ha	000 ha	000 ha	000 ha
Afghanistan	1993	65 209	1 077	274		29 470		34 137	251
Armenia	1996	2 980	322		13	58		2 427	160
Azerbaijan	1988	8 660	918		20	54		7 367	301
Bahrain	1999	69						69	
Bangladesh	1996	14 400	720		232	105	17	11 943	1 383
Bhutan	1990	4 701	2 807	209	8	382	103	1 192	
Brunei Darussalam	1998	577	442		2			83	50
Cambodia	1997	18 104	5 500	3 921	82	2 448	746	4 955	452
China	1996	959 806	110 172	5 616	39 876	34 446		742 633	27 063
Cyprus	1999	926	172			214		539	1
Dem People's Rep. of Korea	1993	12 054	8 210					3 831	13
East Timor	1985	1 489	387	171		388	99	434	10
Gaza Strip		38	-	-	-	-	-	-	-
Georgia	1995	6 970	2 788		200			3 843	139
India	1997	328 759	38 223	25 506		5 190		228 400	31 440
Indonesia	1997	190 457	100 382					80 775	9 300
Iran, Islamic Rep.	1999	163 320	2 488	2 527		1 760	5 626	149 800	1 119
Iraq	1990	43 832	552	237		259	986	41 703	95
Israel	1997	2 106	26		91	46		1 899	44
Japan	1995	37 780	13 382		10 682	1 082		12 506	128
Jordan	2000	8 921		40		1	68	8 784	28
Kazakhstan	1993	271 731	10 470		5	6 152		250 447	4 657
Kuwait	2000	1 782						1 782	
Kyrgyzstan	1993	19 850	786		57			18 337	670
Lao People's Dem. Rep	1989	23 680	11 493	1 663		760	7 417	1 747	600
Lebanon	1996	1 041	20	15		35		954	17
Malaysia	1995	32 975	19 148		91	684	560	12 372	120
Maldives	1999	30	1					29	
Mongolia	1987	156 650	10 481	944		4 896		140 329	
Myanmar	1997	67 658	25 177	10 081	91	10 723	1 196	18 487	1 903
Nepal	1994	14 718	3 201	1 067		1 560		8 472	418
Oman	2000	21 246						21 246	
Pakistan	1990	79 609	1 217	857	93	1 078		73 842	2 522
Philippines	1997	30 000	5 288	104		2 232		22 193	183
Qatar	2000	1 100						1 100	
Republic of Korea	1999	9 926	6 253					3 620	53
Saudi Arabia	1994	214 969	1 100	400		905	300	212 264	
Singapore	1990	62	2					59	1
Sri Lanka	1992	6 561	1 569	368	72	92		4 362	98
Syrian Arab Republic	1992	18 517	164	33		35		18 145	140
Tajikistan	1995	14 310	390		10	330		13 357	223
Thailand	1998	51 312	10 127	2 845				38 117	223
Turkey	1985	77 482	8 041		1 854	10 695		56 373	519
Turkmenistan	1995	48 809	3 742		12			43 238	1 817
United Arab Emirates	2000	8 360						8 360	
Uzbekistan	1995	44 740	1 646		300			39 478	3 316
Viet Nam	1995	33 170	7 312	940		4 951	2 659	16 688	620
West Bank		580	-	-	-	-	-	-	-
Yemen	1993	52 797	11	503		1 277	254	50 752	
Total Asia	**1995**	**3 174 823**	**416 207**	**58 321**	**53 791**	**122 308**	**20 031**	**2 413 470**	**90 077**

Table 5. Forest cover – latest national statistics (cont.)

Country/area	Ref. year	Total area	Forest Closed	Forest Open	Plantation	Other wooded land Shrubs/trees	Other wooded land Forest fallow	Other land	Inland water
	year	000 ha	000 ha	000 ha	000 ha	000 ha	000 ha	000 ha	000 ha
American Samoa	1999	20	12			1	5	2	0
Australia	1992	774 122	155 834		1 043	421 590		189 763	5 892
Cook Islands	1998	23	22					1	
Fiji	1995	1 827	747		93		153	834	
French Polynesia	1991	400	95		10			261	34
Guam	1999	55	21					34	
Kiribati	1997	73	2					71	
Marshall Islands	1999	18					18		
Micronesia	1983	69	30			4	20	15	
Nauru	1993	2						2	
New Caledonia	1993	1 858	372					1 456	30
New Zealand	1996	27 053	6 248		1 542	1 079		17 930	254
Niue	1981	26	6				12	8	
Northern Mariana Isl.	1984	46	14			20		12	
Palau	1985	46	35				2	9	
Papua New Guinea	1996	46 283	30 150	829		585		13 675	1 044
Samoa	1992	283	103		3			176	1
Solomon Islands	1993	2 890	2 208	316				332	34
Tonga	1990	76	4			1	2	66	3
Vanuatu	1993	1 218	442			239	239	298	
Total Oceania	**1992**	**856 388**	**196 345**	**1 145**	**2 691**	**423 519**	**451**	**224 945**	**7 292**
Albania	1995	2 875	928		102			1 710	135
Andorra		45	-	-	-	-	-	-	-
Austria	1994	8 386	3 840			84		4 349	113
Belarus	1994	20 760	7 670		195	1 071		11 812	12
Belgium & Luxemb.	1997	3 310	732			29		2 521	28
Bosnia & Herzegovina	1995	5 113	2 216		57	433		2 394	13
Bulgaria	1995	11 091	2 619		969	314		7 153	36
Croatia	1996	5 654	1 728		47	330		3 487	62
Czech Republic	1995	7 886	2 630					5 098	158
Denmark	1990	4 309	104		341			3 705	66
Estonia	1996	4 510	1 705		305	146		2 071	283
Finland	1994	33 815	21 884			885		7 690	3 356
France	1997	55 150	14 195		961	1 833		38 021	140
Germany	1987	35 698	10 740					24 187	771
Greece	1992	13 196	3 239		120	3 154		6 377	306
Hungary	1996	9 303	1 675		136			7 423	69
Iceland	1998	10 300	18		12	100		9 895	275
Ireland	1996	7 028	1		590			6 298	139
Italy	1995	30 127	9 722		133	985		18 566	721
Latvia	1997	6 460	2 741		143	111		3 210	255
Liechtenstein	1995	15	6					9	
Lithuania	1996	6 521	1 691		284	72		4 211	263
Malta	1996	32						32	
Netherlands	1994	4 084	261		100			3 031	692
Norway	1995	32 388	8 413		300	3 291		18 679	1 705
Poland	2000	32 325	9 008		39			21 395	1 883
Portugal	1995	9 198	2 547		834	84		5 685	48
Republic of Moldova	1997	3 369	321		1	31		2 943	73
Romania	1990	23 839	6 210		91			16 733	805
Russian Federation	1998	1 707 541	833 783		17 340			837 728	18 690
San Marino		6	-	-	-	-	-	-	-
Slovakia	2000	4 901	2 162		15	15		2 616	93
Slovenia	1996	2 025	1 097		1	67		847	13
Spain	1990	50 600	11 731		1 925	12 611		23 678	655
Sweden	1994	44 996	26 561		569	2 980		11 052	3 834
Switzerland	1994	4 129	1 169		4	61		2 721	174
The FYR of Macedonia	1995	2 571	876		30	82		1 555	28

Table 5. Forest cover – latest national statistics (cont.)

Country/area	Ref. year	Total area	Forest Closed	Forest Open	Planta-tion	Other wooded land Shrubs/trees	Other wooded land Forest fallow	Other land	Inland water
	year	000 ha	000 ha	000 ha	000 ha	000 ha	000 ha	000 ha	000 ha
Ukraine	1996	60 370	5 035		4 425	36		48 439	2 435
United Kingdom	2000	24 488	866		1 928			21 366	328
Yugoslavia	1995	10 217	2 855		39	586		6 720	17
Total Europe	**1997**	**2 298 631**	**1 002 979**		**32 036**	**29 484**		**1 195 407**	**38 674**
Antigua and Barbuda	1983	44	8	1		10		25	
Bahamas	1986	1 388	528	32				441	387
Barbados	1998	43	2					41	
Belize	1993	2 296	1 575	19			85	601	16
Bermuda		5	-	-	-	-	-	-	-
British Virgin Islands	1980	15	3	2				10	
Canada	1994	997 061	244 571			173 013		504 513	74 964
Cayman Islands	1998	26							
Costa Rica	1997	5 110	2 058			8		3 040	4
Cuba	1998	11 086	1 981	251	353	21		8 376	104
Dominica	1984	75	52					23	
Dominican Republic	1998	4 873	1 346				683	2 809	35
El Salvador	1990	2 104	101		5		52	1 914	32
Greenland		34 170	-	-	-	-	-	-	-
Grenada	1992	34	5			3		26	
Guadeloupe	1991	171	65					104	2
Guatemala	1999	10 889	2 824			949	2 297	4 773	46
Haiti	1995	2 775	107					2 649	19
Honduras	1995	11 209	3 811	2 157				5 221	20
Jamaica	1997	1 099	293	57	5		638	90	16
Martinique	1998	111	46		2			59	4
Mexico	1993	195 820	33 613	24 835	63	56 836	22 235	53 287	4 951
Montserrat	1983	11	3			1		7	
Netherlands Antilles	1991	80		1			33	46	
Nicaragua	1999	13 000	3 156			240	20	8 724	860
Panama	1998	7 552	3 052					4 391	109
Puerto Rico	1990	895	231			29	8	619	8
Saint Kitts and Nevis	1992	36	4				2	30	
Saint Lucia	1992	62		12		8		41	1
Saint Pierre & Miquelon		23	-	-	-	-	-	-	-
Saint Vincent and Grenadines	1993	39	8			3	2	26	
Trinidad and Tobago	1997	513	250		15			248	
United States	1997	962 908	208 591		16 238	77 464		613 602	47 013
US Virgin Islands	1976	34	14					20	
Total North and Central America	**1995**	**2 265 557**	**508 298**	**27 367**	**16 681**	**308 585**	**26 055**	**1 215 756**	**128 591**
Argentina	1993	278 040	9 000	27 600				237 069	4 371
Bolivia	1993	109 858	47 999	6 589		4 205		49 645	1 420
Brazil	1989	854 740	564 581	1 335		57		279 678	9 089
Chile	1995	75 664	12 298	1 133	2 153	14 670		44 627	783
Colombia	1996	113 892	51 437	14 075	16			38 343	10 021
Ecuador	1992	28 356	10 854	558	72	1 215		14 985	672
Falkland Islands	2000	1 217						1 217	
French Guiana	1990	9 000	7 925					890	185
Guyana	1999	21 498	16 916				3 580	1 002	
Paraguay	1991	40 675	8 354	15 125		2 309	1 077	12 865	945
Peru	1990	128 522	64 204	2 431		10 946	4 864	45 555	522
Suriname	1995	16 327	14 100					1 500	727
Uruguay	1998	17 741	667		170			16 644	260
Venezuela	1995	91 206	49 926		663	7 474	18 682	11 461	3 000
Total South America	**1991**	**1 786 736**	**858 261**	**68 846**	**3 074**	**40 876**	**28 203**	**755 481**	**31 995**
TOTAL WORLD	**1994**	**13 413 257**	**3 334 790**	**444 585**	**112 844**	**1 302 768**	**126 823**	**7 707 198**	**349 357**

Table 6. Forest plantations 2000

Country/area	Total plantation area	Annual planting rate	Acacia	Eucalyptus	Hevea	Tectona	Other broad-leaved	Pinus	Other coniferous	Unspecified
	000 ha	000 ha	000 ha	000 ha	000 ha	000 ha	000 ha	000 ha	000 ha	000 ha
Algeria	718	29	0.4	40			1	48	17	611
Angola	141	0.1	1	113			1	21	4	
Benin	112	1	5	5		15	8		2	78
Botswana	1	0.0		1						
Burkina Faso	67	5								67
Burundi	73	2		26			4	12	32	
Cameroon	80	0.2		8	42		26	4		
Cape Verde	85	5	7	3			74	2		
Central African Rep.	4				1					3
Chad	14	0.3	3	4			7			
Comoros	2	0.1		2						
Congo	83	6		68			10	5		
Côte d'Ivoire	184	5			68	58	58			
Dem Rep of the Congo	97	0.1								97
Djibouti	-	-	-	-	-	-	-	-	-	-
Egypt	72	2	2	8			15	0.1	46	
Equatorial Guinea										
Eritrea	22	4		11						11
Ethiopia	216	2								216
Gabon	36			3	10		23	2		
Gambia	2						2			
Ghana	76	2		19		40	18			
Guinea	25	1								25
Guinea-Bissau	2									2
Kenya	232	2	26	39				53	76	37
Lesotho	14	2		7				6		1
Liberia	119	0.1			110	2	5	2		
Libyan Arab Jamahiriya	168	5								168
Madagascar	350	6	10	163			61	109	7	
Malawi	112	2		26	3		9	74		
Mali	15	1					10			5
Mauritania	25	3	12				12			
Mauritius	13	0.0					1	8	2	2
Morocco	534	10	27	214			43	16	235	
Mozambique	50	1		20				26	4	
Namibia	0.3			0.3						
Niger	73	3	36	4			33			
Nigeria	693	23	2	41	318	74	249	10		
Réunion	3		2						2	
Rwanda	261	2	13	170			5	34	31	8
Saint Helena	2	0.1								2
Sao Tome & Principe	-	-	-	-	-	-	-	-	-	-
Senegal	263	11	32	63			5	129	34	
Seychelles	5									5
Sierra Leone	6				2					4
Somalia	3									3
South Africa	1 554	12	109	606			16	824		
Sudan	641	30								641
Swaziland	161		25	33				102		
Togo	38	1		17		11	10			
Tunisia	202	14		55				88	59	
Uganda	43	1		23				13	6	
United Republic of Tanzania	135		13	3		3	68	35	14	
Western Sahara	-	-	-	-	-	-	-	-	-	-
Zambia	75	2		15				60		
Zimbabwe	141	2	21	13			6	94	7	
Total Africa	**8 036**	**194**	**345**	**1 799**	**573**	**207**	**902**	**1 648**	**578**	**1 985**

Table 6. Forest plantations 2000 (cont.)

Country/area	Total plantation area	Annual planting rate	Acacia	Eucalyptus	Hevea	Tectona	Other broad-leaved	Pinus	Other coniferous	Unspecified
	000 ha	000 ha	000 ha	000 ha	000 ha	000 ha	000 ha	000 ha	000 ha	000 ha
Afghanistan	-	-	-	-	-	-	-	-	-	-
Armenia	13									13
Azerbaijan	20									20
Bahrain	0.4	0.0								0.4
Bangladesh	625	22	32	37	92	144	320			
Bhutan	21	1				3	15	3	1	
Brunei Darussalam	3	0.2	0.4		1	0.1	1		0.2	0.0
Cambodia	90	3		0.1	73	7	10	1		
China	45 083	1 154	129	1 334	592	24	11 468	12 909	17 168	1 459
Cyprus										27
Dem People's Rep. of Korea	-	-	-	-	-	-	-	-	-	-
East Timor	-	-	-	-	-	-	-	-	-	-
Gaza Strip	-	-	-	-	-	-	-	-	-	-
Georgia	200									200
India	32 578	1 509	6404	8 005	560	2 561	11 847	640	2 561	
Indonesia	9 871	271	642	128	3 476	1 470	3 385	770		
Iran, Islamic Rep.	2 284	63					2 136	43		105
Iraq	10	0.3		7			1	1	1	1
Israel	91									91
Japan	10 682									10 682
Jordan	45	1	5	1			2	32	3	2
Kazakhstan	5									5
Kuwait	5	0.1								5
Kyrgyzstan	57									57
Lao People's Dem. Rep	54	6	5	8		14	27			
Lebanon	2							1		1
Malaysia	1 750	35	180	19	1 478	12	12	47		
Maldives	-	-	-	-	-	-	-	-	-	-
Mongolia	-	-	-	-	-	-	-	-	-	-
Myanmar	821	37		71	111	291	333	14		
Nepal	133	5		11			73	33	16	
Oman	1	0.0								1
Pakistan	980	30	196	245			490	49		
Philippines	753	30	49	189	97	38	359	23		
Qatar	1	0.0								1
Republic of Korea	-	-	-	-	-	-	-	-	-	-
Saudi Arabia	4	0.1								4
Singapore	-	-	-	-	-	-	-	-	-	-
Sri Lanka	316	4	47	44	164	4	34	23		
Syrian Arab Republic	229	24								229
Tajikistan	10									10
Thailand	4 920	225	148	443	2 115	836	541	689	148	
Turkey	1 854									1 854
Turkmenistan	12									12
United Arab Emirates	314	0.2								314
Uzbekistan	300									300
Viet Nam	1 711	80	127	452	300	4	504	254	71	
West Bank	-	-	-	-	-	-	-	-	-	-
Yemen	-	-	-	-	-	-	-	-	-	-
Total Asia	**115 847**	**3 500**	**7964**	**10 994**	**9 058**	**5 409**	**31 556**	**15 532**	**19 968**	**15 392**

Table 6. Forest plantations 2000 (cont.)

Country/area	Total plantation area	Annual planting rate	Acacia	Eucalyptus	Hevea	Tectona	Other broad-leaved	Pinus	Other coniferous	Unspecified
	000 ha	000 ha	000 ha	000 ha	000 ha	000 ha	000 ha	000 ha	000 ha	000 ha
American Samoa	0.0	0.0	0.0	0.0			0.0		0.0	
Australia	1 043									1 043
Cook Islands	1	0.1	0.2					1		
Fiji	97	9					47	43		7
French Polynesia	5	0.1					0.3	5		
Guam	0.4	0.0	0.3				0.0		0.0	
Kiribati										
Marshall Islands	-	-	-	-	-	-	-	-	-	-
Micronesia	0.1	0.0	0.0	0.0		0.0	0.1			
Nauru	-	-	-	-	-	-	-		-	-
New Caledonia	10	0.2						10		
New Zealand	1 542									1 542
Niue	0.4	0.0					0.4			
Northern Mariana Isl.	-		-	-	-	-	-	-	-	-
Palau	0.4	0.0	0.1				0.3			
Papua New Guinea	90	4	7	21	20	5	12	14	10	
Samoa	5	1	0.0	0.2		0.1	4			
Solomon Islands	50	1		12		2	36			
Tonga	1	0.0					0.2	0.3		
Vanuatu	3	0.2								3
Total Oceania	**2 848**	**15**	**8**	**33**	**20**	**7**	**101**	**73**	**10**	**2 595**
Albania	102									102
Andorra	-	-	-	-	-	-	-	-	-	-
Austria										
Belarus	195									195
Belgium & Luxembourg										
Bosnia & Herzegovina	57									57
Bulgaria	969									969
Croatia	47									47
Czech Republic										
Denmark	341									341
Estonia	305									305
Finland										
France	961									961
Germany										
Greece	120									120
Hungary	136									136
Iceland	12									12
Ireland	590									590
Italy	133									133
Latvia	143									143
Liechtenstein										
Lithuania	284									284
Malta	0.3									0.3
Netherlands	100									100
Norway	300									300
Poland	39									39
Portugal	834									834
Republic of Moldova	1									1
Romania	91									91
Russian Federation	17 340									17 340
San Marino	-	-	-	-	-	-	-	-	-	-
Slovakia	15							15		
Slovenia	1									1
Spain	1 904									1 904
Sweden	569									569
Switzerland	4									4
The FYR of Macedonia	30									30

Table 6. Forest plantations 2000 (cont.)

Country/area	Total plantation area	Annual planting rate	Acacia	Eucalyptus	Hevea	Tectona	Other broad-leaved	Pinus	Other coniferous	Unspecified
	000 ha	000 ha	000 ha	000 ha	000 ha	000 ha	000 ha	000 ha	000 ha	000 ha
Ukraine	4 425									4 425
United Kingdom	1 928	5								1 928
Yugoslavia	39									39
Total Europe	**32 015**	**5**					**15**			**32 000**
Antigua and Barbuda										
Bahamas	-	-	-	-	-	-	-	-	-	-
Barbados	0.0									0.0
Belize	3	0.1				0.1	2	2		
Bermuda	-	-	-	-	-	-	-	-	-	-
British Virgin Islands	-	-	-	-	-	-	-	-	-	-
Canada										
Cayman Islands	-	-	-	-	-	-	-	-	-	-
Costa Rica	178	11		17		30	112	6	6	6
Cuba	482	26		53			154	207	67	
Dominica	0.1									0.1
Dominican Republic	30	4								30
El Salvador	14	2	0.1	1		5	5	2	1	
Greenland	-	-	-	-	-	-	-	-	-	-
Grenada	0.2						0.2	0.0		0.0
Guadeloupe	4	0.1					4			0.3
Guatemala	133	21		13	41	4	29	37	8	
Haiti	20	1								20
Honduras	48	7						16		32
Jamaica	9						4	4		
Martinique	2	0.0					2			
Mexico	267	35		102	11	8	59	83	5	
Montserrat	-	-	-	-	-	-	-	-	-	-
Netherlands Antilles	-	-	-	-	-	-	-	-	-	-
Nicaragua	46	4		12			1	31		2
Panama	40	3				19	7	14		
Puerto Rico	4			1		0.1	2	0.4		1
Saint Kitts and Nevis										
Saint Lucia	1	0.0								1
Saint Pierre & Miquelon	-	-	-	-	-	-	-	-	-	-
Saint Vincent and Grenadines	0.3	0.0					0.2	0.0		
Trinidad and Tobago	15					9	2	4		
United States	16 238	121						15 033		1 205
US Virgin Islands	-	-	-	-	-	-	-	-	-	-
Total North and Central America	**17 533**	**234**	**0**	**198**	**52**	**76**	**383**	**15 440**	**88**	**1 297**
Argentina	926	126		278			185	463		
Bolivia	46	1		41						5
Brazil	4 982	135		2 964	180	14		1 769	55	
Chile	2 017	85		343			149	1 525		
Colombia	141	7		27		4	21	78	11	
Ecuador	167	4		81	3		15	67		2
Falkland Islands	-	-	-	-	-	-	-	-	-	-
French Guiana	1	0								1
Guyana	12									12
Paraguay	27	2		21			1			4
Peru	640	50		480			128		32	
Suriname	13	0					5	7		
Uruguay	622	50		498			25	100		
Venezuela	863	50		104			69	690		
Total South America	**10 455**	**509**		**4 836**	**183**	**18**	**599**	**4 699**	**98**	**23**
TOTAL WORLD	**186 733**	**4 458**	**8317**	**17 860**	**9 885**	**5 716**	**33 556**	**37 391**	**20 743**	**53 292**

Appendix 3. Global tables 403

Table 7. Volume and biomass in forest

Country/area	Total forest 2000	Volume By area	Volume Total	Biomass By area	Biomass Total	Information source
	000 ha	m³/ha	M m³	t/ha	M t	NI / PI / ES / EX
Algeria	2 145	44	94	75	160	NI
Angola	69 756	39	2 714	54	3 774	NI
Benin	2 650	140	371	195	518	PI
Botswana	12 427	45	560	63	779	NI
Burkina Faso	7 089	10	74	16	113	NI
Burundi	94	110	10	187	18	ES
Cameroon	23 858	135	3 211	131	3 129	PI
Cape Verde	85	83	7	127	11	ES
Central African Republic	22 907	85	1 937	113	2 583	PI/EX
Chad	12 692	11	134	16	205	ES
Comoros	8	60	0	65	1	ES
Congo	22 060	132	2 916	213	4 699	EX
Côte d'Ivoire	7 117	133	948	130	924	PI
Dem. Rep. of the Congo	135 207	133	17 932	225	30 403	NI
Djibouti	6	21	0	46	0	ES
Egypt	72	108	8	106	8	ES
Equatorial Guinea	1 752	93	163	158	277	PI
Eritrea	1 585	23	36	32	50	NI
Ethiopia	4 593	56	259	79	363	PI
Gabon	21 826	128	2 791	137	2 991	ES
Gambia	481	13	6	22	11	NI
Ghana	6 335	49	311	88	556	ES
Guinea	6 929	117	808	114	788	PI
Guinea-Bissau	2 187	19	41	20	44	NI
Kenya	17 096	35	593	48	826	ES
Lesotho	14	34	0	34	0	ES
Liberia	3 481	201	699	196	681	ES
Libyan Arab Jamahiriya	358	14	5	20	7	ES
Madagascar	11 727	114	1 339	194	2 270	NI
Malawi	2 562	103	264	143	365	NI
Mali	13 186	22	289	31	402	PI
Mauritania	317	4	1	6	2	ES
Mauritius	16	88	1	95	2	ES
Morocco	3 025	27	80	41	123	NI
Mozambique	30 601	25	774	55	1 683	NI
Namibia	8 040	7	54	12	94	PI
Niger	1 328	3	4	4	6	PI
Nigeria	13 517	82	1 115	184	2 493	ES
Réunion	71	115	8	160	11	ES
Rwanda	307	110	34	187	58	ES
Saint Helena	2	-	-	-	-	-
Sao Tome and Principe	27	108	3	116	3	NI
Senegal	6 205	31	192	30	187	NI
Seychelles	30	29	1	49	1	ES
Sierra Leone	1 055	143	151	139	147	ES
Somalia	7 515	18	138	26	192	ES
South Africa	8 917	49	437	81	720	EX
Sudan	61 627	9	531	12	740	ES
Swaziland	522	39	20	115	60	NI
Togo	510	92	47	155	79	PI
Tunisia	510	18	9	27	14	NI
Uganda	4 190	133	559	163	681	NI
United Republic of Tanzania	38 811	43	1 676	60	2 333	NI
Western Sahara	152	18	3	59	9	NI
Zambia	31 246	43	1 347	104	3 262	ES
Zimbabwe	19 040	40	765	56	1 065	NI
Total Africa	**649 866**	**72**	**46 472**	**109**	**70 917**	

Table 7. Volume and biomass in forest (cont.)

Country/area	Total forest 2000	Volume By area	Volume Total	Biomass By area	Biomass Total	Information source
	000 ha	m³/ha	M m³	t/ha	M t	NI / PI / ES / EX
Afghanistan	1 351	22	30	27	37	-
Armenia	351	128	45	66	23	-
Azerbaijan	1 094	136	149	105	115	-
Bahrain	n.s.	14	-	14	-	-
Bangladesh	1 334	23	31	39	52	-
Bhutan	3 016	163	492	178	537	-
Brunei Darussalam	442	119	52	205	90	-
Cambodia	9 335	40	376	69	648	-
China	163 480	52	8 437	61	10 038	NI
Cyprus	172	43	7	21	4	-
Dem People's Rep. of Korea	8 210	41	333	25	209	ES
East Timor	507	79	40	136	69	-
Gaza Strip	-	-	-	-	-	-
Georgia	2 988	145	434	97	291	-
India	64 113	43	2 730	73	4 706	NI
Indonesia	104 986	79	8 242	136	14 226	-
Iran, Islamic Rep.	7 299	86	631	149	1 089	-
Iraq	799	29	23	28	22	-
Israel	132	49	6	-	-	-
Japan	24 081	145	3 485	88	2 128	-
Jordan	86	38	3	37	3	-
Kazakhstan	12 148	35	428	18	214	-
Kuwait	5	21	0	21	0	-
Kyrgyzstan	1 003	32	32	-	-	-
Lao People's Dem. Rep	12 561	29	359	31	391	NI
Lebanon	36	23	1	22	1	-
Malaysia	19 292	119	2 288	205	3 949	ES
Maldives	1	-	-	-	-	-
Mongolia	10 645	128	1 359	80	853	NI
Myanmar	34 419	33	1 137	57	1 965	NI
Nepal	3 900	100	391	109	427	PI
Oman	1	17	0	17	0	-
Pakistan	2 361	22	53	27	64	-
Philippines	5 789	66	383	114	661	NI
Qatar	1	13	0	12	0	-
Republic of Korea	6 248	58	362	36	227	NI
Saudi Arabia	1 504	12	18	12	18	-
Singapore	2	119	0	205	0	-
Sri Lanka	1 940	34	66	59	114	-
Syrian Arab Republic	461	29	13	28	13	-
Tajikistan	400	14	6	10	4	-
Thailand	14 762	17	252	29	434	NI
Turkey	10 225	136	1 386	74	754	-
Turkmenistan	3 755	4	14	3	10	-
United Arab Emirates	321	-	-	-	-	-
Uzbekistan	1 969	6	11	-	-	-
Viet Nam	9 819	38	372	66	643	ES
West Bank	-	-	-	-	-	-
Yemen	449	14	6	19	9	-
Total Asia	**547 793**	**63**	**34 506**	**82**	**45 062**	

Appendix 3. Global tables

Table 7. Volume and biomass in forest (cont.)

Country/area	Total forest 2000	Volume By area	Volume Total	Biomass By area	Biomass Total	Information source
	000 ha	m³/ha	M m³	t/ha	M t	NI / PI / ES / EX
American Samoa	12	-	-	-	-	-
Australia	154 539	55	8 506	57	8 840	-
Cook Islands	22	-	-	-	-	-
Fiji	815	-	-	-	-	-
French Polynesia	105	-	-	-	-	-
Guam	21	-	-	-	-	-
Kiribati	28	-	-	-	-	-
Marshall Islands	n.s.	-	-	-	-	-
Micronesia	15	-	-	-	-	-
Nauru	n.s.	-	-	-	-	-
New Caledonia	372	-	-	-	-	-
New Zealand	7 946	125	992	217	1 726	-
Niue	6	-	-	-	-	-
Northern Mariana Isl.	14	-	-	-	-	-
Palau	35	-	-	-	-	-
Papua New Guinea	30 601	34	1 025	58	1 784	NI
Samoa	105	-	-	-	-	-
Solomon Islands	2 536	-	-	-	-	-
Tonga	4	-	-	-	-	-
Vanuatu	447	-	-	-	-	-
Total Oceania	**197 623**	**55**	**10 771**	**64**	**12 640**	
Albania	991	81	80	58	57	-
Andorra	-	0	-	0	-	-
Austria	3 886	286	1 110	250	970	-
Belarus	9 402	153	1 436	80	755	-
Belgium & Luxembourg	728	218	159	101	74	-
Bosnia & Herzegovina	2 273	110	250	-	-	-
Bulgaria	3 690	130	480	76	279	-
Croatia	1 783	201	358	107	190	-
Czech Republic	2 632	260	684	125	329	-
Denmark	455	124	56	58	26	-
Estonia	2 060	156	321	85	175	-
Finland	21 935	89	1 945	50	1 089	NI
France	15 341	191	2 927	92	1 418	-
Germany	10 740	268	2 880	134	1 440	-
Greece	3 599	45	163	25	90	-
Hungary	1 840	174	320	112	207	-
Iceland	31	27	1	17	1	-
Ireland	659	74	49	25	16	-
Italy	10 003	145	1 450	74	742	-
Latvia	2 923	174	509	93	272	-
Liechtenstein	7	254	2	119	1	-
Lithuania	1 994	183	366	99	197	-
Malta	n.s.	232	0	-	-	-
Netherlands	375	160	60	107	40	-
Norway	8 868	89	785	49	432	-
Poland	9 047	213	1 930	94	851	-
Portugal	3 666	82	299	33	120	-
Republic of Moldova	325	128	42	64	21	-
Romania	6 448	213	1 373	124	801	-
Russian Federation	851 392	105	89 136	56	47 423	-
San Marino	-	0	-	0	-	-
Slovakia	2 177	253	552	142	308	-
Slovenia	1 107	283	313	178	197	-
Spain	14 370	44	632	24	347	-
Sweden	27 134	107	2 914	63	1 722	NI
Switzerland	1 199	337	404	165	198	-
The FYR of Macedonia	906	70	63	-	-	-

Table 7. Volume and biomass in forest (cont.)

Country/area	Total forest 2000 000 ha	Volume By area m³/ha	Volume Total M m³	Biomass By area t/ha	Biomass Total M t	Information source NI / PI / ES / EX
Ukraine	9 584	179	1 719	-	-	-
United Kingdom	2 794	128	359	76	213	-
Yugoslavia	2 887	111	321	23	67	-
Total Europe	**1 039 251**	**112**	**116 448**	**59**	**61 070**	
Antigua and Barbuda	9	116	1	210	2	ES
Bahamas	842	-	-	-	-	-
Barbados	2	-	-	-	-	-
Belize	1 348	202	272	211	284	ES
Bermuda	-	-	-	-	-	-
British Virgin Islands	3	-	-	-	-	-
Canada	244 571	120	29 364	83	20 240	-
Cayman Islands	13	-	-	-	-	-
Costa Rica	1 968	211	414	220	433	ES
Cuba	2 348	71	167	114	268	NI
Dominica	46	91	4	166	8	ES
Dominican Republic	1 376	29	40	53	73	ES
El Salvador	121	223	27	202	24	-
Greenland	-	-	-	-	-	-
Grenada	5	83	0	150	1	PI
Guadeloupe	82	-	-	-	-	-
Guatemala	2 850	355	1 012	371	1 057	ES
Haiti	88	28	2	101	9	ES
Honduras	5 383	58	311	105	566	ES
Jamaica	325	82	27	171	56	ES
Martinique	47	5	0	5	0	ES
Mexico	55 205	52	2 871	54	2 981	NI
Montserrat	3	-	-	-	-	-
Netherlands Antilles	1	-	-	-	-	-
Nicaragua	3 278	154	506	161	528	ES
Panama	2 876	308	887	322	926	ES
Puerto Rico	229	-	-	-	-	-
Saint Kitts and Nevis	4	-	-	-	-	-
Saint Lucia	9	190	2	198	2	ES
Saint Pierre & Miquelon	-	-	-	-	-	-
Saint Vincent and Grenadines	6	166	1	173	1	NI
Trinidad and Tobago	259	71	18	129	33	ES
United States	225 993	136	30 838	108	24 428	-
US Virgin Islands	14	-	-	-	-	-
Total North and Central America	**549 304**	**123**	**67 329**	**95**	**52 357**	
Argentina	34 648	25	866	68	2 356	ES
Bolivia	53 068	114	6 050	183	9 711	PI
Brazil	543 905	131	71 252	209	113 676	ES
Chile	15 536	160	2 486	268	4 164	ES
Colombia	49 601	108	5 359	196	9 722	NI
Ecuador	10 557	121	1 275	151	1 594	ES
Falkland Islands	-	-	-	-	-	-
French Guiana	7 926	145	1 151	253	2 003	ES
Guyana	16 879	145	2 451	253	4 264	ES
Paraguay	23 372	34	792	59	1 379	ES
Peru	65 215	158	10 304	245	15 978	NI
Suriname	14 113	145	2 049	253	3 566	ES
Uruguay	1 292	-	-	-	-	-
Venezuela	49 506	134	6 629	233	11 535	ES
Total South America	**885 618**	**125**	**110 826**	**203**	**180 210**	
TOTAL WORLD	**3 869 455**	**100**	**386 352**	**109**	**422 256**	

Table 8. Forest fires 1990-2000

Country/area	Forest area 2000	Number of fires/year in 1990s			Area of fires/year in 1990s		
		Average	Min	Max	Average	Min	Max
	000 ha				*000 ha*	*000 ha*	*000 ha*
Algeria	2 145	-	-	-	-	-	-
Angola	69 756	-	-	-	-	-	-
Benin	2 650	-	-	-	-	-	-
Botswana	12 427	-	-	-	-	-	-
Burkina Faso	7 089	-	-	-	-	-	-
Burundi	94	-	-	-	-	-	-
Cameroon	23 858	-	-	-	-	-	-
Cape Verde	85	-	-	-	-	-	-
Central African Republic	22 907	-	-	-	-	-	-
Chad	12 692	-	-	-	-	-	-
Comoros	8	-	-	-	-	-	-
Congo	22 060	-	-	-	-	-	-
Côte d'Ivoire	7 117	-	-	-	-	-	-
Dem. Rep. of the Congo	135 207	-	-	-	-	-	-
Djibouti	6	-	-	-	-	-	-
Egypt	72	-	-	-	-	-	-
Equatorial Guinea	1 752	-	-	-	-	-	-
Eritrea	1 585	-	-	-	-	-	-
Ethiopia	4 593	-	-	-	-	-	-
Gabon	21 826	-	-	-	-	-	-
Gambia	481	-	-	-	-	-	-
Ghana	6 335	-	-	-	-	-	-
Guinea	6 929	-	-	-	-	-	-
Guinea-Bissau	2 187	-	-	-	-	-	-
Kenya	17 096	-	-	-	-	-	-
Lesotho	14	-	-	-	-	-	-
Liberia	3 481	-	-	-	-	-	-
Libyan Arab Jamahiriya	358	-	-	-	-	-	-
Madagascar	11 727	-	-	-	-	-	-
Malawi	2 562	-	-	-	-	-	-
Mali	13 186	-	-	-	-	-	-
Mauritania	317	-	-	-	-	-	-
Mauritius	16	-	-	-	-	-	-
Morocco	3 025	-	-	-	-	-	-
Mozambique	30 601	-	-	-	-	-	-
Namibia	8 040	-	-	-	-	-	-
Niger	1 328	-	-	-	-	-	-
Nigeria	13 517	-	-	-	-	-	-
Réunion	71	-	-	-	-	-	-
Rwanda	307	-	-	-	-	-	-
Saint Helena	2	-	-	-	-	-	-
Sao Tome and Principe	27	-	-	-	-	-	-
Senegal	6 205	-	-	-	-	-	-
Seychelles	30	-	-	-	-	-	-
Sierra Leone	1 055	-	-	-	-	-	-
Somalia	7 515	-	-	-	-	-	-
South Africa	8 917	-	-	-	-	-	-
Sudan	61 627	-	-	-	-	-	-
Swaziland	522	-	-	-	-	-	-
Togo	510	-	-	-	-	-	-
Tunisia	510	-	-	-	-	-	-
Uganda	4 190	-	-	-	-	-	-
United Republic of Tanzania	38 811	-	-	-	-	-	-
Western Sahara	152	-	-	-	-	-	-
Zambia	31 246	-	-	-	-	-	-
Zimbabwe	19 040	-	-	-	-	-	-
Total Africa	**649 866**						

Table 8. Forest fires 1990-2000 (cont.)

Country/area	Forest area 2000	Number of fires/year in 1990s			Area of fires/year in 1990s		
		Average	Min	Max	Average	Min	Max
	000 ha				000 ha	000 ha	000 ha
Afghanistan	1 351	-	-	-	-	-	-
Armenia	351	7	2	24	0.0	0.0	0.2
Azerbaijan	1 094	5	1	8	0.0	0.0	0.1
Bahrain	n.s.	-	-	-	-	-	-
Bangladesh	1 334	-	-	-	-	-	-
Bhutan	3 016	-	-	-	-	-	-
Brunei Darussalam	442	-	-	-	-	-	-
Cambodia	9 335	-	-	-	-	-	-
China	163 480	-	-	-	-	-	-
Cyprus	172	30	16	64	-	-	-
Dem People's Rep. of Korea	8 210	-	-	-	-	-	-
East Timor	507	-	-	-	-	-	-
Gaza Strip	-	-	-	-	-	-	-
Georgia	2 988	6	1	11	-	-	-
India	64 113	-	-	-	-	-	-
Indonesia	104 986	-	-	-	-	-	-
Iran, Islamic Rep.	7 299	-	-	-	-	-	-
Iraq	799	-	-	-	-	-	-
Israel	132	959	697	1 211	-	-	-
Japan	24 081	3 242	2 262	4 534	2.3	0.2	3.0
Jordan	86	-	-	-	-	-	-
Kazakhstan	12 148	1 017	354	2 257	1.8	0.2	4.3
Kuwait	5	-	-	-	-	-	-
Kyrgyzstan	1 003	-	-	-	-	-	-
Lao People's Dem. Rep	12 561	-	-	-	-	-	-
Lebanon	36	-	-	-	-	-	-
Malaysia	19 292	-	-	-	-	-	-
Maldives	1	-	-	-	-	-	-
Mongolia	10 645	-	-	-	-	-	-
Myanmar	34 419	-	-	-	-	-	-
Nepal	3 900	-	-	-	-	-	-
Oman	1	-	-	-	-	-	-
Pakistan	2 361	-	-	-	-	-	-
Philippines	5 789	-	-	-	-	-	-
Qatar	1	-	-	-	-	-	-
Republic of Korea	6 248	-	-	-	-	-	-
Saudi Arabia	1 504	-	-	-	-	-	-
Singapore	2	-	-	-	-	-	-
Sri Lanka	1 940	-	-	-	-	-	-
Syrian Arab Republic	461	-	-	-	-	-	-
Tajikistan	400	-	-	-	-	-	-
Thailand	14 762	-	-	-	-	-	-
Turkey	10 225	1 973	1 339	3 221	8.5	0.0	20.2
Turkmenistan	3 755	8	2	16	-	-	-
United Arab Emirates	321	-	-	-	-	-	-
Uzbekistan	1 969	-	-	-	-	-	-
Viet Nam	9 819	-	-	-	-	-	-
West Bank	-	-	-	-	-	-	-
Yemen	449	-	-	-	-	-	-
Total Asia	**547 793**						

Appendix 3. Global tables

Table 8. Forest fires 1990-2000 (cont.)

Country/area	Forest area 2000 000 ha	Number of fires/year in 1990s Average	Min	Max	Area of fires/year in 1990s Average 000 ha	Min 000 ha	Max 000 ha
American Samoa	12	-	-	-	-	-	-
Australia	154 539	-	-	-	-	-	-
Cook Islands	22	-	-	-	-	-	-
Fiji	815	-	-	-	-	-	-
French Polynesia	105	-	-	-	-	-	-
Guam	21	-	-	-	-	-	-
Kiribati	28	-	-	-	-	-	-
Marshall Islands	n.s.	-	-	-	-	-	-
Micronesia	15	-	-	-	-	-	-
Nauru	n.s.	-	-	-	-	-	-
New Caledonia	372	-	-	-	-	-	-
New Zealand	7 946	1 503	928	2 198	0.3	0.0	0.7
Niue	6	-	-	-	-	-	-
Northern Mariana Isl.	14	-	-	-	-	-	-
Palau	35	-	-	-	-	-	-
Papua New Guinea	30 601	-	-	-	-	-	-
Samoa	105	-	-	-	-	-	-
Solomon Islands	2 536	-	-	-	-	-	-
Tonga	4	-	-	-	-	-	-
Vanuatu	447	-	-	-	-	-	-
Total Oceania	**197 623**						
Albania	991	406	110	695	0.5	0.0	1.0
Andorra	-	-	-	-	-	-	-
Austria	3 886	114	41	225	0.1	0.0	0.2
Belarus	9 402	3 190	1 466	7 743	4.1	0.0	18.6
Belgium & Luxembourg	728	64	26	185	0.1	0.0	0.8
Bosnia & Herzegovina	2 273	139	104	158	0.8	0.0	1.2
Bulgaria	3 690	413	73	1 196	3.6	0.0	10.1
Croatia	1 783	259	109	372	3.9	0.0	7.0
Czech Republic	2 632	1 671	961	2 586	0.9	0.0	3.5
Denmark	455	7	2	14	0.0	0.0	0.1
Estonia	2 060	233	39	359	0.2	0.0	0.8
Finland	21 935	812	286	1 289	0.8	0.0	1.6
France	15 341	5 415	3 888	7 200	31.5	0.0	56.5
Germany	10 740	1 789	1 237	3 012	1.4	0.0	4.9
Greece	3 599	1 874	858	3 113	19.7	0.0	49.6
Hungary	1 840	-	-	-	-	-	-
Iceland	31	-	-	-	-	-	-
Ireland	659	222	123	721	0.2	0.0	0.3
Italy	10 003	11 470	6 225	15 380	20.5	0.0	44.0
Latvia	2 923	994	582	1 510	0.8	0.0	3.0
Liechtenstein	7	-	-	-	-	-	-
Lithuania	1 994	602	147	1 154	0.3	0.0	0.7
Malta	n.s.	6	1	12	0.0	0.0	0.0
Netherlands	375	81	51	117	0.0	0.0	0.0
Norway	8 868	513	181	976	0.5	0.0	1.4
Poland	9 047	4 792	3 008	9 305	7.0	0.0	33.3
Portugal	3 666	20 019	13 118	29 078	45.8	0.0	98.8
Republic of Moldova	325	22	0	91	0.0	0.0	0.1
Romania	6 448	102	34	187	0.3	0.0	0.7
Russian Federation	851 392	24 649	17 965	32 833	799.9	0.0	1 853.5
San Marino	-	-	-	-	-	-	-
Slovakia	2 177	413	142	674	0.1	0.0	0.2
Slovenia	1 107	81	25	211	0.4	0.0	1.0
Spain	14 370	17 497	12 474	25 827	68.3	0.0	250.4
Sweden	27 134	3 280	1 100	6 240	1.6	0.0	3.3
Switzerland	1 199	104	52	216	0.5	0.0	1.5
The FYR of Macedonia	906	123	18	294	3.4	0.0	10.1

Table 8. Forest fires 1990-2000 (cont.)

Country/area	Forest area 2000	Number of fires/year in 1990s			Area of fires/year in 1990s		
		Average	Min	Max	Average	Min	Max
	000 ha				000 ha	000 ha	000 ha
Ukraine	9 584	4 090	2 309	7 411	21.6	0.1	126.7
United Kingdom	2 794	427	61	906	0.4	0.1	1.0
Yugoslavia	2 887	175	26	313	2.9	1.5	6.9
Total Europe	**1 039 251**						
Antigua and Barbuda	9	-	-	-	-	-	-
Bahamas	842	-	-	-	-	-	-
Barbados	2	-	-	-	-	-	-
Belize	1 348	-	-	-	-	-	-
Bermuda	-	-	-	-	-	-	-
British Virgin Islands	3	-	-	-	-	-	-
Canada	244 571	8 125	5 681	10 267	501.6	143.0	1 239.0
Cayman Islands	13	-	-	-	-	-	-
Costa Rica	1 968	-	-	-	-	-	-
Cuba	2 348	-	-	-	-	-	-
Dominica	46	-	-	-	-	-	-
Dominican Republic	1 376	-	-	-	-	-	-
El Salvador	121	-	-	-	-	-	-
Greenland	-	-	-	-	-	-	-
Grenada	5	-	-	-	-	-	-
Guadeloupe	82	-	-	-	-	-	-
Guatemala	2 850	-	-	-	-	-	-
Haiti	88	-	-	-	-	-	-
Honduras	5 383	-	-	-	-	-	-
Jamaica	325	-	-	-	-	-	-
Martinique	47	-	-	-	-	-	-
Mexico	55 205	7 767	2 829	14 445	66.8	12.4	195.4
Montserrat	3	-	-	-	-	-	-
Netherlands Antilles	1	-	-	-	-	-	-
Nicaragua	3 278	-	-	-	-	-	-
Panama	2 876	-	-	-	-	-	-
Puerto Rico	229	-	-	-	-	-	-
Saint Kitts and Nevis	4	-	-	-	-	-	-
Saint Lucia	9	-	-	-	-	-	-
Saint Pierre & Miquelon	-	-	-	-	-	-	-
Saint Vincent and Grenadines	6	-	-	-	-	-	-
Trinidad and Tobago	259	328	156	764	0.4	0.3	7.2
United States	225 993	108 597	86 660	130 226	-	-	-
US Virgin Islands	14	-	-	-	-	-	-
Total North and Central America	**549 304**						
Argentina	34 648	4 787	343	10 587	465.2	98.4	1 279.0
Bolivia	53 068	-	-	-	-	-	-
Brazil	543 905	-	-	-	-	-	-
Chile	15 536	5 688	4 114	6 830	24.1	3.8	64.1
Colombia	49 601	-	-	-	-	-	-
Ecuador	10 557	-	-	-	-	-	-
Falkland Islands	-	-	-	-	-	-	-
French Guiana	7 926	-	-	-	-	-	-
Guyana	16 879	-	-	-	-	-	-
Paraguay	23 372	-	-	-	-	-	-
Peru	65 215	-	-	-	-	-	-
Suriname	14 113	-	-	-	-	-	-
Uruguay	1 292	-	-	-	-	-	-
Venezuela	49 506	-	-	-	-	-	-
Total South America	**885 618**						
TOTAL WORLD	**3 869 455**						

Table 9. Status and trends in forest management

Country/area	Forest area 2000 *000 ha*	Criteria & indicators for SFM	Area under forest management plans 2000 *000 ha*	2000 %	1990 *000 ha*	1990 %	1980 *000 ha*	1980 %	Forest area certified *000 ha*	Scheme
Algeria	2 145	NE	597	28	-	-	-	-	-	-
Angola	69 756	DZAf/ATO	-	-	-	-	-	-	-	-
Benin	2 650	-	-	-	-	-	-	-	-	-
Botswana	12 427	DZAf	-	-	-	-	-	-	-	-
Burkina Faso	7 089	DZAf	694	10	-	-	-	-	-	-
Burundi	94	-	-	-	-	-	-	-	-	-
Cameroon	23 858	ATO	-	-	-	-	-	-	-	-
Cape Verde	85	DZAf	-	-	-	-	-	-	-	-
Central African Republic	22 907	ATO	269*	n.ap.	-	-	-	-	-	-
Chad	12 692	DZAf	-	-	-	-	-	-	-	-
Comoros	8	-	-	-	-	-	-	-	-	-
Congo	22 060	ATO	-	-	-	-	-	-	-	-
Côte d'Ivoire	7 117	ATO	1 387	19	-	-	1	1	-	-
Dem. Rep. of the Congo	135 207	ATO	-	-	-	-	-	-	-	-
Djibouti	6	NE/DZAf	-	-	-	-	-	-	-	-
Egypt	72	NE	-	-	-	-	-	-	-	-
Equatorial Guinea	1 752	ATO	-	-	-	-	-	-	-	-
Eritrea	1 585	DZAf	-	-	-	-	-	-	-	-
Ethiopia	4 593	DZAf	112	2	-	-	-	-	-	-
Gabon	21 826	ATO	-	-	-	-	-	-	-	-
Gambia	481	DZAf	-	-	-	-	-	-	-	-
Ghana	6 335	ATO	-	-	-	-	1 167	13	-	-
Guinea	6 929	-	112*	n.ap.	-	-	-	-	-	-
Guinea-Bissau	2 187	DZAf	-	-	-	-	-	-	-	-
Kenya	17 096	DZAf	120*	n.ap.	-	-	70	3	-	-
Lesotho	14	DZAf	n.s.	2	-	-	-	-	-	-
Liberia	3 481	ATO	-	-	-	-	-	-	-	-
Libyan Arab Jamahiriya	358	NE	-	-	-	-	-	-	-	-
Madagascar	11 727	-	-	-	-	-	-	-	-	-
Malawi	2 562	DZAf	-	-	-	-	-	-	-	-
Mali	13 186	DZAf	-	-	-	-	-	-	-	-
Mauritania	317	NE/DZAf	-	-	-	-	-	-	-	-
Mauritius	16	DZAf	-	-	-	-	-	-	-	-
Morocco	3 025	NE	-	-	-	-	421	12	-	-
Mozambique	30 601	DZAf	-	-	-	-	-	-	-	-
Namibia	8 040	DZAf	54*	n.ap.	-	-	-	-	54	FSC
Niger	1 328	DZAf	-	-	-	-	-	-	-	-
Nigeria	13 517	ATO	832*	n.ap.	-	-	n.s.	n.s.	-	-
Réunion	71	-	-	-	-	-	2	2	-	-
Rwanda	307	-	-	-	-	-	-	-	-	-
Saint Helena	2	-	-	-	-	-	-	-	-	-
Sao Tome & Principe	27	ATO	-	-	-	-	-	-	-	-
Senegal	6 205	DZAf	-	-	-	-	n.s.	n.s.	-	-
Seychelles	30	DZAf	-	-	-	-	-	-	-	-
Sierra Leone	1 055	-	-	-	-	-	-	-	-	-
Somalia	7 515	NE/DZAf	-	-	-	-	-	-	-	-
South Africa	8 917	DZAf	828*	n.ap.	-	-	-	-	828	FSC
Sudan	61 627	NE/DZAf	-	-	-	-	50	n.s.	-	-
Swaziland	522	DZAf	-	-	-	-	-	-	-	-
Togo	510	ITTO	12	2	-	-	-	-	-	-
Tunisia	510	NE	400	78	-	-	163	38	-	-
Uganda	4 190	DZAf	-	-	-	-	440	7	-	-
United Republic of Tanzania	38 811	DZAf/ATO	-	-	-	-	n.s.	n.s.	-	-
Western Sahara	152	-	-	-	-	-	-	-	-	-
Zambia	31 246	DZAf	-	-	-	-	5	2	-	-
Zimbabwe	19 040	DZAf	92*	n.ap.	-	-	-	-	92	FSC
Total Africa	**649 866**								**974**	

*Partial results only. National figure not available.

Table 9. Status and trends in forest management (cont.)

| Country/area | Forest area 2000 | Criteria & indicators for SFM | Area under forest manangement plans |||||| Forest area certified ||
| | | | 2000 || 1990 || 1980 || | |
	000 ha		000 ha	%	000 ha	%	000 ha	%	000 ha	Scheme
Afghanistan	1 351	NE	-	-	-	-	100	8	-	-
Armenia	351	-	351	100	-	-	-	-	-	-
Azerbaijan	1 094	NE	1 094	100	-	-	-	-	-	-
Bahrain	n.s.	NE	-	-	-	-	-	-	-	-
Bangladesh	1 334	DFAs	1 334	100	-	-	795	75	-	-
Bhutan	3 016	DFAs	699	23	-	-	n.s.	n.s.	-	-
Brunei Darussalam	442	-	-	-	-	-	-	-	-	-
Cambodia	9 335	ITTO	-	-	-	-	-	-	-	-
China	163 480	MON/DFAs/ITTO	-	-	-	-	-	-	-	-
Cyprus	172	NE	172	100	153	100	-	-	-	-
Dem People's Rep. of Korea	8 210	-	-	-	-	-	-	-	-	-
East Timor	507	-	-	-	-	-	-	-	-	-
Gaza Strip	-	-	-	-	-	-	-	-	-	-
Georgia	2 988	EUR	2 438	82	-	-	-	-	-	-
India	64 113	DFAs/ITTO	46 159	72	-	-	31 917	54	-	-
Indonesia	104 986	ITTO	72*	n.ap.	-	-	40	n.s.	72	FSC
Iran, Islamic Rep.	7 299	NE	-	-	-	-	400	11	-	-
Iraq	799	NE	-	-	-	-	-	-	-	-
Israel	132	-	132	100	56	75	-	-	-	-
Japan	24 081	MON	24 081	100	-	-	-	-	3	FSC
Jordan	86	NE	-	-	-	-	-	-	-	-
Kazakhstan	12 148	-	12 148	100	-	-	-	-	-	-
Kuwait	5	NE	-	-	-	-	-	-	-	-
Kyrgyzstan	1 003	NE	1 003	100	-	-	-	-	-	-
Lao People's Dem. Rep	12 561	-	-	-	-	-	-	-	-	-
Lebanon	36	NE	-	-	-	-	-	-	-	-
Malaysia	19 292	ITTO	14 020	73	-	-	2 499	12	55	FSC
Maldives	1	-	-	-	-	-	-	-	-	-
Mongolia	10 645	DFAs	-	-	-	-	-	-	-	-
Myanmar	34 419	DFAs/ITTO	-	-	-	-	3 419	11	-	-
Nepal	3 900	DFAs	1 010	26	-	-	0	0	-	-
Oman	1	NE	-	-	-	-	0	0	-	-
Pakistan	2 361	NE	-	-	-	-	410	16	-	-
Philippines	5 789	ITTO	6 935	120	-	-	-	-	15	FSC
Qatar	1	NE	-	-	-	-	-	-	-	-
Republic of Korea	6 248	MON	4 096	66	-	-	-	-	-	-
Saudi Arabia	1 504	NE	-	-	-	-	-	-	-	-
Singapore	2	-	2	100	-	-	-	-	-	-
Sri Lanka	1 940	DFAs	1 940	100	-	-	-	-	13	FSC
Syrian Arab Republic	461	NE	-	-	-	-	60	32	-	-
Tajikistan	400	NE	400	100	-	-	-	-	-	-
Thailand	14 762	DFAs/ITTO	-	-	-	-	-	-	-	-
Turkey	10 225	NE/EUR	9 954	97	8 812	100	8 856	100	-	-
Turkmenistan	3 755	NE	3 755	100	-	-	-	-	-	-
United Arab Emirates	321	NE	-	-	-	-	-	-	-	-
Uzbekistan	1 969	-	1 969	100	-	-	-	-	-	-
Viet Nam	9 819	-	-	-	-	-	-	-	-	-
West Bank	-	-	-	-	-	-	-	-	-	-
Yemen	449	NE	-	-	-	-	-	-	-	-
Total Asia	**547 793**								**158**	

*Partial results only. National figure not available.

Table 9. Status and trends in forest management (cont.)

Country/area	Forest area 2000	Criteria & indicators for SFM	Area under forest mananagement plans						Forest area certified	
			2000		1990		1980			
	000 ha		*000 ha*	*%*	*000 ha*	*%*	*000 ha*	*%*	*000 ha*	Scheme
American Samoa	12	-	-	-	-	-	-	-	-	-
Australia	154 539	MON	154 539	100	-	-	-	-	-	-
Cook Islands	22	-	-	-	-	-	-	-	-	-
Fiji	815	ITTO	-	-	-	-	-	-	-	-
French Polynesia	105	-	-	-	-	-	-	-	-	-
Guam	21	-	-	-	-	-	-	-	-	-
Kiribati	28	-	-	-	-	-	-	-	-	-
Marshall Islands	n.s.	-	-	-	-	-	-	-	-	-
Micronesia	15	-	-	-	-	-	-	-	-	-
Nauru	n.s.	-	-	-	-	-	-	-	-	-
New Caledonia	372	-	-	-	-	-	-	-	-	-
New Zealand	7 946	MON	6 912	87	-	-	-	-	363	FSC
Niue	6	-	-	-	-	-	-	-	-	-
Northern Mariana Isl.	14	-	-	-	-	-	-	-	-	-
Palau	35	-	-	-	-	-	-	-	-	-
Papua New Guinea	30 601	ITTO	5 341	17	-	-	n.s.	n.s.	4	FSC
Samoa	105	-	-	-	-	-	-	-	-	-
Solomon Islands	2 536	-	43*	n.ap.	-	-	-	-	43	FSC
Tonga	4	-	-	-	-	-	-	-	-	-
Vanuatu	447	ITTO	-	-	-	-	-	-	-	-
Total Oceania	**197 623**								**410**	
Albania	991	EUR	406	41	-	-	1 046	100	-	-
Andorra	-	EUR	n.a.	n.a.	-	-	-	-	-	-
Austria	3 886	EUR	3 886	100	2 135	55	1 489	40	550	PEFC
Belarus	9 402	EUR	7 577	81	-	-	-	-	-	-
Belgium & Luxembourg	728	EUR & EUR	656	90	519	74	310	46	4	FSC
Bosnia & Herzegovina	2 273	EUR	2 007	88	-	-	-	-	-	-
Bulgaria	3 690	EUR	3 690	100	1 213	36	3 600	106	-	-
Croatia	1 783	EUR	1 531	86	-	-	-	-	167	FSC
Czech Republic	2 632	EUR	2 632	100	-	-	-	-	10	FSC
Denmark	455	EUR	455	100	381	82	330	71	n.s.	FSC
Estonia	2 060	EUR	1 125	55	-	-	-	-	-	-
Finland	21 935	EUR	21 900	100	16 392	82	10 578	53	21 900	PEFC
France	15 341	EUR	15 341	100	2 957	21	-	-	1	FSC
Germany	10 740	EUR	10 740	100	6 597	63	6 583	68	3 242	PEFC/FSC
Greece	3 599	EUR	2 009	56	980	39	1 603	64	-	-
Hungary	1 840	EUR	1 840	100	1 674	100	1 612	100	-	-
Iceland	31	EUR	13	42	-	-	-	-	-	-
Ireland	659	EUR	551	84	394	100	298	86	-	-
Italy	10 003	EUR	1 117	11	753	12	-	-	11	FSC
Latvia	2 923	EUR	2 923	100	-	-	-	-	-	-
Liechtenstein	7	EUR	7	100	-	-	-	-	-	-
Lithuania	1 994	EUR	1 938	97	-	-	-	-	-	-
Malta	n.s.	NE	n.s.	100	-	-	-	-	-	-
Netherlands	375	EUR	375	100	254	76	225	77	69	FSC
Norway	8 868	EUR	7 147	81	6 020	69	1 130	15	5 600	PEFC
Poland	9 047	EUR	9 047	100	8 261	95	8 099	94	2 743	FSC
Portugal	3 666	EUR	1 201	33	-	-	448	16	-	-
Republic of Moldova	325	EUR	325	100	-	-	-	-	-	-
Romania	6 448	EUR	6 448	100	6 190	100	5 940	100	-	-
Russian Federation	851 392	MON/EUR	851 392	100	-	-	-	-	33	FSC
San Marino	-	EUR	n.a.	n.a.	-	-	-	-	-	-
Slovakia	2 177	EUR	1 988	91	-	-	-	-	-	-
Slovenia	1 107	EUR	1 107	100	-	-	-	-	-	-
Spain	14 370	EUR	11 694	81	1 588	19	2 007	29	-	-
Sweden	27 134	EUR	27 134	100	16 650	68	14 301	59	11 167	FSC/PEFC
Switzerland	1 199	EUR	1 153	96	646	57	627	67	49	FSC
The FYR of Macedonia	906	-	906	100	-	-	-	-	-	-

*Partial results only. National figure not available.

Table 9. Status and trends in forest management (cont.)

Country/area	Forest area 2000	Criteria & indicators for SFM	Area under forest mananagement plans						Forest area certified	
			2000		1990		1980			
	000 ha		000 ha	%	000 ha	%	000 ha	%	000 ha	Scheme
Ukraine	9 584	EUR	9 584	100	-	-	-	-	203	FSC
United Kingdom	2 794	EUR	2 319	83	946	43	1 505	74	958	FSC
Yugoslavia	2 887	EUR	2 723	94	6 320	76	6 300	69	-	-
Total Europe	**1 039 251**								**46 708**	
Antigua and Barbuda	9	-	-	-	-	-	-	-	-	-
Bahamas	842	-	-	-	-	-	-	-	-	-
Barbados	2	-	-	-	-	-	-	-	-	-
Belize	1 348	LEP	1 000	74	-	-	-	-	96	FSC
Bermuda	-	-	-	-	-	-	-	-	-	-
British Virgin Islands	3	-	-	-	-	-	-	-	-	-
Canada	244 571	MON	173 400	71	-	-	148 087	60	4 360	FSC/CSA/SFI
Cayman Islands	13	-	-	-	-	-	-	-	-	-
Costa Rica	1 968	LEP	116*	n.ap.	-	-	-	-	41	FSC
Cuba	2 348	-	730	31	-	-	200	12	-	-
Dominica	46	-	-	-	-	-	-	-	-	-
Dominican Republic	1 376	-	152	11	-	-	-	-	-	-
El Salvador	121	LEP	-	-	-	-	-	-	-	-
Greenland	-	-	-	-	-	-	-	-	-	-
Grenada	5	-	-	-	-	-	-	-	-	-
Guadeloupe	82	-	28*	n.ap.	-	-	-	-	-	-
Guatemala	2 850	LEP	54	2	-	-	-	-	100	FSC
Haiti	88	-	-	-	-	-	-	-	-	-
Honduras	5 383	LEP	821	15	-	-	58	1	20	FSC
Jamaica	325	-	44	14	-	-	-	-	-	-
Martinique	47	-	10	21	-	-	-	-	-	-
Mexico	55 205	MON	7 100	13	-	-	-	-	169	FSC
Montserrat	3	-	-	-	-	-	-	-	-	-
Netherlands Antilles	1	-	-	-	-	-	-	-	-	-
Nicaragua	3 278	LEP	236	7	-	-	250	6	-	-
Panama	2 876	LEP	20*	n.ap.	-	-	-	-	1	FSC
Puerto Rico	229	-	57	25	-	-	-	-	-	-
Saint Kitts and Nevis	4	-	-	-	-	-	-	-	-	-
Saint Lucia	9	-	-	-	-	-	-	-	-	-
Saint Pierre & Miquelon	-	-	-	-	-	-	-	-	-	-
Saint Vincent and Grenadines	6	-	-	-	-	-	-	-	-	-
Trinidad and Tobago	259	ITTO	120	46	-	-	14	6	-	-
United States	225 993	MON	125 707	56	-	-	86 697	41	26 129	FSC/SFI/ATFP/GT
US Virgin Islands	14	-	-	-	-	-	-	-	-	-
Total North and Central America	**549 304**								**30 916**	
Argentina	34 648	MON	-	-	-	-	-	-	-	-
Bolivia	53 068	TARA	6 900	13	-	-	-	-	885	FSC
Brazil	543 905	TARA	4 000	1	-	-	n.s.	n.s.	666	FSC
Chile	15 536	MON	-	-	-	-	-	-	-	-
Colombia	49 601	TARA	85	n.s	-	-	-	-	-	-
Ecuador	10 557	TARA	14	n.s	-	-	-	-	-	-
Falkland Islands	-	-	-	-	-	-	-	-	-	-
French Guiana	7 926	-	400	5	-	-	-	-	-	-
Guyana	16 879	TARA	4 200	25	-	-	-	-	-	-
Paraguay	23 372	-	3 000	13	-	-	-	-	-	-
Peru	65 215	TARA	1 573	2	-	-	-	-	-	-
Suriname	14 113	TARA	1 568	11	-	-	-	-	-	-
Uruguay	1 292	MON	99	8	-	-	-	-	-	-
Venezuela	49 506	TARA	3 970	8	-	-	-	-	-	-
Total South America	**885 618**								**1 551**	
TOTAL WORLD	**3 869 455**								**80 717**	

*Partial results only. National figure not available.

Table 10. Removals

Country/area	Forest area 2000	Area under timber harvesting scheme				Volume harvested
		Total area 2000	Area actually harvested	Harvesting intensity, low	Harvesting intensity, high	
	000 ha	*000 ha*	*000 ha/year*	*m^3/ha*	*m^3/ha*	*000 m^3 o.b./year*
Algeria	2 145	-	-	-	-	-
Angola	69 756	245	12	-	-	-
Benin	2 650	-	-	-	-	-
Botswana	12 427	-	-	-	-	-
Burkina Faso	7 089	-	-	-	-	-
Burundi	94	-	-	-	-	-
Cameroon	23 858	4 054	338	6	8	-
Cape Verde	85	-	-	-	-	-
Central African Republic	22 907	1 762	65	4	9	-
Chad	12 692	-	-	-	-	-
Comoros	8	-	-	-	-	-
Congo	22 060	-	383	-	-	-
Côte d'Ivoire	7 117	-	604	-	-	-
Dem. Rep. of the Congo	135 207	-	166	-	-	-
Djibouti	6	-	-	-	-	-
Egypt	72	-	-	-	-	-
Equatorial Guinea	1 752	-	45	6	10	-
Eritrea	1 585	-	-	-	-	-
Ethiopia	4 593	-	-	-	-	-
Gabon	21 826	-	378	7	13	-
Gambia	481	-	-	-	-	-
Ghana	6 335	-	39	-	-	-
Guinea	6 929	-	21	-	-	-
Guinea-Bissau	2 187	-	-	-	-	-
Kenya	17 096	-	-	-	-	-
Lesotho	14	-	-	-	-	-
Liberia	3 481	-	32	-	-	-
Libyan Arab Jamahiriya	358	-	-	-	-	-
Madagascar	11 727	-	3	-	-	-
Malawi	2 562	-	-	-	-	-
Mali	13 186	-	-	-	-	-
Mauritania	317	-	-	-	-	-
Mauritius	16	-	-	-	-	-
Morocco	3 025	-	-	-	-	-
Mozambique	30 601	-	22	4	4	-
Namibia	8 040	-	-	-	-	-
Niger	1 328	-	-	-	-	-
Nigeria	13 517	-	1 035	4	12	-
Réunion	71	-	-	-	-	-
Rwanda	307	-	-	-	-	-
Saint Helena	2	-	-	-	-	-
Sao Tome and Principe	27	-	-	-	-	-
Senegal	6 205	-	-	-	-	-
Seychelles	30	-	-	-	-	-
Sierra Leone	1 055	-	-	-	-	-
Somalia	7 515	-	-	-	-	-
South Africa	8 917	-	-	-	-	-
Sudan	61 627	-	-	-	-	-
Swaziland	522	-	-	-	-	-
Togo	510	-	-	-	-	-
Tunisia	510	-	-	-	-	-
Uganda	4 190	-	2	-	-	-
United Republic of Tanzania	38 811	-	49	-	-	-
Western Sahara	152	-	-	-	-	-
Zambia	31 246	-	119	1	1	-
Zimbabwe	19 040	-	-	-	-	-
Total Africa	**649 866**					

Table 10. Removals (cont.)

Country/area	Forest area 2000	Area under timber harvesting scheme				Volume harvested
		Total area 2000	Area actually harvested	Harvesting intensity, low	Harvesting intensity, high	
	000 ha	*000 ha*	*000 ha/year*	*m³/ha*	*m³/ha*	*000 m³ o.b./year*
Afghanistan	1 351	-	-	-	-	-
Armenia	351	-	-	-	-	150
Azerbaijan	1 094	-	-	-	-	60
Bahrain	n.s.	-	-	-	-	-
Bangladesh	1 334	-	-	-	-	-
Bhutan	3 016	-	-	-	-	-
Brunei Darussalam	442	-	-	-	-	-
Cambodia	9 335	6 416	75	5	5	-
China	163 480	-	-	-	-	-
Cyprus	172	-	-	-	-	48
Dem People's Rep. of Korea	8 210	-	-	-	-	-
East Timor	507	-	-	-	-	-
Gaza Strip	-	-	-	-	-	-
Georgia	2 988	-	-	-	-	-
India	64 113	-	3 011	-	-	-
Indonesia	104 986	-	1 840	-	-	-
Iran, Islamic Rep.	7 299	-	-	-	-	-
Iraq	799	-	-	-	-	-
Israel	132	-	-	-	-	-
Japan	24 081	-	-	-	-	-
Jordan	86	-	-	-	-	-
Kazakhstan	12 148	-	-	-	-	1 400
Kuwait	5	-	-	-	-	-
Kyrgyzstan	1 003	-	-	-	-	-
Lao People's Dem. Rep	12 561	-	35	-	-	-
Lebanon	36	-	-	-	-	-
Malaysia	19 292	-	520	-	-	-
Maldives	1	-	-	-	-	-
Mongolia	10 645	-	-	-	-	-
Myanmar	34 419	17 852	411	5	8	-
Nepal	3 900	-	-	-	-	-
Oman	1	-	-	-	-	-
Pakistan	2 361	-	-	-	-	-
Philippines	5 789	-	31	6	23	-
Qatar	1	-	-	-	-	-
Republic of Korea	6 248	-	-	-	-	-
Saudi Arabia	1 504	-	-	-	-	-
Singapore	2	-	-	-	-	-
Sri Lanka	1 940	-	-	-	-	-
Syrian Arab Republic	461	-	-	-	-	-
Tajikistan	400	-	-	-	-	-
Thailand	14 762	1 081	15	-	-	-
Turkey	10 225	-	-	-	-	16 436
Turkmenistan	3 755	-	-	-	-	-
United Arab Emirates	321	-	-	-	-	-
Uzbekistan	1 969	-	-	-	-	-
Viet Nam	9 819	-	109	17	23	-
West Bank	-	-	-	-	-	-
Yemen	449	-	-	-	-	-
Total Asia	**547 793**					

Table 10. Removals (cont.)

Country/area	Forest area 2000	Area under timber harvesting scheme - Total area 2000	Area actually harvested	Harvesting intensity, low	Harvesting intensity, high	Volume harvested
	000 ha	*000 ha*	*000 ha/year*	*m³/ha*	*m³/ha*	*000 m³ o.b./year*
American Samoa	12	-	-	-	-	-
Australia	154 539	-	-	-	-	-
Cook Islands	22	-	-	-	-	-
Fiji	815	-	-	-	-	-
French Polynesia	105	-	-	-	-	-
Guam	21	-	-	-	-	-
Kiribati	28	-	-	-	-	-
Marshall Islands	n.s.	-	-	-	-	-
Micronesia	15	-	-	-	-	-
Nauru	n.s.	-	-	-	-	-
New Caledonia	372	-	-	-	-	-
New Zealand	7 946	-	-	-	-	19 770
Niue	6	-	-	-	-	-
Northern Mariana Isl.	14	-	-	-	-	-
Palau	35	-	-	-	-	-
Papua New Guinea	30 601	1 938	178	17	17	-
Samoa	105	-	-	-	-	-
Solomon Islands	2 536	-	-	-	-	-
Tonga	4	-	-	-	-	-
Vanuatu	447	-	-	-	-	-
Total Oceania	**197 623**					
Albania	991	-	-	-	-	692
Andorra	-	-	-	-	-	-
Austria	3 886	-	-	-	-	17 171
Belarus	9 402	-	-	-	-	9 550
Belgium & Luxembourg	728	-	-	-	-	-
Bosnia & Herzegovina	2 273	-	-	-	-	-
Bulgaria	3 690	-	-	-	-	3 887
Croatia	1 783	-	-	-	-	4 300
Czech Republic	2 632	-	-	-	-	13 140
Denmark	455	-	-	-	-	2 194
Estonia	2 060	-	-	-	-	-
Finland	21 935	-	-	-	-	49 500
France	15 341	-	-	-	-	47 611
Germany	10 740	-	-	-	-	38 867
Greece	3 599	-	-	-	-	2 408
Hungary	1 840	-	-	-	-	5 375
Iceland	31	-	-	-	-	0
Ireland	659	-	-	-	-	2 330
Italy	10 003	-	-	-	-	8 381
Latvia	2 923	-	-	-	-	6 710
Liechtenstein	7	-	-	-	-	14
Lithuania	1 994	-	-	-	-	4 740
Malta	n.s.	-	-	-	-	-
Netherlands	375	-	-	-	-	1 219
Norway	8 868	-	-	-	-	10 880
Poland	9 047	-	-	-	-	26 212
Portugal	3 666	-	-	-	-	11 400
Republic of Moldova	325	-	-	-	-	353
Romania	6 448	-	-	-	-	13 600
Russian Federation	851 392	-	-	-	-	116 200
San Marino	-	-	-	-	-	-
Slovakia	2 177	-	-	-	-	5 600
Slovenia	1 107	-	-	-	-	2 300
Spain	14 370	-	-	-	-	-
Sweden	27 134	-	-	-	-	61 593
Switzerland	1 199	-	-	-	-	6 408
The FYR of Macedonia	906	-	-	-	-	-

Table 10. Removals (cont.)

Country/area	Forest area 2000	Area under timber harvesting scheme				Volume harvested
		Total area 2000	Area actually harvested	Harvesting intensity, low	Harvesting intensity, high	
	000 ha	*000 ha*	*000 ha/year*	*m³/ha*	*m³/ha*	*000 m³ o.b./year*
Ukraine	9 584	-	-	-	-	-
United Kingdom	2 794	-	-	-	-	8 200
Yugoslavia	2 887	-	-	-	-	3 058
Total Europe	**1 039 251**					
Antigua and Barbuda	9	-	-	-	-	-
Bahamas	842	-	-	-	-	-
Barbados	2	-	-	-	-	-
Belize	1 348	-	-	-	-	-
Bermuda	-	-	-	-	-	-
British Virgin Islands	3	-	-	-	-	-
Canada	244 571	-	-	-	-	214 128
Cayman Islands	13	-	-	-	-	-
Costa Rica	1 968	-	-	-	-	-
Cuba	2 348	-	-	-	-	-
Dominica	46	-	-	-	-	-
Dominican Republic	1 376	-	-	-	-	-
El Salvador	121	-	-	-	-	-
Greenland	-	-	-	-	-	-
Grenada	5	-	-	-	-	-
Guadeloupe	82	-	-	-	-	-
Guatemala	2 850	-	27	-	-	-
Haiti	88	-	-	-	-	-
Honduras	5 383	42	3	13	13	-
Jamaica	325	-	-	-	-	-
Martinique	47	-	-	-	-	-
Mexico	55 205	-	28	-	-	-
Montserrat	3	-	-	-	-	-
Netherlands Antilles	1	-	-	-	-	-
Nicaragua	3 278	-	15	-	-	-
Panama	2 876	-	-	-	-	-
Puerto Rico	229	-	-	-	-	-
Saint Kitts and Nevis	4	-	-	-	-	-
Saint Lucia	9	-	-	-	-	-
Saint Pierre & Miquelon	-	-	-	-	-	-
Saint Vincent and Grenadines	6	-	-	-	-	-
Trinidad and Tobago	259	-	-	-	-	-
United States	225 993	-	-	-	-	452 000
US Virgin Islands	14	-	-	-	-	-
Total North and Central America	**549 304**					
Argentina	34 648	-	-	-	-	-
Bolivia	53 068	4 977	137	1	2	-
Brazil	543 905	1 768	53	12	34	-
Chile	15 536	-	-	-	-	-
Colombia	49 601	-	49	16	18	-
Ecuador	10 557	-	202	-	-	-
Falkland Islands	-	-	-	-	-	-
French Guiana	7 926	400	10	6	6	-
Guyana	16 879	3 703	160	4	9	-
Paraguay	23 372	-	427	-	-	-
Peru	65 215	2 014	661	2	2	-
Suriname	14 113	1 711	41	1	6	-
Uruguay	1 292	-	-	-	-	-
Venezuela	49 506	2 125	122	4	13	-
Total South America	**885 618**					
TOTAL WORLD	**3 869 455**					

Table 11. Comparison of forest management areas

Country/area	Forest area 2000	Area under forest management plans	Forest in protected areas - Country report	Forest in protected areas - Global maps	Removal areas - Under schemes	Removal areas - Actually harvested	Forest area certified
	000 ha	%	%	%	%	%	%
Algeria	2 145	28	-	6	-	-	-
Angola	69 756	-	-	3	0.4	0.02	-
Benin	2 650	-	-	32	-	-	-
Botswana	12 427	-	-	26	-	-	-
Burkina Faso	7 089	10	-	11	-	-	-
Burundi	94	-	-	29	-	-	-
Cameroon	23 858	-	-	11	17	1.4	-
Cape Verde	85	-	-	-	-	-	-
Central African Republic	22 907	n.ap.	-	15	8	0.3	-
Chad	12 692	-	-	27	-	-	-
Comoros	8	-	-	-	-	-	-
Congo	22 060	-	-	14	-	1.7	-
Côte d'Ivoire	7 117	19	-	10	-	8.5	-
Dem. Rep. of the Congo	135 207	-	-	9	-	0.1	-
Djibouti	6	-	-	0	-	-	-
Egypt	72	-	-	0	-	-	-
Equatorial Guinea	1 752	-	-	11	-	2.6	-
Eritrea	1 585	-	-	0	-	-	-
Ethiopia	4 593	2	-	15	-	-	-
Gabon	21 826	-	-	16	-	1.7	-
Gambia	481	-	-	3	-	-	-
Ghana	6 335	-	-	9	-	0.6	-
Guinea	6 929	n.ap.	-	5	-	0.3	-
Guinea-Bissau	2 187	-	-	1	-	-	-
Kenya	17 096	n.ap.	-	40	-	-	-
Lesotho	14	2	-	16	-	-	-
Liberia	3 481	-	-	1	-	0.9	-
Libyan Arab Jamahiriya	358	-	-	19	-	-	-
Madagascar	11 727	-	-	4	-	0.03	-
Malawi	2 562	-	-	45	-	-	-
Mali	13 186	-	-	7	-	-	-
Mauritania	317	-	-	3	-	-	-
Mauritius	16	-	-	-	-	-	-
Morocco	3 025	-	-	7	-	-	-
Mozambique	30 601	-	-	7	-	0.1	-
Namibia	8 040	n.ap.	-	5	-	-	0.7
Niger	1 328	-	-	77	-	-	-
Nigeria	13 517	n.ap.	-	7	-	7.7	-
Réunion	71	-	-	-	-	-	-
Rwanda	307	-	-	76	-	-	-
Saint Helena	2	-	-	-	-	-	-
Sao Tome and Principe	27	-	-	-	-	-	-
Senegal	6 205	-	-	16	-	-	-
Seychelles	30	-	-	-	-	-	-
Sierra Leone	1 055	-	-	5	-	-	-
Somalia	7 515	-	-	3	-	-	-
South Africa	8 917	n.ap.	-	7	-	-	9.3
Sudan	61 627	-	-	10	-	-	-
Swaziland	522	-	-	4	-	-	-
Togo	510	2	-	14	-	-	-
Tunisia	510	78	-	4	-	-	-
Uganda	4 190	-	-	18	-	0.04	-
United Republic of Tanzania	38 811	-	-	14	-	0.1	-
Western Sahara	152	-	-	0	-	-	-
Zambia	31 246	-	-	24	-	0.4	-
Zimbabwe	19 040	n.ap.	-	12	-	-	0.5
Total Africa	**649 866**						

Table 11. Comparison of forest management areas (cont.)

Country/area	Forest area 2000	Area under forest management plans	Forest in protected areas - Country report	Forest in protected areas - Global maps	Removal areas - Under schemes	Removal areas - Actually harvested	Forest area certified
	000 ha	%	%	%	%	%	%
Afghanistan	1 351	-	-	0	-	-	-
Armenia	351	100	31	5	-	-	-
Azerbaijan	1 094	100	100	7	-	-	-
Bahrain	n.s.	-	-	-	-	-	-
Bangladesh	1 334	100	-	14	-	-	-
Bhutan	3 016	23	-	25	-	-	-
Brunei Darussalam	442	-	-	22	-	-	-
Cambodia	9 335	-	-	24	69	0.8	-
China	163 480	-	-	3	-	-	-
Cyprus	172	100	100	37	-	-	-
Dem People's Rep. of Korea	8 210	-	-	3	-	-	-
East Timor	507	-	-	3	-	-	-
Gaza Strip	-	-	-	-	-	-	-
Georgia	2 988	82	4	3	-	-	-
India	64 113	72	-	8	-	4.7	-
Indonesia	104 986	n.ap.	-	16	-	1.8	0.1
Iran, Islamic Rep.	7 299	-	-	12	-	-	-
Iraq	799	-	-	0	-	-	-
Israel	132	100	-	63	-	-	-
Japan	24 081	100	7	8	-	-	0.01
Jordan	86	-	-	0	-	-	-
Kazakhstan	12 148	100	100	11	-	-	-
Kuwait	5	-	-	0	-	-	-
Kyrgyzstan	1 003	100	86	10	-	-	-
Lao People's Dem. Rep	12 561	-	-	20	-	0.3	-
Lebanon	36	-	-	0	-	-	-
Malaysia	19 292	73	-	9	-	2.7	0.3
Maldives	1	-	-	-	-	-	-
Mongolia	10 645	-	-	11	-	-	-
Myanmar	34 419	-	-	5	52	1.2	-
Nepal	3 900	26	-	9	-	-	-
Oman	1	-	-	0	-	-	-
Pakistan	2 361	-	-	3	-	-	-
Philippines	5 789	120	-	7	-	0.5	0.3
Qatar	1	-	-	0	-	-	-
Republic of Korea	6 248	66	-	4	-	-	-
Saudi Arabia	1 504	-	-	9	-	-	-
Singapore	2	100	-	-	-	-	-
Sri Lanka	1 940	100	-	18	-	-	0.7
Syrian Arab Republic	461	-	-	0	-	-	-
Tajikistan	400	100	100	1	-	-	-
Thailand	14 762	-	-	23	7	0.1	-
Turkey	10 225	97	2	2	-	-	-
Turkmenistan	3 755	100	3	13	-	-	-
United Arab Emirates	321	-	-	0	-	-	-
Uzbekistan	1 969	100	96	30	-	-	-
Viet Nam	9 819	-	-	6	-	1.1	-
West Bank	-	-	-	-	-	-	-
Yemen	449	-	-	0	-	-	-
Total Asia	**547 793**						

Table 11. Comparison of forest management areas (cont.)

Country/area	Forest area 2000	Area under forest management plans	Forest in protected areas - Country report	Forest in protected areas - Global maps	Removal areas - Under schemes	Removal areas - Actually harvested	Forest area certified
	000 ha	%	%	%	%	%	%
American Samoa	12	-	-	-	-	-	-
Australia	154 539	100	15	13	-	-	-
Cook Islands	22	-	-	-	-	-	-
Fiji	815	-	-	0	-	-	-
French Polynesia	105	-	-	-	-	-	-
Guam	21	-	-	-	-	-	-
Kiribati	28	-	-	-	-	-	-
Marshall Islands	n.s.	-	-	-	-	-	-
Micronesia	15	-	-	-	-	-	-
Nauru	n.s.	-	-	-	-	-	-
New Caledonia	372	-	-	2	-	-	-
New Zealand	7 946	87	21	3	-	-	4.6
Niue	6	-	-	-	-	-	-
Northern Mariana Isl.	14	-	-	-	-	-	-
Palau	35	-	-	-	-	-	-
Papua New Guinea	30 601	17	-	9	6	0.6	0.01
Samoa	105	-	-	-	-	-	-
Solomon Islands	2 536	n.ap.	-	0	-	-	1.7
Tonga	4	-	-	-	-	-	-
Vanuatu	447	-	-	0	-	-	-
Total Oceania	**197 623**						
Albania	991	41	14	2	-	-	-
Andorra	-	n.a.	-	-	-	-	-
Austria	3 886	100	20	22	-	-	14.2
Belarus	9 402	81	9	10	-	-	-
Belgium & Luxembourg	728	90	25	21	-	-	0.5
Bosnia & Herzegovina	2 273	88	-	1	-	-	-
Bulgaria	3 690	100	38	8	-	-	-
Croatia	1 783	86	23	8	-	-	9.4
Czech Republic	2 632	100	25	28	-	-	0.4
Denmark	455	100	21	9	-	-	n.s.
Estonia	2 060	55	9	21	-	-	-
Finland	21 935	100	11	7	-	-	99.8
France	15 341	100	18	17	-	-	0.01
Germany	10 740	100	67	29	-	-	30.2
Greece	3 599	56	29	4	-	-	-
Hungary	1 840	100	20	16	-	-	-
Iceland	31	42	7	7	-	-	-
Ireland	659	84	1	5	-	-	-
Italy	10 003	11	19	11	-	-	0.1
Latvia	2 923	100	16	15	-	-	-
Liechtenstein	7	100	22	-	-	-	-
Lithuania	1 994	97	15	10	-	-	-
Malta	n.s.	100	10	-	-	-	-
Netherlands	375	100	24	9	-	-	18.4
Norway	8 868	81	26	1	-	-	63.1
Poland	9 047	100	16	16	-	-	30.3
Portugal	3 666	33	17	8	-	-	-
Republic of Moldova	325	100	-	4	-	-	-
Romania	6 448	100	7	4	-	-	-
Russian Federation	851 392	100	3	3	-	-	n.s.
San Marino	-	n.a.	-	-	-	-	-
Slovakia	2 177	91	41	29	-	-	-
Slovenia	1 107	100	7	6	-	-	-
Spain	14 370	81	24	17	-	-	-
Sweden	27 134	100	-	8	-	-	41.2
Switzerland	1 199	96	4	12	-	-	4.1
The FYR of Macedonia	906	100	-	5	-	-	-

Table 11. Comparison of forest management areas (cont.)

Country/area	Forest area 2000	Area under forest management plans	Forest in protected areas - Country report	Forest in protected areas - Global maps	Removal areas - Under schemes	Removal areas - Actually harvested	Forest area certified
	000 ha	%	%	%	%	%	%
Ukraine	9 584	100	10	6	-	-	2.1
United Kingdom	2 794	83	32	23	-	-	34.3
Yugoslavia	2 887	94	100	6	-	-	-
Total Europe	**1 039 251**						
Antigua and Barbuda	9	-	-	-	-	-	-
Bahamas	842	-	-	4	-	-	-
Barbados	2	-	-	-	-	-	-
Belize	1 348	74	-	37	-	-	7.1
Bermuda	-	-	-	-	-	-	-
British Virgin Islands	3	-	-	-	-	-	-
Canada	244 571	71	8	5	-	-	1.8
Cayman Islands	13	-	-	-	-	-	-
Costa Rica	1 968	n.ap.	-	36	-	-	2.1
Cuba	2 348	31	-	25	-	-	-
Dominica	46	-	-	-	-	-	-
Dominican Republic	1 376	11	-	15	-	-	-
El Salvador	121	-	-	1	-	-	-
Greenland	-	-	-	-	-	-	-
Grenada	5	-	-	-	-	-	-
Guadeloupe	82	n.ap.	-	-	-	-	-
Guatemala	2 850	2	-	35	-	1.0	3.5
Haiti	88	-	-	1	-	-	-
Honduras	5 383	15	-	5	1	0.05	0.4
Jamaica	325	14	-	11	-	-	-
Martinique	47	21	-	-	-	-	-
Mexico	55 205	13	-	4	-	0.1	0.3
Montserrat	3	-	-	-	-	-	-
Netherlands Antilles	1	-	-	-	-	-	-
Nicaragua	3 278	7	-	23	-	0.5	-
Panama	2 876	n.ap.	-	35	-	-	0.03
Puerto Rico	229	25	-	5	-	-	-
Saint Kitts and Nevis	4	-	-	-	-	-	-
Saint Lucia	9	-	-	-	-	-	-
Saint Pierre & Miquelon	-	-	-	0	-	-	-
Saint Vincent and Grenadines	6	-	-	-	-	-	-
Trinidad and Tobago	259	46	-	-	-	-	-
United States	225 993	56	30	40	-	-	11.6
US Virgin Islands	14	-	-	-	-	-	-
Total North and Central America	**549 304**						
Argentina	34 648	-	-	7	-	-	-
Bolivia	53 068	13	-	31	9	0.3	1.7
Brazil	543 905	1	-	17	0.3	0.01	0.1
Chile	15 536	-	-	14	-	-	-
Colombia	49 601	0.2	-	24	-	0.1	-
Ecuador	10 557	0.1	-	20	-	1.9	-
Falkland Islands	-	-	-	-	-	-	-
French Guiana	7 926	5	-	7	5	0.1	-
Guyana	16 879	25	-	1	22	0.9	-
Paraguay	23 372	13	-	5	-	1.8	-
Peru	65 215	2	-	10	3	1.0	-
Suriname	14 113	11	-	4	12	0.3	-
Uruguay	1 292	8	-	5	-	-	-
Venezuela	49 506	8	-	66	4	0.2	-
Total South America	**885 618**						
TOTAL WORLD	**3 869 455**						

Table 12. Non-wood forest products – major product groups

Country/area	Plant products										Animal products							
	Food	Fodder	Medicines	Medicinal plants	Perfumes, cosmetics	Dying and tanning	Utensils, handicrafts, construction materials	Ornamental	Exudates	Other	Living animals	Honey, beeswax	Bushmeat	Other edible	Hides, skins	Medicines	Colorants	Other non-edible
Africa																		
Algeria							x											
Angola	x			x														
Benin	x			x			x					x	x					
Botswana	x			x									x					
Burkina Faso	x												x					
Burundi				x							x		x					
Cameroon	x	x					x						x					
Cape Verde	-	-	-	-	-	-	-	-	-	-	-	-	-	-	-	-	-	-
Central African Rep.	x	x											x					
Chad	x	x							x									
Comoros	x							x			x	x						
Congo	x			x			x	x				x	x					
Côte d'Ivoire	x						x											
Dem. Rep. Congo	x												x					
Djibouti		x																
Egypt	x			x	x							x						
Equatorial Guinea	x			x			x						x					
Eritrea							x		x									
Ethiopia				x					x			x						
Gabon	x						x						x					
Gambia	x																	
Ghana	x			x			x						x					
Guinea	x			x			x						x					
Guinea-Bissau	x	x																
Kenya		x		x		x			x									
Lesotho	x	x		x								x	x					
Liberia	x												x					
Libyan Arab Jamahiriya	-	-	-	-	-	-	-	-	-	-	-	-	-	-	-	-	-	-
Madagascar	x			x				x			x							
Malawi	x											x	x	x				
Mali	x	x							x			x						
Mauritania	x	x		x					x									
Mauritius	x	x		x				x				x	x					
Morocco	x			x				x										
Mozambique	x			x								x	x					
Namibia	x	x		x								x	x					
Niger	x	x		x					x									
Nigeria	x												x					
Réunion	-	-	-	-	-	-	-	-	-	-	-	-	-	-	-	-	-	-
Rwanda	x			x							x	x						
Saint Helena	-	-	-	-	-	-	-	-	-	-	-	-	-	-	-	-	-	-
Sao Tome and Principe				x														
Senegal	x	x							x		x							
Seychelles	x																	
Sierra Leone	-	-	-	-	-	-	-	-	-	-	-	-	-	-	-	-	-	-
Somalia									x									
South Africa	x	x		x				x					x					
Sudan	x	x		x		x			x			x	x					
Swaziland	x	x		x								x						
Togo	x	x		x		x												
Tunisia	x	x					x											
Uganda												x	x					
United Republic of Tanzania		x		x							x	x						
Western Sahara	-	-	-	-	-	-	-	-	-	-	-	-	-	-	-	-	-	-
Zambia	x	x		x			x					x		x				
Zimbabwe	x	x		x									x	x				

Table 12. Non-wood forest products – major product groups (cont.)

Country/area	Plant products									Animal products						
	Food	Fodder	Medicines	Medicinal plants	Perfumes, cosmetics	Dying and tanning	Utensils, handicrafts, construction materials	Ornamental	Exudates	Other	Living animals	Honey, beeswax	Bushmeat	Other edible	Hides, skins	Medicines	Colorants	Other non-edible
Asia																		
Afghanistan	x			x														
Armenia	x											x						
Azerbaijan				x														
Bahrain	-	-	-	-	-	-	-	-	-	-	-	-	-	-	-	-	-	-
Bangladesh	-	-	-	-	-	-	-	-	-	-	-	-	-	-	-	-	-	-
Bhutan	-	-	-	-	-	-	-	-	-	-	-	-	-	-	-	-	-	-
Brunei Darussalam	-	-	-	-	-	-	-	-	-	-	-	-	-	-	-	-	-	-
Cambodia	-	-	-	-	-	-	-	-	-	-	-	-	-	-	-	-	-	-
China	-	-	-	-	-	-	-	-	-	-	-	-	-	-	-	-	-	-
Cyprus	x											x						
Dem People's Rep. of Korea	-	-	-	-	-	-	-	-	-	-	-	-	-	-	-	-	-	-
East Timor	-	-	-	-	-	-	-	-	-	-	-	-	-	-	-	-	-	-
Gaza Strip	-	-	-	-	-	-	-	-	-	-	-	-	-	-	-	-	-	-
Georgia	-	-	-	-	-	-	-	-	-	-	-	-	-	-	-	-	-	-
India	-	-	-	-	-	-	-	-	-	-	-	-	-	-	-	-	-	-
Indonesia	-	-	-	-	-	-	-	-	-	-	-	-	-	-	-	-	-	-
Iran, Islamic Rep.	x			x	x	x			x			x						
Iraq	-	-	-	-	-	-	-	-	-	-	-	-	-	-	-	-	-	-
Israel	-	-	-	-	-	-	-	-	-	-	-	-	-	-	-	-	-	-
Japan	-	-	-	-	-	-	-	-	-	-	-	-	-	-	-	-	-	-
Jordan	x			x			x					x						
Kazakhstan	x			x														
Kuwait	-	-	-	-	-	-	-	-	-	-	-	-	-	-	-	-	-	-
Kyrgyzstan	-	-	-	-	-	-	-	-	-	-	-	-	-	-	-	-	-	-
Lao People's Dem. Rep	-	-	-	-	-	-	-	-	-	-	-	-	-	-	-	-	-	-
Lebanon	x	x										x						
Malaysia	-	-	-	-	-	-	-	-	-	-	-	-	-	-	-	-	-	-
Maldives	-	-	-	-	-	-	-	-	-	-	-	-	-	-	-	-	-	-
Mongolia	-	-	-	-	-	-	-	-	-	-	-	-	-	-	-	-	-	-
Myanmar	-	-	-	-	-	-	-	-	-	-	-	-	-	-	-	-	-	-
Nepal	-	-	-	-	-	-	-	-	-	-	-	-	-	-	-	-	-	-
Oman				x		x	x											
Pakistan	x			x	x	x	x											
Philippines	-	-	-	-	-	-	-	-	-	-	-	-	-	-	-	-	-	-
Qatar	-	-	-	-	-	-	-	-	-	-	-	-	-	-	-	-	-	-
Republic of Korea	-	-	-	-	-	-	-	-	-	-	-	-	-	-	-	-	-	-
Saudi Arabia		x																
Singapore	-	-	-	-	-	-	-	-	-	-	-	-	-	-	-	-	-	-
Sri Lanka	-	-	-	-	-	-	-	-	-	-	-	-	-	-	-	-	-	-
Syrian Arab Republic	x	x		x		x		x				x	x					
Tajikistan	-	-	-	-	-	-	-	-	-	-	-	-	-	-	-	-	-	-
Thailand	-	-	-	-	-	-	-	-	-	-	-	-	-	-	-	-	-	-
Turkey	x	x		x				x	x			x						
Turkmenistan	-	-	-	-	-	-	-	-	-	-	-	-	-	-	-	-	-	-
United Arab Emirates	-	-	-	-	-	-	-	-	-	-	-	-	-	-	-	-	-	-
Uzbekistan	-	-	-	-	-	-	-	-	-	-	-	-	-	-	-	-	-	-
Viet Nam	-	-	-	-	-	-	-	-	-	-	-	-	-	-	-	-	-	-
West Bank	-	-	-	-	-	-	-	-	-	-	-	-	-	-	-	-	-	-
Yemen	x	x										x						

Appendix 3. Global tables 425

Table 12. Non-wood forest products – major product groups (cont.)

Country/area	Food	Fodder	Medicines	Medicinal plants	Perfumes, cosmetics	Dying and tanning	Utensils, handicrafts, construction materials	Ornamental	Exudates	Other	Living animals	Honey, beeswax	Bushmeat	Other edible	Hides, skins	Medicines	Colorants	Other non-edible
Oceania																		
American Samoa	-	-	-	-	-	-	-	-	-	-	-	-	-	-	-	-	-	-
Australia	-	-	-	-	-	-	-	-	-	-	-	-	-	-	-	-	-	-
Cook Islands	-	-	-	-	-	-	-	-	-	-	-	-	-	-	-	-	-	-
Fiji	-	-	-	-	-	-	-	-	-	-	-	-	-	-	-	-	-	-
French Polynesia	-	-	-	-	-	-	-	-	-	-	-	-	-	-	-	-	-	-
Guam	-	-	-	-	-	-	-	-	-	-	-	-	-	-	-	-	-	-
Kiribati	-	-	-	-	-	-	-	-	-	-	-	-	-	-	-	-	-	-
Marshall Islands	-	-	-	-	-	-	-	-	-	-	-	-	-	-	-	-	-	-
Micronesia	-	-	-	-	-	-	-	-	-	-	-	-	-	-	-	-	-	-
Nauru	-	-	-	-	-	-	-	-	-	-	-	-	-	-	-	-	-	-
New Caledonia	-	-	-	-	-	-	-	-	-	-	-	-	-	-	-	-	-	-
New Zealand	-	-	-	-	-	-	-	-	-	-	-	-	-	-	-	-	-	-
Niue	-	-	-	-	-	-	-	-	-	-	-	-	-	-	-	-	-	-
Northern Mariana Isl.	-	-	-	-	-	-	-	-	-	-	-	-	-	-	-	-	-	-
Palau	-	-	-	-	-	-	-	-	-	-	-	-	-	-	-	-	-	-
Papua New Guinea	-	-	-	-	-	-	-	-	-	-	-	-	-	-	-	-	-	-
Samoa	-	-	-	-	-	-	-	-	-	-	-	-	-	-	-	-	-	-
Solomon Islands	-	-	-	-	-	-	-	-	-	-	-	-	-	-	-	-	-	-
Tonga	-	-	-	-	-	-	-	-	-	-	-	-	-	-	-	-	-	-
Vanuatu	-	-	-	-	-	-	-	-	-	-	-	-	-	-	-	-	-	-
Europe																		
Albania	x			x				x	x									
Andorra	-	-	-	-	-	-	-					-			-			-
Austria	-	-	-	-	-	-	-					-			-			-
Belarus	x			x								x	x					
Belgium & Luxembourg													x					
Bosnia & Herzegovina	-	-	-	-	-	-	-	-	-	-	-	-	-	-	-	-	-	-
Bulgaria	-	-	-	-	-	-	-	-	-	-	-	-	-	-	-	-	-	-
Croatia	-	-	-	-	-	-	-	-	-	-	-	-	-	-	-	-	-	-
Czech Republic	x												x		x			
Denmark								x										
Estonia	x							x					x		x			
Finland	x			x				x					x					
France	x							x				x						
Germany								x				x						
Greece	-	-	-	-	-	-	-	-	-	-	-	-	-	-	-	-	-	-
Hungary												x	x					
Iceland								x										
Ireland								x										
Italy	x						x											
Latvia	-	-	-	-	-	-	-	-	-	-	-	-	-	-	-	-	-	-
Liechtenstein	-	-	-	-	-	-	-	-	-	-	-	-	-	-	-	-	-	-
Lithuania	x			x				x					x		x			
Malta	-	-	-	-	-	-	-	-	-	-	-	-	-	-	-	-	-	-
Netherlands								x					x					
Norway	x											x	x					
Poland	x												x					
Portugal	x						x						x					
Republic of Moldova	x			x								x						
Romania	-	-	-	-	-	-	-	-	-	-	-	-	-	-	-	-	-	-
Russian Federation	x			x								x	x		x			
San Marino	-	-	-	-	-	-	-	-	-	-	-	-	-	-	-	-	-	-
Slovakia	x			x				x				x	x					
Slovenia	x							x				x	x					
Spain	-	-	-	-	-	-	-	-	-	-	-	-	-	-	-	-	-	-
Sweden	x							x					x		x			
Switzerland	x			x				x				x	x		x			
The FYR of Macedonia	-	-	-	-	-	-	-	-	-	-	-	-	-	-	-	-	-	-

Table 12. Non-wood forest products – major product groups (cont.)

Country/area	Plant products									Animal products								
	Food	Fodder	Medicines	Medicinal plants	Perfumes, cosmetics	Dying and tanning	Utensils, handicrafts, construction materials	Ornamental	Exudates	Other	Living animals	Honey, beeswax	Bushmeat	Other edible	Hides, skins	Medicines	Colorants	Other non-edible
---	---	---	---	---	---	---	---	---	---	---	---	---	---	---	---	---	---	---
Ukraine								x										
United Kingdom								x										
Yugoslavia	x			x								x	x					
North and Central America																		
Antigua and Barbuda	-	-	-	-	-	-	-	-	-	-	-	-	-	-	-	-	-	-
Bahamas	-	-	-	-	-	-	-	-	-	-	-	-	-	-	-	-	-	-
Barbados	-	-	-	-	-	-	-	-	-	-	-	-	-	-	-	-	-	-
Belize	x							x										
Bermuda	-	-	-	-	-	-	-	-	-	-	-	-	-	-	-	-	-	-
British Virgin Islands	-	-	-	-	-	-	-	-	-	-	-	-	-	-	-	-	-	-
Canada								x							x			
Cayman Islands	-	-	-	-	-	-	-	-	-	-	-	-	-	-	-	-	-	-
Costa Rica				x		x		x	x									
Cuba	x	x			x	x	x		x			x						
Dominica				x			x											
Dominican Republic				x								x						
El Salvador	x							x				x						
Greenland	-	-	-	-	-	-	-	-	-	-	-	-	-	-	-	-	-	-
Grenada		x					x						x					
Guadeloupe	-	-	-	-	-	-	-	-	-	-	-	-	-	-	-	-	-	-
Guatemala	x			x	x			x	x									
Haiti	x			x								x	x					
Honduras	x			x			x		x	x								
Jamaica	-	-	-	-	-	-	-	-	-	-	-	-	-	-	-	-	-	-
Martinique	-	-	-	-	-	-	-	-	-	-	-	-	-	-	-	-	-	-
Mexico	x			x	x	x	x	x	x			x						
Montserrat	-	-	-	-	-	-	-	-	-	-	-	-	-	-	-	-	-	-
Netherlands Antilles	-	-	-	-	-	-	-	-	-	-	-	-	-	-	-	-	-	-
Nicaragua							x					x			x			
Panama							x											
Puerto Rico	-	-	-	-	-	-	-	-	-	-	-	-	-	-	-	-	-	-
Saint Kitts and Nevis							x											
Saint Lucia	-	-	-	-	-	-	-	-	-	-	-	-	-	-	-	-	-	-
Saint Pierre & Miquelon	-	-	-	-	-	-	-	-	-	-	-	-	-	-	-	-	-	-
Saint Vincent and Grenadines							x											
Trinidad and Tobago	x						x					x	x					
United States	x							x					x					
US Virgin Islands	-	-	-	-	-	-	-	-	-	-	-	-	-	-	-	-	-	-
South America																		
Argentina	x				x	x	x	x	x			x						
Bolivia	x		x	x				x									x	
Brazil	x		x	x	x		x	x				x						
Chile	x		x	x			x	x	x			x	x					
Colombia	x						x	x				x						
Ecuador	x						x	x	x		x							
Falkland Islands	-	-	-	-	-	-	-	-	-	-	-	-	-	-	-	-	-	-
French Guiana	-	-	-	-	-	-	-	-	-	-	-	-	-	-	-	-	-	-
Guyana	x		x		x	x			x			x		x				
Paraguay	x		x	x	x							x						
Peru	x	x	x			x		x				x			x		x	
Suriname	x		x	x	x	x	x		x			x	x					
Uruguay												x						
Venezuela	x						x		x			x						

Table 13. Endangered, endemic species for seven species groups

Country/area	All 7 species groups - Total species	All 7 species groups - Endangered species	Country-endemic endangered species - Total, all 7 species groups	Amphibians	Birds	Ferns	Mammals	Palms	Reptiles	Trees	Total
Africa											
Algeria	523	27	4		1						1
Angola	1387	55	10		7						7
Benin	807	24									
Botswana	933	12									
Burkina Faso	659	10									
Burundi	771	12									
Cameroon	1435	158	28		7	5	2			8	22
Cape Verde	186	9	2		1						1
Central African Rep.	1069	23	2								
Chad	777	20	1								
Comoros	250	18	12		4		1	3		3	11
Congo	1009	52	1							1	1
Côte d'Ivoire	1064	140	17							15	15
Dem. Rep. Congo	1435	125	43		6		2			18	26
Djibouti	475	11	1		1						1
Egypt	639	35	3								
Equatorial Guinea	737	39	3		3	4					7
Eritrea	739	16	2							1	1
Ethiopia	898	80	39		1		4			6	11
Gabon	971	89	24				2			17	19
Gambia	629	9									
Ghana	1073	145	5							5	5
Guinea	803	49	2			1				2	3
Guinea-Bissau	479	12									
Kenya	2283	200	66		5		2			41	48
Lesotho	371	7									
Liberia	1019	76	3		1					1	2
Libyan Arab Jamahiriya	422	16	2								
Madagascar	1579	441	381	16	30	33	149	5		144	377
Malawi	1201	39	6							6	6
Mali	872	26	1							1	1
Mauritania	657	20	1								
Mauritius	306	95	73	9	2	2	3			28	44
Morocco	571	38	5							1	1
Mozambique	1054	78	11		1		1			8	10
Namibia	839	27	2				1				1
Niger	635	16									
Nigeria	1157	151	33		1		1			23	25
Réunion	271	29	6		1				1	2	4
Rwanda	898	18									
Saint Helena	960	48	22			15				6	21
Sao Tome&d Principe	275	61	35		8	8	1			13	30
Senegal	795	33									
Seychelles	359	77	68		7	1	1	6		21	36
Sierra Leone	908	68									
Somalia	1087	48	21		2					2	4
South Africa	1705	124	57	2		3	4	1	1	9	20
Sudan	1334	52	6								
Swaziland	829	15									
Togo	950	21									
Tunisia	481	20									
Uganda	1930	63	7				1			3	4
United Republic of Tanzania	2053	390	230	10	3	3	1			191	208
Western Sahara	220	10			-						
Zambia	1192	32	6		1		1				2
Zimbabwe	1420	37	4							4	4

Table 13. Endangered, endemic and forest occuring for seven species groups (cont.)

Country/area	All 7 species groups		Country-endemic endangered species								
	Total species	Endangered species	Total, all 7 species groups	\multicolumn{8}{c|}{Forest occuring by species group}
				Am-phibians	Birds	Ferns	Mam-mals	Palms	Reptiles	Trees	Total
Asia											
Afghanistan	723	27	1								
Armenia	92	12	1								
Azerbaijan	92	22									
Bahrain	337	2					-				
Bangladesh	1074	75	2					1			1
Bhutan	832	39									
Brunei Darussalam	623	119	14					1		10	11
Cambodia	1071	80	2							2	2
China	4310	402	137		16	10	8	4	1	69	108
Cyprus	415	11	2								
Dem People's Rep. of Korea	579	29	1							1	1
East Timor	-	-	-	-	-	-	-	-	-	-	-
Gaza Strip	-	-	-	-	-	-	-	-	-	-	-
Georgia	132	22									
India	3008	494	210	3	9	6	9	18		128	173
Indonesia	5952	762	370		55	38	50	27		125	295
Iran, Islamic Rep.	1006	45	8								
Iraq	644	21	1								
Israel	610	26	3								
Japan	1351	353	41		5	53	6	2		-	66
Jordan	443	12									
Kazakhstan	453	32	1								
Kuwait	379	6									
Kyrgyzstan	66	13	2								
Lao People's Dem. Rep	1325	85	3					1		1	2
Lebanon	391	12									
Malaysia	3121	966	534		1	5	10	107		358	481
Maldives	128	3					-				
Mongolia	623	26									
Myanmar	2117	139	9		2					4	6
Nepal	1263	63	1								
Oman	560	25	4								
Pakistan	1104	47	3								
Philippines	2097	447	317		70	36	37	40	1	131	315
Qatar	283	3					-				
Republic of Korea	551	25								-	
Saudi Arabia	599	25									
Singapore	567	87	3							5	5
Sri Lanka	1018	390	313		5	44	4	10	2	277	342
Syrian Arab Republic	434	14	1								
Tajikistan	85	16	1								
Thailand	2293	195	17					1	2	8	11
Turkey	737	51	8								
Turkmenistan	117	25									
United Arab Emirates	422	9	1				-				
Uzbekistan	73										
Viet Nam	1999	258	85	1	7	8	4	17		32	69
West Bank	-	-	-	-	-	-	-	-	-	-	-
Yemen	497	79	50							35	35

Appendix 3. Global tables

Table 13. Endangered, endemic and forest occuring for seven species groups (cont.)

Country/area	Total species	Endangered species	Total, all 7 species groups	Amphibians	Birds	Ferns	Mammals	Palms	Reptiles	Trees	Total
Oceania											
American Samoa	255	7									
Australia	2439	300	184	17	19	48	17	26	11	17	155
Cook Islands	136	9	4		3						3
Fiji	455	110	95		6	3	1	19		54	83
French Polynesia	193	87	73		14	1		1		9	25
Guam	162	10									
Kiribati	69	6					-				
Marshall Islands	82	3					-				
Micronesia	135	20	11		5		2	1		2	10
Nauru	22	2	1				-				
New Caledonia	394	268	257		7	3	1	32		57	100
New Zealand	553	90	55		10	7	1	2		13	33
Niue	42	3									
Northern Mariana Isl.	132	14	4		3					1	4
Palau	161	10	1			-				1	1
Papua New Guinea	3372	323	166		11	45	20	2		103	181
Samoa	10	7	6	-	-	-	-	6	-		6
Solomon Islands	798	77	41		6	3	6	13		6	34
Tonga	141	7	2		1					1	2
Vanuatu	269	31	21		5			9		5	19
Europe											
Albania	467	10					-				
Andorra							-				
Austria	593	14									
Belarus	68	8					-				
Belgium & Luxemb.	977	13									
Bosnia & Herzegovina	78	16	1							1	1
Bulgaria	575	26									
Croatia	81	18	1							1	1
Czech Republic	66	15									
Denmark	563	6									
Estonia	486	6									
Finland	555	9									
France	780	37	1								
Germany	693	21	8							6	6
Greece	624	35	6							1	1
Hungary	519	19									
Iceland	352	1									
Ireland	502	6									
Italy	781	30	5	1							1
Latvia	473	10									
Liechtenstein	361	1									
Lithuania	448	9									
Malta	443	2									
Netherlands	599	9									
Norway	585	8									
Poland	595	20	1								
Portugal	628	44	6			5	1			5	11
Republic of Moldova	405	10									
Romania	540	32									
Russian Federation	138	77	3								
San Marino	162										
Slovakia	67	13									
Slovenia	522	15									
Spain	778	66	25		2	1	2		1	10	16
Sweden	606	10									
Switzerland	595	11									
The FYR of Macedonia		14									

Table 13. Endangered, endemic and forest occuring for seven species groups (cont.)

| Country/area | All 7 species groups || Country-endemic endangered species ||||||||
	Total species	Endangered species	Total, all 7 species groups	Am-phibians	Birds	Ferns	Mam-mals	Palms	Reptiles	Trees	Total
Ukraine	60	29	1								
United Kingdom	723	17	9							8	8
Yugoslavia	98	21				-				-	
North and Central America											
Antigua and Barbuda	202	11	1								
Bahamas	100	21	5				1				1
Barbados	221	6	1								
Belize	1163	47	4							4	4
Bermuda	369	12	5			4					4
British Virgin Islands	233	14	1								
Canada	1013	43	2			3					3
Cayman Islands	228	7	2							1	1
Costa Rica	2629	214	56	5	23	3	10			36	77
Cuba	1047	237	185	3	11	1	7			102	124
Dominica	400	17	3	2						1	3
Dominican Republic	479	65	13			3		1	2	6	12
El Salvador	1028	46	5							4	4
Greenland	12	2									
Grenada	352	15	1	1							1
Guadeloupe	462	20	3				1				1
Guatemala	2283	130	19				1	6		9	16
Haiti	494	61	14					1		6	7
Honduras	1849	145	52		1		2	1		36	40
Jamaica	976	264	234	2	3	5	2	1	1	164	178
Martinique	383	17	4		1					2	3
Mexico	4033	362	185	15	11	13	17			65	121
Montserrat	258	11	1						1		1
Netherlands Antilles	353	10	1						1		1
Nicaragua	1832	63	3			1				2	3
Panama	2645	297	129		2	6	5	12		106	131
Puerto Rico	694	91	39	1	2	4			1	21	29
Saint Kitts and Nevis	230	8									
Saint Lucia	320	16	5		2				1		3
Saint Pierre & Miquelon	-	-	-	-	-	-	-	-	-	-	-
Saint Vincent and Grenadines	330	13	4		2				1		3
Trinidad and Tobago	963	13	1		1	1					2
United States	1283	436	292	13	20	13	4	18		147	215
US Virgin Islands	214	25	1								
South America											
Argentina	2311	123	16		1	1					2
Bolivia	3193	152	46		4	1	1	2		24	32
Brazil	3744	621	360		57		27	20		185	289
Chile	798	131	42	2	27	1	3			14	47
Colombia	5133	435	216	27	9	3	23			78	140
Ecuador	4031	303	157	6	6	1	5			127	145
Falkland Islands	205	1	1								
French Guiana	1495	47	10				2	2		6	10
Guyana	1370	50	9							9	9
Paraguay	1313	57	3							1	1
Peru	4247	462	277	13	10		6	4		182	215
Suriname	1453	53	16								
Uruguay	532	16									
Venezuela	3313	158	73	11	7	4	5			38	65

Table 14. Distribution of total forest area by ecological zone

Country/area	Tropical rain forest %	Tropical Moist %	Tropical Dry %	Tropical Shrub %	Tropical Desert %	Tropical Mountain %	Subtropical Humid %	Subtropical Dry %	Subtropical Steppe %	Subtropical Desert %	Subtropical Mountain %	Temperate Oceanic %	Temperate Continental %	Temperate Steppe %	Temperate Desert %	Temperate Mountain %	Boreal Coniferous %	Boreal Tundra %	Boreal Mountain %	Polar %
Africa																				
Algeria					n.s.			96	1		3									
Angola	9	65	25	1	n.s.	n.s.														
Benin	5	66	29																	
Botswana			73	27																
Burkina Faso		9	90	1																
Burundi						100														
Cameroon	81	16	2	n.s.		1														
Cape Verde				100																
Central African Republic	23	53	24																	
Chad		10	88	2																
Comoros		100																		
Congo	95		5																	
Côte d'Ivoire	63	37				n.s.														
Dem. Rep. of the Congo	82	15	n.s.			3														
Djibouti					100															
Egypt										100										
Equatorial Guinea	100																			
Eritrea				75	7	18														
Ethiopia		3	39	30	n.s.	29														
Gabon	99		1																	
Gambia		24	76																	
Ghana	47	32	21																	
Guinea	28	71				1														
Guinea-Bissau	23	77																		
Kenya	1	18	1	28		53														
Lesotho											100									
Liberia	99	1				n.s.														
Libyan Arab Jamahiriya								44	56											
Madagascar	34	9		38		18														
Malawi			48	37		15														
Mali			17	81	3															
Mauritania				100																
Mauritius		100																		
Morocco					n.s.			75	3		22									
Mozambique	1	18	81			n.s.														
Namibia			53	43	1	3														
Niger			99	1																
Nigeria	22	36	38	2		2														
Réunion		100																		
Rwanda						100														
Saint Helena		100																		
Sao Tome and Principe	100																			
Senegal		20	70	10																
Seychelles		100																		
Sierra Leone	40	60				1														
Somalia		1		97	1	1														
South Africa		1	61	3	1	2	15	7			11									
Sudan	7	57	26	9		1														
Swaziland			86								14									
Togo	19	68	12																	
Tunisia								96			4									
Uganda	78	5		1		16														
United Republic of Tanzania	1	18	65	13		3														
Western Sahara				100																
Zambia		49	51			n.s.														
Zimbabwe			99			1														

Table 14. Distribution of total forest area by ecological zone (cont.)

Country/area	Tropical Rain forest %	Tropical Moist %	Tropical Dry %	Tropical Shrub %	Tropical Desert %	Tropical Mountain %	Subtropical Humid %	Subtropical Dry %	Subtropical Steppe %	Subtropical Desert %	Subtropical Mountain %	Temperate Oceanic %	Temperate Continental %	Temperate Steppe %	Temperate Desert %	Temperate Mountain %	Boreal Coniferous %	Boreal Tundra %	Boreal Mountain %	Polar %
Asia																				
Afghanistan									42	2	56			n.s.		n.s.				
Armenia											61		39							
Azerbaijan							8				29		42		3	17				
Bahrain										100										
Bangladesh	63	37																		
Bhutan	14					55					31									
Brunei Darussalam	100					n.s.														
Cambodia	7	16	77																	
China	n.s.	1				3	37				22		17	4	n.s.	9	8		n.s.	
Cyprus								100												
Dem People's Rep. of Korea													100							
East Timor	33	3		59		5														
Gaza Strip																				
Georgia							16				25		11		n.s.	48				
India	13	11	56	9	n.s.	7		n.s.			5									
Indonesia	88	2	n.s.	1		9														
Iran, Islamic Rep.							72		1	1	25	1	1	n.s.						
Iraq									62	36	3									
Israel								62	38											
Japan							40				14	6				40				
Jordan								100												
Kazakhstan													5	41	15	23			17	
Kuwait										100										
Kyrgyzstan															n.s.	100				
Lao People's Dem. Rep	25	25	35			14														
Lebanon								91	9											
Malaysia	94					6														
Maldives	100																			
Mongolia														35	n.s.	64				
Myanmar	35	37	4			24					1									
Nepal	19	21	1			16					42									
Oman					100															
Pakistan		n.s.		1	1				66		31					n.s.				
Philippines	81	10				9														
Qatar										100										
Republic of Korea							15						85							
Saudi Arabia				91							9									
Singapore	100																			
Sri Lanka	18	20	62			n.s.														
Syrian Arab Republic								68	32											
Tajikistan															5	95				
Thailand	23	21	54			2														
Turkey							9	33	1		48	7				1				
Turkmenistan											4			81	14					
United Arab Emirates					100															
Uzbekistan														3	45	52				
Viet Nam	26	37	16	8		10	2				n.s.									
West Bank																				
Yemen					37	63														

Appendix 3. Global tables

Table 14. Distribution of total forest area by ecological zone (cont.)

Country/area	Tropical Rain forest %	Tropical Moist %	Tropical Dry %	Tropical Shrub %	Tropical Desert %	Tropical Mountain %	Subtropical Humid %	Subtropical Dry %	Subtropical Steppe %	Subtropical Desert %	Subtropical Mountain %	Temperate Oceanic %	Temperate Continental %	Temperate Steppe %	Temperate Desert %	Temperate Mountain %	Boreal Coniferous %	Boreal Tundra %	Boreal Mountain %	Polar %
Oceania																				
American Samoa	100																			
Australia	2	39	14				5	6	23	4		4				4				
Cook Islands	100																			
Fiji	100																			
French Polynesia	100																			
Guam	100																			
Kiribati	100																			
Marshall Islands	100																			
Micronesia	100																			
Nauru	100																			
New Caledonia	100																			
New Zealand							51					34				16				
Niue	100																			
Northern Mariana Isl.	100																			
Palau	100																			
Papua New Guinea	80	4	5			11														
Samoa	100																			
Solomon Islands	100																			
Tonga	100																			
Vanuatu	100																			
Europe																				
Albania							76		7			4				13				
Andorra																100				
Austria												7	12			81				
Belarus													100							
Belgium & Luxembourg												92				8				
Bosnia & Herzegovina							19						31			50				
Bulgaria							6						55	n.s.		39				
Croatia							28				n.s.		48			24				
Czech Republic												n.s.	54			46				
Denmark												100								
Estonia													100							
Finland													2				95		3	
France												69				31				
Germany												71	9			20				
Greece							88			9			2			1				
Hungary													100							
Iceland																87	13			
Ireland												93				7				
Italy							75		9		4				12					
Latvia													100							
Liechtenstein																100				
Lithuania													100							
Malta							100													
Netherlands												100								
Norway												6	1				46	n.s.	47	n.s.
Poland												18	72			10				
Portugal							73			8	17				2					
Republic of Moldova													95	5						
Romania													58	2		40				
Russian Federation													9	1	n.s.	4	50	3	33	1
San Marino																				
Slovakia													44			56				
Slovenia							12						40			47				
Spain							65	1		14	10				11					
Sweden												2	26				67		5	
Switzerland													26			74				
The FYR of Macedonia							57			7		18				18				

Table 14. Distribution of total forest area by ecological zone (cont.)

Country/area	Tropical						Subtropical					Temperate					Boreal			
	Rain forest	Moist	Dry	Shrub	Desert	Mountain	Humid	Dry	Steppe	Desert	Mountain	Oceanic	Continental	Steppe	Desert	Mountain	Coniferous	Tundra	Mountain	Polar
	%	%	%	%	%	%	%	%	%	%	%	%	%	%	%	%	%	%	%	%
Ukraine													78	8		14				
United Kingdom												85				2	10		4	
Yugoslavia							16				n.s.		65			19				
North and Central America																				
Antigua and Barbuda	22	43	34																	
Bahamas	29	54				17														
Barbados			100																	
Belize	42	58																		
Bermuda																				
British Virgin Islands	26		74																	
Canada												n.s.	13	n.s.	1	12	40	24	9	1
Cayman Islands		100																		
Costa Rica	61	24	2			13														
Cuba	32	56	4			8														
Dominica	79		21																	
Dominican Republic	59	15				26														
El Salvador	7	66	25			2														
Greenland																				
Grenada	71	25	4																	
Guadeloupe	62	8	30																	
Guatemala	41	40	2	n.s.		17														
Haiti	75	11				14														
Honduras	53	26	11	n.s.		9														
Jamaica	84	16																		
Martinique	68	11	21																	
Mexico	9	32	11			18		4	6		20									
Montserrat	26	56	18																	
Netherlands Antilles		21	7	72																
Nicaragua	74	19	6			1														
Panama	66	30	1			4														
Puerto Rico	93	7																		
Saint Kitts and Nevis	54	45	2																	
Saint Lucia	61	37	2																	
Saint Pierre & Miquelon																	100			
Saint Vincent and Grenadines	56	43	1																	
Trinidad and Tobago	100																			
United States		n.s.					23	1	3	1	10	1	18	1	1	26	5	8		2
US Virgin Islands	44		56																	
South America																				
Argentina	4	22	61			5	3	n.s.	1		1	2		1		2				
Bolivia	32	18	40			10														
Brazil	76	14	8			1	2													
Chile			n.s.	n.s.	n.s.		51	n.s.			4	39		2		5				
Colombia	84	3	2	n.s.		11														
Ecuador	60	4	3	n.s.		33														
Falkland Islands												100								
French Guiana	100																			
Guyana	74	23				4														
Paraguay	1	9	89																	
Peru	86		n.s.	n.s.	n.s.	14														
Suriname	58	42																		
Uruguay							100													
Venezuela	51	19	9	3		18														

Appendix 3. Global tables

Table 15. Forest in protected areas / available for wood supply

Country/area	Forest area 2000	Forest in protected areas		Forest available for wood supply with different distance limits to infrastructure				
		Country report	Global maps	No limit	50 km	30 km	20 km	10 km
	000 ha	000 ha	%	%	%	%	%	%
Algeria	2 145	-	-	6	96	96	96	96	91
Angola	69 756	-	-	3	97	96	94	89	64
Benin	2 650	-	-	32	61	61	61	60	50
Botswana	12 427	-	-	26	75	75	72	67	47
Burkina Faso	7 089	-	-	11	86	86	86	85	74
Burundi	94	-	-	29	79	79	79	79	75
Cameroon	23 858	-	-	11	89	89	86	81	67
Cape Verde	85	-	-	-	-	-	-	-	-
Central African Rep.	22 907	-	-	15	80	76	70	63	45
Chad	12 692	-	-	27	95	94	89	83	61
Comoros	8	-	-	-	100	100	100	100	100
Congo	22 060	-	-	14	86	78	69	60	41
Côte d'Ivoire	7 117	-	-	10	92	92	92	91	81
Dem. Rep. of the Congo	135 207	-	-	9	92	92	90	86	67
Djibouti	6	-	-	0	100	100	100	100	84
Egypt	72	-	-	0	-	-	-	-	-
Equatorial Guinea	1 752	-	-	11	89	89	89	89	83
Eritrea	1 585	-	-	0	100	100	100	100	78
Ethiopia	4 593	-	-	15	86	84	73	59	34
Gabon	21 826	-	-	16	84	84	79	71	51
Gambia	481	-	-	3	98	98	98	98	96
Ghana	6 335	-	-	9	91	91	91	91	85
Guinea	6 929	-	-	5	95	95	95	95	88
Guinea-Bissau	2 187	-	-	1	99	99	99	99	87
Kenya	17 096	-	-	40	61	61	61	56	41
Lesotho	14	-	-	16	87	87	75	47	28
Liberia	3 481	-	-	1	99	99	99	96	84
Libyan Arab Jamahiriya	358	-	-	19	100	100	100	100	100
Madagascar	11 727	-	-	4	98	98	97	94	83
Malawi	2 562	-	-	45	54	54	54	54	47
Mali	13 186	-	-	7	94	94	94	94	82
Mauritania	317	-	-	3	98	98	98	98	76
Mauritius	16	-	-	-	-	-	-	-	-
Morocco	3 025	-	-	7	99	99	99	99	87
Mozambique	30 601	-	-	7	94	94	94	91	66
Namibia	8 040	-	-	5	94	94	94	91	72
Niger	1 328	-	-	77	100	100	100	100	79
Nigeria	13 517	-	-	7	94	93	92	90	78
Réunion	71	-	-	-	-	-	-	-	-
Rwanda	307	-	-	76	100	100	100	100	69
Saint Helena	2	-	-	-	-	-	-	-	-
Sao Tome & Principe	27	-	-	-	-	-	-	-	-
Senegal	6 205	-	-	16	84	84	84	83	75
Seychelles	30	-	-	-	-	-	-	-	-
Sierra Leone	1 055	-	-	5	98	98	98	98	93
Somalia	7 515	-	-	3	100	100	99	88	62
South Africa	8 917	-	-	7	96	96	96	95	81
Sudan	61 627	-	-	10	95	94	88	79	58
Swaziland	522	-	-	4	96	96	96	96	88
Togo	510	-	-	14	86	86	86	86	77
Tunisia	510	-	-	4	95	95	95	95	90
Uganda	4 190	-	-	18	84	84	84	83	74
United Republic of Tanzania	38 811	-	-	14	85	85	81	74	55
Western Sahara	152	-	-	0	-	-	-	-	-
Zambia	31 246	-	-	24	71	71	68	62	42
Zimbabwe	19 040	-	-	12	88	88	88	87	77
Total Africa	**649 866**				**89**	**89**	**86**	**82**	**65**

Table 15. Forest in protected areas / available for wood supply (cont.)

Country/area	Forest area 2000	Forest in protected areas (Country report)		Global maps	Forest available for wood supply with different distance limits to infrastructure				
					No limit	50 km	30 km	20 km	10 km
	000 ha	000 ha	%	%	%	%	%	%	%
Afghanistan	1 351	-	-	0	69	69	68	66	45
Armenia	351	107	31	5	92	92	92	92	89
Azerbaijan	1 094	1 094	100	7	95	95	95	95	87
Bahrain	n.s.	-	-	-	-	-	-	-	-
Bangladesh	1 334	-	-	14	83	83	82	82	76
Bhutan	3 016	-	-	25	52	52	52	52	43
Brunei Darussalam	442	-	-	22	81	81	81	78	62
Cambodia	9 335	-	-	24	78	78	78	77	62
China	163 480	-	-	3	83	83	82	79	64
Cyprus	172	172	100	37	100	100	100	100	100
Dem People's Rep. of Korea	8 210	-	-	3	99	99	99	99	90
East Timor	507	-	-	3	97	97	95	95	91
Gaza Strip	-	-	-	-	-	-	-	-	-
Georgia	2 988	111	4	3	88	88	88	88	81
India	64 113	-	-	8	88	88	88	88	80
Indonesia	104 986	-	-	16	80	65	57	49	34
Iran, Islamic Rep.	7 299	-	-	12	98	98	98	98	82
Iraq	799	-	-	0	100	100	100	92	66
Israel	132	-	-	63	100	100	100	100	100
Japan	24 081	1 758	7	8	84	84	84	84	82
Jordan	86	-	-	0	100	100	100	100	100
Kazakhstan	12 148	12 148	100	11	81	81	81	80	66
Kuwait	5	-	-	0	-	-	-	-	-
Kyrgyzstan	1 003	866	86	10	26	26	26	26	21
Lao People's Dem. Rep	12 561	-	-	20	81	81	81	79	68
Lebanon	36	-	-	0	100	100	100	100	100
Malaysia	19 292	-	-	9	90	79	67	55	36
Maldives	1	-	-	-	-	-	-	-	-
Mongolia	10 645	-	-	11	83	79	71	59	34
Myanmar	34 419	-	-	5	93	93	93	90	78
Nepal	3 900	-	-	9	76	76	76	76	68
Oman	1	-	-	0	100	100	100	100	100
Pakistan	2 361	-	-	3	80	80	79	78	64
Philippines	5 789	-	-	7	95	95	95	91	67
Qatar	1	-	-	0	-	-	-	-	-
Republic of Korea	6 248	-	-	4	95	95	95	95	88
Saudi Arabia	1 504	-	-	9	100	100	91	76	53
Singapore	2	-	-	-	100	100	100	100	100
Sri Lanka	1 940	-	-	18	75	75	75	75	72
Syrian Arab Republic	461	-	-	0	100	100	100	100	99
Tajikistan	400	400	100	1	48	48	48	48	45
Thailand	14 762	-	-	23	72	72	71	68	54
Turkey	10 225	194	2	2	100	100	100	100	95
Turkmenistan	3 755	113	3	13	100	100	100	100	90
United Arab Emirates	321	-	-	0	-	-	-	-	-
Uzbekistan	1 969	1 888	96	30	100	100	100	100	74
Viet Nam	9 819	-	-	6	90	90	90	89	77
West Bank	-	-	-	-	-	-	-	-	-
Yemen	449	-	-	0	100	100	86	80	43
Total Asia	**547 793**				**84**	**81**	**79**	**75**	**63**

Table 15. Forest in protected areas / available for wood supply (cont.)

Country/area	Forest area 2000	Forest in protected areas		Forest available for wood supply with different distance limits to infrastructure				
		Country report	Global maps	No limit	50 km	30 km	20 km	10 km
	000 ha	000 ha	%	%	%	%	%	%
American Samoa	12	-	-	-	-	-	-	-
Australia	154 539	23 335	15	13	90	89	85	80	64
Cook Islands	22	-	-	-	-	-	-	-
Fiji	815	-	-	0	-	-	-	-
French Polynesia	105	-	-	-	-	-	-	-
Guam	21	-	-	-	-	-	-	-
Kiribati	28	-	-	-	-	-	-	-
Marshall Islands	n.s.	-	-	-	-	-	-	-
Micronesia	15	-	-	-	-	-	-	-
Nauru	n.s.	-	-	-	-	-	-	-
New Caledonia	372	-	-	2	97	97	97	96	88
New Zealand	7 946	1 661	21	3	82	82	82	81	73
Niue	6	-	-	-	-	-	-	-
Northern Mariana Isl.	14	-	-	-	-	-	-	-
Palau	35	-	-	-	-	-	-	-
Papua New Guinea	30 601	-	-	9	90	61	46	35	21
Samoa	105	-	-	-	-	-	-	-
Solomon Islands	2 536	-	-	0	100	65	61	57	47
Tonga	4	-	-	-	-	-	-	-
Vanuatu	447	-	-	0	100	87	84	79	68
Total Oceania	**197 623**				**90**	**83**	**77**	**71**	**56**
Albania	991	137	14	2	100	100	100	100	89
Andorra	-	-	-	-	56	56	56	56	56
Austria	3 886	785	20	22	98	98	98	98	94
Belarus	9 402	856	9	10	91	91	91	91	88
Belgium & Luxembourg	728	179	25	21	100	100	100	100	98
Bosnia & Herzegovina	2 273	-	-	1	100	100	100	100	95
Bulgaria	3 690	1 391	38	8	98	98	98	98	92
Croatia	1 783	414	23	8	97	97	97	97	91
Czech Republic	2 632	647	25	28	74	74	74	74	71
Denmark	455	93	21	9	97	97	97	97	96
Estonia	2 060	187	9	21	94	94	94	94	91
Finland	21 935	2 391	11	7	94	94	94	92	78
France	15 341	2 746	18	17	94	94	94	94	92
Germany	10 740	7 207	67	29	99	99	99	99	98
Greece	3 599	1 047	29	4	97	97	97	97	96
Hungary	1 840	368	20	16	85	85	85	85	83
Iceland	31	2	7	7	93	93	93	93	87
Ireland	659	7	1	5	100	100	100	100	94
Italy	10 003	1 881	19	11	99	99	99	99	97
Latvia	2 923	476	16	15	97	97	97	97	96
Liechtenstein	7	2	22	-	100	100	100	100	100
Lithuania	1 994	297	15	10	97	97	97	97	97
Malta	n.s.	n.s.	10	-	-	-	-	-	-
Netherlands	375	89	24	9	100	100	100	100	98
Norway	8 868	2 297	26	1	95	95	95	94	87
Poland	9 047	1 420	16	16	85	85	85	85	84
Portugal	3 666	634	17	8	96	96	96	96	92
Republic of Moldova	325	-	-	4	100	100	100	100	100
Romania	6 448	477	7	4	99	99	99	99	94
Russian Federation	851 392	25 542	3	3	90	74	63	55	39
San Marino	-	-	-	-	-	-	-	-	-
Slovakia	2 177	897	41	29	72	72	72	72	70
Slovenia	1 107	80	7	6	100	100	100	100	94
Spain	14 370	3 420	24	17	89	89	89	89	83
Sweden	27 134	-	-	8	93	93	93	93	88
Switzerland	1 199	43	4	12	80	80	80	80	80
The FYR of Macedonia	906	-	-	5	98	98	98	98	94

Table 15. Forest in protected areas / available for wood supply (cont.)

Country/area	Forest area 2000	Forest in protected areas Country report	Forest in protected areas Country report %	Forest in protected areas Global maps %	Forest available for wood supply No limit %	50 km %	30 km %	20 km %	10 km %
	000 ha	000 ha	%	%	%	%	%	%	%
Ukraine	9 584	987	10	6	95	95	95	95	92
United Kingdom	2 794	897	32	23	78	78	78	78	76
Yugoslavia	2 887	2 887	100	6	99	99	99	99	95
Total Europe	**1 039 251**				**91**	**78**	**70**	**63**	**50**
Antigua and Barbuda	9	-	-	-	100	100	100	100	100
Bahamas	842	-	-	4	100	98	97	96	91
Barbados	2	-	-	-	100	100	100	100	100
Belize	1 348	-	-	37	62	62	61	60	51
Bermuda	-	-	-	-	-	-	-	-	-
British Virgin Islands	3	-	-	-	100	100	100	100	100
Canada	244 571	19 321	8	5	87	64	55	47	33
Cayman Islands	13	-	-	-	99	85	85	85	85
Costa Rica	1 968	-	-	36	67	67	67	65	59
Cuba	2 348	-	-	25	85	85	84	83	78
Dominica	46	-	-	-	84	84	84	84	83
Dominican Republic	1 376	-	-	15	91	91	91	91	86
El Salvador	121	-	-	1	99	99	99	99	99
Greenland	-	-	-	-	-	-	-	-	-
Grenada	5	-	-	-	100	100	100	100	97
Guadeloupe	82	-	-	-	89	89	89	89	89
Guatemala	2 850	-	-	35	76	76	74	70	62
Haiti	88	-	-	1	100	100	100	100	97
Honduras	5 383	-	-	5	88	87	83	80	70
Jamaica	325	-	-	11	88	88	88	88	88
Martinique	47	-	-	-	43	43	43	43	43
Mexico	55 205	-	-	4	93	93	91	88	71
Montserrat	3	-	-	-	100	100	100	100	100
Netherlands Antilles	1	-	-	-	100	100	100	100	100
Nicaragua	3 278	-	-	23	74	74	72	68	57
Panama	2 876	-	-	35	70	68	55	42	27
Puerto Rico	229	-	-	5	97	97	97	97	97
Saint Kitts and Nevis	4	-	-	-	100	100	100	100	100
Saint Lucia	9	-	-	-	100	100	100	100	100
Saint Pierre & Miquelon	-	-	0	-	100	100	100	100	91
Saint Vincent and Grenadines	6	-	-	-	100	100	100	100	96
Trinidad and Tobago	259	-	-	-	100	100	100	100	100
United States	225 993	66 668	30	40	59	58	56	55	49
US Virgin Islands	14	-	-	-	100	100	100	100	100
Total North and Central America	**549 304**				**77**	**66**	**61**	**56**	**45**
Argentina	34 648	-	-	7	96	96	96	93	79
Bolivia	53 068	-	-	31	81	80	74	64	43
Brazil	543 905	-	-	17	85	44	36	31	22
Chile	15 536	-	-	14	88	83	78	74	63
Colombia	49 601	-	-	24	75	45	35	29	20
Ecuador	10 557	-	-	20	79	68	58	50	34
Falkland Islands	-	-	-	-	100	100	77	38	23
French Guiana	7 926	-	-	7	100	84	59	43	23
Guyana	16 879	-	-	1	100	78	58	44	25
Paraguay	23 372	-	-	5	95	95	90	80	57
Peru	65 215	-	-	10	90	42	27	19	10
Suriname	14 113	-	-	4	96	49	31	22	12
Uruguay	1 292	-	-	5	97	97	97	97	78
Venezuela	49 506	-	-	66	39	25	23	21	16
Total South America	**885 618**				**84**	**54**	**46**	**40**	**29**
TOTAL WORLD	**3 869 455**				**86**	**73**	**67**	**62**	**49**

Table 16. FRA 2000 country interaction

Country/area	FRA 2000 country correspondent	In-country assign-ment	Workshops and meetings	Country report
Algeria	Directeur Général des Forêts, Ministère de l'Agriculture	Yes		
Angola	Zola Alfonso, Deputy Director General IDF, Institute of Forestry Development, Ministry of Agriculture	Yes	Yes	WP47, EC-FAO
Benin	Sylla Alioune, Directeur des forêts et des ressources naturelles, Ministére du Développement Rural	Yes	Yes	EC-FAO
Botswana	Kujinga K.K., Chief Forestry, Range Ecology and Beekeeping, Ministry of Agriculture, Department of Crop Production and Forestry	Yes	Yes	WP47, EC-FAO
Burkina Faso	Dakkar Djiri, Ministère de l' Environnement et de l' Eau	Yes		
Burundi	Ntamagendero Liberata, Directeur-adjoint des Forêts, Ministère de l' Agriculture et de l' Élevage	Yes		
Cameroon	Kameni Foteu R., Chef de la cellule d´etudes et de planification forestière and Conseiller Technique auprès du Ministère des eaux et forêts, Ministère de l'Environnement et des forêts		Yes	EC-FAO
Cape Verde	Carvalho Leao, Directeur National des Forêts		Yes	EC-FAO
Central African Republic	Mbitikon Raymond, Chargé de mission, Cabinet du Ministre du Tourisme Chargé de l' Environnement	Yes	Yes	EC-FAO
Chad	Hassan Mahamat Ali, Dircteur des forets, Ministère de l'Environnement et du Tourisme	Yes	Yes	EC-FAO
Comoros	Directeur du Developpement Forestier, Miinistère de la Production			
Congo	Bouetoukadilamio V., Ministère de l'Agriculture, de l' Élevage, des Eaux et Forêts et de la Pêche	Yes	Yes	EC-FAO
Côte d'Ivoire	Nzoré Kadja, Directeur Général adjoint des Forêts, Ministère de l'Agriculture et des Ressources Animales	Yes	Yes	EC-FAO
Dem. Rep. of the Congo	Vangu-Lutete Clément	Yes		
Djibouti	Chef du Service de l' Agriculture, Ministère de l' Agriculture			
Egypt	Riad Mamdouh, Undersecretary of State for Afforestation and Environment, Ministry of Agriculture and Land Reclamation			
Equatorial Guinea	Obama Carlos Eyi, Director General de Economia Forestal, Ministerio de Agricultura y Alimentacion		Yes	EC-FAO
Eritrea	Iyassu Mebrahtu, Director General, Crop and Land Resources Department, Ministry of Agriculture		Yes	EC-FAO
Ethiopia	Head, Forestry and Wildlife Dept., Ministry of Natural Resources and Environtal Protection	Yes	Yes	EC-FAO
Gabon	Nyar-Ollame Pierre, Coordinnateur national PAFN	Yes		
Gambia	Bojang Lamin, Senior Forestry Officer in Charge of Technical Unit of Forestry Department, State Department of the Presidency and Natural Resources	Yes		
Ghana	Tuffuor Kwabena, Senior official, NFAP Focal Point	Yes		
Guinea	Directeur National des forêts et de la faune, Ministère de l'Agriculture		Yes	EC-FAO
Guinea-Bissau	Diombera Kaoussou, Chef de la Division des Études et de la Planification Forestière, Directeur National du projet GCP/GBS/023/NET	Yes		
Kenya	Muita Daniel W., Forest Department	Yes	Yes	EC-FAO
Lesotho	Chief Forestry Officer, Forestry Division, Ministry of Agriculture	Yes	Yes	WP47.EC-FAO
Liberia	Kaydea Shad G., Senior official: Managing Director, Forestry Department Authority			
Libyan Arab Jamahiriya	Bourjini Salah, Senior official: Resident Representative, UNDP			
Madagascar	Henri Finoama, Directeur des Eaux et Forêts, Ministére du Developpement Rural	Yes	Yes	EC-FAO
Malawi	Chipompha N.W.S., Senior official: Deputy Chief Forestry Officer	Yes	Yes	WP47, EC-FAO
Mali	Berthé Yafon, Directeur des Forêts, Ministère de l'Environnement	Yes	Yes	EC-FAO
Mauritania	Ibrahim Sall, Coordinnateur du PMLCD	Yes		
Mauritius				
Morocco	Ministére de l' Agriculture			
Mozambique	Adamo, Senior official: Director of Forestry, National Directorate of Forestry and Wildlife (DNFFB)	Yes	Yes	WP47, EC-FAO

Table 16. FRA 2000 country interaction (cont.)

Country/area	FRA 2000 country correspondent	In-country assign-ment	Workshops and meetings	Country report
Namibia	Chakamga Moses, Principal Forester, National Forest Inventory Project (component of Namibia-Finland Forestry Programme) Ministry of Environment and Tourism	Yes	Yes	WP47, EC-FAO
Niger	Adamou Abdou, Ministère de l' Hydraulique et de l' Environnement	Yes	Yes	EC-FAO
Nigeria	Okenyi I.I., Senior official, Forestry Management, Evaluation and Coordinating Unit (FORMECU)	Yes	Yes	EC-FAO
Réunion				
Rwanda	Directeur des Forests, Ministère de l' Agriculture et de l' Élevage	Yes	Yes	EC-FAO
Saint Helena				
Sao Tome and Principe	Directeur du Developpement Technique, Ministére de l' Agriculture			
Senegal	Dieng Ndiawar, Conseiller technique, Parc forestier de Hann, Ministère de l'Environnement et de la Protection de la Nature	Yes		
Seychelles	Directeur de la Foret, Division de l' Environnement			
Sierra Leone	Palmer Prince, Senior Official: Deputy Chief Conservator of Forests, Ministry of Agriculture and Forestry			
Somalia	Gammadid Ismail Deria, Senior official: Director of Forestry, National Range Agency			EC-FAO
South Africa	Mondlane Stephen, Deputy Director, International Forestry, Department of Water Affairs and Forestry	Yes	Yes	WP47, EC-FAO
Sudan	Abdel Nour Hassal Oman, General Manager, Sudan National Forests Cooperation, Ministry of Agriculture and Natural Resources	Yes	Yes	EC-FAO
Swaziland	Gamedze Solomon T., (SADC Forestry Contact Point) Senior Forestry Officer, Forestry Section, Ministery of Agriculture and Cooperatives	Yes	Yes	WP47, EC-FAO
Togo	Kodjo M. Tengue, Coordinator of PAFN	Yes		
Tunisia	Directeur General des Forests, Ministére de l'Agriculture	Yes		
Uganda	Kanabahita Charlotte, Forest Officer, Ministry of Water, Lands and Environment		Yes	EC-FAO
United Republic of Tanzania	Yonazi R. P., Forestry and Beekeeping Division, Ministery of Tourism, Natural Resources and Environment	Yes	Yes	EC-FAO
Western Sahara				
Zambia	Akepelwa, Chief conservator of Forests, Forest Department, Ministery of Land and Natural Resources	Yes	Yes	WP47, EC-FAO
Zimbabwe	Nyoni J., Policy Review Coordinator, Forestry Commission, Old Mutual Centre	Yes	Yes	WP47, EC-FAO
Total Africa		**36**	**30**	
Afghanistan	General President, Forestry and Range Department Ministry of Agriculture and Land Reform			
Armenia	Ter-Ghazaryan Karen, Deputy Executive Director, Forest Research and Experimental Centre, Ministry of Nature Protection			UNECE
Azerbaijan	Amirov Faik Acad ogly, Director, Forestry Scientific, Research and Project Development Institute, Forestry and Industry Association, (AzerbLes)			UNECE
Bahrain	Director, Agriculture Department, Ministry of Commerce and Agriculture			
Bangladesh	Ahmad Ishtiaq Uddin, DCF, Bangladesh Forest Department	Yes	Yes	WP45,15
Bhutan	Dhital D.B., Head Forest Resources Development Division, Bhutan Forest Department	Yes	Yes	WP45,14
Brunei Darussalam	Tuan Haji Abd. Rahman bin Hj. Chuchu , Director, Forestry Department Headquarters, Ministry of Industry and Primary Resources	Yes		
Cambodia	Syphan Ouk, Deputy Director-General, Department of Forestry and Wildlife			
China	Chen Xuefeng, Deputy Division Director, Division of Inventory, Department of Forest Resources, State Forestry Administration (SFA)	Yes		
Cyprus	Kourtellarides I., Department of Forests, Ministry of Agriculture, Natural Resources and Environment			UNECE
Dem People's Rep. of Korea	Administrator, Forestry Administration			
East Timor				

Table 16. FRA 2000 country interaction (cont.)

Country/area	FRA 2000 country correspondent	In-country assign-ment	Workshops and meetings	Country report
Gaza Strip				
Georgia	Kandelaki Teimuraz E., Doctor, Vice-Chairman of the State Department of Forest Management			UNECE
India	Pandey Devendra, Director, Forest Survey of India	Yes	Yes	WP45
Indonesia	Sumantri Ishak, Director of Inventory Division, Directorate General of Inventory and Landuse Planning (INTAG), Ministry of Forestry	Yes		
Iran, Islamic Rep.	Shirazi M.H., Director-General of International Affairs Bureau, Forest and Range Organization			
Iraq	Director General of Forestry, Ministry of Agriculture and Irrigation, Ministry of Agriculture and Irrigation			
Israel	Sapir Gil, Forest Resources Unit, Research and Development Authority			UNECE
Japan	Amano Masahiro, Director, Resource Planning Section, Forestry and Forest Products Research Institute (FFPRI)			UNECE
Jordan	Director General of Forest and Range, Ministry of Agriculture			
Kazakhstan	Musataev Murat, Vice-Minister, Ministry of Nature Resources and Environment Protection			UNECE
Kuwait	Director, National Parks and Afforestation, General Authority for Agriculture and Fisheries			
Kyrgyzstan	Venglovsky Bronislav I., Director, Institute of Forestry and Nut-trees Management, Academy of Science of the Kyrghyz Republic			UNECE
Lao People's Dem. Rep	Panzer Kersten F. Director Natural Science Department of Forestry	Yes		
Lebanon	Chef du Département des forêts, Ministère de l'Agriculture			
Malaysia	Hooi Chiew Thang, Assistant Director General, Forestry Department	Yes		
Maldives	Director-General, Ministry of Fisheries and Agriculture			
Mongolia	Director, Forest Office, Ministry of Nature and Environment	Yes		
Myanmar	Thin Kyau, Director General, Forest Department, Ministry of Forestry		Yes	WP45
Nepal	Karki Indra, Director General, Department of Forestry, Ministry of Forests and Soil Conservation	Yes	Yes	WP45,16
Oman	Director General of Agriculture, Ministry of Agriculture and Fisheries			
Pakistan	Jan Abeedulah, Additional Secretary and Inspector General of Forests, Ministry of Food Agriculture and Livestock		Yes	WP45
Philippines	Malvas or Ms. Mayumi Ma. Quintos Jose D., Director, Forest Managment Bureau, Department of Environment and Natural Resources			
Qatar	His Excellency The Minister, Ministry of Industry and Agriculture			
Republic of Korea				
Saudi Arabia	Director-General, Range and Forestry Department			
Singapore		Yes		
Sri Lanka	Ariyadasa K.P., DCF, Sri Lanka Forest Department	Yes	Yes	WP45,17
Syrian Arab Republic	Farouk El Ahmed, Director of Forests, Ministry of Agriculture and Agrarian Reform+D145			
Tajikistan	Avsalov Gaidulo A., Director-General, Forestry Production Association			UNECE
Thailand	Charuppat Thongchai, Chief, Royal Forest Department Remote Sensing and Forest Mapping Sub-division, Forest Management Division	Yes		
Turkey	Us Ulvi, Head of Research Planning and Coordination Department, General Directorate of Forestry, Ministry of Forestry			UNECE
Turkmenistan	Baigeldyev Batyr Artykovitch, Head, Reforestation Department, Ministry of the Use of Natural Resources and Environmental Protection			UNECE
United Arab Emirates	Deputy Minister, Ministry of Agriculture and Fisheries			
Uzbekistan				UNECE
Viet Nam	Sau, Director or Dr. Nguyen Huy Phon, Deputy Director, Forest Inventory and Planning Institute (FIPI)	Yes		
West Bank				
Yemen	Director of Forestry Department, Ministry of Agriculture and Fisheries	Yes		
Total Asia		**15**	**7**	

Table 16. FRA 2000 country interaction (cont.)

Country/area	FRA 2000 country correspondent	In-country assign-ment	Workshops and meetings	Country report
American Samoa	Markstein Robert, Director of Agriculture		Yes	WP51
Australia	Hnatiuk Roger, Senior Principal Research Scientist, Forest Section, Bureau of Resource Sciences			UNECE
Cook Islands	Tangianau Otheniel, Chief Executive, Ministry of Outer Islands Development		Yes	WP51
Fiji	Swaarup Ram, Conservator of Forests, Department of Forestry		Yes	WP51
French Polynesia			Yes	WP51
Guam	Limtiaco David, Head Forestry and Soil Resources, Department of Agriculture			
Kiribati	Ubaiitoi Ioane, Agroforestry Officer, Division of Agriculture		Yes	WP51
Marshall Islands				
Micronesia			Yes	WP51
Nauru				
New Caledonia	Trimari Bernard, Directeur de l'Agriculture et de la Forêt			
New Zealand	Barton James P., Senior Policy Analyst, Agriculture and Forestry Statistical Information, Policy Information Team, MAF Policy, Ministry of Agricultrue and Forestry			UNECE
Niue	Utalo Shiela, Forestry Officer		Yes	WP51
Northern Mariana Isl.				
Palau				
Papua New Guinea	Pouru Kanawi, Managing Director, PNG Forest Authority	Yes	Yes	WP51
Samoa	Iakopo Malaki, Assistant Director, Department of Agriculture, Fisheries and Forests		Yes	WP51
Solomon Islands	Loliano Ed, Forestry Department, Ministry of Forests, Environment and Conservation			
Tonga	Faka'osi Tevita, Director, Forestry and Conservation		Yes	WP51
Vanuatu	Nimoho Feke Pedro, Acting Director, Ministry of Agriculture, Livestock, Forestry and Fisheries		Yes	WP51
Total Oceania		**16**	**18**	
Albania	Karadumi Spiro, Institute of Forest and Pasture Researches		Yes	UNECE
Andorra			Yes	
Austria	Knieling Albert, Deputy Director, Forestry Department, International Forest Policy Division, Federal Ministry of Agriculture and Forestry		Yes	UNECE
Belarus	Kuzmenkov Mikhail V., Head of the Forestry Department, Ministry for Forestry		Yes	UNECE
Belgium & Luxembourg	Laurent Christian, Attaché, Ministere de la Région Wallone, Direction Générale des Ressources Naturelles et de l'Environnement, respectively Wagner Marc, Chef du Service de l´Aménagement des Bois et de l´Economie Forestière		Yes	UNECE
Bosnia & Herzegovina	Melic Frank, Vice-Minister, Ministry of Agriculture, Water Management and Forestry		Yes	UNECE
Bulgaria	Anguelov Ilija Petrov, Chief of Department of Forest Arrangement, Ministry for Agriculture, Forestry and Land Reform		Yes	UNECE
Croatia	Bilandzija Jela, World Bank Project Coordinator, Ministry of Agriculture and Forestry		Yes	UNECE
Czech Republic	Stransky Vaclav, Ministry of Agriculture of the Czech Republic, Forestry Department		Yes	UNECE
Denmark	Dralle Kim, Head of Section, Ministry of Environment and Energy, National Forest and Nature Agency		Yes	UNECE
Estonia	Viilup Ulo, Director, Estonian Forest Survey Centre		Yes	UNECE
Finland	Tomppo Erkki, Professor, Finnish Forest Research Institute		Yes	UNECE
France	Wencelius François, Director, National Forest Inventory (IFN), Inventaire Forestier National		Yes	UNECE
Germany	Lohner Peter, Deputy Head, Section 531 Federal Ministry of Food, Agriculture and Forestry		Yes	UNECE
Greece	Vogiatzis Stephanos, Section of Forest Research, Ministry of Agriculture, General Secretariat of Forests and Natural Environment		Yes	UNECE
Hungary	Csoka Peter, Director-General, State Forest Service, Ministry of Agriculture and Regional Development		Yes	UNECE
Iceland	Eysteinsson Thröstur, Deputy Director, Iceland Forestry Service (IFS)		Yes	UNECE
Ireland	Coggins Karl, Executive Officer, Forest Service, Department of Marine and Natural Resources		Yes	UNECE

Appendix 3. Global tables 443

Table 16. FRA 2000 country interaction (cont.)

Country/area	FRA 2000 country correspondent	In-country assignment	Workshops and meetings	Country report
Italy	Cavalensi Roberto, Ispettore Forestale, Funzionario Ufficio di Statistica Forestale (Div.III), Corpo Forestale dello Stato, Ministero delle Risorse Agricole Alimentari e Forestali		Yes	UNECE
Latvia	Bisenieks Janis, Senior Specialist, State Forest Service, Ministry of Agriculture		Yes	UNECE
Liechtenstein	Näscher Felix, Landesforstamt (National Office for Forestry), Ministry for Environment		Yes	UNECE
Lithuania	Kuliesis Andrius, Director (Direktorius) Habil.dr., Forest Inventory and Management Institute		Yes	UNECE
Malta	Borg Joseph, Principal Agricultural Officer (PAO), Ministry of Agriculture and Fisheries, Programmes and Initiatives for Director of Agriculture, Department of Agriculture		Yes	UNECE
Netherlands	Daamen Win P., Stichting Bosdata		Yes	UNECE
Norway	Tomter Stein Michael, Research Officer, Norwegian Institute of Land Inventory (NIJOS)		Yes	UNECE
Poland	Smykala Jerzy, Deputy Director, Forest Research Institute		Yes	UNECE
Portugal	Pinheiro Duarte Maria Odete, Head of National Forestry Inventory and Statistics Division, Direccío Geral Das florestas		Yes	UNECE
Republic of Moldova	Galupa Dimitru, Deputy Director-General, State Forestry, Association		Yes	UNECE
Romania	Zaharescu Claudiu, Expert, Ministry of Waters, Forests and Environment Protection, Department of Forests		Yes	UNECE
Russian Federation	Filipchuk Andrew, Deputy Director, All Russian Scientific Research and Information Center for Forest Resources		Yes	UNECE
San Marino			Yes	
Slovakia	Gecovic Miroslav, Head of Department of Foreign Relations and Information Technology, Forest Research Institute		Yes	UNECE
Slovenia	Hocevar Milan, Professor, Slovenian Forestry Institute		Yes	UNECE
Spain	Lopez Jose Solano, Direccíon General Conservacíon de la Naturaleza, Ministero de Medio Ambiente		Yes	UNECE
Sweden	Svensson S., National Board of Forestry		Yes	UNECE
Switzerland	Brassel Peter, Dr., Swiss Federal Institute for Forest,Snow and Landscape Research, Institut Fédéral de Recherches sur la Forêt, la Neige et le Paysage		Yes	UNECE
The FYR of Macedonia	Trendafilov Aleksandar, Minister´s Assistant, Ministry of Agriculture Forestry and Water Economy, Forestry Department		Yes	UNECE
Ukraine	Torosov Artijom S., Dr. Deputy Director of Economy, Chief of Laboratory of Economy, Ukrainian Scientiific Research Institute of Forestry and Forest Amelioration		Yes	UNECE
United Kingdom	Gillam Simon, Head of Statistics, Forestry Commission		Yes	UNECE
Yugoslavia	Medarevic Milan, Associate Professor, Dr. Sc., Faculty of Forestry of Belgrade University		Yes	UNECE
Total Europe		**0**	**40**	
Antigua and Barbuda	Mc Ronnie Henry, Foresrty Officer, Ministry of Agriculture, Fisheries, Lands and Housing, Temple and Nevis Streets			
Bahamas	Russel Cristopher C., Forest Officer, Department of Lands and Surveys		Yes	EC-FAO
Barbados	Jones Nigel, Soil Conservation Unit – Ministry of Agriculture and Rural Development		Yes	EC-FAO
Belize	Chun Angel V., Forest Management Officer, Ministry of Natural Resources	Yes	Yes	WP10,40,52 EC-FAO
Bermuda				
British Virgin Islands				
Canada	Boulter David W.K., Director, Economics ans Statistical Services, Canadian Forest Service			UNECE
Cayman Islands				
Costa Rica	Rojas Luis, Director General, Sistema Nacional de Areas de Conservación	Yes	Yes	WP10,36,52
Cuba	Nieto Lara Marcos, Direccion de Relaciones Internacionales Area Forestal, Ministerio de Agricultura	Yes	Yes	EC-FAO
Dominica	Colmore Christian, Director of Parks, Forestry and Wildlife, Ministry of Agriculture, Botanical Gardens		Yes	EC-FAO
Dominican Republic	Manon Rossi Bernabé, Presidente de la Comisión Técnica Forestal, CONATEF		Yes	EC-FAO
El Salvador	Olano Julio Alberto, Director General, Dirección General de Recursos Naturales Renovables, MAG	Yes	Yes	WP10,37,52

Table 16. FRA 2000 country interaction (cont.)

Country/area	FRA 2000 country correspondent	In-country assignment	Workshops and meetings	Country report
Greenland				
Grenada	Rolax Frederick, Chief Forestry Officer, Ministry of Agriculture		Yes	EC-FAO
Guadeloupe				
Guatemala	Cabrera Claudio, Gerente General, Instituto Nacional de Bosques	Yes	Yes	WP10,13,52
Haiti	Ogé Jean Pierre-Louis, Head Chief, Service de Forêts, Ministère de l'Agriculture et Développement Rural		Yes	EC-FAO
Honduras	Martinez Salomón, Subgerente, Administración Forestal del Estado AFE-COHDEFOR	Yes	Yes	WP10,44,52
Jamaica	Headley Marilyn, Conservation of Forests, Forestry Department, Ministry of Agriculture and Mining		Yes	EC-FAO
Martinique				
Mexico	Varela Hernández Sergio, Director del Inventario Nacional de recursos Naturales, Dirección General Forestal, Subsecretaría de Recursos Naturales, Secretaría de Medio Ambiente, Recursos Naturales y Pesca (SEMARNAP)	Yes	Yes	WP10,35,52
Montserrat				
Netherlands Antilles				
Nicaragua	Montalbán Alvaro, Director Ejecutivo, Instituto Nacional Forestal (INAFOR)	Yes	Yes	WP10,34,52
Panama	Vargas Lombardo Carlos, Director Nacional de Administración Forestal, Autoridad Nacional del Ambiente	Yes	Yes	WP10,41,52
Puerto Rico				
Saint Kitts and Nevis	Mills Henry, Ministry of Agriculture, Lands, Housing and Development		Yes	EC-FAO
Saint Lucia	James Brian, Ministry of Agriculture, Lands, Fisheries and Forestry		Yes	EC-FAO
Saint Pierre & Miquelon				
Saint Vincent and Grenadines	Weeks Nigel, Ministry of Agriculture, Industry and Labour, Forestry Division		Yes	EC-FAO
Trinidad and Tobago	Faizool Sheriff, Ministry of Agriculture, Land and marine Resources		Yes	EC-FAO
United States	Smith Brad, Associate Branch Chief, Forest Inventory Research, International Resource Assessment Liaison, USDA Forest Service (FIERR)		Yes	UNECE
US Virgin Islands				
Total North and Central America		**9**	**21**	
Argentina	Merenson Carlos E., Director, Direc. Recursos Forestales Nativos, Secretaría Recursos Naturales y Ambiente Humano	Yes		WP52
Bolivia	Alborta Rodolfo, Director General Forestal y Silvicultura, Ministerio de Agricultura, Ganaderia y Desarrollo Forestal	Yes		WP52
Brazil	Prado Antonio Carlos, Director, Departamento de Formulaçao de Politicas e Programas Ambientales, Ministério de Neio Ambiente, dos Recursos Hídricos e da Amazonia Legal	Yes		WP52
Chile	Guerra M. Guillermo, Gerente de Desarrollo y Fomento Forestal, Corporación Nacional Forestal (CONAF)	Yes		WP52
Colombia	Otavo Rodriguez Edgar, Coordinador Grupo Bosques y Plantaciones Forestales, Ministerio del Medio Ambiente	Yes	Yes	WP10,43,52
Ecuador	Thiel Hans, Director Forestal, Ministerio de Medio Ambiente	Yes	Yes	WP10,38,52
Falkland Islands				
French Guiana				
Guyana	Marshall Godfrey, Head of the Forest Resource Management Division, Ministry of Agriculture		Yes	EC-FAO
Paraguay	Rodas Manuel, Director del servicio Forestal Nacional, Subsecretaría de Estado de Recursos Naturales y medio Ambiente, Ministerio de Agricultura y Ganadería	Yes		WP52
Peru	Morisaki Antonio, Director General Forestal, Instituto Nacional de Recursos Naturales (IRENA)	Yes		WP52
Suriname	Playfair Maureen, Head of Planning, Surinam Forest Service, Ministry of Natural Resources		Yes	EC-FAO
Uruguay	Ligrone Atilio, Ministrio de Ganaderia, Agricultura y Pesca	Yes		
Venezuela	Mendoza Samuel, Director General, Dirección General Sectorial de Recursos Forestales, Ministerio del Ambiente y de los Recursos Naturales Renovables (MARNR)	Yes	Yes	WP10,39,52
Total South America		**10**	**5**	
TOTAL WORLD		**86**	**121**	

Appendix 4. FRA 2000 publications

FRA 2000 results and documents are available on the FAO Forestry Web site at www.fao.org/forestry/fo/fra/index.jsp (FRA subject page) and www.fao.org/forestry/fo/country/nav_world.jsp (FAO Forestry country profiles). This appendix lists the FRA Working Papers to date and gives an introduction to the contents of the FAO Forestry country profiles.

FRA WORKING PAPERS

The below list contains the FRA Working Papers as of 1 August 2001. The papers are available on line at www.fao.org/forestry/fo/fra/index.jsp. Papers can also be requested by e-mail to fra@fao.org, or by ordinary mail to FAO, Forestry Department, FRA Programme, Viale delle Terme di Caracalla, 00100 Rome, Italy.

E/F/S/P refers to the languages English, French, Spanish and Portuguese.

1998
1. FRA 2000 Terms and Definitions (18 pp. - E/F/S/P)
2. FRA 2000 Guidelines for assessments in tropical and subtropical countries (43 pp. - E/F/S/P)

1999
3. The status of the forest resources assessment in the South-Asian subregion and the country capacity building needs. Proceedings of the GCP/RAS/162/JPN regional workshop held in Dehradun, India, 8-12 June 1998. (186 pp. – E)
4. Volume/Biomass Special Study: georeferenced forest volume data for Latin America (93 pp. - E)
5. Volume/Biomass Special Study: georeferenced forest volume data for Asia and Tropical Oceania (102 pp. - E)
6. Country Maps for the Forestry Department website (21 pp. - E)
7. Forest Resources Information System (FORIS) – Concepts and Status Report (20 pp. E)
8. Remote Sensing and Forest Monitoring in FRA 2000 and beyond. (22 pp. - E)
9. Volume/Biomass special Study: Georeferenced Forest Volume Data for Tropical Africa (97 pp. – E)
10. Memorias del Taller sobre el Programa de Evaluación de los Recursos Forestales en once Países Latinoamericanos (pp. 194 - S)
11. Non-wood forest Products study for Mexico, Cuba and South America (draft for comments) (82 pp. - E)
12. Annotated bibliography on Forest cover change – Nepal (59 pp. – E)
13. Annotated bibliography on Forest cover change – Guatemala (66 pp. – E)
14. Forest Resources of Bhutan – Country Report (80 pp. – E)
15. Forest Resources of Bangladesh – Country Report (93 pp. – E)
16. Forest Resources of Nepal – Country Report (78 pp. – E)
17. Forest Resources of Sri Lanka – Country Report (77 pp. – E)
18. Forest plantation resource in developing countries (75 pp. – E)
19. Global forest cover map (14 pp. – E)
20. A concept and strategy for ecological zoning for the global FRA 2000 (23 pp. – E)

2000

21. Planning and information needs assessment for forest fires component (32 pp. – E)
22. Evaluación de los productos forestales no madereros en América Central (102 pp. – S)
23. Forest resources documentation, archiving and research for the Global FRA 2000 (77 pp. – E)
24. Maintenance of Country Texts on the FAO Forestry Department Website (25 pp. – E)
25. Field documentation of forest cover changes for the Global FRA 2000 (40 pp. – E)
26. FRA 2000 Global Ecological Zones Mapping Workshop Report Cambridge, 28-30 July 1999 (53 pp. - E)
27. Tropical Deforestation Literature:Geographical and Historical Patterns in the Availability of Information and the Analysis of Causes (17 pp. – E)
28. World Forest Survey – Concept Paper (30 pp. – E)
29. Forest cover mapping and monitoring with NOAA-AVHRR and other coarse spatial resolution sensors (42 pp. E)
30. Web Page Editorial Guidelines (22 pp. – E)
31. Assessing state & change in Global Forest Cover: 2000 and beyond (15 pp. – E)
32. Rationale & methodology for Global Forest Survey (60 pp. – E)
33. On definitions of forest and forest change (13 pp.- E)
34. Bibliografia comentada. Cambios en la cobertura forestal: Nicaragua (51 pp. – S)
35. Bibliografia comentada. Cambios en la cobertura forestal: México (35 pp. – S)
36. Bibliografia comentada. Cambios en la cobertura forestal: Costa Rica (55 pp. – S)
37. Bibliografia comentada. Cambios en la cobertura forestal: El Salvador (35 pp. – S)
38. Bibliografia comentada. Cambios en la cobertura forestal: Ecuador (47 pp. – S)
39. Bibliografia comentada. Cambios en la cobertura forestal: Venezuela (32 pp. – S)
40. Annotated bibliography. Forest Cover Change: Belize (36 pp. – E)
41. Bibliografia comentada. Cambios en la cobertura forestal: Panamà (32 pp. – S)
42. Proceedings of the FAO Expert Consultation to Review FRA 2000 Methodology for Regional and Global Forest Change Assessment. Rome 6-10 March 2000.
43. Bibliografia comentada. Cambios en la cobertura forestal: Colombia (32 pp. – E)
44. Bibliografia comentada. Cambios en la cobertura forestal: Honduras (42 pp. – E)

2001

45. Proceedings of South Asian Regional Workshop on Planning, Database and Networking for Sustainable Forest Management, Thimpu, Bhutan, 23 – 26 May 2000 (263pp. – E)
46. Global Forest Survey – Field Site Specification and Guidelines
47. Proceedings from regional workshop on forestry information services. Stellenbosch, South Africa. 12-17 february 2001 (69 pp. – E)
48. Forest cover assessment in the Argentinean regions of the Monte and Espinal (E)
49. Pan tropical survey of forest cover changes 1980-2000 (E)
50. Global forests cover mapping – final report
51. FRA 2000 Data collection for Pacific Region – FAO wshop – Apia, Samoa
52. Causas y tendencias de la deforestación en América Latina (S)
53. Forest occurring species of conservation concern: Review of status of information for FRA 2000 (E)
54. Assessing Forest Integrity and Naturalness in Relation to Biodiversity. Forest Resources Assessment Programme (E)
55. Global forest fire assessment 1990-2000 (E)
56. Global ecological zoning – final report (E)
57. Ecofloristic zone mapping (E)
58. FRA 2000 Project Processes (E)

WEB COUNTRY PROFILES

FRA 2000 hectaress produced large quantities of information by country, some of which is summarized in this report mainly in the form of country statistics (see Appendix 3). The main publication of country information is, however, made on the FAO Forestry website, in the Forestry country profiles. These country profiles aim at a comprehensive presentation of the forest sector of each country, which includes also other aspects than those covered by FRA 2000. A large proportion of the currently (1 July 2001) published information is however derived from FRA 2000 work.

Currently, the FAO Forestry country profiles contain more than 20 000 published pages covering 213 countries, four languages and up to 30 unique pages by country. The majority of these pages are related to FRA 2000 and include:

- Country statistics for a range of subjects, including documentation of estimates;
- Country maps of forest cover, ecological zones and protected areas – derived from FRA 2000 global maps;
- Narratives of geography, natural woody vegetation, plantations, forest management, protected areas removals, non-wood forest products.

Appendix 5. Authors by chapter

Chapter	Principal author(s)	Contact for further information
1	Peter Holmgren	peter.holmgren@fao.org
2	Mohamed Saket	mohamed.saket@fao.org
3	Jim Carle	jim.carle@fao.org
4	Michelle Gauthier	fra@fao.org
5	Peter Holmgren and Christel Palmberg	christel.palmberg@fao.org
6	Mette L. Wilkie	mette.loyche-wilkie@fao.org
7	Peter Holmgren	peter.holmgren@fao.org
8	Gillian Allard and Bob Mutch	gillian.allard@fao.org
9	Dan Altrell	dan.altrell@fao.org
10	Laura Russo, Paul Vantomme, François Ndeckere-Ziangba and Sven Walter	paul.vantomme@fao.org
11	FRA-team	fra@fao.org
12	James Space	
13	Mohamed Saket	mohamed.saket@fao.org
14	Isabelle Amsallem	fra@fao.org
15	Isabelle Amsallem	fra@fao.org
16	Alberto Del Lungo	alberto.dellungo@fao.org
17	Mohamed Saket	mohamed.saket@fao.org
18	Hivy Ortiz Chour	fra@fao.org
19	FRA-Team	fra@fao.org
20	James Space	
21	Talat Abdel-Hamid Omran	fra@fao.org
22	Hivy Ortiz-Chour	hivy.ortizchour@fao.org
23	Kailash Govil	fra@fao.org
24	Jonas Cedergren	fra@fao.org
25	Jonas Cedergren and Hivy Ortiz-Chour	hivy.ortizchour@fao.org
26	FRA team	fra@fao.org
27	James Space	fra@fao.org
28	Tomas Thuresson	fra@fao.org
29	Tim Peck	fra@fao.org
30	Tim Peck	fra@fao.org
31	Tim Peck	fra@fao.org
32	FRA team	fra@fao.org
33	James Space	fra@fao.org
34	James Space	fra@fao.org
35	Hivy Ortiz-Chour	hivy.ortizchour@fao.org
36	Hivy Ortiz-Chour	hivy.ortizchour@fao.org
37	FRA team	fra@fao.org
38	James Space	fra@fao.org
39	Chris Brown	fra@fao.org
40	James Space	fra@fao.org
41	FRA team	fra@fao.org
42	James Space	fra@fao.org
43	Jorge Malleux	jorge.malleux@fao.org
44	Jorge Malleux	jorge.malleux@fao.org
45	Robert Davis	robert.davis@fao.org
46	Anne Branthomme, Sören Holm and Ingemar Eriksson	anne.branthomme@fao.org, ingemar.eriksson@fao.org
47	Robert Davis	robert.davis@fao.org
48	Peter Holmgren	peter.holmgren@fao.org

Appendix 6. Earlier global assessments

At the first session of the Conference of FAO in the autumn of 1945, the need for up-to-date information on the forest resources of the world was fully recognized and it was recommended that an inventory should be undertaken as soon as possible. In May 1946 the Forestry and Forest Products Division was founded and work was immediately initiated on FAO's first worldwide assessment of forests (FAO 1948). After reviewing the results of the assessment in 1947, the sixth session of the FAO Conference in 1951 recommended that the Organization "maintain a permanent capability to provide information on the state of forest resources worldwide on a continuing basis" (FAO 1951). Since that time, various other regional and global surveys have been conducted every five to ten years. Each has taken a somewhat different form.

Statistics released by FAO on world forest cover from 1948 through 1963 were largely collected through questionnaires sent to the countries. The assessments since 1980 have taken a more solid technical form, being based on the analysis of country references supported by expert judgements, remote sensing and statistical modelling. FRA 2000 is the most comprehensive in terms of the number of references used and information analysed on forest cover, forest state, forest services and non-wood forest products (NWFP). FRA 2000 is also notable for applying for the first time a single technical definition of forest at the global level, based on 10 percent crown cover density.

Statistics from the different assessments are difficult to use for comparative purposes, owing to changes in baseline information, methods and definitions. However, better correlations can be achieved for time series in many countries for certain assessments, especially with information generated since 1980, when reporting parameters stabilized. Consistent definitions were applied for developing countries for subsequent assessments.

FAO'S GLOBAL AND REGIONAL ASSESSMENTS 1946-1997

Forest resources of the world (1948)

In 1946, the year following the founding of FAO, a first global survey was conducted by the Organization; it was published in 1948 as *Forest resources of the world* (FAO 1948). Initially, a questionnaire was sent to all countries, of which 101 responded, representing about 66 percent of the world's forests. Parameters included in the survey were forest area (total and productive), types of forest by accessibility, growth and fellings.

One of the noteworthy conclusions of the first world forest inventory report was that:

> All these investigations made valuable additions to our knowledge, but all suffered from certain fundamental difficulties. Most important of these were the lack of reliable forest inventory information which existed and still exists in many countries, and the lack of commonly accepted definitions of some of the more important forestry terms. Hence, to the weakness of some of the quantitative estimates there was added doubt as to the real meanings of some of the qualitative descriptions (FAO 1948).

This statement remains largely true today, over 50 years later. While technical and scientific advances have greatly increased the potential to improve the information base in countries, many still lack the training, institutional and financial resources to conduct periodic assessments.

Major findings
- Total forest cover (global): 4.0 billion hectares
- Net forest change (global): not reported

World forest inventories (1953, 1958 and 1963)

World forest inventories including all countries were carried out on three occasions during the 1950s and 1960s. Lanly (1983) describes these various inventories:

> ...126 countries and territories replied to the 1953 questionnaire representing about 73 percent of the world forest area. The picture was completed by information from the replies to the 1947 questionnaire for 10 other countries (representing 3 percent of the total world forested area) and official statistics for the remaining 57 countries, representing 24 percent of the world forest area. The results were published by FAO in 1955 under the title *World forest resources –results of the inventory undertaken in 1953 by the Forestry Division of FAO.*

> The 1958 inventory of the FAO World Forest Inventory published in 1960 (*World forest inventory 1958 – the third in the quinquennial series compiled by the Forestry and Forest Products Division of FAO*) utilized the replies of the 143 countries or territories, representing the 88 percent of the world forest area,

complemented by the replies to the 1953 questionnaire for 13 countries (2 percent) and to the 1947 questionnaire for 5 countries (3 percent). Necessary changes and precisions introduced in the definition of some concepts, more precise definitions of forests and changes in such concepts as forest-in-use and accessible forests affected comparability with the previous inventory. However, changes in area and other forest characteristics during the 1953-58 period were, for several countries, either reported directly from them or could be derived by comparison of the replies to both questionnaires (changes in area of permanent forests, in management status in forests-in-use, increase in accessible areas and in forest-in-use, afforested area between 1953 and 1957, etc.).

The *World Forest Inventory 1963* published by FAO in 1965 witnessed a slightly lower rate of response of (105 compared to 130), "at least partly accounted for by temporary strains on administration in countries gaining their independence" as was reported in the document. Again comparability with the former enquiries was limited, and as pointed out by the authors of the report, "large differences for some countries (between the results of the 1958 and 1963 enquires) resulted more from better knowledge about the forests, or stricter application of definitions, than from effective changes in the forest resources".

The main parameters assessed during the 1963 World Forest Inventory were forest area (total, productive, and protected), ownership, management status, composition (softwoods and hardwoods), growing stock and removals (FAO 1966).

Major findings (1963)
- Total forest cover (global): 3.8 billion hectares
- Net forest change: not reported

Regional forest resources assessments (1970s)

During the 1970s FAO carried out no global surveys. Instead a series of regional assessments were made with the intention that each would be more regionally appropriate and specific. Beginning in the late 1960s FAO sent out questionnaires to all industrialized countries. The results were published in 1976 as *Forest resources of the European Region* (FAO 1976b). Questionnaires were also sent to Latin America and Asia and the results were published in *Forest resources in Asia and the Far East Region* (FAO 1976c) and *Appraisal of forest resources of the Latin American Region* (FAO 1976a). A similar questionnaire was sent to African countries by the Department of Forest Survey of the Swedish Royal College of Forestry and published in *Forest resources of Africa – an approach to international forest resources appraisal, Part I: Country descriptions* (Persson 1975) and *Part II: Regional analyses* (Persson 1977).

According to Lanly (1983), the regional assessments of the developing areas had the following main features in common:

- they were based only in part upon questionnaires, the rest of the information having been collected in another form, in particular through travels to countries of the region concerned;
- they included more qualitative information (descriptions of forest types, indication of species planted, quotations of figures on volumes and other stand characteristics extracted from inventory reports, etc.) than the World Forest Inventory assessments, which were essentially statistical;
- in addition to regional statistical tables, country notes were prepared regrouping all quantitative information selected for each country;
- since the information provided was not limited to the replies to the questionnaires, the draft country notes were sent back to the national forest institutions for their comments and suggested amendments.

Although FAO did not compile the regional findings into a global synthesis, a global survey was done outside FAO and published in *World forest resources – review of the world's forest resources in the early 1970's* (Persson 1974). Finally, another FAO study, *Attempt at an assessment of the world's tropical moist forests* (Sommer 1976), provided a summary on findings on the forest situation in all tropical moist forests.

FRA 1980

FRA 1980 covered 97 percent of the land area of developing countries or 76 tropical countries: 36 in Africa, 16 in Asia and 23 in Latin America and the Carribean. FRA 1980 was distinguished by many features. Its breadth was the greatest to date, and in many cases remains unmatched by the present assessment. It is also notable as the first assessment to use a technical definition of forests, in which measurable parameters were indicated – notably 10 percent canopy cover density, minimum tree height of 7 m, and 10 ha as the minimum area for defining a forest. Previous assessments had relatively broad definitions which could be interpreted quite differently by different countries. The consistent definition provided parameters useful in adjusting country information to a common standard. An adjustment in time was also made using expert opinion to project the information to common reference years of 1976, 1980, 1981 and 1985.

FRA 1980 relied extensively on existing documentation from countries to formulate its estimates on forest cover (state and change), plantation resources and timber volume. Existing information from multiple sources in the countries was gathered and analysed. Dialogues with national and international experts on information utility and reliability helped to firm up the estimates for the countries. The assessment noted that information was abundant but hard to locate and synthesize in the coherent manner needed for a consistent global survey.

Extended narratives, explanatory text and qualitative information complemented the statistical data set. During the tenure of FRA 1980, FAO was conducting extensive work on forest inventories in tropical countries. Roughly one project existed for every two to three countries, and FAO experts employed in the projects provided valuable input to the 1980 assessment results.

In major forested areas where existing information was lacking, the assessment conducted manual interpretations of satellite imagery (1:1 000 000 scale). This was done for six Latin American countries, two African countries, two Asian countries and portions of two other Asian countries. The interpretations covered about 70 to 99 percent of these countries, with 55 satellite images used.

The final documentation for FRA 1980 included three volumes of country briefs (one for each developing country region) (FAO 1981a, FAO 1981b, FAO 1981c), three regional summaries and a condensed main report, published as an FAO Forestry Paper (FAO 1982). While the findings were not global, FRA 1980 was used again in 1988 to make an interim global assessment.

Major findings
- Total forest area (tropical developing countries only) 1980: 2.1 billion hectares (natural forests and plantations)
- Net forest change (tropical developing countries only) 1981-1985: -10.2 million hectares per year
- Net forest change (global): not reported

Interim assessment 1988

An interim report on the state of forest resources in the developing countries (FAO 1988) provided information on 129 developing countries (53 more than FRA 1980) as well as the industrialized countries. The report provided information on the state of the forests at the year 1980 and the changes over the period 1981-1985. Definitions varied between the industrialized and the developing countries, specifically in regard to canopy closure thresholds for forests, which were set at 20 percent for industrialized countries and 10 percent for developing countries. Information for the industrialized countries was collected by UNECE/FAO in Geneva, which drew on the report *The forest resources of the ECE region (Europe, the USSR, North America)* (UNECE/FAO 1985). Parameters also varied for the two groups of countries, so that a global synthesis of core elements was needed, in order to achieve a uniform global data set.

The elements of the global synthesis included forest, operable forest, inoperable forest, other wooded lands, broad-leaved forest and coniferous forest.

Major findings
- Total forest area (global) 1980: 3.6 billion hectares
- Net forest change (tropical developing countries) 1981-1985: -11.4 million hectares per year
- Net forest change (global): not reported

FRA 1990

FRA 1990 (FAO 1995) covered all developing and industrialized countries and was distinguished by two innovations: the development and use of a computerized "deforestation model" which was applied to the developing country data for projecting the forest area statistics to a common reference year; and an independent pan-tropical remote sensing survey of forest change based on high-resolution remote sensing data.

FRA 1990 sought to improve estimates by eliminating the bias of expert opinions in the assessments through a statistical model to predict forest cover loss (and thereby deforestation rates). The model was based on forest cover change derived from the few comparable multi-date assessments available. Deforestation rates were then regressed against independent variables to determine the rate of forest loss relative to changes in population densities within specific ecological zones. Forest cover change rates were obtained by applying the model to existing baseline statistics available for the countries.

The advantages of the 1990 method were the near-uniformity achieved by applying the model equally to almost all developing countries and the ability to streamline production of statistics using computer routines.[60] The disadvantages of the 1990 method were the low number of variables used in the deforestation algorithm and the low number of observations used to construct the model, introducing a relatively high random error (low precision) in country estimates.

Because of the many uncertainties involved in working with existing national data, FRA 1990 implemented a remote sensing survey to provide a quality-controlled set of statistics on forest resources and to complement the survey based on country information. The use of statistical sampling combined with a uniform data source (satellite imagery) and common data collection methods made this approach an important tool for providing a set of statistics to compare with the country data.

The survey relied on statistical sampling (10 percent) of the world's tropical forests through 117 sample units distributed throughout the tropics to produce estimates of the state and changes of tropical forest at the regional, ecological and pan-tropical levels (but not at the national level). Each of the sample units consisted of three multi-date Landsat satellite images which provided the raw material for producing statistics on forest and other land cover changes from 1980 to 1990 and later to 2000.

FAO used an interdependent manual interpretation of satellite scenes at a scale of 1:250 000, conducted by local professionals where possible, and internationally experienced professionals in other areas. Multi-date image interpretations were manually registered to one another. Ground information was incorporated into about 50 percent of the interpretations. In some areas, ground truthing was not necessary owing to the high and consistent amount of forest. In other locations, especially where the composition of the landscape was highly differentiated, ground truthing was found to be highly valuable.

The principal output of the remote sensing survey was the change matrix, which illustrated and quantified how the forest and landscape change over time. The forest and land cover classification scheme of the remote sensing survey was linked closely to the FRA classes established for global reporting by countries.

Different definitions of forests for developing and industrialized countries again limited the utility of the final global synthesis, as did the absence of change information on forests in industrialized countries. Only changes in the area of forest combined with other wooded lands were assessed. (The definition of forest was set at 20 percent crown cover density for industrialized countries and 10 percent for developing countries.)

The assessment covered the parameters of volume, biomass, annual harvesting (tropics) and plantations. Brief summaries were also made on conservation, forest management and biological diversity. The country briefs prominent in FRA 1980 were unfortunately discontinued.

Major findings
- Total forest area (global) 1990: 3.4 billion hectares
- Net forest change (tropical developing countries) 1980-1990: -13.6 million hectares per year
- Net forest change (global) 1980-1990: -9.9 million hectares per year (forest and other wooded lands combined)

Interim 1995 assessment

An interim 1995 assessment was published in *State of the World's Forests 1997* (FAO 1997). This report published new statistics on forest cover state and change for all countries with a reference year of 1995, and a change interval from 1991-1995. The definition of forest varied between the industrialized and the developing countries; canopy closure thresholds were set at 20 percent for industrialized countries and 10 percent for developing countries.

The baseline information set for the assessment with only a minimum of updates was drawn from the FRA 1990 data set and had an average reference year of only 1983. Although FAO contacted all developing countries and requested their latest inventory reports, updated information was only submitted and used for Brazil, Bolivia, Cambodia, Côte d'Ivoire, Guinea-Bissau, Mexico, Papua New Guinea, the Philippines and Sierra Leone.

The FRA 1990 deforestation model was used for adjusting developing country statistics to a standard reference year (1991 and 1995). No adjustments to standard reference years were made for the industrialized country statistics. Consequently, the industrialized and developing country data were not harmonized in terms of their definitions or reference year.

---

[60] Two different models were used – one for the tropics and one for subtropical areas. Other differences among countries included lack of baseline data in some countries, lack of a uniform ecological map and lack of comparable multi-date observations.

Major findings
- Total forest area (global) 1995: 3.4 billion hectares
- Net forest change (tropical developing countries) 1990-1995: -12.7 million hectares per year
- Net forest change (global): -11.3 million hectares per year (total forests)

BIBLIOGRAPHY

FAO. 1948. *Forest resources of the world.* Washington, DC.

FAO. 1951. *Report of the sixth session of the Conference of FAO.* Rome.

FAO. 1960. *World forest inventory 1958 – the third in the sequential series compiled by the Forestry and Forest Division of FAO.* Rome.

FAO. 1966. *World forest inventory 1963.* Rome.

FAO. 1976a. *Appraisal of the forest resources in the Latin American Region.* Document presented at the 12th session of the Latin American Forestry Commission, Havana, Cuba, February 1976. FO:LACF/76.

FAO. 1976b. *Forest resources in the European Region.* Rome.

FAO. 1976c. *Forest resources in the Asia and Far-East Region.* Rome.

FAO. 1981a. *Tropical Forest Resources Assessment project (in the framework of GEMS) – Forest resources of tropical Asia.* Rome.

FAO. 1981b. *Los recursos forestales de la America tropical.* Rome.

FAO. 1981c. *Forest resources of tropical Africa.* Rome.

FAO. 1982. *Tropical forest resource,* by J.P. Lanly. FAO Forestry Paper No. 30. Rome.

FAO. 1988. *An interim report on the state of forest resources in the developing countries.* Rome.

FAO. 1995. *Forest resources assessment 1990 – global synthesis.* FAO Forestry Paper No. 124. Rome.

FAO. 1997. *State of the World's Forests 1997.* Rome.

Lanly, J.P. 1983. Assessment of forest resources of the tropics. *Commonwealth Forestry Review,* 44 (6): 287-318.

Lanly, J.P. 1988. *An interim report on the state of the forest resources in the developing countries.* FO: Misc/88/7. Rome, FAO.

Persson, R. 1974. *World forest resources – review of the world's forest resources in the early 1970's.* Department of Forest Survey, Reports and Dissertations No. 17. Stockholm, Royal College of Forestry.

Persson, R. 1975. *Forest resources of Africa – an approach to international forest resource appraisals, Part I: Country appraisals.* Department of Forest Survey, Reports and Dissertations No. 18. Stockholm, Royal College of Forestry.

Persson, R. 1977. *Forest resources of Africa – an approach to international forest resource appraisals, Part II: Regional analyses.* Department of Forest Survey, Reports and Dissertations No. 22. Stockholm, Sweden, Royal College of Forestry.

Sommer, A. 1976. Attempt at an assessment of the world's tropical moist forests. *Unasylva,* 112-113: 5-27.

Space, J. 1997. *Strategic plan, Global Forest Resources Assessment 2000.* Rome, FAO. (unpublished).

UNECE/FAO. 1985. *The forest resources of the ECE region (Europe, the USSR, North America).* Geneva.

Index of geographic names

A

Accra 105
Adelaide 260
Afghanistan 87, 150, 151, 157, 158, 159, 160
Africa ... xxv, 1, 3, 6, 9, 12, 25, 36, 55, 56, 66, 67,
75, 76, 83, 101, 305, 307, 310, 314, 315,
316, 328, 329
Ain Draham 109
Agadir Basin 106
Alaska 236, 237
Albania 211, 212, 213, 214, 216
Algeria 68, 84, 85, 106, 109, 110, 111, 112
Alps 69, 188, 190, 204
Altai 153, 154, 164, 193
Alto Paraná Region 90
Alto Uruguay Region 90
Amapà 281
Amazon Basin 65, 281, 282, 283, 287
Amazon Region 70, 75, 76, 88, 90, 282, 289
Amazon River 281
America 75
American Samoa 271, 272, 273, 274, 275, 276
Amu Darya delta 165
Amur River 190, 194
Andamans 170
Andean Region 70
Andes 281, 283, 284, 285, 287, 288, 294
Andiroba 90
Andorra 211, 212, 214
Angara River 192
Angola . 67, 83, 104, 105, 133, 134, 135, 136, 137
Annamitic Range 146, 148
Antigua and Barbuda 83, 249, 250
Apennines 187, 188
Apia 273
Appalachian Mountains 227, 228, 229
Arabian Peninsula 148
Arctic Circle 190, 192, 217
Ardennes 206
Argentina 10, 70, 75, 90, 89, 91, 283, 284, 293,
294, 295
Arizona 231
Armenia 157, 158, 159, 161
Aruba 249
Asia .... xxiv, xxvi, 3, 6, 9, 23, 55, 66, 67, 75, 76,
85, 305, 307, 310, 314, 315, 328, 329, 345
Asia-Pacific Region 36
Astrocaryum spp 92
Asturia 188
Atacama Desert 285, 294
Atlantic Ocean 69, 109, 203, 211, 226, 227,
243, 250
Atlantic-Indian Basin 133
Atlas Mountains 107

Australia .... xxii, 6, 11, 53, 56, 65, 69, 70, 83,
165, 256, 257, 258, 259, 260, 261,
263, 265, 266, 267, 268
Australian Alps 263
Austria .... xxiv, 69, 203, 204, 206, 207
Awara 93
Azerbaijan 157, 158, 159, 160, 161

B

Bahamas 83, 249, 250, 251
Bahrain 157, 158, 159, 160
Bali 174
Balkan Peninsula 213
Balkans 189
Baltic 197, 200
Bangladesh 18, 32, 33, 42, 43, 86, 87, 167,
168, 169, 171
Barbados 83, 249, 250, 251
Belarus 69, 186, 190, 217, 218, 219, 220, 222
Belgium 55, 203, 206, 207
Belgium and Luxembourg 204
Belize 83, 92, 243, 244, 247, 251
Belo Horizonte 70
Bengal 170
Benin 83, 115, 116, 118
Benin City 84
Bermuda 249, 250
Bhabar 170
Bhutan 86, 87, 167, 168, 169, 171
Bishan Mountain 152
Black Sea 148, 149, 151, 211
Bolivia 57, 59, 78, 88, 90, 89, 91, 283, 284,
287, 288, 289, 290, 294
Bonaire 249
Borneo 146
Bosnia 55
Bosnia and Herzegovina 211, 212, 213, 214
Botswana 83, 133, 134, 135, 136
Brahmaputra 145
Brazil .... xxiv, 10, 26, 27, 32, 35, 41, 57, 59, 65,
70, 71, 78, 88, 90, 91, 235, 281,
282, 283, 284, 287, 288, 289, 290
Brazilian Amazon 91
Brazilian Shield 282
British Columbia 236
British Isles 188, 190
British Virgin Islands 250
Brunei 18
Brunei Darussalam 173, 174, 175, 177
Bucharest 189
Bulgaria 211, 212, 213, 214, 216
Burkina Faso 83, 84, 115, 116, 117, 118, 119
Burma 147
Burundi 83, 121, 122, 123, 125

C

Cabinda 105
Cabo Delgado 135
California 230, 231
Cambodia 86, 146, 147, 173, 174, 175, 177
Cambridge 328
Cameroon 53, 58, 83, 84, 105, 121, 122, 124, 125
Canada 2, 53, 56, 57, 58, 59, 67, 69, 75,
226, 235, 236, 237, 238, 239, 240
Canarium spp 146
Cantabrica 188
Cantabrican Mountains 190
Cape Camorin 170
Cape of Good Hope 135
Cape region 107, 133
Cape Verde 83, 139, 140
Caribbean 8, 55, 57, 69, 83, 92, 224, 249, 283
Caroline Islands 258
Carpathians 190
Carpentaria Gulf 259
Cascade Mountains 227, 228, 230
Caspian Sea 148, 149, 151, 157, 163
Cauca River 283
Caucasus 189, 190
Caucasus Mountains 148, 189
Cayman Islands 250
Central Africa 8, 121
Central African Republic 67, 83, 102, 121,
122, 123, 124
Central America 41, 56, 65, 66, 69, 75, 76,
83, 91, 224, 329
Central American xxiv
Central Asia 65, 144, 163
Central Europe 186, 203, 204
Central European xxiv
Central Yakutia 192
Chaco 283, 285
Chad 83, 85, 115, 116, 117, 118, 119
Chamba 170
Changbaishan 152
Chao Phraya River 147
Charkov 189
Chernobyl 221
Chersky Range 194
Chiapas 65
Chiapas Sierra Madre 231
Chico 285
Chihuahuan Desert 243
Chile 28, 70, 75, 87, 90, 89, 91, 285,
293, 294, 295, 296
Chilean Norte Chico 285
Chimanimani 133
China .... xxiv, 10, 26, 27, 35, 37, 38, 66, 67,
86, 87, 88, 146, 147, 148, 149, 150,
151, 152, 153, 154, 163, 179, 180,
181, 182, 221, 274
Chukotka 192
Coast Range 227, 228, 230

Cobar Peneplain 261
Colchis 149
Colombia 41, 90, 89, 281, 283, 287, 288, 289, 290
Colombian Andes 284
Colorado Plateau 228, 230
Columbia 283
Columbia-Snake River Plateaus 228
Comoros 83, 139, 140
Conacaste 92
Congo 67, 83, 122, 315
Congo Basin 103, 122, 124
Cook Islands 271, 272, 274
Cordillera Cantabrica 188
Coromandel Coast 147
Corsica 188
Costa Rica 32, 37, 41, 83, 91, 92, 232,
243, 244, 246, 247
Côte d'Ivoire 83, 116
Crete 187, 188
Crimean Mountains 189
Croatia 68, 211, 212, 213, 214
Cuba 83, 93, 249, 250, 251, 252
Cumaru 90
Cunas 245
Curaçao 249
Cyprus 52, 68, 83, 157, 158, 159, 161
Czech Republic 203, 204, 205, 207

D

Dalmatia 187
Danube 190
Danzig 188
Darjeeling 170
Darling Riverine Plain 261
Darussalam 173
Daxinganling 154
Deccan Plateau 147, 170
Democratic People's Republic of Korea. 179, 180, 182
Democratic Republic of Congo 121, 122, 124, 125, 134
Denmark 188, 203, 204, 206, 207
Desna 190
Dinaric Alps 190
Dja National Park 124
Djebel El Ghorra 109
Djibouti 127, 128, 130
Dominica 83, 249, 250
Dominican Republic 83, 93, 249, 250, 251, 252
Drakensberg 107
Dzanga-Ndoki National Park 124

E

East Africa 8, 83, 102, 127
East Asia 144, 179
East Timor 173, 174, 177
Ecuador 89, 91, 281, 283, 284, 287, 288, 289

Index of geographic names 459

Egypt 84, 85, 109, 110, 111, 112
El Salvador 83, 91, 92, 232, 243, 244, 246, 247
Elbe 190
English Channel 69
Equator 65, 257, 283
Equatorial Guinea 83, 121, 122, 123
Erfurt 188
Eritrea 83, 84, 85, 127, 128, 129, 130
Estonia 69, 197, 198
Ethiopia 35, 65, 67, 83, 84, 85, 127, 128, 129, 130
Ethiopian highlands 105
Europe . 3, 25, 52, 56, 66, 68, 75, 78, 83, 212, 329
European Community 88
Euskal 188
Evenkija 192
Everglades 232
Eyre 261

F

Falkland Islands (Malvinas) 294
False beech 266
Far East 218
Federated States of Micronesia 271, 273, 274, 276
Fennoscandia 191
Fiji 33, 258, 271, 272, 273, 274, 275, 276
Finland 57, 59, 190, 191, 197, 198
Florida 237
Former Soviet Union 329
Fouta Djalon 105
France 43, 68, 188, 203, 204, 206, 207
French Guiana 288, 289, 291
French Polynesia 271, 272, 274, 275

G

Gabon 75, 78, 83, 121, 122, 124
Galicia 188
Gambia 83, 115, 116, 117, 118
Ganges Delta 145
Ganges Valley 147
Gangetic Plains 170
Gansu 153
Garagum 164
Garifonas 245
Gaza 159
Gaza Strip 157, 158, 161
Georgia 55, 157, 158, 159, 160, 161
Germany . 57, 59, 67, 69, 188, 203, 204, 206, 207
Ghana 53, 58, 83, 84, 85, 105, 115, 116, 118
Gippsland 262
Great African Plateau 104
Great Basin 228
Great Dividing Range 261
Great Lakes 69, 227
Greater Antilles 249
Greater Xingan Range 154
Greece 68, 187, 188, 211, 212, 213
Grenada 83, 92, 93, 249, 250, 251
Guadeloupe 249, 250

Guam 271, 272, 274, 276
Guanacaste 232
Guararema 283
Guatemala .... xxiv, 41, 57, 59, 83, 91, 92,
232, 243, 244, 246
Guayaquil Gulf 283
Guiana Shield 282, 284
Guinea 83, 84, 104, 116
Guinea-Bissau 83, 115, 116
Guineo-Congolian Basin 104
Gujarat 170
Gulf Coast 230
Gulf of Bengal 147
Gulf of Mexico 232
Gurue 133
Guyana 53, 58, 83, 90, 93, 251,
287, 288, 289, 291

H

Hainan Island 146, 147
Haiti 41, 83, 249, 250, 251, 252
Haryana State 41
Hawaii 236, 237
Heilongjiang River 154
Helsinki 189
Herzegovina 55
High Atlas 109
Highveld region 107
Himachal Pradesh 152, 170
Himalayas 148, 150, 170
Hinoki 181
Hispaniola 249
Hoggar 105
Hokkaido 153
Hokkaido Region 179
Honduras 41, 83, 92, 232, 243, 244
Hong Kong 87, 88
Honshu 153, 179
Hudson Bay 225
Hudson Plain 225
Hummocky 192
Hungary 203, 204, 205, 206, 207

I

Iberian Peninsula 187, 188, 213, 215
Iceland 190, 192, 193, 197, 198
Idaho 69
Iguaçu 281
India .... xxiv, 10, 26, 27, 31, 32, 33, 35, 38,
41, 43, 52, 86, 87, 146, 147, 167,
168, 169, 170, 171
Indian Ocean 127, 133, 151
Indigirka River 192
Indochina 147, 148
Indo-Malaya 258
Indonesia .... xxiv, 10, 18, 26, 27, 32, 33, 35, 36,
38, 53, 58, 86, 87, 173, 174, 176, 177, 257
Indonesian Archipelago 68

Indus River 151, 170
Inner Mongolia 67
Insular East Africa 83
Insular Southeast Asia 67
Iran 26, 27, 83, 151, 157, 158, 159, 160, 161
Iran (Islamic Republic of) 158
Iraq 157, 158, 159, 160
Ireland 203, 204, 206, 207
Irian Jaya 257
Irtish River 191
Islands 271
Israel 52, 68, 157, 158, 159, 161
Italy 56, 211, 212, 213, 214, 215
Ivory Coast 115, 116, 119

J

Jamaica 83, 93, 249, 250, 252
Japan .... xxiv, 26, 27, 35, 83, 88, 148, 149, 152,
153, 165, 179, 180, 181, 182, 221, 274
Japanese Archipelago 180
Jarrah 261, 262
Java 174, 175
Jordan 83, 157, 158, 159, 160
Jordan-Arava Rift Valley 150
Jos Plateau 105
Jura 190

K

Kalahari 105
Kalahari Desert 133, 134
Kalahari-Highveld phytoregion 137
Kalimantan 68, 173
Kamchatka River 194
Karnataka Coast 170
Karoo-Namib phytoregion 137
Karri 261
Kashmir 151, 170
Kazakhstan 163, 164, 165
Kenya 11, 41, 83, 85, 105, 127, 128, 129
Kerala 32, 170
Kerkira 187
Khangai 153, 164
Khangai Mountains 154
Khasi 170
Khentii 164
Kimberley Plateau 265
Kiribati 271, 272, 274
Kivu 106
Kivu ridge 105
Kola Peninsula 192
Kolyma River 192
Korat Plateau 147
Korea 274
Korean Peninsula 148, 152, 180
Kufa 93
Kusnetsky Ala-Tau 193
Kuwait 157, 158, 159, 160
Kwahu District 84

Kyrgyzstan 163, 164, 165

L

La Pampa 70
Labrador 225
Ladakh 170
Lahol 170
Lake Baikal 193, 221
Länder 207
Lao People's Dem. Rep. 174
Lao People's Democratic Republic 86, 87, 146,
147, 148
Latin America 6, 9, 13, 55, 305, 307, 310, 314,
315
Latvia 69, 197, 198
Lebanon 83, 151, 157, 158, 160
Lena River 192
Les Landes 204
Lesotho 83, 133, 134, 135
Lesser Antilles 249
Lezhou Peninsula 146
Liberia 83, 84, 116, 119
Libya 109, 110, 112
Libyan Arab Jamahiriya 110
Liechtenstein 203, 204, 207
Lithuania 69, 197, 198
Lofty Block 261
Luxembourg 55, 203, 206, 207

M

Madagascar 10, 35, 83, 84, 85, 104, 105,
133, 134, 135, 136
Madhya Pradesh 170
Madras 170
Madrone 230
Mahableswar 170
Maharashtra 170
Malawi 83, 133, 134, 135, 136
Malay Archipelago 145
Malaysia 36, 53, 58, 86, 87, 88, 146, 148,
173, 174, 177
Malaysian Peninsula 148
Maldives 167, 168, 169, 170, 171
Mali 83, 85, 115, 116, 117, 119
Malta 211, 212
Maluku 173
Mandalay 147
Mandara Plateau 105
Marrakech Basin 106
Marri 261
Marshall Islands 271, 272, 274
Martinique 249, 250, 251
Massif Central 190
Massive kauri 260
Mauritania 115, 116, 117, 119
Mauritius 83, 139, 140
Mayas 245
Mediterranean Sea 68, 109, 187, 203, 211

Index of geographic names

Mekong River 147
Melanesia 257, 258, 272
Melanesian archipelago 257, 258
Mesquite 230
Mexico 8, 57, 59, 65, 69, 91, 231,
              232, 235, 238, 243, 244, 245, 246
Micronesia 257, 272
Micronesian archipelago 257
Middle East 150, 221
Mirzachol 165
Mississippi River 229
Mizquitos 245
Moldova 222
Mongolia 67, 153, 154, 163, 164, 165
Mongolian poplar 192
Montana 69
Montserrat 249, 250
Morocco 41, 67, 68, 84, 85, 106, 109, 110, 111, 112
Mozambique .. 11, 67, 83, 133, 134, 135, 136, 137
Mulga Lands 261
Murray-Darling 261
Myanmar 33, 53, 58, 78, 86, 87, 145, 146,
                 147, 148, 173, 174, 176, 177
Mysore 170

N

Namibia 83, 84, 133, 134, 135, 136
Nanling Mountains 149
Naracoorte Coastal Plain 261
Narrow-leaved red mallee 262
Nauru 271, 272, 274
Near East 83, 329
Nepal 86, 87, 152, 167, 168, 169, 170, 171
Netherlands 203, 204, 206, 207
Netherlands Antilles 249, 250
New Caledonia 258, 271, 272, 273, 274, 275
New England Tablelands 263
New Guinea 145, 146
New Ireland 275
New South Wales 260, 261, 267
New Zealand 28, 32, 37, 56, 69, 70, 83, 165,
        256, 257, 260, 262, 265, 266, 267, 272, 274
Newfoundland 226
Nibi 93
Nicaragua 83, 92, 232, 243, 244, 245, 247
Niger 83, 85, 116, 117, 118, 119
Nigeria 83, 84, 85, 115, 116, 117, 118
Nilgiri 170
Niue 271, 272, 274, 275
Non Tropical South America 280, 293
North Africa 83, 102, 109
North African Sahara 109
North America 55, 56, 66, 69, 75, 76, 78,
                     83, 209, 224, 235, 236, 328, 329
North and Central America 3, 25
North Europe 197
North Island 260, 265
North Sea 69

Northern Europe 186
Northern Mariana Islands 271, 272, 273, 274, 276
Northern pine 236
Northern Viet Nam 146
Norway 57, 59, 188, 190, 193, 197, 198, 200, 205
Novgorod 189
Nusa Tenggara 174

O

Oceania 3, 8, 25, 55, 56, 65, 69, 76, 256, 329
Oimjakon Upland 194
Okhotsk sea 194
Okoumé 123
Old Crow Basin 227
Olympus 188
Oman 157, 158, 159, 160
Ontario 236
Oregon 69
Orinoco 287
Orissa 170
Oslo 189
Other Africa 102
Outeniekwaberge 107

P

Pacific Coast Mountains 228
Pacific Islands 257, 259
Pacific Ocean 243, 271, 284, 285
Pakistan 33, 86, 87, 150, 151, 167, 168,
                                   169, 170, 171
Palau 271, 272, 273, 274
Palembang 68
Panama 83, 91, 92, 233, 243, 244, 245
Papua New Guinea 56, 75, 86, 147, 257,
                             259, 272, 273, 274, 275
Paraguay 90, 89, 91, 283, 287, 288, 289, 294
Paraná 287
Parana River 281, 283
Patagonia 70, 294
Patagonian Andes 285
Pechora River 192
Peleponnesus 188
Penibética 188
Peninsular Malaysia 86, 176, 177
Perth 260, 265
Peru 35, 41, 88, 90, 89, 91, 282, 283,
                              284, 287, 288, 289, 290
Petén 92
Philippines 36, 37, 86, 87, 145,
                        146, 148, 173, 174, 176, 177, 274
Pindus 188
Po Plain 188
Podkamennaja Tunguska River 192
Poland 57, 59, 68, 69, 190, 203, 204, 206
Polynesia 272
Polynesian archipelagos 257
Pomeroon Basin 93
Portugal 68, 188, 211, 212, 214

Prairie Parkland 230
Pripet 190
Puerto Rico 249, 250, 253
Punjab 170, 171
Pyrenees 188, 190, 204

Q

Qatar 157, 158, 159
Qinling Range 149
Quebec 236
Queensland 257, 260, 261
Quillay 90, 89
Qyzylqum 165
Qyzylqum Desert 163

R

Radom-Katowice 68
Rajasthan 170
Red River 146
Red River Plain 147
Red Sea 109, 127, 159
Republic of Congo 121, 122, 124, 125
Republic of Korea 87, 179, 180, 181, 182
Republic of Moldova 186, 217, 218, 220
Réunion 139 140
Rhine 190
Rhodope Mountains 190
Rhone Basin 187
Rio Grande do Sul 284
River redgum 261
Riverina 261
Rocky Mountains 228, 230, 231
Romania 211, 212, 213, 214, 216
Roughleaf dogwood 230
Russian Far East 66
Russian Federation .... xxii, 11, 26, 27, 56, 58,
68, 78, 163, 185, 186, 190, 193,
194, 217, 218, 219, 220, 221, 235
Rwanda 83, 84, 121, 122, 123, 124, 125, 128

S

Sabah 174, 176, 177
Sahara 106
Sahel Region 115
Saint Helena 133, 134, 135
Saint Kitts and Nevis 83, 249, 250
Saint Lucia 83, 249, 250, 251
Saint Vincent and the Grenadines 83, 249, 250
Sakhalin Island 194
Salair Range 193
Salt Lake City 328
Salta 283
Samoa 258, 271, 272, 273, 274, 275, 276
San Juan 231
San Marino 55, 211, 212, 214
San Vincent and Grenadines 251
Sangre de pozo 231

Sangre real 231
Santa Lucia 93
Sao Tome and Principe 139, 140
Sarawak 174, 176, 177
Sardinia 188
Saudi Arabia 83, 157, 158, 159, 160
Savai'i 274
Save 190
Scandinavia 188, 189
Scotland 190
Scots pine 197, 205, 213, 218
Scottish Highlands 188, 192, 193
Seliyon 231
Senegal 83, 85, 104, 115, 116, 117, 119
Seychelles 83, 139, 140
Shanxi 153
Sheesham 169
Siberia 163, 191, 192, 193, 218
Siberian Plateau 191, 193
Sichuan 153
Sichuan Basin 149
Sierra Leone 83, 115, 116, 119
Sierra Madre 231, 232
Sierra Nevada 228, 230
Sikhote-Alin 194
Sikkim 170
Sind 170
Singapore 87, 88, 173, 174, 176, 177
Sistema Berico 188
Sistema Central 188
Slovakia 203, 204, 205, 207
Slovenia 211, 212, 213
Sofia 189
Solomon Islands 271, 272, 273, 274, 275, 276
Somalia 83, 85, 127, 128, 129, 130
Sonora Desert 243
Sous 107
South Africa 8, 67, 83, 133, 134, 135, 137
South America .... xxiv, 3, 12, 23, 55, 57, 65,
70, 74, 76, 88, 328, 329, 345
South Asia 8, 144, 167
South East Asia 173
South Island 262, 263, 265
South Korea 148
South Punjab 170
South Sumatra 68
South Western Slopes 261
Southeast Asia 36, 65
South-East Asia 8, 144
Southeastern Europe 69
Southeastern Highlands 263
Southern Africa 8, 102, 133
Southern Baltic Sea 69
Southern Brigalow Belt 261
Southern Europe 186, 211, 212
Southern Hemisphere 32
Southern Viet Nam 146
Southwest Islands 179
Spain 68, 92, 187, 211, 212, 214
Sri Lanka 33, 37, 86, 88, 145, 146,

147, 167, 168, 169, 171
State of Roraima 70
Straight of Magellan 293
Stringybark 262, 263
Sudan 6, 9, 35, 41, 83, 85, 105,
127, 128, 129, 130, 315
Sugi..181
Sulawesi...173
Sumatra.....................................68, 146, 148, 173
Suriname 75, 83, 93, 287, 288, 289, 290, 291
Swaziland...........................33, 83, 133, 135, 136
Sweden. 57, 59, 185, 188, 189, 190, 197, 198, 200
Switzerland 69, 203, 204, 205, 207
Syria...83, 151, 159, 160
Syrian Arab Republic.......................157, 158, 159

T

Table Mountain..106
Tahiti..275
Taimir 192
Taiwan 88, 149, 152, 274
Tajikistan 163, 164, 165, 166
Talysh Mountains 148, 190
Tamarack 225, 226, 227, 237
Tanoak 230, 231
Tanzania...........67, 83, 84, 85, 127, 128, 129, 134
Tapia 106
Tara...91
Tashkent...165
Tasmania...............................262, 263, 265, 267
Tasmanian Highlands 263
Thailand .... xxiv, 10, 26, 27, 33, 36, 37, 38, 68,
86, 87, 88, 147, 148, 173, 174
The former Yugoslav Republic of Macedonia 211,
212 ,213, 214, 216
Tiama 123
Tibet..148, 152
Tibetan Plateau 153
Togo..83, 116
Tonga 271, 272, 273, 274
Trinidad 32
Trinidad and Tobago.............83, 93, 249, 250, 251
Tropic of Capricorn 283
Tropical America 52
Tropical South America............................280, 287
Tunisia 83, 84, 85, 106, 109, 110, 111, 112
Turkey..........68, 83, 150, 151, 157, 158, 159, 160
Turkmenistan 163, 164, 165
Tuva...193

U

Ufa 189
Uganda................................83, 127, 128, 129, 130
Ukraine ...... xxiv, 26, 27, 186, 217, 218, 220, 222
Ukrania 221
United Arab Emirates 157, 158, 159
United Kingdom 43, 69, 203, 204, 206, 207
United Republic of Tanzania 128

United States..235, 236
United States of America... xxiv, 5, 10, 27, 32, 38,
53, 56, 57, 58, 59, 65, 67, 68, 69, 71,
75, 79, 87, 88, 227, 228, 232, 235, 236,
237, 238, 239, 240
United States Virgin Islands............................250
Upolu 274
Upper Assam 170
Ural Mountains........................ 189, 191, 192, 193
Urals 190, 219
Uruguay 35, 75, 90, 89, 284, 293, 294, 296
USSR 53
Utah 228
Uttar Pradesh 170
Uzbekistan 163, 164, 165

V

Valparaiso..284
Vanuatu 258, 271, 272, 273, 274, 275, 276
Vekhojansky Range... 194
Venezuela 90, 89, 282, 283, 284,
287, 288, 289, 290
Vetiver..92
Victoria...262, 263, 267
Vienna 188
Viet Nam 37, 78, 86, 87, 88, 145,
147, 148, 173, 174, 177
Virgin Islands 249
Volga 190
Volta River 115

W

Wallace line 145, 146
West Africa...8, 83, 102
West Asia 144, 157
West Bank 157, 158, 159, 161
West Indies 249, 250
West New Britain 275
Western Ghats 170
Western Sahara... 109, 110
Western Sierra Madre......................................230
White Sea 191
Windhoek Mountain...106
Wyoming Basin..228

X

Y

Yakutia 193
Yamoussoukro..116
Yangtze River...148, 149
Yellow Loess Plateau 153
Yemen 158, 159, 161

Yenisey River 191, 192
Yorell 262
York Block 261
Yucatan Peninsula 232
Yugoslavia 55, 211, 212, 213
Yukagir Upland 194
Yungui Plateau 149
Yushan Mountain 152

Z

Zagros 158
Zambia 6, 9, 28, 67, 77, 83, 84,
133, 134, 135, 136, 137, 316
Zealand's Southern Alps 263
Zimbabwe 10, 28, 83, 85, 133, 134, 135, 136
Zululand Republic of South Africa 32

Index of botanical names

A

Abies
- alba 188, 190
- amabilis 228
- balsamea 225, 226, 228, 229
- borisii-regis 188, 190
- cephalonica 188
- chensiensis 149
- cilicica 151
- concolor 230
- ernestii 149
- faberi 152
- fargesii 149
- faxoniana 153
- firma 150
- georgei 152
- holophylla 152
- kawakamii 152
- lasiocarpa 227, 228
- lasiocarpa var. arizonica 231
- magnifica 230
- nephrolepis 152, 153, 193
- nordmanniana 190
- pinsapo 188
- sachalinensis 154, 194
- sibirica 190, 191, 193
- spectabilis 151, 152
- webbiana 151

Abies spp. 148, 152, 190, 213, 218, 243

Acacia 23, 157, 174, 262, 267
- albida 105
- aneura 261
- auriculiformis 31
- caffra 105
- cambagei 261
- caven 283, 284, 285
- davyi 105
- dudgeoni 105
- farnesiana 85
- gourmaensis 105
- gummifera 107
- harpophylla 259, 261
- karroo 105
- loderi 261
- luederitzii 105
- macrostachya 105
- mangium 174
- mearnsii 31, 174
- melanoxylon 262, 263
- modesta 150
- nilotica 105, 129
- pendula 261
- senegal 83, 84, 85, 128, 129
- seyal 85
- shirleyi 259
- tortilis 85

Acacia spp. 32, 105, 112, 134, 147, 148

Acer
- campestre 190
- formosum 152
- insigne 151
- macrophyllum 227, 228
- mono 152, 153
- monspessulanum 150
- negundo 228
- oblongum 149
- pseudoplatanus 190
- saccharum 226, 227
- tegmentosum 152
- ukurunduense 152

Acer spp. 33, 150, 152, 153, 154, 194, 229,
.... 232, 236
Achras zapota 232
Acmena smithii 260
Acrostichum aureum 104
Adansonia digitata 105, 135
Adansonia spp. 105
Adina cordifolia 147
Adina spp. 147
Aesculus spp. 227
Aetoxicon punctatum 285
African oil palm 24, 36
Afrormosia spp. 34
Afzelia
- africana 104
- quanzensis 134
- xylocarpa 147

Agarwood 86
Agathis australis 260, 266
Ailanthus altissima 153
Albizia
- amara 147
- macrophylla 149

Albizia spp. 140
Alchornea bogotensis 284
Alder 191, 192, 193, 194, 225
Aleppo pine 150, 213
Aleurites spp. 88
Alfa 110
Algarrobo 283
Allspice tree 231
Almond 151
Alnus
- barbata 150
- glutinosa 110, 190
- incana 225
- japonica 153

maximowiczii 154
rubra 228
subcordata 150
Alnus spp. 191, 236
Alpine ash 263
Altingia obovata 147
Amburana spp. 283
American beech 227, 229
American elm 230
American sycamore 230
Amesiodendron chinense 147
Ampelocera hottlei 231
Amygdalus
arabica 150
cf. *communis* 151
korshinskyi 150
kuramica 151
Amygdalus spp. 151
Amyris balsamifera 92
Anacardiaceae 104, 146
Anacardium
excelsum 232
occidentale 105
Anamirta cocculus 86
Anastrabe integerrima 106
Aniba rosaeodora 90, 92
Aningeria
adolfi-fredrici 106
altissima 104
robusta 104
Aningeria spp. 34
Annatto 90
Annona spp. 92
Annonaceae 281
Anodopetalum biglandulosum 262
Anogeissus leiocarpus 105
Anogeissus spp. 105
Anono 92
Antidesma venosum 105
Apache pine 231
Apodytes spp. 107
Aponogeton spp. 85
Apuleia leiocarpa 283
Aquifoliaceae 147
Aquilaria spp. 86, 88
Araliaceae 104
Araucaria
angustifolia 31, 32, 90, 284
araucana 90, 91, 285
cunninghamii 31
Araucaria spp. 146, 148, 260, 284
Araucariaceae 258
Arbutus 229
andrachne 150
menziesii 227, 229, 230, 231
unedo 110
Ardisia granatensis 282
Argan tree 85
Argania spinosa 85
Argania spp. 107

Argyrodendron
actinophyllum 260
trifoliolatum 260
Argyrodendron spp. 260
Arundinaria alpina 106
Ash 33, 189, 197, 237
Ashe juniper 230
Aspidosperma
peroba 283
polyneuron 283
Aspidosperma spp. 283
Assam 170
Aspen 191, 218, 231, 237
Astrocaryum
maripa 93
segregatum 93
Astronium
graveolens 232
urundeuva 283
Astronium spp. 283
Atalaya natalensis 106
Ateleia guaraya 283
Atherosperma moschatum 262
Athrotaxis selaginoides 262
Athyrium pycnosorum 153
Atriplex spp. 112
Attalea
cohune 92
funifera 91
Aucoumea klaineana 123
Aucuba japonica 150
Austrobaileya spp. 258
Avicennia
africana 104
alba 146
marina 105, 146, 147, 259
nitida 104, 282
officinalis 146
tomentosa 282
Avicennia spp. 258

B

Backhousia spp. 258
Bactris gasipaes 90, 92
Balanites aegyptiaca 85
Baldcypress 229
Balfourodendron riedlianum 283
Balkan oak 190
Balsam fir 225, 226, 228, 229
Balsam poplar 225, 227, 228
Bamboo 24, 81, 86, 87, 88, 91, 92, 93, 106,
147, 149, 169, 194, 233
Bambusa vulgaris 93
Banyan figs 258
Baobab 105
Basswood 227, 228, 229
Bathiaea spp. 105
Bauhinia spp. 134

Beech 188, 190, 197, 204, 205, 213, 237, 260, 262, 285
Befaria spp. 284
Beilschmiedia natalensis 106
Beilschmiedia spp. 260, 284
Bellota 92
Berlinia spp. 105
Bertholletia excelsa 88
Betula
    albo-sinensis 149, 152, 154
    alleghaniensis 228, 229
    cajanderi 194
    costata 152
    davurica 152
    ermanii 154, 194
    kajanderi 194
    papyrifera 225, 226, 227
    pendula 190, 191
    platyphylla 152, 153, 154, 193
    pubescens 188
    pubescens subsp. *czerepanovii* 192, 193
    utilis 151, 154
Betula spp. 152, 163, 191, 197, 218, 236
Big-leaf maple 227, 228
Bignoniaceae 282
Birch 191, 192, 193, 194, 197, 218, 229, 236, 237
Bird-cherry trees 193
Bishop pine 230
Bitter oak 190
Bitternut hickory 227
Bixa orellana 90
Black ash 226
Black cottonwood 227, 228
Black locust 205
Black oak 227, 229
Black spruce 225, 226, 227
Blackbox 261
Blackjack oak 230
Blackwood 263
Blepharocarya spp. 258
Blue oak 230
Blue pine 170
Boldo 90
Boldoa fragrans 90
Bombacaceae 281
Bombacopsis quinata 232
Bombax
    aquaticum 282
    munguba 282
Boswellia papyrifera 84, 85
Bowdichia spp. 282
Box 261, 263
Box-elder 228
Brachychiton discolor 260
Brachylaena uniflora 106
Brachystegia
    floribunda 105
    glaberrima 105
    laurentii 103

    longifolia 105
    spiciformis 105
    taxifolia 105
    utilis 105
    wangermeeana 105
Brachystegia spp. 105, 134
Breadnut 231
Brigalow 259
Bristlecone pine 231
Brosimum alicastrum 92, 231
Brosimum spp. 232
Bruguiera
    conjugata 147
    cylindrica 146, 147
    gymnorhiza 259
    gymnorrhiza 105, 146
Brunellia
    comocladifolia 284
    occidentalis 284
Brunellia spp. 284
Brysonima coriacea 93
Bubbia spp. 258
Buckeye 227
Buckinghamia spp. 258
Bulnesia arborea 283
Bur oak 228
Burkea africana 134
Burseraceae 281
Byrsonima crassifolia 232

C

Cabralea spp. 283
Cactaceae 283
Caesalpinia
    coriaria 283
    pulcherrima 92
    spinosa 89, 91
Caesalpinia spp. 283
Caimito 92
Calahuala 92
Calamus manan 86
Calamus spp. 86
California laurel 231
California live oak 231
California red fir 231
Calligonum comosum 112
Callitris glauca 259, 261
Callitris spp. 261, 267
Calocedrus decurrens 231
Calophyllum
    brasiliense 232, 282
    inophyllum 140, 170
Calophyllum spp. 258
Calycophyllum spruceanum 282
Calycotome villosa 110
Campnosperma spp. 258
Candlewood 92
Canyon live oak 231
Caoba 231

Caobina.. 231
Capparis
   coccolobifolia 283
   zeylanica 147
Capparis spp. 283
Carapa guianensis 232, 282
Carapa spp. 90
Cardwellia sp. 258
Carex middemdorfii 194
Carludovica palmata 89, 91, 92, 93
Carpinus
   betulus 150, 151, 189, 190, 197
   orientalis 150
Carya
   cordiformis 227
   illinoiensis 229
   ovata 227
   ovata 237
Carya spp. 229
Caryocar brasiliense 282
Cassia 88
Cassia spp. 86
Castanea
   mollissima 153
   sativa 150
Castanopsis
   carlesii 149
   chrysophylla 231
   cuspidata 149
   eyrei 149
   fargesii 149
   hystrix 149
   kawakamii 149
   kusanoi 149
   lamontii 149
   sclerophylla 149
   uraiana 149
Castanopsis spp. 148, 149
Castanospermum spp. 258
Castilla
   elastica 231
   tunu 231
   ulei 282
Casuarina
   equisetifolia 31
   junghuhniana 31, 148
Casuarina spp. 32, 140, 146, 157, 267
Cat's claw 90, 89
Catalpa bungei 153
Cedar 151, 193, 237
Cedrela
   fissilis 282, 283, 284
   mexicana 232
   odorata 92, 245, 282
Cedrela spp. 246, 283
Cedro 92, 245
Cedrus
   atlantica 107, 111
   deodara 151, 152
   libani 151

Ceiba pentandra 88, 282
Celery top pine 262
Celtis
   laevigata 229
   mildbraedii 104
   sinensis 149
   spinosa 283
Ceratonia siliqua 41, 110
Ceratonia spp. 150
Ceratopetalum apetalum 260
Cercidium australe 90
Cercidium praecox 283
Cercis siliquastrum 150
Ceriops
   decandra 146
   tagal 105, 147, 259
Chamaecyparis
   formosensis 152
   nootkatensis 228
   obtusa 182
   obtusa var. *formosana* 152
Cherry 33
Chihuahuan pine 231
Chilean hazelnut 90
Chilghoza pine 87
Chir pine 170
Chlorophora
   excelsa 104
   tinctoria 232
Chlorophora spp. 34
Chloroxylon swietenia 147
Chorophora spp. 34
Chosenia arbutifolia 154, 192
Chrysophyllum
   cainito 92
   gorungosanum 106
   perpulchrum 104
Chukrasia tabularis 147
Chusquea spp. 91
Cinchona cuatrecasasii 284
Cinchona spp. 83, 84, 90
Cinnamomum
   camphora 86, 149
   chekiangense 149
Cinnamomum spp. 149
Citronella 92
Citronella spp. 86
Cliffortia spp. 106
Clusia spp. 93, 284
Clusiaceae 258, 281
Coachwood 260
Coccus lacca 88
Coco-de-Mer palm 140
Coconut 38
Coconut palm 24, 36, 275
Cocos
   comosa 283
   nucifera 24, 36, 38, 275
Cocothrinax barbadensis 93
Cohune 92

Cola
- *acuminata* 84
- *gigantea* 104
- *greenwayi* 106
- *natalensis* 106

Colophospermum mopane 105, 135
Combretaceae 147
Combretum spp. 105, 134
Commiphora
- *harveyi* 106
- *myrrha* 85

Commiphora spp. 85, 105, 148
Common hazel 92
Congo mallee 262
Conocarpus erectus 282
Conostegia polyandra 282
Copaiba 90
Copaifera spp. 90
Copai-yé wood 231
Copernicia
- *cerifera* 283
- *prunifera* 90

Coprosma virescens 262
Cordia
- *alliodora* 31, 229, 231, 232, 275
- *bicolor* 231
- *caffra* 106

Cordia alliodora 229
Cordia bicolor 229
Corkbark fir 231
Cornus drummondii 230
Cornus spp. 232
Corylus avellana 189
Corymbia maculata 260
Cottonwood 228, 229
Couepia
- *longipendula* 90
- *polyandra* 92

Couma spp. 90
Crataegus aronia 150
Croton draconoides 90
Cryptocarya chinensis 149
Cryptocarya spp. 146, 284
Cryptomeria japonica 181
Cryptomeria spp. 140
Cuajada 231
Cumbillo 231
Cunninghamia lanceolata 149, 152, 181
Cunoniaceae 258
Cupressaceae 258, 260
Cupressus
- *chengii* 154
- *funebris* 152
- *lusitanica* 31
- *macrocarpa* 230
- *sempervirens* 188

Cupressus spp. 237
Curatella americana 232, 282
Cyathea spp. 85
Cybistax donnell-smithii 232

Cycas thouarsii 85
Cyclobalanopsis
- *acuta* 150
- *gilva* 149
- *glauca* 149
- *myrsinaefolia* 149, 150
- *salicina* 149
- *stenophylloides* 152

Cyclobalanopsis spp. 149
Cymbopogon citratus 92
Cynometra alexandri 103
Cypress 181, 188, 229, 230, 237

D

Dacrycarpus spp. 260
Dacrydium
- *cupressinum* 262, 266

Dacrydium spp. 148, 260
Daemonorops spp. 86
Dalbergia
- *hupeana* 149
- *sissoo* 31, 33, 169

Dalbergia spp. 33, 34, 105, 232
Damnacanthus indicus 150
Darlingia spp. 258
Dendrocalamus spp. 87
Dendrocalamus strictus 147
Dendropanax arboreus 231
Desmoncus sp. 92
Dialium guianense 231, 232
Dicoryphe spp. 106
Didiereaceae 105
Dillenia
- *pentagyna* 152
- *turbinata* 147

Dillenia spp. 147, 258
Dilleniaceae 146
Dioscorea deltoidea 86
Dioscorea spp. 84
Diospyros
- *abyssinica* 105, 106
- *hainanensis* 147
- *inhacaensis* 106
- *kaki* 153
- *lotus* 150

Diospyros spp. 147
Dipterocarpaceae 145, 147, 258
Dipterocarpus
- *intricatus* 147
- *obtusifolius* 147
- *tuberculatus* 147

Dipterocarpus spp. 34, 146
Dipteronia sinensis 154
Dipteryx panamensis 232
Discaria toumatou 262
Dodonea viscosa 150
Doryphora sassafras 260
Doryphora spp. 258
Douglas fir 188, 204, 206, 227, 229, 230,

.... 231, 236
Dracophyllum traversi... 263
Dragon's blood.. 90
Drimys winteri... 284, 285
Dry ash.. 262
Dryobalanops spp.. 146
Dryopteris crassirhizoma.. 153
Drypetes
    australasica.. 260
    gerrardii... 106
Duschekia kamschatika.. 194
Dwarf birch... 227
Dwarf pine... 192, 193, 194, 218
Dypterix odorata.. 90
Dysoxylum binectariferum.. 147

E

Eastern hemlock... 227, 228, 229
Eastern white cedar... 226
Ebenaceae... 104, 146
Elaeis
    guineensis.. 24
Elaeis guineensis... 36, 38, 84
Elaeocarpaceae.. 104
Elaeocarpus
    holopetalus... 262
    japonica.. 149
Elaeocarpus spp... 258
Elm... 163, 208, 229, 237
Empetrum sibiricum... 194
Endospermum spp... 258
Engelhardtia
    roxburghiana... 147, 149
Engelhardtia spp.. 148
Englemann spruce.. 229, 231
Entandophragma cylindricum....................................... 34
Entandophragma utile.. 34
Entandrophragma
    angolense... 123
    cylindricum... 123
Entandrophragma spp... 103
Enterolobium cyclocarpum................................... 92, 232
Ephedra spp... 86
Erica spp.. 106
Ericaceae... 232
Erythrina berteroana... 92
Eschweilera calyculata... 232
Eschweilera spp.. 284
Eucalypt................................. 177, 259, 260, 262, 263, 267
Eucalyptus.. 148
    acmenioides... 260
    alba.. 261
    albens... 261, 263
    baxteri... 261
    blakelyi... 261, 263
    botryoides... 262
    brevifolia... 259
    caliginosa.. 263
    calophylla... 261
    camaldulensis.. 31, 261
    crebra.. 260, 261
    cypellocarpa.. 262
    dalrympleana... 263
    deglupta.. 31
    delegatensis... 263
    dichromophloia... 259
    diversicolor... 261
    diversifolia.. 262
    dives.. 262
    drepanophylla.. 259
    dumosa.. 262
    eremophila.. 262
    fastigata.. 262
    fibrosa.. 260, 261
    foecunda... 262
    globulus... 31, 212
    gomphocephala.. 261
    gracilis.. 262
    grandifolia.. 259
    grandis.. 31, 32
    gummifera... 262
    incrassata.. 262
    intermedia.. 259, 260
    jacksonii.. 261
    laevopinea... 263
    largiflorens.. 261
    leptophleba... 259
    leucoxylon... 261
    maculata... 260
    marginata... 261, 262
    melanophloia... 259, 260
    melliodora.. 261, 263
    microcarpa.. 261
    microcorys.. 260
    microneuro.. 259
    miniata.. 259
    nitida... 262
    normantonensis... 259
    nova-anglica.. 263
    obliqua.. 261, 262
    odorata... 261
    oleosa... 262
    pellita... 259
    pilularis.. 260
    populnea... 261
    pruinosa.. 259
    radiata.. 262
    regnans... 263
    robusta... 31
    saligna... 31, 260
    salmonophloia... 262
    setosa.. 259
    sideroxylon.. 261
    sieberi... 262
    socialis.. 262
    tectifica... 259
    tereticornis.. 259, 260
    terminalis.. 259
    tessellaris... 259, 260

Index of botanical names

tetrodonta 259
urophylla 31, 32
viminalis 261, 262, 263
wandoo 262
Eucalyptus spp. .... xxiv, 23, 25, 31, 32, 85, 92,
 157, 160, 181, 237, 259, 266, 289, 294
Eucryphia cordifolia 285
Eugenia jambos 93
Euphorbiaceae 104, 146, 281
Eupomatia spp. 259
Eurya spp. 152
Euterpe
oleracea 90, 93, 282
precatoria 90
Excoecaria agallocha 146

F

Fagaceae 147, 148
Fagetea hyrcanica 151
Fagus
americana 229
crenata 153, 154
grandifolia 197, 227
orientalis 151
sylvatica 188, 190
sylvatica subsp. *moesiaca* 190
sylvatica subsp. *orientalis* 150, 190
sylvatica subsp. *sylvatica* 190
Fagus spp. 213, 237
Faidherbia albida 85
Faurea saligna 105
Ficus microcarpa 149
Ficus spp. 85, 258, 283
Fir 151, 152, 170, 188, 190, 191, 193,
 .... 194, 213, 218, 236, 237
Fitzroya cupressoides 285
Flindersia spp. 258, 260
Foothill pine 230
Fraxinus
angustifolia 190
chinensis 153
excelsior 189, 190, 197
latifolia 227
mandshurica 152, 194
nigra 226
oxycarpa 151
pennsylvanica 229
rhynchophylla 152
xanthoxyloides 151
Fraxinus spp. 33, 237

G

Galbulimima spp. 259
Gallesia gorazema 283
Garcinia spp. 84
Garry oak 229
Garuga floribunda 147
Gevuina avellana 90, 89

Gevuina spp. 92
Giant mallee 262
Giant sequoia 231
Gidgee 261
Gilbertiodendron dewevrei 103
Ginkgo biloba 152
Gironniera subaequalis 147
Gmelina arborea 31, 32
Gmelina spp. 258
Gnetum africanum 84, 85
Gnetum spp. 84
Golden chinkapin 231
Green ash 229
Greybox 261
Griselinia litoralis 263
Guabo 92
Guadua angustifolia 91
Guarea
cedrata 103
thompsonii 103
Guarea spp. 232
Gum 229, 237
Guttiferae 104
Gynoxys spp. 284

H

Haloxylon ammodendron 163
Hard maple 33
Harpagophytum
procumbens 84
zeyheri 84
Harpagophytum spp. 84
Hemlock 237
Heritiera parvifolia 147
Heritiera spp. 34
Heteropsis flexuosa 93
Heteropsis spp. 91
Hevea brasiliensis 24 34, 36, 38, 90
Hevea spp. .... xxiv, 27, 35, 174
Hibiscus tiliaceus 170
Hickory 227, 228, 229, 237
Holoptelea grandis 104
Homalium hainanensis 147
Hopea hainanensis 147
Hopea spp. 146
Hornbeam 188, 190, 197
Hule 231
Hura crepitans 283
Hydnocarpus hainanense 147
Hydrangea spp. 152
Hymenaea courbaril 90, 232
Hymenaea spp. 282
Hymenolobium spp. 282

I

Idiospermum sp. 258

Ilex
- *aquifolium* 188
- *paraguariensis* 90, 284
- *purpurea* 149
- *rotunda* 149

Ilex spp. 152
Imperata cylindrica 119
Incense cedar 231
Inga spp. 92
Interior live oak 231
Intsia bijuga 147
Ironbark 260, 261
Irvingia gabonensis 84
Isoberlinia doka 105
Isoberlinia spp. 105, 134

J

Jacaranda spp. 33
Jack pine 225, 226, 228
Jatropha curcas 139
Jeffrey pine 230, 231
Jessenia bataua 90
Jippi jappa 93
Jubaea chilensis 91, 284
Juglandaceae 148
Juglans
- *cinerea* 227
- *mandshurica* 152, 194
- *nigra* 227

Julbernardia seretii 103
Julbernardia spp. 105, 134
Juniper 151, 188, 228, 230, 231
Juniperus
- *excelsa* 151, 188
- *foetidissima* 188
- *phoenicea* 150
- *polycarpos* 188
- *procera* 148
- *seravschanica* 151
- *thurifera* 107, 188

Juniperus ashei 230
Juniperus spp. 151, 228, 230
Kalopanax septemlobus 153
Khaya
- *grandifolia* 104
- *senegalensis* 85

Khaya spp. 34
Kielmeyera coriacea 282
Kimberly 259
King Billy pine 262
Knightia spp. 260
Korean cedar pine 193
Korean willow 192, 194
Korthalsia spp. 86

L

Lagarostrobos franklinii 262
Lagerstroemia spp. 147
Laguncularia racemosa 104, 282
Lance wood 259
Larch 191, 192, 193, 194, 206, 213, 218, 236
Larix
- *cajanderi* 192, 194
- *gmelini* 154, 192, 193
- *gmelinii* 192
- *kurilensis* 194
- *laricina* 225, 226
- *principis-rupprechtii* 153

Larix sibirica 163, 190, 191, 192
Larix spp. 181, 213, 218, 236
Latanier 93
Lauraceae 104, 147, 148, 232, 233, 260
Laurea spp. 105
Laurel 229, 231
Laurelia serrata 285
Laurelia spp. 260
Laurus nobilis 84, 150
Lecythis spp. 232
Leguminosae 103, 146, 147, 281, 283
Leopoldina piassaba 91
Leptospermum ericoides 262
Lerp mallee 262
Leucadendron argenteum 106
Leucaena
- *leucocephala* 31, 139

Leucospermum spp. 106
Libocedrus bidwillii 263
Libocedrus spp. 148, 260
Licania spp. 284
Limba 123
Limber pine 231
Liquidambar
- *formosana* 147, 149
- *styraciflua* 228

Liriodendron tulipifera 227
Litchi chinensis 147
Lithocarpus
- *amygdalifolius* 149
- *brevicaudatus* 149
- *densiflorus* 230, 231
- *fenzelianus* 147
- *glabra* 149
- *ternaticupula* 149

Lithocarpus spp. 148, 149
Lithraea caustica 284
Litsea spp. 260
Live oak 230
Loblolly pine 229, 230
Lodgepole pine 225, 226, 227, 229, 231

M

Machaerium spp. 282
Machilus thunbergii 149
Macrolobium acaciaefolium 282
Madhuca hainanensis 147
Magnolia 229
    grandiflora 229
    virginiana 229
Magnoliaceae 148
Mahogany 33, 244
Mallee 261
Malpighiaceae 104
Mangifera indica 105
Manglietia hainanensis 147
Mangrove 105, 119, 145, 146, 147, 149, 169,
    170, 232, 249, 258, 259, 267, 274, 282
Manicaria saccifera 93, 282
Manicole palm 93
Manilkara
    bidentata 93
    concolor 106
    hexandra 147
    huberi 90
    zapota 92
Manilkara spp. 232
Manna gum 263
Mansonia altissima 104
Mansonia spp. 34
Manteco 231
Maple 194, 229, 236
Maprounea africana 105
Maranthes
    glabra 103
    polyandra 105
Maranthes spp. 258
Maria 231
Maripa 93
Maritime 212, 213
Maritime pine 204, 212
Marquesia macroura 105
Marquesia spp. 105
Masaquilla 231
Masica 231
Mauria sessiliflora 231
Mauritia flexuosa 90
Mauritiella pacifica 282
Maximiliana caribea 93
Melaleuca
    dealbata 259
    leucadendra 259
    minutifolia 259
    viridiflora 259

Lodoicea maldivica 140
Lophira lanceolata 84, 105
Lophira spp. 34
Lovoa spp. 34
Lovoa trichilioides 103
Lumnitzera spp. 258

Melaleuca spp. 267
Meliaceae 123, 147, 260
Melinis minutiflora 275
Metrosideros
    collina 258
    umbellata 263
Metroxylon spp. 86, 88
Michelia balansae 147
Michelsonia microphylla 103
Miconia calvescens 275
Miconia spp. 284
Millettia thonningii 105
Mimosa spp. 283
Mimosaceae 147
Miombo 134
Monotes kerstingii 105
Monterey cypress 230
Monterey pine 230
Montrouziera spp. 258
Mopane 105
Moraceae 281
Morchella spp. 87
Mountain ash 263
Mountain hemlock 228, 229, 231
Mountain white gum 263
Muraltia spp. 106
Musgravea spp. 258
Musk rose 90
Myall 261
Myrciaria dubia 90
Myrica
    californica 231
    tomentosa 194
Myristica fragrans 92
Myroxylon balsamum 90, 92
Myrtaceae 104, 147, 148, 258, 260
Myrtle 262, 263
Myrtoideae 258
Myrtus communis 110, 150

N

Nardostachys jatamansi 86
Nectandra spp. 233
Nelia 261
Neolitsea sericea 150
Nestegis spp. 260
Nipa fruticans 146
Norway spruce 197, 205, 218
Nothofagus
    antarctica 285
    betuloides 285
    cunninghamii 262, 263
    dombeyi 284, 285
    fusca 262
    menziesii 262
    nitida 285
    obliqua 284, 285
    procera 284, 285
    pumilio 285

solandri 262
solandri var. *cliffortiodes* 266
solandri var. *cliffortioides* 263
truncata 262
Nothofagus spp. 148, 258, 260, 262, 266, 284, 285
Nutmeg tree 92
Nypa fruticans 86
Nyssa aquatica 229

O

Oak 33, 148, 150, 151, 160, 187, 188, 190,
197, 205, 209, 213, 227, 229, 230, 231,
233, 236
Ochna
afzelii 105
schweinfurthiana 105
Ochroma lagopus 232
Ocotea
architectorum 284
pretiosa 90, 91, 92
Ocotea rodiaei 34
Ocotea spp. 107, 233
Oder 190
Oil palm 38
Ojushte 92
Olea
africana 148
capensis 106, 107
chrysophylla 148
cuspidata 150
europaea 110, 150, 187
Olearia
ilicifolia 263
lineata 263
Oncosperma spp. 86
Opuntia ficus-indica 112
Orbignya phalerata 90, 89
Orchid 85
Oregon ash 227
Oregon white oak 227
Ormosia balansae 147
Ostrearia spp. 258
Ostrya spp. 232

P

Pacific bayberry 231
Pacific cedar 228
Pacific madrone 227, 231
Pacific silver fir 228
Padus
asiatica 193
maackii 193
Paleto 231
Palm 47, 88, 90, 91, 92, 231, 282, 283, 284
Palma cana 93
Palma chonga 92
Panicum maximum 119

Paper birch 225
Paque 231
Parapiptadenia spp. 284
Parinari spp. 258
Parkia bicolor 103
Parkia spp. 282
Parkinsonia aculeata 112
Paulownia fortunei 153
Paulownia spp. 153, 181
Pecan 229
Pejibaye palm 92
Pemphis acidula 140, 170
Pentacme siamensis 147
Peppermint 262, 263
Pericopsis
angolensis 134
elata 103
Perm 189
Persea lingue 285
Persea spp. 233
Petersianthus macrocarpus 103
Peumus boldus 90, 284
Phillyrea spp. 150
Phoebe
porosa 284
sheareri 149
Phoebe spp. 149, 233
Phyllocladus aspleniifolius 262
Phyllocladus spp. 148, 260
Phyllostachys
bambusoides 149, 153
edulis 149
glauca 153
heteroclada 149
mannii 149
nidularis 149
nigra var. *henonis* 149
propinqua 153
vivax 153
Phyllostachys spp. 87, 149
Phytelephas seemannii 92
Phytelephas spp. 89, 91
Piassaba 91
Picea
abies 188, 189, 190, 191, 197, 218
ajanensis 194
ajsnensis 193
asperata 153
balfouriana 152
brachytyla 153
complanata 149, 152, 153
engelmannii 229, 231
glauca 225, 226, 227, 229
glehnii 154
jezoensis 154
jezoensis var. *microsperma* 152
koraiensis 152
likiangensis 152
linzhiensis 152
mariana 225, 226

meyeri ... 153
morinda ... 151
neoveitchii ... 149
obovata ... 190, 191, 192, 193
omorika ... 190
orientalis ... 190
rubens ... 228, 229
sitchensis ... 229
smithiana ... 151
wilsonii ... 153
Picea spp. ... 152, 188, 190, 213, 218, 236
Pilgerodendron uvifera ... 285
Pimenta dioica ... 231
Pine ... 23, 87, 88, 90, 151, 160, 175, 181, 188, 190, 192, 193, 205, 213, 229, 230, 231, 236, 237, 245, 246, 262

Pinus
   albicaulis ... 229, 231
   armandii ... 149, 153, 154
   ayacahuite ... 244
   banksiana ... 225, 226, 228
   brutia ... 150
   bungeana ... 149, 153
   caribae ... 244
   caribaea ... 89
   caribaea ... 90
   caribaea var. *caribaea* ... 31
   caribaea var. *hondurensis* ... 31, 32
   cembra ... 87
   cembra var *sibirica* ... 163
   contorta ... 198, 225, 226, 227, 229, 231
   densiflora ... 150, 152
   echinata ... 229
   edulis ... 228
   elliottii ... 32, 90, 230
   engelmannii ... 231
   excelsa ... 151, 152
   flexilis ... 231
   gerardiana ... 87, 151
   griffithii ... 152
   halepensis ... 84, 106, 107, 110, 111, 150, 213
   henryi ... 149
   jeffreyi ... 230
   koraiensis ... 87, 152, 193, 194
   lambertiana ... 230
   latteri ... 147
   leiophylla var. *chihuahuana* ... 231
   longaeva ... 231
   massoniana ... 149, 152
   merkusii ... 147
   montezumae ... 244
   monticola ... 229
   mugo ... 190
   muricata ... 230
   nigra ... 151, 188
   oocarpa ... 31, 92, 244
   patula ... 31, 33
   pinaster ... 106, 107, 111, 204, 212
   pinea ... 87, 150
   ponderosa ... 229, 230
   pseudostrobus ... 232
   pumila ... 154, 192, 193, 194, 218
   radiata ... 31, 32, 90, 212, 230, 266, 295
   resinosa ... 226, 227
   roxburghii ... 148, 151, 152, 170
   sabiniana ... 230
   sibirica ... 154, 191, 193, 218
   silvestris ... 163
   strobus ... 226, 227
   strobus ... 236
   sylvestris ... 190, 191, 193, 197, 205, 213, 218
   sylvestris var. *mongolica* ... 154
   sylvetris var. *sylvestriformis* ... 152
   tabulaeformis ... 149, 152, 153, 154
   taeda ... 229, 230
   taiwanensis ... 149
   torreyana ... 230
   wallichiana ... 151, 170
   yunnanensis ... 152
Pinus spp. ... xxiv, 23, 25, 31, 87, 90, 111, 148, 181, 236, 243, 289
Pinyon pine ... 228, 230, 231
Piper guineense ... 84
Piptadenia flava ... 283
Piptadenia inaequalis ... 282
Pistachio ... 151
Pistacia
   atlantica ... 107, 150, 151
   chinensis ... 149
   lentiscus ... 110, 150
   palaestina ... 150
Pithecellobium
   saman ... 232, 283
   unguis-cati ... 283
Pito ... 92
Placospermum sp. ... 258
Platanus
   occidentalis ... 230
   orientalis ... 151
Platonia insignis ... 90
Platycladus orientalis ... 149, 152, 153
Platymiscium spp. ... 232
Plectocomia spp. ... 86
Podocarpaceae ... 258
Podocarpus
   ferruginea ... 262
   halii ... 263
   imbricata ... 147
   latifolius ... 106
   macrophyllus ... 150
   nagi ... 150
   nubigena ... 285
   oleifolius ... 284
   totara ... 266
Podocarpus spp. ... 85, 107, 148, 260, 284
Pogostomon cablin ... 86
Polypodium aureum ... 92
Polypodium spp. ... 92
Pometia spp. ... 146
Ponderosa pine ... 229, 230

Poplar 157, 163, 190, 193, 204, 205, 261
Populus
    alba 190
    balsamifera 225, 226, 227, 228
    cathayana 153, 154
    davidiana 152, 153, 154
    deltoides 229
    maximoviczii 193
    nigra 190
    purdomii 154
    suaveolens 154, 192
    tremula 190, 191, 218
    tremuloides .. 197, 225, 226, 227, 228, 229, 231
    trichocarpa 227, 228
Populus spp. 151, 152, 163, 181, 204, 228, 295
Post oak 230
Potentilla fruticosa 154
Pouteria
    izabalensis 231
    sapota 92
Prioria copaifera 232
Prosopis
    alba 283
    chilensis 91, 283
    juliflora 112
    juliflora 139
    nigra 283
    pallida 91
    tamarugo 91
Prosopis pallida 89
Prosopis spp. 230, 283, 285
Protea madiensis 105
Protea spp. 106
Proteaceae 147, 258
Protium macgregorii 147
Prunus
    africana 84, 85, 106
    avium 110
Prunus spp. 33, 232
Pseudolmedia cf. *spurea* 231
Pseudotsuga menziesii 227, 229, 230, 266
Pseudotsuga spp. 188, 204 236
Pterocarpus
    officinalis 231
Pterocarpus spp. 105, 283
Pterocarya
    pterocarpa 150
    rhoifolia 153
Pterospermum heterophyllum 147
Pterospermum spp. 147
Puya spp. 285
Pyrus bovei 150

Q

Quaking aspen 197, 225
Qualea spp. 282
Quebracho colorado 89, 91
Quercus
    acutissima 149, 150, 152
    afares 111
    agrifolia 231
    alba 227
    aliena 152
    aliena var. *acuteserrata* 149
    baloot 151
    baronii 153
    boissieri 151
    calliprinos 150
    castaneifolia 150, 151
    cerris 151, 189, 190
    chrysolepis 231
    coccifera 110, 150
    copeyensis 233
    costaricensis 233
    crispula 154
    dentata 149, 152
    dilatata 151
    douglasii 230
    fabrei 149
    faginea 106, 111, 188
    frainetto 190
    garryana 227, 229
    glandulifera 149
    hartwissiana 150
    iberica 150
    ilex 106, 107, 110, 111, 187
    imeretina 150
    infectoria 150
    ithaburensis 150
    laurifolia 229
    liaotungensis 149, 152, 153
    libani 151
    lobata 230
    macrocarpa 228
    marilandica 230
    mongolica 152, 154, 194
    mongolica var. *grosseserrata* 153
    myrtifolia 229
    oleoides 232
    persica 151
    petraea 188, 189, 190
    pubescens 188, 189
    pyrenaica 188
    robur 188, 190
    rubra 227, 229
    seemanni 233
    semecarpifolia 151
    serrata 150, 152
    stellata 230
    suber 84, 85, 106, 110, 111, 112
    variabilis 149, 152
    velutina 227
    virginiana 229, 230
    wislizeni 231
Quercus spp. 33, 190, 197, 204, 213, 229, 230, 232, 243
Quillaja saponaria 90, 89, 284
Quintinia acutifolia 262

R

Radiata pine 212
Raphia taedigera 282
Rattan 81, 86, 87, 92
Ravenala madagascariensis 104
Red alder 228, 229
Red fir 230
Red oak 227, 229
Red pine 226, 227
Red spruce 228, 229
Red tingle 261
Redwood 230, 262
Restio spp. 106
Rhamnaceae 285
Rhamnus palaestina 150
Rhamnus spp. 284
Rhizophora
    apiculata 146, 147
    brevistyla 282
    harrisonii 104
    mangle 104, 282
    mucronata 105, 146, 147
    racemosa 104
    stylosa 259
Rhizophora spp. 259
Rhizophoraceae 258
Rhododendron campanulatum 151
Rhododendron spp. 190
Robinia pseudoacacia 153
Rosa moschata 90
Rose apple 93
Rosewood 33, 90, 92
Rosmarinus officinalis 84, 85, 110
Rubber .... xxiv, 27, 38, 174
Rubberwood 36
Rubiaceae 104
Rustia occidentalis 282
Rutaceae 147, 260

S

Sabal mauritiiformis 93
Sabal umbraculifera 93
Sabina chinensis 153
Sago 86
Sago palm 88
Sajan 193
Salix
    alba 190
    fragilis 190
    schwerin 192
    udensis 192
    viminalis 89, 91
Salix spp. .. 151, 152, 154, 191, 228, 232, 295, 236
Salmon gum 262
Sambucus spp. 232
Sandalwood 86, 88
Santalum spp. 86, 88
Sapelli 123
Sapindaceae 146
Sapodilla 92
Sapotaceae 258
Sasa kurilensis 194
Sassafras 90, 92
Sassafras randaiense 152
Schefflera octophylla 147
Schima spp. 148
Schinopsis spp. 91, 283
Schizomeria ovata 260
Schizomeria spp. 258
Sclerocarya birrea 135
Sequoia sempervirens 230
Sequoiadendron giganteum 231
Serruria spp. 106
Sessile oak 190
Shagbark hickory 227
Shorea
    obtusa 147
    robusta 145, 147, 152, 169
    talura 147
Shorea spp. 34, 88, 146
Shortleaf pine 229
Siberian cedar pine 191, 193
Siberian larch 191, 192
Siberian spruce 191, 192
Siberian stone
pine 218
Silver fir 205
Silverleaf ironbark 259
Simaroubaceae 260
Sindora
    cochinchinensis 147
    glabra 147
Sitka spruce 206, 229
Siwalik 170
Slash pine 230
Smilax spp. 92
Sombrerete 231
Sonneratia
    acida 147
    alba 105, 146
    caseolaris 146, 259
Sonneratia spp. 258
Sophora
    japonica 152
    microphylla 263
Sorbus aucuparia 190
Southern pine 237
Southern sassafras 262
Sphagnum spp. 194
Sphenostemon spp. 258
Spondias mombin 93
Spruce 188, 189, 190, 191, 192, 193, 194, 213,
                                                         218, 229, 236, 237
Sterculia spp. 88, 104
Sterculiaceae 281
Stipa tenacissima 110
Styrax spp. 86
Subalpine fir 227, 228, 229

Sugar maple 226, 227, 229
Sugar pine 230
Sugarberry 229
Suriana maritima 170
Swartzia spp. 282
Sweetgum 228, 229
Sweetia panamensis 232
Swertia chirayta 86
Swietenia
    humilis 232, 244
    macrophylla 31, 33, 231, 244, 275
    mahagoni 244
Swietenia spp. 33, 34, 246
Symphonia globulifera 231
Symplocos pichindensis 284
Symplocos spp. 152
Syzigium guineense ssp. afromontanum 106
Syzygium spp. 146, 260

T

Tabebuia
    chrysantha 232
    pentaphylla 232
Tabebuia spp. 284
Tachiglia spp. 282
Tagua 89
Tagua palm 91
Tall messmate 262
Tall sand mallee 262
Tamarindus indica 105
Tamarix spp. 151
Tambourissa spp. 106
Taxaceae 258
Taxodium distichum 229
Taxus cuspidata 152
Teak 32, 33, 147, 169, 174, 175, 176
Tectona grandis 31, 32, 33, 147, 169, 174
Terminalia
    amazonia 231
    calamansanai 258
    catappa 152
    chiriquensis 232
    glaucescens 105
    ivorensis 31
    superba 31, 123
Terminalia spp. 34, 146, 147, 258
Tetraclinis articulata 106
Tetrameles nudiflora 147, 152
Theaceae 147
Theobroma grandiflorum 90
Thuja
    koraiensis 152
    occidentalis 226, 228
Thuya
    plicata 227, 229, 230
    standishii 154
Thuya spp. 236
Thymus spp. 84, 85

Tieghemella spp. 34
Tilia
    americana 227, 228, 229
    amurensis 152, 194
    chinensis 154
    cordata 190
    japonica 153
    miqueliana 149
Tilia spp. 152, 153
Tina spp. 106
Toona sinensis 153
Torrey pine 230
Torreya nucifera 150
Toxicodendron vernicifluum 149, 153
Tree fern 231
Trembling aspen 226, 227, 228, 229
Tremella fuciformis 88
Tridax procumbens 92
Trilepisium madagascariense 104
Triplaris surinamensis 282
Triplochiton scleroxylon 104
Tristania laurina 260
Trochetia boutoniana 85
Trochodendron aralioides 152
Trochonanhus comphoratus 148
Tsaratanana 133
Tsuga
    canadensis 227, 229
    chinensis 152, 153
    diversifolia 154
    dumosa 153
    heterophylla 227, 228, 230, 236
    mertensiana 228, 229, 231
    sieboldii 150
Tsuga spp. 148
Tuart 261
Tulip-tree 227
Tulipwood 33

U

Uapaca
    bojeri 106
    kirkiana 134
    togoensis 105
Ulmus
    alata 229
    americana 227, 230
    campestris 151
    davidiana var. japonica 152, 154
    glabra 190
    laciniata 153, 194
    laevis 190
    minor 190
    parvifolia 149
Ulmus spp. 152, 163, 208, 229, 237
Umbellularia californica 231
Uncaria tomentosa 90, 89, 91

V

Varillo 231
Vatica hainanensis 147
Verbenaceae 147
Vetiveria zizanioides 92
Virola
    koschnyi 231
Virola spp. 232
Vistula 190
Vitellaria paradoxa 84
Vitex spp. 147, 232
Vochysia hondurensis 231

W

Walnut 33, 227
Wandoo 262
Warburgia salutaris 84, 85
Water tupelo 229
Weinmannia
    balbisiana 284
    racemosa 262, 263
Weinmannia spp. 106, 284
Western hemlock 227, 228, 230, 236
Western red cedar 227, 229, 230, 236
Western white pine 229, 231
Wheatbelt 262
White birch 226, 227
White cedar 228
White elm 227
White fir 230
White mallee 262
White oak 227
White pine 228, 229
White spruce 225, 226, 227, 229
Whitebark pine 229, 231
Willow 190, 191, 227, 228
Winged elm 229

X

Xylia kerrii 147
Xylocarpus
    granatum 147
    obovatus 105
Xylocarpus spp. 258
Xylopia aethiopica 85
Xymalos monospora 106

Y

Yellow birch 228, 229
Yellow pine 229
Yellow poplar 237
Yerba de toro 92

Z

Zapote 92
Zapotillo 92
Zarzaparilla 92
Zelkova
    carpinifolia 150
    schneideriana 149
    sinica 153
Ziziphus jujuba 150, 153
Ziziphus lotus 107

FAO TECHNICAL PAPERS

FAO FORESTRY PAPERS

1	Forest utilization contracts on public land, 1977 (E F S)	32	Classification and definitions of forest products, 1982 (Ar/E/F/S)
2	Planning forest roads and harvesting systems, 1977 (E F S)	33	Logging of mountain forests, 1982 (E F S)
3	World list of forestry schools, 1977 (E/F/S)	34	Fruit-bearing forest trees, 1982 (E F S)
3 Rev.1	World list of forestry schools, 1981 (E/F/S)	35	Forestry in China, 1982 (C E)
3 Rev.2	World list of forestry schools, 1986 (E/F/S)	36	Basic technology in forest operations, 1982 (E F S)
4/1	World pulp and paper demand, supply and trade 1– Vol. 1, 1977 (E F S)	37	Conservation and development of tropical forest resources, 1982 (E F S)
4/2	World pulp and paper demand, supply and trade – Vol. 2, 1977 (E F S)	38	Forest products prices 1962-1981, 1982 (E/F/S)
		39	Frame saw manual, 1982 (E)
5	The marketing of tropical wood in South America, 1976 (E S)	40	Circular saw manual, 1983 (E)
6	National parks planning, 1976 (E F S**)	41	Simple technologies for charcoal making, 1983 (E F S)
7	Forestry for local community development, 1978 (Ar E F S)	42	Fuelwood supplies in the developing countries, 1983 (Ar E F S)
8	Establishment techniques for forest plantations, 1978 (Ar C E* F S)	43	Forest revenue systems in developing countries, 1983 (E F S)
9	Wood chips – production, handling, transport, 1976 (C E S)	44/1	Food and fruit-bearing forest species – 1. Examples from eastern Africa, 1983 (E F S)
10/1	Assessment of logging costs from forest inventories in the tropics – 1. Principles and methodology, 1978 (E F S)	44/2	Food and fruit-bearing forest species – 2. Examples from southeastern Asia, 1984 (E F S)
10/2	Assessment of logging costs from forest inventories in the tropics – 2. Data collection and calculations, 1978 (E F S)	44/3	Food and fruit-bearing forest species – 3. Examples from Latin America, 1986 (E S)
		45	Establishing pulp and paper mills, 1983 (E)
11	Savanna afforestation in Africa, 1977 (E F)	46	Forest products prices 1963-1982, 1983 (E/F/S)
12	China: forestry support for agriculture, 1978 (E)	47	Technical forestry education – design and implementation, 1984 (E F S)
13	Forest products prices 1960-1977, 1979 (E/F/S)	48	Land evaluation for forestry, 1984 (C E F S)
14	Mountain forest roads and harvesting, 1979 (E)	49	Wood extraction with oxen and agricultural tractors, 1986 (E F S)
14 Rev.1	Logging and transport in steep terrain, 1985 (E)	50	Changes in shifting cultivation in Africa, 1984 (E F)
15	AGRIS forestry – world catalogue of information and documentation services, 1979 (E/F/S)	50/1	Changes in shifting cultivation in Africa – seven case-studies, 1985 (E)
16	China: integrated wood processing industries, 1979 (E F S)	51/1	Studies on the volume and yield of tropical forest stands – 1. Dry forest formations, 1989 (E F)
17	Economic analysis of forestry projects, 1979 (E F S)	52/1	Cost estimating in sawmilling industries: guidelines, 1984 (E)
17 Sup.1	Economic analysis of forestry projects: case studies, 1979 (E S)	52/2	Field manual on cost estimation in sawmilling industries, 1985 (E)
17 Sup.2	Economic analysis of forestry projects: readings, 1980 (C E)	53	Intensive multiple-use forest management in Kerala, 1984 (E F S)
18	Forest products prices 1960-1978, 1980 (E/F/S)	54	Planificación del desarrollo forestal, 1984 (S)
19/1	Pulping and paper-making properties of fast-growing plantation wood species – Vol. 1, 1980 (E)	55	Intensive multiple-use forest management in the tropics, 1985 (E F S)
19/2	Pulping and paper-making properties of fast-growing plantation wood species – Vol. 2, 1980 (E)	56	Breeding poplars for disease resistance, 1985 (E)
20	Forest tree improvement, 1985 (C E F S)	57	Coconut wood – Processing and use, 1985 (E S)
20/2	A guide to forest seed handling, 1985 (E S)	58	Sawdoctoring manual, 1985 (E S)
21	Impact on soils of fast-growing species in lowland humid tropics, 1980 (E F S)	59	The ecological effects of eucalyptus, 1985 (C E F S)
22/1	Forest volume estimation and yield prediction – Vol. 1. Volume estimation, 1980 (C E F S)	60	Monitoring and evaluation of participatory forestry projects, 1985 (E F S)
22/2	Forest volume estimation and yield prediction – Vol. 2. Yield prediction, 1980 (C E F S)	61	Forest products prices 1965-1984, 1985 (E/F/S)
		62	World list of institutions engaged in forestry and forest products research, 1985 (E/F/S)
23	Forest products prices 1961-1980, 1981 (E/F/S)	63	Industrial charcoal making, 1985 (E)
24	Cable logging systems, 1981 (C E)	64	Tree growing by rural people, 1985 (Ar E F S)
25	Public forestry administrations in Latin America, 1981 (E)	65	Forest legislation in selected African countries, 1986 (E F)
26	Forestry and rural development, 1981 (E F S)	66	Forestry extension organization, 1986 (C E S)
27	Manual of forest inventory, 1981 (E F)	67	Some medicinal forest plants of Africa and Latin America, 1986 (E)
28	Small and medium sawmills in developing countries, 1981 (E S)	68	Appropriate forest industries, 1986 (E)
29	World forest products, demand and supply 1990 and 2000, 1982 (E F S)	69	Management of forest industries, 1986 (E)
30	Tropical forest resources, 1982 (E F S)	70	Wildland fire management terminology, 1986 (E/F/S)
31	Appropriate technology in forestry, 1982 (E)	71	World compendium of forestry and forest products research institutions, 1986 (E/F/S)
		72	Wood gas as engine fuel, 1986 (E S)

73	Forest products: world outlook projections 1985-2000, 1986 (E/F/S)	115	Forestry policies of selected countries in Asia and the Pacific, 1993 (E)
74	Guidelines for forestry information processing, 1986 (E)	116	Les panneaux à base de bois, 1993 (F)
75	Monitoring and evaluation of social forestry in India – an operational guide, 1986 (E)	117	Mangrove forest management guidelines, 1994 (E)
76	Wood preservation manual, 1986 (E)	118	Biotechnology in forest tree improvement, 1994 (E)
77	Databook on endangered tree and shrub species and provenances, 1986 (E)	119	Number not assigned
		120	Decline and dieback of trees and forests – A global overview, 1994 (E)
78	Appropriate wood harvesting in plantation forests, 1987 (E)	121	Ecology and rural education – Manual for rural teachers, 1995 (E S)
79	Small-scale forest-based processing enterprises, 1987 (E F S)	122	Readings in sustainable forest management, 1994 (E F S)
80	Forestry extension methods, 1987 (E)	123	Forestry education – New trends and prospects, 1994 (E F S)
81	Guidelines for forest policy formulation, 1987 (C E)	124	Forest resources assessment 1990 – Global synthesis, 1995 (E F S)
82	Forest products prices 1967-1986, 1988 (E/F/S)	125	Forest products prices 1973-1992, 1995 (E F S)
83	Trade in forest products: a study of the barriers faced by the developing countries, 1988 (E)	126	Climate change, forests and forest management – An overview, 1995 (E F S)
84	Forest products: World outlook projections – Product and country tables 1987-2000, 1988 (E/F/S)	127	Valuing forests: context, issues and guidelines, 1995 (E F S)
85	Forestry extension curricula, 1988 (E/F/S)	128	Forest resources assessment 1990 – Tropical forest plantation resources, 1995 (E)
86	Forestry policies in Europe, 1988 (E)	129	Environmental impact assessment and environmental auditing in the pulp and paper industry, 1996 (E)
87	Small-scale harvesting operations of wood and non-wood forest products involving rural people, 1988 (E F S)	130	Forest resources assessment 1990 – Survey of tropical forest cover and study of change processes, 1996 (E)
88	Management of tropical moist forests in Africa, 1989 (E F P)	131	Ecología y enseñanza rural – Nociones ambientales básicas para profesores rurales y extensionistas, 1996 (S)
89	Review of forest management systems of tropical Asia, 1989 (E)	132	Forestry policies of selected countries in Africa, 1996 (E/F)
90	Forestry and food security, 1989 (Ar E S)	133	Forest codes of practice – Contributing to environmentally sound forest operations, 1996 (E)
91	Design manual on basic wood harvesting technology, 1989 (E F S) (Published only as FAO Training Series, No. 18)	134	Estimating biomass and biomass change of tropical forests – A primer, 1997 (E)
92	Forestry policies in Europe – An analysis, 1989 (E)	135	Guidelines for the management of tropical forests – 1. The production of wood, 1998 (E S)
93	Energy conservation in the mechanical forest industries, 1990 (E S)	136	Managing forests as common property, 1998 (E)
94	Manual on sawmill operational maintenance, 1990 (E)	137/1	Forestry policies in the Caribbean – Volume 1: Proceedings of the Expert Consultation, 1998 (E)
95	Forest products prices 1969-1988, 1990 (E/F/S)	137/2	Forestry policies in the Caribbean – Volume 2: Reports of 28 selected countries and territories, 1998 (E)
96	Planning and managing forestry research: guidelines for managers, 1990 (E)	138	FAO Meeting on Public Policies Affecting Forest Fires, 2001 (E F S)
97	Non-wood forest products: the way ahead, 1991 (E S)	139	Governance principles for concessions and contracts in public forests, 2001 (E)
98	Timber plantations in the humid tropics of Africa, 1993 (E F)	140	Global Forest Resources Assessment 2000 - Main report, 2001 (E)
99	Cost control in forest harvesting and road construction, 1992 (E)		
100	Introduction to ergonomics in forestry in developing countries, 1992 (E F I)		
101	Management and conservation of closed forests in tropical America, 1993 (E F P S)		
102	Research management in forestry, 1992 (E F S)		
103	Mixed and pure forest plantations in the tropics and subtropics, 1992 (E F S)		
104	Forest products prices 1971-1990, 1992 (E/F/S)		
105	Compendium of pulp and paper training and research institutions, 1992 (E)		
106	Economic assessment of forestry project impacts, 1992 (E/F)		
107	Conservation of genetic resources in tropical forest management – Principles and concepts, 1993 (E/F/S)		
108	A decade of wood energy activities within the Nairobi Programme of Action, 1993 (E)		
109	Directory of forestry research organizations, 1993 (E)		
110	Proceedings of the Meeting of Experts on Forestry Research, 1993 (E/F/S)		
111	Forestry policies in the Near East region – Analysis and synthesis, 1993 (E)		
112	Forest resources assessment 1990 – Tropical countries, 1993 (E)		
113	*Ex situ* storage of seeds, pollen and *in vitro* cultures of perennial woody plant species, 1993 (E)		
114	Assessing forestry project impacts: issues and strategies, 1993 (E F S)		

Availability: October 2001

Ar	–	Arabic	Multil –	Multilingual
C	–	Chinese	*	Out of print
E	–	English	**	In preparation
I	–	Italian		
F	–	French		
P	–	Portuguese		
S	–	Spanish		

The FAO Technical Papers are available through the authorized FAO Sales Agents or directly from Sales and Marketing Group, FAO, Viale delle Terme di Caracalla, 00100 Rome, Italy.